CONSERVATION AND THE GENETICS OF POPULATIONS

D1490389

Companion website

This book has a companion website at:

www.wiley.com/go/allendorf/populations

with additional resources

CONSERVATION AND THE GENETICS OF POPULATIONS

Second Edition

Fred W. Allendorf

Division of Biological Sciences, University of Montana, Missoula, MT 59812, USA

Gordon Luikart

Flathead Lake Biological Station, University of Montana, Polson, MT 59860, USA

Sally N. Aitken

Department of Forest Sciences, University of British Columbia, Vancouver, British Columbia, Canada V6T 1Z4

With illustrations by Agostinho Antunes

WILEY-BLACKWELL

A John Wiley & Sons, Ltd., Publication

This edition first published 2013 © 2013 by Fred W. Allendorf, Gordon Luikart and Sally N. Aitken.

Blackwell Publishing was acquired by John Wiley & Sons in February 2007. Blackwell's publishing program has been merged with Wiley's global Scientific, Technical and Medical business to form Wiley-Blackwell.

Registered office: John Wiley & Sons, Ltd, The Atrium, Southern Gate, Chichester, West Sussex, PO19 8SQ, UK

Editorial offices: 9600 Garsington Road, Oxford, OX4 2DQ, UK
The Atrium, Southern Gate, Chichester, West Sussex, PO19 8SQ, UK
111 River Street, Hoboken, NJ 07030-5774, USA

For details of our global editorial offices, for customer services and for information about how to apply for permission to reuse the copyright material in this book please see our website at www.wiley.com/wiley-blackwell.

The right of the author to be identified as the author of this work has been asserted in accordance with the UK Copyright, Designs and Patents Act 1988.

All rights reserved. No part of this publication may be reproduced, stored in a retrieval system, or transmitted, in any form or by any means, electronic, mechanical, photocopying, recording or otherwise, except as permitted by the UK Copyright, Designs and Patents Act 1988, without the prior permission of the publisher.

Designations used by companies to distinguish their products are often claimed as trademarks. All brand names and product names used in this book are trade names, service marks, trademarks or registered trademarks of their respective owners. The publisher is not associated with any product or vendor mentioned in this book. This publication is designed to provide accurate and authoritative information in regard to the subject matter covered. It is sold on the understanding that the publisher is not engaged in rendering professional services. If professional advice or other expert assistance is required, the services of a competent professional should be sought.

Library of Congress Cataloging-in-Publication Data
Allendorf, Frederick William.
   Conservation and the genetics of populations / Fred W. Allendorf, Gordon Luikart, Sally N. Aitken; with illustrations by Agostinho Antunes. – 2nd ed.
       p. cm.
   Includes bibliographical references and index.
   ISBN 978-0-470-67146-7 (cloth) – ISBN 978-0-470-67145-0 (pbk.)   1. Biodiversity conservation.   2. Population genetics.   3. Evolutionary genetics.   I. Luikart, Gordon.   II. Aitken, Sally N.   III. Title.
   QH75.A42 2013
   333.95'16–dc23
                    2012016197

A catalogue record for this book is available from the British Library.

Wiley also publishes its books in a variety of electronic formats. Some content that appears in print may not be available in electronic books.

Cover image: © Lee Rentz Photography. Clark's nutcracker feeding on whitebark pine. Whitebark pine seeds are a crucial food resource for many animal species (e.g., Clark's nutcracker and grizzly bears). Whitebark pine currently are threatened by an introduced pathogen (Chapter 20 Invasive Species) and by climate change (Chapter 21 Climate Change).

Cover design by: www.simonlevyassociates.co.uk

Set in 9/11 pt Photina by Toppan Best-set Premedia Limited

Printed in Singapore by C.O.S. Printers Pte Ltd

4   2017

Contents

Companion website

This book has a companion website at:

www.wiley.com/go/allendorf/populations

with additional resources

Guest Box authors

C. Scott Baker, Marine Mammal Institute and Department of Fisheries and Wildlife, Oregon State University, Newport, Oregon, USA (scott.baker@oregonstate.edu). Chapter 22.

Louis Bernatchez, Département de biologie, Université Laval, Quebec, Canada (Louis.Bernatchez@bio.ulaval.ca). Chapter 4.

David Coates, Department of Environment and Conservation, Western Australia, Australia (Dave.Coates@dec.wa.gov.au). Chapter 16.

David W. Coltman, Department of Biological Sciences, University of Alberta, Edmonton, Alberta, Canada (dcoltman@ualberta.ca). Chapter 11.

William A. Cresko, Institute of Ecology and Evolution, Department of Biology, University of Oregon, Eugene, Oregon, USA (wcresko@uoregon.edu). Chapter 8.

James F. Crow, Laboratory of Genetics, University of Wisconsin, Madison, Wisconsin, USA. Appendix (1916–2012).

Chris J. Foote, Fisheries and Aquaculture Department, Malaspina University College, Nainamo, British Columbia, Canada (Chris.Foote@viu.ca). Chapter 2.

Steven J. Franks, Department of Biology, Fordham University, Bronx, New York, USA (franks@fordham.edu). Chapter 21.

Birgita D. Hansen, Arthur Rylah Institute for Environmental Research, Department of Sustainability and Environment, Victoria, Australia (birgita.hansen@dse.vic.gov.au). Chapter 5.

Paul A. Hohenlohe, Institute for Bioinformatics and Evolutionary Studies, University of Idaho, Moscow, Idaho, USA (hohenlohe@uidaho.edu). Chapter 8.

Menna E. Jones, School of Zoology, University of Tasmania, Hobart, Tasmania, Australia (Menna.Jones@utas.edu.au). Chapter 6.

Lukas F. Keller, Institute of Evolutionary Biology and Environmental Studies, University of Zurich, Switzerland (lukas.keller@ieu.uzh.ch). Chapter 13.

Robert C. Lacy, Department of Conservation Science, Chicago Zoological Society, Brookfield, Illinois, USA (rlacy@ix.netcom.com). Chapter 19.

Guðrún Marteinsdóttir, Institute of Biology, University of Iceland, Sturlugata, Iceland (runam@hi.is). Chapter 18.

Craig R. Miller, Departments of Mathematics and Biological Sciences, University of Idaho, Moscow, Idaho, USA (crmiller@uidaho.edu). Chapter 7.

L. Scott Mills, Wildlife Biology Program, University of Montana, Missoula, Montana, USA (lscott.mills@umontana.edu). Chapter 1.

B.G. Murray, School of Biological Sciences, University of Auckland, Auckland, New Zealand (b.murray@auckland.ac.nz). Chapter 14.

Michael W. Nachman, Department of Ecology and Evolutionary Biology, University of Arizona, Tucson, Arizona, USA (nachman@u.arizona.edu). Chapter 12.

M. Pickup, Department of Ecology and Evolutionary Biology, University of Toronto, Toronto, Ontario, Canada (melinda.pickup@utoronto.ca). Chapter 14.

Loren H. Rieseberg, University of British Columbia, Department of Botany, Vancouver, British Columbia, Canada, and Indiana University, Department of Biology, Bloomington, Indiana, USA (lriesebe@mail.ubc.ca). Chapter 17.

Michael K. Schwartz, Rocky Mountain Research Station, US Forest Service, Missoula, Montana, USA (mkschwartz@fs.fed.us). Chapter 9.

Richard Shine, School of Biological Sciences, University of Sydney, New South Wales, Australia (rick.shine@sydney.edu.au). Chapter 20.

Michael E. Soulé, 212 Colorado Ave, Paonia, Colorado, USA (msoule36@gmail.com). Chapter 1.

Paul Sunnucks, School of Biological Sciences and Australian Centre for Biodiversity, Monash University, Victoria, Australia (Paul.Sunnucks@monash.edu). Chapter 5.

Jody M. Tucker, Sequoia National Forest, US Forest Service, Porterville, California, USA, and Wildlife Biology Program, University of Montana, Missoula, Montana, USA (jtucker@fs.fed.us). Chapter 9.

Elaina M. Tuttle, Department of Biology, Indiana State University, Terre Haute, Indiana, USA (elaina.tuttle@indstate.edu). Chapter 3.

Robert C. Vrijenhoek, Monterey Bay Aquarium Research Institute, Moss Landing, California, USA (vrijen@mbari.org). Chapter 15.

Lisette P. Waits, Fish and Wildlife Sciences, University of Idaho, Moscow, Idaho, USA (lwaits@uidaho.edu). Chapter 7.

Robin S. Waples, Northwest Fisheries Science Center, Seattle, Washington, USA (Robin.Waples@noaa.gov). Chapter 10.

Andrew Young, Centre for Plant Biodiversity Research, CSIRO Plant Industry, Canberra, Australian Capital Territory, Australia (andrew.young@csiro.au). Chapter 14.

Preface to the second edition

Truth is a creation, not a discovery.

R.H. Blyth

This second edition is an updated and expanded version of our book published in 2007. The need for applying genetics to problems in conservation has continued to increase. In addition, many important technological and conceptual developments have changed the field of conservation genetics over the least five years. We have added new chapters on climate change and on genetic effects of harvest (e.g., hunting and fishing), and have extensively revised all other chapters. In an effect to restrict the size of this new edition, we have moved the end of the chapter questions to the book's website. A new edition is needed to keep pace with the changes, to remain a useful overview and synthesis, and to help advance the field. Nearly one-third of the over 1800 references in this book were published after the first edition.

We are excited to have added Sally Aitken as a coauthor. Sally was a reviewer of our first edition for Wiley-Blackwell. Sally's review and suggested changes were so helpful that she should have been an author of that edition!

As we said in the Preface of the first edition, this book is not an argument for the importance of genetics in conservation. Rather, it is designed to provide the reader with the appropriate background and conceptual understanding to apply genetics to problems in conservation. The primary current causes of extinction are anthropogenic changes that affect ecological characteristics of populations (habitat loss, fragmentation, introduced species, etc.). However, genetic information and principles can be invaluable in developing conservation plans for species threatened with such effects. Phil Hedrick (2007) suggested in his review of the first edition that we should have made more of a sales pitch for the successes of conservation genetics, rather than our very even-handed approach. We do not agree. We are advocates for conservation, not genetics.

We have strived for a balance between theory, empirical data, and statistical analysis in this text. Empirical population genetics depends upon this balance (see Figure i.1). The effort to measure and understand the evolutionary significance of genetic variation in natural populations began with the rediscovery of Mendel's principles in the early 1900s. For many years, empirical observation of genetic variation lagged far behind the development of sophisticated theory by some of the greatest minds of the twentieth century (e.g., J.B.S. Haldane, R.A. Fisher, and Sewall Wright). Today it is relatively easy to obtain and analyze enormous amounts of information on genetic variation (e.g., markers or genome sequences) in any species.

There are a wide variety of computer programs available to analyze data and estimate parameters of interest. However, the ease of collecting and analyzing data has led to an unfortunate and potentially dangerous reduction in the emphasis on understanding theory in the training of population and conservation geneticists. Understanding theory remains crucial for correctly interpreting outputs from computer programs and statistical analyses. For example, the most powerful software programs that estimate important parameters, such as effective population size (Chapter 7) and gametic disequilibrium (Chapter 10), are not useful if their assumptions and limitations are not understood. We are still disturbed when we read statements in the literature that the loci studied are not linked because they are not in linkage (gametic) disequilibrium.

The tools being used by molecular population geneticists are changing rapidly. Deciding which techniques

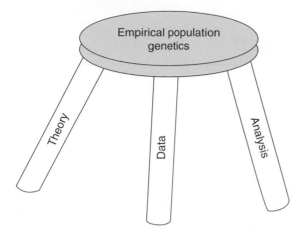

Figure i.1 The application of population genetics to understand genetic variation in natural populations relies upon a balance of understanding the appropriate theory, collecting appropriate data, and understanding its analysis.

to include in Chapter 4 was difficult. We have added new genomics (and other 'omics') information and still include some techniques that are no longer in use (e.g., minisatellites) because they are crucial for understanding previous literature. On the other hand, we have not included older techniques that are now known to provide data that are not reliable (e.g., RAPDs).

ACKNOWLEDGMENTS

We would like to thank again everyone acknowledged in the Preface to the previous edition of this book. FWA thanks the Hatfield Marine Science Center and the Hawai'i Institute of Marine Biology for hosting his extended visits to work on this edition. Chapter 18 is based partially on work supported by the US National Science Foundation, Grant DEB 074218 to FWA and GL. In addition, we give special thanks to the Minitab Corporation for providing software through their Author Assistance Program, Paul Sunnucks for his many helpful comments on the previous edition, Nils Ryman for help with the Appendix, and Ian Jamieson for help with Table 13.1. In addition, we thank the many colleagues who have helped us by providing comments, information, unpublished data, and answers to questions associated with this edition: Steve Amish, Peter Beerli, Kurt Benirschke, Des Cooper, Rob Cowie, Kirsten Dale, Pam Diggle, Suzanne Edmands, Zac Forsman, Ned Friedman, Oscar Gaggiotti Roxanne Haverkort, Paul Hohenlohe, Marty Kardos, Meng-Hua Li, Ian MacLachlan, Sierra McLane, Juha Merilä, Gordon Orians, Barb Taylor, Mark Tanaka, Dave Towns, C. Susannah Tysor, and Sam Yeaman.

DEDICATION

We dedicate this book to James F. Crow, who died just before his 96th birthday as we were putting the finishing touches on this book. Jim was a great scientist and wonderful human being whose contributions were enormous. We were honored that he agreed to write a Guest Box for the Statistical Appendix. Over the years, we always appreciated Jim's clear and useful reviews of our papers when they were sent to him. FWA got to know Jim in the last few years while his daughter was attending graduate school in Madison. My most vivid memory of talking with Jim in his tiny office was finding Sewall Wright's National Medal of Science haphazardly placed on a filing cabinet behind the office door.

12 February 2012

Fred W. Allendorf
Gordon Luikart
Sally N. Aitken

Preface to the first edition

The many beings are numberless; I vow to save them all.

Traditional Zen vow

The one process now going on that will take millions of years to correct is the loss of genetic and species diversity by the destruction of natural habitats. This is the folly our descendants are least likely to forgive us.

Edward O. Wilson, 1984

This book is about applying the concepts and tools of genetics to problems in conservation. Our guiding principle in writing has been to provide the conceptual basis for understanding the genetics of biological problems in conservation. We have not attempted to review the extensive and ever growing literature in this area. Rather we have tried to explain the underlying concepts and to provide enough clear examples and key citations for further consideration. We also have strived to provide enough background so that students can read and understand the primary literature.

Our primary intended audience is broadly trained biologists who are interested in understanding the principles of conservation genetics and applying them to a wide range of particular issues in conservation. This includes advanced undergraduate and graduate students in biological sciences or resource management, as well as biologists working in conservation biology for management agencies. The treatment is intermediate and requires a basic understanding of ecology and genetics.

This book is not an argument for the importance of genetics in conservation. Rather, it is designed to provide the reader with the appropriate background to determine how genetic information may be useful in any specific case. The primary current causes of extinction are anthropogenic changes that affect ecological characteristics of populations (habitat loss, fragmentation, introduced species, etc.). However, genetic information and principles can be invaluable in developing conservation plans for species threatened with such effects.

The usefulness of genetic tools and concepts in conservation of biological diversity is continually expanding as new molecular technologies, statistical methods, and computer programs are being developed at an increasing rate. Conservation genetics and molecular ecology are under explosive growth, and this growth is likely to continue for the foreseeable future. Indeed we have recently entered the age of genomics. New laboratory and computational technologies for generating and analyzing molecular genetic data are emerging at a rapid pace.

There are several excellent texts in population genetics available (e.g., Hedrick 2005, Hartl and Clark 1997, Halliburton 2004). Those texts concentrate on questions related to the central focus of population and evolutionary genetics, which is to understand the processes and mechanisms by which evolutionary changes occur. There is substantial overlap between those texts and this book. However, the theme underlying this book is the application of an understanding of the genetics of natural populations to conservation.

We have endeavored to present a balanced view of theory and data. The first four chapters (Part I) provide an overview of the study of genetic variation in natural populations of plants and animals. The middle eight chapters (Part II) provide the basic principles of population genetics theory with an emphasis on concepts especially relevant for problems in conservation. The final eight chapters (Part III) synthesize these principles and apply them to a variety of topics in conservation.

We emphasize the interpretation and understanding of genetic data to answer biological questions in conservation. Discussion questions and problems are included at the end of each chapter to engage the reader in understanding the material. We believe well written problems and questions are an invaluable tool in learning the information presented in the book. These problems feature analysis of real data from populations, conceptual theoretical questions, and the use of computer simulations. A web site contains example datasets and software programs for illustrating population genetic processes and for teaching methods for data analysis.

We have also included a comprehensive glossary. Words included in the glossary are bolded the first time that they are used in the text. Many of the disagreements and long-standing controversies in population and conservation genetics result from people using the same words to mean different things. It is important to define and use words precisely.

We have asked many of our colleagues to write guest boxes that present their own work in conservation genetics. Each chapter contains a guest box that provides further consideration of the topics of that chapter. These boxes provide the reader with a broader voice in conservation genetics, as well as familiarity with recent case study examples, and some of the major contributors to the literature in conservation genetics.

The contents of this book have been influenced by lecture notes and courses in population genetics taken by the senior author from Bob Costantino and Joe Felsenstein. We also thank Fred Utter for his contagious passion to uncover and describe genetic variation in natural populations. This book began as a series of notes for a course in conservation genetics that the senior author began while on sabbatical at the University of Oregon in 1993. About one-quarter of the chapters were completed within those first six months. However, as the demands of other obligations took over, progress slowed to a near standstill. The majority of this book has been written by the authors in close collaboration over the last two years.

ACKNOWLEDGMENTS

We are grateful to the students in the Conservation Genetics course at the University of Montana over many years who have made writing this book enjoyable by their enthusiasm and comments. Earlier versions of this text were used in courses at the University of Oregon in 1993 and at the University of Minnesota in 1997; we are also grateful to those students for their encouragement and comments. We also gratefully acknowledge the University of Montana and Victoria University of Wellington for their support. Much appreciated support was also provided by Pierre Taberlet and the CNRS Laboratoire d'Ecologie Alpine. This book could not have been completed without the loving support of our wives Michel and Shannon.

We thank the many people from Blackwell Publishing for their encouragement, support, and help throughout the process of completing this book. We are grateful to the authors of the Guest Boxes who quickly replied to our many inquiries. We are indebted to Sally Aitken for her extremely thorough and helpful review of the entire book, John Powell for his excellent help with the glossary, and to Kea Allendorf for her help with the literature cited. We also thank many colleagues who have helped us by providing comments, information, unpublished data, and answers to questions: Teri Allendorf, Agostinho Antunes, Jon Ballou, Mark Beaumont, Albano Beja-Pereira, Steve Beissinger, Pierre Berthier, Giorgio Bertorelle, Matt Boyer, Brian Bowen, Ron Burton, Chris Cole, Kirsten Dale, Charlie Daugherty, Sandie Degnan, Dawson Dunning, Norm Ellstrand, Dick Frankham, Chris Funk, Oscar Gaggiotti, Neil Gemmell, John Gilliespie, Mary Jo Godt, Dave Goulson, Ed Guerrant, Bengt Hansson, Sue Haig, Kim Hastings, Phil Hedrick, Kelly Hildner, Rod Hitchmough, Denver Holt, Jeff Hutchings, Mike Ivie, Mike Johnson, Carrie Kappel, Joshua Kohn, Peter Lesica, Laura Lundquist, Shujin Luo, Lisa Meffert, Don Merton, Scott Mills, Andy Overall, Jim Patton, Rod Peakall, Robert Pitman, Kristina Ramstad, Reg Reisenbichler, Bruce Rieman, Pete Ritchie, Bruce Rittenhouse, Buce Robertson, Rob Robichaux, Nils Ryman, Mike Schwartz, Jim Seeb, Brad Shaffer, Pedro Silva, Paul Spruell, Paul Sunnucks, David Tallmon, Kathy Traylor-Holzer, Randal Voss, Hartmut Walter, Robin Waples, and Andrew Whiteley.

August 2005

Fred W. Allendorf
Gordon Luikart

List of symbols

This list includes mathematical symbols with definitions, and references to the primary chapters in which they are used. There is some duplication of usage which reflects the general usage in the literature. However, the specific meaning should be apparent from the context and chapter.

Symbol	Definition	Chapter
	Latin symbols	
A	number of alleles at a locus	3, 4, 5, 6
B	number of lethal equivalents per gamete	13
c	probability of recolonization	15
D	Jost's measure of differentiation	9
D	Nei's genetic distance	9, 16
D	coefficient of gametic disequilibrium	10, 17
D'	standardized measure of gametic disequilibrium	10
D_B	gametic disequilibrium caused by population subdivision	10
D_C	composite measure of gametic disequilibrium	10
D_{ST}	the proportion of total heterozygosity due to genetic divergence between subpopulations	20
d^2	the squared difference in number of repeat units between the two microsatellite alleles in an individual	10
E	probability of an event	A
e	probability of extinction of a subpopulation	15
f	inbreeding coefficient	6
F	pedigree inbreeding coefficient	10, 13, A
F	proportion by which heterozygosity is reduced relative to heterozygosity in a random mating population	9
F_{ij}	coefficient of coancestry	9
F_{IS}	departure from Hardy-Weinberg proportions within local demes or subpopulations	9, 13
F_{IT}	overall departure from Hardy-Weinberg proportions	9
F_k	temporal variance in allele frequencies	A
F_{SR}	proportion of the total differentiation due to differences between subpopulations within regions	9
F_{ST}	proportion of genetic variation due to differences among populations	3, 9, 12, 13

Symbol	Definition	Chapter
$F2_{ST}$	F_{ST} value using the frequency of the most common allele and all other allele frequencies binned together	9
G	generation interval	7, 20
G_i	frequency of gamete i	10
G_{ST}	F_{ST} extended for 3 or more alleles	9
h	gene diversity, computationally equivalent to H_e, especially useful for haploid marker systems	4, 7
h	degree of dominance of an allele	12
h	heterozygosity	6
H_A	alternative hypothesis	A
H_B	broad-sense heritability	11
H_e	expected proportion of heterozygotes	3, 4, A
H_o	observed heterozygosity	3, 9, A
H_o	null hypothesis	A
H_N	narrow-sense heritability	10
H_S	mean expected heterozygosity	3, 12, 15
H_T	total genetic variation	3, 9, 15
K	equilibrium size of a population	6
k	number of gametes contributed by an individual to the next generation	7
k	number of populations	16, A
L	number of loci	10
m	proportion of migrants	9, 15, 16
mk	mean kinship	22
MP	match probability	20
n	sample size	A
n	number of subpopulations	9
N	population size	6, 7, 8, A
N_C	census population size	7, 14, 19, A
N_C	proportion of individuals that reproduce in captivity	19
N_W	proportion of individuals that reproduce in the wild	19
N_e	effective population size	7, 11, 12, 14, 19, A
N_{eI}	inbreeding effective population size	7
N_{eV}	variance effective population size	7
N_f	number of females in a population	6, 7, 9
N_m	number of males in a population	6, 7, 9
NS	Wright's neighborhood size	9
P	proportion of loci that are polymorphic	3, 5
P	probability of an event	5, A
p	frequency of allele A_1 (or A)	5, 6, 8
p	proportion of patches occupied in a metapopulation	15
PE	probability of paternity exclusion; average probability of excluding (as father) a randomly sampled non-father	22
PI_{av}	average probability of identity	22
q	frequency of allele A_2 (or a)	5, 8
q	proportion of an individual's genome that comes from one parental population in an admixed population	17

Symbol	Definition	Chapter
Q	probability of an individual's genotype originating from each population	16
Q_{ST}	proportion of total genetic variation for a phenotypic trait due to genetic differentiation among populations (analogous to F_{ST})	11
r	frequency of allele A_3	5
r	correlation coefficient	A
r	rate of recombination	10
r	intrinsic population growth rate	6, 14
r_A	correlation between two traits	10
R	correlation coefficient between alleles at two loci	10
R	response to selection	11, 19
R	number of recaptured individuals	14
$R(g)$	allelic richness in a sample of g genes	5
R_{ST}	analog to F_{ST} that accounts for differences in length of microsatellite alleles	9, 16
S	selection differential	11, 19
S	effects of inbreeding on the probability of survival	13
S	selfing rate	9
s	selection coefficient (intensity of selection)	8, 9, 12
s_x	standard deviation	A
s_x^2	sample variance	A
t	number of generations	6
V_A	proportion of phenotypic variability due to additive genetic differences between individuals	11
V_D	proportion of phenotypic variability due to dominance effects (interactions between alleles)	11
V_E	proportion of phenotypic variability due to environmental differences between individuals	2, 11
V_G	proportion of phenotypic variability due to genetic differences between individuals	2, 11
V_I	proportion of phenotypic variability due to epistatic effects	11
V_k	variance of the number of offspring contributed to the next generation	7
V_m	increase in additive genetic variation per generation due to mutation	12, 14
V_P	total phenotypic variability for a trait	2
W	absolute fitness	8
w	relative fitness	8

Greek symbols

Symbol	Definition	Chapter
α	probability of a Type I error	A
β	probability of a Type II error	A
Δ	change in value from one generation to the next	6, 7, 8
δ	proportional reduction in fitness due to selfing	13
Φ_{ST}	analogous to F_{ST} but incorporates genealogical relationships among alleles	16
λ	the factor by which the population size increases each time unit	14
π	probability of an event	A
σ_x^2	population variance	A
μ	population mean	A
μ	neutral mutation rate	12
χ^2	chi-square statistic	5, 22, A

Symbol	Definition	Chapter
	Other symbols	
x	number of observations or times that an event occurs	A
$\bar{x}$	sample mean	A
$\bar{x}^2$	sample variance (2nd moment)	A
$\bar{x}^3$	skewness of sample distribution (3rd moment)	A
$\hat{x}$	estimate of parameter x	A
x'	value of parameter x in the next generation	6, 8, 9, 10
x^*	equilibrium value of parameter x	8, 9, 12

PART I

INTRODUCTION

CHAPTER 1

Introduction

One-horned rhinoceros, Section 5.3

We are at a critical juncture for the conservation and study of biological diversity: such an opportunity will never occur again. Understanding and maintaining that diversity is the key to humanity's continued prosperous and stable existence on Earth.

US National Science Board Committee on Global Biodiversity (1989)

The extinction of species, each one a pilgrim of four billion years of evolution, is an irreversible loss. The ending of the lines of so many creatures with whom we have traveled this far is an occasion of profound sorrow and grief. Death can be accepted and to some degree transformed. But the loss of lineages and all their future young is not something to accept. It must be rigorously and intelligently resisted.

Gary Snyder (1990)

Chapter Contents

Conservation and the Genetics of Populations, Second Edition. Fred W. Allendorf, Gordon Luikart, and Sally N. Aitken.
© 2013 Fred W. Allendorf, Gordon Luikart and Sally N. Aitken. Published 2013 by Blackwell Publishing Ltd.

We are living in a time of unprecedented extinctions (Myers and Knoll 2001, Stuart *et al.* 2010, Barnosky *et al.* 2011). Current extinction rates have been estimated to be 50–500 times background rates and are increasing; an estimated 3000–30,000 species go extinct annually (Woodruff 2001). Projected extinction rates vary from 5–25% of the world's species by 2015 or 2020. Approximately 23% of mammals, 12% of birds, 42% of turtles and tortoises, 32% of amphibians, 34% of fish, and 9–34% of major plant taxa are threatened with extinction over the next few decades (IUCN 2001, Baillie *et al.* 2004). Over 50% of animal species are considered to be critically endangered, endangered, or vulnerable to extinction (Baillie *et al.* 2004). A recent assessment of the status of the world's vertebrates, based on the International Union for the Conservation of Nature (IUCN) Red List, concluded that approximately 20% are classified as Threatened. This figure is increasing primarily because of agricultural expansion, logging, overexploitation, and invasive introduced species (Hoffmann *et al.* 2010).

The true picture is much worse than this because the conservation status of most of the world's species remains poorly known. Recent estimates indicate that less than 30% of the world's arthropod species have been described (Hamilton *et al.* 2010). Less than 5% of the world's described animal species have been evaluated for the IUCN Red List. Few invertebrate groups have been evaluated, and the evaluations that have been done have tended to focus on molluscs and crustaceans. Among the insects, only the swallowtail butterflies, dragonflies, and damselflies have received much attention.

Conservation biology poses perhaps the most difficult and important questions ever faced by science (Pimm *et al.* 2001). The problems are difficult because they are so complex and cannot be approached by the reductionist methods that have worked so well in other areas of science. Moreover, solutions to these problems require a major readjustment of our social and political systems. There are no more important scientific challenges because these problems threaten the continued existence of our species and the future of the biosphere itself.

1.1 GENETICS AND CIVILIZATION

Genetics has a long history of application to human concerns. The domestication of animals and cultivation of plants is thought to have been perhaps the key step in the development of civilization (Diamond 1997). Early peoples directed genetic change in domestic and agricultural species to suit their needs. It has been estimated that the dog was domesticated over 15,000 years ago, followed by goats and sheep around 10,000 years ago (Darlington 1969, Zeder and Hess 2000). Wheat and barley were the first crops to be domesticated in the Old World approximately 10,000 years ago; beans, squash, and maize were domesticated in the New World at about the same time (Darlington 1969, Kingsbury 2009).

The initial genetic changes brought about by cultivation and domestication were not due to intentional selection but apparently were inadvertent and inherent in cultivation itself. Genetic change under domestication was later accelerated by thousands of years of purposeful selection as animals and crops were selected to be more productive or to be used for new purposes. This process became formalized in the discipline of agricultural genetics after the rediscovery of Mendel's principles at the beginning of the 20th century.

The 'success' of these efforts can be seen everywhere. Humans have transformed much of the landscape of our planet into croplands and pasture to support the over 7 billion humans alive today. It has been estimated that 35% of the Earth's ice-free land surface is now occupied by crops and pasture (Foley *et al.* 2007), and that 24% of the primary terrestrial productivity is used by humans (Haberl *et al.* 2007). Recently, however, we have begun to understand the cost at which this success has been achieved. The replacement of wilderness by human-exploited environments is causing the rapidly accelerating loss of species and ecosystems throughout the world. The continued growth of the human population and their direct and indirect effects on environments imperils a large proportion of the wild species that now remain.

Aldo Leopold inspired a generation of biologists to recognize that the actions of humans are embedded into an ecological network that should not be ignored (Meine 1998). The organized actions of humans are controlled by sociopolitical systems that operate into the future on a timescale of a few years at most. All too often our systems of conservation are based on the economic interests of humans in the immediate future. We tend to disregard, and often mistreat, elements that lack economic value but that are essential to the stability of the ecosystems upon which our lives and the future of our children depend.

In 1974, Otto Frankel published a landmark paper entitled 'Genetic conservation: our evolutionary responsibility', which set out conservation priorities:

First, . . . we should get to know much more about the structure and dynamics of natural populations and communities. . . . Second, even now the geneticist can play a part in injecting genetic considerations into the planning of reserves of any kind. . . . Finally, reinforcing the grounds for nature conservation with an evolutionary perspective may help to give conservation a permanence which a utilitarian, and even an ecological grounding, fail to provide in men's minds.

Frankel, an agricultural plant geneticist, came to the same conclusions as Leopold, a wildlife biologist, by a very different path. In Frankel's view, we cannot anticipate the future world in which humans will live in a century or two. Therefore, it is our responsibility to "keep evolutionary options open". It is time to apply our understanding of genetics to conserving the natural ecosystems that are threatened by human civilization.

1.2 WHAT SHOULD WE CONSERVE?

Conservation can be viewed as an attempt to protect the genetic diversity that has been produced by evolution over the previous 3.5 billion years on our planet (Eisner *et al.* 1995). Genetic diversity is one of three forms of biodiversity recognized by the IUCN as deserving conservation, along with species and ecosystem diversity (www.cbd.int, McNeely *et al.* 1990). Unfortunately, genetics has been generally ignored by the member countries in their National Biodiversity Strategy and Action Plans developed to implement the Convention on Biological Diversity (CBD) (Laikre *et al.* 2010a).

We can consider the implications of the relationship between genetic diversity and conservation at many levels: genes, individuals, populations, varieties, subspecies, species, genera, and so on. Genetic diversity provides a retrospective view of evolutionary lineages of taxa (phylogenetics), a snapshot of the current genetic structure within and among populations (population and ecological genetics), and a glimpse ahead to the future evolutionary potential of populations and species (evolutionary biology).

1.2.1 Phylogenetic diversity

The amount of genetic divergence based upon **phylogenetic** relationships is often considered when setting conservation priorities for different species (Mace *et al.* 2003, Avise 2008). For example, the United States Fish and Wildlife Service (USFWS) assigns priority for listing under the Endangered Species Act (ESA) of the United States on the basis of "taxonomic distinctiveness" (USFWS 1983). Species of a **monotypic** genus receive the highest priority. The tuatara raises several important issues about assigning conservation value and allocating our conservation efforts based upon taxonomic distinctiveness (Example 1.1).

Faith (2008) recommends integrating evolutionary processes into conservation decision-making by considering phylogenetic diversity. Faith provides an approach that goes beyond earlier recommendations that species that are taxonomically distinct deserve greater conservation priority. He argues that the phylogenetic diversity approach provides two ways to consider maximizing biodiversity. First, considering phylogeny as a product of evolutionary process enables the interpretation of diversity patterns to maximize biodiversity for future evolutionary change. Second, phylogenetic diversity also provides a way to better infer biodiversity patterns for poorly described taxa when used in conjunction with information about geographic distribution.

Vane-Wright *et al.* (1991) presented a method for assigning conservation value on the basis of phylogenetic relationships. This system is based upon the information content of the topology of a particular phylogenetic hierarchy. Each extant species is assigned an index of taxonomic distinctness that is inversely proportional to the number branching points to other extant lineages. May (1990) has estimated that the tuatara (Example 1.1) represents between 0.3 and 7% of the taxonomic distinctness, or perhaps we could say genetic information, among reptiles. This is equivalent to saying that each of the two tuatara species is equivalent to approximately 10 to 200 of the 'average' reptile species. Crozier and Kusmierski (1994) developed an approach to setting conservation priorities based upon phylogenetic relationships and genetic divergence among taxa. Faith (2002) has presented a method for quantifying biodiversity for the purpose of identifying conservation priorities that considers phylogenetic diversity both between and within species.

Example 1.1   The tuatara: a living fossil

The tuatara is a lizard-like reptile that is the remnant of a taxonomic group that flourished over 200 million years ago during the Triassic Period (Figure 1.1). Tuatara are now confined to some 30 small islands off the coast of New Zealand (Daugherty *et al.* 1990). Three species of tuatara were recognized in the 19th century. One of these species is now extinct. A second species, *Sphenodon guntheri*, was ignored by legislation designed to protect the tuatara which 'lumped' all extant tuatara into a single species, *S. punctatus*.

Daugherty *et al.* (1990) reported allozyme and morphological differences from 24 of the 30 islands on which tuatara are thought to remain. These studies support the status of *S. guntheri* as a distinct species and indicate that fewer than 300 individuals of this species remain on a single island, North Brother Island in Cook Strait. Another population of *S. guntheri* became extinct earlier in this century. Daugherty *et al.* (1990) argued that not all tuatara populations are of equal conservation value. As the last remaining population of a distinct species, the tuatara on North Brother Island represent a greater proportion of the genetic diversity remaining in the genus *Sphenodon* and deserve special recognition and protection. However, recent results with other molecular techniques indicate that the tuatara on North Brother Island probably do not warrant recognition as a distinct species (Hay *et al.* 2010, Example 16.3).

On a larger taxonomic scale, how should we value the tuatara relative to other species of reptiles? Tuatara species are the last remaining representatives of the Sphenodontida, one of four extant orders of reptiles (tuatara, snakes and lizards, alligators and crocodiles, and tortoises and turtles). In contrast, there are approximately 5000 species in the Squamata, the speciose order that contains lizards and snakes.

One position is that conservation priorities should regard all species as equally valuable. This position would equate the two tuatara species with any two species of reptiles. Another position is that we should take phylogenetic diversity into account in assigning conservation priorities. The extreme phylogenetic position is that we should assign equal conservation value to each major sister group in a phylogeny. According to this position, tuatara would be weighed equally with the over 5000 species of other snakes and lizards. Some intermediate between these two positions seems most reasonable.

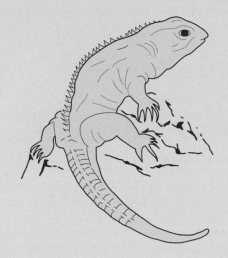

Figure 1.1   Adult male tuatara.

There is great appeal to placing conservation emphasis on distinct evolutionary lineages with few living relatives. Living fossils, such as the tuatara, ginkgo (Royer *et al.* 2003), or the coelacanth (Thompson 1991), represent important pieces in the jigsaw puzzle of evolution. Such species are relics that are representatives of taxonomic groups that once flourished. Study of the primitive morphology, physiology, and behavior of living fossils can be extremely important in understanding evolution. For example, tuatara morphology has hardly changed in nearly 150 million years.

Among the many primitive features of the tuatara is a rudimentary third, or pineal, eye on the top of the head.

Tuatara represent an important ancestral outgroup for understanding vertebrate evolution. For example, a recent study has used genomic information from tuatara to reconstruct and understand the evolution of 18 human retroposon elements (Lowe *et al.* 2010). Most of these elements were quickly inactivated early in the mammalian lineage, and thus study of other mammals provides little insight into these elements in humans. These authors conclude that species with historically low population sizes (such as tuatara) are more likely to maintain ancient mobile elements for long periods of time with little change. Thus, these species are indispensable in understanding the evolutionary origin of functional elements in the human genome.

In contrast, others have argued that our conservation strategies and priorities should be based primarily upon conserving the evolutionary process rather than preserving only those pieces of the evolutionary puzzle that are of interest to humans (Erwin 1991). Those species that will be valued most highly under the schemes that weigh phylogenetic distinctness are those that may be considered evolutionary failures. Evolution occurs by changes within a single evolutionary lineage (**anagenesis**) and the branching of a single evolutionary lineage into multiple lineages (**cladogenesis**). Conservation of primitive, nonradiating taxa is not likely to be beneficial to the protection of the evolutionary process and the environmental systems that are likely to generate future evolutionary diversity (Erwin 1991).

Figure 1.2 illustrates the phylogenetic relations among seven hypothetical species (from Erwin 1991) Species A and B are phylogenetically distinct taxa that are endemic to small geographic areas (e.g., tuataras in New Zealand). Such lineages carry information about past evolutionary events, but they are relatively unlikely to be sources of future evolution. In contrast, the stem resulting in species C, D, E, and F is relatively likely to be a source of future anagenesis and cladogenesis. In addition, species such as C, D, E, and F may be widespread, and therefore are not likely to be the object of conservation efforts.

The problem is more complex than just identifying species with high conservation value; we must take a broader view and consider the habitats and environments where our conservation efforts could be concentrated. Conservation emphasis on phylogenetically distinct species will lead to protection of environments that are not likely to contribute to future evolution (e.g., small islands along the coast of New Zealand). In contrast, geographic areas that are the center of evolutionary activity for diverse taxonomic groups could be identified and targeted for long-term protection.

Recovery from our current extinction crisis should be a central concern of conservation (Myers *et al.* 2000). It is important to maintain the potential for the generation of future biodiversity. We should identify and protect contemporary hotspots of evolutionary radiation and the functional taxonomic group from which tomorrow's biodiversity is likely to originate. In addition, we should protect those phylogenetically distinct species that are of special value for our understanding of biological diversity and the evolutionary process. These species are also potentially valuable for future evolution of biodiversity because of their combination of unusual phenotypic characteristics that may give rise to a future evolutionary radiation. Isaac *et al.* (2007) have proposed using an index that combines both evolutionary distinctiveness and IUCN Red List categories to set conservation priorities.

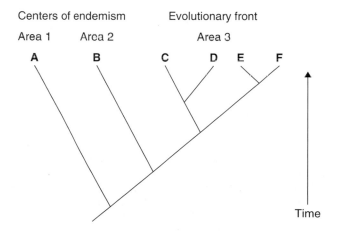

Figure 1.2 Hypothetical phylogeny of seven species. Redrawn from Erwin (1991).

1.2.2 Populations, species, or ecosystems?

A related, and sometimes impassioned, dichotomy between protecting centers of biodiversity or phylogenetically distinct species is the dichotomy between emphasis on species conservation or on the conservation of habitat or ecosystems (Soulé and Mills 1992, Armsworth *et al.* 2007). Conservation efforts to date have emphasized the concerns of individual species. For example, in the US the Endangered Species Act (ESA) has been the legal engine behind much of the conservation efforts. However, it is frustrating to see enormous resources being spent on a few high profile species when little is spent on less charismatic taxa or in preventing environmental deterioration that would benefit many species. It is clear that a more comprehensive and proactive conservation strategy emphasizing protection of habitat and ecosystems, rather than species, is needed. Some have advocated a shift from saving things, the products of evolution (species, communities, or ecosystems), to saving the underlying processes of evolution "that underlie a dynamic biodiversity at all levels" (Templeton *et al.* 2001).

It has been argued that more concern about extinction should be focused on the extinction of genetically distinct populations, and less on the extinction of species (Hughes *et al.* 1997, Hobbs and Mooney 1998). The conservation of many distinct populations is required to maximize evolutionary potential of a species and to minimize the long-term extinction risks of a species. In addition, a population focus would also help to prevent costly and desperate 'last-minute' conservation programs that occur when only one or two small populations of a species remain. The first attempt to estimate the rate of population extinction worldwide was published by Hughes *et al.* (1997). They estimated that tens of millions of local populations that are genetically distinct go extinct each year. Approximately 16 million of the world's three billion genetically distinct natural populations go extinct each year in tropical forests alone.

Luck *et al.* (2003) have considered the effect of population diversity on the functioning of ecosystems and so-called **ecosystem services**. They argue that the relationship between biodiversity and human wellbeing is primarily a function of the diversity of populations within species. They have also proposed a new approach for describing population diversity that considers the value of groups of individuals to the services that they provide.

Ceballos and Ehrlich (2002) have compared the historical and current distributions of 173 declining mammal species from throughout the world. Their data included all of the terrestrial mammals of Australia and subsets of terrestrial mammals from other continents. Nearly 75% of all species they included have lost over 50% of their total geographic range. Approximately 22% of all Australian species are declining, and they estimated that over 10% of all Australian terrestrial mammal populations have been extirpated since the 19th century. These estimates, however, all assume that population extirpation is proportional to loss of range area rather than defining populations using genetic criteria.

The amount of genetic variation within a population may also play an important ecosystem role in the relationships among species in some functional groups and ecosystems. Clark (2010) has found that intraspecific genetic variation within forest trees in the southeastern US allows higher species diversity. Recent results in **community genetics** suggest that individual alleles within some species can affect community diversity and composition (Crutsinger *et al.* 2006). For example, alleles at tannin loci in cottonwood trees affect palatability and decay rate of leaves, which in turn influences abundance of soil microbes, fungi, and arboreal insects and birds (Whitham *et al.* 2008). Genetic variation in the bark characteristics of a **foundation species** (Tasmanian blue gum tree) has been found to affect the abundance and distribution of insects, birds, and marsupials. Loss or restoration of such alleles to populations could thus influence community diversity and ecosystem function (Whitham *et al.* 2008).

Conservation requires a balanced approach that is based upon habitat protection which also takes into account the natural history and viability of individual species. Consider Chinook salmon in the Snake River basin of Idaho, which are listed under the ESA. These fish spend their first two years of life in small mountain rivers and streams far from the ocean. They then migrate over 1500 km downstream through the Snake and Columbia Rivers and enter the Pacific Ocean. There they spend two or more years ranging as far north as the coast of Alaska before they return to spawn in their natal freshwater streams. There is no single ecosystem that encompasses these fish, other than the biosphere itself. Protection of this species requires a combination of habitat measures and management actions that take into account the complex life-history of these fish.

1.3 HOW SHOULD WE CONSERVE BIODIVERSITY?

Extinction is a **demographic** process: the failure of one generation to replace itself with a subsequent generation. Demography is of primary importance in managing populations for conservation (Lacy 1988, Lande 1988). Populations are subject to uncontrollable **stochastic** demographic factors as they become smaller. It is possible to estimate the expected mean and variance of a population's time to extinction if one has an understanding of a population's demography and environment (Goodman 1987, Belovsky 1987, Lande 1988).

There are two main types of threats causing extinction: **deterministic** and **stochastic** threats (Caughley 1994). Deterministic threats are habitat destruction, pollution, overexploitation, species translocation, and global climate change. Stochastic threats are random changes in genetic, demographic or environmental factors. Genetic stochasticity is random genetic change (drift) and increased inbreeding (Shaffer 1981). Genetic stochasticity leads to loss of genetic variation (including beneficial alleles) and increase in frequency of harmful alleles. An example of demographic stochasticity is random variation in sex ratios, for example producing only male offspring. Environmental stochasticity is simply random environmental variation, such as the occasional occurrence of several harsh winters in a row. In a sense, the effects of small population size are both deterministic and stochastic. We know that genetic drift in small populations is likely to have harmful effects, and the smaller the population,

the greater the probability of such effects. However, the effects of small population size are stochastic because we cannot predict what traits will be affected.

Under some conditions, extinction is likely to be influenced by genetic factors. Small populations are also subject to genetic stochasticity, which can lead to loss of genetic variation through genetic drift. The 'inbreeding effect of small populations' (see Box 1.1) is likely to lead to a reduction in the fecundity and viability of individuals in small populations. For example, Frankel and Soulé (1981, p. 68) suggested that a 10% decrease in genetic variation due to the inbreeding effect of small populations is likely to cause a 10–25% reduction in reproductive performance of a population. This in turn is likely to cause a further reduction in population size, and thereby reduce a population's ability to persist (Gilpin and Soulé 1986). This has come to be known as the **extinction vortex** (see Figure 14.2).

Some have argued that genetic concerns can be ignored when projecting the viability of small populations because they are in much greater danger of extinction by purely demographic stochastic effects (Lande 1988, Pimm *et al.* 1988, Caughley 1994, Frankham 2003, Sarre and Georges 2009). It has been argued that such small populations are not likely to persist long enough to be affected by inbreeding depression, and that efforts to reduce demographic stochasticity will also reduce the loss of genetic variation. The disagreement over whether or not genetics should be considered in demographic predictions of population persistence has been unfortunate and

---

Box 1.1   What is an 'inbred' population?

The term 'inbred population' is used in the literature to mean two very different things (Chapter 13, Templeton and Read 1994). In the conservation literature, 'inbred population' is often used to refer to a small population in which mating between related individuals occurs because after a few generations, all individuals in a small population will be related. Thus, matings between related individuals (inbreeding) will occur in small populations even if they are random mating (panmictic). This has been called the 'inbreeding effect of small populations' (see Chapter 6).

Formally in population genetics, an 'inbred' population is one in which there is a tendency for related

individuals to mate with one another. For example, many extremely large populations of pine trees are inbred because of their spatial structure (see Section 9.2). Nearby trees tend to be related to one another because of limited seed dispersal, and nearby trees also tend to fertilize each other because of wind pollination. Therefore, a population of pine trees with millions of individuals may still be 'inbred'.

Population genetics is a complex field. The incorrect, ambiguous, or careless use of words can sometimes result in unnecessary confusion. We have made an effort throughout this book to use words precisely and carefully.

misleading. Extinction is a demographic process that is likely to be influenced by genetic effects under some circumstances. The important issue is to determine under what conditions genetic concerns are likely to influence population persistence (Nunney and Campbell 1993).

Perhaps most importantly, we need to recognize when management recommendations based upon demographic and genetic considerations may be in conflict with each other. For example, small populations face a variety of genetic and demographic effects that threaten their existence. Management plans aim to increase the population size as soon as possible to avoid the problems associated with small populations. However, efforts to maximize growth rate may actually increase the rate of loss of genetic variation by relying on the exceptional reproductive success of a few individuals (see Example 19.1, Caughley 1994).

Ryman and Laikre (1991) considered what they termed **supportive breeding** in which a portion of wild parents are brought into captivity for reproduction and their offspring are released back into the natural habitat where they mix with wild conspecifics. Programs similar to this are carried out in a number of species to increase population size and thereby temper stochastic demographic effects (e.g., Blanchet *et al.* 2008). Under some circumstances, supportive breeding may reduce effective population size and cause a drastic reduction in genetic heterozygosity (Ryman 1994).

Genetic information also can provide valuable insight into the demographic structure and history of a population (Escudero *et al.* 2003). Estimation of the number of unique genotypes can be used to estimate total population size in populations that are difficult to census (Luikart *et al.* 2010). Many demographic models assume a single random mating population. Examination of the distribution of genetic variation over the distribution of a species can identify what geographic units can be considered separate demographic units. Consider the simple example of a population of trout found within a single small lake for which it would seem appropriate to consider these fish a single demographic unit. However, under some circumstances the trout in a single small lake can actually represent two or more separate reproductive (and demographic) groups with little or no exchange between them (e.g., Ryman *et al.* 1979).

The issue of population persistence is a multidisciplinary problem that involves many aspects of the biology of the populations involved (Lacy 2000b). A similar statement can be made about most of the issues we are faced with in conservation biology. We can only resolve these problems by an integrated approach that incorporates demography and genetics, as well as other biological considerations that are likely to be critical for a particular problem (e.g., behavior, physiology, interspecific interactions, as well as habitat loss and environmental change).

1.4 APPLICATIONS OF GENETICS TO CONSERVATION

Darwin (1896) was the first to consider the importance of genetics in the persistence of natural populations. He expressed concern that deer in British nature parks may be subject to loss of vigor because of their small population size and isolation. Voipio (1950) presented the first comprehensive consideration of the application of population genetics to the management of natural populations. He was primarily concerned with the effects of genetic drift in game populations that were reduced in size by trapping or hunting and fragmented by habitat loss.

The modern concern for genetics in conservation began around 1970 when Sir Otto Frankel (Frankel 1970) began to raise the alarm about the loss of primitive crop varieties and their replacement by genetically uniform cultivars (see Guest Box 1). It is not surprising that these initial considerations of conservation genetics dealt with species that were used directly as resources by humans. Conserving the genetic resources of wild relatives of agricultural species remains an important area of conservation genetics (Maxted 2003, Hanotte *et al.* 2010). A surprisingly modern view of the importance and role of genetics in conservation was written by J.C. Greig in 1979. This interesting paper emphasized the importance of maintaining the integrity of local population units.

The application of genetics to conservation in a more general context did not blossom until around 1980, when three books established the foundation for applying the principles of genetics to conservation of biodiversity (Soulé and Wilcox 1980, Frankel and Soulé 1981, Schonewald-Cox *et al.* 1983). Today conservation genetics is a well-established discipline, with its own journals (*Conservation Genetics* and *Conservation Genetics Resources*) and two textbooks, including this one and Frankham *et al.* (2010).

Maintenance of biodiversity primarily depends upon the protection of the environment and maintenance of habitat. Nevertheless, genetics has played an important and diverse role in conservation biology in the last few years. Nearly 10% of the articles published in the journal *Conservation Biology* since its inception in 1988 have "genetic" or "genetics" in their title. Probably at least as many other articles deal with largely genetic concerns but do not have the term in their title. Thus, some 15% of the articles published in *Conservation Biology* have genetics as a major focus.

The subject matter of papers published on conservation genetics is extremely broad. However, most of articles dealing with conservation and genetics fit into one of the five broad categories below:

1 Management and reintroduction of captive populations, and the restoration of biological communities.
2 Description and identification of individuals, genetic population structure, kin relationships, and taxonomic relationships.
3 Detection and prediction of the effects of habitat loss, fragmentation, and isolation.
4 Detection and prediction of the effects of hybridization and introgression.
5 Understanding the relationships between adaptation or fitness and genetic characters of individuals or populations.

These topics are listed in order of increasing complexity and decreasing uniformity of agreement among conservation geneticists. Although the appropriateness of captive breeding in conservation has been controversial (Snyder *et al.* 1996, Adamski and Witkowski 2007, Fraser 2008), procedures for genetic management of captive populations are well developed with relatively little controversy. However, the relationship between specific genetic types and fitness or adaptation has been a particularly vexing issue in evolutionary and conservation genetics. Nevertheless, studies have shown that natural selection can bring about rapid genetic changes in populations that may have important implications for conservation (Stockwell *et al.* 2003).

Invasive species are recognized as one of the top two threats to global biodiversity (Chapter 20). Studies of genetic diversity and the potential for rapid evolution of invasive species may provide useful insights into what causes species to become invasive (Lee and Gelembiuk 2008). More information about the genetics and evolution of invasive species or native species in invaded communities, as well as their interactions, may lead to predictions of the relative susceptibility of ecosystems to invasion, identification of key alien species, and predictions of the subsequent effects of removal.

Recent advances in molecular genetics, including sequencing of the entire genomes of many species, have revolutionized applications of genetics (e.g., medicine, forestry, and agriculture). For example, it has been suggested that genetic engineering should be considered as a conservation genetics technique (Adams *et al.* 2002). Many native trees in the northern temperate zone have been devastated by introduced diseases for which little or no genetic resistance exists (e.g., European and North American elms, and the North American chestnut. Adams *et al.* (2002) suggested that transfer of resistance genes by genetic modification is perhaps the only available method for preventing the loss of important tree species. Transgenic trees have been developed for both American elm and American chestnut, and are now being tested for stable resistance to Dutch elm disease and chestnut blight (Newhouse *et al.* 2007). The use of genetic engineering to improve crop plants has been very controversial. There no doubt will continue to be a lively debate in the near future about the use of these procedures to prevent the extinction of natural populations.

The loss of key tree species is likely to affect many other species as well. For example, whitebark pine is currently one of the two most important food resources for grizzly bears in the Yellowstone National Park ecosystem (Mattson and Merrill 2002). However, virtually all of the whitebark pine in this region is projected to be **extirpated** because of an exotic pathogen (Mattson *et al.* 2001), and with predicted geographic shifts in the climatic niche-based habitat of this species in the next century (Warwell *et al.* 2007).

There are a variety of efforts around the world to store samples of DNA libraries, frozen cells, gametes, and seeds that could yield DNA (Frozen Ark Project, Millennium Seed Bank Project, Svalbard Global Seed Vault, Ryder *et al.* 2000). The hope is that these resources would at least provide complete genome sequences of species that might become extinct in the not-distant future. These sequences could be invaluable for reconstructing evolutionary relationships, understanding how specific genes arose to encode proteins that perform specialized functions, and how the regulation of genes has evolved. In some cases (e.g., seed banks), these resources could be used to recover apparent extinct species.

1.5 THE FUTURE

Genetics is likely to play even a greater role in conservation biology in the future (Primmer 2009, Ouborg *et al.* 2010, Avise 2010, Frankham 2010). We will soon have complete genome sequences from thousands of species, as well as many individuals within species (Haussler *et al.* 2009). This coming explosion of information will transform our understanding of the amount, distribution, and functional significance of genetic variation in natural populations (Amato *et al.* 2009, Allendorf *et al.* 2010). Now is a crucial time to explore the potential implications of this information revolution for conservation genetics, as well as to recognize limitations in applying genomic tools to conservation issues. The ability to examine hundreds or thousands of genetic markers with relative ease will make it possible to answer many important questions in conservation that have been intractable until now (Figure 1.3).

As in other areas of genetics, model organisms have played an important research role in conservation genetics (Frankham 1999). Many important theoretical issues in conservation biology cannot be answered by research on threatened species (e.g., how much gene flow is required to prevent the inbreeding effects of small population size?). Such empirical questions are often best resolved in species that can be raised in captivity in large numbers with a rapid generation interval (e.g., the fruit fly *Drosophila*, the guppy, deer mouse, and the fruit-fly equivalent in plants, *Arabidopsis*). The genome sequencing of plant and animal model and agricultural species have been crucial for transferring genomic tools to wild populations of other

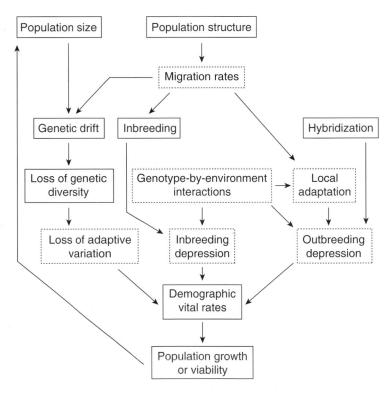

Figure 1.3 Schematic diagram of interacting factors in the conservation of natural populations. Traditional conservation genetics, using neutral markers, provides direct estimates of some of these factors (outlined by solid lines). Conservation genomics can address a wider range of factors (outlined by dashed lines). It also promises more precise estimates of neutral processes and understanding of the specific genetic basis of all of these factors. For instance, traditional conservation genetics can estimate overall migration rates or inbreeding coefficients, while genomic tools can assess gene flow rates specific to adaptive loci or founder-specific inbreeding coefficients. From Allendorf *et al.* (2010).

species. Such laboratory investigations can also provide excellent training opportunities for students. We have tried to provide a balance of examples from model and threatened species. Nevertheless, where possible we have chosen examples from threatened species, even though many of the principles were first demonstrated with model species.

This is an exciting time to be interested in the genetics of natural populations. Molecular techniques make it possible to detect genetic variation in any species of interest, not just those that can be bred and studied in the laboratory (Wayne and Morin 2004, Allendorf *et al.* 2010). However, interpretation of this explosion of data requires a solid understanding of population genetics theory (see Preface, Figure i.1). This book is meant to provide a thorough examination of our understanding of the genetic variation in natural populations. Based upon that foundation, we will consider the application of this understanding to the many problems faced by conservation biologists, with the hope that our more informed actions can make a difference.

Guest Box 1   The role of genetics in conservation
L. Scott Mills and Michael E. Soulé

Until recently, most conservationists ignored genetics and most geneticists ignored the biodiversity catastrophe. It was agricultural geneticists, led by Sir Otto Frankel (1974), who began to sound an alarm about the disappearance of thousands of land races – crop varieties coaxed over thousands of years to adapt to local soils, climates, and pests. Frankel challenged geneticists to help promote an "evolutionary ethic" focused on maintaining evolutionary potential and food security in a rapidly changing world.

Frankel's pioneering thought inspired the first international conference on conservation biology in 1978. It brought together ecologists and evolutionary geneticists to consider how their fields could help slow the extinction crisis. Some of the chapters in the proceedings (Soulé and Wilcox 1980) foreshadowed population viability analysis and the interactions of demography and genetics in small populations (the extinction vortex). Several subsequent books (Frankel and Soulé 1981, Schonewald-Cox *et al.* 1983, Soulé 1987a) consolidated the role of genetic thinking in nature conservation.

Thus, topics such as inbreeding depression and loss of heterozygosity were prominent since the beginning of the modern discipline of conservation biology, but like inbred relatives, they were conveniently forgotten at the end of the 20th century. Why? Fashion. Following the human proclivity to champion simple, singular solutions to complex problems, a series of papers on population viability in the late 1980s and early 1990s argued that – compared with demographic and environmental accidents – inbreeding and loss of genetic variation were trivial contributors to extinction risk in small populations. Eventually, however, this swing in scientific fashion was arrested by the friction of real world complexity.

Thanks to the work of F_1 and F_2 conservation geneticists, it is now clear that inbreeding depression can increase population vulnerability by interacting with random environmental variation, not to mention deterministic factors including habitat degradation, new diseases, and invasive exotics. Like virtually all dualisms, the genetic versus non-genetic battles abated in the face of the overwhelming evidence for the relevance of both.

Genetic approaches have become prominent in other areas of conservation biology as well. These include: (1) the use of genetic markers in forensic investigations concerned with wildlife and endangered species; (2) genetic analyses of hybridization and invasive species; (3) noninvasive genetic estimation of population size and connectivity; and (4) studies of taxonomic affiliation and distance. And, of course, the overwhelming evidence of rapid climate change has renewed interest in the genetic basis for adaptation as presaged by Frankel 30 years ago. Nowadays, genetics is an equal partner with ecology, systematics, physiology, epidemiology, and behavior in conservation, and both conservation and genetics are enriched by this pluralism.

CHAPTER 2

Phenotypic variation in natural populations

Western terrestrial garter snake, Section 2.3

Few persons consider how largely and universally all animals are varying. We know however, that in every generation, if we would examine all the individuals of any common species, we should find considerable differences, not only in size and color, but in the form and proportions of all the parts and organs of the body.

Alfred Russel Wallace (1892, p. 57)

It would be of great interest to determine the critical factors controlling the variability of each species, and to know why some species are so much more variable than others.

David Lack (1947, p. 94)

Chapter Contents

Conservation and the Genetics of Populations, Second Edition. Fred W. Allendorf, Gordon Luikart, and Sally N. Aitken.
© 2013 Fred W. Allendorf, Gordon Luikart and Sally N. Aitken. Published 2013 by Blackwell Publishing Ltd.

Genetics has been defined as the study of differences among individuals (Sturtevant and Beadle 1939). If all of the individuals within a species were identical, we could still study and describe their morphology, physiology, behavior, and so on. However, geneticists would be out of work. Genetics and the study of inheritance are based upon comparing the similarity of parents and their progeny relative to the similarity among unrelated individuals within populations or species.

Variability among individuals is also essential for adaptive evolutionary change. **Natural selection** cannot operate unless there are **phenotypic** differences between individuals. Transformation of individual variation within populations into differences between populations, or species, by the process of natural selection is the basis for adaptive evolutionary change described by Charles Darwin almost 150 years ago (Darwin 1859). Nevertheless, there is surprisingly little in Darwin's extensive writings about the extent and pattern of differences between individuals in natural populations. Instead, Darwin relied heavily on examples from animal breeding and the success of artificial selection to argue for the potential of evolutionary change by natural selection (Ghiselin 1969).

Alfred Russel Wallace (the co-founder of the principle of natural selection) was perhaps the first biologist to emphasize the extent and importance of variability within natural populations (Figure 2.1). Wallace felt that "Mr. Darwin himself did not fully recognise the enormous amount of variability that actually exists" (Wallace 1923, p. 82). Wallace concluded that for morphological measurements, individuals commonly varied by up to 25% of the mean value; that is, from 5 to 10% of the individuals within a population differ from the population mean by 10 to 25% (Wallace 1923, p. 81). This was in opposition to the commonly held view of naturalists in the 19th century that individual variation was comparatively rare in nature.

Mendel's classic work was an attempt to understand the similarity of parents and progeny for traits that varied in natural populations. The original motivation for Mendel's work was to test a theory of evolution developed by his botany professor (Unger 1852) who proposed that "variants arise in natural populations which in turn give rise to varieties and subspecies until finally the most distinct of them reach species level" (Mayr 1982, p. 711). The importance of this inspiration can be seen in the following quote from Mendel's original paper (1865): "this appears, however, to be the only right way by which we can finally reach the solu-

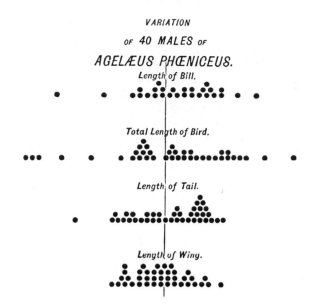

Figure 2.1 Diagram from Alfred Russel Wallace of variation in body dimensions of 40 red-winged blackbirds in the United States. From Wallace (1923, p. 64).

tion of a question the importance of which cannot be overestimated in connection with the history of the evolution of organic forms".

Population genetics was limited to the study of species that could be studied experimentally in the laboratory for most of the 20th century. Experimental population genetics was dominated by studies that dealt with *Drosophila* fruitflies until the mid-1960s because of the difficulty in determining the genetic basis of phenotypic differences between individuals (Lewontin 1974). *Drosophila* that differed phenotypically in natural populations could be brought into the laboratory for detailed analysis of the genetic differences underlying phenotypic differences. Similar studies were not possible for species with long generation times that could not be raised in captivity in large numbers. However, population genetics underwent an upheaval in the 1960s when biochemical techniques allowed genetic variation to be studied directly in natural populations of any organism. Lewontin (1974) provided an excellent and highly readable account of the state of population genetics at the beginning of the molecular revolution.

Molecular techniques today make it possible to study differences in the DNA sequence of any species. The complete genomes of many species have been sequenced and plans are under way to obtain the complete genome sequences of 10,000 vertebrate species (Haussler *et al.* 2009). However, even this level of detail will not provide sufficient information to understand the significance of genetic variation in natural populations (Allendorf *et al.* 2010). Adaptive evolutionary change within populations consists of gradual changes in morphology, life history, physiology, and behavior. Such traits are usually affected by a combination of many genes and the environment, so that it is difficult to identify single genes that contribute to the genetic differences between individuals for many of the phenotypic traits that are of interest.

This difficulty has been described as a paradox in the study of the genetics of natural populations (Lewontin 1974). We are interested in the phenotype of those characteristics for which genetic differences at individual loci have only a slight phenotypic effect relative to the contributions of other loci and the environment. "What we can measure is by definition uninteresting and what we are interested in is by definition unmeasurable" (Lewontin 1974, p. 23). This paradox can only be resolved by a multidisciplinary approach that combines molecular biology, developmental biology, and population genetics, so that we can understand the developmental processes that connect the genotype and the phenotype (e.g., Lewontin 1999, Clegg and Durbin 2000).

Much effort has been devoted recently to using recently developed genomic techniques to search for genome-wide associations in humans, in order to detect the genetic basis of complex traits, particularly disease, using large samples of individuals and genetic markers. The results have not been as productive as anticipated. While many **candidate genes** have been identified, a large proportion of the heritability generally has remained unexplained (Frazer *et al.* 2009). However, the search for ecologically relevant genomic variation underlying important phenotypic traits in wild plant and animal populations is starting to have some success, and has identified some key genes underlying phenotypic variation in multiple species (Nadeau and Jiggins 2010).

The more complex and distant the connection between the genotype and the phenotype, the more difficult it is to determine the genetic basis of observed phenotypic differences. For example, human behavior is influenced by many genes and an extremely complex developmental process that continues to be influenced by the environment throughout the lifetime of an individual (Figure 2.2). It is relatively easy to identify genetic differences in the structural proteins, enzymes, and hormones involved in behavior because they are direct expressions of DNA sequences in the genotype. We also know that these genetic differences do result in differences in behavior between individuals; for example, a mutation in the enzyme monoamine oxidase *A* apparently results in a tendency for aggressive and violent behavior in humans (Morell 1993, McDermott *et al.* 2009). However, it is extremely difficult in general to identify the specific differences in the genotype that are responsible for differences in behavior.

We are faced with a dilemma in our study of the genetic basis of phenotypic variation in natural populations. We can start with the genotype (the bottom of Figure 2.2) and find genetic differences between individuals at specific loci; however, it is difficult to relate those differences to the phenotypic differences between individuals that are of interest. The alternative is to start with the phenotype of interest (the top of Figure 2.2), and determine whether the phenotypic differences do have a genetic basis; however, it is usually extremely difficult to identify which specific genes contribute to those phenotypic differences (Mackay 2001). Genomic technologies are now available that can potentially "bridge the chasm between genotype and phenotype" (Brenner 2000), but they remain financially out of reach for most species of conservation interest.

There is accumulating evidence that the environment can impose effects that are transmitted from generation to generation and generate heritable variation for traits by external influences on the genome (Bonduriansky and Day 2009). So-called **nongenetic inheritance** or **epigenetics** includes any effect on offspring phenotype brought about by factors other than DNA sequences from parents or more remote ancestors. Nongenetic inheritance includes mechanisms whereby the environment in which an ancestor's genes reside can influence development in descendants. We will briefly introduce this topic and consider whether it is likely to have influence on conservation biology.

In this chapter, we consider the amount and pattern of phenotypic variation in natural populations. We introduce approaches and methodology used to understand the genetic, and nongenetic, basis of phenotypic variation. In the next two chapters, we will examine genetic variation directly in chromosomes

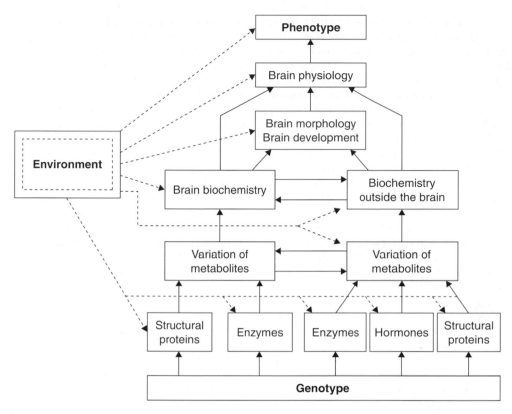

Figure 2.2 Diagrammatic representation of interconnections between genotype and environment resulting in expression of the phenotype of human. Redrawn from Vogel and Motulsky (1986).

and molecules and consider how it relates to evolution and conservation.

2.1 COLOR PATTERN

Mendel chose to study the inheritance of seven characters that had clearly distinguishable forms without intermediates: tall versus dwarf plants, violet versus white flowers, green versus yellow pods, and so on. His revolutionary success depended upon the selection of qualitative traits in which the variation could be classified into discrete categories, rather than quantitative traits in which individuals vary continuously (e.g., weight, height).

Variation is the essence of genetics. However, the presence of discrete polymorphisms in species has sometimes been problematic for naturalists and taxonomists. For example, the king coat color pattern in cheetahs was first described as a cheetah-leopard hybrid (van Aarde and van Dyk 1986). Later, animals with this pattern were recognized as a new species of cheetah. It was then suggested that this coat pattern was a genetic polymorphism within cheetahs. Inheritance results with captive cheetahs eventually confirmed that this phenotype results from a recessive allele at an **autosomal** locus (van Aarde and van Dyk 1986).

Rare color phenotypes often attract wide interest from the public. For example, a photojournalist published a picture of a white-phase black bear near Juneau, Alaska, in the summer of 2002. In response to public concerns, the Alaska Board of Game ordered an emergency closure of hunting on all "white phase" black bears in the Juneau area during the 2002 hunting season. White-phases of species like the tiger are often maintained in zoos because of public interest. A single, recessive autosomal allele is responsible for the white color. However, this allele also causes abnormal vision in tigers (Thornton 1978). This is an example of **pleiotropy**, which occurs when a single gene affects two distinct phenotypic characteristics.

The Spirit Bear, or Kermode bear, is a white phase of the black bear caused by a single gene that presents an interesting case of pleiotropy (Hedrick and Ritland 2011) (Figure 2.3). The white morph occurs at low frequencies along the coast of British Columbia and Alaska (Ritland *et al.* 2001). However, the Kermode bear is at frequencies as high as 40% on some islands off the coast of British Columbia. These bears have been protected from hunting since 1925 (Ritland *et al.* 2001). Klinka and Reimchen (2009) have reported that white bears are more successful at capturing salmon during the day. Experiments indicate that salmon were twice as evasive to black as the white morph during the day, but both morphs were equally successful capturing salmon at night.

Discrete color polymorphisms are widespread in plants and animals. For example, a recent review of color and pattern polymorphisms in anurans (frogs and toads) cites polymorphisms in 225 species (Hoffman and Blouin 2000). However, surprisingly little work has been done to describe the genetic basis of these polymorphisms or their adaptive significance. Hoffman and Blouin (2000) reported that the mode of inheritance has been described in only 26 species, but

conclusively demonstrated in only two! Nevertheless, available results suggest that, in general, color pattern polymorphisms are highly heritable in anurans.

Color pattern polymorphisms have been described in many bird species (Hoekstra and Price 2004). Polymorphism in this context can be considered to be the occurrence of two or more discrete, genetically based phenotypes in a population in which the frequency of the rarest type is greater than 1% (Hoffman and Blouin 2000). For example, red and gray morphs of the eastern screech owl occur throughout its range (VanCamp and Henny 1975). This polymorphism has been recognized since 1874 when it was realized that red and gray birds were conspecific and that the types were independent of age, sex, or season of the year.

The inheritance of this phenotypic polymorphism has been studied by observing progeny produced by different mating types in a population in northern Ohio (VanCamp and Henny 1975). Matings between gray owls produced all gray progeny. The simplest explanation of this observation is a single locus with two alleles, and the red allele (R) is dominant to the gray allele (r). Under this model, gray owls are homozygous rr, and red owls are either homozygous RR or heterozygous Rr. Homozygous RR red owls are expected to produce all red progeny regardless of the genotype of their mate. One-half of the progeny between heterozygous Rr owls and gray birds (rr) are expected to be red and one-half are expected to be gray. We cannot predict the expected progeny from matings involving red birds without knowing the frequency of homozygous RR and heterozygous Rr birds in the population. Progeny frequencies from red parents in Table 2.1 are compatible with most red birds being heterozygous Rr; this is expected because the red morph is relatively rare in northern Ohio based upon the number of families with red parents in Table 2.1. We will take

Figure 2.3 A white-phase Kermode fishing next to a black bear. From Ritland *et al.* (2001).

Table 2.1 Inheritance of color polymorphism in eastern screech owls from northern Ohio (VanCamp and Henny 1975).

Mating	Number of families	Progeny Red	Gray
Red × red	8	23	5
Red × gray	46	68	63
Gray × gray	135	0	439

another look at these results in Chapter 5 after we have considered estimating genotypic frequencies in natural populations.

A series of papers on flower color polymorphism in the morning glory provides a model system for connecting adaptation with the developmental and molecular basis of phenotypic variation (reviewed in Clegg and Durbin 2000). Flower color variation in this species is determined primarily by allelic variation at four loci that affect flux through the flavonoid biosynthetic pathway. Perhaps the most surprising finding is that almost all of the mutations that determine the color polymorphism are the result of the insertion of mobile elements that are called **transposons**. In addition, the gene that is most clearly subject to natural selection is not a structural gene that encodes a protein, but is rather a regulatory gene that determines the floral distribution of pigmentation (Clegg and Durbin 2003).

Flower color can have a great effect on pollinator visits. Different types of pollinators are attracted to different colors of flowers. Bradshaw and Schemske (2003) bred lines of two species of monkeyflowers that had substitutions for a single locus (yellow upper, *YUP*) controlling the presence or absence of yellow carotenoid pigments. They found that a change in color of monkeyflowers resulting from a *YUP* allele substitution resulted in a near-wholesale shift in pollinators from bumblebees to hummingbirds.

We have entered an exciting new era where for the first time it has become possible to identify the genes responsible for color polymorphisms, and these genes often play a similar role in multiple species. A recent series of papers has shown that a single gene, melanocortin-1 receptor (*MC1R*), is responsible for color polymorphism in a variety of birds and mammals, including the Spirit Bear (Example 2.1, Guest Box 12, Majerus and Mundy 2003, Mundy *et al.* 2004). This same approach has been used to detect color polymorphisms in extinct mammoths and Neandertals (Rompler *et al.* 2006, Lalueza-Fox *et al.* 2007). Field studies of natural selection, combined with study of genetic variation at *MC1R*, will eventually lead to understanding of the roles of selection and mutation in generating similarities and differences between populations and species (Hoekstra 2006).

Example 2.1 Plumage polymorphism in the Arctic skua

A color polymorphism in Arctic skua (or parasitic jaeger) was the subject of a long-term population genetic study by O'Donald (1987). Three color phases (pale, intermediate, and dark) occur throughout the range of the Arctic skua. Some birds have pale plumage with a white neck and body, while other birds have a dark brown head and body. O'Donald and Davis (1959, 1975) classified monogamous breeding adults and their chicks (normally two per brood) on Fair Isle, Scotland, as either pale or melanic (dark or intermediate). The following results were obtained through 1951–1958:

Parental types	Chicks	
	Pale	**Melanic**
Pale × Pale	29	0
Pale × Melanic	52	86
Melanic × Melanic	25	240
Total	106	326

These results suggested that this color polymorphism was controlled by a single Mendelian locus in which the dark allele is dominant to the pale.

This prediction was confirmed by Mundy *et al.* (2004) who sequenced *MC1R* as a candidate locus responsible for this polymorphism. They found that a single amino acid substitution at amino acid 230 ($Arg^{230} \rightarrow His^{230}$) correlates perfectly with this polymorphism (Figure 2.4). All melanic birds were either heterozygous Arg^{230}/His^{230} or homozygous His^{230}/His^{230}; darker birds were more likely to be homozygous. All pale birds were homozygous Arg^{230}/Arg^{230}. A similar substitution at this site is associated with melanism in rock pocket mice (see Figure 8.13 and Guest Box 12).

Sequence analysis of *MC1R* indicates that the ancestral Arctic skuas were pale and the melanic form is the derived trait. Comparison of the amount of sequence divergence between alleles suggests that the His^{230} mutation arose during the Pleistocene approximately 340,000 years ago.

The frequencies of these phenotypes show a latitudinal cline, with the pale birds being more common

(Continued)

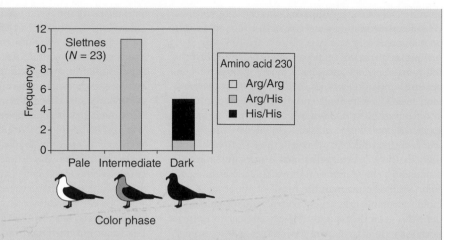

Figure 2.4 Association between genotypes at *MC1R* and color phases of the Arctic skua. From Mundy *et al.* (2004).

northward in the circumpolar breeding area. This polymorphism appears to be maintained by a combination of natural and sexual selection. Melanic individuals find mates more quickly than pale individuals and benefit from increased reproductive effect associated with earlier breeding (O'Donald 1987). Apparently, pale individuals are favored by natural selection at higher latitudes.

It has been proposed that color polymorphisms could play an important role in the ecology and conservation of species and populations (Forsman and Åberg 2008, Forsman *et al.* 2008). Forsman and Åberg (2008) found that species of Australian lizards and snakes with color polymorphisms use a greater diversity of habitats, have larger ranges, and are less likely to be listed as threatened. Forsman *et al.* (2008) have predicted that populations with color polymorphisms are less vulnerable to extirpation when facing population declines (see Chapter 14) and are more likely to be successful invasive species (see Chapter 20).

2.2 MORPHOLOGY

Morphological variation is everywhere. Plants and animals within the same population differ in size, shape, and numbers of body parts. However, there are serious difficulties with using morphological traits to understand patterns of genetic variation. The biggest problem is that variation in morphological traits is caused by both genetic and environmental differences among individuals. Therefore, variability in morphological traits cannot be used to estimate the amount of genetic variation within populations or the amount of genetic divergence between populations. In fact, we will see in Section 2.5 that morphological differences between individuals in different populations may actually be misleading in terms of genetic differences between populations.

Size traits can be correlated with other important life history traits. For example, in blue-eyed Mary, a small winter annual, there is considerable variation among populations in flower size. Populations of small-flowered plants are more likely to be self-pollinating rather than insect pollinated, with anthers shedding pollen before flowers open (Elle *et al.* 2010). They also flower earlier, and are found in drier climates than large-flowered populations where the period favorable for growth is shorter.

Observation of changes in morphology over time has produced some interesting results. For example, pink salmon on the west coast of North America tended to become smaller at sexual maturity between 1950 and 1974 (Figure 2.5). Pink salmon have an

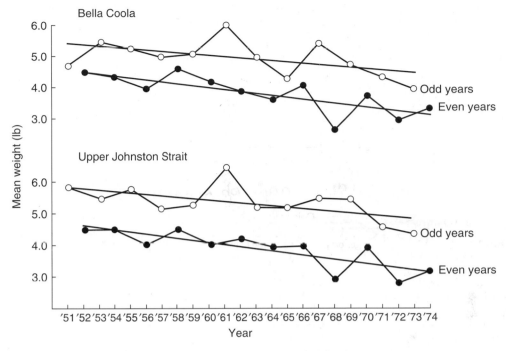

Figure 2.5 Decrease in size of pink salmon caught in two rivers in British Columbia, Canada, between 1950 and 1974. Two lines are drawn for each river: one for the salmon caught in odd-numbered years, the other for even years. Redrawn from Ricker (1981).

unusual life history in that all individuals become sexually mature and return from the ocean to spawn in freshwater at two years of age, and all fish die after spawning (Heard 1991). Therefore, pink salmon within a particular stream comprise separate odd- and even-year populations that are reproductively isolated from each other (Aspinwall 1974). Both the odd- and even-year populations of pink salmon have become smaller over this time period. This effect is thought to be largely due to the selective effects of fishing for larger individuals (see Chapter 18).

Most phenotypic differences between individuals within populations have *both* genetic and environmental causes. Geneticists often represent this distinction by partitioning the total phenotypic variability for a trait (V_P) within a population into two components:

$$V_P = V_G + V_E \qquad (2.1)$$

where V_G is the proportion of phenotypic variability due to genetic differences between individuals, and V_E is the proportion due to environmental differences. The **heritability** of a trait is defined as the proportion of the total phenotypic variation that has a genetic basis (V_G/V_P). The greater the heritability of a trait, the more phenotypic differences among individuals within a population are due to genetic differences among individuals. Equation 2.1 is an extreme simplification of complex interactions which are considered in more detail in Chapter 11.

One of the first attempts to tease apart genetic and environmental influences on morphological variation in a natural population was by Punnett (of Punnett square fame) in 1904. He obtained a number of velvet belly sharks from the coast of Norway to study the development of the limbs in vertebrates. The velvet belly is a small, round-bodied, viviparous shark that is common along the European continental shelf. Punnett counted the total number of vertebrae in 25 adult females that each carried from 2–14 fully developed young (Figure 2.6). He estimated the correlation between vertebrae number in females and their young

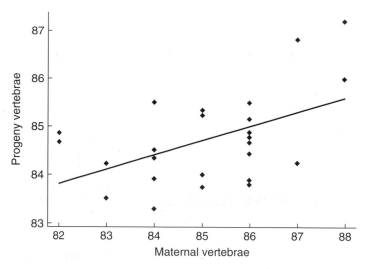

Figure 2.6 Regression of the mean number of total vertebrae in sharks before birth on the total number of vertebrae in their mothers ($P < 0.01$). Data from Punnett (1904).

to test the inheritance of this morphological character. He assumed that the similarity between females and their progeny would be due to inheritance since the females and their young developed in different environments.

Punnett (1904) concluded "the values of these correlations are sufficiently large to prove that the number of units in a primary linear meristic series is not solely due to the individual environment but is a characteristic transmitted from generation to generation". In fact, approximately 25% of the total variation in progeny vertebrae number can be attributed to the effect of their mothers ($r = 0.504$; $P < 0.01$). We will take another look at these data in Chapter 11 when we consider the genetic basis of morphological variation in more detail (see also Tave 1984).

Some phenotypic variation can be attributed to neither genetic nor environmental differences among individuals. Bilateral characters of an organism may differ in size, shape, or number. Take, for example, the number of gill rakers in fish species. The left and right branchial arches of the same individual usually have the same number of gill rakers (Figure 2.7). However, some individuals are asymmetric, that is, they have different numbers of gill rakers on the left and right sides. **Fluctuating asymmetry** of such bilateral traits occurs when most individuals are symmetric

and there is no tendency for the left or right side to be greater in asymmetric individuals (Palmer and Strobeck 1986).

What is the source of such fluctuating asymmetry? The cells on the left and right sides are genetically identical, and it seems unreasonable to attribute such variability to environmental differences between the left and right side of the developing embryo. Fluctuating asymmetry is thought to be the result of the inability of individuals to control and integrate development, so that random physiological differences occur during development and result in asymmetry (Palmer and Strobeck 1997, Leamy and Klingenberg 2005). That is, fluctuating asymmetry is a measure of developmental noise – random molecular events (Lewontin 2000). Mather (1953) ascribed the regulation or suppression of these chance physiological differences to the genotypic stabilization of development, and proposed that developmental stability could be measured by fluctuating asymmetry. Thus, increased 'noise' or accidents during development (i.e., decreased developmental stability) will result in greater fluctuating bilateral asymmetry. The amount of fluctuating asymmetry in populations may be a useful measure of stress resulting from either genetic or environmental causes in natural populations (Leary and Allendorf 1989, Clarke 1993, Zakharov 2001).

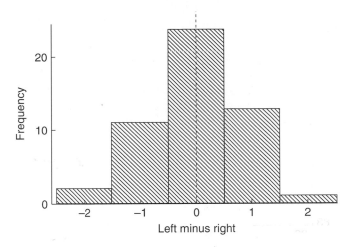

Figure 2.7 Fluctuating asymmetry for gill rakers on lower first branchial arches from a randomly mating population of rainbow trout. Redrawn from Leary and Allendorf (1989).

2.3 BEHAVIOR

Behavior is another aspect of the phenotype and thus will be affected by natural selection and other evolutionary processes, just as any phenotypic characteristic will be. Human behavioral genetic research has been controversial over the years because of concerns that the results from behavioral genetic studies might be used to stigmatize individuals or groups of people (Ridley 2004). Genetically based differences in behavior are of special interest in conservation because many behavioral differences are of importance for local adaptation and because captive breeding programs often result in changes in behavior because of adaptation to captivity (Caro 2007, Moore *et al.* 2008).

Most research in behavioral genetics has used laboratory species such as mice and *Drosophila*. These studies have focused on determining the genetic, neurological, and molecular basis of differences in behavior among individuals. *Drosophila* behavioral geneticists are especially creative in naming genes affecting behavior; they called a gene *couch potato* that is associated with reduced activity in adults (Bellen *et al.* 1992).

The extent to which genetic factors are involved in differences in bird migratory behavior has been studied systematically over the last 20 years in the blackcap, a common warbler of western Europe (Berthold 1991, Berthold and Helbig 1992). Selective breeding for the tendency to migrate has shown that the tendency to migrate itself is inherited and is based upon a multilo-

cus system with a threshold for expression (Figure 2.8). In addition, differences between geographical populations in the direction of migration are also genetically influenced (Figure 2.9).

The importance of genetically based differences in behavior for adaptation to local conditions has been shown by an elegant series of experiments with the western terrestrial garter snake (Arnold 1981). This garter snake occurs in a wide variety of habitats throughout the west from Baja to British Columbia, and occurs as far east as South Dakota. Arnold compared the diets of snakes living in the foggy and wet coastal climate of California and the drier, high-elevation inland areas of that state. As hard as it may be to believe, the major prey of coastal snakes is the banana slug; in contrast, banana slugs do not occur at the inland sites.

Arnold captured pregnant females from both locations and raised the young snakes in isolation away from their littermates and mother to remove this possible environmental influence on its behavior. The young snakes were offered a small chunk of freshly thawed banana slug. Native coastal snakes usually ate the slugs; inland snakes did not (Figure 2.10). Hybrid snakes between the coastal and inland sites were intermediate in slug-eating proclivity. These results confirm that the difference between populations in slug-eating behavior has a strong genetic component.

Studies with several salmon and trout species have demonstrated innate differences in migratory behavior

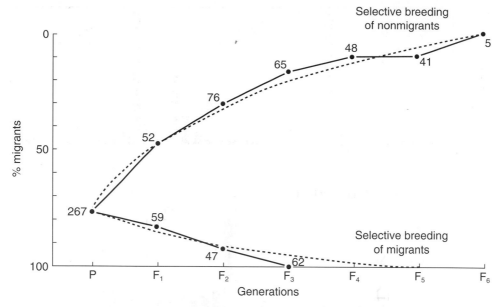

Figure 2.8 Results of two-way selective breeding experiment for migratory behavior with blackcaps from a partially migratory Mediterranean population. From Berthold and Helbig (1992).

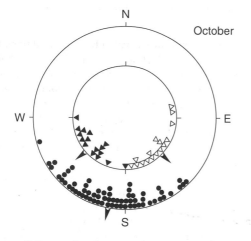

Figure 2.9 Results of tests in orientation cages allowing blackcaps to choose direction of migration. Solid and open triangles represent birds from Germany and Austria, respectively. The difference in direction of migration corresponds to geographic differences between these populations. The solid circles show the orientation of hybrids experimentally produced between the two parental groups. From Berthold and Helbig (1992).

that correspond to specializations in movement from spawning and incubation habitat in streams to lakes favorable for feeding and growth (reviewed in Allendorf and Waples 1996, de Leaniz *et al.* 2007). Fry emerging from lake outlet streams typically migrate upstream upon emergence, and fry from inlet streams typically migrate downstream. Differences in compass orientation behavior of newly emerged sockeye salmon correspond to movements to feeding areas.

Kaya (1991) has shown that behavioral local adaptations have evolved in Arctic grayling in just a few generations under strong selection. Arctic grayling from the Big Hole River have been planted in lakes throughout Montana, US, over the last 50 years. Arctic grayling from mountain lakes emerge as fry from gravel in streams and immediately migrate into a nearby lake. Fry from adults that spawn upstream from the lake innately swim downstream after emerging. Conversely, fry from adults that spawn downstream innately swim upstream after emerging. Fry that swim in the wrong direction upon emergence would be expected to have greatly reduced survival chances. Such differences between upstream and downstream spawning populations have apparently arisen by

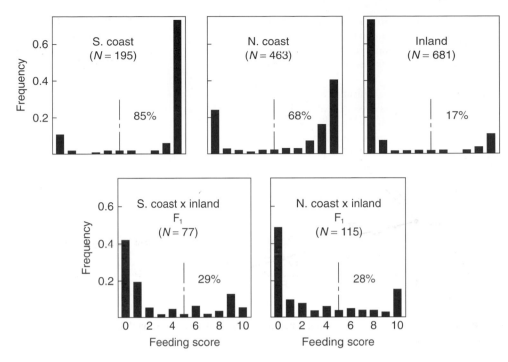

Figure 2.10 Response of newborn garter snakes to availability of pieces of banana slugs as food. Snakes from coastal populations tend to have a high slug-feeding score. A score of 10 indicates that a snake ate a piece of slug on each of the 10 days of the experiment. Inland snakes rarely ate a piece of slug on even one day. From Arnold (1981).

natural selection since the introduction of these populations.

There is a very interesting pleiotropic behavioral effect of the *MC1R* gene that we considered in Section 2.1 (Ducrest *et al.* 2008). Individuals with the darker *MC1R* allele in many species tend to be more aggressive, more sexually active, and more resistant to stress than lighter individuals. This results from effects of *MC1R* within the meloncortin system which produces a variety of peptide hormones that act as neurocrine and endocrine factors.

2.4 PHENOLOGY

Many environments fluctuate seasonally between more and less favorable conditions, and for plants and animals living in those environments, timing is critical. The timing of biological events, particularly in relation to climatic or other environmental variation, is called **phenology**. Species have evolved genetic responses to environmental cues to use favorable cli-

matic periods and avoid typically poor conditions for growth and reproduction. Genetically based differences in phenology are of special interest for conservation since these traits are often locally adapted because of the environmental differences among populations. Humans have used phenological events, such as bud break or flowering date, or the arrival of migratory birds, to track seasons and initiate agricultural activities for millennia.

Plants need to germinate, grow, flower, and produce seeds while temperatures and moisture levels are favorable, and avoid active growth when conditions are sufficiently cold, hot or dry enough to cause damage. **Annual plants** need to synchronize their growth and reproduction with local climatic conditions to complete their lifecycle in a short period of time, and the timing of events is under both genetic and environmental control. The heritability of flowering time in populations of the annual field mustard plants in southern California is estimated to be one third to one half of the total phenotypic variation in flowering time (Franks *et al.* 2007). Successive recent droughts resulted in an average advance of the date of flowering by up to a week, showing a genetic response to selection. The capacity for adaptation to new climates will, in part, be determined by the extent of genetic variation for phenological events.

Perennial plants that grow for multiple years in temperate and boreal environments need to cease growth, develop dormancy, shed leaves (if deciduous), and cold-acclimate prior to freezing events. These events are particularly well studied in forest trees. The initiation of growth in spring for many woody species requires a period of low temperatures in winter (chilling requirement), followed by a period of warm temperatures (heat sum requirement) signaling the arrival of spring. The timing of bud break in spring is highly heritable; a given individual will be consistently early or late in bud break from year to year compared with others (Howe *et al.* 2003), although the average date in any given spring will be dependent upon temperature accumulation that year.

Genetic variation has been documented in both chilling and heat sum requirements for bud break in Douglas-fir (Campbell and Sugano 1993). Growth cessation, bud set, leaf abscission, and dormancy in higher latitude perennial plants are usually triggered by photoperiod rather than temperature and have somewhat lower heritabilities than growth initiation (Howe *et al.* 2003). When night length exceeds the critical genetic threshold for an individual, it is detected by phytochrome genes, which initiate a cascade of events involved in preparing for winter. For example, variation at a single nucleotide position in the *phytochrome B2* locus is associated with phenotypic variation in bud set timing in populations of European aspen (Figure 2.11, Ingvarsson *et al.* 2008). Genetic control of flowering time in many plants also involves the detection of photoperiod by phytochrome genes. Depending on the species, flowering can be initiated in response to either long or short days.

Animals also have genetically determined responses to environmental cues indicating seasonality. Key phenological traits include timing of migration, egg laying, fertility, molting, and hibernation. Reproductive synchrony among individuals of the same species is important for reproductive success. Timing of long-distance spring and fall migration is critical to avoid arriving when climatic conditions are unfavorable or when food is unavailable (Coppack and Both 2002). The hibernation period of nonmigratory animals similarly needs to be synchronized with local climate and

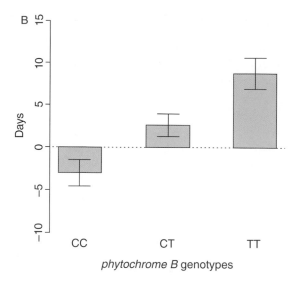

Figure 2.11   Association between genotypes at a single nucleotide polymorphism at *phytochrome B* and the date of bud set in the Eurasian aspen. The effect is displayed as the deviation (± SE) from the mean number of days to bud set. From Ingvarsson *et al.* (2008).

availability of resources. As in plants, photoperiod plays a key role in many animals in triggering phenological events.

Many organisms have endogenous molecular timekeeping systems that control circadian rhythms, and these are also involved in the sensing of seasons. 'Clock genes' have been identified in plants and animals that involve light and cold signaling, and mediate both diurnal and seasonal responses (Resco *et al.* 2009). O'Malley *et al.* (2010) found that variation at a clock gene corresponds to latitudinal variation in reproductive timing among species of Pacific salmon along the west coast of North America. Those species (chum and Chinook salmon) that have strong phenotypic latitudinal clines in both spawn timing and age at spawning had the strongest allele frequency clines at *clock*. No allele frequency cline was found in coho salmon, which displays no phenotypic clines in spawn timing or age at spawning. Moderate phenotypic and allelic clines were found in pink salmon.

Genetic variation within animal species exists for many phenological traits. We saw in Section 2.3 that the tendency to migrate and the direction of migration are both under genetic control in the blackcap. In addition, studies have found that the timing of fall migration in the blackcap has a substantial genetic component (Pulido *et al.* 2001). Selection experiments for later migration in the same captive blackcap population have delayed the onset of migration activity by an average of one week over two generations, demonstrating that considerable genetic variation exists for timing of migration.

Differences in reproductive phenology can create reproductive barriers within species. The rockhopper penguin in subantarctic and subtropical waters comprises two geographically and genetically distinct groups, considered subspecies or sibling species. These groups differ by two months in breeding phenology, reflecting water temperature differences rather than physical distances between their respective habitats (Jouventin *et al.* 2006). Similarly, the fragrant orchid in Europe has **sympatric** populations of early- and late-flowering individuals that are considered different subspecies, and their reproductive phenological differences allow little gene flow between them (Soliva and Widmer 1999).

In addition to adapting phenology to the abiotic environment, the timing of lifecycle events needs to correspond to that of conspecific individuals and **mutualist species**. Flowering time needs to be not only synchronous with other individuals of the same species to allow for cross-pollination, but also needs to correspond with the availability of animal pollinators for successful fertilization. The maturation of fruit should also synchronize with the life cycles of seed dispersers. Growth phenology (e.g., the timing of bud break and 'leafing out' in spring) can also affect the impact of herbivorous insects on host plants. The phenology of forest tree caterpillars is locally synchronized with that of host trees, and under genetic control in both herbivore and host (Van Asch and Visser 2007). The date on which red squirrels in the Yukon, Canada, give birth varies both genetically and with environment, particularly with the cone abundance of white spruce (Berteaux *et al.* 2004).

Climate change has the potential to disrupt phenological synchrony between plant and animal species and their biotic and abiotic environments (Parmesan 2006). Timing of these events is now being tracked as one measure of the extent of recent climate change (Chapter 21). While phenological traits that are dependent on temperature cues may adjust to new climates without genetic changes, those dependent on photoperiod will need to adapt genetically as photoperiods remain constant but temperatures change. We will explore this more in Chapter 21.

2.5 DIFFERENCES AMONG POPULATIONS

Populations from different geographical areas are detectably different for many phenotypic attributes in almost all species. Gradual changes in phenotypes across geographic or environment gradients are found in many species. However, there is no simple way to determine whether such a **cline** for a particular phenotype results from genetic or environmental differences between populations. The most common way to test for genetic differences between populations is to eliminate environmental differences by raising individuals under identical environmental differences in a so-called common-garden experiment. That is, by making V_E in equation 2.1 equal to zero, any remaining phenotypic differences must be due to genetic differences between individuals.

The classic common-garden experiments were conducted with altitudinal forms of yarrow plants along an altitudinal gradient from the coast of central California to over 3000 m altitude in the Sierra Nevada Mountains (Clausen *et al.* 1948). Individual plants were cloned into genetically identical individuals by cutting them into pieces and rooting the cuttings. The clones were then raised at three different altitudes (Figure 2.12). Phenotypic differences among plants from different altitudes persisted when the plants were grown in common locations at each of the altitudes (Figure 2.12). Coastal plants had poor survival at high altitude, but grew much faster than high-altitude plants when grown at sea-level. A contemporary example of a similar pattern of phenotypic and genetic divergence along an environmental gradient is presented in Example 2.2.

Transplant and common-garden experiments are much more difficult with animals than with plants for several obvious reasons. However, James and her colleagues have partitioned clinal variation in size and shape of the red-winged blackbird into genetic and environmental components by conducting transplantation experiments (reviewed in James 1991). Eggs were transplanted between nests in northern and southern Florida, and between nests in Colorado and Minnesota. A surprisingly high proportion of the regional differences in morphology were explained by

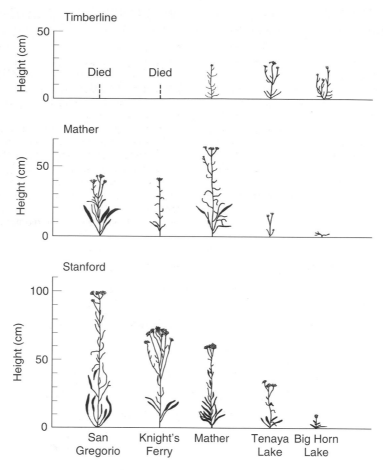

Figure 2.12   Representative clones of yarrow plants originating from five different altitudinal locations grown at three altitudes: 30 m above sea-level at Stanford, 1200 m above sea-level at Mather, and 3000 m above sea-level at Timberline. The San Gregorio clone was from a coastal population, and the Big Horn Lake clone was from the highest altitude site (over 3000 m); the other three clones were from an altitudinal gradient between these two extremes. From Clausen *et al.* (1948), redrawn from Strickberger (2000).

the locality in which eggs developed (James 1983, 1991).

James (1991) has reviewed experimental studies of geographic variation in bird species. She found a remarkably consistent pattern of intraspecific variation in size in breeding populations of North American bird species. Individuals from warm humid climates tend to be smaller than birds from increasingly cooler and drier regions. In addition, birds from regions with greater humidity tend to have more darkly colored feathers. The consistent patterns in body size and coloration among many species suggest that these patterns are adaptations that have evolved by natural selection in response to differential selection in different environments.

For example, there is some evidence that the color polymorphism in Eastern screech owls that we considered earlier affects the survival and reproductive success of individuals. The frequency of red owls increases from north (less than 20% red) to south (approximately 80% red; Pyle 1997). VanCamp and Henny (1975) found evidence that red owls suffered relatively greater mortality than gray owls during severe winter conditions in Ohio, and suggested that this may be due to metabolic

Example 2.2 Adaptive gradient in Sitka spruce

Sitka spruce has a large geographic range along the Pacific Coast of North America from northern California to Alaska, spanning a wide spectrum of climatic conditions. Populations in the southern portion of the range occupy relatively warm, wet habitats with long favorable growing seasons. There, Sitka spruce trees face interspecific competition from some of the other tallest tree species in the world, including coast redwood and Douglas-fir. In these environments, competition for light results in strong selection for rapid height growth. In the northern portion of the range, temperatures are considerably lower, and the growing season length between late spring and early

fall frosts is relatively short. In general, trees cannot withstand temperatures much below freezing without injury during active growth, but can tolerate subfreezing temperatures during the dormant period when tissues are cold acclimated.

Seedlings grown from seed collected in populations of Sitka spruce across the species range were planted in a common-garden experiment in Vancouver, British Columbia (Mimura and Aitken 2007). The results show strong local adaptation, with a tradeoff among populations between height growth and adaptation to low temperatures (Figure 2.13). Trees from the southern portion of the species range did not set bud until late

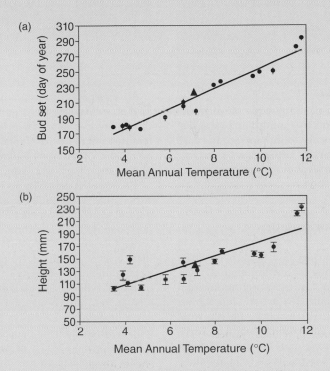

Figure 2.13 Phenotypic clines for (a) date of bud set and (b) height growth for Sitka spruce populations along a gradient of population mean annual temperature (data from Mimura and Aitken 2007). Seedlings were grown in a common-garden in Vancouver, British Columbia. Mean annual temperature for populations of Sitka spruce along the western coast of North America is strongly correlated with latitude, as the available frost-free period for growth decreases from south to north. The triangles indicate values for the Prince Rupert. This population currently has a mean annual temperature of 7.1°C, but is predicted to warm to 9.8 to 10.8°C by the 2080s, resulting in maladaptation assuming that populations are currently adapted.

(Continued)

fall, while those from northern Alaska provenances set bud in July or August (Figure 2.13a), and achieved much greater cold hardiness. This large difference in growing season length translated into a large difference in total height, with trees from California reaching sizes that were over twice as tall as trees from Alaska (Figure 2.13b).

Holliday and others have investigated the genomic basis of these population differences in bud set and cold acclimation phenotypes. Populations from California and Alaska differed in the extent to which genes were expressed during fall cold acclimation for approximately 300 of 22,000 genes studied (Holliday *et al.* 2008). Seedlings in populations from across the species range were genotyped for 768 single nucleotide polymorphisms (SNPs; see Section 4.2.3) in over 200 nuclear genes, and phenotyped for timing of bud set and for cold hardiness in artificial freeze tests (Holliday *et al.* 2010). A total of 35 SNPs in 28 genes were significantly associated with phenotypes for bud set, cold hardiness, or both. The genes included a homolog to phytochrome A (phyA), which detects photoperiod in plants, and several downstream genes involved in light signal transduction. Fourteen of these SNPs associated with phenotype also had significant genetic clines with climatic variables for population locations.

From this work it is clear that the synchronization of growth and dormancy with local climate in woody plants is genetically complex, involving many genes. It is also now possible to unravel these complex genetics through combining traditional common gardens with genomic methods. For conservation purposes, the greatest contribution of this type of study may be the identification of promising candidate genes that can be focused on for the development of potentially useful adaptive markers for species that cannot be studied in common gardens.

differences between red and gray birds (Mosher and Henny 1976). A similar north–south clinal pattern of red and gray morphs has also been reported in ruffed grouse; Gullion and Marshall (1968) reported that the red morph has lower survival during extreme winter conditions than the gray morph.

2.5.1 Countergradient variation

Countergradient variation is a pattern in which genetic influences counteract environmental influences, so that phenotypic change along an environmental gradient is minimized (Conover and Schultz 1995, see Guest Box 2). For example, Berven *et al.* (1979) used transplant and common-garden experiments in the laboratory to examine the genetic basis in life history traits of green frogs. In the wild, montane tadpoles experience lower temperatures; they grow and develop slowly and are larger at metamorphosis than are lowland tadpoles that develop at higher temperatures. Egg masses collected from high- and low-altitude populations were cultured side-by-side in the laboratory at temperatures that mimic developmental conditions at high and low altitude (18°C and 28°C). The differences observed between low- and high-altitude frogs raised under common conditions in the laboratory for some traits were *opposite* in direction to the differences observed in nature. That is, at low (montane-like) temperatures, lowland tadpoles grew even slower than, took longer to complete metamorphosis, and were larger than, montane tadpoles.

A reversal of naturally occurring phenotypic differences under common environments may occur when natural selection favors development of a similar phenotype in different environments. Consider the developmental rate in a frog or fish species and assume that there is some optimal developmental rate. Individuals from populations occurring naturally at colder temperatures will be selected for relatively fast developmental rate to compensate for the reduction in developmental rate caused by lower temperatures. Individuals in the lower temperature environment may still develop more slowly in nature. However, if grown at the same temperature, the individuals from the colder environment will develop more quickly. This will result in countergradient variation (Figure 2.14).

Therefore, phenotypic differences between populations observed in the wild are not a reliable indicator of genetic differences between populations without additional information. In some cases, all of the phenotypic differences between populations may result from environmental conditions. And, even if genetic differences do exist, they actually may be in the oppo-

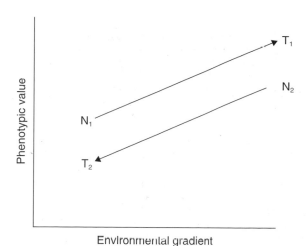

Figure 2.14 Countergradient variation. The end-points of the lines represent outcomes of a reciprocal transplant experiment: N_1 and N_2 are the native phenotypes of each population in its home environment. T_1 and T_2 are the phenotypes when transplanted to the other environment. Redrawn from Conover and Schultz (1995).

site direction of the observed phenotypic differences between populations.

Differences among populations in the amount of total phenotypic variation within populations can also be misleading. Using the relationship represented by equation 2.1, we would expect a positive association between V_P and V_G. That is, if V_E is constant, then greater total phenotypic variability (V_P) in a population would be indicative of greater genetic variability (V_G). However, assuming that V_E is constant is a very poor assumption, because different populations are subject to different environmental conditions. In addition, the reduction in genetic variation associated with small population size can sometimes decrease developmental stability and thereby increase total phenotypic variability in populations (Leary and Allendorf 1989).

Thus, it is not appropriate to use the amount of total phenotypic variability (V_P) in separate populations to detect differences in the amount of genetic variation between populations. The relationship between V_P and V_G is not straightforward. It differs for different traits within a single population and also depends on the history of the population. The genetic analysis of polygenic phenotypic variation is considered in more detail in Chapter 11.

2.6 NONGENETIC INHERITANCE

Nongenetic inheritance is an apparent oxymoron. Nevertheless, it has been used to describe phenotypic variation that is transmitted from parent to offspring by mechanisms other than changes in DNA sequences (Richards *et al.* 2010). **Epigenetics** is the study of phenotypic effects of changes in gene expression caused by mechanisms other than changes in the underlying DNA sequence, such as inherited changes to DNA structure and packaging including DNA methylation and histone modifications.

In Norway spruce, for example, temperature during seed development and maturation has a significant effect on the phenotypes of offspring (Johnsen *et al.* 2005a,b). Mother trees were cloned through grafting, planted in both cold and warm environments, and pollinated by hand with pollen from the same fathers. Seedlings grown from seed produced in warmer environments had phenotypes more like those typical of populations from warmer climates than their full siblings produced in colder climates. They had later bud set timing and cessation of growth at the end of summer, and delayed development of cold hardiness. The potential importance of this type of epigenetic response for rapid nongenetic response to climate change is discussed further in Chapter 21.

There is increasing evidence that epigenetic processes might be important following some hybridization events (Salmon *et al.* 2005, Bossdorf *et al.* 2008). Therefore, an epigenetics perspective might be important for understanding the effects of hybridization and predicting outbreeding depression (Chapter 17).

In addition, epigenetic effects might be an important source of phenotypic variation for invasive species. Richards *et al.* (2008) have shown that the invasive Japanese knotweed, which has little variation in DNA sequence, maintains substantial phenotypic variation even under controlled environmental conditions. Epigenetic effects associated with this phenotypic variation might enhance their ability to invade novel environments. This could be a partial explanation for the recognized paradox of the ability of invasive species that have lost genetic variation during a bottleneck associated with their introduction to adapt to new environmental conditions.

It has been suggested that understanding genetic change in populations requires the incorporation of cultural inheritance: the part of the phenotypic variation that is inherited socially or learned from others

(Danchin *et al.* 2011). This concept has potential implications with captive breeding, and also in small populations. Animals raised in captivity often fare poorly in the wild because they have not learned behaviors needed to be successful in the wild (Moore *et al.* 2008). Moreover, inappropriate imprinting of young cross-fostered by similar species or humans has also been a problem with captive rearing.

Cultural drift has been defined as the loss of culturally transmitted traits in small populations that is analogous to genetic drift (Koerper and Stickel 1980). Under some circumstances, this process could reduce the viability of small populations. For example, humans colonized the island of Tasmania approximately 35,000 years ago and became isolated from humans on mainland Australia by rising sea-level (Henrich 2004). This small, isolated population apparently lost a number of skills through cultural drift over this time period: manufacture of bone tools, cold-weather clothing, fishhooks, hafted tools, fishing spears, barbed spears, nets, bows, arrows, and boomerangs.

Some have suggested that nongenetically transmitted factors could increase in frequency within populations, even if they reduce individual and population-mean fitness, because the rate of spread is a function of transmission probability, as well as selection (Freedberg and Wade 2001, Bonduriansky and Day 2009). These effects could affect population viability and pose some interesting questions for their possible relevance to conservation.

Guest Box 2   Looks can be deceiving: countergradient variation in secondary sexual color in sympatric morphs of sockeye salmon
Chris J. Foote

Sockeye salmon and kokanee are respectively the anadromous (sea-going; physically large) and non-anadromous (lake-dwelling; small) morphs of sockeye salmon (Figure 2.15). Both morphs occur throughout the native range of the species in the North Pacific drainages of North America and Asia. The morphs are polyphyletic, with one morph, likely sockeye, having given rise to the other on numerous independent occasions throughout their range (Taylor *et al.* 1996). The morphs occur together or separately in lakes (where sockeye typically spend their first year of life), but wherever they occur sympatrically, even where they spawn together, they are always genetically distinct in a wide array of molecular and physical traits (Taylor *et al.* 1996, Wood and Foote 1996). This reproductive isolation appears to result from significant size-related, pre-zygotic isolation coupled with the large selective differences between marine and lacustrine environments (Wood and Foote 1996).

On viewing sockeye and kokanee on the breeding grounds, one is struck by the large size difference between them (sockeye can be 2–3 times the length and 20–30 times the weight of kokanee), and by their shared striking bright red breeding color (Figure 2.15, Craig and Foote 2001). However, with respect to genetic differentiation, looks can be deceiving. The size difference between the morphs results largely from differences in food availability in the marine versus lacustrine environments, with only a slight genetic difference in growth evident between the morphs when grown in a common environment (Wood and Foote 1996).

In contrast, their similarity in breeding color masks substantial polygenic differentiation in the mechanism by which they produce their red body color. The progeny of sockeye that rear in freshwater throughout their life cannot turn red at maturity like kokanee; rather, they turn green (Craig and Foote 2001). Therefore, as sockeye repeatedly gave rise to kokanee over the last 10,000 years, they did so by at first producing a green freshwater morph that over time changed genetically to converge on the ancestral red breeding color.

The convergence in breeding color in sockeye and kokanee is an example of countergradient variation. Kokanee, which live in carotenoid-poor lake environments, are three times more efficient in utilizing carotenoids than sockeye, which live in a carotenoid-rich marine environment (Craig and Foote 2001). Interestingly, the selective force for the re-emergence of red in kokanee appears to be inherited from ancestral sockeye. Sockeye possess a very strong, and apparently innate, preference for red

Figure 2.15 Photograph of sockeye salmon (large) and kokanee (small) on the breeding grounds in Dust Creek a tributary of Takla Lake, British Columbia. Both kokanee and sockeye salmon have bright red bodies and olive green heads. See Color Plate 1. Photo by Chris Foote.

mates (Foote *et al.* 2004), a preference that is shared by kokanee. This pre-existing bias appears to have independently driven the evolution of red breeding colour in kokanee throughout their distribution. This contrasts with other examples of countergradient variation, where natural selection, and not **sexual selection**, is thought to be the driving selective force (Conover and Shultz 1995).

CHAPTER 3

Genetic variation in natural populations: chromosomes and proteins

Sumatran orangutan, Section 3.1.7

The empirical study of population genetics has always begun with and centered around the characterization of the genetic variation in populations.

Richard C. Lewontin (1974, p. 16)

Nevertheless, if populations with unrecognized intraspecific chromosome variation are crossed, progeny fitness losses will range from partial to complete sterility, and reintroductions and population augmentation of rare plants may fail.

Paul M. Severns and Aaron Liston (2008)

Conservation and the Genetics of Populations, Second Edition. Fred W. Allendorf, Gordon Luikart, and Sally N. Aitken.
© 2013 Fred W. Allendorf, Gordon Luikart and Sally N. Aitken. Published 2013 by Blackwell Publishing Ltd.

Genetic variation is the raw material of evolution. Change in the genetic composition of populations and species is the basis of evolutionary change. In Chapter 2, we examined phenotypic variation in natural populations. In this chapter, we will examine the genetic basis of this phenotypic variation by examining genetic differences between individuals in their chromosomes and proteins. In Chapter 4, we will examine variation in DNA sequences. This order, from the chromosomes that are visible under a light microscope down to the study of molecules, reflects the historical sequence of study of natural populations.

In our consideration of conservation, we are concerned with genetic variation at two fundamentally different hierarchical levels:

1 Genetic differences among individuals within local populations.
2 Genetic differences among populations within the same species.

The amount of genetic variation within a population provides insight into the demographic structure and evolutionary history of a population. For example, lack of genetic variation may indicate that a population has gone through a recent dramatic reduction in population size. Genetic divergence among populations is indicative of the amount of genetic exchange that has occurred among populations, and can play an important role in the conservation and management of species. For example, genetic analysis of North Pacific minke whales has shown that some 20–40% of the whale meat sold in Korean and Japanese markets comes from a protected and genetically isolated population of minke whales in the Sea of Japan (Baker *et al.* 2000, see Guest Box 22).

Population geneticists struggled throughout most of the 20th century to measure genetic variation in natural populations (Table 3.1). Before the advent of biochemical and molecular techniques, genetic variation could only be examined by bringing individuals into the laboratory and using experimental matings. The fruit fly (*Drosophila*) was the workhorse of empirical population genetics during this time because of its short generation time and ease of laboratory culture. For example, 41% of the papers (9 of 22) in the first volume of the journal *Evolution* published in 1947 had *Drosophila* in the title; approximately 5% of the papers (16 of 299) in the volume of *Evolution* published in the year 2010 had *Drosophila* in the title.

The tools to examine chromosomal and allozyme variation have been in use for many years. Their utility

Table 3.1 Historical overview of primary methods used to study genetic variation in natural populations.

Time period	Primary techniques
1900–1970	Laboratory matings and chromosomes
1970s	Protein electrophoresis (allozymes)
1980s	Mitochondrial DNA
1990s	Nuclear DNA
2000s	Genomics
2010s	Metagenomics

has been eclipsed by powerful new techniques that we will consider in the next chapter, which allow direct examination of genetic variation in entire genomes! Nevertheless, these 'old' tools and the information that they continue to provide are valuable.

The technique of protein electrophoresis is fading away and is being replaced by techniques that examine genetic variation in the DNA that encodes the proteins studied by allozyme electrophoresis (Utter 2005). However, a surprisingly large number of studies of genetic variation in natural populations using allozymes continue to be published every year. This is especially true for species for which it is more difficult to obtain funding. Unfortunately, this includes many species of interest for conservation, especially plants.

Study of DNA sequences, however, cannot replace examination of chromosomes. We anticipate a rejuvenation of chromosomal studies in evolutionary and conservation genetics as new technologies are developed that allow rapid examination of chromosomal differences between individuals (de Jong 2003, Hoffmann and Rieseberg 2008).

3.1 CHROMOSOMES

Surprisingly little emphasis has been placed on chromosomal variability in conservation genetics (Benirschke and Kumamoto 1991, Robinson and Elder 1993, Severns and Liston 2008). This is unfortunate because heterozygosity for chromosomal differences often causes reduction in fertility (Nachman and Searle 1995, Rieseberg 2001, Severns and Liston 2008). For example, the common cross between a female horse with 64 chromosomes and a male donkey with 62 chromosomes produces a sterile mule that has 63

Example 3.1   Cryptic chromosomal species in the graceful tarplant

The graceful tarplant is a classic example of the importance of chromosomal differentiation between populations for conservation and management. Clausen (1951) described the karyotype of plants from four populations of this species ranging from Alder Springs in northern California to a population near San Diego, California. These populations can hardly be distinguished morphologically and live in similar habitats. Plants from all of these populations had a haploid set of four chromosomes ($n = 4$). However, the size and shape of these four chromosomes differed among populations. Experimental crossings revealed that matings between individuals in different populations either failed to produce F_1 individuals or the F_1 individuals were sterile.

Clausen (1951) concluded that these populations were distinct species because of their chromosomal characteristics and infertility. Nevertheless, he felt that it would be "impractical" to classify them as taxonomic species because of their morphologic similarity and lack of ecological distinctness. These populations are still classified as the same species today. Nevertheless, for purposes of conservation, each of the chromosomally distinct populations should be treated as separate species because of their reproductive isolation. Translocations of individuals among populations could have serious harmful effects because of reduction in fertility or the production of sterile hybrids.

---

chromosomes. Some captive breeding programs unwittingly have hybridized individuals that are morphologically similar but have distinct chromosomal complements: orangutans (Ryder and Chemnick 1993); gazelles (Ryder 1987); and dik-diks (Ryder *et al.* 1989). Similarly, translocation or reintroduction programs may cause problems if individuals are translocated among chromosomally distinct groups (e.g., Example 3.1).

The possible occurrence of hybridization in captivity of individuals from chromosomally distinct populations is much more common than expected because small, isolated populations have a greater rate of chromosomal evolution than common widespread taxa (Lande 1979). Thus, the very demographic characteristics that make a species a likely candidate for captive breeding are the same characteristics that favor the evolution of chromosomal differences between groups. For example, extensive chromosomal variability has been reported in South American primates (Matayoshi *et al.* 1987).

The direct examination of genetic variation in natural population began with the description of differences in chromosomes between individuals. One of the first reports of differences in the chromosomes of individuals within populations was by Stevens (1908) who described different numbers of **supernumerary chromosomes** in beetles (White 1973). For many years, study of chromosomal variation in natural populations was dominated by the work of

Theodosius Dobzhansky and his colleagues on *Drosophila* (Dobzhansky 1970) because of the presence of giant polytene chromosomes in salivary glands (Painter 1933). However, the study of chromosomes in other species lagged far behind. For example, until 1956 it was thought that humans had 48, rather than 46 chromosomes in each cell. It is amazing that the complete human genome was sequenced within 50 years of the development of the technical ability to even count the number of human chromosomes!

3.1.1 Karyotypes

A **karyotype** is the characteristic chromosome complement of a cell, individual, or species. Chromosomes in the karyotype of a species are usually arranged beginning with the largest chromosome (Figure 3.1). The large number of **microchromosomes** in this karyotype is typical for many bird species (Shields 1982). Evidence indicates that bird microchromosomes are essential, unlike the supernumerary chromosomes discussed later in this section (Shields 1982).

Chromosomes of eukaryotic cells consist of DNA and associated proteins. Each chromosome consists of a single highly folded and condensed molecule of DNA. Some large chromosomes would be several centimeters long if they were stretched out – thousands of times longer than a cell nucleus. The DNA in a chromosome is coiled again and again and is tightly packed around

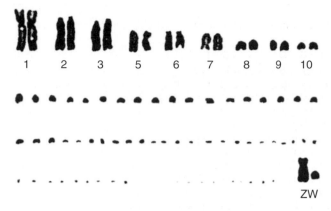

Figure 3.1 Karyotype ($2n = 84$) of a female cardinal. From Bass (1979).

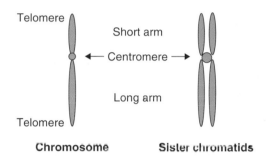

Figure 3.2 Diagram of an unreplicated chromosome and a chromosome that has replicated into two identical sister chromatids that are joined at the centromere.

histone proteins. Chromosomes are generally thin and difficult to observe, even with a microscope. Before cell division (mitosis or meiosis), however, they condense into thick structures that are readily seen with a light microscope. This is the stage that we usually observe chromosomes (Figure 3.1). The chromosomes right before cell division have already replicated so that each chromosome consists of two identical sister chromatids (Figure 3.2).

Chromosomes function as the vehicles of inheritance during the processes of **mitosis** and **meiosis**. Mitosis is the separation of the sister chromatids of replicated chromosomes during somatic cell division to produce two genetically identical cells. Meiosis is the pairing of and separation of homologous replicated chromosomes during the division of sex cells to produce gametes.

Certain physical characteristics and landmarks are used to describe and differentiate among chromosomes. The first is size; the chromosomes are numbered from the largest to the smallest. The **centromere** appears as a constricted region and serves as the attachment point for spindle microtubules that are the filaments responsible for chromosomal movement during cell division (Figure 3.2). The centromere divides a chromosome into two arms. Chromosomes in which the centromere occurs approximately in the middle are called **metacentric**. In **acrocentric** chromosomes, the centromere occurs near one end of the chromosome. **Chromosomal satellites** are small chromosomal segments separated from the main body of the chromosome by a secondary constriction (see chromosome-13 in Figure 3.3). Staining techniques have been developed that differentially stain different regions of a chromosome to help distinguish chromosomes that have similar size and centromere location (Figure 3.3). Recent developments of so-called chromosome painting allow determination of regions of shared common ancestry (**homology**) between populations and species (Ferguson-Smith and Trifonov 2007).

3.1.2 Sex chromosomes

Many groups of animals and some plant taxa have evolved sex-specific chromosomes that are involved in the process of sex determination (see Rice 1996 for an excellent review). In mammals, females are homogametic XX and males are heterogametic XY. The chromosomes that do not differ between the sexes are called

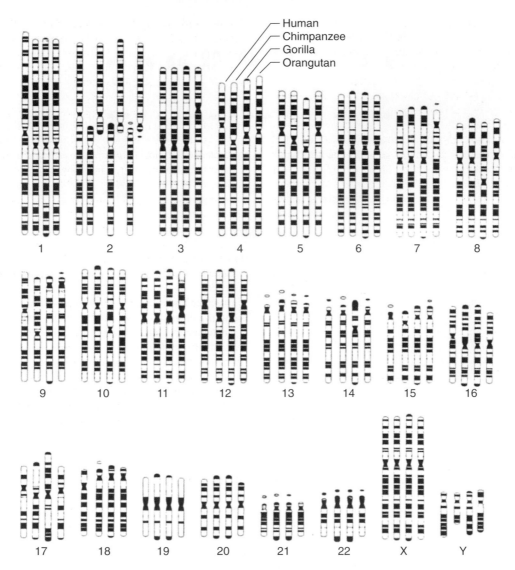

Human
Chimpanzee
Gorilla
Orangutan

1   2   3   4   5   6   7   8

9   10   11   12   13   14   15   16

17   18   19   20   21   22   X   Y

Figure 3.3 Karyotypes and chromosomal banding patterns of humans, chimpanzees, gorillas, and orangutans. From Strickberger (2000), modified from Yunis and Prakash (1982).

autosomes. For example, there are 28 pairs of chromosomes in the karyotype of African elephants ($2n = 56$; Houck *et al.* 2001); thus, each African elephant has 54 autosomes and two **sex chromosomes**. The heterogametic sex is reversed in birds: males are homogametic ZZ and females are heterogametic ZW (Figure 3.1). Note that the XY and ZW notations are strictly arbitrary and are used to indicate which sex is homogametic. For example, Lepidopterans (butterflies and moths) are ZZ/ZW; this indicates that the males are the homogametic sex.

The heterogametic sex differs between species in some taxa (Charlesworth 1991). Some fish species are XX/XY, some are ZZ/ZW, some do not have detectable sex chromosomes, and a few species even have more than two sex chromosomes (Devlin and Nagahama 2002).

Over 95% of plant species are hermaphrodites and therefore do not have sex-determining chromosomes or sex-determining loci within chromosomes (Charlesworth 2002, Heslop-Harrison and Schwarzacher 2011). However, both XX/XY and ZZ/ZW sex determination systems occur in dioecious plant species which have separate male and female individuals. Sex chromosomes appear to have evolved rather recently in plant species. There are no examples of ancient sex chromosomes that are shared among large taxonomic groups of plants, such as the XY system of mammals or the ZW system of birds. While some plant species, like white campion, have two morphologically distinguishable **heteromorphic** sex chromosomes, others like papaya have indistinguishable nonheteromorphic sex chromosomes or chromosomal regions, and some are at earlier stages in the evolution of dioecy with only sex-determining loci (Heslop-Harrison and Schwarzacher 2011).

Heteromorphic sex chromosomes can provide useful markers for conservation. The sex of individuals can be determined by karyotypic examination. However, many other far easier procedures can be used to sex individuals by their sex chromosomes complement. For example, one of the two X chromosomes in most mammal species is inactivated and forms a darkly coiling structure (a Barr body) that can be readily detected with a light microscope in epithelial cells scraped from the inside of the mouth of females but not males (White 1973). We will see in the next chapter that DNA sequences specific to one of the sex chromosomes can be used in many taxa to identify the sex of individuals.

3.1.3 Polyploidy

Most animal species contain two sets of chromosomes and therefore are diploid ($2n$) for most of their lifecycles. The eggs and sperm of animals are haploid and contain only one set of chromosomes ($1n$). Some species, however, are polyploid because they possess more than two sets of chromosomes: triploids ($3n$), tetraploids ($4n$), pentaploids ($5n$), hexaploids ($6n$), and even greater number of chromosome sets. Polyploidy is relatively rare in animals, but it does occur in invertebrates, fishes, amphibians, and lizards (White 1973). Polyploidy is common in plants and is a major mechanism of speciation (Stebbins 1950, Soltis and Soltis 1989). Estimates of the proportion of plant species

that arose through polyploidy range from 2–4% (Otto and Whitton 2000) to 20–40% (Stebbins 1938). In plants, polyploidy can lead to the evolution of asexual reproduction through apomixis, the production of seed without meiosis, resulting in embryos that are clones of maternal plants (Whitton et al. 2008).

Perhaps the most interesting cases of polyploidy occur when diploid and tetraploid forms of the same taxon are both in sympatry (see Guest Box 14). For example, both diploid (*Hyla chrysoscelis*) and tetraploid (*H. versicolor*) forms of gray tree frogs occur throughout the central US (Ptacek et al. 1994). Reproductive isolation between diploids and tetraploids is maintained by call recognition; the larger cells of the tetraploid males result in a lower calling frequency that is recognized by the females (Bogart 1980). Hybridization between diploids and tetraploids does occur and results in triploid progeny that are not fertile (Gerhardt et al. 1994).

Fireweed provides another example of reproductive isolation between diploids and polyploids. The diploids and tetraploids largely avoid mating through ecological specialization resulting in spatial isolation, differences in flowering phenology, and different pollinators (Husband and Sabara 2004). Survival was as high in triploid offspring as it was for diploids and tetraploids; however, pollen production and viability were significantly less in triploids. The overall relative fitnesses of diploids, triploids and tetraploids were 1, 0.09, and 0.61, respectively.

A thorough treatment of polyploidy is beyond the scope of this chapter. Nevertheless, examination of ploidy levels is an important taxonomic tool when describing units of conservation in some plant taxa (Chapter 16). Flow cytometry is a valuable method that can be used to evaluate ploidy level quickly in individual plants (Doležel and Bartoš 2005).

3.1.4 Numbers of chromosomes

Many closely related species have different numbers of chromosomes, such as the horse and donkey discussed in Section 3.1. For example, a haploid set of human chromosomes has $n = 23$ chromosomes while the extant species that are most closely related to humans all have $n = 24$ chromosomes (Figure 3.3). This difference is due to a fusion of two chromosomes to form a single chromosome (human chromosome 2) that occurred sometime in the human evolutionary lineage following its separation from the ancestor of the other species. This is an example of a **Robertsonian fusion**, which we will consider in Section 3.1.8.

Chromosome numbers have been found to evolve very slowly in some taxa. For example, the approximately 100 or so cetaceans have either $2n = 42$ or 44 chromosomes (O'Brien *et al.* 2006, Benirschke and Kumamoto 1991). Most species of conifer trees in the pine family (Pinaceae) have $2n = 24$, except for Douglas-fir which has $2n = 26$ (Krutovsky *et al.* 2004). In contrast, chromosome numbers have diverged rather rapidly in other taxa. For example, the $2n$ number in horses (genus *Equus*) varies from $2n = 32$ to 66 (Table 3.2). The Indian muntjac has a karyotype that is extremely divergent from other species in the same genus (Figure 3.4). In the next few sections, we will consider the types of chromosome rearrangements that bring about karyotypic changes among species (Ferguson-Smith and Trifonov 2007).

Table 3.2 Characteristic chromosome numbers of some living members of the horse family (White 1973, Richard *et al.* 2001, O'Brien *et al.* 2006).

Species		2n
Przewalski's horse	*Equus przewalski*	66
Domestic horse	*E. caballus*	64
Donkey	*E. asinus*	62
Kulan	*E. hemionus*	54
Grevy's zebra	*E. grevyi*	46
Burchell's zebra	*E. burchelli*	44
Mountain zebra	*E. zebra*	32

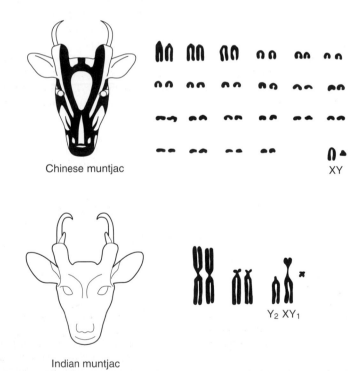

Figure 3.4 The Chinese and Indian muntjac and their karyotypes (from Strickberger 2000). The Indian muntjac has the lowest known chromosome number of any mammal. First-generation hybrids ($2n = 27$) created in captivity between these species are viable (Wang and Lan 2000). The two Y-chromosomes in this species have resulted from a centric fusion between an autosome and the sex chromosomes (White 1973). From Strickberger (2000).

3.1.5 Supernumerary chromosomes

Supernumerary chromosomes (also called B chromosomes) are not needed for normal development and vary in number in many plant and animal species. They are usually small, lack functional genes (**heterochromatic**), and do not pair and segregate during meiosis (White 1973, Jones 1991). In general, the presence or absence of B chromosomes does not affect the phenotype or the fitness of individuals (Battaglia 1964). It is thought that B chromosomes are "parasitic" genetic elements that do not play a role in adaptation (Jones 1991). B chromosomes have been reported in many species of higher plants (Müntzing 1966). In animals, B chromosomes have been described in many invertebrates, but they are rarer in vertebrates. However, Green (1991) has described extensive polymorphism for B chromosomes in populations of the Pacific giant salamander along the west coast of North America.

3.1.6 Chromosomal size

In many species, differences in size between homologous chromosomes have been detected. In most of the cases it appears that the 'extra' region is due to a heterochromatic segment that does not contain functional genes (White 1973, p. 306). These extra heterochro-matic regions resulting in size differences between homologous chromosomes are analogous to supernumerary chromosomes, except that they are inherited in a Mendelian manner. Heterochromatic differences in chromosomal size seem to be extremely common in several species of South American primates (Matayoshi *et al.* 1987).

3.1.7 Inversions

Inversions are segments of chromosomes that have been turned around so that the gene sequence has been reversed. There has been recent renewed interest in inversions because of the ability to detect them with comparative genomics (Hoffmann and Rieseberg 2008, Kirkpatrick 2010). For example, comparison of karyotypes identified only nine inversions that distinguish humans and chimpanzees; comparison of their complete genome sequences has revealed over 1000 inversions (Feuk *et al.* 2005). Lowry and Willis (2010) have detected a widespread chromosomal inversion polymorphism in the yellow monkeyflower that contributes to major life-history differences, local adaptation, and reproductive isolation (see Example 3.2).

Inversions are produced by two chromosomal breaks and a rejoining with the internal piece inverted (Figure 3.5). An inversion is **paracentric** if both breaks are

---

Example 3.2 A widespread chromosomal inversion in the yellow monkeyflower

Lowry and Willis (2010) described a widespread chromosomal inversion in the yellow monkeyflower, which has become an important model species for the integration of ecological and genomic studies. One arrangement of the inverted region is found in an ecotype of this species that lives in habitats characterized by reduced soil water availability in the summer and has an annual life history. The other arrangement lives in habitats with high year-round soil moisture and has a perennial life history. The inversion influences morphological and flowering time differences between these ecotypes throughout its range in western North America.

Field experiments were performed by breeding plants to reciprocally swap the alternative chromosomal arrangements between the annual and perennial genetic backgrounds, to investigate how polymor-phism contributes to adaptation and reproductive isolation. Late-flowering coastal perennial plants failed to flower before the onset of the hot seasonal summer drought in the inland habitat. Inland annual plants were at a disadvantage in the coastal habitat because they invested more resources in reproduction instead of growth and thus failed to take advantage of year-round soil moisture and cool foggy conditions.

Thus, this inversion polymorphism contributes to local adaptation, the annual versus perennial life-history polymorphism, and reproductive isolating barriers. These results indicate that adaptation to local environments can drive the spread of chromosomal inversions and promote reproductive isolation. It will be important to discover in the future how common such polymorphisms are in other plant species.

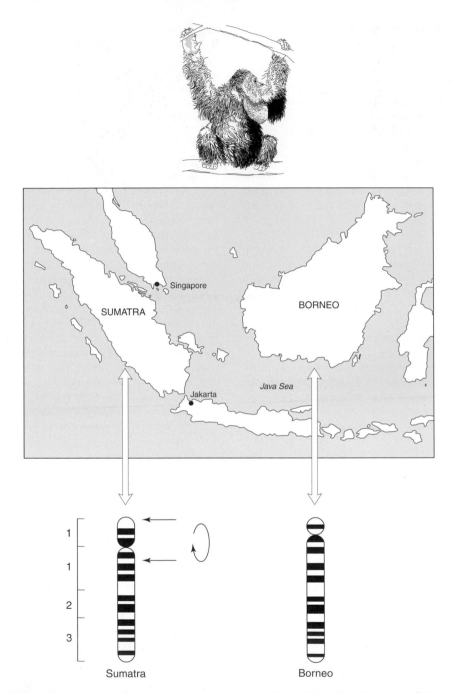

Figure 3.5   Pericentric inversion in chromosome-2 of two subspecies of orangutans from Sumatra and Borneo (chromosomes from Seuanez 1986).

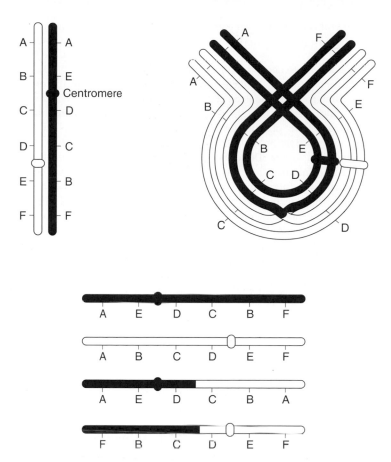

Figure 3.6 Crossing-over in a heterozygote for a pericentric inversion. The two chromosomes are shown in the upper left. The pairing configuration and crossing-over between two nonsister chromatids is shown in the upper right. The resulting products of meiosis are shown below. Only the two top chromosomes have complete sets of genes; these are noncrossover chromosomes that have the same sequences as the two original chromosomes. From Dobzhansky *et al.* (1977).

situated on the same side of the centromere, and **pericentric** if the two breaks are on opposite side of the centromere.

Heterozygosity for inversions is often associated with reduced fertility. **Recombination** (crossing over) within inversions produces aneuploid gametes that form inviable zygotes (Figure 3.6). The allelic combinations at different loci within inversion loops will tend to stay together because of the low rate of successful recombination within inversions. In situations where several loci within an inversion affect the same trait, the loci are collectively referred to as a supergene, with allelic combinations across loci within them acting as alleles (see Guest Box 3). Examples of phenotypes controlled by supergenes include shell color and pattern in

snails (Ford 1971), mimicry in butterflies (Turner 1985), and flower structure loci in *Primula* (Kurian and Richards 1997).

Inversions are exceptionally common in some taxonomic groups because they do not have the usual harmful effects of producing inviable zygotes (White 1973, p. 241). For example, crossing-over and recombination does not occur in male dipteran (two-winged) flies (e.g., *Drosophila*). Therefore, lethal chromatids will only be produced in females. However, there seems to be a meiotic mechanism in these species so that chromatids without a centromere or with two centromeres pass into the polar body rather than into the egg nucleus so that fertility is not reduced. Chromosomes with a single centromere having deficiencies or duplications which are produced in heterozygotes for pericentric inversions (Figure 3.6) are just as likely to be passed into the egg nucleus as normal chromosomes.

Paracentric inversions are difficult to detect because they do not change the relative position of the centromere on the chromosome. They can only be detected by examination of meiotic pairing or by using some technique that allows visualization of the genic sequence on the chromosome such as in polytenic chromosomes of *Drosophila* and other Dipterans. Several chromosome-staining techniques that reveal banding patterns were discovered in the early 1970s (Figure 3.3, Comings 1978). These techniques have been extremely helpful in identifying homologous chromosomes in karyotypes, and for detecting chromosomal rearrangements such as paracentric inversions. However, relatively few species have been studied with these techniques, so we know little about the frequency of paracentric inversions in natural populations.

Pericentric inversions are more readily detected than paracentric inversions because they change the relative position on the centromere. Two frequent pericentric inversions have been described in orangutans (*Pongo pygmaeus*). Orangs from Borneo (*P. p. pygmaeus*) and Sumatra (*P. p. abelii*) are fixed for different forms of an inversion at chromosome-2 (Figure 3.5, Seuanez 1986, Ryder and Chemnick 1993). All wild captured orangs have been homozygous for these two chromosomal types, while over a third of all captive-born orangs have been heterozygous (Table 3.3). A pericentric inversion of chromosome-9 is polymorphic in both subspecies (Table 3.3). The persistence of the polymorphism in chromosome-9 for this period of time is surprising. Divergence in proteins and mitochondrial DNA

sequences between Bornean and Sumatran orangutans support the chromosomal evidence that these two subspecies have been isolated for at least a million years (Ryder and Chemnick 1993). Two groups of authors proposed in 1996 that the orangutan of Borneo and Sumatra should be recognized as separate species on the basis of these chromosomal differences and molecular genetic divergence (Xu and Arnason 1996, Zhi et al. 1996). However, Muir et al. (2000) have argued that the pattern of genetic divergence among orangutans is complex and is not adequately described by a simple Sumatra–Borneo split.

Chromosomal polymorphisms seem to be unusually common in some bird species (Shields 1982). Figure 3.1 shows the karyotype of a cardinal that is heterozygous for a pericentric inversion of chromosome-5 (Bass 1979). There is evidence that suggests that such chromosomal polymorphisms may be associated with important differences in morphology and behavior among individuals (see Guest Box 3). For example, Rising and Shields (1980) have described pericentric inversion polymorphisms in slate-colored juncos that are associated with morphological differences in bill size and appendage size. They suggest that this polymorphism is associated with habitat partitioning during the winter when they live and forage in flocks.

3.1.8 Translocations

A translocation is a chromosomal rearrangement in which part of a chromosome becomes attached to a different chromosome. Reciprocal or mutual translocations result from a break in each of two nonhomologous chromosomes and an exchange of chromosomal sections. In general, polymorphisms for translocations are rare in natural populations because of infertility problems in heterozygotes.

Robertsonian translocations are a special case of translocations in which the break points occur very close to either the centromere or the **telomere**. A Robertsonian fusion occurs when a break occurs in each of two acrocentric chromosomes near their centromeres, and the two chromosomes join to form a single metacentric chromosome. Robertsonian fissions also occur where this process is reversed. Robertsonian polymorphisms can be relatively frequent in natural populations because the translocation involves the entire chromosome arm and balanced gametes are usually produced by heterozygotes (Searle 1986, Nachman and Searle 1995).

Table 3.3 Chromosomal inversion polymorphisms in the orangutan. The inversion in chromosome-2 distinguishes the Sumatran (*S*) and Bornean (*B*) subspecies. The two inversion types in chromosome-9 (*C* and *R*) are polymorphic in both subspecies. From Ryder and Chemnick (1993).

	Chromosome 2			Chromosome 9		
	BB	SB	SS	CC	CR	RR
Wild-born	51	0	41	67	22	3
Zoo-born	90	44	82	71	34	3

Table 3.4 Litter sizes produced by mice heterozygous for Robertsonian translocations characteristic of three different chromosomal races (AA, POS, and UV). From Hauffe and Searle (1998).

Female	Male	No. litters	Litter size
AA	AA (control)	17	6.7 ± 0.8
AA	(AA × POS)	16	4.1 ± 0.4
AA	(AA × UV)	18	2.6 ± 0.3
AA	(UV × POS)	19	3.8 ± 0.3
AA (control)	AA	18	6.8 ± 0.4
(AA × POS)	AA	7	1.0 ± 0
(AA × UV)	AA	10	3.1 ± 0.6
(POS × UV)	AA	11	4.0 ± 0.5

The Western European house mouse has an exceptionally variable karyotype (Piálek *et al.* 2005). The rate of evolution of Robertsonian changes in this species is nearly 100 times greater than most other mammals (Nachman and Searle 1995). Over 40 chromosomal races of this species have been described in Europe and North Africa on the basis of Robertsonian translocations and whole-arm reciprocal translocations (Hauffe and Searle 1998, Piálek *et al.* 2005). As expected, hybrids between three of these races have reduced fertility. On average, the litter size of crosses with one hybrid parent is 44% less than crosses between two parents from the AA race (Table 3.4).

3.1.9 Chromosomal variation and conservation

The main concern of chromosomal differences for conservation has been as a possible source of reduced fitness in individuals resulting from crosses between different populations (**outbreeding depression**, Frankham *et al.* 2011). As we have seen, chromosomal heterozygotes often have reduced fitness. We expect greater rates of chromosomal evolution by genetic drift in small, isolated populations in the case of reduced fitness of heterozygotes (Lande 1979, see Section 8.2.3). Thus, we should be most concerned about outbreeding depression because of chromosomal rearrangements in taxa that tend to have small, fragmented populations (e.g., rodents and primates).

Plants tend to have greater rates of chromosomal rearrangements than animals (Hoffmann and Rieseberg 2008). Chromosomal evolution appears to be highest in annual plants, probably because they are prone to dramatic fluctuations in population size (Harrison *et al.* 2000). Second, many plant species can reproduce by *selfing*, and this greatly increases the probability of chromosomal rearrangements becoming fixed. In addition, differences in ploidy among plant populations are another possible source of reduced fitness of chromosomal heterozygotes resulting from matings between populations.

Recent studies that include both chromosomal and genomic analyses have shown that chromosomal rearrangements can contribute to local adaptation (Example 3.2). In addition, the reduced recombination in inversion regions can be favored because it allows locally adapted alleles to remain together and be inherited as a unit (Kirkpatrick and Barton 2006). Finally, genes located within inversions have been found to be associated with traits involved in adaptation to a changing environment (Hoffmann and Rieseberg 2008). For example, several *Drosophila* studies on chromosomal inversion polymorphisms have found adaptive shifts in response to the unprecedented increases in temperature in the past 30 years (Hoffmann and Rieseberg 2008).

3.2 PROTEIN ELECTROPHORESIS

The first major advance in our understanding of genetic variation in natural populations began in the mid-1960s with the advent of protein electrophoresis (Powell 1994). The detection of variation in amino acid sequences of proteins by electrophoresis allowed an immediate assessment of genetic variation in a wide variety of species (Lewontin 1974). There is a direct relationship between genes (DNA base pair sequences) and proteins (amino acid sequences). Proteins have an electrical charge and migrate in an electrical field at different rates depending upon their charge, size, and shape. A single amino acid substitution can affect migration rate and thus can be detected by electrophoresis. Moreover, the genomes of all animals (from elephants to *Drosophila*), all plants (from sequoias to the Furbish lousewort), and all microbes (from *E. coli* to HIV) encode proteins. Empirical population genetics has become universal, as genetic variation has been described in natural populations of thousands of species in the last 50 years.

Figure 3.7 outlines the procedures in the gel electrophoresis of enzymes (see also May 1998). There are

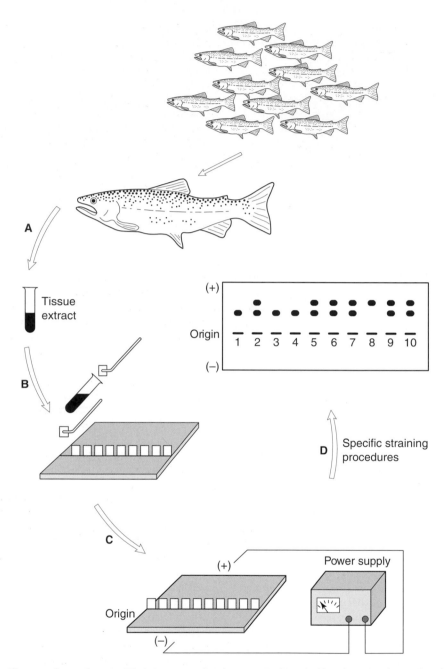

Figure 3.7 Allozyme electrophoresis. (A) A tissue sample is homogenized in a buffer solution and centrifuged. (B) The supernatant liquid is placed in the gel with filter paper inserts. (C) Proteins migrate at different rates in the gel because of differences in their charge, size, or shape. (D) Specific enzymes are visualized in the gel by biochemical staining procedures. From Utter *et al.* (1987).

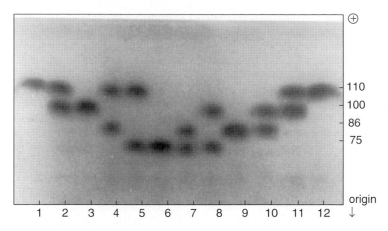

Figure 3.8 Gel electrophoresis of the enzyme aconitate dehydratase in livers of 12 Chinook salmon. The relative mobilities of the allozymes encoded by four alleles at this locus are on the right. The genotypes of all 12 individuals are (1) *110/110*, (2) *100/110*, (3) *100/100*, (4) *110/86*, (5) *110/75*, (6) *75/75*, (7) *86/75*, (8) *100/75*, (9) *86/86*, (10) *100/86*, (11) *110/100*, and (12) *110/110*. From Utter *et al.* (1987).

two fundamental steps to electrophoresis. The first is to separate proteins with different mobilities in some kind of a supporting medium (usually a gel of starch or polyacrylamide). However, most tissues contain proteins encoded by hundreds of different genes. The second step of the process, therefore, is to locate the presence of specific proteins. This step is usually accomplished by taking advantage of the specific catalytic activity of different enzymes. Specific enzymes can be located by staining gels with a chemical solution containing the substrate specific for the enzyme to be assayed, and a salt that reacts with the product of the reaction catalyzed by the enzyme and produces a visible product.

We can stain different gels for many enzymes and thus examine genetic variation at many protein loci. Figure 3.8 shows variation at the enzyme aconitate dehydrogenase in Chinook salmon from a sample from the Columbia River of North America. Alleles are generally identified by their relative migration distances in the gel. Thus, the *86* allele migrates approximately 86% as far as the common allele, *100*.

3.2.1 Strengths and limitations of protein electrophoresis

Protein markers were the workhorse for describing the genetic structure of natural populations for 40 years (Lewontin 1991). They were the first genetic markers used to estimate the amount of genetic variation in natural populations. In addition, they provided initial glimpses into the reproductive behavior of wild species populations.

The strengths of protein electrophoresis are many. First, genetic variation at a large number of nuclear loci can be studied with relative ease, speed, and low cost. In addition, the genetic basis for variation of protein loci can often be inferred directly from electrophoretic patterns because of the codominant expression of isozyme loci, the constant number of subunits for the same enzyme in different species, and consistent patterns of tissue-specific expression of different loci. Third, it is relatively easy for different laboratories to examine the same loci and use identical allelic designations so that datasets from different laboratories can be combined (White and Shaklee 1991).

Protein electrophoresis has several weaknesses. First, it can examine only a specific set of genes within the total genome, those that code for water-soluble enzymes. In addition, this technique cannot detect genetic changes that do not affect the amino acid sequence of a protein subunit. Thus, silent substitutions within codons or genetic changes in noncoding regions within genes cannot be detected with protein electrophoresis. Finally, this technique usually requires that multiple tissues be taken for analysis and stored in ultra-cold freezers. Thus, individuals to be analyzed often must be sacrificed and the samples must be treated with care and stored under proper refrigeration. Techniques that examine DNA directly using the **polymerase chain reaction** (PCR) do not require lethal sampling and can often use old (or even ancient) specimens that have not been carefully stored.

Protein electrophoresis is still a good tool for some questions. For example, so-called cryptic species occur in many groups of invertebrates (Gouws *et al.* 2004). Protein electrophoresis is the quickest and best initial method for detecting cryptic species in a sample of individuals from an unknown taxonomic group (e.g., Close and Gouws 2007). Individuals from different species will generally be fixed for different alleles at some loci. The absence of heterozygotes at these loci would suggest the presence of two reproductively isolated genetic divergent groups (Ayala and Powell 1972). As we will see in the next chapter, PCR-based DNA techniques generally either require prior genetic information or they rely upon techniques in which heterozygotes cannot be distinguished from some homozygotes.

3.3 GENETIC VARIATION WITHIN NATURAL POPULATIONS

Protein electrophoresis often provides more meaningful comparisons of the amount of genetic variation between different species than more recent techniques. With protein electrophoresis, the same or a similar suite of loci are used to study genetic variation in different species. In addition, all available loci are screened, even those that are monomorphic. This is in contrast with microsatellite loci, which we will consider in the next chapter. Generally, many microsatellite loci are screened and then only those that are variable in the target species are chosen for analysis. This selection process results in an **ascertainment bias**, so that the amount of variation found in different species does not provide a meaningful comparison. For example, assume we select a group of microsatellites for use in a particular species because they are polymorphic. A comparison between the amount of genetic variation at these loci in our target species and a closely related one is biased because we selected these loci to use because they were polymorphic in our target species (Morin *et al.* 2004).

The most commonly used measure to compare the amount of genetic variation within different populations is **heterozygosity**. At a single locus, heterozygosity is the proportion of individuals that are heterozygous; heterozygosity (H) ranges between zero and one. Two different measures of heterozygosity are used. H_o is the observed proportion of heterozygotes. For example, 7 of the 12 individuals in Figure 3.8 are heterozygotes, so H_o at this locus in this sample is 0.58 ($7/12 = 0.58$). H_e is the expected proportion of heterozygotes if the population is mating at random; the estimation of H_e is discussed in detail in Chapter 5. H_e provides a better measure than H_o to compare the relative amount of variation in different populations, as long as the populations are mating at random (Nei 1977). Protein electrophoresis is commonly used to estimate heterozygosity at many loci in a population by calculating the mean heterozygosities averaged over all loci.

Another measure often used is **polymorphism**, or the proportion of loci that are genetically variable (P). The likelihood of detecting genetic variation at a locus increases as more individuals are sampled from a population. This dependence on sample size is partially avoided by setting an arbitrary limit for the frequency of the most common allele. We use the criterion that the most common allele must have a frequency of 0.99 or less.

3.3.1 Data from natural populations

Harris (1966) was one of the first to describe protein heterozygosity at multiple loci (Table 3.5). He described genetic variation in the human population in England and found three of ten loci to be polymorphic ($P = 0.30$). On the average, individuals were heterozygous at approximately 10% of all loci examined ($H_e = 0.097$). These initial results were strikingly similar to more extensive studies using many more loci and larger sample sizes of human populations (Nei 1975).

Table 3.5 Genetic variation at ten protein loci in humans (Harris 1966). H_e is the expected proportion of heterozygotes assuming random mating (see Chapter 5). H_o is the observed proportion of heterozygotes in the sample.

Locus	Allele frequency 1	2	3	H_e	H_o
AP	0.600	0.360	0.040	0.509	0.510
PGM-1	0.760	0.240	–	0.365	0.360
AK	0.950	0.050	–	0.095	0.100
7 loci	1.000	–	–	0.000	0.000
Total				0.097	0.097

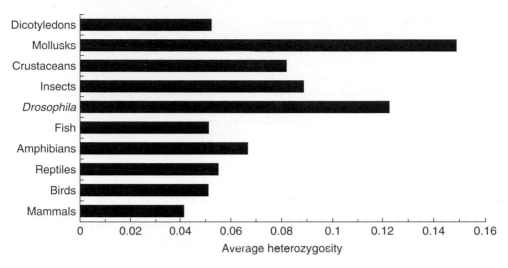

Figure 3.9 Average heterozygosities (H_S) at allozymes loci from major taxa. Modified from Gillespie (1992); data are from Nevo *et al.* (1984).

Nevo *et al.* (1984) summarized the results of protein electrophoresis surveys of some 1111 species! Average heterozygosities for major taxonomic groups are shown in Figure 3.9. Different species sometimes have enormous differences in the amount of genetic variation they possess (Table 3.6). For example, red pine and ponderosa pine are closely related species with vastly different amounts of genetic variation (Table 3.6, Example 11.2). Differences between species in amounts of genetic variation can have important significance. Remember – evolutionary change cannot occur unless there is genetic variation present and this can have important implications for conservation.

Differences between species in their amount of genetic variation need to be carefully interpreted. Species with lower variation are not necessarily more vulnerable to extinction (Hedrick *et al.* 1996). For example, initial studies indicated that the cheetah has much less genetic variation than other large cats (O'Brien *et al.* 1983, 1985). This finding of low allozyme variation in the cheetah led to the conclusion that the cheetah is vulnerable to extinction because of its lack of genetic variation (Table 3.6). However, the equilibrium genetic variation among species is expected to vary, largely because of differences in long-term effective population size. A species with low genetic variation does not necessarily suffer reduced fitness or is necessarily less able to become adapted to future environmental conditions (see Chapter 12).

Table 3.6 Summary of genetic variation demonstrating the range of genetic variation found in different species (Nevo *et al.* 1984). H_S is the mean expected heterozygosity (H_e) over all loci for all populations examined. P is the proportion of loci that is variable.

Species	No. loci	P(%)	H_e
Alligator	44	7	0.016
American toad	14	34	0.116
Cheetah	52	0	0.000
Humans	107	47	0.125
Moose	23	9	0.018
Polar bear	29	2	0.000
Roundworm	21	29	0.027
Gilia	13	52	0.106
Ponderosa pine	35	83	0.180
Red pine	35	3	0.007
Salsify	21	9	0.026
Yellow evening primrose	20	25	0.028

On the other hand, low genetic variation in a species might indicate a recent reduction in population size, and there are many reasons to expect that the loss of variation and increase in inbreeding associated with such a reduced population size (bottleneck) does potentially indicate vulnerability to extinction. First, a recent reduction bottleneck might indicate demographic instability that is not obvious from contemporary population size alone. Second, a species that has gone through a bottleneck severe enough to erode detectable molecular genetic variation might suffer from fixation of detrimental alleles, resulting in reduced fitness that might increase vulnerability to extinction. Finally, loss of genetic variation caused by the bottleneck may limit the ability of the population to evolve and adapt. The more recent a bottleneck has been, the more we would expect the bottleneck to influence the future of a species.

Thus, low genetic variation in itself does not indicate a conservation concern. However, a recent loss of genetic variation is a concern. We will see in later chapters (Chapters 6 and 12) that there are ways to detect recent losses of genetic variation.

Returning to the cheetah, recent studies suggest that the lower genetic variation results from a severe population bottleneck that occurred at the end of the last ice age, some 10,000 years ago (Marker *et al.* 2008). Since that time, cheetah populations have regained genetic variation at markers with higher mutation rates (see Chapter 12). As we will see, allozymes recover very slowly from bottlenecks because of their low mutation rate (Chapter 12). The low genetic variation in cheetahs appears to be associated with relatively low levels of normal spermatozoa and an increased susceptibility to infectious disease agents (Marker *et al.* 2008).

3.4 GENETIC DIVERGENCE AMONG POPULATIONS

The total amount of genetic variation within a species (H_T) can be partitioned into genetic differences among individuals within a single population (H_S) and genetic differences among different populations (Nei 1977). The proportion of total genetic variation that is due to differences among populations is generally represented by F_{ST}

$$F_{ST} = 1 - \frac{H_S}{H_T}$$

The meaning and estimation of F_{ST} is considered in detail in Chapter 9. For now, we use F_{ST} simply as an indicator of the amount of genetic divergence between populations in different taxa: the greater the value of F_{ST}, the greater the relative divergence among populations.

Ward *et al.* (1992) and Hamrick and Godt (1990, 1996) have summarized estimated values of these parameters in animal and plant species (Table 3.7). Some interesting patterns emerge from Table 3.7. First, invertebrates tend to have greater mean expected heterozygosity within populations (H_S) than vertebrates. This reflects the tendency for local populations of invertebrates to be larger than vertebrates because of the relationship that species with smaller body size tend to have larger population size (Cotgreave 1993). An analogous pattern is seen for plants. Species with a

Table 3.7 Comparison of H_T, H_S, and F_{ST} for different major taxa of animals (Ward *et al.* 1992) and plants classified by their geographic range (Hamrick and Godt 1990, 1996).

Taxa	H_T	H_S	F_{ST}	No. of species
Amphibians	0.136	0.094	0.315	33
Birds	0.059	0.054	0.076	16
Fish	0.067	0.054	0.135	79
Mammals	0.078	0.054	0.242	57
Reptiles	0.124	0.090	0.258	22
Crustaceans	0.088	0.063	0.169	19
Insects	0.138	0.122	0.097	46
Mollusks	0.157	0.121	0.263	44
Endemic plants	0.096	0.063	0.248	100
Regional plants	0.150	0.118	0.216	180
Widespread plants	0.202	0.159	0.210	85

wider range, and therefore greater total population size, have greater total mean heterozygosity (H_T) than endemic plants that have a more narrow range (Table 3.7, Example 3.3). We will examine the relationship between population size and genetic variation in Chap-

ters 6 and 12. Annual plant species, which complete their lifecycle within a year, generally have lower heterozygosity and greater divergence among populations than long-lived perennial plants, particularly woody perennials such as trees.

Example 3.3   Do rare plants have less genetic variation?

There are several reasons to expect that rare species with restricted geographic distributions will have less genetic variation. First, loss of genetic variation caused by chance events (e.g., genetic and the founder effect, Chapter 6) will be greater in smaller populations. In addition, species with restricted geographic distributions will occur in a limited number of environments and therefore be less affected by natural selection to exist under different environmental conditions.

   Karron (1991) provided a very interesting test of this expectation by comparing the amount of genetic variation at allozyme loci in closely related species. He compared congeneric species from ten genera in which both locally **endemic** and widespread species were present. One to four species of each type (rare and widespread) was used in each genus. In nine of ten cases, the widespread species had a greater number of average alleles per locus (Figure 3.10).

   These data support the prediction that rare species contain less genetic variation than widespread species. On average, widespread species tend to have greater molecular genetic variation than rare endemic species with a limited distribution in nine of ten genera. However, this relationship is not so simple. The amount of genetic variation in a species will be profoundly affected by the history of a species, as well as its current condition. For example, some very common and widespread species have little genetic variation because they may have gone through a recent population bottleneck (the red pine, Example 11.2). Rare endemic species should have relatively little genetic variation unless their current rareness is recent and they historically were more widespread. Thus, 'rareness' should be used cautiously as a predictor of the amount of genetic variation within individual species.

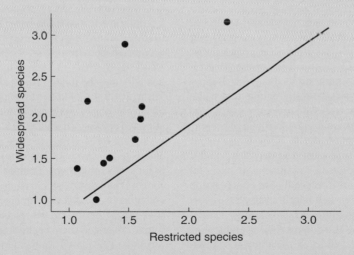

Figure 3.10   Comparison of the mean number of alleles per allozyme locus in widespread and restricted species of plants in ten genera. One to four species of each type was compared within each genus. The solid line shows equal mean values. In nine of ten cases, the mean of the widespread species was greater than the mean of the restricted species. From Karron (1991).

In addition, those taxa that we would expect to have greater ability for movement and exchange among populations have less genetic divergence among populations. For example, bird species have the same mean amount of genetic variation within populations as fish and mammals, but they have much less genetic divergence among populations. This difference reflects the greater ability of birds for exchange among geographi-cally isolated populations because of flight. Plant species that are wind pollinated, or have wind-dispersed seeds, generally have less genetic divergence among populations than species that are insect pollinated or have animal-dispersed seeds (Hamrick and Godt 1996). We will consider the relationship between exchange among populations and genetic divergence in Chapter 9.

Guest Box 3   Chromosomal polymorphism in the white-throated sparrow
E. M. Tuttle

Chromosomal inversions reduce the rate of recom-bination, often resulting in genotypes that have distinct evolutionary histories (see Section 3.1.7). The white-throated sparrow, a songbird that breeds throughout the boreal forests of northeastern United States and Canada, is an unusual case illus-trating this concept. This species is polymorphic, and both sexes occur as either tan or white morphs based on the color of their crown stripes (Figure 3.11, Lowther 1961).

Behavioral and ecological differences exhibited by the two morphs are striking. White males are promiscuous and invest heavily in securing extra matings through song, aggression, and territorial intrusion at the apparent expense of mate-guarding and paternal care (Tuttle 2003). Tan males invest heavily in monogamy (Tuttle 2003) and high levels of parental care (Knapton and Falls 1983). Behav-ioral and genetic data show that while only white males are promiscuous, white males are also 'cuck-olded' more than tan males and lose a large propor-tion of their own paternity to other white males (Tuttle 2003). White and tan females exhibit similar tradeoffs between investment in parental effort and investment in reproductive effort (Tuttle 2003), although their behavioral strategies are less clear. The morphs mate disassortatively (Tuttle 2003), maintaining the polymorphism and result-ing in pair types that differ in the amount of bipa-rental care they provide (Knapton and Falls 1983). The two disassortative pair types differ in other key behavioral and ecological attributes, such as terri-tory placement (Formica and Tuttle 2009).

The presence or absence of a complex series of inversions on chromosome-2 determines these plumage and behavioral morphs (Thorneycroft 1975). Birds with the genotype, ZAL2/ZAL2 are monogamous tan morphs, whereas birds with the heterozygous genotype, ZAL2^m/ZAL2, are pro-miscuous white morphs; homozygous ZAL2^m/ZAL2^m birds are rarely found (<0.05%; Tuttle, unpublished data). The rearrangement covers over 86% of chromosome-2 (approximately 104 Mb), thereby limiting recombination in heterozygotes (Romanov *et al.* 2009, Huynh *et al.* 2011). Since the first inversion of ZAL2^m originated approximately 2.3 million years ago (Thomas *et al.* 2008), it is likely that mutations that alter gene function have since accumulated within that chromosomal region, resulting in the differences in morphology and behavior we see today (Tuttle 2003, Formica and Tuttle 2009). An additional rearrangement located on chromosome 3 may also influence morphic vari-ation via epistasis (Romanov *et al.* 2009).

Current research has focused on identifying can-didate genes whose function may have been altered by this rearrangement. Comparative mapping has shown that sparrow chromosome 2 is mainly orthologous to chicken chromosome 3 (Thomas *et al.* 2008). Specific genes affected by the chromo-somal inversion are beginning to be identified through BAC-based mapping and comparative sequence analysis (Huynh *et al.* 2011, Romanov *et al.* 2011).

The evolutionary origins of this chromosomal inversion are of important interest. Thus far, compar-

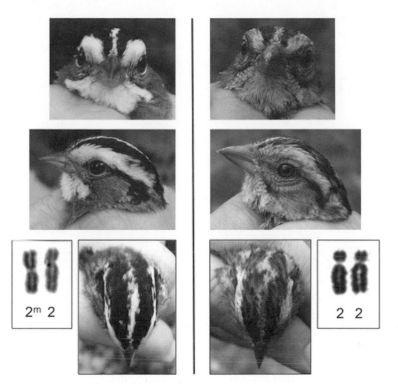

Figure 3.11 White and tan plumage morphs of the white-throated sparrow. Morph of an individual is absolutely associated with the presence (ZAL2^m/ZAL2 = white) or absence (ZAL2/ZAL2 = tan) of a chromosomal rearrangement. Photo by Elaina Tuttle. See Color Plate 2.

ative sequence analysis suggests that the three other North American congeneric species (white-crowned sparrow, Harris's sparrow, and golden-crowned sparrow) share at least part of the inversion with the white-throated sparrow; in contrast, the only South American *Zonotrichia* species (rufous-collared sparrow) does not (Romanov *et al.* 2009). Thorneycroft (1975) originally predicted that the tan morph was the ancestral form because homozygotes for ZAL2^m rarely occur. However, sequence analyses suggest that the white morph is likely the ancestral form and that the tan morph is derived (Romanov *et al.* 2009).

Chromosomal rearrangements, such as those exhibited by the white-throated sparrow, are important considerations for evolution, as gene clusters often function together enabling rapid adaptation (Crombach and Hogeweg 2007). Chromosomal rearrangements can initiate both micro- and macro-evolutionary processes (Hoffmann *et al.* 2004), and so it is important that we understand how they influence trait expression. Morphs of the white-throated sparrow provide a comparison of chromosomal types that occur as a polymorphism within the same population, so that factors affecting white and tan birds can be associated with specific genes and genotypes. They are, in a sense, analogous to natural 'inbred lines' with regard to chromosome 2, showing recombination at all other areas of the genome except chromosome 2 (and probably chromosome 3). Future research is sure to establish firmly the link between genes and behavior in this species, thereby revealing the evolutionary bases of phenotypic variation.

CHAPTER 4

Genetic variation in natural populations: DNA

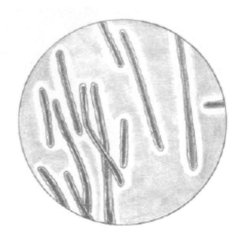

Thermus aquaticus ("Taq"), Box 4.1

The most direct, but unfortunately not the most useful, approach to the phylogeny of recent animals is through their genetics. The stream of heredity makes phylogeny; in a sense, it is phylogeny. Complete genetic analysis would provide the most priceless data for the mapping of this stream.

George Gaylord Simpson (1945, p. 5)

Just as the polymerase chain reaction leveled the genetic playing field at the end of the 20th century by providing easy access to the genes of all organisms, so the 21st century promises to sweep away the technological privileges of classic model organisms and democratize genomic exploration.

Camille Bonneaud *et al*. (2008)

Chapter Contents

Conservation and the Genetics of Populations, Second Edition. Fred W. Allendorf, Gordon Luikart, and Sally N. Aitken.
© 2013 Fred W. Allendorf, Gordon Luikart and Sally N. Aitken. Published 2013 by Blackwell Publishing Ltd.

Today we are able to perform "complete genetic analysis" in ways that George Gaylord Simpson could not imagine in 1945. Remember, Watson and Crick (1953) described the structure of DNA as the hereditary material eight years after the above quote from Simpson.

The polymerase chain reaction (PCR), described in Section 4.1.2, and technological advances in high-throughput DNA sequencing have revolutionized our ability to study genetic variation in wild populations beyond our wildest dreams. Who could imagine a few years ago that we would have nearly complete genome sequence information for Neandertals from three fossil bones that are nearly 40,000 years old (Green *et al.* 2010)! Moreover, usable DNA samples from extant species can be found in a variety of amazing sources: feces, hair left on trees, host blood in ticks, a single pollen grain, and even in the breath of dolphins (Matsuki *et al.* 2007, Frère *et al.* 2010).

Constantly changing methods are used for detecting variation in DNA sequences in natural populations (Figure 4.1). We will discuss the primary approaches that are used to study this variation. Sunnucks (2000) and Schlötterer (2004) have reviewed the principle methods for DNA analysis and their advantages and disadvantages. It is important to remember that there is no universal 'best' technique. The best technique to examine genetic variation depends upon both the question being asked and the extent and type of genetic information available for the species of concern. The toolkit of a molecular geneticist is analogous to the toolbox of a carpenter. Whether a hammer or power screwdriver is the best tool depends on whether you are trying to drive in a nail or set a screw.

This chapter provides a conceptual overview of the primary techniques employed to study genetic variation in natural populations. Our emphasis is on the nature of the genetic information produced by each technique and how it can be used in conservation genetics. Some of these techniques are no longer in common use. Our primary criterion for inclusion in this chapter is that an understanding of the technique is needed to read the conservation genetics literature. Detailed descriptions of the techniques and procedures can be found in the original papers. Many of these techniques are described in Hoelzel (1998); we are not aware of a more recent technical description of these techniques in one source. Millar *et al.* (2008) provided a very helpful overview of the application of new sequencing techniques to evolutionary questions.

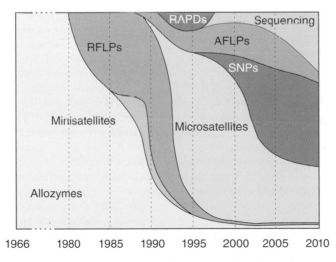

Figure 4.1 Subjective view of the changing relative popularity of major molecular markers in conservation genetics. The horizontal axis indicates time. At each time point, the vertical axis corresponds to the relative use of molecular markers. AFLP, amplified fragment length polymorphism; RAPD, randomly amplified polymorphic DNA; RFLP, restriction fragment length polymorphism; SNP, single nucleotide polymorphism. Modified and extended from Schlötterer (2004).

4.1 MITOCHONDRIAL AND CHLOROPLAST ORGANELLE DNA

Animal and plant cells have circular DNA in organelles called mitochondria, and plants also have circular DNA in organelles called chloroplasts, in addition to nuclear DNA in chromosomes. These molecules are relatively small, are usually inherited from a single parent, and usually undergo no recombination. The first studies of DNA variation in natural populations examined animal mitochondrial DNA (mtDNA) because it is small (approximately 17,000 base pairs in vertebrates and many other animals), relatively easy to isolate from genomic DNA, and occurs in thousands of copies per cell. These characteristics allowed investigators to separate mtDNA from nuclear DNA by ultracentrifugation.

In 1979, two independent groups published the first reports of genetic variation in DNA from natural populations. Avise *et al.* (1979a,b) used **restriction enzyme** analysis of mtDNA to describe sequence variation and the genetic population structure of mice and pocket gophers. Avise (1986) provided an overview of that early work. Brown and Wright (1979) used the maternal inheritance of mtDNA to determine the sex of two lizard species that originally hybridized to produce **parthenogenetic** species. A paper by Brown *et al.* (1979) compared the rate of evolution of mtDNA and nuclear DNA in primates. This latter work was done in collaboration with Allan C. Wilson, whose lab became a center for the study of the evolution of mtDNA (Wilson *et al.* 1985).

Several characteristics of animal mtDNA make it especially valuable for certain applications in understanding patterns of genetic variation. Mitochondrial DNA is haploid and maternally inherited in most species. That is, a progeny generally inherits a single mtDNA genotype from its mother (Figure 4.2). There are thousands of mtDNA molecules in an egg, but relatively few in sperm. In addition, mitochondria from the sperm are actively destroyed once they are inside the egg. There are many exceptions to strict maternal inheritance. For example, there is evidence of some incorporation of male mitochondria ('paternal leakage') in species that generally show maternal inheritance, such as mice (Gyllensten *et al.* 1991) and humans (Awadalla *et al.* 1999). In addition, some species (e.g., many mussels) show doubly uniparental inheritance of mitochondrial DNA in which there are separate maternally and paternally inherited mtDNA molecules

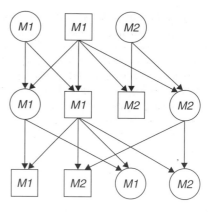

Figure 4.2   Pedigree showing maternal inheritance of two mtDNA genotypes: *M1* and *M2*. By convention in pedigrees, males are represented by squares and females are represented by circles. Each progeny inherits the mtDNA of its mother.

(Sutherland *et al.* 1998). Both maternal leakage and mutation can lead to **heteroplasmy** (the presence of more than one organelle genotype within a cell or individual).

Mitochondrial DNA molecules are especially valuable for reconstructing phylogeny because there is generally no recombination between mtDNA molecules. Unlike nuclear DNA, the historical genealogical record of descent is not 'shuffled' by recombination between different mtDNA lineages during gamete production, as occurs in nuclear DNA during meiosis. Recombination between lineages is unlikely because mtDNA generally occurs in only one lineage per individual (one haploid genome) since the male gamete does not contribute mtDNA to the zygote. However, there is some evidence for rare recombination events in animal mtDNA (Slate and Gemmell 2004, Ujvari *et al.* 2007). Thus, the mtDNA of a species can be considered a single non-recombining genealogical unit with multiple alleles or haplotypes (Avise 2004).

The lack of recombination, which makes mtDNA especially valuable in reconstructing phylogenies, reduces its value for describing genetic population structure within species. The primary problem is that the entire mtDNA genome acts as a single locus because there is no recombination. As we will see in Chapters 6 and 9, there can be substantial differences between loci in the patterns of genetic variation just by chance alone. In addition, there is increasing evidence that mtDNA is affected by natural selection, and therefore might not accurately reflect the demographic or evolutionary history and processes within a species (Ballard and Whitlock 2004, Dowling *et al.* 2008, Galtier *et al.* 2009, Balloux 2010). Finally, maternal inheritance makes it especially inappropriate to be used as the sole source of genetic knowledge for describing units of conservation since patterns of divergence are not influenced by genetic contributions of males (see Example 9.3).

Plant mitochondrial DNA has been less useful than animal mtDNA for genetic studies of natural populations primarily because of a lower mutation rate resulting in low variation (Clegg 1990, Powell 1994, Petit *et al.* 2005). In addition, in some cases rearrangements within the plant mtDNA genome and the transfer of so-called promiscuous DNA between the nuclear and chloroplast genomes make it difficult to use so-called universal primers (see Section 4.1.2) with plant mtDNA (Kubo and Mikami 2007). There is also some recombination for mitochondrial and chloroplast molecules in many plant species (Mackenzie and McIntosh 1999, McCauley and Ellis 2008).

DNA from plant chloroplasts (cpDNA) has proven very useful for phylogeographic studies in recent years (Petit *et al.* 2005), with the identification of regions of cpDNA that are variable across a wide range of taxa (Provan *et al.* 2001, Ebert and Peakall 2009, see Section 4.2.1). Most plant species have largely or completely maternal inheritance of cpDNA, but conifers in the pine family have largely paternal inheritance (Petit *et al.* 2005). Within some plant species, inheritance can be biparental. For example, McCauley *et al.* (2005) found primarily maternal inheritance (96% of all individuals) in *Silene vulgaris*, a gynodioecious species with some bisexual individuals and some plants producing only female flowers. Despite this high rate of maternal inheritance, the cumulative genetic effects of paternal leakage were evident as heteroplasmy was detected in over 20% of all individuals.

Chloroplast DNA markers have also played a key role in detecting hybridization and introgression between plant species, and revealing cases in which one species has taken on the chloroplast genotype of another, in a process called chloroplast capture. For example, the European oaks *Quercus petraea* and *Q. robur* are morphologically distinct, yet share chloroplast haplotypes in sympatric populations (Petit *et al.* 1997).

4.1.1 Restriction endonucleases and RFLPs

The discovery of restriction **endonucleases** (restriction enzymes) in 1968 (Meselson and Yuan 1968) marked the beginning of the era of genetic engineering (i.e., the cutting and splicing together of DNA fragments from different chromosomes or organisms). Restriction endonucleases are enzymes in bacteria that provide a protective function by cleaving foreign DNA from intracellular viral pathogens (bacteriophages) harmful to the bacteria. The bacterial DNA is protected from cleavage because it is methylated. Nearly 1000 different restriction enzymes have been described from bacteria.

The most commonly used restriction endonuclease is *Eco*RI from the bacterium *Escherichia coli*. *Eco*RI cleaves a specific six base sequence: GAATTC (and the reverse compliment CTTAAG). The cleavage is uneven such that each strand is left with an overhang of AATT as follows:

5′-XXXXXXXX**GAATTC**XXXXXXX-3′
3′-XXXXXXXX**CTTAAG**XXXXXXX-5′

$$\downarrow$$

5′-XXXXXXXXG   **AATTC**XXXXXXX-3′
3′-XXXXXXXXX**CTTAA**   GXXXXXXX-5′

where the overhang is in bold and the Xs represent the sequence flanking the restriction site sequence.

Sequence differences between individuals can produce different results when we cut DNA molecules with restriction enzymes. Within the same segment of DNA, some individuals might have only one restriction site, while others might have two or three. A circular DNA molecule (such as mitochondrial DNA) with one restriction site will yield one linear DNA fragment after cleavage (Figure 4.3). If two cleavage sites exist, then two linear DNA fragments are produced from the cleavage (Figure 4.3). We can visualize the number of fragments using gel electrophoresis to separate them by length; short fragments migrate faster than long ones (Figure 4.3).

This is the basis of the **restriction fragment length polymorphism** (RFLP) technique for detecting DNA polymorphisms. Restriction site polymorphisms are usually generated by a single nucleotide

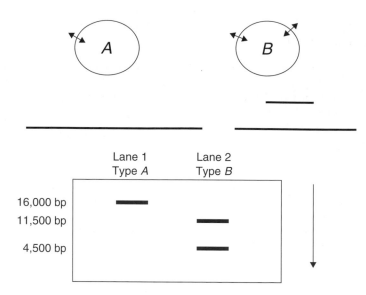

Figure 4.3   Hypothetical examination of sequence differences in mtDNA revealed by restriction enzyme analysis. Type *A* has only one cleavage site (arrow) which produces a single linear fragment of 16,000 base pairs; type *B* has two cleavage sites which produce two linear fragments of 11,500 and 4500 base pairs. Electrophoresis of the digested products results in the pattern shown. The DNA fragments move in the direction indicated by the arrow, and the smaller fragments migrate faster.

substitution in the restriction site (e.g., from GA*A*TTC to GA*T*TTC). This causes the loss of the restriction site in the individual because the enzyme will no longer cleave the individual's DNA. Thus a restriction site polymorphism is detectable as an RFLP following digestion of the molecule with a restriction enzyme and gel electrophoresis.

Each restriction enzyme cuts a different specific DNA sequence (usually four, six, or eight base pairs in length). For example, *Taq*1 cuts at TCGA. More than 600 different enzymes are commercially available. Thus we can easily study polymorphism across a mtDNA molecule by using a large number of different restriction enzymes.

Figure 4.4 shows RFLP variation in the mtDNA molecule of two subspecies of cutthroat trout digested by two restriction enzymes (*Bgl*I and *Bgl*II) that each recognize six base pair sequences. There are three cut sites for *Bgl*I in the W (westslope cutthroat trout) haplotype; there is an additional cut site in the Y (Yellowstone cutthroat trout) haplotype so that the largest fragment in the W haplotype is cut into two smaller pieces. The Y haplotype also has an additional cut site for *Bgl*II resulting in one more fragment than in the W haplotype. RFLP analysis is also useful for studies of nuclear genes following PCR amplification of a gene fragment (see Section 4.2.2).

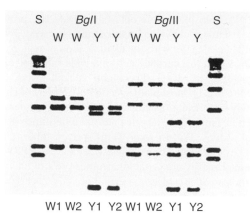

Figure 4.4   Restriction fragment polymorphism of mtDNA in cutthroat trout digested by two restriction enzymes (*Bgl*I and *Bgl*II). The W lanes are two (W1 and W2) westslope cutthroat trout and the Y lanes are two (Y1 and Y2) Yellowstone cutthroat trout. The S lanes are size standards. From Forbes (1990).

4.1.2 Polymerase chain reaction

Detection and screening of mtDNA polymorphism is now conducted using PCR (polymerase chain reaction, Box 4.1), followed by restriction enzyme analysis or by direct sequencing of the PCR product. For conducting

Box 4.1   Polymerase chain reaction

The polymerase chain reaction (PCR) can generate millions of copies of a specific target DNA sequence in about 3 hours, even when starting from small DNA quantities (e.g., one target DNA molecule!). Millions of copies are necessary to facilitate analysis of DNA sequence variation. PCR involves the following three steps conducted in a small (0.5 ml) plastic tube in a thermocycling machine: (a) denature (make single stranded) a DNA sample from an individual by heating the DNA to 95°C; (b) cool the sample to about 60°C to allow hybridization (i.e., annealing) of a primer (i.e., a DNA fragment of approximately 20 bp) to each flanking region of the target sequence; (c) reheat slightly (72°C) to facilitate extension of the single strand into a double-stranded DNA by the enzyme *Taq* polymer-

ase. These three steps are repeated 30–40 times until millions of copies result (Figure 4.5).

PCR was developed by Kary Mullis and his colleagues at Cetus Corporation in California in the 1980s. They used a DNA polymerase enzyme from the heat-stable organism *Thermus aquaticus* (thus the name '*Taq*'). This bacterium was originally obtained from hot springs of Yellowstone National Park in Montana and Wyoming. PCR has revolutionized modern biology and has widespread applications in the areas of genomics, population genetics, forensics, medical diagnostics, and gene expression analysis. Mullis was awarded the Nobel Prize in Chemistry in 1993 for his contributions to the development of PCR.

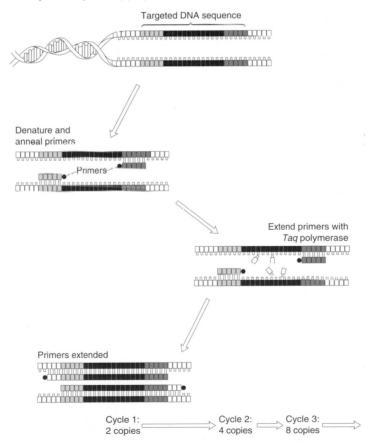

Figure 4.5   The main steps of the polymerase chain reaction (PCR) for amplifying specific DNA sequences: denaturing of the double-stranded template DNA, annealing of primers flanking the target sequence, and extension from each primer by *Taq* polymerase to add nucleotides across the target sequence and generate a double-stranded DNA molecule.

PCR, 'universal' primers are available for both mtDNA (Kocher *et al.* 1989) and chloroplast DNA (Taberlet *et al.* 1991). These primers will amplify a specific sequence (e.g., the cytochrome *b* gene) across a wide range of taxa. This universality has facilitated the accumulation of many DNA studies since the 1980s.

4.2 SINGLE-COPY NUCLEAR LOCI

There are a variety of techniques available to study variation at individual nuclear loci. Allozyme electrophoresis (Section 3.2) was the first such technique, but the techniques described here detect more variation than allozymes because they examine the DNA itself, rather than the protein product. We will describe a few of the most popular techniques that are currently in use.

4.2.1 Microsatellites

Microsatellites have become the most widely used DNA marker in population genetics for genome mapping, molecular ecology, and conservation studies. Microsatellite DNA markers were first discovered in the 1980s (Schlötterer 1998, Takezaki 2010). They are also called VNTRs (variable number of tandem repeats) or SSRs (simple sequence repeats) and consist of tandem repeats of a short sequence motif of 1 to 6 nucleotides (e.g., "CGTCGTCGTCGTCGT" which can be represented by $(CGT)^n$ where $n = 5$). The number of repeats at a polymorphic locus generally ranges from approximately 5 to 100. PCR primers are designed to hybridize to the conserved DNA sequences flanking the variable repeat units (Example 4.1). Microsatellite PCR products are generally between 75 and 300 bp long, depending

---

Example 4.1 Modified GenBank sequence database entry for the *Lla71CA* locus in the hairy-nosed wombat (Figure 4.6). The primers in the sequence at the bottom have been capitalized and the dinucleotide repeat region (CA) is shown in bold. The **n**'s in the sequence are base pairs that could not be resolved in the sequencing process.

```
1: AF185107. Lasiorhinus latif

LOCUS           AF185107 310 bp DNA linear MAM 01-JAN-2000

DEFINITION   Lasiorhinus latifrons microsatellite Lla71CA sequence.
  AUTHORS    Beheregaray,L.B., Sunnucks,P., Alpers,D.L. and Taylor,A.C.
  TITLE      Microsatellite loci for the hairy-nosed wombats
                 (Lasiorhinus krefftii and Lasiorhinus latifrons)
  JOURNAL    Unpublished
  AUTHORS    Taylor,A.C.
   JOURNAL     Submitted (31-AUG-1999) Biological Sciences, Monash
                 University, Wellington Rd., Clayton, VIC 3168, Australia
FEATURES               Location/Qualifiers
     source            1..310
     repeat_region     109..154
                       /rpt_type=tandem
                       /rpt_unit=ca

BASE COUNT        99 a     94 c     42 g     68 t     7 others

ORIGIN
    1 gngctcggnn cccctggatc acagaatcta aatctgagca tctcagAATG AGAAGGTATC
   61 TCCAGGataa ccannnccct ctacctaaac aagaattcca ctcccctaca cacacacaca
  121 cacacacaca cacacacaca cacacacaca cacactcaat agacccaaca agtggaatgt
  181 cacacagcct ttggggnagg tggggggatat acttCCTATG ACATAGCCTA TACCacttct
  241 gaatagtaac tttcctatcc ataaatctaa aacctacttc ccactctttt ctgctagttc
  301 tataatctgg
```

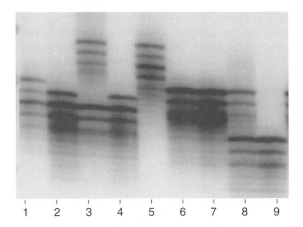

Figure 4.6 Microsatellite variation in the hairy-nosed wombat The sub-bands or stutter bands occur as a result of 'slippage' during PCR amplification. Allele sizes for each individual are: 1 (*187/191*), 2 (*187/189*), 3 (*187/199*), 4 (*187/189*), 5 (*195/199*), 6 (*191/191*), 7 (*191/191*), 8, (*183/191*), 9 (*183/183*). From Taylor *et al.* (1994).

on the locus and the location of the primers. Microsatellites are usually amplified using PCR and alleles are visualized using gel electrophoresis to separate fragments on the basis of differences in length resulting from different numbers of tandem repeats (Figure 4.6).

More than one microsatellite locus can be PCR amplified from a single tube and then identified separately on a sequencing gel using different colors of fluorescent dyes for each locus (Figure 4.7). Such multiplexes can greatly improve the rate of genotyping many individuals at many loci and lower costs. Nevertheless, the development of appropriate multiplex conditions can be time-consuming and often involves redesign of PCR primers. The choice of developing multiplex conditions is a tradeoff between the time taken to develop the conditions and the time saved by running fewer gels.

The main advantage of microsatellites is that they are usually highly polymorphic, even in small populations and endangered species (e.g., polar bears or cheetahs). This high polymorphism results from a high mutation

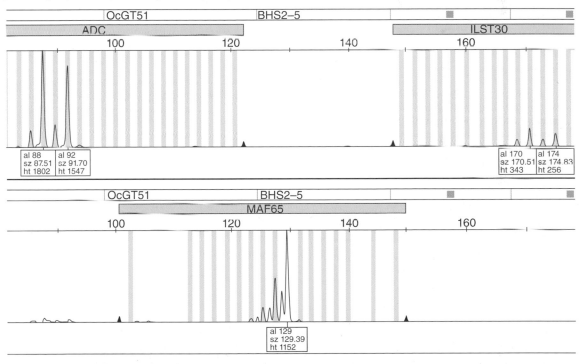

Figure 4.7 Three loci from a single microsatellite multiplex in bighorn sheep. Two fluorescent dyes are shown on different rows, with the fragment length (peak size) in base pairs along the x-axis and the fluorescent intensity (peak height) along the y-axis. Locus names are in grey boxes above the peaks, while allele length (al), peak size (sz), and peak height (ht) are in the boxes below the peaks. This individual is heterozygous at locus *ADC* (*88/92*) and locus *ILST30* (*170/174*), but homozygous at locus *MAF65* (*129/129*). From M. Kardos (unpublished).

rate due primarily to slippage during DNA replication (Ellegren 2000a, Chapter 12). A microsatellite mutation usually results in a change in the number of repeats (usually an increase or decrease of one repeat unit). The rate of mutation is typically around one mutation in every 1000 or 10,000 meioses (10^{-3} or 10^{-4} per generation). The high mutation rate of microsatellites (see Chapter 12) results in greater heterozygosity and allelic diversity at microsatellites than allozymes. Nevertheless, the relative amount of genetic variation for different marker types in different populations is expected to be the same (Box 4.2).

Box 4.2   Genetic variation in natural populations

The multiplicity of techniques presented in this chapter makes it possible to detect and study genetic variation in any species of choice. Some genetic variation has been discovered in virtually every species that has been studied. The Wollemi pine is a fascinating exception to this rule (see Example 4.2).

As we mentioned at the beginning of this chapter, there is no single 'best' technique to study variation in natural populations. The most appropriate technique to be used in a particular study depends on the question that is being asked. Generally, the relative amount of genetic variation detected by different techniques within a population or species is concordant. It is often informative to use more than one kind of marker. For example, using both mtDNA and nuclear markers allow assessment of female versus male-mediated gene flow (see Section 9.6).

Substantial differences in the amount of genetic variation can occur even between different populations within the same species. From a conservation perspective, such intraspecific differences are more meaningful than differences between species because they could indicate recent reductions in genetic variation caused by human actions.

Table 4.1 shows differences in the amount of genetic variation found between different population samples of brown bears from North America as detected with allozymes, microsatellites, and mtDNA. There are substantial differences in the absolute amount of variation in different marker types, but all three marker types demonstrate the identical relative pattern of variation: Alaska/Canada > NCDE > YE > Kodiak Island. Thus, the same population genetic processes appear to be affecting all three marker types in a similar fashion.

Not surprisingly, the isolated population of bears on Kodiak Island has the least amount of genetic variation. There are approximately 3000 bears on this island that have been isolated for approximately some 5000 to 10,000 years. More surprising is the substantially lower genetic variation in bears from the Yellowstone ecosystem in comparison with their nearest neighboring population in the Northern Continental Divide Ecosystem. The Yellowstone population has been isolated for less than 100 years. It is important to determine whether or not this reduced variation in Yellowstone bears is historical or results from a bottleneck associated with recent human activities (Guest Box 7).

Table 4.1   Summary of genetic variation in brown bears from four regions of North America.

Sample	Allozymes		Microsatellites		mtDNA	
	H_e	$\bar{A}$	H_e	$\bar{A}$	h	A
Alaska/Canada	0.032	1.2	0.763	7.5	0.689	5
Kodiak Island	0.000	1.0	0.265	2.1	0.000	1
NCDE	0.014	1.1	0.702	6.8	0.611	5
YE	0.008	1.1	0.554	4.4	0.240	3

The allozyme (34 loci) data are from K.L. Knudsen *et al.* (unpublished); the microsatellite (8 loci) and mtDNA data are from Waits *et al.* (1998). The allozyme samples for Alaska/Canada are from the Western Brooks Range in Alaska, and the microsatellite and mtDNA samples for this sample are from Kluane National Park, Canada. H_e is the mean expected heterozygosity (see Section 3.3), $\bar{A}$ is the average number of alleles observed, and h is gene diversity. h is computationally equivalent to H_e, but is termed gene diversity because mtDNA is haploid so that individuals are not heterozygous (Nei 1987, p. 177).
NCDE, Northern Continental Divide Ecosystem (including Glacier National Park); YE, Yellowstone Ecosystem (including Yellowstone National Park).

Example 4.2   The Wollemi pine: coming soon to a garden near you?

The discovery of this tree in 1994 has been described as the botanical find of the century. At the time of discovery, the Wollemi pine was thought to have been extinct for over 100 million years; there are no other extant species in this genus (Jones *et al.* 1995). There are currently less than 100 individuals known to exist in a 'secret' and inaccessible canyon in Wollemi National Park, 150 km west of Sydney, Australia (Hogbin *et al.* 2000).

An initial study of 12 allozyme loci and 800 AFLP fragments failed to reveal any genetic variation (Hogbin *et al.* 2000). A study of 20 microsatellite loci also failed to detect any genetic variation in this species (Peakall *et al.* 2003). The exceptionally low genetic variation in this species combined with its known susceptibility to exotic fungal pathogens provides strong justification for current policies of strict control of access and the secrecy of their location.

The Wollemi pine reproduces both by sexual reproduction and asexual coppicing in which additional stems grow from the base of the tree. Some individual trees are more than 500 years old, and there are indications that coppicing can result in the longevity of a plant greatly exceeding the age of individual trunks (Peakall *et al.* 2003). It is possible that **genets** are thousands of years old.

Wollemi pine became available as a horticultural plant in 2006 (Figure 4.8). The plant is distinct in appearance and somewhat resembles its close relative the Norfolk Island pine, which is a popular ornamental tree throughout the world.

Figure 4.8   Wollemi pine in a pot. Photographer J. Plaza, Royal Botanic Gardens, Sydney, Australia.

Microsatellite primer pairs developed in one species can sometimes be used in closely related species because primer sites are generally highly conserved. For example, about 50% of primers designed from cattle will work in wild sheep and goats that diverged approximately 20 million years ago (Maudet *et al.* 2001). This is an enormous advantage because over 3500 microsatellites have been mapped in cattle; thus cattle primers can be tested to find polymorphic markers across the genome of any ungulate, without the time and cost of cloning and mapping for each new species. Similar transfer of markers is possible for wild canids, felids, primates, salmonid fishes, and galliform birds because genome maps are available with microsatellites. However, there may be some downward bias in estimates of genetic diversity for microsatellite primers transferred across species, particularly when a relatively small number of loci are selected for development based on high polymorphism in the focal species as they may be fixed or have relatively low polymorphism in a congener.

Microsatellite primer sets for thousands of species can be found at several websites. *Molecular Ecology Resources* has a website that contains all primers published in that journal plus many others. The United States National Center for Biotechnology Information (NCBI) maintains one of the primary websites for sequence information: http://www.ncbi.nih.gov/. This resource was established in 1988 (Wheeler *et al.* 2000). Example 4.1 shows the GenBank sequence database entry for the microsatellite locus shown in Figure 4.6.

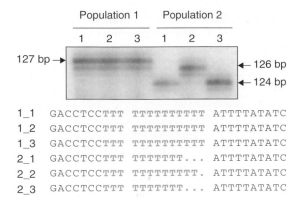

	Population 1			Population 2		
	1	2	3	1	2	3

127 bp →

← 126 bp
← 124 bp

1_1	GACCTCCTTT	TTTTTTTTTT	ATTTTATATC
1_2	GACCTCCTTT	TTTTTTTTTT	ATTTTATATC
1_3	GACCTCCTTT	TTTTTTTTTT	ATTTTATATC
2_1	GACCTCCTTT	TTTTTTT...	ATTTTATATC
2_2	GACCTCCTTT	TTTTTTTT.	ATTTTATATC
2_3	GACCTCCTTT	TTTTTTT...	ATTTTATATC

Figure 4.9 Chloroplast microsatellite polymorphism in six individuals of the leguminous tree *Caesalpinia echinata* from two populations. The three individuals from Population 1 all have the *127* allele; the sequence of these individuals shown below the gel indicates that they have 13 copies of the (T) mononucleotide repeat. Two different alleles are present in the three individuals from Population 2. From Provan *et al.* (2001).

4.2.1.1 Chloroplast microsatellites

An unusual type of microsatellite has been found to occur in the genome of chloroplasts (Provan *et al.* 2001). Chloroplast microsatellites are usually mononucleotide repeats and often have less than 15 repeats (Figure 4.9). These markers have proven exceptionally useful in the study of a wide variety of plants (Petit *et al.* 2005). Furthermore, the uniparental inheritance of chloroplasts (usually maternal in angiosperms and paternal in most conifers) make them useful for distinguishing the relative contributions of seed and pollen flow to the genetic structure of natural populations by comparing nuclear and chloroplast markers in angiosperms, or mitochondrial and chloroplast markers in conifers. By using cpDNA SSRs and mtDNA markers based on insertions or deletions in introns, Gerardi *et al.* (2010) were able to determine the number and distribution of glacial refugia in black spruce. They concluded that postglacial genetic contact among refugial lineages during recolonization from pollen-based but not seed-based gene flow has increased genetic diversity within populations and potentially increased adaptive capacity in this species.

4.2.2 PCR of protein-coding loci

Sometimes we are interested in studying variation at specific genes that are known to have effects on impor-

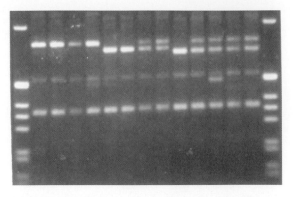

Figure 4.10 Length polymorphism in an intron for a growth hormone gene (*GH-1*) in coho salmon. The lanes on the ends are size standards. There are three alleles at this locus that differ by the number of copies of a 31-base pair repeat. The repeat occurs 11 times in the *a* allele, 9 times in *b*, and 8 in *c*. The genotypes from left to right are *a*/*a*, *a*/*a*, *a*/*a*, *a*/*a*, *b*/*b*, *b*/*b*, *a*/*b*, *a*/*b*, *c*/*c*, *a*/*c*, *a*/*c*, *a*/*c*, and *a*/*c*. From S.H. Forbes (unpublished).

tant phenotypes (**candidate genes**) in natural populations. If appropriate sequence information is available, PCR primers can be designed to detect genetic variation at these candidate loci. Those regions that actually encode proteins (**exons**) tend to be much less variable than the intervening noncoding regions (**introns**). PCR primers can be designed using exon sequences that will produce a PCR product that consists primarily of the more variable introns. These have been called EPIC PCR-markers (exon-priming intron-crossing) or expressed sequence tag polymorphisms (ESTPs). Figure 4.10 shows a polymorphism in an intron from a growth-hormone gene (*GH-1*) in coho salmon. There are three alleles at this locus that differ by the number of copies of a 31-base pair repeat (Forbes *et al.* 1994). The repeat occurs 11 times in the *a* allele, 9 times in *b*, and 8 in *c*.

4.2.3 Single nucleotide polymorphisms

Single nucleotide polymorphisms (**SNPs**) are the most abundant type of polymorphism in the genome, with one occurring about every 200–500 bp in many wild animal populations (Brumfield *et al.* 2003, Morin *et al.* 2004). For example, a G and a C might exist in different individuals at a particular nucleotide position within a population (or within a heterozygous individual). Because the mutation rate at a single base pair is low (about 10^{-8} changes per nucleotide per generation), SNPs usually consist of only two alleles. Thus, SNPs are usually bi-allelic markers. Transitions are a replacement of a purine with a purine (G⇔A) or a pyrimidine with a pyrimidine (C⇔T). Transversions are a replacement of a purine with a pyrimidine (A or G ⇔ C or T). Even though there are twice as many possible transversions as transitions, SNPs in most species tend to be transitions. This is both because of the physical and chemical nature of the mutation process (transition mutations are more common than transversions), and because transversions in coding regions are more likely to cause an amino-acid substitution than transitions and be subject to selection.

SNPs are useful markers for describing genetic variation in natural populations. For example, Akey et al. (2002) described allele frequencies at over 26,000 SNPs in three human populations. There are now over ten million SNPs available in humans. Two randomly chosen humans will differ at up to several million single nucleotide sites over their entire genomes. On average, there is one SNP in humans every 300 bp. SNPs are even more common in many other species because humans arose recently in evolutionary terms from relatively few founders, and thus have somewhat limited genomic variation.

The marker types discussed above are largely selectively neutral, not affecting phenotypes or fitness, but very useful for determining genetic relationships among individuals, gene flow, population structure, and demographic history. SNPs in non-coding regions of genes (i.e., introns), or in intergenic regions are also likely to be selectively neutral markers. However, SNPs in coding regions are more likely to have a phenotypic effect and thus affect fitness. Within coding regions, third base-pair substitutions can be synonymous or silent, resulting in no change in amino-acid sequence. This means that SNPs are powerful markers for separating the effects of history and demographics from natural selection on a genome-wide basis (Luikart et al. 2003); however, it also means that their analysis can be complex as different evolutionary forces can be acting on different SNPs (Allendorf et al. 2010). Selectively neutral and non-neutral markers provide different yet complementary information on population structure, demographics, and adaptation, and require different analytical approaches, as we will discuss in later chapters.

The first step in using SNPs is to identify single nucleotides that are polymorphic using a discovery panel of a small number of individuals from several populations (Seeb et al. 2009). This is most commonly done through DNA sequencing (see Section 4.4). SNPs and some other types of polymorphisms can also be identified through ecotilling (targeted induced local lesions in genomes) (Comai et al. 2004). DNA from a reference individual is hybridized with that of a target individual, and the enzyme CEL1 is used to cut strands at SNPs where base pairs are mismatched. This technique was used to discover SNPs in black cottonwood populations (Gilchrist et al. 2006).

Once SNPs have been discovered, many techniques exist for SNP genotyping. The increase in interest in SNPs (Figure 4.1) has been driven by development of a diverse range of inexpensive SNP genotyping methods (Perkel 2008). For example hundreds of SNPs now can be genotyped quickly using **quantitative (q) PCR** SNP chip microarrays. An advantage of **qPCR** assays is they allow analysis of samples with degraded or low quantities of DNA such as from noninvasive, historical, or environmental DNA samples (Section 22.1). For example, Campbell and Narum (2009) used 29 SNP qPCR assays to genotype old carcasses of Chinook salmon found decaying along river banks. The average genotyping success was 79% for SNPs but only 24% for microsatellites which had longer **amplicons** (PCR products) (Figure 4.11). Hundreds of thousands of SNPs can be genotyped by hybridization of DNA to SNP chip hybridization microarrays (Syvänen 2005). For example, vonHoldt et al. (2010) used a SNP genotyping microarray developed for the domestic dog to assay variation at ~48,000 loci for the Great Lakes wolf and red wolf. Using an analysis across all 38 canid autosomes they suggested that these canids are admixed varieties derived from gray wolves and coyotes, respectively. The recent interspecific admixture could complicate decisions regarding endangered species restoration and protection (Chapter 17). Many other SNP genotyping methods are available (Seeb et al. 2011).

Ascertainment bias is a crucial issue in many applications of both microsatellites and SNPs (Morin et al. 2004). Ascertainment bias results from the selection of loci from an unrepresentative sample of individuals, or using a particular method, which yields loci that are not representative of the spectrum of allele frequencies in a population. For example, if few individuals are used for SNP discovery (e.g., via DNA sequencing), then SNP loci with rare alleles may be underrepresented, and future genotyping studies using those SNPs could reveal a (false) deficit of rare alleles

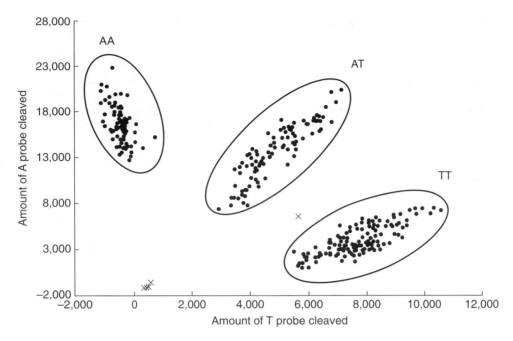

Figure 4.11 SNP genotyping assay in Chinook salmon (locus *OtspP450*). Each filled circle represents an individual fish whose genotype is determined by its position with respect to the two axes, which indicate the amount of each allele-specific probe (adenine [A] or thymine [T]) cleaved during the course of the assay. The ellipses indicate clusters of single genotypes. The Xs represent unreadable samples (due to air bubbles, failed PCR, etc.). From Smith *et al.* (2005a).

(e.g., false bottleneck signature). Ascertainment bias has the potential to introduce a systematic bias in estimates of variation within and among populations. The protocol used to identify SNPs for a study must be recorded in detail, including the number and origin of individuals screened, to enable ascertainment bias to be assessed and potentially corrected.

SNPs are in the process of replacing microsatellites as the marker of choice for many applications in conservation genetics (Figure 4.1, Seeb *et al.* 2011). One big advantage of SNPs is that it is much easier to standardize the scoring of genotypes when more than one laboratory is studying the same species (Stokstad 2010). SNPs will be especially useful for studies involving partially degraded DNA (from noninvasive and ancient DNA samples) because they are short and thus can be PCR-amplified from DNA fragments of less than 50 bases (PCR primers flanking SNPs must each be 20 bases or longer, depending on the genotyping system used). The time it takes until SNPs become more popular in conservation depends on the speed with which new technologies become available to permit rapid and inexpensive genotyping of SNPs in many individuals.

4.2.4 Sex-linked markers

Genetic markers in sex-determining regions can be especially valuable in understanding genetic variation in natural populations of animals. For example, markers that are specific to the sex-determining Y or W chromosomes can be used to identify the sex of individuals in species in which it is difficult to identify sex phenotypically, as in many bird species (Ellegren 2000b). In addition, Y-chromosome markers, like mtDNA, are especially useful for phylogenetic reconstruction because Y-chromosome DNA is haploid and non-recombining in mammals.

Mammals can be sexed using PCR amplification of Y chromosome fragments, or by coamplification of a homologous sequence on both the Y and X that are subsequently discriminated by size, restriction-enzyme cleavage of diagnostic sites, or by sequencing (Fernando and Melnick 2001). Similar molecular sexing techniques exist for birds (see Example 4.3) and other taxa (e.g., amphibians). The W chromosome of birds has conserved sequences not found on the Z, allowing nearly universal avian sexing PCR techniques (e.g.,

Example 4.3 Sexing and the detection of cryptic species of extinct moa

Moa were massive flightless ratites endemic to New Zealand that weighed up to 250 kg (Worthy and Holdaway 2002). All ten or so species in six genera of moa quickly became extinct within 100–200 years after Pacific Islanders colonized New Zealand approximately 700 years ago. Moa were the only wingless birds that lacked even vestigial wings, which all other ratites have. They were the dominant herbivores in New Zealand forest, shrubland, and subalpine areas. Before the arrival of humans, moa were hunted only by a single predator, the enormous Haast's eagle.

The taxonomy of moa has long been problematic, despite an extensive fossil record. Three moa species in the genus *Dinornis* that differed markedly in size were found throughout both major islands in New Zealand. Comparison of mtDNA sequences from **subfossil** remains indicated that the three species were genetically indistinguishable from each other within both the North and South Islands (Bunce *et al.* 2003, Huynen *et al.* 2003). However, the mtDNA genotypes of birds from each island were extremely different.

Sexing with a W-linked marker specific to females (Figure 4.12) indicated that the three previously described species morphological forms actually represented just one species on the North Island and one species on the South Island. Thus, rather than three species widespread on both islands, there were actually just two. The size of individuals differed dramatically according to sex and habitat. Females were much larger than males. The largest females were about 280% the weight and 150% the height of the largest males.

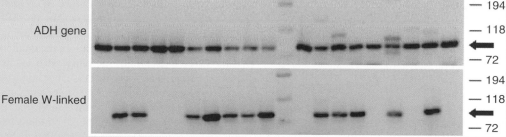

Figure 4.12 Subfossil moa samples screened for a short (85 bp, arrows) W-specific marker and the autosomal alcohol dehydrogenase (ADH) gene using PCR and gel electrophoresis. Those individuals with the W-linked fragment are females and those without are males. ADH PCR products were used to make sure the sample had sufficient quality DNA for reliable detection of the W-linked fragment. Note that those individuals that lack the W-linked fragment (males) do show an ADH gene PCR product. Gels from Bunce *et al.* (2003).

Huynen *et al.* 2002). Some plants also have sex-linked sequences (Korpelainen 2002). Robertson and Gemmell (2006) provide a needed caution and guidelines for using these techniques.

4.3 MULTIPLE LOCUS TECHNIQUES

Multiple locus techniques assay many anonymous locations across the genome simultaneously with a single PCR reaction (Bruford *et al.* 1998). The advantage of these techniques is that many loci can be examined readily with little or no information about sequences from the genome. The major disadvantages are that it is generally difficult to associate individual bands with particular loci, and that alternative alleles are either dominant or recessive. Unlike codominant loci, we usually cannot resolve between a heterozygote (with one band) and the homozygote 'dominant' type (also with one band, but two copies of it). Consequently, we cannot compute individual (observed) heterozygosity to test for Hardy-Weinberg proportions. In addition, these markers are bi-allelic and thus provide less information per locus than the more polymorphic microsatellites. It can require 5–10 times more of these loci to provide the same information as multiple allelic microsatellite loci (e.g., Waits *et al.* 2001). However, more loci (10–25) can be analyzed per PCR and per gel lane using some of these techniques, compared with microsatellites (5–10 loci per lane using fluorescent labels).

The RAPD method (randomly amplified polymorphic DNA) involved PCR amplification using short (usually 10 nucleotides) arbitrary primer sequences (Welsh and McClelland 1990). There was a brief crazy period in the 1990s when, in desperation, people thought that it might be useful to amplify poor, irreproducible, unidentified bands that often did not even come from the target organism, but then microsatellites came along, so most people stopped thinking this way (Paul Sunnucks, personal communication).

4.3.1 Minisatellites

Minisatellites are tandem repeats of a sequence motif that is approximately 20 to several hundred nucleotides long – much longer than *micro*satellite motifs. Minisatellites were first discovered by Jeffreys *et al.*

(1985), and used for DNA 'fingerprinting' in human forensics cases. They were used in wildlife populations soon after, for example, to study paternity and detect extra-pair copulations in birds thought to be monogamous, but they are rarely used today.

The high polymorphism of minisatellites made them most useful for interindividual studies such as parentage analysis and individual identification before microsatellites became widely available. The minisatellite genotyping technique was the first to be referred to as DNA fingerprinting because individuals possess unique minisatellite signatures (Jeffreys *et al.* 1985). Alleles are identified by the number of tandem repeats of the sequence motif. Disadvantages include difficulty in determining allelic relationships (identifying alleles that belong to one locus) because most minisatellite typing systems reveal bands (alleles) from many loci that are all visualized together in one gel lane. Also, the repeat motifs are long, so they cannot be studied in samples of partially degraded DNA generally containing small fragments of only 100–300 bases.

4.3.2 AFLPs and ISSRs

The AFLP technique uses PCR to generate DNA fingerprints (i.e., multilocus band profiles) based on many anonymous locations across the genome. These 'fingerprints' are generated by selective PCR amplification of DNA fragments produced by cleaving genomic DNA (Figure 4.13). This technique was named AFLP because it resembles the RFLP technique (Vos *et al.* 1995). However, these authors said that AFLP is not an acronym for amplified fragment length polymorphism because it does not detect length polymorphisms. The main advantage of the AFLP method is that many polymorphic markers can be developed quickly for most species, even if no sequence information exists for the species. In addition, the markers generally provide a broad sampling of the genome. Characterizing AFLPs is faster, less labor intensive, and provides more information than other commonly used techniques. Foll *et al.* (2010) have developed a Bayesian statistical treatment of AFLP banding patterns that allows a more accurate use of these markers for understanding genetic population structure.

Inter-simple sequence repeat (ISSR) markers generate a large number of DNA fragments from a single PCR. ISSR primers are based upon the simple sequence repeats found in microsatellites. Bands are generated

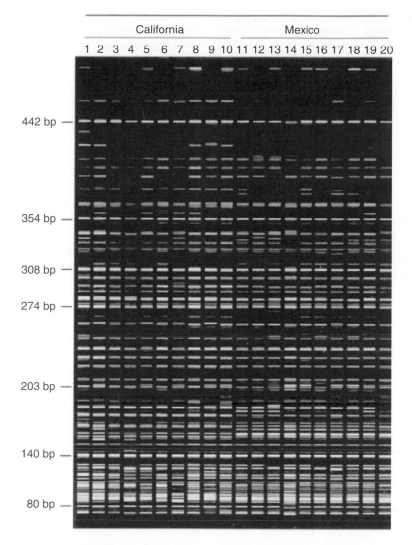

Figure 4.13   AFLP gel showing geographic differences among populations of the house finch. Lanes 1–10 are birds from California, and lanes 11–20 are birds from Mexico. From Wang *et al.* (2003).

by a single-primer PCR reaction where the primer is a repetition of a di-, tri- or tetranucleotide and the amplified region is a portion of genome between two identical microsatellite primers with an opposite orientation on the DNA strand. These primer sequences are broadly distributed on the genome. Therefore, the ISSR-PCR technique allows one quickly to screen a wide part of the genome without prior DNA sequence knowledge.

4.4 GENOMIC TOOLS AND MARKERS

The explosion in genomic methods, both in sequencing technology and in analytical 'bioinformatics' computing methods, is having a pronounced impact on population and conservation genetics. While many people think of genomics in terms of whole genome sequencing, the primary impacts for conservation genetics are the potential for far greater numbers of marker loci, rapid and inexpensive genotyping methods, and in the capacity for studying both selectively neutral and adaptive variation. Genomics research to date has largely focused on applying genomic techniques to model organisms, but advances for these model species have direct applications to populations of related wild species. Only now are we starting to see the application of these tools to inform conservation of threatened species and populations.

4.4.1 High-throughput sequencing

DNA sequencing methods were first developed independently by Walter Gilbert and Frederick Sanger. Gilbert and Sanger, along with Paul Berg, were awarded the Nobel Prize in Chemistry in 1980. Sanger and his colleagues used their own sequencing method to determine the complete nucleotide sequence of the bacteriophage *fX174*, the first genome ever completely sequenced.

The first application of DNA sequencing to the study of genetic variation in natural populations was by Kreitman (1983) who published the DNA sequences of 11 alleles at the alcohol dehydrogenase locus from *Drosophila melanogaster*. Initial studies of DNA variation were technically involved and time-consuming, so that it was expensive and difficult to sample large numbers of individuals from natural populations. However, the advent of the polymerase chain reaction in the mid-1980s removed these obstacles.

DNA sequencing is becoming more common as the process becomes less expensive and more automated. The complete mtDNA sequences (each just over 15,000 base pairs) for 53 humans from diverse origins were published over ten years ago (e.g., Ingman *et al.* 2000). More recently, Roach *et al.* (2010) analyzed the complete nuclear genome sequence (over 3 billion base pairs per genome) of two parents and their two children, who suffered from two clinical recessive disorders. They identified four candidate genes for both of these Mendelian disorders using family-based genome analysis.

In the past decade, new high-throughput sequencing methods have replaced Sanger sequencing and have revolutionized the study of genetic variation. Large-scale sequencing is becoming an accessible tool for the study of natural populations. The global genome sequencing capacity in the year 2001 was 2000 base pairs per second. As of 2011, one state-of-the-art

sequencing machine can generate 6000 base pairs of data per second! As a result, the management, analysis and archiving of genomic data through bioinformatics techniques has become a greater challenge than the sequencing itself.

While whole genome sequences still lie beyond available financial resources and provide more data than needed for most conservation questions (Allendorf *et al.* 2010), the genomic technological revolution has yielded three major benefits for conservation genetics in terms of DNA markers (Primmer 2009). First, it has made the identification and development of primers for traditional selectively neutral molecular markers, particularly microsatellites, easier. Secondly, it has facilitated the development and genotyping of new marker types in large numbers, most notably single-nucleotide polymorphisms (SNPs, see RADs below). Finally, it has allowed for the development of markers for functional polymorphisms that facilitate the study of variation directly relevant for adaptation. The availability of genome sequences for model species and associated genetic resources has greatly facilitated the development of genomic tools for related species of conservation concern, like the dog genome for population studies of wild canids such as wolves and foxes (Gray *et al.* 2009, Example 4.4).

4.4.2 Inferences from sequence data

DNA sequence data can be analyzed to provide several useful types of information that traditional genetic markers cannot. First, the distribution of frequencies of polymorphisms segregating across the regions sequenced, called the site frequency spectrum (SFS), can be used to make inferences about the demographic history of populations, and the past occurrence of bottlenecks (Charlesworth *et al.* 2003). For example, Holliday *et al.* (2010) resequenced Sitka spruce trees from six populations spanning the species range for 153 genes, and found evidence to support recent bottlenecks in northern but not southern populations, a strong demographic signature resulting from sequential postglacial recolonization. Secondly, local deviations from the genome-wide SFS can provide information about portions of the genome that are under selection, and these can be used to identify potential candidate genes for future study of adaptation (Kreitman 2000). In a whole genome resequencing study of the domestic chicken, Rubin *et al.* (2010)

Example 4.4   Use of genomic tools from model species for wild populations

The whole genome sequences of farmed animals including the cow, sheep, chicken, dog, and Atlantic salmon have provided a plethora of SNPs, microsatellite markers and genomic information for the study of wild populations of their relatives. For example, Poissant *et al.* (2009) tested approximately 600 mapped microsatellite primer pairs from domestic sheep on both bighorn sheep and mountain goats. Of the markers tested, 247 were polymorphic in bighorn sheep and 149 were polymorphic in mountain goats, increasing the number of markers available for population genetics in these wild species greatly.

SNPs are the most abundant type of genetic marker, and they have been highly transferable to closely related species. As most SNPs are bi-allelic, individual SNPs are less informative than individual microsatellite loci with several to many alleles; however, the sheer number of SNPs available, and the development of 'SNP chips', makes genotyping many loci simultaneously relatively easy. vonHoldt *et al.* (2011) used a ~48,000 SNP chip developed for the domestic dog to assess among- and within-species genetic diversity in several species of wolves and coyotes. The proportion of these SNPs that were polymorphic ranged from 7% in the Ethiopian wolf to 97% in the gray wolf.

The power in numbers of SNPs is well illustrated in a case using applying cow SNPs to the endangered European bison (Tokarska *et al.* 2009). This bison went through a severe bottleneck in the last century, was extirpated in the wild, and genetic variation in animals maintained in captivity was low. The Lowland Population of bison, with a current population size of ~1800, was founded by only four bulls and three cows, with two animals estimated to be responsible for 80% of the current gene pool. The bovine SNP chip containing 52,968 SNPs was used to identify 960 polymorphic SNPs in the European bison. These SNPs were then tested against 17 microsatellite loci for determining paternity of 92 offspring in a reintroduced Polish bison population. Using 50 of the most polymorphic SNPs, or 90 SNPs chosen at random, identified fathers with 95% confidence in at least 50% of the cases. In contrast, the microsatellite loci could only identify fathers with high confidence for two of the offspring.

identified several regions with a molecular signature suggesting that **selective sweeps** had occurred, resulting in low variation surrounding a recent mutation that has been favored by selection. They also identified over seven million single nucleotide polymorphisms in the chicken genome!

DNA nuclear or organelle sequence data can also be used to distinguish cryptic species that are morphologically indistinguishable but are reproductively isolated (see Section 22.1). Hebert *et al.* (2004) used partial sequence data for the mitochondrial gene COI1 to identify at least ten cryptic species of the neotropical skipper butterfly which are indistinguishable as adults, and to relate sequence variation to differences in caterpillar morphology and host plants. Sequence data can be used in combination with other types of markers and phenotypic trait variation to clarify taxonomic relationships and identify cryptic variation. Toews and Irwin (2008) combined data from mitochondrial gene sequences, AFLP variation, and recorded bird songs to identify cryptic eastern and western species of winter wrens previously considered to be a single interbreeding species.

4.4.3 EST sequencing applications

Expressed sequence tags (ESTs) are one of the most abundant types of genomic data and target that small fraction of the genome containing genes that are transcribed. These data are obtained through extracting and reverse transcribing bulk mRNA into cDNA libraries (Nagaraj *et al.* 2007). cDNA libraries are then sequenced to obtain single-read sequences of genes being expressed in the tissues that were sampled. PCR primers can then be developed from flanking sequences to resequence genes in multiple individuals and populations. The untranslated regions (UTRs) of ESTs can contain microsatellite sequences useful as highly polymorphic markers for a wide range of applications as described above in Section 4.2.

4.4.4 SNP discovery and genotyping by sequencing

The discovery of SNPs for nonmodel species has in the past required a large effort and investment, or genomic resources for closely-related model species (see Example 4.4). A number of recently developed methods using high-throughput sequencing now facilitate the discovery of SNPs distributed across and representative of the genome in a cost-effective manner (Davey *et al.* 2011). These methods are starting to be widely applied in wild populations (see Seeb *et al.* 2011 and references therein). It is now feasible to discover thousands of SNPs in nonmodel species quickly and at reasonable cost, even if no prior genomic information is available. Thousands of SNPs can be discovered and subsequently genotyped for the same cost required to discover and genotype only 10–20 microsatellite markers.

RADs

An inexpensive way to identify SNPs is through restriction-site associated DNA (RAD) tags, also called RAD-seq (Baird *et al.* 2008). DNA is fragmented with restriction enzymes, and DNA adjacent to the restriction sites is sequenced to identify SNPs. RAD tags are anonymous markers, like AFLPs, meaning their location within the genome is unknown, but unlike AFLPs, they are codominant. This approach was used to discover thousands of SNPs in westslope cutthroat trout populations (Hohenlohe *et al.* 2011). RAD tags were also used to conduct a genome-wide study of variation in several thousand SNPs in sticklebacks and to confirm the repeated parallel evolution of freshwater sticklebacks from marine ancestral populations (Guest Box 8). In addition to confirming the multiple origins of the freshwater forms of sticklebacks, they were able to identify genomic regions that appear to have been under selection in the transition from a marine to a lake environment, and candidate genes within those regions.

RRLs

It is not feasible, nor is it necessary, to sequence the whole genome of every individual sampled to obtain genotypic data. Reduced-representation libraries (RRLs) are a way of obtaining representative sequences from across the genome. Genomic DNA is extracted from multiple individuals, cut with a restriction enzyme, and pooled (Davey *et al.* 2011). Fragments are selected based on size and then sequenced. The resulting sequences can then be mapped onto a whole-genome sequence if one is available, but for most species of conservation interest, the sequences from the fragments are assembled, and SNPs in the population or species are identified based on variation between homologous fragments. van Bers *et al.* (2010) obtained over 16 million short sequence reads from 550,000 loci to discover 20,000 novel SNPs in the great tit for use in mapping genes associated with phenotypes.

CRoPS

The complexity reduction of polymorphic sequences (CRoPS) approach is another way to accomplish reduced-representation sequencing, but uses AFLP fragments as a starting point for sequencing (Davey *et al.* 2011). This was the first method to label DNA from different individuals by tagging fragments with a short DNA 'barcode' sequence. In this way, SNP polymorphisms in the discovery panel can be assigned to the individuals from which they came. The CRoPS approach was successfully used by Mammadov *et al.* (2010) to identify SNPs in maize, and by Gompert *et al.* (2010) to determine population structure in *Lycaeides* butterflies.

Exome sequencing

Exome sequencing (also known as targeted exome capture) allows selective sequencing of coding region of the genome. It is a cheaper but efficient alternative to whole genome sequencing and can be downscaled to sequence only tens to hundreds of genes (Hodges *et al.* 2007). The technique uses exon sequences as probes anchored on a microarray in order to capture (i.e., hybridize with) exons from genomic DNA of the study individual. The targeting of the entire exome or any number of genes is not possible with other approaches such as transcriptomics (Section 4.5), which detects only expressed genes. Cosart *et al.* (2011) used exon capture to sequence 16,131 exons from 2570 genes in a single bison, and discovered 2400 polymorphic (heterozygous) sites in a single resequencing experiment. This one experiment yielded sequences from 203 candidate genes and from genes evenly spaced across all chromosomes for use in genome-wide association studies.

4.5 TRANSCRIPTOMICS

Genetic variation exists in regulatory regions as well as in protein-coding DNA sequences. One way to study this variation is through assays of levels of gene expression. Genomic tools have been developed to study variation in levels of expression of many genes simultaneously through quantifying levels of mRNA present in different tissues or individuals. cDNA microarrays and oligonucleotide microarrays are used for this purpose. Thousands or tens of thousands of different short DNA fragments are spotted onto a glass slide or other template, and cDNA from the individuals being studied, labeled with fluorescent dyes or other markers, is hybridized with that array. The intensity of fluorescence provides a quantification of the relative expression levels of targeted genes. Results are often validated with more precise estimates of RNA levels using real-time PCR for a subset of genes.

Levels of gene expression can be viewed as phenotypes as they are the joint product of both genetic and environmental variation (Hansen 2010). If genetic differences in gene expression are to be determined, then individuals need to be reared in a common environment. Information on gene expression differences among populations can be used to complement data on neutral genetic markers and adaptive traits for circumscribing conservation units. For example, Tymchuk *et al.* (2010) quantified gene expression for populations of Atlantic salmon in and around the Bay of Fundy, Newfoundland, using a 16,000 gene cDNA microarray. They found consistent year-to-year population differences in the expression of 389 genes when fish were reared in common environments. Population differentiation for gene expression was stronger and patterns were somewhat different than those observed for seven microsatellite loci.

Gene expression differences among populations or between environments can also be used to identify candidate genes that may be involved in adaptation (see Guest Box 4). Gene expression levels for specific stress-related genes can also be used as biomarkers for health or response to environmental toxins for a range of animal species. While these assays typically investigate environmental rather than genetic sources of variation, they are another example of the use of genomic tools in conservation, and may be used in the future to monitor population health. For example, Chapman *et al.* (2011) characterized the transcriptomic response of eastern oysters to water temperature, pH, salinity, oxygen, and pollutant levels.

4.6 OTHER 'OMICS' AND THE FUTURE

Several other 'omic' tools exist to bridge the gap between genomes and phenotypes beyond transcrip-

tomics (Kristensen *et al.* 2010). The proteome includes all of the proteins and the metabolome is the entire set of small-molecule metabolites found in a cell, tissue, or organism. These 'omics' require highly specialized and instrument-intensive chemical analysis to identify molecules present and quantify their concentrations. These technologies are already being used to study genetic diversity and develop tools for plant breeding in some crop plants and their wild progenitors. Mensack *et al.* (2010) were able to distinguish among cultivars, and between Central American and South American, two centers of diversity of the common bean using transcriptomics, proteomics, and metabolomics. Proteomics and metabolomics to date have had little use or impact in conservation genetics; however, they may find some limited, highly specific applications for particular populations or species in the future, or provide biomarkers for monitoring population health.

4.6.1 Metagenomics

Metagenomics can be used to describe the diversity and relative abundance of taxonomic groups present within a single sequencing experiment (DeLong 2009). These techniques have been applied primarily to microbes; samples are collected from such environments as seawater, soil, or an animal's gut, and subjected to high-throughput sequencing. Further, analysis of the functional groups of genes and their relative abundance, without requiring knowledge of which organism each sequence fragment came from, can provide a functional metabolic profile of the microbial community (Dinsdale *et al.* 2008).

The application of metagenomics to conservation is still in its early stages, but a few specific areas show promise for the future. First, functional metagenomics of microbial communities opens a novel perspective on ecosystem processes such as nutrient and energy flux. While some studies have made comparisons across a broad scale of biomes, similar comparative approaches may identify aspects of ecosystem function across sites within a habitat (Dinsdale *et al.* 2008). Microarrays targeting microbial genes related to bioremediation have identified functional differences in microbial community composition across sites with varying levels of contamination. Waldron *et al.* (2009) used such an array with over 2000 probes and found significant effects on both overall diversity and functional composition of microbial communities in groundwater from a variety of environmental factors, including heavy metal contamination. The development of similar arrays for marine and other systems promises a standardized platform for high-throughput monitoring of functional microbial diversity, with implications for ecosystem processes affecting species of conservation concern.

The second potential application of metagenomics to conservation is in the assessment of physiological condition of individual organisms. For instance, Vega Thurber *et al.* (2009) have found numerous shifts in the **endosymbiont** community of corals in response to multiple stressors such as reduced pH, increased nutrients, and increased temperature. Such shifts in

the endosymbiont community could serve as indicators of reef health, and they could also suggest mechanisms by which coral condition affects other taxa in the reef ecosystem. Finally, a large-scale study used metagenomic techniques on fecal samples to catalog 3.3 million microbial genomes in the human gut fauna, and found significant differences in the microbial metagenome between healthy individuals and those with two types of inflammatory bowel disease (Qin *et al.* 2010). It may be possible in the future to apply metagenomic techniques to fecal samples from wildlife species in order to assess physiological state, such as starvation stress.

Guest Box 4   Rapid evolutionary changes of gene expression in domesticated Atlantic salmon and its consequences for the conservation of wild populations
Louis Bernatchez

Since the 1970s, Atlantic salmon have responded successfully to intense artificial selection aimed at improving growth rates, as well as other traits of production interest in aquaculture (Gjoen and Bentsen 1997). Genetically based phenotypic changes not specifically selected for have also resulted from such selection programs, including increased fat content in flesh and poorer performance in the wild, as well as physiological, morphological, and behavioural changes (e.g., Rye and Gjerde 1996, Fleming *et al.* 2000). As a consequence, escaped farmed salmon represent an important threat for natural populations (McGinnity *et al.* 2003). Indeed, fugitive farmed salmon are thought to greatly enhance the risk of extinction of wild populations through ecological interactions, as well as by spreading diseases. In addition, there is ample evidence that farmed salmon successfully hybridize with wild populations. A pressing question is, therefore, to what extent such interbreeding will alter the genetic integrity of wild Atlantic salmon populations.

It has been proposed for more than 30 years that changes in gene regulation might play a crucial role in driving rapid evolutionary changes as a response to selection (King and Wilson 1975). The development of microarray technologies, allowing the simultaneous detection of expression modulations at thousands of genes, offers a powerful means of

assessing the importance of evolutionary change in gene regulation involved in population divergence and adaptation. We compared gene transcription profiles by means of a 3557-gene cDNA microarray in the progeny of farmed with wild progeny from the same river of origin, grown in controlled conditions, both in Norway and Canada (Roberge *et al.* 2006). We showed that 5 to 7 generations of artificial selection sufficed to cause heritable changes in transcription for genes representing numerous functional classes between farmed and wild salmon (Figure 4.14). Thus, the average magnitude of the differences in levels of expression was 25% and 18% for at least 1.4% and 1.7% of the expressed genes in juvenile salmon from Norway and Canada, respectively. This suggested that at the scale of the whole genome, artificial selection probably led to rapid evolutionary change in gene expression for at least several hundreds of genes. Moreover, genes showing expression differences in both farmed strains (16%) exhibited parallel changes in Canada and Norway.

In a follow-up study that used a microarray this time containing 17,328 cDNA features, we compared patterns of gene transcription between pure and introgressed wild salmon populations from Norway (Roberge *et al.* 2008). This revealed substantial gene misregulation in introgressed fish. For

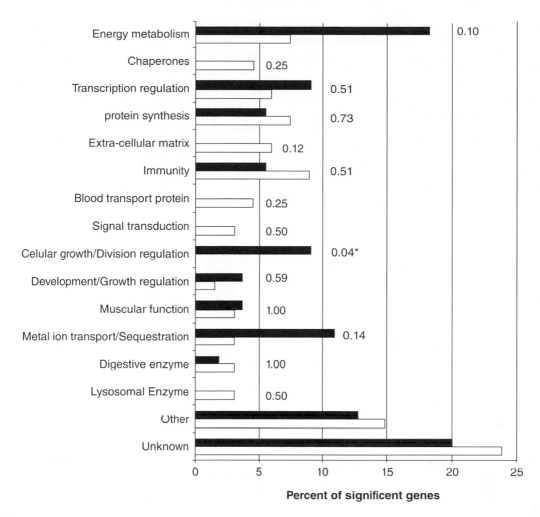

Figure 4.14 Distribution of all genes showing significant expression differences between farmed and wild salmon originating from the St John River, Canada (black), and from the River Namsen, Norway (white), in each of 16 functional classes. The numbers to the right of the bars are the *P* values testing for the significance of the differences in proportion between the Canadian and Norwegian systems. From Roberge *et al.* (2006).

example, over 6% of the detected genes exhibited highly significantly different transcription levels, and the range and average magnitude of those differences was strikingly higher than differences we had observed between pure farmed and wild strains. Thus, most differences resulted from nonadditive gene interactions. A third study using the same microarray, but this time performed on different populations from Canada, revealed population-specific gene expression responses to hybridization between farmed and wild Atlantic salmon (Normandeau *et al.* 2009). In particular, altered biological processes in introgressed relative to pure wild salmon differed between populations both in number and in the type of biological functions that were impacted.

These three studies show that interbreeding of fugitive farmed and wild salmon can substantially

(Continued)

modify the regulatory control of gene transcription in wild salmon populations, resulting in potentially detrimental effects on the survival of these populations. This further stresses the urgent need to reduce considerably the number of escaped farmed salmon and their reproduction in the wild. Moreover, since the consequences of introgression of farm genetic material on gene expression depend upon population-specific genetic architectures, the results of these studies highlight the need to evaluate impacts of farm-wild genetic interactions at the population scale.

The development of next-generation sequencing technologies will greatly enhance the integration of gene expression studies in conservation. When carried out on cDNA synthesized from messenger RNA by reverse transcriptase, such methods allow accurate and rapid gene expression analysis (RNA-seq) of the whole transcriptome, without the need for investing in the costly development of microarrays (Torres *et al.* 2008). Moreover, they enable simultaneous transcript quantification and gene dis-

covery, even for nonmodel organisms for which sequenced genomes do not exist (Goetz and MacKenzie 2008). RNA-seq can also measure allele-specific expression rather than total gene expression, thereby offering insight into regulatory variation associated with the expression of particular phenotypes (Jeukens *et al.* 2010). Among many potential applications in conservation, such methods will allow more thorough investigations of association between genome-wide patterns of gene expression and phenotypic variation among locally adapted populations. They will also allow acute measurements of rapid adaptive responses (or lack thereof) to environmental change or to quantify plastic responses in gene regulation at the scale of the whole genome in the face of environmental stressors (pollutants, climate change, etc.). Clearly, the full integration of such regulatory studies will be crucial for finding causal relationships between genetic variation, phenotypes and environment, to predict future dynamics of selectively important variation and potential for adaptation to new conditions.

PART II

MECHANISMS OF EVOLUTIONARY CHANGE

CHAPTER 5

Random mating populations: Hardy-Weinberg principle

Leadbeater's possum, Guest Box 5

In a sexual population, each genotype is unique, never to recur. The life expectancy of a genotype is a single generation. In contrast, the population of genes endures.

James F. Crow (2001)

Today, the Hardy-Weinberg Law stands as a kind of Newton's First Law (bodies remain in their state of rest or uniform motion in a straight line, except insofar as acted upon by external forces) for evolution: gene frequencies in a population do not alter from generation to generation in the absence of migration, selection, statistical fluctuation, mutation, etc.

Robert M. May (2004)

Chapter Contents

Conservation and the Genetics of Populations, Second Edition. Fred W. Allendorf, Gordon Luikart, and Sally N. Aitken.
© 2013 Fred W. Allendorf, Gordon Luikart and Sally N. Aitken. Published 2013 by Blackwell Publishing Ltd.

A description of genetic variation by itself, as in chapters 3 and 4, will not help us understand the evolution and conservation of populations. We need to develop the theoretical expectations of the effects of Mendelian inheritance in natural populations in order to understand the influence of natural selection, small population size, and other evolutionary factors that affect the persistence of populations and species. The strength of population genetics is the rich foundation of theoretical expectations that allows us to test the predictions of hypotheses to explain the patterns of genetic variation found in natural populations (Provine 2001).

This chapter introduces the structure of the basic models used to understand the genetics of populations (Crow 2001). In later chapters, we will explore expected changes in allele and genotype frequencies in the presence of such evolutionary factors as natural selection or mutation. In this chapter, we will focus on the relationship between allele frequencies and genotype frequencies. In addition, we will examine techniques for estimating allele frequencies and for testing observed genotypic proportions with those expected.

We will use a series of models to consider the pattern of genetic variation in natural populations and to understand the mechanisms that produce evolutionary change. Models allow us to simplify the complexity of the world around us. Models may be either conceptual or mathematical. Conceptual models allow us to simplify the world so that we can represent reality with words and in our thoughts. Mathematical models allow us to specify the relationship between empirical quantities that we can measure and parameters that we specify in our biological theory. These models are essential in understanding the factors that affect genetic change in natural populations, and in predicting the effects of human actions on natural populations.

In addition, models are very helpful in a variety of additional ways:

1 Models make us define the parameters that need to be considered.
2 Models allow us to test hypotheses.
3 Models allow us to generalize results.
4 Models allow us to predict how a system will operate in the future.

The use of models in biology is sometimes criticized because genetic and ecological systems are complex, and simple models ignore many important properties of these systems. This criticism has some validity. Nevertheless, it is impossible to think without using models because reality is too complex (de Brabandere and Iny

2010). Our brains receive information and process this information in order to construct a mental model of how things work. We are often not aware of the models that our brains are using to interpret reality. Construction of mathematical models, such as we will be using in population genetics, forces us to explicitly define the assumptions and parameters that we need to include in order to interpret observed patterns of genetic change in populations.

As a general rule of thumb, models that we develop to understand natural populations should be as simple as possible. That is, a hypothesis or model should not be any more complicated than necessary (**Ockham's razor**). There are several reasons for this. First, hypotheses and models are scientifically useful only if they can be tested and rejected. Simpler models are easier to reject, and, therefore, are more useful. Second, simple models are likely to be more general and therefore more applicable to a wider number of situations.

5.1 HARDY-WEINBERG PRINCIPLE

We will begin with the simplest model of population genetics: a random mating population in which no factors are present to cause genetic change from generation to generation. This model is based upon the fundamental framework of Mendelian segregation for diploid organisms that are reproducing sexually in combination with fundamental principles of probability (Box 5.1). These same principles apply to virtually all species, from elephants to pine trees to violets. We will make the following assumptions in constructing this model:

1 **Random mating.** "Random mating obviously does not mean promiscuity; it simply means . . . that in the choice of mates . . . there is neither preference for nor aversion to the union of persons similar or dissimilar with respect to a given trait or gene" (Wallace and Dobzhansky 1959).

A population can be random mating with regard to most loci, but be mating nonrandomly with regard to other loci that influence mate choice. For example, snow geese are commonly white, but there is a blue phase which is caused by a dominant allele at the *MC1R* locus (see Section 2.2, Mundy *et al.* 2004). Snow geese prefer mates who have the same coloration as the parents that reared them (Cooke 1987). Thus, snow geese show positive **assortative mating** with regard to *MC1R*, but they mate at

---

Box 5.1 Probability

Genetics is a science of probabilities. Mendelian inheritance itself is based upon probability. We cannot know for certain which allele will be placed into a gamete produced by a heterozygote, but we know that there is a one-half probability that each of the two alleles will be transmitted. This is an example of a random, or **stochastic**, event. There are a few simple rules of probability that we will use to understand the extension of Mendelian genetics to populations.

The **probability** (P) of an event is the number of times the event will occur (a) divided by the total number of possible events (n):

$P = a/n$

For example, a die has six faces that are equally likely to land up if the die is tossed. Thus, the probability of throwing any particular number is one-sixth:

$P = a/n = 1/6$

We often are interested in combining the probabilities of different events. There are two different rules that we will use to combine probabilities.

The **product rule** states that the probability of two or more independent events occurring simultaneously is equal to the product of their individual probabilities. For example, what is the probability of throwing a total

of 12 with a pair of dice? This can only occur by a six landing up on the first die and also on the second die. According to the product rule:

$P = 1/6 \times 1/6 = 1/36$

The **sum rule** states that the probability of two or more mutually exclusive events occurring is equal to the sum of their individual probabilities. For example, what is the probability of throwing either a five or six with a die? According to the sum rule:

$P = 1/6 + 1/6 = 2/6 = 1/3$

In many situations, we need to use both of these rules to compute a probability. For example, what is the probability of throwing a total of seven with a pair of dice?

Solution: There are six mutually exclusive ways that we can throw seven with two dice: 1 + 6, 2 + 5, 3 + 4, 4 + 3, 5 + 2, and 6 + 1. As we saw in the example for the product rule, each of these combinations has a probability of $1/6 \times 1/6 = 1/36$ of occurring. They are all mutually exclusive so we can use the sum rule. Therefore, the probability of throwing a seven is:

$1/36 + 1/36 + 1/36 + 1/36 + 1/36 + 1/36 = 6/36 - 1/6$

---

random with regard to the rest of their genome. See Section 9.1.2 for another example of positive assortative mating at *MC1R*.

2 **No mutation.** We assume that the genetic information is transmitted from parent to progeny (i.e., from generation to generation) without change. Mutations provide the genetic variability that is our primary concern in genetics. Nevertheless, mutation rates are generally quite small and are only important in population genetics from a long-term perspective, generally hundreds or thousands of generations. We will not consider the effects of mutations on changes in allele frequencies in detail, since in conservation genetics we are more concerned with factors that can influence populations in a more immediate timeframe.

3 **Large population size.** Many of the theoretical models that we will consider assume an infinite

population size. This assumption may effectively be correct in some populations of insects or plants. However, it is obviously not true for many of the populations of concern in conservation genetics. Nevertheless, we will initially consider the ideal large population in order to develop the basic concepts of population genetics, and we will then consider the effects of small population size in later chapters.

4 **No natural selection.** We will assume that there is no differential survival or reproduction of individuals with different genotypes (that is, no natural selection). Again, this assumption will not be true at all loci in any real population, but it is necessary that we initially make this assumption in order to develop many of the basic concepts of population genetics. We will consider the effects of natural selection in later chapters.

5 No immigration. We will assume that we are dealing with a single isolated population. We will later consider multiple populations in which gene flow between populations is brought about through exchange of individuals.

There are two important consequences of these assumptions. First, the population will not evolve. Mendelian inheritance has no inherent tendency to favor any one allele. Therefore, allele and genotype frequencies will remain constant from generation to generation. This is known as the Hardy-Weinberg equilibrium. In the next few chapters we will explore the consequences of relaxing these assumptions on changes in allele frequency from generation to generation. We will not be able to consider all possibilities. However, our goal is to develop an intuitive understanding of the effects of each of these evolutionary factors.

The second important outcome of the above assumptions is that genotype frequencies will be in binomial (Hardy-Weinberg) proportions. That is, **genotypic** frequencies after one generation of random mating will be a binomial function of **allele** frequencies. It is important to distinguish between the two primary ways in which we will describe the genetic characteristics of populations at individual loci: allele (gene) frequencies and genotypic frequencies.

The Hardy-Weinberg principle greatly simplifies the task of describing the genetic characteristics of populations; it allows us to describe a population by the frequencies of the alleles at a locus rather than by the many different genotypes that can occur at a single diploid locus. This simplification becomes especially important when we consider multiple loci. For example, there are 59,049 different genotypes possible at just ten loci that each has just two alleles. We can describe this tremendous genotypic variability by specifying only ten allele frequencies if the populations is in Hardy-Weinberg proportions.

This principle was first described by G.H. Hardy (1908), a famous English mathematician, and independently by a German physician Wilhelm Weinberg (1908). The principle was actually first used by an American geneticist W.E. Castle (1903) in a description of the effects of natural selection against recessive alleles. However, this aspect of the paper by Castle was not recognized until nearly 60 years later (Li 1967). A detailed and interesting history of the development of population genetics is provided by Provine (2001). There is great irony in our use of Hardy's name to describe a fundamental principle that has been of great practical value in medical genetics and now in our efforts to conserve biodiversity (Edwards 2008). Hardy (1967) saw himself as a "pure" mathematician whose work had no practical relevance: "I have never done anything 'useful'. No discovery of mine has made, or is likely to make, directly or indirectly, for good or ill, the least difference to the amenity of the world."

5.2 HARDY-WEINBERG PROPORTIONS

We will first consider a single locus with two alleles (A and a) in a population such that the population consists of the following numbers of each genotype

AA	Aa	aa	Total
N_{11}	N_{12}	N_{22}	N

Each homozygote (AA or aa) contains two copies of the same allele, while each heterozygote (Aa) contains one copy of each allele. Therefore, the allele frequencies are:

$$p = freq(A) = \frac{(2N_{11} + N_{12})}{2N}$$
$$q = freq(a) = \frac{(N_{12} + 2N_{22})}{2N} \tag{5.1}$$

where $p + q = 1.0$.

Our assumption of random mating will result in random union of gametes to form zygotes. Thus, the frequency of any particular combination of gametes from the parents will be equal to the product of the frequencies of those gametes, which are the allele frequencies. This is shown graphically in Figure 5.1. Thus, the expected genotypic proportions are predicted by the binomial expansion:

$$(p+q)^2 = p^2 + 2pq + q^2$$
$$\quad\quad AA \quad Aa \quad aa \tag{5.2}$$

These proportions will be reached in one generation, providing all of the above assumptions are met and allele frequencies are equal in males and females. Additionally, these genotypic frequencies will be maintained forever, as long as these assumptions hold.

The Hardy-Weinberg principle can readily be extended to more than two alleles with two simple rules:

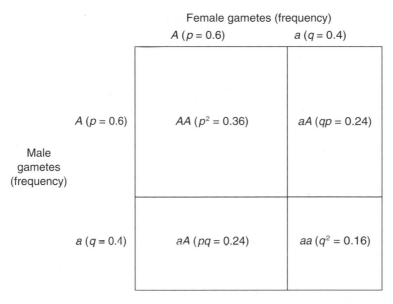

Female gametes (frequency)

$A\,(p = 0.6)$ $a\,(q = 0.4)$

Male gametes (frequency)

$A\,(p = 0.6)$ $AA\,(p^2 = 0.36)$ $aA\,(qp = 0.24)$

$a\,(q = 0.4)$ $aA\,(pq = 0.24)$ $aa\,(q^2 = 0.16)$

Figure 5.1 Hardy-Weinberg proportions at a locus with two alleles (A and a) generated by random union of gametes produced by females and males. The area of each rectangle is proportional to the genotypic frequencies.

1 the expected frequency of homozygotes for any allele is the square of the frequency of that allele, and

2 the expected frequency of any heterozygote is twice the product of the frequency of the two alleles present in the heterozygote.

In the case of three alleles the following genotypic frequencies are expected:

$$p = \text{freq}(A_1)$$
$$q = \text{freq}(A_2)$$
$$r = \text{freq}(A_3)$$

and

$$(p+q+r)^2 = \underset{A_1A_1}{p^2} + \underset{A_1A_2}{2pq} + \underset{A_2A_2}{q^2} + \underset{A_1A_3}{2pr} + \underset{A_2A_3}{2qr} + \underset{A_3A_3}{r^2}$$
(5.3)

5.3 TESTING FOR HARDY-WEINBERG PROPORTIONS

Genotypic frequencies of samples from natural populations can be tested readily to see whether they conform to expected Hardy-Weinberg proportions. However, there are a profusion of papers that discuss the some-

times hidden intricacies of testing for goodness-of-fit to Hardy-Weinberg proportions (Fairbairn and Roff 1980). Lessios (1992) has provided an interesting and valuable review of this literature.

Many papers in the literature refer to these analyses as testing for "Hardy-Weinberg equilibrium", rather than the more precise description of testing for "Hardy-Weinberg proportions". This is potentially confusing because populations whose genotypic proportions are in expected Hardy-Weinberg proportions will not be in Hardy-Weinberg equilibrium (no mutation, large population size, etc.). In general, genotypic proportions will be as expected under the Hardy-Weinberg model as long as the population is randomly mating. The effects of mutation, small population size, natural selection, and immigration are generally too small to cause genotypic proportions to deviate from Hardy-Weinberg expectations.

Dinerstein and McCracken (1990) described genetic variation using allozyme electrophoresis at ten variable loci in a population of one-horned rhinoceros from the Chitwan Valley of Nepal. The following numbers of each genotype were detected at a lactate dehydrogenase (*LDH*) locus with two alleles (*100* and *125*). Do these values differ from what we expect with Hardy-Weinberg proportions?

100/100	100/125	125/125	Total
$N_{11} = 5$	$N_{12} = 12$	$N_{22} = 6$	$N = 23$

We first need to estimate the allele frequencies in this sample. We do not know the true allele frequencies in this population, which consisted of some 400 animals at the time of sampling. However, we can estimate the allele frequency in this population based upon the sample of 23 individuals. The estimate of the allele frequency of the *100* allele obtained from this sample will be designated as $\hat{p}$ (called *p* hat) to designate that it is an estimate rather than the true value.

$$\hat{p} = \frac{2N_{11} + N_{12}}{2N} = \frac{10 + 12}{46} = 0.478$$

and

$$\hat{q} = \frac{N_{12} + 2N_{22}}{2N} = \frac{12 + 12}{46} = 0.522$$

We now can estimate the expected number of each genotype in our sample of 23 individual genotypes assuming Hardy-Weinberg proportions:

	100/100	**100/125**	**125/125**
Observed	5	12	6
Expected	($\hat{p}^2 N = 5.3$)	($2\hat{p}\hat{q}N = 11.5$)	($\hat{q}^2 N = 6.3$)

The agreement between observed and expected genotypic proportions in this case is very good. In fact, this is the closest fit possible in a sample of 23 individuals from a population with the estimated allele frequencies. Therefore, we would conclude that there is no indication that the genotype frequencies at this locus are not in Hardy-Weinberg proportions.

The chi-square method provides a statistical test to determine whether the deviation between observed genotypic and expected Hardy-Weinberg proportions is greater than we would expect by chance alone. We first calculate the chi-square value for each of the genotypes and sum them into a single value:

$$\chi^2 = \sum \frac{(OBSERVED - EXPECTED)^2}{EXPECTED}$$
$$= \frac{(5 - 5.3)^2}{5.3} + \frac{(12 - 11.5)^2}{11.5} + \frac{(6 - 6.3)^2}{6.3}$$
$$= 0.02 + 0.02 + 0.01 = 0.05$$

The χ^2 value becomes increasingly greater as the difference between the observed and expected values becomes greater.

The computed χ^2 value is then tested by comparing it with a set of values (Table 5.1) calculated under the assumption that the null hypothesis we are testing is correct; in this case, our null hypothesis is that the population from which the samples was drawn is in Hardy-Weinberg proportions. We need one additional value to apply the chi-square test, the **degrees of freedom**. In using the chi-square test for Hardy-Weinberg proportions, the number of degrees of freedom is equal to the number of possible genotypes minus the number of alleles.

Number of alleles	Number of genotypes	Degrees of freedom
2	3	1
3	6	3
4	10	6
5	15	10

By convention, if the probability estimated by a statistical test is less than 0.05, then the difference between the observed and expected values is said to be significant. We can see in Table 5.1 that the chi-square value with one degree of freedom must be greater than 3.84 before we would conclude that the deviation between observed and expected proportions is greater than we would expect by chance with one degree of freedom. Our estimated χ^2 value of 0.05 for the *LDH* locus in the one-horned rhino is much smaller than this. Therefore, we would accept the null hypothesis that the population from which this sample was drawn was in Hardy-Weinberg proportions at this locus. See Example 5.1 for a situation where the null hypothesis of Hardy-Weinberg proportions can be rejected. An example of the chi-square test for Hardy-Weinberg proportions in the case of three alleles is given in Example 5.2.

5.3.1 Small sample sizes or many alleles

Sample sizes in conservation genetics are often smaller than our statistical advisors recommend because of the limitations imposed by working with rare species. The chi-square test is only an approximation of the actual probability distribution, and the approximation becomes poor when expected numbers are small. The

Table 5.1 Critical values of the chi-square distribution for up to 5 degrees of freedom (d.f.). The proportions in the table (corresponding to $\alpha = 0.05, 0.01$, etc.) represent the area to the right of the critical value of chi-square given in the table, as shown in the figure below. The null hypothesis is usually not rejected unless the probability associated with the calculated chi-square is less than 0.05.

Table of chi-square values

Degrees of freedom	Probability (P)					
	0.90	0.50	0.10	0.05	0.01	0.001
1	0.02	0.46	2.71	3.84	6.64	10.83
2	0.21	1.39	4.60	5.99	9.21	13.82
3	0.58	2.37	6.25	7.82	11.34	16.27
4	1.06	3.86	7.78	9.49	13.28	18.47
5	1.61	14.35	9.24	11.07	15.09	20.52

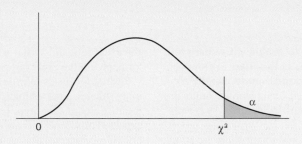

Example 5.1 Test for Hardy-Weinberg proportions

Leary *et al.* (1993b) reported the following genotype frequencies at an allozyme locus (*mlDHP-1*) in a sample of bull trout from the Clark Fork River in Idaho.

Genotype	Observed	Expected	Chi-square
100/100	1	$(\hat{p}^2N = 5.8)$	3.97
100/75	22	$(2\hat{p}\hat{q}N = 12.5)$	7.22
75/75	2	$(\hat{q}^2N = 6.8)$	3.38
Total	25	(25.1)	14.58

Estimated frequency of $100 = \hat{p} = [(2 \times 1) + 22]/50$
$= 0.480$

Estimated frequency of $75 = \hat{q} = [22 + (2 \times 2)]/50$
$= 0.520$

Degrees of freedom = 1

The calculated χ^2 of 14.58 is greater than the critical value for $P < 0.001$ with 1 d.f. of 10.83 (Table 5.1). Therefore, the probability of getting such a large deviation by chance alone is less than 0.001. Therefore we would reject the null hypothesis that the sampled population was in Hardy-Weinberg proportions at this locus.

There is a significant excess of heterozygotes in this sample of bull trout. We will return to this example in the next chapter to see the probable cause of this large deviation from Hardy-Weinberg proportions (see Example 6.4).

Example 5.2 Test for Hardy-Weinberg proportions at a locus (aminoacylase-1, *Acy-1*) with three alleles in the Polish brown hare (Hartl et al. 1992)

Genotype	Observed	Expected	Chi-square
100/100	4	$(\hat{p}^2 N = 2.1)$	1.72
100/81	6	$(2\hat{p}\hat{q}N = 9.7)$	1.41
81/81	14	$(\hat{q}^2 N = 11.0)$	0.82
100/66	4	$(2\hat{p}\hat{r}N = 4.0)$	0.00
81/66	7	$(2\hat{q}\hat{r}N = 9.2)$	0.53
66/66	3	$(\hat{r}^2 N = 1.9)$	0.64
Total	38	(37.9)	5.12

Estimated frequency of 100 $= \hat{p} = [(2 \times 4) + 6 + 4]/76$
$= 0.237$

Estimated frequency of 81 $= \hat{q} = [6 + (2 \times 14) + 7]/76$
$= 0.539$

Estimated frequency of 66 $= \hat{r} = [4 + 7 + (2 \times 3)]/76$
$= 0.224$

There are 6 genotypic classes and two independent allele frequencies at a locus with 3 alleles.

Degrees of freedom $= 6 - 3 = 3$

The calculated χ^2 of 5.12 is less than the critical value with 3 d.f. of 7.82 (Table 5.1). Therefore, we accept the null hypothesis that the sampled population was in Hardy-Weinberg proportions at this locus.

usual rule-of-thumb is not to use the chi-square test when any expected number is less than 5. However, some have argued that this rule is unnecessarily conservative and have suggested using smaller limits on expected values (3 by Cochran 1954, and 1 by Lewontin and Felsenstein 1965).

In addition, there is a systematic bias in small samples because of the discreteness of the possible numbers of genotypes. Levene (1949) has shown that in a finite sample of N individuals, the heterozygotes are increased by a fraction of $1/(2N - 1)$ and homozygotes are correspondingly decreased (Crow and Kimura 1970, pp. 55–56). For example, if only one copy of a rare allele is detected in a sample, then the only genotype containing the rare allele must be heterozygous. The simple binomial Hardy-Weinberg proportions will predict that some fraction of the sample is expected to be homozygous for the rare allele; however, this is impossible because there is only one copy of the allele in the sample. Levene's correction will adjust for this bias.

Exact tests provide a method to overcome the limitation of small expected numbers with the chi-square test (Fisher 1935). Exact tests are performed by determining the probabilities of all possible samples, assuming that the null hypothesis is true (Example 5.3). The probability of the observed distribution is then added to the sum of all less probable but possible sample outcomes. Weir (1996, pp. 98–101) described the use of

the exact test, and Vithayasai (1973) presented tables for applying the exact test with two alleles.

Testing for Hardy-Weinberg proportions at loci with many alleles, such as microsatellite loci, is a problem because many genotypes will have extremely low expected numbers. There are $A(A - 1)/2$ heterozygotes and A homozygotes at a locus with A alleles. Therefore, there is the following possible number of genotypes at a locus in a population with A alleles:

$$\frac{A(A-1)}{2} + A = \frac{A(A+1)}{2} \tag{5.4}$$

For example, Olsen *et al.* (2000) found an average of 23 alleles at eight microsatellite loci in pink salmon, compared with an average of 2.3 alleles at 24 polymorphic allozyme loci in the same population. There are a total of 279 genotypes possible with 23 alleles at a single locus (equation 5.4). Exact tests for Hardy-Weinberg proportions are possible with more than two alleles (Louis and Dempster 1987, Engels 2009). However, the number of possible genotypes increases very quickly with more than two alleles, and computation time becomes prohibitive. Engels (2009) has recently provided software that uses a likelihood ratio approach for exact Hardy-Weinberg tests which is much faster that previous approaches.

Nearly exact tests are generally used to analyze data from natural populations using computer-based per-

Example 5.3 Exact test for Hardy-Weinberg proportions (Weir 1996)

We would reject the null hypothesis of Hardy-Weinberg proportions in the table below because our calculated chi-square value is greater than 3.84. However, an exact test indicates that we would expect to get a deviation as great or greater than the one we observed some 8% (0.082) of the time. Therefore, we would not reject the null hypothesis in this case using the exact test.

Possible samples					
100/ 100	*100/ 80*	*80/ 80*	Probability	Cumulative probability	χ^2
30	1	9	0.0000	0.0000	34.67
29	3	8	0.0000	0.0000	25.15
28	5	7	0.0001	0.0001	17.16
27	7	6	0.0023	0.0024	10.69
26	9	5	0.0205	0.0229	5.74
21	**19**	**0**	**0.0594**	**0.0823**	**3.88**
25	11	4	0.0970	0.1793	2.32
22	17	1	0.2308	0.4101	1.20
24	13	3	0.2488	0.6589	0.42
23	15	2	0.3411	1.0000	0.05

Genotypes				
100/100	*100/80*	*80/80*		
21	19	0	$\hat{p} = 0.763$	$\chi^2 = 3.88$
(23.3)	(14.5)	(2.3)		

There are 10 possible samples of 40 individuals that would provide us with the same allele frequency estimates. We can calculate the exact probabilities for each of these possibilities if the sampled population is in Hardy-Weinberg proportions using the binomial distribution as shown below:

In practice, exact tests are performed using computer programs because calculating the exact binomial probabilities is extremely complicated and time-consuming.

mutation or randomization testing. In the case of Example 5.3, a computer program would randomize genotypes by sampling, or creating, 40 diploid individuals from a pool of 61 copies of the *100* allele and 19 copies of the *80* allele. A chi-square value is then calculated for a thousand or more of these randomized datasets and its value compared with the statistic obtained from the observed dataset. The proportion of chi-square values from the randomized datasets that give a value as large or larger than the observed provides an unbiased estimation of the proportion of distributions expected to be as bad or worse fit than the observed if the null hypothesis is true.

5.3.2 Multiple simultaneous tests

In most studies of natural populations, multiple loci are examined from several populations resulting in multiple tests for Hardy-Weinberg proportions (see Guest Box 5). For example, if we examine ten loci in ten population samples, 100 tests of Hardy-Weinberg proportions will be performed. If all of these loci are in

Hardy-Weinberg proportions (that is, our null hypothesis is true at all loci in all populations), we expect to find five significant tests if we use the 5% significance level. Thus, simply applying the statistical procedure presented here would result in rejection of the null hypothesis of Hardy-Weinberg proportions approximately five times when our null hypothesis is true. This is called a type-I error. (A type-II error occurs when a false null hypothesis is accepted.)

There are a variety of approaches that can be used to treat this problem (see Rice 1989). One common approach is to use the so-called Bonferroni correction in which the significance level (say 5%) is adjusted by dividing it by the number of tests performed (Cooper 1968). Therefore in the case of 100 tests, we would use the adjusted nominal level of $0.05/100 = 0.0005$. The critical chi-square value for $P = 0.0005$ with 1 degree of freedom is 12.1. That is, we expect a chi-square value greater than 12.1 with one degree of freedom less than 0.0005 of the time if our null hypothesis is correct. Thus, we would reject the null hypothesis for a particular locus only if our calculated chi-square value was greater than 12.1. This procedure is known to be conservative and results in a loss of statistical power to detect multiple deviations from the null hypothesis. A procedure known as the sequential Bonferroni can be used to increase power to detect more than one deviation from the null hypothesis (Rice 1989).

It is also extremely important to examine the data to detect possible patterns for those loci that do not conform to Hardy-Weinberg proportions. For example, let's say that eight of our 100 tests have probability values less than 5%; this value is not much greater than our expectation of five. If the eight cases are spread fairly evenly among samples and loci, and none of the individual probability values are less than 0.0005 obtained from the Bonferroni correction, then it is reasonable not to reject the null hypothesis that these samples are in Hardy-Weinberg proportions at these loci.

However, we may reach a different conclusion if all eight of the deviations from Hardy-Weinberg proportions occurred in the same sample, and all the deviations were in the same direction (e.g., a deficit of heterozygotes). This would suggest that this particular sample was taken from a population that was not in Hardy-Weinberg proportions. Perhaps this sample was collected from a group that consisted of two separate random mating populations (Wahlund effect; see Guest Box 5 and Chapter 9).

Another possibility is that all eight deviations from Hardy-Weinberg proportions occurred at the same locus in eight different population samples, and all the deviations were in the same direction (e.g., a deficit of heterozygotes). This would suggest that there is something unusual about this particular locus. For example, the presence of a null allele (see Section 5.4.2) would result in a tendency for a deficit of heterozygotes.

5.4 ESTIMATION OF ALLELE FREQUENCIES

So far we have estimated allele frequencies when the number of copies of each allele in a sample can be counted directly from the genotypic frequencies. However, sometimes we cannot identify the alleles in every individual in a sample. The Hardy-Weinberg principle can be used to estimate allele frequencies at loci in which there is not a unique relationship between genotypes and phenotypes. We will consider two such situations that are often encountered in analyzing data from natural populations.

5.4.1 Recessive alleles

There are many cases in which heterozygotes cannot be distinguished from one of the homozygotes. For example, color polymorphisms and metabolic disorders in many organisms are caused by recessive alleles. The frequency of recessive alleles can be estimated if we assume Hardy-Weinberg proportions.

$$\hat{q} = \sqrt{\frac{N_{22}}{N}} \tag{5.5}$$

Dozier and Allen (1942) described differences in coat color in the muskrat of North America. A dark phase, the so-called blue muskrat, is generally rare relative to the ordinary brown form, but occurs at high frequencies along the Atlantic coast between New Jersey and North Carolina. Breeding studies (Dozier 1948) have shown that the blue phase is caused by an allele (*b*) which is recessive to the brown allele (*B*). A total of 9895 adult muskrats were trapped on the Blackwater National Wildlife Refuge, Maryland, in 1941. The blue muskrat occurred at a frequency of 0.536 in this sample. If we assume Hardy-Weinberg proportions (equation 5.2), then:

$$\hat{q}^2 = 0.536$$

and taking the square root of both sides of this relationship:

$$\hat{q} = \sqrt{0.536} = 0.732$$

The estimated frequency of the *B* allele is $\{1 - 0.732\} = 0.268$.

Example 5.4 demonstrates how the genotypic proportions in a population for a recessive allele can be examined for Hardy-Weinberg proportions.

5.4.2 Null alleles

Null alleles at protein coding loci are alleles that do not produce a detectable protein product; null alleles at microsatellite loci are alleles that do not produce a detectable PCR amplification product. Null alleles at allozyme loci result from alleles that produce either no protein product, or a protein product that is enzymati-

---

Example 5.4   Color polymorphism in the eastern screech owl

We concluded in Chapter 2 based on the data below from Table 2.1 that the red morph of the eastern screech owl is caused by a dominant allele (R) at a single locus with two alleles; gray owls are homozygous for the recessive allele (rr).

Mating	Number of families	Progeny Red	Progeny Gray
Red × red	8	23	5
Red × gray	46	68	63
Gray × gray	135	0	439

We can estimate the frequency of the r allele by assuming that this population is in Hardy-Weinberg proportions and the total progeny observed is representative of the entire population:

$$\hat{q}^2 = (507/598) = 0.847$$

$$\hat{q} = \sqrt{0.847} = 0.921$$

and

$$\hat{p} = 1 - \hat{q} = 0.079$$

We can also check to see whether the progeny produced by matings between two red birds is close to what we would expect if this population was in Hardy-Weinberg proportions. Remember that red birds may be either homozygous (RR) or heterozygous (Rr). What proportion of gray progeny do we expect to be produced by matings between two red parents?

Three things must occur for a progeny to be gray: (1) the mother must be heterozygous (Rr); (2) the father must be heterozygous (Rr); and (3) the progeny must receive the recessive allele (r) from both parents:

$$\text{Prob(progeny } rr) = \text{Prob(mother } Rr) \times \text{Prob(father } Rr)$$
$$\times 0.25$$

The proportion of red birds in the population who are expected to be heterozygous is the proportion of heterozygotes divided by the total proportion of red birds:

$$\text{Prob(parental bird } (Rr) = (2pq) / (p^2 + 2pq) = 0.959$$

Therefore, the expected proportion of gray progeny is

$$0.959 \times 0.959 \times 0.25 = 0.230$$

This is fairly close to the observed proportion of 0.178 (5/28). Thus, we would conclude that this population appears to be in Hardy-Weinberg proportions at this locus.

---

cally non-functional (Foltz 1986). Null alleles at microsatellite loci result from nucleotide substitutions that prevent the primers from binding (Brookfield 1996). Heterozygotes for a null allele and another allele appear to be homozygotes on a gel. The presence of null alleles results in an apparent excess of homozygotes relative to Hardy-Weinberg proportions. Brookfield (1996) discusses the estimation of null allele frequencies in the case of more than three alleles. Kalinowski and Taper (2006) have presented a maximum likelihood approach to estimate the frequency of null alleles at microsatellite loci.

The familiar ABO blood group locus in humans presents a parallel situation to the case of a null allele in which all genotypes cannot be distinguished. In this case, the I^A and I^B alleles are codominant, but the I^O allele is recessive (i.e., null). This results in the following relationship between genotypes and phenotypes (blood types):

Genotypes	Blood types	Expected frequency	Observed number
$I^A I^A$, $I^A I^O$	A	$p^2 + 2pr$	N_A
$I^B I^B$, $I^B I^O$	B	$q^2 + 2qr$	N_B
$I^A I^B$	AB	$2pq$	N_{AB}
$I^O I^O$	O	r^2	N_O

where p, q, and r are the frequencies of the I^A, I^B, and I^O alleles.

We can estimate allele frequencies at this locus by the expectation maximization (EM) algorithm which finds the allele frequencies that maximize the probability of obtaining the observed data from a sample of a population assumed to be in Hardy-Weinberg proportions (Dempster *et al.* 1977). This is an example of a

maximum likelihood estimate (see Section A5), which has many desirable statistical properties (Fu and Li 1993).

We could estimate the frequency p directly, as in Example 5.2, if we knew how many individuals in our sample with blood type A were $I^A I^A$ and how many were $I^A I^O$:

$$\hat{p} = \frac{2N_{AA} + N_{AO} + N_{AB}}{2N} \qquad (5.6)$$

where N is the total number of individuals. However, we cannot distinguish the phenotypes of the $I^A I^A$ and $I^A I^O$ genotypes. The EM algorithm solves this ambiguity with a technique known as gene counting. We start with guesses of the allele frequencies, and use them to calculate the expected frequencies of all genotypes (step E of the EM algorithm), assuming Hardy-Weinberg proportions. Then, we use these genotypic frequencies to obtain new estimates of the allele frequencies, using maximum likelihood (step M). We then use these new allele frequency estimates in a new E step, and so forth, in an iterative fashion, until the values converge.

We first guess the three allele frequencies (remember $p + q + r = 1.0$). The next step is to use these guesses to calculate the expected genotype frequencies assuming Hardy-Weinberg proportions. We next use gene counting to estimate the allele frequencies from these genotypic frequencies. The count of the I^A alleles is twice the number of $I^A I^A$ genotypes plus the number of $I^A I^O$ genotypes. We expect p^2 of the total individuals with blood type A ($p^2 + 2pr$) to be homozygous $I^A I^A$, and $2pr$ of them to be heterozygous $I^A I^O$. These counts are then divided by the total number of genes in the sample ($2N$) to estimate the frequency of the I^A allele, as we did in equation 5.6. A similar calculation is performed for the I^B allele with the following result:

$$\hat{p} = \frac{\left[2\left(\frac{p^2}{p^2 + 2pr}\right)N_A + \left(\frac{2pr}{p^2 + 2pr}\right)N_A + N_{AB}\right]}{2N}$$

$$= \frac{2\left(\frac{p+r}{p+2r}\right)N_A + N_{AB}}{2N} \qquad (5.7)$$

$$\hat{q} = \frac{2\left(\frac{q+r}{q+2r}\right)N_B + N_{AB}}{2N}$$

$$\hat{r} = 1 - \hat{p} - \hat{q}$$

These equations produce new estimates of p, q, and r that can be substituted into the right hand side of equations 5.7 to produce new estimates of p, q, and r. This iterative procedure is continued until the estimates converge: that is, until the estimated values on the left side are nearly equal to the values substituted into the right side.

5.5 SEX-LINKED LOCI

So far we have only considered autosomal loci in which there are no differences between males and females. However, the genotypes of genes on sex chromosomes will often differ between males and females. The most familiar situation is genes on the X chromosome of mammals (and *Drosophila*) in which females are homogametic XX and males are heterogametic XY. In this case, genotype frequencies for females conform to the Hardy-Weinberg principle. However, the Y chromosome is largely void of genes so that males will have only one gene copy, and the genotype frequency in males will be equal to the allele frequencies. The situation is reversed in bird species: females are heterogametic ZW and males are homogametic ZZ (Ellegren 2000b). In this case, genotype frequencies for the ZZ males conform to the Hardy-Weinberg principle, and the genotype frequency in the ZW females will be equal to the allele frequencies (Figure 5.2).

Phenotypes resulting from rare recessive X-linked alleles will be much more common in males than in females because q^2 will always be less than q. The most familiar case of this is X-linked red–green color blindness in humans in which approximately 8% of males of northern European origin (including the senior author of this book!) lack a pigment in the retina of their eye so they do not perceive colors as most people (Deeb 2005). In this case, $q = 0.08$ and therefore we expect the frequency of color-blindness in females to be $q^2 = (0.08)^2 = 0.0064$. Thus, we expect more than 10 times as many red–green color-blind males than females in this case.

A variety of other mechanisms for sex determination occur in other animals and plants (Bull 1983, see Section 3.1.2). Many plant species possess either XY or ZW systems. The use of XY or ZW indicates which sex is heterogametic (Charlesworth 2002). The sex chromosomes are identified as XY in species in which males are heterogametic, and ZW in species in which females are heterogametic. Many reptiles have a ZW system

| | Z-bearing eggs | | W-bearing eggs |
	Z^A (p)	Z^a (q)	W
Z^A (p)	Z^AZ^A (p^2)	Z^AZ^a (pq)	Z^AW (p)
Z^a (q)	Z^AZ^a (pq)	Z^aZ^a (q^2)	Z^aW (q)
	Males		Females

Sperm

Figure 5.2   Expected genotypic proportions with random mating for a Z-linked locus with two alleles (*A* and *a*).

(e.g., all snakes, Graves and Shetty 2001). A wide variety of genetic sex determination systems are found in fish species (Devlin and Nagahama 2002) and in invertebrates. Some species have no detectable genetic mechanism for sex determination. For example, sex is determined by the temperature in which eggs are incubated in some reptile species (Graves and Shetty 2001).

5.5.1 Pseudoautosomal inheritance

The classic XY system of mammals and *Drosophila* (with the Y chromosome being largely devoid of functional genes) taught in introductory genetic classes has been over-generalized. A broader taxonomic view suggests that mammals and *Drosophila* are exceptions and that both sex chromosomes contain many functional genes across a wide variety of animal taxa. The sex chromosomes of many species have so-called pseudoautosomal regions in which functional genes are present on both the X and Y, or Z and W, chromosomes.

Morizot *et al.* (1987) found that functional genes for the creatine kinase enzyme locus are present on both the Z and W chromosomes of Harris' hawk. Recent genome mapping efforts support these early results. For example, three of six loci found to be sex-linked in the Siberian jay were present on both the Z and W chromosomes (Jaari *et al.* 2009). Wright and Richards (1983) found that two of 12 allozyme loci that they mapped in the leopard frog were sex-linked and that two functional gene copies of both loci are found in XY males. Functional copies of a peptidase locus are present on both the Z and W chromosomes in the sala-

mander *Pleurodeles waltlii* (Dournon *et al.* 1988). Two allozyme loci in rainbow trout have functional alleles on both X and Y chromosomes (Allendorf *et al.* 1994).

Differences in allele frequencies between the males and females for genes found on both sex chromosomes will result in an excess of heterozygotes in comparison with expected Hardy-Weinberg proportions in the heterogametic sex (Clark 1988, Allendorf *et al.* 1994). This excess of heterozygotes can persist for many generations if the locus is closely linked to the sex determining locus. These regions happen to be quite small in those species for which we are most familiar with sex-linked inheritance (e.g., humans and *Drosophila*). However, these pseudoautosomal regions comprise a large proportion of the sex chromosomes in many taxa (e.g., salmonid fishes and birds). Detecting such loci in population genetic studies will become much more frequent in the future as more and more loci are examined with genomic techniques. For example, a recent genetic linkage map of the Siberian jay found that three of six sex-linked loci are pseudoautosomal (Jaari *et al.* 2009). This linkage map included a total of 117 loci, so that nearly 3% of genome-wide loci were pseudoautosomal.

Differences in allele frequency between the sex chromosomes will result in an excess of heterozygotes compared with expected Hardy-Weinberg proportions in the heterogametic sex for pseudoautosomal loci (Clark 1988, Allendorf *et al.* 1994). For example, Berlocher (1984) observed the following genotypic frequencies at a sex-linked allozyme locus (*Pgm*) with two alleles (*100* and *82*) in the walnut husk fly for which males are XY and females are XX:

		100/100	100/82	82/82
Females	XX	25	0	0
Males	XY	4	21	0

Based on these data, the *100* allele is fixed on the X chromosome because the *82* allele is not present in females. We can estimate allele frequencies on the Y chromosome by removing the *100* allele on the X chromosome in each male. Both the *100* and *82* alleles are on the Y chromosome; the *82* allele is at an estimated frequency of 0.84 (21/25) on the Y chromosome.

5.6 ESTIMATION OF GENETIC VARIATION

We are often interested in comparing the amount of genetic variation in different populations. For example, we saw in Table 4.1 that brown bears from Kodiak Island and Yellowstone National Park had less genetic variation than other populations for allozymes, microsatellites, and mtDNA. In addition, comparisons of the amount of genetic variation in a single population sampled at different times can provide evidence for loss of genetic variation because of population isolation and fragmentation due to habitat loss or other causes. In this section we will consider measures that have been used to compare the amount of genetic variation.

5.6.1 Heterozygosity

The average expected (Hardy-Weinberg) heterozygosity at n loci within a population is the best general measure of genetic variation:

$$H_e = 1 - \sum_{i=1}^{n} p_i^2 \qquad (5.8)$$

It is easier to calculate one minus the expected homozygosity, as in equation 5.8, than summing over all heterozygotes because there are fewer homozygous than heterozygous genotypes with three or more alleles. Nei (1987) has referred to this measure as gene diversity, and pointed out that it can be thought of as either the average proportion of heterozygotes per locus in a randomly mating population, or the expected proportion of heterozygous loci in a randomly chosen individual. Gorman and Renzi (1979) have shown that estimates of H_e are generally insensitive to sample size, and that even a few individuals are sufficient for estimating H_e if a large number of loci are examined. In general, comparisons of H_e among populations are not valid unless a large number of loci are examined.

There are a variety of characteristics of average heterozygosity that make it valuable for measuring genetic variation. It can be used for genes of different ploidy levels (e.g., haploid organelles) and in organisms with different reproductive systems. We will see in later chapters that there is considerable theory available to predict the effects of reduced population size on heterozygosity (Chapter 6), that average heterozygosity is a good measure of the expected response of a population to natural selection (Chapter 11), and that it can also provide an estimate of individual inbreeding coefficients (Chapter 14).

5.6.2 Allelic richness

The total number of alleles at a locus has also been used as a measure of genetic variation (see Table 4.1). This is a valuable complementary measure of genetic variation because it is more sensitive to the loss of genetic variation because of small population size than heterozygosity, and it is an important measure of the long-term evolutionary potential of populations (see Section 6.4, Allendorf 1986).

The major drawback of the number of alleles is that, unlike heterozygosity, it is highly dependent on sample size. That is, there will be a tendency to detect more alleles at a locus as sample sizes increase. Therefore, comparisons between samples are not meaningful unless samples sizes are similar because of the presence of many low-frequency alleles in natural populations. This problem can be avoided by using **allelic richness**, which is a measure of allelic diversity that takes into account sample size (El Mousadik and Petit 1996). This measure uses a **rarefaction** method to estimate allelic richness at a locus for a fixed sample size, usually the smallest sample size if a series of populations are sampled (see Petit *et al.* 1998). Allelic richness can be denoted by $R(g)$, where g is the number of genes sampled.

The **effective number of alleles** is sometimes used to describe genetic variation at a locus. However, this parameter provides no more information about the number of alleles present at a locus than does hetero-zygosity. The effective number of alleles is the number of alleles that, if equally frequent, would result in the observed heterozygosity or homozygosity. It is computed as $A_e = 1/\sum p_i^2$ where p_i is the frequency of the ith allele. For example, consider two loci that both have an H_e of 0.50. The first locus has two equally frequent alleles $(p = q = 0.5)$, and the second locus has five alleles at frequencies of 0.68, 0.17, 0.05, 0.05, and 0.05. Both of these loci will have the same value of $A_e = 2$.

5.6.3 Proportion of polymorphic loci

The proportion of loci that are polymorphic (P) in a population has been used to compare the amount of variation between populations and species at allozyme loci (see Table 3.5). Strictly speaking, a locus is polymorphic if it contains more than one allele. However, generally some standard definition is used to avoid problems associated with comparisons of samples that are different sizes: that is, the larger the sample, the more likely we are to detect a rare allele. A locus is usually considered to be polymorphic if the frequency of the most common allele is less than either 0.95 or 0.99 (Nei 1987). The 0.99 standard has been used most often, but it is not reasonable to use this definition unless all sample sizes are greater than 50 (which is often not the case).

This measure of variation is of limited value. In some circumstances it can provide a useful measure of another aspect of genetic variation that is not provided by heterozygosity or allelic richness. It has been most valuable in studies of allozyme loci with large sample sizes in which many loci are studied, and many of the loci are monomorphic. However, it is of much less value in studies of highly variable loci (e.g., microsatellites) in which most loci are polymorphic in most populations. In addition, microsatellite loci are often selected to be studied because they are highly polymorphic in preliminary analysis.

Guest Box 5   Null alleles and Bonferroni 'abuse': treasure your exceptions (and so get it right for Leadbeater's possum)
Paul Sunnucks and Birgita D. Hansen

Nevertheless if I may throw out a word of counsel to beginners, it is: Treasure your exceptions! When there are none, the work gets so dull that no one cares to carry it further. Keep them always uncovered and in sight. Exceptions are like the rough brickwork of a growing building which tells that there is more to come and shows where the next construction is to be.

William Bateson (1912, p. 21)

The simplest model of population genetics, Hardy-Weinberg equilibrium, can be a very useful practical tool. However, it is important to avoid some common mistakes when testing for Hardy-Weinberg proportions.

When loci show a significant deviation from Hardy-Weinberg proportions in the direction of homozygous excess, researchers often infer that these loci have null alleles (see Section 5.4.2), and so should be removed from the dataset. Such pruning of data can do more harm than good. What matters is *why* the loci show homozygous excess. Three very common causes are presence of multiple demes (the Wahlund effect, Section 9.1.1), sex-linkage (Section 5.5), and null alleles (Section 5.4.2). But of these, only null alleles might present a compelling case for removing loci. Importantly, the presence of the Wahlund effect usually says something important about spatial population structure, and by removing the loci that show homozygous excess, we may inadvertently compromise our ability to detect one of the main population features we are looking for.

Before removing any loci, it is important to demonstrate that they really do have null alleles. The best approach is to test for Mendelian inheritance of null alleles at loci showing an excess of homozygotes. Unfortunately, this is generally not possible

(Continued)

for many species. Nulls at a locus are also implicated when good DNA samples fail to amplify. An effective way to detect such putative homozygous null genotypes is within multiplex PCRs in which other loci register PCR success. Another alternative is to ensure sufficient DNA quality, and then re-PCR individuals that have not amplified. This kind of quality control should be done early to prevent wasted work and resources.

Research on a highly endangered species serves as an excellent example of the importance of checking basic population genetic statistics and identifying the causes of deviation from Hardy-Weinberg proportions. The Leadbeater's possum is an arboreal marsupial, endemic to the State of Victoria, Australia, and thought to be extinct until its redis-

covery in 1961. In 1986, a small population was found at the 4.6 km^2 Yellingbo Nature Conservation Reserve (Smales 1994). Population genetics have played an important role in setting conservation priorities for this species, including showing the Yellingbo animals to be the last known representatives of a historically differentiated lowland, swamp-dwelling lineage (Hansen and Taylor 2008, Hansen *et al.* 2009).

Let us take a dataset of 85 Yellingbo individuals that were adults when first captured in 1996–2001 and screened for 20 microsatellite loci (Hansen *et al.* 2005). If this sample is considered one 'population', six loci show significant, and mostly very strong, homozygous excess (Table 5.2, as indicated by F_{IS}, the proportional excess of homozygotes; see

Table 5.2 Hardy-Weinberg analysis for 20 microsatellite loci in 85 adult Leadbeater's possums sampled at Yellingbo Nature Conservation Reserve. H_o is the proportion of individuals that were heterozygous, and H_e is that proportion expected under Hardy-Weinberg proportions. F_{IS} is the proportional excess of homozygotes relative to Hardy-Weinberg proportions. Loci are listed in order from the lowest to the greatest excess of homozygotes. F_{ST} is a measure of divergence between the two putative demes classified on the basis of genotypic similarity. F-statistic values significantly different than zero are in bold. F_{IS} values marked * would be disregarded under 'Bonferroni abuse', as described in the text.

Locus	No. alleles	H_o	H_e	F_{IS}	F_{ST}
GL39	2	0.341	0.300	−0.138	0.063
GL7	3	0.624	0.551	−0.133	0.245
GL6	2	0.435	0.392	−0.112	0.000
GL95	3	0.624	0.568	−0.099	**0.022**
GL35	3	0.600	0.548	−0.096	**0.106**
GL27B	5	0.682	0.634	−0.076	**0.117**
GL5A	3	0.635	0.602	−0.056	**0.351**
DT1	4	0.541	0.537	−0.008	**0.301**
GL33	3	0.518	0.521	**0.007***	**0.199**
GL1	3	0.494	0.505	0.022	**0.059**
GL42	4	0.624	0.650	0.040	**0.165**
GL28	2	0.471	0.491	0.042	**0.311**
GL13	3	0.576	0.623	0.076	**0.130**
GL44	5	0.588	0.651	0.097	**0.348**
GL24	2	0.424	0.491	0.139	**0.383**
GL110	5	0.482	0.563	**0.144***	**0.295**
GL19B	6	0.659	0.788	**0.165**	**0.190**
GL38	4	0.376	0.468	**0.196***	**0.335**
GL83	6	0.471	0.599	**0.215**	**0.339**
GL4	4	0.506	0.648	**0.220***	**0.419**
Overall	3.6	0.533	0.556	**0.041**	**0.236**

Section 9.1). Under a common, but incorrect approach, those loci would be assumed to have null alleles and be pruned from the dataset. But these loci do not have problematic null alleles. Failed PCR reactions were repeated, and there were no putative homozygous nulls. Also in other populations of the species, the same loci do not persistently show homozygous excess, and there is no evidence of X-linkage (Hansen *et al.* 2005).

Instead, the six loci show homozygous excess because Yellingbo, rather surprisingly given its small size relative to possum dispersal distances, contains two demes (Hansen and Taylor 2008, Hansen *et al.* 2009). This subdivision can be revealed using the genotypic clustering computer program STRUCTURE (see Figure 16.12), which shows two fairly distinct groups of genotypes, one in the north and one in the south of Yellingbo. We can calculate F_{ST} (a measure of differentiation, see Section 9.1) for each locus between the two demes. As expected, if the excess of homozygotes is caused by subdivision, those loci with the greatest differentiation between demes tend to show the greatest excess of homozygotes (Table 5.2).

If we erroneously assume that null alleles rather than a Wahlund effect cause the homozygous excess, and remove the offending loci, we will tend to remove the most useful loci for detecting population subdivision. On the other hand, if we split data into two demes and retest for Hardy-Weinberg proportions, much of the homozygous excess disappears; there are no significant cases for the southern deme, and only three (of reduced intensity) for the northern deme. Thus by mistaking Wahlund effect for null alleles and attempting to correct for nulls,

we would reduce or even remove our ability to find a potentially important, unanticipated population split in a highly endangered species.

Another common mistake is the misapplication of sequential Bonferroni correction (Section 5.3.2). For Leadbeater's possum, six localities have been sampled, and 20 loci are available. Correcting for multiple tests for Hardy-Weinberg proportions at 6 locations $\times$ 20 loci = 120 tests. So an individual result remains significant ($P \leq 0.05$) only if its uncorrected P value is $\leq 0.05 / (1 + N - i)$, where N is the number of tests (120 here), and i is the rank of the result among the 120. Under these conditions, the most significant result must jump a very substantial hurdle ($P \leq 0.00042$) to be significant after correction. The number of independent tests depends upon the question that we are asking. Correcting for this table of 120 tests does not ask the right questions. Does a particular locus show deviations from Hardy-Weinberg proportions? Is a particular population panmictic? In the former case, we should make one table for each locus (20 tables of six tests each, so the most significant result must be $P \leq 0.0083$ to survive the correction). In the second case, we should make one table per location (six tables of 20 tests, demands only $P \leq 0.0025$).

In fact, we argue that it is probably even more sensible to examine the matrix of locus $\times$ population homozygous deviations for patterns: are certain loci or locations over-represented in homozygous excess? In the case outlined here, four of the six loci that showed strong homozygous excess at Yellingbo would have been ruled nonsignificant by 'Bonferroni abuse' in a table of 120 tests, yet these results are meaningful and important.

CHAPTER 6

Small populations and genetic drift

Land snail, Example 6.2

The race is not always to the swift, nor the battle to the strong, for time and chance happens to us all.
Ecclesiastes 9:11

. . . the conservationist is faced with the ultimate sampling problem – how to preserve genetic variability and evolutionary flexibility in the face of diminishing space and with very limited economic resources. Inevitably we are concerned with the genetics and evolution of small populations, and with establishing practical guidelines for the practicing conservation biologist.
Sir Otto H. Frankel and Michael E. Soulé (1981, p. 31)

Chapter Contents

Conservation and the Genetics of Populations, Second Edition. Fred W. Allendorf, Gordon Luikart, and Sally N. Aitken.
© 2013 Fred W. Allendorf, Gordon Luikart and Sally N. Aitken. Published 2013 by Blackwell Publishing Ltd.

Genetic change will not occur in populations if all the assumptions of the Hardy-Weinberg equilibrium are met (see Section 5.1). In this and the next several chapters, we will see what happens when the assumptions of Hardy-Weinberg equilibrium are violated. In this chapter, we will examine what happens when we violate the assumption of infinite population size: that is, what will be the effect on allele and genotype frequencies when population size (N) is finite?

All natural populations are finite, so genetic drift will occur in all natural populations, even very large ones. For example, consider a new mutation that increases fitness, which occurs in an extremely large population of insects that numbers in the millions. Whether or not the single copy of this advantageous mutation is lost from this population will be determined primarily by the sampling process that determines which alleles are transmitted to the next generation. For example, if the individual with the mutation does not reproduce, the new allele will be lost immediately. Even if the individual with the mutation produces two progeny, there is a 25% chance, based on Mendelian segregation, that the mutation will be lost. Thus, the fate of a rare allele in an extremely large population will be determined primarily by genetic drift.

Understanding genetic drift and its effects is extremely important for conservation. Fragmentation and isolation due to habitat loss and modification has reduced the population size of many species of plants and animals throughout the world. We will see in future chapters how genetic drift is expected to affect genetic variation in these populations. More importantly, we will consider how genetic drift may reduce the fitness of individuals in these populations and limit the evolutionary potential of these populations to evolve by natural selection.

6.1 GENETIC DRIFT

Genetic drift is random change in allele frequencies from generation to generation because of sampling error. That is, the finite number of genes transmitted to progeny will be an imperfect sample of the allele frequencies in the parents (Figure 6.1). The mathematical treatment of genetic drift began with R.A. Fisher (1930) and Sewall Wright (1931) who independently considered the effects of binomial sampling in small populations. This model is therefore often referred to as the Wright-Fisher model or Fisher-Wright model.

However, Fisher and Wright strongly disagreed on the importance of drift in bringing about evolutionary change (Crow 2010). Genetic drift is sometimes called the 'Sewall Wright effect' in recognition that the importance of drift in evolution was largely introduced by Wright's arguments.

It is often helpful to consider extreme situations in order to understand the expected effects of relaxing assumptions on models. Consider the example of a plant species capable of self-fertilization with a constant population size of $N = 1$, consisting of a single individual of genotype Aa; the allele frequency in this generation is 0.5. We cannot predict what the allele frequency will be in the next generation because the genotype of the single individual in the next generation will depend upon which alleles are transmitted via the chance elements of Mendelian inheritance. However, we do know that the allele frequency in the next generation will be 0.0, 0.50, or 1.0, because the only three possible genotypes are AA, Aa, or aa. Based upon Mendelian expectations, there is a 50% probability that the frequency of the A allele will be either zero or one in the next generation.

Genetic drift is an example of a **stochastic** process in which the actual outcome cannot be predicted because it is affected by random elements (chance). Tossing a coin is one example of a stochastic process. One-half of the time, we expect a head to result, and one-half of the time we expect a tail. However, we do not know what the outcome of any specific coin-toss will be. We can mimic or simulate the effects of genetic drift by using a series of coin-tosses. Consider a population initially consisting of two heterozygous (Aa) individuals, one male and one female. Heterozygotes are expected to transmit the A and a alleles with equal probability to each gamete. A coin is tossed to specify which allele is transmitted by heterozygotes; an outcome of head (H) represents an A allele; and a tail (T) represents an a. No coin toss is needed for homozygous individuals since they will always transmit the same allele.

The results of one such simulation using these rules are shown in Table 6.1 and Figure 6.2. In the first generation, the female transmitted the A allele to both progeny because both coin tosses resulted in heads. The male transmitted an A to his daughter and an a to his son because the coin tosses resulted in a head and then a tail. Thus, the allele frequency (p) changed from 0.5 in the initial generation to 0.75 in the first generation, and the expected heterozygosity in the population

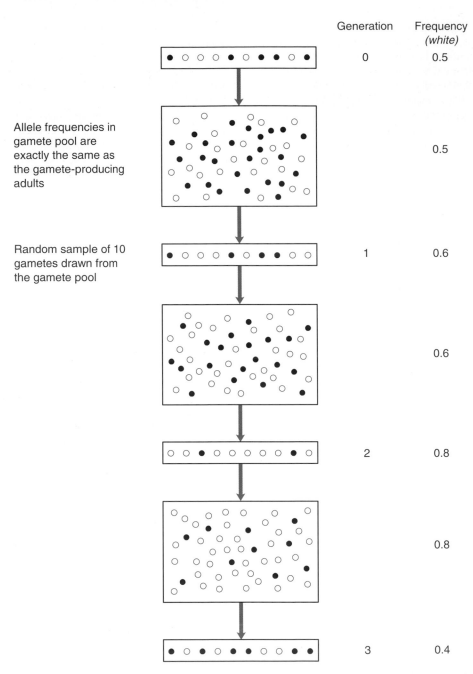

Figure 6.1 Random sampling of gametes resulting in genetic drift in a population. Allele frequencies in the gamete pools (large boxes) are assumed to reflect exactly the allele frequencies in the adults of the parental generation (small boxes). The allele frequencies fluctuate from generation to generation because the population size is finite ($N = 5$). From Graur and Li (2000).

Table 6.1 Simulation of genetic drift by coin-tossing in a population of one female and one male over seven generations. A coin is tossed twice to specify which alleles are transmitted by heterozygotes; an outcome of head (H) represents an A allele, and a tail (T) represents an a. The first toss represents the allele transmitted to the female in the next generation and the second toss the male (as shown in Figure 6.2). p is the frequency of the A allele. The observed and expected heterozygosities (assuming Hardy–Weinberg proportions) are also shown.

Generation	Mother	Father	p	H_o	H_e
0	Aa (HH)	Aa (HT)	0.50	1.000	0.500
1	AA	Aa (TT)	0.75	0.500	0.375
2	Aa (TH)	Aa (HT)	0.50	1.000	0.500
3	aA (HH)	Aa (TH)	0.50	1.000	0.500
4	Aa (TT)	AA	0.75	0.500	0.375
5	aA (HT)	aA (TT)	0.50	1.000	0.500
6	Aa (TT)	aa	0.25	0.500	0.375
7	aa	aa	0.00	0.000	0.000

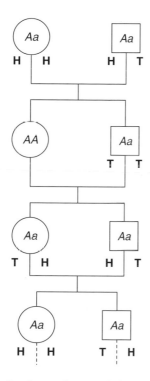

Figure 6.2 Simulation of genetic drift in a population consisting of a single female (circle) and male (square) in each generation. A coin is tossed twice to simulate the two gametes produced by each heterozygote. A head (**H**) indicates that the A allele is transmitted and a tail (**T**) indicates the a allele. Homozygotes always transmit the allele for which they are homozygous.

changed as well. This process is continued until the seventh generation when both individuals become homozygous for the a allele, and, thus, no further gene frequency changes can occur.

Table 6.1 shows one of many possible outcomes of genetic drift in a population with two individuals. However, we are nearly certain to get a different result if we start over again. In addition, it would be helpful to simulate the effects of genetic drift in larger populations. In principle, this can be done by tossing a coin; however, it quickly becomes extremely time-consuming.

A better way to simulate genetic drift is through computer simulations. Computational methods are available to produce a random number that is uniformly distributed between zero and one. This random number can be used to determine which allele is transmitted by a heterozygote. For example, if the random number is in the range of 0.0 to 0.5, we can specify that the A allele is transmitted; similarly, a random number in the range of 0.5 to 1.0 would specify an a allele. Models such as this are often referred to as Monte Carlo simulations in reference to the gambling tables in Monte Carlo. Figure 6.3 shows changes in allele frequencies in three populations of different sizes as simulated with a computer. The smaller the population size, the greater are the changes in allele frequency due to drift (compare N of 10 and 200).

The sampling process that we have examined here has two primary effects on the genetic composition of small populations:

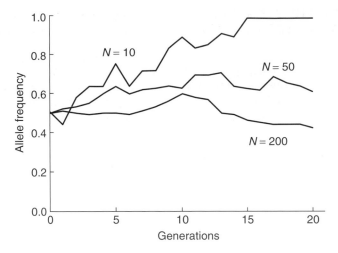

Figure 6.3 Results of computer simulations of changes in allele frequency by genetic drift for each of three population sizes (N) with an initial allele frequency of 0.5.

1 Allele frequencies will change.

2 Genetic variation will be lost.

We can model genetic changes in small populations either by changes in allele frequencies or increases in homozygosity caused by inbreeding. As allele frequencies change because of genetic drift, heterozygosity is expected to decrease (and homozygosity increase). For example, heterozygosity became zero in generation 16 with N of 10 because only one allele remained in the population. Once such a **fixation** of one allele or another occurs, it is permanent; only mutation (see Chapter 12) or gene flow (see Chapter 9) from another population can introduce new alleles. We will consider the effects of genetic drift on both allele frequencies and genetic variation in the next two sections.

6.2 CHANGES IN ALLELE FREQUENCY

We cannot predict the direction of change in allele frequencies from generation to generation because genetic drift is a random process. The frequency of any allele is equally likely to increase or decrease from one generation to the next because of genetic drift. Although we cannot predict the direction of change in allele frequency, we can describe the expected magnitude of change. In general, the smaller the population,

the greater the change in allele frequency we might expect (Figure 6.3).

The change in allele frequencies from one generation to the next because of genetic drift is a problem in sampling. A finite sample of gametes is drawn from the parental generation to produce the next generation. Both the sampling of gametes and the coin toss can be described by the binomial sampling distribution (see Section A3.3.2). The variance of change in allele frequency from one generation to the next is thus the binomial sampling variance:

$$V_q = \frac{pq}{2N}$$

Given that the current allele frequency is p with a population size of N, there is approximately a 95% probability that the allele frequency in the next generation will be in the interval:

$$p' = p \pm 2\sqrt{\frac{(pq)}{(2N)}} \tag{6.1}$$

For example, with an allele frequency of 0.50 and an N of 10, the allele frequency in the next generation will be in the interval 0.28 to 0.72 with 95% probability (equation 6.1). In contrast, with a p of 0.5 and an N of 200, this interval is only 0.45–0.55.

6.3 LOSS OF GENETIC VARIATION: THE INBREEDING EFFECT OF SMALL POPULATIONS

Genetic drift is expected to cause a loss of genetic variation from generation to generation. **Inbreeding** occurs when related individuals mate with one another. Inbreeding is one consequence of small population size; see Chapter 13 for a detailed consideration of inbreeding. For example, in an animal species with $N = 2$, the parents in each generation will be full-sibs (that is, brother and sister). Matings between relatives will cause an increase in homozygosity. The **inbreeding coefficient** (f) is the probability that the two alleles at a locus within an individual are identical by descent (that is, identical because they are derived from the same allele in a common ancestor in a previous generation). We will consider several different inbreeding coefficients that have specialized meaning (e.g., F_{IS}, F_{ST}, etc. in Chapter 9 and F in Chapter 13). We will use the general inbreeding coefficient f in this chapter, as defined above, along with its counterpart heterozygosity (h), which is equal to $1 - f$.

In general, the increase in homozygosity due to genetic drift will occur at the following rate per generation:

$$\Delta f = \frac{1}{2N} \tag{6.2}$$

This effect was first discussed by Gregor Mendel who pointed out that only half of the progeny of a heterozygous self-fertilizing plant will be heterozygous; one-quarter will be homozygous for one allele and the remaining one-quarter will be homozygous for the other allele (Table 6.2). This is as predicted by equation 6.2 ($N = 1$, ($\Delta f = 0.50$).

We have seen that the expected rate of loss of heterozygosity per generation is $\Delta f = 1/2N$; therefore, after t generations

$$f_t = 1 - \left(1 - \frac{1}{2N}\right)^t \tag{6.3}$$

f_t is the expected increase in homozygosity at generation t and is known by a variety of names (e.g., **autozygosity**, **fixation index**, or the inbreeding coefficient) depending upon the context in which it is used.

It is often more convenient to keep track of the amount of variation remaining in a population using h (heterozygosity), where:

$$f = 1 - h \tag{6.4}$$

Therefore, the expected decline in h per generation is:

$$\Delta h = -\frac{1}{2N} \tag{6.5}$$

so that after one generation:

$$h_{t+1} = \left(1 - \frac{1}{2N}\right)h_t \tag{6.6}$$

Table 6.2 This table appeared in Mendel's original classic paper in 1865. He was considering the expected genotypic ratios in subsequent generations from a single hybrid (i.e., heterozygous) individual that reproduced by self-fertilization. He assumed that each plant in each generation (Gen) had four offspring. The homozygosity (Homo) and heterozygosity (Het) columns did not appear in the original paper.

Gen	AA	Aa	aa	Ratio AA:Aa:aa	Homo	Het
1	1	2	1	1:2:1	0.500	0.500
2	6	4	6	3:2:3	0.750	0.250
3	28	8	28	7:2:7	0.875	0.125
4	120	16	120	15:2:15	0.938	0.062
5	496	32	496	31:2:31	0.939	0.031
n				$2^n-1:2:2^n-1$	$1-(1/2)^n$	$(1/2)^n$

Example 6.1   Bottleneck in the Mauritius kestrel

Kestrels on the Indian Ocean island of Mauritius went through a bottleneck of one female and one male in 1974 (Nichols *et al.* 2001). The population had fewer than 10 birds throughout the 1970s, and there were fewer than 50 birds in this population for many years because of the widespread use of pesticides from 1940 to 1960. However, this population grew to nearly 500 birds by the mid-1990s. Nichols *et al.* (2001) examined the loss in genetic variation in this population at 10 microsatellite loci by comparing living birds to 26 ancestral birds from museum skins that were up to 170 years old. The heterozygosity of the restored population was 0.099 compared to heterozygosity in the ancestral birds of 0.231.

The amount of heterozygosity expected to remain in Mauritius kestrels after one generation of a bottleneck of $N = 2$ can be estimated with equation 6.6:

$$\left(1-\frac{1}{2N}\right)h_t = \left(1-\frac{1}{4}\right)(0.231) = 0.173$$

We can use expression 6.7 to see whether the amount of heterozygosity in the restored population of Mauritius kestrels is approximately the same as we would expect after a bottleneck of two individuals for three generations:

$$\left(1-\frac{1}{2N}\right)^t h_o = (0.75)^3(0.231) = 0.097$$

The actual bottleneck in Mauritius kestrels was almost certainly longer than three generations, with more birds than two in each generation. However, the expressions in this chapter all assume discrete generations and cannot be applied directly to species such as the Mauritius kestrels that have overlapping generations.

The heterozygosity after t generations can be found by:

$$h_t = \left(1-\frac{1}{2N}\right)^t h_0 \tag{6.7}$$

where h_o is the initial heterozygosity. Example 6.1 shows how these expressions can be used to predict the effects of population **bottlenecks**.

Figure 6.4 shows this effect at a locus with two alleles and an initial frequency of 0.5 in a series of computer simulations of eight populations that consist of 20 individuals each. These 20 individuals possess 40 gene copies at any given locus. Forty gametes must be drawn from these 40 parental gene copies to form the next generation. The genotype of any one selected gamete does not affect the probability of the next gamete that is drawn; this is similar to a coin toss where one outcome does not affect the probability of the next toss.

Two of the eight populations simulated became fixed for the A allele and one became fixed for the a allele. Both of the alleles were retained by five of the populations after 20 generations. The heterozygosity in each of the populations is shown in Figure 6.5. There are large differences among populations in the decline in

heterozygosity over time. Nevertheless, the mean decline in heterozygosity for all eight populations is very close to that predicted with equation 6.6.

The heterozygosity at any single locus with two alleles is equally likely to increase or decrease from one generation to the next (except in the case of maximum heterozygosity when the allele frequencies are at 0.5). This may seem counterintuitive in view of equation 6.6, which describes a monotonic decline in heterozygosity. Heterozygosity at a locus with two alleles is at a maximum when the two alleles are equally frequent ($p = q = 0.5$; Figure 6.6). The frequency of any particular allele is equally likely to increase or decrease due to genetic drift. Thus, heterozygosity will increase if the allele frequency drifts towards 0.5, and it will decrease if the allele frequency drifts toward 0 or 1. However, the expected net loss is greater than the net gain in heterozygosity in each generation by $1/2N$.

6.4 LOSS OF ALLELIC DIVERSITY

We have so far measured the loss of genetic variation caused by small population size by looking at the expected reduction in heterozygosity (h). There are other ways to

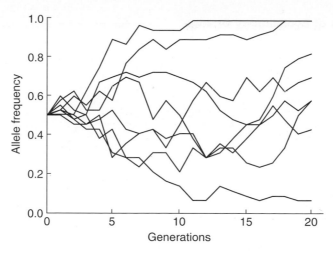

Figure 6.4 Computer simulations of genetic drift at a locus having two alleles with initial frequencies of 0.5 in eight populations of 20 individuals each.

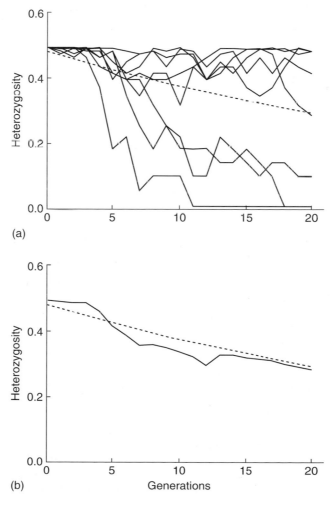

Figure 6.5 (a) Expected heterozygosities ($2pq$) in the eight populations ($N = 20$) undergoing genetic drift as shown in Figure 6.4. The dashed line shows the expected change in heterozygosity using equation 6.6. (b) Mean heterozygosity for all loci (solid line) and the expected heterozygosity using equation 6.6 (dashed line).

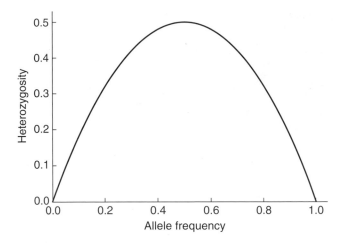

Figure 6.6 Expected heterozygosity ($2pq$) at a locus with two alleles, as a function of allele frequency.

measure genetic variation and its loss. A second important measure of genetic variation is allelic diversity or the number of alleles present at a locus (A). There are advantages and disadvantages to both of these measures.

Heterozygosity has been widely used because it is proportional to the amount of **genetic variance** at a locus, and it lends itself readily to theoretical considerations of the effects of finite population size on genetic variation. In addition, the expected reduction in heterozygosity because of genetic drift is independent of the number of alleles present. Finally, estimates of heterozygosity from empirical data are relatively insensitive to sample size, whereas estimates of the number of alleles in a population are strongly dependent upon sample size. Therefore, comparisons of heterozygosities in different species or populations are generally more meaningful than comparisons of the number of alleles detected.

Nevertheless, heterozygosity has the disadvantage of being relatively insensitive to the effects of bottlenecks (Allendorf 1986). The difference between heterozygosity and A is greatest with extreme bottlenecks (Figure 6.7). For example, a population with two individuals is expected to lose only 25% ($1/2N = 25\%$) of its heterozygosity. Thus, 75% of the heterozygosity in a population will be retained even through such an extreme bottleneck. However, two individuals can possess a maximum of four different alleles. Thus, considerably more of the allelic variation may be lost during a bottleneck if there are many alleles present at a locus.

The effect of a bottleneck on the number of alleles present is more complicated than the effect on heterozygosity because it is dependent upon the number and frequencies of alleles present (Allendorf 1986). The probability of an allele being lost during a bottleneck of size N is:

$$(1-p)^{2N} \tag{6.8}$$

where p is the frequency of the allele. This is the probability of sampling all of the gametes to create the next generation ($2N$) without selecting at least one copy of the allele in question. Rare alleles (say $p < 0.05$) are especially susceptible to loss during a bottleneck. However, the loss of rare, potentially important, alleles will have little effect on heterozygosity. For example, an allele at a frequency of 0.01 has a 60% chance of being lost following a bottleneck of 25 individuals (expression 6.8). Figure 6.8 shows the probability of the loss of rare alleles during a bottleneck of N individuals.

In general, if a population is reduced to N individuals for one generation then the expected total number of alleles (A') remaining is:

$$E(A') = A - \sum_{j=1}^{A}(1-p_j)^{2N} \tag{6.9}$$

where A is the initial number of alleles and p_j is the frequency of the jth allele. For example, consider a locus with two alleles at frequencies of 0.9 and 0.1 and a bottleneck of just two individuals. In this case:

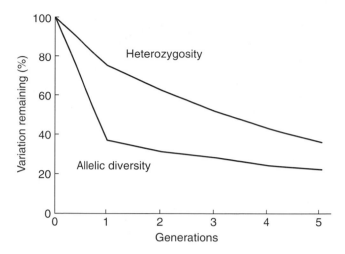

Figure 6.7 Simulated loss of heterozygosity and allelic diversity at eight microsatellite loci during a bottleneck of two individuals for five generations. The initial allele frequencies are from a population of brown bears from the Western Brooks Range of Alaska. Redrawn from Luikart and Cornuet (1998).

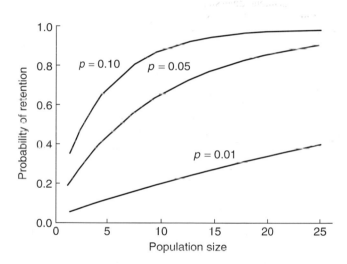

Figure 6.8 Probability of retaining a rare allele ($p = 0.01$, 0.05, or 0.10) after a bottleneck of size N for a single generation (expression 6.8).

$$E(A') = 2 - (1 - 0.9)^4 - (1 - 0.1)^4 = 1.34$$

Thus, on average, we expect to lose one of these two alleles nearly two-thirds of the time. In contrast, there is a much greater expected probability of retaining both alleles at a locus with two alleles if the two alleles are equally frequent:

$$E(A') = 2 - (1 - 0.5)^4 - (1 - 0.5)^4 = 1.88$$

Thus, the expected loss of alleles during a bottleneck depends upon the number and frequencies of the alleles present. This is in contrast to heterozygosity, which is lost at a rate of $1/2N$, regardless of the current heterozygosity.

The loss of alleles during a bottleneck will have a drastic effect on the overall genotypic diversity of a population. As we saw in Section 5.3.1, the number of genotypes grows very quickly as the number of alleles increases. For example, the number of possible genotypes at a locus with 2, 5, and 10 alleles is 3, 15, and 55, respectively. Thus, the loss of alleles during a bottleneck will greatly reduce the genotypic diversity in a population.

6.5 FOUNDER EFFECT

The founding of a new population by a small number of individuals will cause abrupt changes in allele frequency and loss of genetic variation (Example 6.2). Such severe bottlenecks in population size are a special case of genetic drift. Perhaps surprisingly, however, even extremely small bottlenecks have relatively little effect on heterozygosity. For example, with sexual species the smallest possible bottleneck is $N = 2$. Even in this extreme case, the population will only lose 25% of its heterozygosity (see equation 6.5). Stated in another way, just two individuals randomly selected from any population, regardless of size, will contain 75% of the total heterozygosity in the original population. We can also use equation 6.5 to estimate the size of the founding population if we know how much heterozygosity has been lost through the founding bottleneck.

A laboratory experiment with guppies demonstrates this effect clearly (Nakajima *et al.* 1991). Sixteen separate **subpopulations** were derived from a large random mating laboratory colony of guppies by mating a female with a single male. After four generations, each of these subpopulations contained more than 500 individuals. Approximately 45 fish were then sampled from each subpopulation and genotyped at two protein loci that were polymorphic in the original colony (Table 6.3). The mean heterozygosity at both loci in these 16 subpopulations was 0.358, compared with heterozygosity in the original colony of 0.482. Thus, the mean heterozygosity in the subpopulations, following a bottleneck of two individuals, was 26% lower than in the population from which the subpopulations were founded. This agrees very closely with equation 6.5 which predicts a 25% reduction following a bottleneck of two individuals.

The total amount of heterozygosity lost during a bottleneck depends upon how long it takes the population to return to a 'large' size (Nei *et al.* 1975). That is, species such as guppies, in which individual females may produce 50 or so progeny, may quickly attain large enough population sizes following a bottleneck so that little further variation is lost following the initial bottleneck. However, species with lower population growth rates may persist at small population sizes for many generations, during which heterozygosity is further eroded.

The growth rate of a population following a bottleneck can be modeled using the so-called logistic growth equation, which describes the size of a population after t generations based upon the initial population size (N_0), the intrinsic growth rate (r), and the equilibrium size of the population (K):

$$N(t) = \frac{K}{1 + be^{-rt}} \tag{6.10}$$

The constant e is the base of the natural logarithms (approximately 2.72), and b is a constant equal to $(K - N_0)/N_0$.

We can estimate the total expected loss in heterozygosity in the guppy example depending upon the rate of population growth of the subpopulations. The initial size of the subpopulations (N_o) was 2, and we assume the equilibrium size (K) was 500. We can then examine three different intrinsic growth rates (r): 1.0, 0.5, and 0.2. An r of 1.0 indicates that population size is increasing by a factor of 2.72 (e) each generation when population size is far below K. Similarly, r values of 0.5 and 0.2 indicate growth rates of 1.65 and 1.22 at small population sizes, respectively. A detailed discussion of use of the logistic equation to describe population growth can be found in chapters 15 and 16 of Ricklefs and Miller (2000).

Equation 6.10 can be used to predict the expected population size each generation following the bottleneck. We expect heterozygosity to be eroded at a rate of $1/2N$ in each of these generations. Figure 6.10 shows the expected loss in heterozygosity in our guppy example for ten generations following the bottleneck. As expected, populations having a relatively high growth rate ($r = 1.0$) will lose little heterozygosity following the initial bottleneck. However, heterozygosity is expected to continue to erode even ten generations following the bottleneck in populations with the

Example 6.2 Effects of founding events on allelic diversity in a snail

The land snail *Theba pisana* was introduced from Europe into Western Australia in the 1890s. A colony was founded in 1925 on Rottnest Island with animals taken from the mainland population near Perth. Johnson (1988) reported the allele frequencies at 25 allozyme loci. Figure 6.9 shows the loss of rare alleles caused by the bottleneck associated with the founding of a population in Perth on the mainland and in the second bottleneck associated with the founding of the population on Rottnest Island. The height of each bar represents the number of alleles in that sample that had the frequency specified on the x-axis. For example, there were eight alleles that had a frequency of less than 0.05 in the founding French population.

However, there were no alleles in either of the two Australian populations at a frequency of less than 0.05.

The distribution of allele frequencies, such as plotted in Figure 6.9, can be used to detect bottlenecks even when data are not available from the pre-bottlenecked population (Luikart *et al.* 1998). Rare alleles (frequency less than 0.05) are expected to be common in samples from populations that have not been bottlenecked in their recent history, such as observed in the French sample (see Chapter 12). The complete absence of such rare alleles in Australia would have suggested that these samples came from recently bottlenecked populations, even if the French sample was not available for comparison.

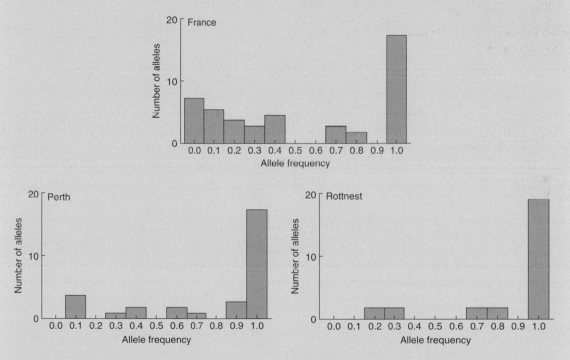

Figure 6.9 Effects of bottlenecks on the number of rare alleles at 25 allozyme loci in the land snail *Theba pisana* which was introduced from Europe into Western Australia in the 1890s. Data from Johnson (1988).

Table 6.3 Allele frequencies (p) and heterozygosities (h) at two loci in 16 subpopulations of guppies founded by a single female and male. H_e is the mean heterozygosity at the two loci. Data from Nakajima *et al.* (1991).

Subpopulation	AAT-1		PGM-1		
	p	h	p	h	H_e
1	0.521	0.499	0.677	0.437	0.468
2	0.738	0.387	0.600	0.480	0.433
3	0.377	0.470	0.131	0.227	0.349
4	0.915	0.156	0.939	0.114	0.135
5	0.645	0.458	0.638	0.461	0.460
6	0.571	0.490	0.548	0.495	0.492
7	0.946	0.102	0.833	0.278	0.190
8	0.174	0.287	0.341	0.449	0.368
9	0.617	0.473	0.500	0.500	0.486
10	0.820	0.295	0.640	0.461	0.378
11	0.667	0.444	0.917	0.152	0.298
12	0.219	0.342	0.531	0.498	0.420
13	1.000	0	0.838	0.272	0.136
14	0.250	0.375	0.853	0.251	0.313
15	0.375	0.469	0.740	0.385	0.427
16	0.152	0.258	0.582	0.486	0.372
—	—	—	—	—	—
Average	0.562	0.344	0.644	0.372	0.358
Original colony	0.581	0.487	0.605	0.478	0.482

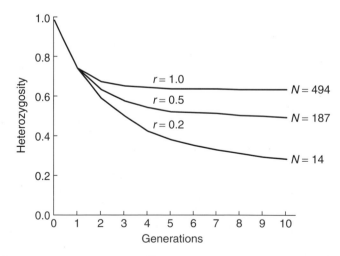

Figure 6.10 Expected heterozygosities in three subpopulations of guppies going through a bottleneck of two individuals and growing at different rates (r) according to the logistic growth equation (equation 6.10). N is the expected population size for each subpopulation in the tenth generation.

slowest growth rate. In general, bottlenecks will have a greater and more long-lasting effect on the loss of genetic variation in species with smaller intrinsic growth rates (e.g., large mammals) than species with high intrinsic growth rates (e.g., insects).

Founder events and population bottlenecks will have a greater effect on the number of alleles in a population than on heterozygosity (see Figure 6.7). Some classes of loci in vertebrates have been found to have many nearly equally frequent alleles. For example, Gibbs *et al.* (1991) have described 37 alleles at the hypervariable major histocompatibility (*MHC*) locus in a sample of 77 adult blackbirds from Ontario (Figure 6.11). As we have seen, two birds chosen at random from this population are expected to contain 75% of the heterozygosity. However, two birds can at best possess four of the 37 different *MHC* alleles. Thus, at least 33 of the 37 detected alleles (89%) will be lost in a bottleneck of two individuals. Thus, bottlenecks of short duration may have little effect on heterozygosity but will reduce severely the number of alleles present at some loci (Example 6.3).

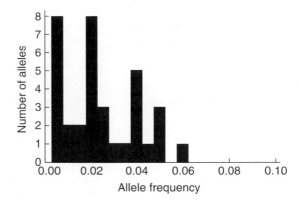

Figure 6.11 Distribution of frequencies of alleles at the highly variable *MHC* locus detected in red-wing blackbirds. The height of each bar represents the number of alleles that had the frequency specified on the *x*-axis. Thus, just over 20% of the 37 alleles had frequencies of less than 1% in the sample. Compare this with Figure 6.9 which is the equivalent plot of allele frequencies at much less variable allozyme loci in a land snail; note the difference in the scale of the allele frequency axis. Data from Gibbs *et al.* (1991).

Example 6.3 *Founding events in the Laysan finch*

The Laysan finch is an endangered Hawaiian honeycreeper found on several islands in the Pacific Ocean (Tarr *et al.* 1998). The species underwent a bottleneck of approximately 100 birds on Laysan Island after the introduction of rabbits in the early 1900s (Figure 6.12). The population recovered rapidly after eradication of the rabbits and has fluctuated around a mean of 10,000 birds since 1968. In 1967, the US Fish and Wildlife Service translocated 108 finches to Southeast Island, one of several small islets approximately 300 km northwest of Laysan that comprise Pearl and Hermes Reef (PHR). The translocated population declined to 30–50 birds and then rapidly increased to some 500 birds on Southeast Island. Several smaller populations have since become established in other islets within PHR. Two birds colonized Grass Island in 1968 and six more finches were moved to this islet in 1970. The population of birds on Grass Island has fluctuated between 20 and 50 birds. In 1973, a pair of finches founded a population on North Island. The

population on North Island has fluctuated between 30 and 550 birds.

Tarr *et al.* (1998) assayed variation at nine microsatellite loci to examine the effects of the founder events and small population sizes in these four populations (Table 6.4). Their empirical results are in close agreement with theoretical expectations. All three newly founded populations have fewer alleles than the founding population on Laysan. The average heterozygosity on Southeast Island is approximately 8% less than on Laysan Island; the heterozygosities on the two other islands is approximately 30% less than the original founding population on Laysan.

However, heterozygosities at four of the nine loci are actually greater in the post-bottleneck population on Southeast Island than on Laysan; see the discussion in Section 6.3. Thus, it is important to examine many loci to detect and quantify the effects of bottlenecks in populations on heterozygosity.

(Continued)

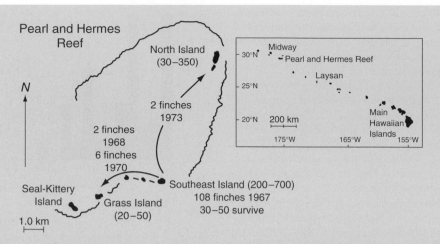

Figure 6.12 Map of Pearl and Hermes Reef and colonization history of the three finch populations. Ranges of population size on the islands are shown in parentheses. The inset is the Hawaiian Archipelago. Redrawn from Tarr *et al.* (1998).

Table 6.4 Numbers of alleles (*A*) and observed heterozygosities (Het) at nine microsatellite loci in Laysan finches in four island populations. The number in parentheses after the name of the island is the sample size. The populations on the other three islands were all founded from birds from Laysan. Data from Tarr *et al.* (1998).

	Laysan (44)		Southeast (43)		North (43)		Grass (36)	
Locus	*A*	Het	*A*	Het	*A*	Het	*A*	Het
Tc.3A2C	2	0.558	2	0.535	2	0.535	2	0.528
Tc.4A4E	2	0.386	2	0.605	2	0.209	2	0.556
Tc.5A1B	3	0.372	3	0.233	1	0	2	0.583
Tc.5A5A	3	0.409	2	0.071	2	0.372	2	0.278
Tc.1A4D	3	0.659	3	0.744	3	0.698	2	0.528
Tc.11B1C	3	0.636	3	0.674	3	0.628	3	0.194
Tc.11B2E	3	0.614	3	0.488	1	0	2	0.500
Tc.11B4E	4	0.614	4	0.628	2	0.256	3	0.444
Tc.12B5E	5	0.568	4	0.442	3	0.372	1	0
All loci	3.11	0.535	2.89	0.491	2.11	0.341	2.11	0.401
		100%		92%		64%		75%

6.6 GENOTYPIC PROPORTIONS IN SMALL POPULATIONS

We saw in the guppy example (Table 6.3) that the separation of a large random mating population into a number of subpopulations can cause a reduction in heterozygosity, and a corresponding increase in homozygosity. However, genotypes within each subpopulation will be in Hardy-Weinberg proportions as long as random mating occurs within the subpopulations. It may seem paradoxical that heterozygosity is decreased in small populations, but the subpopulations themselves remain in Hardy-Weinberg proportions. The explanation is that the reduction in heterozygosity is caused by changes in allele frequency from one generation to the next, while Hardy-Weinberg genotypic proportions will occur in any one generation as long as mating is random (see Section 5.3).

In fact, there is actually a tendency for an *excess* of heterozygotes in small populations of animals and plants with separate sexes (see Example 6.4). Different allele frequencies in the two sexes will cause an excess in heterozygotes relative to Hardy-Weinberg proportions (Robertson 1965, Kirby 1975, Brown 1979). An extreme example of this is that of hybrids produced by males from one strain (or species) and females from another, so that all progeny are heterozygous at any loci where the two strains differ. In this case, however, genotypic proportions will return to Hardy-Weinberg proportions in the next generation.

In small populations, allele frequencies are likely to differ between the sexes just due to chance. On average, the frequency of heterozygotes in the progeny population will exceed Hardy-Weinberg expectations by a proportion of

$$\frac{1}{8N_m} + \frac{1}{8N_f} \tag{6.11}$$

where N_m and N_f are the numbers of male and female parents (Robertson 1965). This result holds regardless of the number of alleles at the locus concerned.

Let us consider the extreme case of a population with one female and one male ($N = 2$) and two alleles (Table 6.6). There are six possible types of matings. Mating between identical homozygotes (either *AA* or *aa*) will produce monomorphic progeny. Progeny produced by matings between two heterozygotes will result in expected Hardy-Weinberg proportions. However, the other three matings will result in an excess of heterozygotes. The extreme case is a mating between opposite homozygotes which will produce

---

Example 6.4 Small populations of bull trout

The expected excess of heterozygotes in small populations can sometimes be used to detect populations with a small number of individuals. For example, Leary *et al.* (1993b) examined bull trout from a hatchery population at four polymorphic allozyme loci and found a strong tendency for an excess of heterozygotes (Table 6.5, see Example 5.1). On average, there was a 38% excess of heterozygotes at four polymorphic loci; a 25% excess of heterozygotes would be expected if all of the progeny came from just two individuals (expression 6.8).

Further examination of genotype frequencies suggests that these fish were produced by a very small number of parents. The exceptionally high proportion of heterozygotes (0.88) at *mIDHP-1* suggests that most fish came from a single pair mating between individuals homozygous for the two different alleles at this locus; all progeny from such a mating will be heterozygous at the locus. Allele frequencies at the four polymorphic loci also support the inference that most of these fish resulted from a single pair mating. The only allele frequencies possible in a full-sib family are 0.00, 0.25, 0.50, 0.75, and 1.00 because two parents possess four copies of each gene (Table 6.5); the allele frequencies at all four loci are near these values.

When we asked about the source of these fish after this genetic evaluation, we were told that the fish sampled were produced from at most three wild females and two wild males that were taken from the Clark Fork River in Idaho.

Table 6.5 Observed (and expected) genotypic proportions in bull trout sampled from a hatchery population. $\hat{p}(1)$ is the estimated frequency of the *1* allele. F_{IS} equals $[1 - (H_o/H_e)]$ and is a measure of the deficit of heterozygotes observed relative to the expected Hardy-Weinberg proportions (see Chapter 9). A negative F_{IS} indicates an excess of heterozygotes. Data from Leary *et al.* (1993b).

Locus	Genotype 11	12	22	$\hat{p}(1)$	F_{IS}
GPI-A	10 (12.2)	15 (10.5)	0 (2.2)	0.700	−0.43*
IDDH	24 (24.1)	1 (1.0)	0 (0.0)	0.980	0.00
mIDHP-1	1 (5.8)	22 (12.5)	2 (6.8)	0.480	−0.76***
IDHP-1	12 (13.7)	13 (9.6)	0 (1.7)	0.740	−0.35
Mean					−0.38

*$P < 0.05$, ***$P < 0.001$.

Table 6.6 Expected Mendelian genotypic proportions at a locus with two alleles in a population with a single female and a single male. F_{IS} is a measure of the deficit of heterozygotes observed relative to the expected Hardy-Weinberg proportions (see Chapter 9). A negative F_{IS} indicates an excess of heterozygotes.

Mating	AA	Aa	aa	Freq(A)	F_{IS}
$AA \times AA$	1.00	0	0	1.00	undefined
$AA \times Aa$	0.50	0.50	0	0.75	−0.33
$AA \times aa$	0	1.00	0	0.50	−1.00
$Aa \times Aa$	0.25	0.50	0.25	0.50	0.00
$Aa \times aa$	0	0.50	0.50	0.25	−0.33
$aa \times aa$	0	0	1.00	0.00	undefined

all heterozygous progeny. On the average, there will be a 25% excess of heterozygotes in populations produced by a single male and a single female (expression 6.11).

With more than two alleles, there will be a deficit of each homozygote and an overall excess of heterozygotes. However, some heterozygous genotypes may be less frequent than expected by Hardy-Weinberg proportions, despite the overall excess of heterozygotes.

6.7 FITNESS EFFECTS OF GENETIC DRIFT

We have considered in some detail how genetic drift is expected to affect allele frequencies and reduce the amount of genetic variation in small populations. We will now preview the effect that this loss of genetic variation is expected to have on the population itself (see Box 6.1 and Guest Box 6). How will the loss of genetic variation expected in small populations affect the capability of a population to persist and evolve? We will take a more in-depth look at these effects in later chapters.

6.7.1 Changes in allele frequency

Large changes in allele frequency from one generation to the next are likely in small populations due to chance. This effect may cause an increase in frequency of alleles that have harmful effects. Such deleterious

alleles are continually introduced by mutation but are kept at low frequencies by natural selection. Moreover, most of these harmful alleles are recessive, so their harmful effects on the phenotype are only expressed in homozygotes. It is estimated that every individual in a population harbors several of these harmful recessive alleles in a heterozygous condition without any phenotypic effects (Chapter 13).

Let us consider the possible effect of a population bottleneck of two individuals. As we have seen, most rare alleles will be lost in such a small bottleneck. However, any allele for which one of the two founders is heterozygous will be found in the new population at a frequency of 25%. Thus, some rare deleterious alleles present in the founders will jump in frequency to 25%. Of course, at most loci the two founders will not carry a harmful allele. However, every individual carries harmful alleles at some loci. Therefore, we cannot predict which particular harmful alleles will increase in frequency following a bottleneck, but we can predict that several harmful alleles that were rare in the original population will be found at much higher frequencies. And if the bottleneck persists for several generations, these harmful alleles may become more frequent in the new population.

This effect is commonly seen in domestic animals such as dogs in which breeds often originated from a small number of founders (Cruz *et al.* 2008, vonHoldt *et al.* 2010). Different dog breeds usually have some characteristic genetic abnormality that is much more common within the breed than in the species as a whole (Hutt 1979). For example, different kinds of hemolytic anemia are common in several dog breeds (e.g., basenjis, beagles, and Alaskan malamutes).

Dalmatians were originally developed from a few founders that were selected for their running ability and distinctive spotting pattern. Dalmatians are susceptible to kidney stones because they excrete exceptionally high amounts of uric acid in their urine. This difference is due to a recessive allele at a single locus (Trimble and Keeler 1938). Apparently, one of the principal founders of this breed carried this recessive allele, and it subsequently drifted to high frequency in this breed.

6.7.2 Loss of allelic diversity

We have seen in Section 6.4 that genetic drift will have a much greater effect on the allelic diversity of a popu-

Box 6.1 Population bottlenecks and decreased hatching success in endangered birds

Many populations have undergone severe bottlenecks because of reduced population size and increased fragmentation caused by habitat loss. As we have seen in this chapter, small populations are vulnerable to genetic drift and inbreeding effects. Increased hatching failure is a common result of inbreeding in birds. Heber and Briskie (2010) have recently shown that bottlenecks in endangered birds have a major effect on hatching success.

Hatching failure is generally less than 10% in non-inbred bird populations. Hatching failure is sometimes more than 50% in some inbred populations. Given the increase of inbreeding in smaller populations, Heber and Briskie tested whether hatching success decreases in bird populations that have gone through a population bottleneck.

They summarized rates of hatching failure in 51 threatened bird species from 31 families. The bottleneck size ranged from 4 to 20,000 individuals. They estimated hatching failure as the proportion of eggs incubated to term that failed to hatch, excluding failure due to desertion, predation, or adverse weather. Under this definition, eggs that failed to hatch were either infertile or died during embryonic development, both of which are thought to increase as inbreeding increases (Jamieson and Ryan 2000).

Heber and Briskie found a substantial increase in hatching failure associated with smaller population bottlenecks ($P < 0.001$; Figure 6.13). The exact threshold of population size below which inbreeding depression is likely to cause a problem varies among species and traits. Nevertheless, hatching failure was greater than 10% in all species that went through a bottleneck of less than 150 individuals.

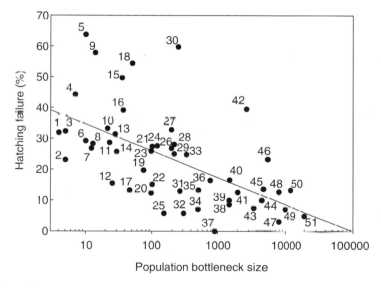

Figure 6.13 Effect of bottleneck size (smallest number of individuals recorded in the population) and percentage hatching failure in 51 bird species. Hatching failure is plotted on a linear scale and bottleneck size is plotted on a logarithmic scale, although both were log transformed in analyses. Redrawn from Heber and Briskie (2010).

lation than on heterozygosity if there are many alleles present at a locus. Evidence from many species indicates that loci associated with disease resistance often have many alleles (Clarke 1979). The best example of this is the major histocompatibility complex (MHC) in vertebrates (Edwards and Hedrick 1998). The MHC in humans consists of over 100 linked genes on chromosome-6 (Vogel and Motulsky 1986). Many alleles occur at all of these loci; for example, there are ten or more nearly equally frequent alleles at the A locus and 15 or more at the B locus.

MHC molecules assist in the triggering of the immune response to disease organisms. Individuals heterozygous at MHC loci are relatively more resistant to a wider array of pathogens than are homozygotes (see Hughes 1991). Most vertebrate species that have been studied have been found to harbor an amazing number of MHC alleles (Figure 6.11, Hughes 1991). Thus, the loss of allelic diversity at MHC loci is likely to render small populations of vertebrates much more susceptible to disease epidemics (Paterson et al. 1998, Gutierrez-Espeleta et al. 2001).

6.7.3 Inbreeding depression

The harmful effects of inbreeding have been known for a long time. Experiments with plants by Darwin and others demonstrated that loss of vigor generally accompanied continued selfing, and that crossing different lines maintained by selfing restored the lost vigor. Livestock breeders also generally accepted that continued inbreeding within a herd or flock could lead to a general deterioration that could be restored by outcrossing. The first published experimental reports of the effects of inbreeding in animals were with rats (Crampe 1883, Ritzema-Bos 1894).

The implication of these results for wild populations did not go unnoticed by Darwin. It occurred to him that deer kept in British parks might be affected by isolation and "long-continued close interbreeding". He was especially concerned because he was aware that the effects of inbreeding may go unnoticed because they accumulate slowly. Darwin inquired about this effect and received the following response from an experienced gamekeeper:

"... the constant breeding in-and-in is sure to tell to the disadvantage of the whole herd, though it may take a long time to prove it; moreover, when we find, as is very constantly the case, that the introduction of fresh blood has been of the greatest use to deer, both by improving their size and appearance, and particularly by being of service in removing the taint of 'rickback' if not other diseases, to which deer are sometime subject when the blood has not been changed, there can, I think, be no doubt but that a judicious cross with a good stock is of the greatest consequence, and is indeed essential, sooner or later, to the prosperity of every well-ordered park." (Darwin 1896, p. 99).

Despite Darwin's concern and warning, these early lessons from agriculture were largely ignored by those responsible for the management of wild populations of game and by captive breeding programs of zoos for nearly 100 years (see Voipio 1950 for an exception).

A seminal paper in 1979 by Kathy Ralls and her colleagues had a dramatic effect on the application of genetics to the management of wild and captive populations of animals. They used zoo pedigrees of 12 species of mammals to show that individuals from matings between related individuals tended to show reduced survival relative to progeny produced by matings between unrelated parents. The pedigree inbreeding coefficient (F) is the expected increase in homozygosity for inbred individuals; it is also the expected decrease in heterozygosity throughout the genome of inbred individuals (Chapter 13). One of us (FWA) can clearly remember being excitedly questioned in the hallway by our departmental mammalogist who had just received his weekly issue of Science and could not believe the data of Ralls and her colleagues. Subsequent studies (Ralls and Ballou 1983, Ballou 1997, Lacy 1997) have supported their original conclusions (Figure 6.14).

Inbreeding depression results from both increased homozygosity and reduced heterozygosity (see Section 13.4). That is, a greater number of deleterious recessive alleles will be expressed in inbred individuals because of their increased homozygosity. In addition, fitness of inbred individuals will be reduced at loci at which the heterozygotes have a selective advantage over all homozygous types (heterozygous advantage or overdominance). Both of these mechanisms are likely to contribute to inbreeding depression, but it is thought that increased expression of deleterious recessive alleles is the more important mechanism (Charlesworth and Willis 2009). Inbreeding depression is considered in detail in Chapter 13.

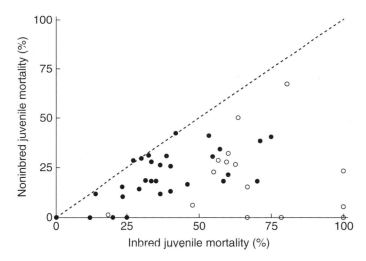

Figure 6.14 Effects of inbreeding on juvenile mortality in 44 captive populations of mammals (16 ungulates, 16 primates, and 12 small mammals). The line shows equal mortality in inbred and non-inbred progeny. The preponderance of points below the line (42 of 44, 95%) indicates that inbreeding generally increased juvenile mortality. The open circles indicate populations in which juvenile mortality of inbred and non-inbred individuals was significantly different ($P < 0.05$; exact test). Data from Ralls and Ballou (1983).

Guest Box 6 Reduced genetic variation and the emergence of an extinction-threatening disease in the Tasmanian devil
Menna E. Jones

Reduced genetic variation is often associated with reduced ability to respond to new pathogens (O'Brien and Evermann 1988). Immune recognition of foreign cells (tumors and pathogens), a function of highly variable genes of the major histocompatibility complex (MHC), is impeded by genetic loss and selective sweeps that result from strong directional selection during major disease epidemics (see Section 6.7.2) (Belov 2011, Radwan *et al.* 2010). Loss of genetic diversity and the emergence of infectious diseases are major conservation threats to wildlife worldwide (see Section 20.6) (O'Brien and Evermann 1988, Daszak *et al.* 2000). Devil facial tumor disease (a contagious cancer) is threatening the Tasmanian devil, the largest carnivorous marsupial, with extinction.

The Tasmanian devil, once widely distributed across mainland Australia (Dawson 1982), was extirpated from its mainland range some 3500 years ago, around the time of the arrival of the dingo (Brown 2006). Now restricted to Tasmania,

genetic diversity of the devil is low (Jones, M.E. *et al.* 2004, Miller *et al.* 2011). The patterns of allelic diversity indicate a genetic bottleneck prior to or at the time of separation from the mainland 13,000 years ago (Jones, M.E. *et al.* 2004). Allelic variation is particularly low at the MHC genes, indicative of a possible earlier selective sweep (Siddle *et al.* 2010).

Contagious cancers are rare in nature (Murchison 2009). They represent a major evolutionary step beyond the vast majority of cancers that arise, spread and die within a single host (McCallum and Jones 2012). Contagious cancers involve the infectious spread of live tumor cell lines between hosts by intimate injurious contact. The evolution of transmissibility is associated with either low genetic diversity or host immunosuppression for evasion of immune recognition (Murchison 2009, McCallum and Jones 2012). Transmissibility of both known natural cases (devil facial tumor disease and canine transmissible venereal tumor) is thought to have originated in genetically restricted populations

(*Continued*)

(Murgia *et al.* 2006, Pearse and Swift 2006, Rebbeck *et al.* 2009). Laboratory and medical cases of transmissible tumors are associated with extreme inbreeding (e.g., laboratory populations of Syrian golden hamster maintained for medical research) (Brindley and Banfield 1961), medical immunosuppression for organ transplants, and an immature immune system in the case of maternal-fetal transmission of leukemia (Kaufmann *et al.* 2002, Tolar and Neglia 2003, Sala-Torra *et al.* 2006).

Devil facial tumor disease is a contagious tumor cell line spread between individuals by injurious biting which facilitates the rapid transfer of live tumor cells to within the new host. Most bites and most primary tumors are around the face and inside the mouth, but the tumor metastasizes and is consistently fatal in 3–6 months. Detected in 1996, the disease has spread to most of the devil's range and caused up to 95% local declines, leading to endangered listing (Figure 6.15).

Managing for the recovery of wild devil populations is challenging. The disease epidemic threatens further genetic loss. Populations will be severely reduced for decades. A selective sweep is anticipated via extreme mortality and strong directional selection for traits that confer disease resistance (rejection or growth inhibition of tumors), reduced aggression (to reduce exposure), and increased juvenile growth rates and smaller adult body size to facilitate early reproduction (precocial breeding has increased 16-fold in diseased populations) (Jones *et al.* 2008). Increased inbreeding and genetic structuring in devils is evident in just 2–3 generations after disease arrival, as females reduce dispersal in response to increased food resources associated with population decline (Lachish *et al.* 2011). Genetic rescue, through translocations to mix the divergent eastern and northwestern Tasmanian population genotypes, will increase the species' genetic diversity, with future benefits for fitness, disease resilience, and adaptive potential.

Conservation options for wildlife threatened by contagious cancer are complicated by evolutionary interactions between the tumor and its host. Selection should favor increased resistance or tolerance in the host as well as reduced virulence of the tumor, which would allow increased tumor transmission and increased host persistence (Raberg *et al.* 2009, Carval and Ferriere 2010).

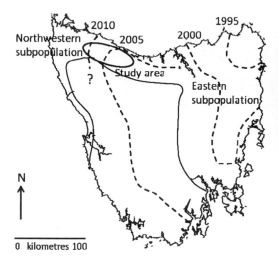

Figure 6.15 Timing of devil facial tumor disease (broken lines) of the Tasmanian devil in Tasmania. The spatial extent of the disease is unknown in the inaccessible wilderness of southwest Tasmania (indicated by ?) where devil populations are sparse. The region where the variant patterns of the disease epidemic are being studied in the transition zone between host genetic subpopulations is shown with an oval.

There is some hope for an alternative epidemic outcome for the Tasmanian devil in patterns of much lower disease prevalence, reduced demographic effects, and longer individual survival times at the current disease-front in northwest Tasmania (Hamede *et al.* in review). The westward spreading tumor, which is of eastern origin and is virtually genetically identical to that found in eastern devils, is encountering different host genotypes for the first time (Siddle *et al.* 2010). If there are resistant host genotypes, the deliberate introduction of these into declined populations could enable devil recovery. If, however, this new epidemic pattern reflects less virulent strains of the tumor, mixing different devil genotypes with different tumor strains would have unpredictable outcomes. In this case, translocating the tumor may be a more effective management strategy. Genetics and evolution of both host and pathogen likely underlie the field observations. The recent origin of Tasmanian devil facial tumor provides a rare opportunity to observe rapid evolution in real time.

CHAPTER 7

Effective population size

Medium ground finch, Example 7.2

Effective population size is whatever must be substituted in the formula (1/2N) to describe the actual loss in heterozygosity.

Sewall Wright (1969)

Effective population size (N_e) is one of the most fundamental evolutionary parameters of biological systems, and it affects many processes that are relevant to biological conservation.

Robin S. Waples (2002)

Chapter Contents

Conservation and the Genetics of Populations, Second Edition. Fred W. Allendorf, Gordon Luikart, and Sally N. Aitken.
© 2013 Fred W. Allendorf, Gordon Luikart and Sally N. Aitken. Published 2013 by Blackwell Publishing Ltd.

We saw in the previous chapter that we expect heterozygosity to be lost at a rate of $1/2N$ in finite populations (equation 6.5). However, this expectation holds only under conditions that rarely apply to real populations. For example, such factors as the number of individuals of reproductive age rather than the total of all ages, the sex ratio, and differences in reproductive success among individuals must considered. Thus, the actual number of adult individuals in a natural population (census size, N_C) is not sufficient for predicting the rate of genetic drift. We will use the concept of effective population size to deal with the discrepancy between the demographic size and population size relevant to the rate of genetic drift in natural populations.

Perhaps the most important assumption of our model of genetic drift has been the absence of natural selection. That is, we have assumed that the genotypes under consideration do not affect the fitness (survival and reproductive success) of individuals. We would not be concerned with the retention of genetic variation in small populations if the assumption of genetic neutrality were true for all loci in the genome. However, the assumption of neutrality, and the use of neutral loci, allows us to predict the effects of finite population size with great generality. In Chapter 8, we will consider the effects of incorporating natural selection into our basic models of genetic drift.

7.1 CONCEPT OF EFFECTIVE POPULATION SIZE

Our consideration in the previous chapter of genetic drift dealt only with 'ideal' populations. **Effective population size** (N_e) is the size of the **ideal** (Wright-Fisher) population (N) that will result in the same amount of genetic drift as in the actual population being considered. The basic ideal population consists of "N diploid individuals reconstituted each generation from a random sample of $2N$ gametes" (Wright 1939, p. 298). In an ideal population, individuals produce both female and male gametes (**monoecy**) and self-fertilization is possible. Under these conditions, heterozygosity will decrease exactly by $1/2N$ per generation.

We can see this by considering an ideal population of N individuals (say 10) in which each individual is heterozygous for two unique alleles (Figure 7.1). All of these 10 individuals will contribute equally to the gamete pool that is sampled to create each individual in the next generation. Thus, each allele will be at a

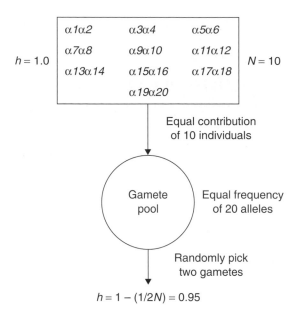

Figure 7.1   Diagram of reduction in heterozygosity (h) in an ideal population consisting of 10 individuals that are each heterozygous for two unique alleles ($h = 1$). Two gametes are picked from the gamete pool to create each individual in the next generation. Let's say the first gamete chosen is $\alpha 15$. This individual will be heterozygous unless the next gamete sampled is also $\alpha 15$. What is the probability that the next gamete sampled is $\alpha 15$? This is the frequency of the $\alpha 15$ allele in the gamete pool, which is $1/2N = 0.05$ because of the equal contribution of individuals to the gamete pool. Therefore, the expected heterozygosity of each individual in the next generation is $1 - (1/2N) = 0.95$.

frequency of $1/2N = 0.05$ in the gamete pool. A new individual will only be homozygous if the same allele is present in both gametes. For the purposes of our calculations, it does not matter which allele is sampled first because all alleles are equally frequent. Let us say the first gamete chosen is $\alpha 15$. This individual will be homozygous only if the next gamete sampled is also $\alpha 15$. What is the probability that the next gamete sampled is $\alpha 15$? This probability is simply the frequency of the $\alpha 15$ allele in the gamete pool, which is $1/2N = 0.05$ because all 20 alleles (2×10) are at equal frequency in the gamete pool (see Figure 7.1). Therefore, the expected homozygosity is $1/2N$, and the expected heterozygosity of each individual in the next generation is $1 - (1/2N) = 0.95$.

This conceptual model becomes more complicated if self-fertilization is prevented, or if the population is **dioecious**. In these two cases, the decrease in heterozygosity due to sampling individuals from the gamete pool will skip a generation because both gametes in an individual cannot come from the same parent. Nevertheless, the mean rate of loss per generation over many generations is similar in this case; heterozygosity is lost at a rate more closely approximated by $1/(2N + 1)$ (Wright 1931, Crow and Denniston 1988). The difference between these two expectations, $1/2N$ and $1/(2N + 1)$, is usually ignored because the difference is unimportant except when N is very small.

For our general purposes, the ideal population consists of a constant number of N diploid individuals ($N/2$ females and $N/2$ males) in which all parents have an equal probability of being the parent of any individual progeny. We will consider the effects of violating the following assumptions of such idealized populations on the rate of genetic drift:

1 Equal numbers of males and females.
2 All individuals have an equal probability of contributing an offspring to the next generation.
3 Constant population size.
4 Non-overlapping (discrete) generations.

We have examined two expected effects of genetic drift: changes in allele frequency (Section 6.2) and a decrease in heterozygosity (Section 6.3). Thus, there are at least two possible measures of the effective population size (i.e., the rate of genetic change due to drift). First, the 'variance effective number' (N_{eV}) is whatever must be substituted in equation 6.1 to predict the expected changes in allele frequency; second, the 'inbreeding effective number' (N_{eI}) is whatever must be substituted in equation 6.5 to predict the expected reduction in heterozygosity (Example 7.1). Crow (1954) and Ewens (1982) have described effective population numbers that predict the expected rate of decay of the proportion of polymorphic loci (P). We will only consider the first two kinds of effective population size (N_{eV} and N_{eI}) because they have more relevance for understanding the loss of genetic variation in populations.

Crow and Denniston (1988) have clarified the distinction between these two measures of effective population size. In many cases, a population has nearly the same effective population size for either measure. Specifically, their values are identical in constant size populations in which the age and sex distributions are unchanging. We will first consider N_e to be the inbreed-

ing effective population size under different circumstances because this number is most widely used, and we will then consider when these two numbers will differ.

7.2 UNEQUAL SEX RATIO

Populations often have unequal numbers of males and females contributing to the next generation. The two sexes, however, contribute an equal number of genes to the next generation regardless of the total of males and females in the population. Therefore, the amount of genetic drift attributable to the two sexes must be considered separately. Consider the extreme case of one male mating with 100 females. In this case, all progeny will be half-sibs because they share the same father. In general, the rarer sex is going to have a much greater effect on genetic drift so that the effective population size will seldom be much greater than twice the size of the rarer sex.

What is the size of the ideal population that will lose heterozygosity at the same rate as the population we are considering, which has different numbers of females and males? We saw in Section 6.3 that the increase in homozygosity due to genetic drift is caused by an individual being homozygous because its two gene copies were derived from a common ancestor in a previous generation. The inbreeding effective population size in a monoecious population in which selfing is permitted may be defined as the reciprocal of the probability that two uniting gametes come from the same parent. With separate sexes, or if selfing is not permitted, uniting gametes must come from different parents; thus, the effective population size is the probability that two uniting gametes come from the same grandparent.

The probability that the two uniting gametes in an individual came from a male grandparent is 1/4. (One-half of the time uniting gametes will come from a grandmother and a grandfather, and 1/4 of the time both gametes will come from a grandmother.) Given that both gametes come from a grandfather, the probability that both come from the *same* male is $1/N_m$, where N_m is the number of males in the grandparental generation. Thus, the combined probability that both uniting gametes come from the same grandfather is $(1/4 \times 1/N_m) = 1/4N_m$. The same probabilities hold for grandmothers. Thus, the combined probability of uniting gametes coming from the same grandparent is:

Example 7.1 Effective population size of grizzly bears

Harris and Allendorf (1989) estimated the effective population size of grizzly bear populations using computer simulations based upon the life history characteristics (survival, age at first reproduction, litter size, etc.). They estimated N_{el} by comparing the loss of heterozygosity in the simulated populations to that expected in ideal populations of $N = 100$ (Figure 7.2).

Over a wide range of conditions, the effective population size was approximately 25% of the actual population size. However, this method, like most, does not account for all factors that might reduce N_e (e.g., heritability of fitness or high variance in female reproductive success).

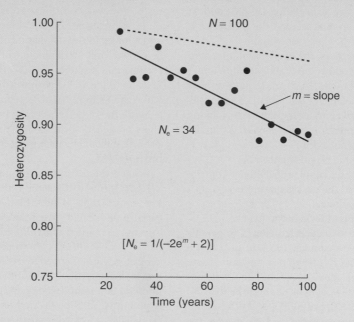

Figure 7.2 Estimation of effective size of grizzly bear populations ($N = 100$) by computer simulation. The dashed line shows the expected decline in heterozygosity over 10 generations (100 years) in an ideal population using equation 6.7. The points are the mean heterozygosity of cubs born that year. The solid line shows the decline in heterozygosity in a simulated population. The decline in heterozygosity in the simulated population is equal to that expected in an ideal population of 34 bears; thus, $N_e = 34$. m is the slope of the regression of the log of heterozygosity on time. From Harris and Allendorf (1989).

$$\frac{1}{N_e} = \frac{1}{4N_f} + \frac{1}{4N_m} \qquad (7.1)$$

This is more commonly represented by solving for N_e with the following result:

$$N_e = \frac{4N_f N_m}{N_f + N_m} \qquad (7.2)$$

As we would expect, if there are equal numbers of males and females ($N_f = N_m = 0.5\,N$), then this expression reduces to $N_e = N$.

Melampy and Howe (1977) described skewed sex ratios in the tropical tree *Triplaris americana* from four sites in Costa Rica. We can use equation 7.2 to predict what effect the observed excess of females would have on the effective population size (Table 7.1). There is a substantial reduction in N_e of this tree only at site 4, where males comprise approximately 20% of the population.

In general, a skewed sex ratio will not have a large effect on the N_e/N ratio unless there is a great excess of one sex or the other. Figure 7.3 shows this for a hypothetical population with a total of 100 individuals. N_e is maximum (100) when there is an equal number of males and females, but declines as the sex ratio departs from 50:50. However, small departures from 50:50 have little effect on N_e. The dashed lines in this figure show that the N_e/N ratio will only be reduced by half if the least common sex is less than 15% of the total population. In the most extreme situations, N_e will be approximately four times the rarer sex:

N_f	N_m	N_e
1000	1	4.0
1000	2	8.0
1000	3	12.0
1000	4	15.9
1000	5	19.9

Some populations of ungulates in which males are more likely to be hunted can have highly skewed sex ratios. For example, males comprised less than 1% of

Table 7.1 Sex ratios of the tropical tree *Triplaris americana* at four study sites in Costa Rica (Melampy and Howe 1977). N_C is the census population size, which in this case is the number of trees present at a site. Ne is estimated using equation 7.2.

Sites	Females	Males	N_C	N_e	N_e/N_C
1	61	41	102	98.1	0.96
2	58	42	100	97.4	0.97
3	56	44	100	98.6	0.99
4	47	12	59	38.2	0.65

all adult elk in the Elkhorn Mountains of Montana in 1985 (Lamb 2010).

7.3 NONRANDOM NUMBER OF PROGENY

Our model of an ideal population assumes that all individuals have an equal probability of contributing progeny to the next generation. That is, a random sample of $2N$ gametes is drawn from a population of N diploid individuals. In real populations, parents seldom have an equal chance of contributing progeny because they differ in fertility and in the survival of their progeny. The variation among parents results in a greater proportion of the next generation coming from

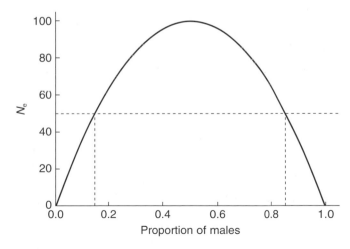

Figure 7.3 The effect of sex ratio on effective population size for a population with a total of 100 males and females using equation 7.2. The dashed lines indicate the sex ratios at which N_e will be reduced by half because of a skewed sex ratio.

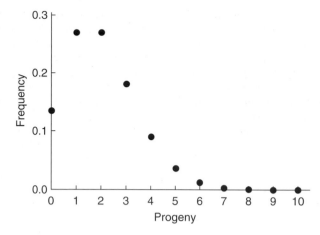

Figure 7.4 Expected frequency of number of progeny per individual in a large stable population in which the mean number of progeny per individual is 2 and all individuals have equal probability of reproducing.

a smaller number of parents. Thus, the effective population size is reduced.

It is somewhat surprising just how much variation in reproductive success there is, even when all individuals have equal probability of reproducing as in the ideal population. Figure 7.4 shows the expected frequency of progeny number in a very large stable population in which the mean number of progeny is two and all individuals have equal probability of reproducing. Take, for example, a stable population of 20 individuals (10 males and 10 females). On average, each individual will have two progeny. However, approximately 12% of all individuals will not contribute any progeny. Consider that the probability of any male *not* fathering a particular child in this population is $(0.90 = 9/10)$. Therefore, the probability of a male not contributing any of the 20 progeny is $(0.90)^{20}$, or approximately 12%. The same statistical reasoning applies for females as well. Thus, on the average, two or three of the 20 individuals in this population will not contribute any genes to the next generation, while one of the 20 individuals is expected to produce five or more progeny.

We can adjust for nonrandom progeny contribution following Wright (1939). Consider N individuals that contribute varying numbers of gametes (k) to the next generation of the same size (N) so that the mean number of gametes contributed per individual is $\bar{k} = 2$. The variance of the number of gametes contributed to the next generation is:

$$V_k = \frac{\sum_{i=1}^{N}(k_i - 2)^2}{N} \tag{7.3}$$

The proportion of cases in which two random gametes will come from the same parent is

$$\frac{\sum_{i=1}^{N} k_i(k_i - 1)}{2N(2N-1)} = \frac{2 + V_k}{4N - 2} \tag{7.4}$$

As we saw in the previous section, the effective population size may be defined as the reciprocal of the probability that two gametes come from the same parent. Thus, we may write the effective population size as

$$N_e = \frac{4N - 2}{2 + V_k} \tag{7.5}$$

Random variation of k will produce a distribution that approximates a Poisson distribution. A Poisson distribution has a mean equal to the variance; thus, $V_k = \bar{k} = 2$ and $N_e = N$ for the idealized population (see Section A3.3.3). However, as the variability in reproductive success among parents (V_k) increases, the effective population size decreases. An interesting result is that the effective population size will be larger than the actual population size if $V_k < 2$. In the extreme where each parent produces exactly two progeny,

$N_e = 2N - 1$. Thus, in captive breeding where we can control reproduction, we may nearly double the effective population size by making sure that all individuals contribute equal numbers of progeny.

This potential near-doubling of effective population size occurs because there are two sources of genetic drift: reproductive differences among individuals, and Mendelian segregation in heterozygotes. These two sources contribute equally to genetic drift. Thus, eliminating differences in reproductive success will approximately double the effective population size. Unfortunately, there is no way to eliminate the second source of genetic drift (Mendelian segregation), except by nonsexual reproduction (cloning, etc.).

The following example considers three hypothetical populations of constant size $N = 10$ with extreme differences in individual reproductive success (Table 7.2). Each population consists of five pairs of mates. In Population A, only one pair of mates reproduces successfully. In Population B, each of the five pairs produces two offspring so that there is no variance in reproductive success. There is an intermediate amount of variability in reproductive success in Population C.

We can estimate N_e of each of these populations using equations 7.3 and 7.5, as shown in Table 7.3. Thus, Population B is expected to lose only approximately 3% ($1/2N_e = 0.026$) of its heterozygosity per generation, while populations A and C are expected to

lose 24% and 5%, respectively. There are very few examples in natural populations where the lifetime reproductive success of individuals is known so that N_e can be estimated using this approach (Example 7.2).

Equation 7.5 assumes that the variance in progeny number is the same in males and females. However, the variation in progeny number among parents is likely to be different for males and females. For many animal species, the variance of progeny number in males is expected to be larger than that for females. For example, according to the 1990 Guinness Book of World Records, the greatest number of children produced by a human mother is 69; in great contrast, the last Sharifian Emperor of Morocco is estimated to have fathered some 1400 children! The current use of sperm donors can also result in males with many progeny. Some

Table 7.3 Estimation of effective population size for three hypothetical populations with high, low, and intermediate variability in family size using equation 7.5.

	$\Sigma(k_i - \bar{k})^2$	V_k	N_e	$1/2N_e$
Population A	160	16	2.11	0.237
Population B	0	0	19.00	0.026
Population C	20	2	9.50	0.053

Table 7.2 Estimation of effective population size in three hypothetical populations of constant size $N = 10$ with extreme differences in individual reproductive success. Each population consists of five pairs of mates. In Population A, only one pair of mates reproduces successfully. In Population B, each of the five pairs produces two offspring so that there is no variance in reproductive success. There is an intermediate amount of variability in reproductive success in Population C.

	A			B			C		
i	k_i	$k_i - \bar{k}$	$(k_i - \bar{k})^2$	k_i	$k_i - \bar{k}$	$(k_i - \bar{k})^2$	k_i	$k_i - \bar{k}$	$(k_i - \bar{k})^2$
1	10	8	64	2	0	0	0	−2	4
2	10	8	64	2	0	0	0	−2	4
3	0	−2	4	2	0	0	3	1	1
4	0	−2	4	2	0	0	3	1	1
5	0	−2	4	2	0	0	2	0	0
6	0	−2	4	2	0	0	2	0	0
7	0	−2	4	2	0	0	1	−1	1
8	0	−2	4	2	0	0	1	−1	1
9	0	−2	4	2	0	0	4	2	4
10	0	−2	4	2	0	0	4	2	4
			160			0			20

Example 7.2   Effective population size of Darwin's finches

Grant and Grant (1992a) reported the lifetime reproductive success of two species of Darwin's ground finches on Daphne Major, Galápagos: the cactus finch and the medium ground finch. They followed survival and lifetime reproductive success of four cohorts born in the years 1975–1978. Figure 7.5 shows the lifetime reproductive success of the 1975 cohort for both species. The variance in reproductive success for both species was much greater than expected in ideal population (see Figure 7.4). Over one-half of the birds in both species did not produce any recruits to the next generation, and several birds produced eight or more recruits. Eighteen cactus finches produced 33 recruits ($\bar{k} = 1.83$) distributed with a variance (V_k) of 6.74; 65 medium ground finches produced 102 recruits ($\bar{k} = 1.57$) distributed with a variance (V_k) of 7.12. The average number of breeding birds (census population sizes) for these years was approximately 94 cactus finches and 197 medium ground finches. The estimated N_e based on these data is 38 cactus finches and 60 medium ground finches. Thus, the N_e/N_C ratio for these two species is 38/94 = 0.40 and 60/197 = 0.30, respectively.

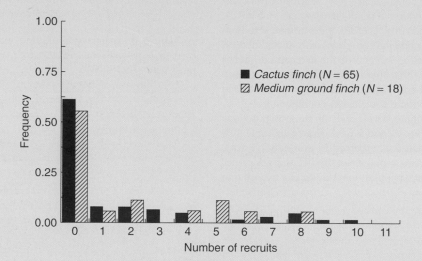

Figure 7.5   Lifetime reproductive success of the 1975 cohort of the cactus finch and medium ground finches on Isla Daphne Major, Galápagos. The x-axis shows the number of recruits (progeny that breed) produced. Thus, over 50% of the breeding birds for both species did not produce any progeny that lived to breed. From Grant and Grant (1992a).

sperm donors have apparently fathered hundreds of progeny (Romm 2011).

We can take such differences between the sexes into account as shown:

$$N_e = \frac{8N - 4}{V_{km} + V_{kf} + 4} \tag{7.6}$$

The estimation of effective population size with non-random progeny number becomes much more complex

if we relax our assumption of constant population size. In the case of separate sexes, the following equation may be used:

$$N_e = \frac{N_{t-2}\bar{k} - 2}{\bar{k} - 1 + \dfrac{V_k}{\bar{k}}} \tag{7.7}$$

where N_{t-2} is N in the grandparental generation (Crow and Denniston 1988).

7.4 FLUCTUATING POPULATION SIZE

Natural populations sometimes fluctuate greatly in size. The rate of loss of heterozygosity ($1/2N$) is proportional to the reciprocal of population size ($1/N$). Thus, generations with small population sizes will dominate the effect on loss of heterozygosity. This is analogous to the sex with the smallest population size dominating the effect on loss of heterozygosity (see Section 7.2). Therefore, the average population size is a poor metric for the loss of heterozygosity over many generations.

For example, consider three generations for a population that goes though a severe bottleneck, say $N_1 = 100$, $N_2 = 2$, and $N_3 = 100$. A very small proportion of the heterozygosity will be lost in generations 1 and 3 ($1/200 = 0.5\%$); however, 25% of the heterozygosity will be lost in the second generation. The exact heterozygosity remaining after these three generations can be found as shown:

$$h = \left(1 - \frac{1}{200}\right)\left(1 - \frac{1}{4}\right)\left(1 - \frac{1}{200}\right) = 0.743$$

The average population size over these three generations is $(100 + 2 + 100)/3 = 67.3$. Using equation 6.7 we would expect to lose only approximately 2% of the heterozygosity over three generations with a population size of 67.3, rather than the 25.7% heterozygosity that is actually lost.

We can estimate the effective population size over these three generations by using the mean of the reciprocal of population size ($1/N$) in successive generations, rather than the mean of N itself. This is known as the harmonic mean. Thus:

$$\frac{1}{N_e} = \frac{1}{t}\left(\frac{1}{N_1} + \frac{1}{N_2} + \frac{1}{N_3} + \ldots + \frac{1}{N_t}\right) \qquad (7.8)$$

After a little algebra, this becomes:

$$N_e = \frac{t}{\sum\left(\frac{1}{N_i}\right)} \qquad (7.9)$$

Generations with the smallest N have the greatest effect. A single generation of small population size may cause a large reduction in genetic variation. A rapid expansion in numbers does not affect the previous loss of genetic variation; it merely reduces the current rate of loss. This is known as the 'bottleneck' effect as discussed in Section 6.3.

We can use equation 7.9 to predict the expected loss of heterozygosity in the example that we began this section with:

$$N_e = \frac{3}{\left(\dfrac{1}{100} + \dfrac{1}{2} + \dfrac{1}{100}\right)} = 5.77$$

We expect to lose 23.8% of the heterozygosity in a population where $N_e = 5.77$ over three generations (equation 6.7). This is very close to the exact value of 25.7% that we calculated previously.

7.5 OVERLAPPING GENERATIONS

We have so far only considered populations with discrete generations. However, most species have overlapping generations. Hill (1979) has shown that the effective number in the case of overlapping generations is the same as that for discrete-generation populations having the same variance in lifetime progeny numbers and the same number of individuals entering the population each generation. Thus, the presence of overlapping generations itself does not have a major effect on N_e. However, this result assumes a constant population size and a stable age distribution. Crow and Denniston (1988) concluded that Hill's results are approximately correct for populations that are growing or contracting, as long as the age distribution is fairly stable. Waples *et al.* (2011) have provided a lucid overview of this problem.

On the other hand, some biological aspects of overlapping generations can have a major effect on N_e (Nunney 2002). For example, N_e is likely to be reduced in polygamous species in which individuals reproduce over many years. In this case, the variance in reproductive success can be greatly increased if the same individuals tend to be relatively successful over many years (Example 7.3). In contrast, the presence of seed banks or diapausing eggs of freshwater crustaceans can greatly reduce the loss of heterozygosity over time, and thereby increase N_e (Nunney 2002).

There are no expressions available to correct for the effects of overlapping generations on N_e, as we have used in previous sections for unequal sex ratios, nonrandom number of progeny, or fluctuations in population size (although see Nunney 2002 and Engen *et al.*

Example 7.3 Reduction of effective population size in red-wing blackbirds because of the reproductive success of a single male for many years

Occasionally a truly superior individual graces a population.

Beletsky and Orians (1989)

A long-term study of reproduction of red-wing black-birds in the Columbia National Wildlife Refuge in central Washington demonstrates the potential for N_e to be greatly reduced in polygamous species with overlapping generations (Beletsky and Orians 1989). Males in this population held breeding territories on average for only 2.1 years. Half of all male breeders held territories for just a single year, and annual adult male mortality was approximately 40%.

The male known as RYB-AR was banded as a nonterritorial subadult during the 1977 spring breeding season. He first acquired a breeding territory in 1978 and held the same breeding territory through 1988 over 11 consecutive years. The mean annual harem size of RYB-AR was almost double that of other males. Harem size is strongly correlated with reproductive success in this population. Over his lifetime, RYB-AR produced 176 fledged young; this is 17 times greater than the average for males in this population.

RYB-AR fathered 4.2% of the total progeny in this population over the 11 years that he bred. Over these years, there was a total of nearly 400 breeding males in this population! As Beletsky and Orians concluded, even if the offspring of RYB-AR are genetically no better than average, he is sure to become a direct ancestor of many individuals in future generations. And, if his exceptional reproductive success is partially inherited by his descendants, his contributions to future generations will even be greater.

2007). The best way to estimate N_e in populations with overlapping generations is with demographic population computer simulations that incorporate genetic change over time (Example 7.1).

7.6 VARIANCE EFFECTIVE POPULATION SIZE

The two measures of effective population size (N_{eI} and N_{eV}) differ when the population size is changing. In general, the inbreeding effective population size (N_{eI}) is more related to the number of parents since it is based on the probability of two gametes coming from the same parent. The variance effective population size (N_{eV}) is more related to the number of progeny since it is based on the number of gametes contributed, rather than the number of parents (Crow and Kimura 1970, p. 361).

Consider the extreme of two parents that have a very large number of progeny. In this case, the allele frequencies in the progeny will be an accurate reflection of the allele frequencies in the parents; therefore, N_{eV} will be nearly infinite. However, all the progeny will be full-sibs and thus their progeny will show the reduction in homozygosity expected in matings between full-sibs; thus, N_{eI} is very small. In the other extreme, if each parent has exactly one offspring, then there will be no tendency for inbreeding in the populations, and, therefore, N_{eI} will be infinite. However, N_{eV} will be small.

Therefore, if a population is growing, the inbreeding effective number is usually less than the variance effective number (Waples 2002). If the population size is decreasing, the reverse is true. In the long run, these two effects will tend to cancel each other and the two effective numbers will be roughly the same (Crow and Kimura 1970, Crow and Denniston 1988).

7.7 CYTOPLASMIC GENES

Genetic variation in cytoplasmic gene systems (e.g., mitochondria and chloroplasts) has come under active investigation in the last few years because of advances in techniques to analyze differences in DNA sequences. We will consider mitochondrial DNA (mtDNA) because so much is known about genetic variation of this molecule. The principles we will consider also apply to genetic variation in chloroplast DNA (cpDNA). However, cpDNA is paternally inherited in some plants (Harris and Ingram 1991).

There are three major differences between mitochondrial and nuclear genes that are relevant for this comparison:

1 Individuals usually possess many mitochondria that share a single predominant mtDNA sequence. That is, individuals are effectively haploid for a single mtDNA type.

2 Individuals inherit their mtDNA genotype from their mother.

3 There is no recombination between mtDNA molecules.

The effective population size for mtDNA is generally smaller than that for diploid nuclear genes because each individual has only one haplotype (allele) and uniparental inheritance (Birky *et al.* 1983).

For purposes of comparison, we will use *h* to compare genetic drift at mtDNA with nuclear genes. It might seem inappropriate to use *h* as a measure of variation for mtDNA since it is haploid and individuals therefore cannot be heterozygous for mtDNA. Nevertheless, *h* in this context is a valuable measure of the variation present within a population (Nei 1987, p. 177). It can be thought of as the probability that two randomly sampled individuals from a population will have the same mtDNA genotype, and has also been called gene diversity (Nei 1987, p. 177).

The probability of sampling the same mtDNA haplotype in two consecutive gametes is $1/N_f$, where N_f is the number of females in the population. And since $N_f = 0.5N$:

$$\Delta h = -\frac{1}{N_f} = -\frac{1}{0.5N} \tag{7.10}$$

In the case of a $1:1$ sex ratio, there are four times as many nuclear genes as mitochondrial genes (N_f):

$$\frac{N_e(nuc)}{N_e(mt)} = \frac{2(N_f + N_m)}{N_f} = 4 \tag{7.11}$$

In general, drift is more important and bottlenecks have greater effects for genes in mtDNA than for nuclear genes because of the generally smaller N_e (Example 7.4). Figure 7.7 shows the relative loss of variation during a bottleneck of a single generation for a nuclear and mitochondrial gene based upon equation 7.11.

Things become more interesting with an unequal sex ratio. If there are more females than males in a population, then the N_e for mtDNA can actually be greater than the N_e for nuclear genes. If we use equation 7.2 for N_e for nuclear genes, then the ratio between the effective number of nuclear genes to the effective number of mitochondrial genes is:

$$\frac{N_e(nuc)}{N_e(mt)} = \frac{\frac{2(4N_fN_m)}{(N_f + N_m)}}{N_f}$$

Example 7.4 Effects of a bottleneck in the Australian spotted mountain trout

Ovenden and White (1990) demonstrated that genetic variation at mtDNA is much more sensitive to bottlenecks than nuclear variation in the southern Australian spotted mountain trout from Tasmania (Figure 7.6). These fish spawn in fresh water, and the larvae are immediately washed to sea where they grow and develop. The juvenile fish reenter fresh water the following spring where they remain until they spawn. Landlocked populations of spotted mountain trout also occur in isolated lakes that were formed by the retreat of glaciers some 3000–7000 years ago.

Ovenden and White found 58 mtDNA genotypes identified by the presence or absence of restriction sites in 150 fish collected from 14 coastal streams. There is evidence of substantial exchange of individuals among the 14 coastal stream populations. In con-

trast, they found only two mtDNA genotypes in 66 fish collected from landlocked populations in isolated lakes. However, the lake populations and coastal populations had nearly identical heterozygosities at 22 allozyme loci (Table 7.4). As expected, the allelic diversity of the lake populations was smaller than the coastal populations.

The reduced genetic variation at mtDNA in the landlocked populations is apparently due to a bottleneck associated with their founding and continued isolation. Ovenden and White suggested that the founding bottleneck may have been exacerbated by natural selection for the landlocked life-history in these populations. Regardless of the mechanism, the reduced N_e of the landlocked populations has had a dramatic effect on genetic variation at mtDNA but virtually no effect on nuclear heterozygosity.

(Continued)

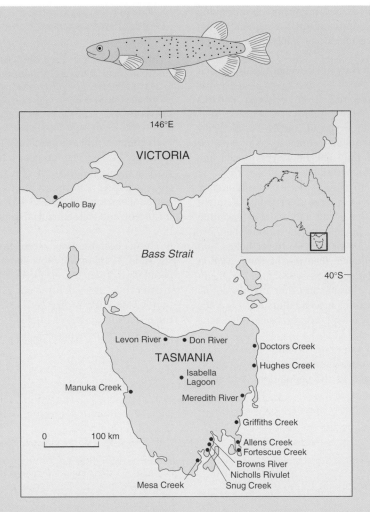

Figure 7.6 Map of southern Australian spotted mountain trout populations (from Ovenden and White 1990). Allens Creek and Fortescue Creek are coastal populations that exchange migrants with each other and other coastal populations. Isabella Lagoon is an isolated landlocked population. Redrawn from Ovenden and White (1990).

Table 7.4 Expected heterozygosity (H_e), diversity (h) at mtDNA, and average number of alleles ($\bar{A}$) per locus in three populations of the southern Australian spotted mountain trout from Tasmania at 22 allozyme loci and mtDNA (see Example 7.4). The Allens Creek and Fortescue Creek populations are coastal populations that are connected by substantial exchange of individuals. The Isabella Lagoon population is an isolated landlocked population.

Sample	Nuclear loci		mtDNA	
	$\bar{A}$	H_e	$\bar{A}$	h
Allens Creek	1.9	0.123	28	0.946
Fortescue Creek	1.9	0.111	25	0.922
Isabella Lagoon	1.3	0.104	2	0.038

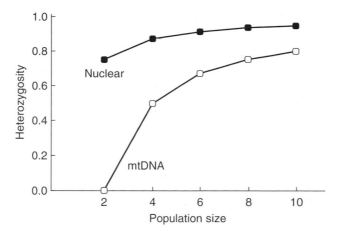

Figure 7.7 Amount of heterozygosity or diversity remaining after a bottleneck of a single generation for a nuclear and mitochondrial gene with equal numbers of males and females. For example, there is no mitochondrial variation left after a bottleneck of two individuals because only one female is present. In contrast, 75% of the nuclear heterozygosity will remain after a bottleneck of two individuals (see equation 6.5).

which, after a little bit of algebra, becomes:

$$\frac{8N_m}{(N_f + N_m)} \tag{7.12}$$

This expression will be less than one if there are more than seven times as many females as males. Therefore, the N_e for mtDNA will be less than the N_e for nuclear genes unless there are at least seven times as many females as males.

We have assumed so far in this section that the variance in reproductive success is equal in males and females. As we saw in Section 7.2, this is often not true. This will decrease the difference in effective population size between nuclear and mitochondrial genes in the many species for which there is much greater variance in reproductive success in males than in females.

7.8 GENE GENEALOGIES, THE COALESCENT, AND LINEAGE SORTING

So far we have described genetic changes in populations due to genetic drift by changes in allele frequencies from generation to generation. There is an alternative approach to study the loss of genetic variation in populations that can be seen most easily for the case of mtDNA in which each individual receives the mtDNA haplotype of its mother. We can trace the transmission of mtDNA haplotypes over many generations. That is, we can trace the genealogy of the mtDNA genotype of each individual in a population. We can see in the example shown in Figure 7.8 that only two of the original 18 haplotypes remain in a population after just 20 generations due to a process called stochastic lineage sorting.

The gene genealogy approach can also be applied to nuclear genes, although it is somewhat more complex because of diploidy and recombination. The recent development of the application of genealogical data to the study of population-level genetic processes is the major advance in population genetics theory in the last 50 years (Hudson 1990, Fu and Li 1999, Schaal and Olsen 2000, Wakeley 2009). This development has been based upon two primary advances, one technical and one conceptual. The technological advance is the collection of DNA sequence data that allow tracing and reconstructing gene genealogies. The conceptual advance that has contributed to the theory to interpret these results is called 'coalescent theory' (see Appendix, Section A10).

Lineage sorting, as in Figure 7.8, will eventually lead to the condition where all alleles in a population are derived from (i.e., coalesce to) a single common ancestral allele. Therefore, the time to coalescence is expected to be shorter for smaller populations. In fact, the mean time to coalescence is approximately equal to N_e for

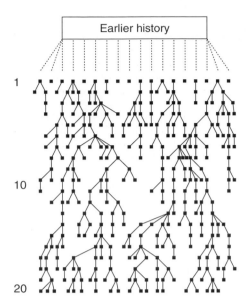

Figure 7.8   The allelic lineage sorting process of mtDNA haplotypes in a population over 20 generations. Each node represents an individual female and branches lead to daughters. The tree was generated by assuming a random distribution of female progeny with a mean of one daughter per female. From Avise (1994).

mtDNA, and is four times as long for a nuclear gene (Felsenstein 2011, p. 369). Coalescent theory provides a powerful framework to study the effects of genetic drift, natural selection, mutation, and gene flow in natural populations (Rosenberg and Nordborg 2002, Cenik and Wakeley 2010).

The coalescent approach can be used to study effective population size over relatively long periods of time. Cenik and Wakeley (2010) have defined the coalescent effective size and applied it to simulated populations of Pacific salmon. Although this application might not improve empirical estimates of effective population size, it is likely to be helpful in understanding how the rate of loss of genetic variation over long periods of time is affected by fluctuations in population size and gene flow.

7.9 LIMITATIONS OF EFFECTIVE POPULATION SIZE

Effective population size can be used to predict the expected rate of loss of heterozygosity or change in allele frequencies resulting from genetic drift. In practice, however, we generally need to know the rate of genetic drift in order to estimate effective population size. Thus, effective population size is perhaps best thought of as a standard, or unit of measure, rather than as a predictor of the loss of heterozygosity. That is, if we know the rate of change in allele frequency or the rate of loss of heterozygosity in a given population, we can use those observed rates to estimate effective population size (see Examples 7.1 and 7.4, and Guest Box 7). We will consider the estimation of effective population size in more detail in chapters 10 and 14.

Perhaps the greatest value of effective population size is heuristic. That is, we can better our understanding of genetic drift by comparing the effects of different violations of the assumptions of ideal populations on N_e (e.g., Figure 7.3). For example, Tanaka *et al.* (2009) have estimated N_e under different management regimes in order to compare the effects of different measures to control population size in overabundant koala populations. Similarly, in applying the concept of effective population size to managing populations, certain specific effective population sizes are often used as benchmarks. For example, it has been suggested that an N_e of at least 50 is necessary to avoid serious loss of genetic variation in the short term (Soulé 1980, Allendorf and Ryman 2002).

7.9.1 Allelic diversity and N_e

We have considered two measures of the loss of genetic variation in small populations: heterozygosity and allelic diversity. By definition, the inbreeding N_e is an estimate of the rate of loss of heterozygosity, but it is not a good indicator of the loss of allelic diversity within populations. That is, two populations that go through a bottleneck of the same N_e may lose very different amounts of allelic diversity. This difference is greatest when the bottleneck is caused by an extremely skewed sex ratio. Bottlenecks generally have a greater effect on allelic diversity than on heterozygosity. However, a population with an extremely skewed sex ratio may experience a substantial reduction in heterozygosity with very little loss of allelic diversity.

The duration of a bottleneck (intense versus diffuse) will also affect heterozygosity and allelic diversity differently (England *et al.* 2003). Consider two populations that fluctuate in size over several generations with the same N_e, and therefore the same loss of heterozygosity. A brief but very small bottleneck (intense) will cause substantial loss of allelic diversity. However, a diffuse bottleneck spread over several generations can result in the same loss of heterozygosity, but will cause a much smaller reduction in allelic diversity.

In summary, populations that experience the same rate of decline of heterozygosity can experience very different rates of loss of allelic diversity. Therefore, we must consider more than just N_e when considering the rate of loss of genetic variation in populations.

7.9.2 Generation interval

In conservation, we are usually concerned with the loss of genetic variation over some specified number of years in developing policies. For example, according to the IUCN, species are considered to be 'vulnerable' if they have greater than a 10% probability of extinction within 100 years (see Table 14.1). The rate of loss of genetic variation through calendar time (e.g., years) depends upon both N_e and mean generation interval (G) because $1/(2N_e)$ is the expected rate of loss *per* generation. Therefore, it is necessary to consider both G and N_e when predicting the expected rate of decline of heterozygosity in natural populations. There are many estimates of N_e in the literature, but very few include estimates of G, which are needed to predict the rate of loss of heterozygosity in calendar time.

There is some confusion in the literature about how to estimate generation interval. The generation interval is the average age of parents (Felsenstein 1971, Hill 1979). Generation interval is not the age of first reproduction nor is it the average age of reproduction if individuals of different ages produce different numbers of offspring; see Table 7.5 for an example of estimating the generation interval.

It is especially important to estimate the generation interval when comparing the effects of different management schemes on the rate of loss of heterozygosity, because conditions that reduce N_e often lengthen the generation interval. For example, Ryman *et al.* (1981) found that different harvest regimes for moose in Sweden can have strong effects on both effective population size and generation interval (see Section 18.1). Populations with smaller N_e tended to lose heterozygosity at a slower rate over calendar time because those effects of hunting that reduced N_e (e.g., harvesting young animals) also tended to increase the generation

Table 7.5 Example of estimation of generation interval ($\hat{G}$) in a hypothetical demographically stable population of sockeye salmon, which die after spawning. The mean age of adult females at sexual maturity is 4.680. However, the mean generation interval of females (4.742) is estimated by using the mean number of eggs produced by females of different ages to estimate the proportion of progeny produced by females of different ages. We assume that males of all ages are equally reproductively successful. The generation interval in this population (4.441) is the mean of the generation interval in females (4.742) and males (4.140).

| Age | Females | | | Male Adults |
	Adults	Eggs	Progeny	
3	0.010	2500	0.007	0.230
4	0.310	2825	0.255	0.510
5	0.670	3712	0.726	0.210
6	0.010	4000	0.012	0.060
Mean	4.680	–	4.742	4.140

interval (see Figure 18.2). That is, hunted populations with relatively smaller N_e and a longer generation interval would lose genetic variation over calendar time (not generations) more slowly than some populations with large N_e and shorter generation interval.

Generation interval was not considered in the koala example above (Tanaka *et al.* 2009), and it is possible that some strategies producing larger values of N_e might actually lose heterozygosity at a faster rate over calendar time than strategies resulting in smaller values of N_e. In fact, the strategy recommended to increase N_e was to administer contraception to all female koalas beyond a particular age; this strategy is likely to reduce the generation interval, and actually increase the rate of loss of heterozygosity over calendar time for a given N_e!

The inverse relationship between N_e and generation interval is often also true for differences between species. For example, Keall *et al.* (2001) estimated the census population size of five species of reptiles on North Brother Island in Cook Strait, New Zealand (Table 7.6). The generation interval for these five species was estimated based upon their life history (age of first reproduction, longevity, etc.; C.H. Daugherty, personal communication). As expected, the species with larger body size (e.g., tuatara) have smaller population sizes and longer generation intervals. The loss of heterozygosity over calendar time is strikingly similar in these five species, although they have very different population sizes. In general, species with larger body size (e.g., elephants) will tend to have smaller population sizes but longer generation intervals than species with smaller body size (e.g., mice), and these effects will tend to counteract each other.

7.10 EFFECTIVE POPULATION SIZE IN NATURAL POPULATIONS

The ratio of effective to census population size (N_e/N_C) in natural populations is of general importance for the conservation of populations. Census size is generally much easier to estimate than N_e. Therefore, establishing a general relationship between N_C and N_e would allow us to predict the rate of loss of genetic variation in a wide variety of species (Waples 2002).

The actual value N_e/N_C in a particular population or species will differ greatly depending upon demography and life history. For example, Hedgecock (1994) has argued that the high fecundities and high mortalities in early life history stages of many marine organisms can lead to exceptionally high variability in mortality in different families. Thus, N_C may be many orders of magnitude greater than N_e in some populations (Hauser and Carvalho 2008) (see Example 7.5).

Frankham (1995) provided a comprehensive review of estimates of effective population size in over 100 species of animals and plants. He concluded that estimates of N_e/N_C averaged approximately 10% in natural populations for studies in which the effects of unequal sex-ratio, variance in reproductive success, and fluctuations in population size were included (Figure 7.10). However, Waples (2002) concluded that Frankham (1995) overestimated the contribution of temporal changes by computing the N_e/N_C ratio as a harmonic mean divided by an arithmetic mean. The empirical estimates of N_e that do not include the effect of temporal changes (Frankham 1995) suggest that 20% of the adult population size is perhaps a better general value to use for N_e for many species. Palstra and Ruzzante (2008) reviewed estimates of N_e published since Frankham's (1995) review and found a median N_e/N_C of 0.14.

Table 7.6 Expected loss in heterozygosity after 1000 years for five species of reptiles on North Brother Island in Cook Strait, New Zealand. N_e estimates for each species are assumed to be 20% of the estimated census size (Keall *et al.* 2001). The estimated generation interval (G) was then used to calculate the number of generations in 1000 years (t) in order to predict the proportion of heterozygosity (H_e) remaining after 1000 years using equation 6.3.

Species	N_c	N_e	G (years)	t	H_e
Tuatara	350	70	50	20	0.866
Duvaucel's gecko	1440	288	15	67	0.890
Common gecko	3738	747	5	200	0.875
Spotted skink	3400	680	5	200	0.863
Common skink	4930	986	5	200	0.904

Example 7.5 Effective population size in a marine fish

A comparison of genetic variation at seven microsatellite loci in New Zealand snapper from the Tasman Bay has shown that the N_e may be four orders of magnitude smaller than N_C in this population (Hauser et al. 2002). Collection of a time series of scale samples began in 1950 just after commencement of a commercial fishery on this population. Genetic variation, as measured by both the number of alleles and heterozygosity, declined between 1950 and 1998 in this population (Figure 7.9).

The estimated N_e over this time period in this population based on the reduction in heterozygosity and temporal changes in allele frequency are 46 and 176, respectively. The minimum estimated population size during this period was 3.3 million fish in 1985; thus, the N_e/N_C is on the order of 0.0001! These results support the conclusion of Hedgecock (1994) that the N_e/N_C ratio may be very small in a variety of marine species. This suggests that even very large exploited marine fish populations may be in danger of losing substantial genetic variation.

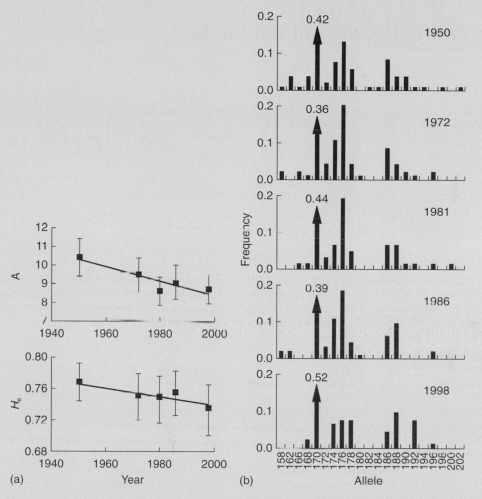

Figure 7.9 Loss of genetic variation of New Zealand snapper from Tasman Bay. The left side shows the decline in the number of alleles (A) and expected heterozygosity (H_e) at seven microsatellite loci. The right side shows loss of alleles at the *GA2B* locus between 1950 and 1998. The frequency of the most common allele is shown above the arrow. From Hauser *et al.* (2002).

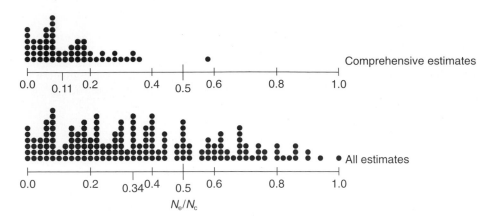

Figure 7.10 Distribution of estimates of N_e/N_c in natural populations. Comprehensive estimates that include unequal sex-ratio, variance in reproductive success, and fluctuations in population size are above, and estimates that included only one or two of these effects are below. The means of the estimates (0.11 and 0.34) are indicated below each line. From Frankham (1995).

Guest Box 7   Estimation of effective population size in Yellowstone grizzly bears
Craig R. Miller and Lisette P. Waits

Grizzly bears have been extirpated from over 99% of their historical range south of the Canadian border (Allendorf and Servheen 1986). During the last century, bears of the Yellowstone ecosystem became isolated from bears in Canada and northern Montana (Figure 7.11). Further, at least 220 bear mortalities occurred between 1967 and 1972 resulting from garbage dump closures and the removal of bears habituated to garbage (Craighead *et al.* 1995). Assessments of genetic variation with allozymes, mtDNA, and nuclear microsatellite DNA all indicated that the Yellowstone population has significantly lower variability than all other North American mainland populations (see Table 4.1).

In the modern Yellowstone population, estimated heterozygosity at eight nuclear microsatellite loci is approximately 20% lower than in the nearby Glacier population (Paetkau *et al.* 1998). If we assume that Yellowstone bears historically had the same heterozygosity as Glacier bears, we can estimate N_e in Yellowstone since isolation using the expectation that heterozygosity declines at a rate of $1/2N_e$ per generation (recall $H_t = H_0(1 - 1/2N_e)^t$, and equation 6.7). With approximately eight generations since the time of isolation (i.e., $t = 8$), this

implies an N_e of only 22. This is depicted in Figure 7.11 as hypothesis B. More troubling is the possibility that most of the postulated decline in heterozygosity occurred following dump closure, implying a very small N_e and a rapid increase in the rate of inbreeding (hypothesis C).

If N_e in Yellowstone has been this small, then genetic drift, inbreeding, and loss of quantitative genetic variation may reduce the population's viability (see Chapter 14). We estimated the effective population size during the 20th century to distinguish among hypotheses B and C, and a third possibility that variation in Yellowstone was historically low and that N_e has remained moderate across the last century (hypothesis A) (Miller and Waits 2003).

DNA was extracted from museum specimens (bones, teeth, and skins) taken from the periods 1910–20 and 1960–70, and individuals were genotyped at the same eight loci as above. We used the changes in allele frequency over time to estimate the harmonic mean N_e using maximum likelihood. For both the periods 1915–65 and 1965–95, the estimated N_e is approximately 80 (95% confidence interval approximately 50–150). Allelic diversity has declined significantly ($P < 0.05$), but only

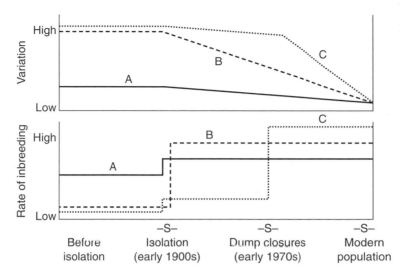

Figure 7.11   Three potential hypotheses (A–C) explaining the low level of genetic diversity observed in the modern Yellowstone grizzly bear population. Genetic samples taken at times indicated by '-S-' are used to resolve among hypotheses. From Miller and Waits (2003).

slightly (Miller and Waits 2003). Estimates of population size in Yellowstone between the 1960s and 1990s are surrounded by large uncertainty, but a summary suggests a harmonic mean N of around 280 individuals. Combining these values yields an estimate of $N_e/N_c = 27\%$; this is similar to an estimate of 25% obtained from a simulation approach (see Example 7.1).

Hence it appears that hypothesis A is supported most by our data. The lower genetic variation in Yellowstone bears appears to pre-date the decline of population size in the 20th century. With N_e apparently near 80, we argue that the need for gene flow into Yellowstone is not pressing yet. Management should focus mostly on habitat protection, restoring natural connectivity, and limiting human-caused mortalities. If natural connectivity cannot be achieved within a few additional generations (20–30 years), we recommend the translocation of a small number of individuals into Yellowstone (or perhaps artificial insemination using nonresident males, if such technology becomes available for bears). Translocations might be warranted sooner if population vital rates (e.g., survival and reproduction) decline substantially. This example illustrates the usefulness of temporally spaced samples and historical museum specimens to estimate N_e and provide information of great relevance to conservation.

CHAPTER 8

Natural selection

Rewardless orchid, Example 8.3

I have called this principle, by which each slight variation, if useful, is preserved, by the term Natural Selection, in order to mark its relation to man's power of selection.

Charles Darwin (1859, p. 61)

Then comes the question, Why do some live rather than others? If all the individuals of each species were exactly alike in every respect, we could only say it is a matter of chance. But they are not alike. We find that they vary in many different ways. Some are stronger, some swifter, some hardier in constitution, some more cunning.

Alfred Russel Wallace (1923, p. 11)

Chapter Contents

Conservation and the Genetics of Populations, Second Edition. Fred W. Allendorf, Gordon Luikart, and Sally N. Aitken.
© 2013 Fred W. Allendorf, Gordon Luikart and Sally N. Aitken. Published 2013 by Blackwell Publishing Ltd.

We have so far assumed that different genotypes have an equal probability of surviving and passing on their alleles to future generations. That is, we have assumed that natural selection is not operating. If this assumption were true in real populations, we would not be concerned with genetic variation in conservation because genetic changes would not affect a population's longevity or its evolutionary future. However, as we saw in Chapter 6, there is ample evidence that the genetic changes that occur when a population goes through a bottleneck often result in increased frequencies of alleles that reduce an individual's probability of surviving to reproduce.

In addition, some alleles and genotypes affect survival and reproductive success under different environmental conditions. Remember the white Kermode bear from Section 2.1 that has an advantage in coastal populations of black bears because it is more successful at fishing for salmon. Genetic differences between local adapted populations can be important for continued persistence of populations. In addition, individuals that are moved by human action between populations or environments may not be genetically suited to survive and reproduce in their new surroundings. And perhaps worse from a conservation perspective, gene flow caused by such translocations can reduce the adaptation of local populations.

For example, many native species of legumes (*Gastrolobium* and *Oxylobium*) in Western Australia naturally synthesize large concentrations of fluoroacetate, which is the active ingredient in 1080 (a poison used to remove mammalian pests; King *et al.* 1978). Native marsupials in Western Australia are resistant to 1080 because they have been eating plants that contain fluoroacetate for thousands of years. Therefore, 1080 does not kill native mammals in Western Australia, which means it can be used as a specific poison for introduced foxes and feral cats that are a serious problem. However, members of the same 1080-resistant mammal species (e.g., brush-tailed possums) that occur to the east beyond the range of the fluoroacetate-producing legumes are susceptible to 1080 poisoning. Therefore, translocating brush-tailed possums into Western Australia from eastern populations might not be successful because the introduced individuals would not be 'adapted' to consume the local vegetation.

Many of the best examples of local adaptation are from plant species because it is possible to do reciprocal transplantations and measure components of fitness (Joshi *et al.* 2001, Hall *et al.* 2010). Nagy and Rice

(1997) performed reciprocal transplant experiments with coastal and inland California populations of the native annual *Gilia capitata*. They compared performance for four traits: seedling emergence, early vegetative size (leaf length), probability of surviving to flowering, and number of inflorescences. Native plants significantly outperformed non-natives for all characters except leaf length. Figure 8.1 shows the results for the proportion of plants that survived to flowering. On average, the native inland plants had over twice the rate of survival compared with non-native plants grown on the inland site; the native coastal plants had 5–10 times greater survival rates compared with non-native plants grown on the coastal site.

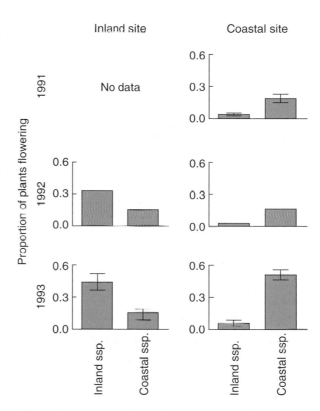

Figure 8.1 Reciprocal transplant experiment with *Gilia capitata* showing local adaptation for the proportion of plants that survived to flowering. The native subspecies had significantly greater survival than the non-native subspecies in each of the five experiments. For example, approximately 45% of the seeds from the inland subspecies survived to flowering in 1993 at the inland site; however, only some 15% of the seeds from the non-native subspecies survived to flowering in the same experiment. From Nagy and Rice (1997).

The adaptive significance of the vast genetic variation that we can now detect using the techniques of biochemical and molecular genetics has long been controversial (Lewontin 1974, Gillespie 1992, Mitton 1997, Nei 2005). Most of the models that we use to interpret data and predict effects in natural populations assume selective neutrality. This is not done because we believe that all genetic variation is neutral. Rather, neutrality is assumed because we sometimes have no choice if we want to use the rich theory of population genetics to interpret data and make predictions, as most models assume the absence of natural selection. Moreover, allele frequency distributions at neutral loci are more useful than at adaptive loci in describing population sizes and exchange among populations (see Section 9.7). Estimating the strength of natural selection in the wild has proven to be very difficult (Hendry 2005). Nevertheless, recent developments using high-throughput sequencing provide great potential for detecting natural selection and adaptation in a variety of organisms across the entire genome (see Guest Box 8).

In this chapter, we consider the effects of natural selection on allele and genotype frequencies. Sewall Wright developed powerful theoretical models that allow us to predict the effects of small populations on genetic variability. Most of these models assume selective neutrality. For example, we have seen that heterozygosity will be lost at a rate of $1/2N$ per generation in the ideal population (equation 6.5). What is the expected rate of loss of heterozygosity if the genetic variability is affected by natural selection? The answer depends upon the pattern and intensity of natural selection in operation. Since it is so difficult to estimate fitness in natural populations, we generally cannot predict the expected rate of loss of heterozygosity, unless we ignore the effects of natural selection. And worse yet, since natural selection acts differently on each locus, there is not one answer, but rather there is a different answer for each of the perhaps thousands of loci affected by selection within a population.

8.1 FITNESS

Natural selection is the differential success of genotypes in contributing to the next generation. In the simplest conceptual model, there are two major life-history components that bring about selective differences between genotypes: **viability** and **fertility**. Viability is the probability of survival to reproductive

age, and fertility is the average number of offspring per individual that survive to reproductive maturity for a particular genotype.

The effect of natural selection on genotypes is measured by fitness. Fitness is the average number of offspring produced by individuals of a particular genotype. Fitness can be calculated as the product of viability and fertility, as defined above, and we can define fitness for a diallelic locus as shown:

Genotype	Viability	Fertility	Fitness
AA	v_{11}	f_{11}	$(v_{11})(f_{11}) = W_{11}$
Aa	v_{12}	f_{12}	$(v_{12})(f_{12}) = W_{12}$
aa	v_{22}	f_{22}	$(v_{22})(f_{22}) = W_{22}$

These are absolute fitnesses based on the total number of expected progeny from each genotype. It is often convenient to use relative fitnesses to predict genetic changes caused by natural selection. Relative fitnesses are estimated by the ratios of absolute fitnesses. For example, in the data below, fitnesses have been standardized by dividing by the fitness of the genotype with the highest fitness (AA). Thus, the relative fitness of heterozygotes is 0.67 because, on average, heterozygotes have 0.67 times as many progeny as AA individuals ($1.80/2.70 = 0.67$).

Genotype	Viability	Fertility	Absolute fitness	Relative fitness
AA	0.90	3.00	2.70	1.00
Aa	0.90	2.00	1.80	0.67
aa	0.45	2.00	0.90	0.33

8.2 SINGLE LOCUS WITH TWO ALLELES

We will begin by modeling changes caused by differential survival (viability selection) in the simple case of a single locus with two alleles. Consider a single diallelic locus with differential reproductive success in a large random mating population in which all of the other assumptions of the Hardy-Weinberg model are valid. We would expect the following result after one generation of selection:

Genotype	Zygote frequency	Relative fitness	Frequency after selection
AA	p^2	w_{11}	$(p^2 w_{11})/\bar{w}$
Aa	$2pq$	w_{12}	$(2pq w_{12})/\bar{w}$
aa	q^2	w_{22}	$(q^2 w_{22})/\bar{w}$

where $\bar{w}$ is used to normalize the frequencies following selection so that they sum to one. This is the average fitness of the population, and it is the fitness of each genotype weighted by its frequency.

$$\bar{w} = p^2 w_{11} + 2pq w_{12} + q^2 w_{22} \tag{8.1}$$

After one generation of selection the frequency of the A allele is:

$$p' = \frac{p^2 w_{11}}{\bar{w}} + \left(\frac{1}{2}\right)\left(\frac{2pq w_{12}}{\bar{w}}\right) = \frac{p^2 w_{11} + pq w_{12}}{\bar{w}} \tag{8.2}$$

and, similarly, the frequency of the a allele is:

$$q' = \frac{pq w_{12} + q^2 w_{22}}{\bar{w}} \tag{8.3}$$

It is often convenient to predict the change in allele frequency from generation to generation, Δp, caused by selection. We get the following result if we solve for Δp in the current case:

$$\Delta p = \frac{pq}{\bar{w}}[p(w_{11} - w_{12}) + q(w_{12} - w_{22})] \tag{8.4}$$

We can see that the magnitude and direction of change in allele frequency are both dependent on the fitnesses of the genotypes and the allele frequency.

Equation 8.4 can be used to predict the expected change in allele frequency after one generation of selection for any array of fitnesses. The allele frequency in the following generation will be:

$$p' = p + \Delta p \tag{8.5}$$

We will use this model to study the dynamics of selection for three basic modes of natural selection with constant fitnesses:

1 Directional selection.
2 Heterozygous advantage (overdominance).
3 Heterozygous disadvantage (underdominance).

8.2.1 Directional selection

Directional selection occurs when one allele is always at a selective advantage. The advantageous allele under directional selection may either be dominant, intermediate, or recessive to the alternative allele as shown below:

Dominant	$w_{11} = w_{12} > w_{22}$
Intermediate	$w_{11} > w_{12} > w_{22}$
Recessive	$w_{11} > w_{12} = w_{22}$

The advantageous allele will increase in frequency and will be ultimately fixed by natural selection under all three modes of directional selection (Figure 8.2). Thus, the eventual or equilibrium outcome is independent of the dominance of the advantageous allele. However, the rate of change of allele frequency does depend on dominance relationships as well as the intensity of selection. For example, selection on a recessive allele is ineffective when the recessive allele is rare because most of the copies of that allele occur in heterozygotes and are therefore 'hidden' from selection.

8.2.2 Heterozygous advantage (overdominance)

Heterozygous advantage occurs when the heterozygote has the greatest fitness:

$$w_{11} < w_{12} > w_{22}$$

This mode of selection is expected to maintain both alleles in the population as a **stable equilibrium**. This pattern of selection is often called overdominance. In the case of dominance, the phenotype of the heterozygote is equal to the phenotype of one of the homozygotes. In overdominance, the phenotype (i.e., fitness) of the heterozygote is greater than the phenotype of either of the homozygotes (Example 8.1).

Let us examine the simple case of heterozygous advantage in which the two homozygotes have equal fitness:

	AA	Aa	aa
Fitness	$1-s$	1.0	$1-s$

where s (the selection coefficient) is greater than 0 and less than or equal to 1. We can examine the dynamics

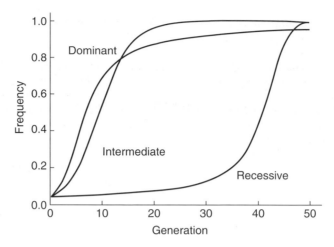

Figure 8.2 Change in allele frequency under directional selection when the homozygote for the favored allele has twice the fitness of the homozygote for the unfavored allele (1.00 vs. 0.50). The heterozygote can have the same fitness as the favored allele (1.00, dominant), the same fitness as the unfavored allele (0.50, recessive), or has intermediate fitness (0.75). The initial frequency of the favored allele is 0.03.

Example 8.1 Natural selection at an allozyme locus

Patarnello and Battaglia (1992) have described an example of heterozygous advantage at a locus encoding the enzyme glucosephosphate isomerase (GPI) in a copepod (*Gammarus insensibilis*) that lives in the Lagoon of Venice. Individuals were collected in the wild, acclimated in the laboratory at room temperature, and then held at a high temperature (27°C) for 36 hours. Individuals with different genotypes differed significantly in their survival at this temperature (Table 8.1; $P < 0.005$). Heterozygotes survived better than either of the homozygotes.

A persistent problem in measuring fitnesses of individual genotypes is whether any observed differences are due to the locus under investigation or to other loci that are linked to that locus (Eanes 1987). *In vitro* measurements show that heterozygotes at the *GPI* locus in *Gammarus insensibilis* have greater enzyme activity than either homozygote over a wide range of temperatures. In addition, the *80/80* homozygote has the greatest mortality and the lowest enzyme activity.

Table 8.1 Differential survival of *GPI* genotypes in the copepod *Gammarus insensibilis* held in the laboratory for 36 hours at high temperature (27°C). From Patarnello and Battaglia (1992).

	Genotype		
	100/100	*100/80*	*80/80*
Alive	48	90	12
Dead	47	53	27
Total	95	143	39
Relative survival	0.803	1.000	0.490

Patarnello and Battaglia (1992) have argued that the observed differences are caused by the *GPI* genotype on the basis of these enzyme kinetic properties and other considerations.

of this case of selection by plotting the values of Δp as a function of allele frequency (Figure 8.3). When p is less than 0.5, selection will increase p, and when p is greater than 0.5, selection will decrease p. Thus, 0.5 is a stable equilibrium; that is, when p is perturbed from 0.5, it will return to that value.

Any overdominant fitness set will produce a stable intermediate equilibrium allele frequency (p^*). However, the value of p^* depends upon the relative fitnesses of the homozygotes. If we solve equation 8.4 for $\Delta p = 0$ we get the following result:

$$p^* = \frac{w_{12} - w_{22}}{2w_{12} - w_{11} - w_{22}} \tag{8.6}$$

Thus, the equilibrium allele frequency will be near 0.5 if the two homozygotes have nearly equal fitnesses.

However, if one homozygote has a great advantage over the other, that allele will be much more frequent at equilibrium. Heterozygous advantage was once thought to be the major mechanism maintaining genetic variation in natural populations (Lewontin 1974). However, there have been relatively few examples of heterozygous advantage found at individual loci in natural populations (see Examples 8.1 and 8.2).

8.2.3 Heterozygous disadvantage (underdominance)

Underdominance occurs when the heterozygote is least fit:

$$w_{11} > w_{12} < w_{22}$$

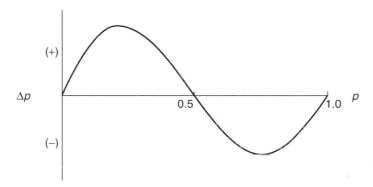

Figure 8.3 Expected change in allele frequency (Δp) as a function of allele frequency (p) in the case of heterozygous advantage when the homozygotes have equal fitness.

Example 8.2 Heterozygous advantage for a color polymorphism in the common buzzard

The European common buzzard has a plumage polymorphism controlled by a single locus (Krüger et al. 2001, Boerner and Krüger 2009). Observations of 162 offspring and their parents indicated that dark brown individuals and light colored individuals at this locus are alternative homozygotes, and that heterozygotes are intermediate in color (Krüger et al. 2001). There is assortative mating at this locus in that individuals are more likely to mate with individuals who have the same plumage pattern as their mother.

A long-term study of lifetime reproductive success revealed that heterozygotes tended to show greater annual survival and marked differences in aggression, habitat preference, and parasite load (Boerner and Krüger 2009). Overall, the lifetime reproductive success of heterozygotes was nearly twice that of either homozygote, averaged over females and males (Figure 8.4).

(Continued)

Figure 8.4   Mean lifetime reproductive success (+ standard error) in three color morphs (above) of male and female European common buzzards. From Boerner and Krüger (2009). Numbers above error bars are the sample sizes. Photo by Oliver Krüger. See Color Plate 3.

An examination of Δp as a function of p reveals that underdominance will produce what is called an **unstable equilibrium** (Figure 8.5). The p^* valuc is found using the same formula as for overdominance (see equation 8.6). However, this equilibrium is unstable because allele frequencies will tend to move away from the equilibrium value once they are perturbed. Underdominance, therefore, is not a mode of selection that will maintain genetic variation in natural populations.

We saw in Chapter 3 that heterozygotes for chromosomal rearrangements often have reduced fertility because they produce unbalanced or aneuploid gametes. Foster *et al.* (1972) examined the behavior of translocations in population cages of *Drosophila melanogaster*. They set up cages in which homozygotes for the chromosomal rearrangements had equal fitness; in this case, the unstable equilibrium frequency is expected to be 0.5 (see equation 8.6). As predicted by this analysis, the populations quickly went to fixation for whichever chromosomal type became most frequent in the early generations because of genetic drift (Figure 8.6).

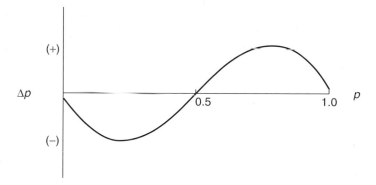

Figure 8.5 Expected change in allele frequency (Δp) as a function of allele frequency (p) in the case of heterozygous disadvantage.

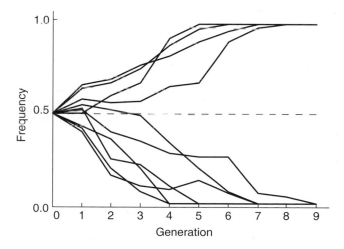

Figure 8.6 Population cage results with *Drosophila melanogaster* showing the change in frequency of a chromosomal translocation in ten populations when the two homozygotes have equal fitness that is approximately twice that of heterozygotes. Populations were founded by 20 individuals, and population sizes fluctuated between 100 and 400 flies. From Foster *et al.* (1972).

8.2.4 Selection and Hardy-Weinberg proportions

The absence of departures from Hardy-Weinberg proportions is sometimes taken as evidence that a particular locus is not affected by natural selection. However, this interpretation is incorrect for several reasons. First, differences in fecundity will not affect Hardy-Weinberg proportions. Thus, only differential survival can be detected by testing for Hardy-Weinberg proportions. Second, even strong differences in survival may not cause departures from Hardy-Weinberg proportions. For example, Lewontin and Cockerham (1959) have shown that at a locus with two alleles, differential survival will not cause a departure from Hardy-Weinberg proportions if the product of the fitnesses of the two homozygotes is equal to the square of the fitness of the heterozygotes. Finally, the goodness-of-fit test for Hardy-Weinberg proportions has little power to detect departures from Hardy-Weinberg proportions caused by differential survival (Lessios 1992).

8.3 MULTIPLE ALLELES

Analysis of the effects of natural selection becomes more complex when there are more than two alleles at a locus because the number of genotypes increases dramatically with a modest increase in the number of alleles; remember, there are 55 possible genotypes with just 10 alleles at a single locus (see equation 5.4). Nevertheless, our model of selection can be readily extended to three alleles (A_1, A_2, and A_3):

Genotype	A_1A_1	A_1A_2	A_2A_2	A_1A_3	A_2A_3	A_3A_3
Fitness	w_{11}	w_{12}	w_{22}	w_{13}	w_{23}	w_{33}
Frequency	p^2	$2pq$	q^2	$2pr$	$2qr$	r^2

The average fitness of the population is:

$$\bar{w} = p^2w_{11} + 2pqw_{12} + q^2w_{22} + 2prw_{13} + 2qrw_{23} + r^2w_{33}$$
$$(8.7)$$

And the expected allele frequencies the next generation are

$$p' = \frac{(p^2w_{11} + pqw_{12} + prw_{13})}{\bar{w}}$$

$$q' = \frac{(pqw_{12} + q^2w_{22} + qrw_{23})}{\bar{w}} \qquad (8.8)$$

$$r' = \frac{(prw_{13} + qrw_{23} + r^2w_{33})}{\bar{w}}$$

We can find any equilibria that exist for a particular set of fitnesses by setting $p' = p = p^*$ and solving these equations. The following conditions emerge after a bit of math.

$$p^* = \frac{z_1}{z} \text{ where } z_1 = (w_{12} - w_{22})(w_{13} - w_{33})$$
$$- (w_{12} - w_{23})(w_{13} - w_{23})$$

$$q^* = \frac{z_2}{z} \text{ where } z_2 = (w_{23} - w_{33})(w_{12} - w_{11})$$
$$- (w_{23} - w_{13})(w_{12} - w_{13})$$

$$r^* = \frac{z_3}{z} \text{ where } z_3 = (w_{13} - w_{11})(w_{23} - w_{22})$$
$$- (w_{13} - w_{12})(w_{23} - w_{12})$$

where:

$$z = z_1 + z_2 + z_3 \qquad (8.9)$$

If these equations give negative values for the allele frequencies, that means there is no three-allele equilibrium (i.e., at least one allele will be lost due to selection). The equilibrium will be stable if the equilibrium is a maximum for average fitness (see equation 8.6) and will be unstable if it is a minimum for average fitness. In general, a three-allele equilibrium will be stable if z_1, z_2, and z_3 are greater than zero and:

$$(w_{11} + w_{22}) < (2w_{13}) \qquad (8.10)$$

There are no simple rules for a locus with three alleles as there are for a diallelic locus. However, the following statements may be helpful.

1 There is at most one stable equilibrium for two or more alleles.
2 A stable equilibrium will be globally stable; that is, it will be reached from any starting point containing all three alleles.
3 If a stable polymorphism exists, the mean fitness of the population exceeds that of any homozygote. If such a homozygote existed, it would become fixed in the population.

4 Heterozygous advantage (i.e., all heterozygotes have greater fitness than all homozygotes) is neither necessary nor sufficient for a stable polymorphism.

The dynamics of selection acting on three alleles can be shown by plotting the trajectories of allele frequencies on triangular coordinate paper. Figure 8.7 shows allele frequency change when all of the homozygotes have a fitness of 0.9 and all heterozygotes have a fitness of 1.0. In this case, we would expect a stable equilibrium to occur when all three alleles are equally frequent at a frequency of 0.33.

Templeton (1982) has described a very interesting set of fitnesses for three alleles at the human ß-chain hemoglobin locus (Table 8.2). Figure 8.8 shows the expected trajectories of gene frequencies at this locus. The stable two-allele polymorphism with the A and S alleles is a familiar example of heterozygous advantage (using equation 8.6, $p^* = 0.89$). However, the fitness of the homozygotes for the C allele is greater than the AS heterozygotes. Nevertheless, the C allele will be selected

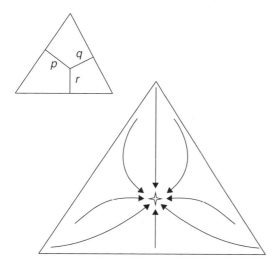

Figure 8.7 Expected trajectory of allele frequency change in the case of heterozygous advantage with three alleles plotted on triangular coordinate paper. All homozygotes have a fitness of 0.9, and all heterozygotes have a fitness of 1.0. As shown in the upper left, allele frequencies are represented by the relative lengths of the three perpendicular lines from any point to the three sides of the triangle.

Table 8.2   Estimated relative fitness at the ß-hemoglobin locus in West African human populations (Templeton 1982).

Genotype	Fitness	Phenotype
AA	0.9	Malarial susceptibility
AS	1.0	Malarial resistance
SS	0.2	Sickle-cell anemia
AC	0.9	Malarial susceptibility
SC	0.7	Malarial susceptibility
CC	1.3	Superior malarial resistance

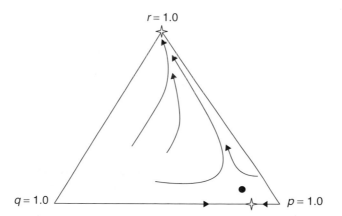

Figure 8.8   Expected allele frequency trajectories for the fitnesses of the hemoglobin locus shown in Table 8.2. There are two stable equilibria indicated by stars at the top of the triangle ($r = 1.0$) and towards the bottom right ($r = 0.0$) of the triangle. The equilibrium indicated by the shaded circle at the bottom right is unstable. $p = $ freq(A); $q = $ freq(S); $r = $ freq(C).

against when it is rare because the AC and SC geno-types both have relatively low fitnesses. Thus, the C allele will be removed by selection from a population if it occurs as a new mutation in a population with the A and S alleles present. The only way the C allele can successfully invade a population is if it increases in frequency through genetic drift so that the CC geno-type becomes frequent enough to outweigh the disad-vantage of the C allele when heterozygous. However, recent data have suggested that the C allele heterozy-gotes do not have decreased fitness (Modiano et al. 2001), and, therefore, would be expected to replace the A and S alleles in malarial regions.

8.3.1 Heterozygous advantage and multiple alleles

Overdominance was thought to be the major mecha-nism maintaining genetic variation in natural popula-tions at the time when most empirical evidence suggested that most polymorphic loci had two primary alleles (see discussion in Lewontin 1974, pp. 23–31). However, molecular techniques quickly revealed that many alleles exist at most loci in natural populations. For example, Singh et al. (1976) discovered 37 different alleles at a locus coding for xanthine dehydrogenase in a sample of 73 individuals collected from 12 natural populations of *Drosophila pseudoobscura*.

Can overdominance maintain many alleles at a single locus? This question was approached in a classic paper by Lewontin et al. (1978). They estimated the proportion of randomly chosen fitness sets that would maintain all alleles through overdominance. For a locus with two alleles, heterozygous advantage is both necessary and sufficient to maintain both alleles. If fit-nesses are selected at random, the heterozygotes will have the greatest fitness one-third of the time because there are three genotypes (Table 8.3). However, it becomes increasingly unlikely that all heterozygotes will have greater fitness than all homozygotes as the number of alleles increases. In addition, heterozygous advantage (i.e., all heterozygotes have greater fitness than all homozygotes) is neither necessary nor suffi-cient to maintain an A-allele polymorphism when A is greater than two. In fact, fitness sets capable of main-taining an A-allele polymorphism quickly become extremely unlikely as A increases (Table 8.3). For example, heterozygous advantage is sufficient to always produce a stable polymorphism with two alleles. However, only 34% of all fitness sets with four alleles

Table 8.3 Proportion of randomly chosen fitness sets that maintain all A alleles in a stable equilibrium (Lewontin et al. 1978). The third column shows the proportion of fitness sets expected to maintain all A alleles considering only those fitness sets in which all het-erozygotes have greater fitness than all homozygotes.

A	All fitness sets	Heterozygous advantage
2	0.33	1.00
3	0.04	0.71
4	0.0024	0.34
5	0.00006	0.10
6	0	0.01

in which all heterozygotes have greater fitness than all homozygotes will maintain all four alleles. Thus, over-dominance with constant fitness is not an effective mechanism for maintaining many alleles at individual loci in natural populations (Kimura 1983).

Spencer and Marks (1993) have revisited this issue using a different approach. Rather than randomly assigning fitness as done by Lewontin et al. (1978), they simulated evolution by allowing new mutations with randomly assigned fitnesses to occur within a large population and then determined how many alleles could be maintained in the population by viabil-ity selection. They found that up to 38 alleles were sometimes maintained by selection in their simulated populations. In general, they found many more alleles could be maintained by this type of selection than pre-dicted by Lewontin et al. (1978).

Spencer and Marks (1993) argued that their approach, which examines how a polymorphism may be constructed by evolution, is a complementary approach to understanding evolutionary dynamics when used along with traditional models that focus only on conditions that maintain equilibrium. Never-theless, the conclusions of Lewontin et al. (1978) are still likely to be valid, even if the approach of Spencer and Marks (1993) is more realistic. One major draw-back of the results of Spencer and Marks (1993) is that their models do not include genetic drift, and, as we will see in Section 8.5, heterozygous advantage is only effective in maintaining alleles that are relatively common in a population at equilibrium.

Hedrick (2002) has considered the maintenance of many alleles at a single locus by 'balancing selection' at the MHC locus. He then assumed resistance to pathogens is conferred by specific alleles and the action of each allele is dominant. He concluded that this model of selection could maintain stable multiple allele polymorphisms, even in the absence of any intrinsic heterozygous advantage, because heterozygotes will have higher fitness in the presence of multiple pathogens.

8.4 FREQUENCY-DEPENDENT SELECTION

We have so far assumed that fitnesses are constant. However, fitnesses are not likely to be constant in natural populations (Kojima 1971). Fitnesses are likely to change under different environmental conditions. Fitnesses may also change when allele frequencies change; this is called frequency-dependent selection. This type of selection is a potentially powerful mechanism for maintaining genetic variation in natural populations (Clarke and Partridge 1988).

8.4.1 Two alleles

Let us begin with the simple case where the fitness of a genotype is a direct function of its frequency. For example:

$$\begin{array}{cccc} & AA & Aa & aa \\ \text{Fitness} & 1-p^2 & 1-2pq & 1-q^2 \end{array} \qquad (8.11)$$

With this model of selection, a genotype becomes less fit as it becomes more common in a population. The change in allele frequency at any value of p can be calculated with equation 8.4. We can predict the expected effects of this pattern of selection by examination of the plot of Δp versus allele frequency; we will get the same plot as Figure 8.3. In this case, there is an equilibrium at $p^* = 0.5$ where Δp is zero. Is this equilibrium stable or unstable? When p is less than 0.5, $w_{11} > w_{22}$ and therefore p will increase; when p is greater than 0.5, $w_{11} < w_{22}$ and p will decrease. This is a stable equilibrium.

Note that the homozygotes have a fitness of 0.75, and the heterozygote has a fitness of 0.5 at equilibrium. Therefore, this is a stable polymorphism in which the heterozygote has a disadvantage at equilibrium. We can see that our rules for understanding the effects of selection with constant fitnesses are not likely to be helpful in understanding the effects of frequency-dependent selection. In general, frequency-dependent

selection will produce a stable polymorphism whenever the rare phenotype has a selective advantage. However, there is no general rule about the relative fitnesses at equilibrium.

8.4.2 Frequency-dependent selection in nature

Frequency-dependent selection is an important mechanism for maintaining genetic variation in natural populations. You are encouraged to read the review by Clarke (1979); additional references on frequency-dependent selection can be found in a collection of papers edited by Clarke and Partridge (1988). Frequency-dependent selection often results from mechanisms of sexual selection, predation and disease, and ecological competition (see Example 8.3).

8.4.3 Self-incompatibility locus in plants

In contrast to heterozygous advantage, frequency dependent selection can be extremely powerful for maintaining multiple alleles. The self-incompatibility locus (S) of many flowering plants is an extreme example of this (see Guest Box 14) (Wright 1965a, Vieira and Charlesworth 2002, Castric and Vekemans 2004). In the simplest system, pollen grains can only fertilize plants that do not have the same S allele as carried by the pollen. Homozygotes cannot be produced at this locus, and at least three alleles must be present at this locus.

The expected equilibrium with three alleles will be a frequency of 0.33 for each allele because fitnesses are equivalent for all three alleles. At equilibrium, any pollen grain will be able to fertilize one-third of the plants in the population (Table 8.4). However, a fourth allele produced by mutation (S_4) would have a great selective advantage because it will be able to fertilize every plant in the population. Thus, we would expect the fourth allele to increase in frequency until it reaches a frequency equal to the other three alleles.

Any new mutation at the S locus is expected to have an initial selective advantage because of its rarity regardless of the existing number of alleles. However, we also would expect rare alleles to be susceptible to loss because of genetic drift. Therefore, the total number of S alleles will be at equilibrium between mutation and genetic drift.

Emerson (1939, 1940) described 45 nearly equal-frequency S alleles in a narrow endemic plant

Example 8.3   Frequency-dependent selection in an orchid

Gigord *et al.* (2001) have presented an elegant example of frequency-dependent selection in the rewardless orchid *Dactylorhiza sambucina*. This species has a dramatic flower-color polymorphism; both yellow- and purple-flowered individuals occur throughout the range of the species in Europe. Rewardless orchids do not provide any reward to insect pollinators, and are usually pollinated by newly emerged insects that are naïve. Laboratory experiments showed that bumblebees tend to sample different color morphs in alternation, because visiting an empty flower increases the probability of switching to a different color morph. This behavior results in rare morphs being proportionately overvisited. This was confirmed in an experiment that demonstrated that whichever color morph is rare has a selective advantage in natural populations (Figure 8.9).

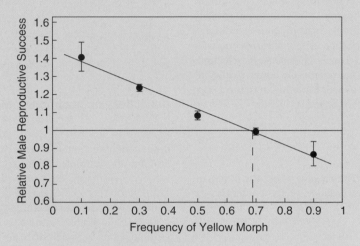

Figure 8.9   Frequency-dependent selection in the orchid *Dactylorhiza sambucina*. Relative male reproductive success of the yellow morph increases as the frequency of the yellow morph decreases. Male reproductive success was estimated by the average proportion of pollinia (mass of fused pollen produced by many orchids) removed from plants by insect pollinators. The horizontal line corresponds to equal reproductive success between the two morphs. The intersection between the regression line and the horizontal line is the value of predicted morph frequencies at equilibrium (represented by vertical dotted lines). From Gigord *et al.* (2001).

Table 8.4   Genotypes possible at the self-incompatibility locus (*S*) in species of flowering plants with three alleles.

Parental genotypes		Progeny frequencies		
Ovule	Pollen	S_1S_2	S_1S_3	S_2S_3
S_1S_2	S_3	0.00	0.50	0.50
S_1S_3	S_2	0.50	0.00	0.50
S_2S_3	S_1	0.50	0.50	0.00

(*Oenothera organensis*) that occurs in an area of approximately 50 km² in the Organ Mountains, New Mexico. Emerson originally thought that the total population size of this species was approximately 500 individuals. More recent surveys indicate that the total population size may be as great as 5000 individuals (Levin *et al.* 1979). Regardless of the actual population size, this is an enormous amount of variability at a single locus. As expected because of its small population size, this species has very little genetic variation at other loci as measured by protein electrophoresis (Levin *et al.* 1979).

8.4.4 Complementary sex determination (*csd*) locus in invertebrates

Nearly 15% of all species of invertebrates have a haplodiploid mechanism of sex determination in which females are diploid and males are haploid (e.g., ants, bees, and wasps) (Crozier 1971, Cook and Crozier 1995). Sex is determined in most haplodiploid species by genotypes at the *csd* locus, which results in frequency-dependent selection similar to the *S* locus in plants (Table 8.5, Hedrick *et al.* 2006). Heterozygotes

at *csd* develop into normal females, and haploid hemizygotes develop into normal males. However, diploid homozygotes generally are either inviable or develop into sterile males (although see Elias *et al.* 2009). Thus, rare alleles are advantageous because they are less likely to produce inviable or sterile males which are homozygotes.

Large populations commonly have 10–20 *csd* alleles, and, therefore, produce very few diploid males (Zayed and Packer 2005). However, genetic drift in small populations reduces allelic diversity at *csd* and increases the proportion of diploid males produced (Figure 8.10). The increase in diploid males reduces the number of females produced in the population and can decrease the population growth rate (Hedrick *et al.* 2006). We will see in Chapter 14 that this can increase the probability of extinction in small populations.

8.5 NATURAL SELECTION IN SMALL POPULATIONS

What happens when we combine the effects of genetic drift and natural selection? More specifically, what are the effects of finite population size on the models of natural selection that we have just considered? There are two general effects of adding genetic drift to these models. First, natural selection becomes less effective because the random changes caused by drift can swamp

Table 8.5 Complementary sex determination system at the *csd* locus found in haplodiploid species. All females are heterozygous at *csd* and therefore place two alleles in their gametes with equal frequency. Females control release of sperm to produce either haploid or diploid progeny. From Hedrick *et al.* (2006).

Male gametes	Female gametes	
	½ A₁	½ A₂
½ A₁	¼ A₁A₁ (diploid males)	¼ A₁A₂ (females)
½ no fertilization	¼ A₁ (haploid males)	¼ A₂ (haploid males)

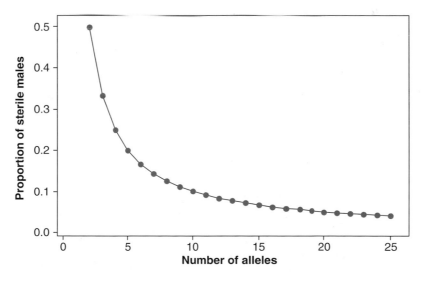

Figure 8.10 Proportion of sterile or inviable males that are produced as a function of the number of equally frequent alleles at the *csd* locus with haplodiploid sex determination. Redrawn from Cook and Crozier (1995).

the effects of increased survival or fertility. Second, the effects of natural selection become less predictable.

As a general rule-of-thumb, changes in allele frequency are determined primarily by genetic drift rather than by natural selection when the product of the effective population size and the selection coefficient ($N_e s$) is less than one (Li 1978). Thus, a deleterious allele that reduces fitness by 5% will act as if it were selectively neutral in a population with an N_e of 20 ($20 \times 0.05 = 1.00$) or less. The results of our models of selection are deterministic so that we always get the same result if we begin with the same fitnesses and the same initial allele frequency. However, the stochasticity due to genetic drift makes it more difficult to predict what the effects of natural selection will be on allele frequencies (see Figure 8.6).

8.5.1 Directional selection

Genetic drift will make directional selection less effective. This may be harmful in small populations in two ways. First, the effects of random genetic drift can outweigh the effects of natural selection, so that alleles that have a selective advantage may be lost in small populations. Second, alleles that are at a selective disadvantage may go to fixation in small populations through genetic drift. This will increase the **genetic load** of a population (see Section 13.6).

Wright (1931, p. 157) first suggested that small populations would continue to decline in vigor slowly over time because of the accumulation of deleterious mutations that natural selection would not be effective in removing, because of the overpowering effects of genetic drift. A number of theoretical papers have considered the expected rate and importance of this effect for population persistence (Lynch and Gabriel 1990, Gabriel and Bürger 1994, Lande 1995). As deleterious mutations accumulate, population size may decrease further and thereby accelerate the rate of accumulation of deleterious mutations. This feedback process has been termed **mutational meltdown** (Lynch et al. 1993).

8.5.2 Underdominance and drift

Most chromosomal rearrangements (translocations, inversions, etc.) cause reduced fertility in heterozygotes because of the production of aneuploid gametes. Homozygotes for such chromosomal mutations, however, may have increased fitness. Thus, chromo-somal mutations generally fit a pattern of underdominance and will always be initially selected against, regardless of their selective advantage when homozygous. However, we know that chromosomal rearrangements are sometimes incorporated into populations and species. In fact, rearrangements are thought to be an important factor in reproductive isolation and speciation.

How can we reconcile our theory with our knowledge from natural populations? That is, how can chromosomal rearrangements be incorporated into a population when they will always be initially selected against? The answer is, of course, genetic drift. If random changes in allele frequency perturb the population across the threshold of the unstable p^*, then natural selection will act to 'fix' the chromosomal rearrangement. Thus, we would expect faster rates of chromosomal evolution in species with small local deme sizes.

In fact, it has been proposed that the rapid rate of chromosomal evolution and speciation in mammals is due to their social structuring and reduced local deme sizes (Wilson et al. 1975). A paper by Lande (1979) examined the theoretical relationship between local deme sizes and rates of chromosomal evolution. As discussed in Chapter 3, chromosomal variability is of special importance for conservation because the demographic characteristics that make a species a likely candidate for being threatened are the same characteristics that favor the evolution of chromosomal differences between groups. Therefore, reintroduction or translocation programs may reduce the average fitness of a population if individuals are exchanged among chromosomally distinct groups.

8.5.3 Heterozygous advantage and drift

We have seen that heterozygous advantage in a two-allele system will always produce a stable polymorphism with infinite population size. However, overdominance may actually accelerate the loss of genetic variation in finite populations if the equilibrium allele frequency is near 0 or 1 (Robertson 1962).

Consider the following fitness set:

	AA	Aa	aa
Fitness	$1-s_1$	1	$1-s_2$

The following equilibrium allele frequency results if we substitute these fitness values into equation 8.6:

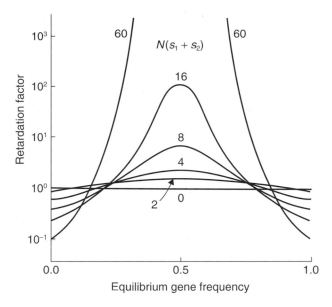

Figure 8.11 Relative effectiveness of heterozygous advantage to maintain polymorphism. The retardation factor is the reciprocal of the rate of decay of genetic variation relative to the neutral case, and N is N_e. Values less than one (i.e., 10^0) indicate a more rapid rate of loss of genetic variation than expected with selective neutrality. Thus, even strong natural selection (e.g., $N(s_1 + s_2) = 60$) is not effective at maintaining a polymorphism if the equilibrium allele frequency is less than 0.20 or greater than 0.80. From Crow and Kimura (1970).

$$p^* = \frac{s_2}{s_1 + s_2} \tag{8.12}$$

$N_e(s_1 + s_2)$ is used as a measure of the effectiveness of selection here; the effectiveness of selection increases as effective population size (N_e) increases and the intensity of selection ($s_1 + s_2$) increases. For example, $N_e(s_1 + s_2)$ will equal 60 when $(s_1 + s_2) = 0.2$ and $N_e = 300$ or $(s_1 + s_2) = 0.4$ and $N_e = 150$. When the equilibrium allele frequency (p^*) is less than 0.2 or greater than 0.8, this mode of selection will actually lose genetic variation more quickly than the neutral case ($s_1 = s_2 = 0$), unless selection is very strong or the population size is very large (Figure 8.11). Thus, heterozygous advantage is only effective at maintaining fairly common alleles (frequency > 0.2). This is similar to our conclusion in Section 8.3.1.

8.6 NATURAL SELECTION AND CONSERVATION

An understanding of natural selection is important for the management and conservation of populations. We saw in Chapter 1 that the size of pink salmon on the west coast of North America has declined dramatically in just 25 years, apparently in response to selective fishing pressure. There are many examples of rapid responses to selection in wild populations in response to human harvest (Darimont et al. 2009, Allendorf and Hard 2009). Rapid genetic change in response to such strong selection has been called 'contemporary evolution' (Stockwell et al. 2003). However, this term is misleading because evolution is more than just change by natural selection. Thus, loss of genetic variation caused by genetic drift or increase in genetic variation caused by hybridization would also represent contemporary evolution. Stockwell et al. (2003) have reviewed the potential importance of short-term responses to natural selection in conservation biology. In addition, adaptation to captive conditions is a major concern for captive breeding programs of plants and animals (Ford 2002, Frankham 2008) (see Chapter 19).

Evidence for natural selection on morphological traits is widespread in natural populations (Example 8.4). However, detecting the effects of natural selection at individual loci has proven to be a very difficult

Example 8.4   Intense natural selection on cliff swallows during winter storm

Brown and Brown (1998) reported dramatic selective effects of body size on survival of the cliff swallows in a population from the Great Plains of North America (Figure 8.12). Cliff swallows in these areas are sometimes exposed to periods of cold weather in late spring that reduce the availability of food. Substantial mortality generally results if the cold spell lasts four or more days. A once in a hundred years six-day cold spell occurred in 1996 that killed approximately 50% of the cliff swallows in southwestern Nebraska.

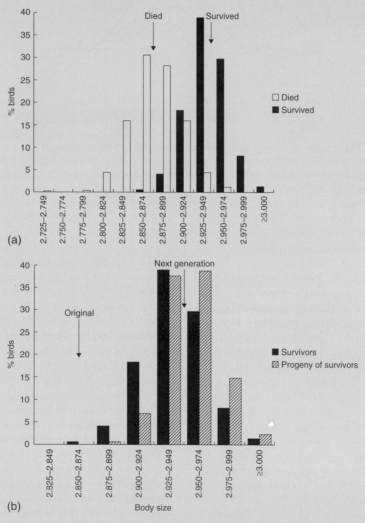

Figure 8.12   Intense natural selection on cliff swallows during a harsh winter storm. The upper figure shows that larger birds were much more likely to survive the storm than smaller birds. Body size is a multivariate measure that includes wing length, tail length, tarsus length, and culmen length and width. The lower figure shows that adult progeny in the next generation were much larger than the mean of the population before the storm event. Thus, natural selection increased the size of this population. Arrows indicate means. From Brown and Brown (1998).

Comparison of survivors and dead birds revealed that larger birds were much more likely to survive (Figure 8.12). Mortality patterns did not differ in males and females, but older birds were less likely to survive. Morphology did not differ with age. Non-survivors were not in poorer condition before the storm, suggesting that selection acted on size and not condition. Larger birds apparently were favored in extreme cold weather due to the thermal advantage of larger size and the ability to store more body fat.

Examination of the adult progeny of the survivors indicated that mean body size of the population responded to the selective event caused by the storm. The body size of progeny was significantly greater than the body size of the population before the storm (Figure 8.12). Thus, body size had high heritability (see Section 2.2 and Chapter 11).

problem ever since the discovery of widespread molecular polymorphisms in natural populations (Lewontin 1974, Watt 1995).

Nachman *et al.* (2003) have presented an elegant example of the action of natural selection on an individual locus resulting in local adaptation. Rock pocket mice are generally light-colored and match the color of the rocks on which they live. However, mice that live on dark lava are dark-colored (melanic), and this concealing coloration provides protection from predation (Figure 8.13). These authors examined several candidate loci that were known to result in changes in pigmentation in other species. They found mutations in *MC1R* (see Section 2.1) that were responsible for the dark coloration in one population of lava-dwelling mice that were melanic. However, they found no evidence of mutations at this locus in another melanic population. Thus, the similar adaptation of dark coloration apparently has evolved by different genetic mechanisms in different populations (see Guest Box 12).

There is substantial evidence for natural selection acting on MHC loci in many species (Edwards and Hedrick 1998). Nevertheless, how this information should be applied in a conservation perspective has been controversial. Hughes (1991) recommended that "all captive breeding programs for endangered vertebrate species should be designed with the preservation of MHC allelic diversity as their main goal". There are a variety of potential problems associated with following this recommendation (Gilpin and Wills 1991, Miller and Hedrick 1991, Vrijenhoek and Leberg 1991). The primary problem is that "selecting" individuals on the basis of their MHC genotype could reduce genetic variation throughout the rest of the genome (Lacy 2000a). We will revisit these issues in

later chapters when we consider the identification of units of conservation (Chapter 16) and captive breeding (Chapter 19).

Frequency-dependent selection has special importance for conservation because of the many functionally distinct alleles that are maintained by frequency-dependent selection at some loci. We have seen that allelic diversity is much more affected by bottlenecks than is heterozygosity (Section 6.4). Reinartz and Les (1994) concluded that some one-third of the remaining 14 natural populations of *Aster furcatus* in Wisconsin, USA, had reduced seed sets because of a diminished number of self-incompatability S-alleles. Young *et al.* (2000a) have considered the effect of loss of allelic variation at the S-locus on the viability of small populations. In addition, frequency-dependent selection probably contributes to the large number of alleles present at some loci associated with disease resistance (e.g., MHC; Section 6.7). Thus, the loss of allelic diversity caused by bottlenecks is likely to make small populations more susceptible to epidemics (Hedrick 2003).

Many local adaptations of native populations will be difficult to detect because they will only be manifest during periodic episodes of extreme environmental conditions, such as winter storms (Example 8.4), drought, or fire (Gutschick and BassiriRad 2003). Wiens (1977) has argued that short-term studies of fitness and other population characteristics are of limited value because of the importance of "ecological crunches" in variable environments. For example, Rieman and Clayton (1997) suggested that the complex life histories (e.g., mixed migratory behaviors) of bull trout are adaptations to periodic disturbances such as fire that may occur only every 25–100 years.

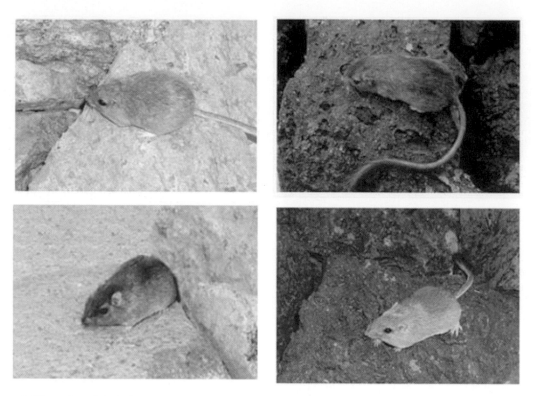

Figure 8.13 Light and dark phenotype of rock pocket mice on light-colored rocks and dark lava. From Nachman *et al.* (2003). See Guest Box 12 and Color Plate 4.

Guest Box 8 Natural selection across the genome of the threespine stickleback fish
Paul A. Hohenlohe and William A. Cresko

Natural selection affects allele frequencies and genetic diversity at loci under selection. Because genes are arrayed along chromosomes, the effects of selection on single loci spread outward across the local genomic neighborhood. With modern, high-throughput DNA sequencing approaches, biologists can now scan across genomes at a fine scale to detect signatures of selection. These signatures can be diverse. Extremes in the distributions of nucleotide diversity, genetic differentiation among populations, allele frequency spectrum, and even the correlation among neighboring loci (gametic disequilibrium) can all indicate past or ongoing responses to natural selection (Hohenlohe *et al.* 2010b).

The threespine stickleback fish provides a model of rapid, parallel adaptation in which to identify signatures of selection. The ancestral oceanic form

has repeatedly colonized freshwater habitats across the Northern Hemisphere, and each time has evolved a suite of physiological, morphological, and behavioral adaptations to fresh water. We used a high-throughput sequencing technique to genotype thousands of single nucleotide polymorphisms (SNPs) in two oceanic and three independently derived, postglacial freshwater populations in Alaska (Hohenlohe *et al.* 2010a). We asked the following question: is parallel phenotypic evolution the result of genetic evolution at the same sets of genes across the genome?

The answer appears to be yes (Figure 8.14). Each of the freshwater populations showed elevated genetic differentiation from the oceanic ancestor at many of the same genomic regions. These greater values result from divergent selection between habitats because allele frequencies have diverged more rapidly in these regions compared with the rest of the genome, where the rate of divergence is governed primarily by genetic drift since founding of the freshwater populations. Selection may have acted on just one or two genes in each region, but the extent of elevated population differentiation – the signature of selection – covers dozens of genes in each region (Hohenlohe *et al.* 2010a).

Other observations indicate that this rapid adaptation resulted from standing genetic variation, not from new mutations in each freshwater habitat. Genetic diversity is not particularly reduced at these regions in fresh water, as one would expect from a rapid selective sweep of a new mutation (Hohenlohe *et al.* 2010a). Gene flow from freshwater populations appears sufficient to maintain freshwater-adapted alleles at low frequency in the ocean (Schluter and Conte 2009). Moreover, some of these freshwater alleles appear to be in gametic disequilibrium with each other in the ocean, main-

tained as a coadapted gene complex that facilitates extremely rapid adaptation in newly colonized freshwater habitats (Hohenlohe *et al.* 2012). Thus a complete picture of the genomics of adaptation emerges only when one looks at multiple genomic signatures – population differentiation, genetic diversity, and gametic disequilibrium – in addition to ecology and demographics.

This type of population genomics approach to natural selection can address many questions important to conservation: is local adaptation the result of just a few genes and genomic regions, or many? Do populations in similar habitats exhibit signatures of selection on the same genomic regions? What aspects of genetic variation and genomic architecture are important for local adaptation? From a practical standpoint, understanding the genetic basis of adaptation can help predict the ability of populations to respond to environmental change and inform conservation decisions about captive breeding and restoration.

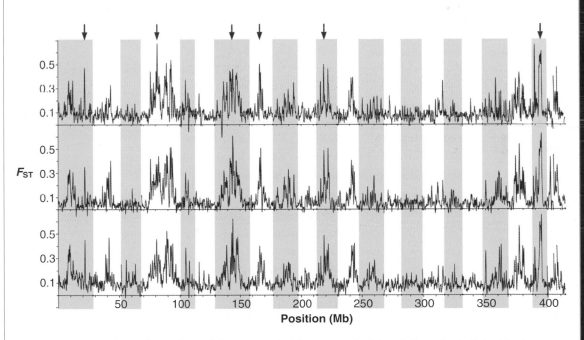

Figure 8.14 Genome-wide population differentiation (F_{ST}) between each of three independently derived freshwater populations of threespine stickleback and the ancestral oceanic population. Continuous distributions were estimated by smoothing across over 45,000 SNPs. The 22 chromosomes are arrayed along the *x*-axis, indicated by alternating gray shading. Arrows indicate examples of elevated differentiation at the same genomic regions in each of the three freshwater populations. From Hohenlohe *et al.* (2010a).

CHAPTER 9

Population
subdivision

Grevillea barklyana, Example 9.1

There is abundant geographical variation in both morphology and gene frequency in most species. The extent of geographic variation results from a balance of forces tending to produce local genetic differentiation and forces tending to produce genetic homogeneity.

Montgomery Slatkin (1987)

*The term "species" includes any subspecies of fish or wildlife or plants, and any **distinct population segment** of any species of vertebrate fish or wildlife which breeds when mature.*

US Endangered Species Act of 1973

Chapter Contents

Conservation and the Genetics of Populations, Second Edition. Fred W. Allendorf, Gordon Luikart, and Sally N. Aitken.
© 2013 Fred W. Allendorf, Gordon Luikart and Sally N. Aitken. Published 2013 by Blackwell Publishing Ltd.

So far we have considered only random mating (i.e., **panmictic**) populations. Natural populations of most species are subdivided or 'structured' into separate local random mating units that are called **demes**. The subdivision of a species into separate subpopulations means that genetic variation within species exists at two primary levels:

1 Genetic variation within local populations.
2 Genetic diversity between local populations.

We saw in Chapter 3 that there are large differences between species in the proportion of total genetic variation that is due to differences among populations (F_{ST}). For example, Schwartz *et al.* (2002) found very little genetic divergence ($F_{ST} = 0.033$) at nine microsatellite loci among 17 Canada lynx population samples collected from northern Alaska to central Montana (over 3100 km). However, other species of vertebrates, including carnivores, can be highly structured over a relatively short geographic distance (Figure 9.1). Spruell *et al.* (2003) found 20 times this amount of genetic divergence among bull trout populations within the Pacific Northwest of the US ($F_{ST} = 0.659$). Even separate spawning populations of bull trout just a few kilometers apart within a small tributary of Lake Pend Oreille in Idaho had twice the amount of genetic divergence ($F_{ST} = 0.063$) than the widespread population samples of lynx (Spruell *et al.* 1999b).

Understanding the patterns and extent of genetic divergence among populations is crucial for protecting species and developing effective conservation plans (see Guest Box 9). For example, translocation of animals or plants to supplement suppressed populations may have harmful effects if the translocated individuals are genetically different from the recipient population (Frankham *et al.* 2011). In addition, developing priorities for the conservation of a species requires an understanding of adaptive genetic differentiation among populations. Perhaps most importantly, an understanding of genetic population structure is essential for identifying units to be conserved. For example, as stated in the quote at the beginning of the chapter, distinct populations can be listed under the US Endangered Species Act (ESA) and receive the same protection as biological species. Similar provisions also exist in endangered species legislation in Australia and Canada (see Box 16.1).

The use of the terms **migration** and **dispersal** is somewhat confusing. The classic population genetics literature uses migration synonymously with gene flow to refer to the movement of individuals or gametes (e.g., pollen) from one genetic population to another (i.e., genetic exchange among breeding groups). This exchange is generally referred to as dispersal in the ecology literature. Migration in the ecology literature

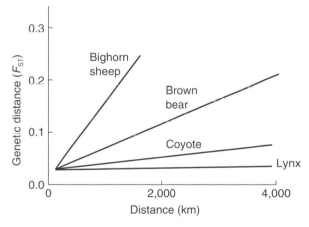

Figure 9.1 General relationship between geographic distance and genetic distance for four species of mammals. Lynx and coyotes show little genetic differentiation over thousands of kilometers; wolves (not shown) are similar to coyotes in this respect. However, less mobile species have significant differences in allele frequencies between populations over only a few hundreds of kilometers. Bighorn sheep, for example, live on mountain tops and tend not to disperse across deep valleys and forests that often separate mountain ranges. Modified from Forbes and Hogg (1999); additional unpublished data from M.K. Schwartz.

refers to movement of individuals during their lifetime from one geographic region to another. For example, anadromous salmon undertake long migrations from their natal stream to the ocean, where they feed for several years before migrating to their natal streams for reproduction. In the genetic sense, migration of salmon refers to an individual returning to a spawning population other than its natal population.

In this chapter, we will consider populations that are subdivided into a series of partially isolated **subpopulations** that are connected by some amount of genetic exchange (migration). We will first consider how genetic variation is distributed at neutral loci within subdivided populations because of the effects of two opposing processes: gene flow and genetic drift. We will next consider the effects of natural selection on the distribution of genetic variation within species. Finally, we will consider the application of this analysis to the observed distribution of genetic variation in natural populations.

9.1 *F*-STATISTICS

The oldest and most widely used metrics of genetic differentiation are *F*-statistics. Sewall Wright (1931, 1951) developed a conceptual and mathematical framework to describe the distribution of genetic variation within a species that used a series of inbreeding coefficients: F_{IS}, F_{ST}, and F_{IT}. Holsinger and Weir (2009) provide an insightful recent review of the application of *F*-statistics to describe and understand genetic population structure.

F_{IS} is a measure of departure from Hardy-Weinberg (HW) proportions within local demes or subpopulations. The term 'subpopulation' as used in this context is the same as the common use of the term 'local population'. As we have seen, F_{ST} is a measure of allele frequency divergence among demes or subpopulations; and F_{IT} is a measure of the overall departure from HW proportions in the entire base population (or species) due to both nonrandom mating within local subpopulations (F_{IS}), and allele frequency divergence among subpopulations (F_{ST}).

In general, inbreeding is the tendency for mates to be more closely related than two individuals drawn at random from the population. It is crucial to define inbreeding relative to some clearly specified base population, which may be the species as a whole, or some specific geographical collection of subpopulations

forming a 'population'. For example, using the entire species as the base population, a mating between two individuals within a local population will produce apparently 'inbred' progeny because individuals from the same local populations are likely to have shared a more recent common ancestor than two individuals chosen at random from throughout the range of a species. As we will see, F_{ST} is a measure of this type of inbreeding.

These parameters were initially defined by Wright for loci with just two alleles. They were extended to three or more alleles by Nei (1977), who used the parameters G_{IS}, G_{ST}, and G_{IT} in what he termed the analysis of gene diversity. *F*- and *G*- are often used interchangeably in the literature – see Chakraborty and Leimar (1987) for a comprehensive discussion of *F*-and *G*-statistics.

F-statistics are a measure of the deficit of heterozygotes relative to expected HW proportions in the specified base population. That is, *F* is the proportion by which heterozygosity is reduced relative to heterozygosity in a random mating population with the same allele frequencies:

$$F = 1 - (H_o/H_e) \tag{9.1}$$

where H_o is the observed proportion of heterozygotes and H_e is the expected HW proportion of heterozygotes.

F_{IS} is a measure of departure from HW proportions within local subpopulations:

$$F_{IS} = 1 - (H_o/H_S) \tag{9.2}$$

where H_o is the observed heterozygosity averaged over all subpopulations, and H_S is the expected heterozygosity averaged over all subpopulations. F_{IS} will be positive if there is a deficit of heterozygotes and negative if there is an excess of heterozygotes. Inbreeding within local populations, such as self-pollinating, will cause a deficit of heterozygotes (Example 9.1). As we saw in Section 6.6 a small effective population size can cause an excess of heterozygotes and result in negative F_{IS} values.

F_{ST} is a measure of genetic divergence among subpopulations

$$F_{ST} = 1 - (H_S/H_T) \tag{9.3}$$

where H_T is the expected HW heterozygosity if the entire base population were panmictic (Example 9.2).

Example 9.1   Selfing in an Australian shrub

Ayre *et al.* (1994) studied genetic variation in the rare Australian shrub *Grevillea barklyana* which reproduces by both selfing and outcrossing. They found a significant ($P < 0.001$) deficit of heterozygotes at the *Gpi* locus in sample of progeny from one of their four populations:

We can estimate the proportion of selfing that can explain these results by estimating S using equation 9.7:

$$\hat{S} = \frac{2F_{IS}}{(1 + F_{IS})} = 0.60$$

This results in an estimated 60% of the progeny in this population being produced by selfing and the remaining 40% by random mating, if we assume that the deficit of heterozygotes is caused entirely by selfing. Similar estimates of selfing were found at both of the allozyme loci studied in these four populations. Inbreeding due to selfing and biparental inbreeding cannot be distinguished based on single locus genotypes, but multilocus analysis can distinguish between selfing and biparental inbreeding as causes of an excess of homozygotes (see Ritland 2002).

Genotypes

A1/A1	A1/A2	A2/A2		
112	43	31	$\hat{p} = 0.718$	$\hat{F}_{IS} = 0.429$
(95.9)	(75.3)	(14.8)		

H_T is the expected HW proportion of heterozygotes using the allele frequencies averaged over all subpopulations. F_{ST} ranges from zero, when all subpopulations have equal allele frequencies, to one, when all the subpopulations are fixed for different alleles. F_{ST} is sometimes called the '**fixation index**'.

F_{IT} is a measure of the total departure from HW proportions which includes departures from HW proportions within local populations and divergence among populations:

$$F_{IT} = 1 - (H_o / H_T)$$

These three F-statistics are related by the expression

$$F_{IT} = F_{IS} + F_{ST} - (F_{IS})(F_{ST}) \tag{9.4}$$

This approach will be used in this chapter to describe the effects of population subdivision on the genetic structure of populations.

9.1.1 The Wahlund effect

The deficit of heterozygotes relative to HW proportions caused by the subdivision of a population into separate demes is often referred to as the '**Wahlund effect**' (see Guest Box 5). For example, a large deficit of heterozygotes was found at many loci when brown trout cap-

tured in Lake Bunnersjöarna (Example 9.2) were initially analyzed without knowledge of the two separate subpopulations (Ryman *et al.* 1979). Wahlund was a Swedish geneticist who first described this effect in 1928. He analyzed the excess of homozygotes and deficit of heterozygotes in terms of the variance of allele frequencies among S subpopulations:

$$Var(q) = \frac{1}{S} \sum (q_i - \bar{q})^2 \tag{9.5}$$

When $Var(q) = 0$, all subpopulations have the same allele frequencies and the population is in HW proportions. As $Var(q)$ increases, the allele frequency differences among subpopulations increase and the deficit of heterozygotes increases. In fact:

$$F_{ST} = \frac{Var(q)}{pq} \tag{9.6}$$

so that we can express the genotypic array of the population in terms of either F_{ST} or $Var(q)$:

Genotype	HW	Wright	Wahlund
AA	p^2	$p^2 + pqF_{ST}$	$p^2 + Var(q)$
Aa	$2pq$	$2pq - 2pqF_{ST}$	$2pq - 2Var(q)$
aa	q^2	$q^2 + pqF_{ST}$	$q^2 + Var(q)$

Example 9.2 The Wahlund effect in a lake of brown trout

The approach of Nei (1977) can be used to compute F-statistics with genotypic data from natural populations. For example, two nearly equal-sized demes of brown trout occurred in Lake Bunnersjöarna in northern Sweden (Ryman *et al.* 1979). One deme spawned in the inlet, and the other deme spawned in the outlet. The fish spent almost all of their life in the lake itself rather than in the inlet and outlet streams. These two demes were nearly fixed for two alleles (*100* and *null*) at the *LDH-A2* locus. Genotype frequencies for a hypothetical sample taken from the lake itself of 100 individuals made up of exactly 50 individuals from each deme are shown below:

	100/100	*100/null*	*null/null*	Total	$\hat{p}$	2pq
Inlet deme	50	0	0	50	1.000	0.000
Outlet deme	1	13	36	50	0.150	0.255
Lake sample	51	13	36	100	0.575	0.489
(expected)	(33.1)	(48.9)	(18.1)			

The mean expected heterozygosity within these two demes (H_S) is 0.128 (the mean of 0.000 and 0.255). Thus, the value of F_{ST} for this population at this locus is 0.738:

$$F_{ST} = 1 - (H_S/H_T) = 1 - (0.128/0.489) = 0.738$$

That is, the heterozygosity of the sample of fish from the lake is approximately 73% lower than we would expect if this population was panmictic.

These two approaches for describing the genotypic effects of population subdivision (Wright and Wahlund) are analogous to the two ways we modeled genetic drift in Chapter 6: either an increase in homozygosity or a change in allele frequency.

The Wahlund effect can readily be extended to more than two alleles (Nei 1965). However, the variance in frequencies will generally differ for different alleles. The frequency of particular heterozygotes may be greater or less than expected with HW proportions. Nevertheless, there will always be an overall deficit of heterozygotes due to the Wahlund effect.

9.1.2 When is F_{IS} not zero?

Generally the first step in analyzing genotypic data from a natural population is to test for HW proportions (see Guest Box 5). As we have seen, F_{IS} is a measure of departure from expected HW proportions. A positive value indicates an excess of homozygotes, and a negative value indicates a deficit of homozygotes. Interpreting the causes of an observed excess or deficit of homozygotes can be difficult.

The most general cause of an excess of homozygotes is nonrandom mating or population subdivision. In the case of the Wahlund effect, the presence of multiple demes within a single population sample will produce an excess of homozygotes at all loci for which the demes differ in allele frequency.

Inbreeding within a single local population will produce a similar genotypic effect. That is, the tendency for related individuals to mate will also produce an excess of homozygotes. Perhaps the simplest example of this is a plant with a mixed mating system that reproduces by both self-pollination and outcrossing. Assume that a proportion S (i.e., the selfing rate) of the matings in a population are the result of selfing and the remainder $(1 - S)$ result from random mating. The equilibrium value of F_{IS} in this case will be

$$F_{IS}^* = \frac{S}{(2-S)} \tag{9.7}$$

For example, consider a population in which half of the progeny are produced by selfing and half by outcrossing ($S = 0.5$). In this case, F_{IS} will be 0.33 (see Example 9.1).

Assortative mating can also cause an excess or deficit of heterozygotes. For example, the white Spirit Bear that we considered in Chapter 2 displays positive assortative mating. There is an average deficit of heterozygotes of 36% at the *MC1R* locus responsible for this color polymorphism in black bears in three populations ($F_{IS} = 0.360$; Hedrick and Ritland 2011). In contrast, 10 microsatellite loci were all in Hardy-Weinberg proportions in these populations. This excess of homozygotes at least partially results from the tendency for bears to mate with individuals having the same phenotype as their mother, resulting in positive assortative mating.

Null alleles that cannot be detected using a particular assay are another possible source of an excess of homozygotes (see Section 5.4.2 for a description of null alleles at allozyme and microsatellite loci). Heterozygotes for a null allele and another allele appear to be homozygotes on a gel, and thus will result in an apparent excess of homozygotes.

Perhaps the best way to discriminate between nonrandom mating (either inbreeding within a deme, or unknowingly including multiple populations in a single sample) or a null allele to explain an excess of homozygotes is to examine whether the effect appears to be locus-specific or population-specific. All loci that differ in allele frequency between demes will have a tendency to show an excess of homozygotes. Assume you examine 10 loci in 10 different population samples ($10 \times 10 = 100$ total tests), and that you detect a significant ($P < 0.05$) excess of homozygotes for 12 tests. If eight of the 12 deviations are in a single population, this would suggest that this population sample consisted of more than one deme. In contrast, a homozygote excess due to a null allele should be locus-specific. In the same example as above, if eight of the deviations were at just one of the 10 loci, this would suggest that a null allele was present at appreciable frequency at that locus.

It may also be possible to discriminate between inbreeding versus including multiple populations in a single sample (the Wahlund effect) to explain an observed excess of homozygotes caused by nonrandom mating. Inbreeding will reduce the frequency of all heterozygotes equally (e.g., equation 9.7). However, as discussed in the previous section, some heterozygotes will be in excess and some will be in deficit in the case of more than two alleles when more than two subpopulations are unknowingly sampled.

A deficit of homozygotes (excess of heterozygotes) may also occur under some circumstances. We saw in Section 6.6 that we expect a slight excess of heterozygotes in small randomly mating populations. Natural selection may also cause an excess of heterozygotes, if heterozygotes have a greater probability of surviving than homozygotes (see Section 8.2 and Table 8.1). However, the survival advantage of heterozygotes has to be very great to have a detectable effect on genotypic proportions.

9.2 SPATIAL PATTERNS OF RELATEDNESS WITHIN LOCAL POPULATIONS

Genotypes are sometimes distributed spatially nonrandomly within local populations. This will be especially true for plant species and for animal species that are not mobile (e.g., many marine invertebrates). In these cases, understanding genetic population structure requires an understanding of the distribution of genotypes on a small geographic scale. For example, if genotypes within a local population are spatially structured, then within-population variation may be underestimated if samples are not spatially well-distributed across the population.

9.2.1 Effects of dispersal distance and population density on patterns of relatedness

The spatial genetic structure (SGS) within a local population depends on both the average dispersal distance of genes, and population density. To illustrate this, imagine two mature populations of a tree species (Figure 9.2). One population has relatively low density (few, widely spaced individuals; Figure 9.2a), while the other has relatively high density (Figure 9.2b). Most gravity-dispersed tree seed falls within approximately two tree heights of the maternal parent. In the lower-density stand, the seed shadows of individual mother trees will overlap relatively little, so there will be a relatively high probability that two seedlings that germinate and grow within that seed shadow are related. In contrast, in the higher-density population, many seed shadows will overlap, and the probability of adjacent seedlings being related is lower. At pairwise distances exceeding the seed shadow width, the probability of relatedness is low. If dispersal increases (i.e., seed shadows increase in diameter), then the SGS will

a. Lower-density population

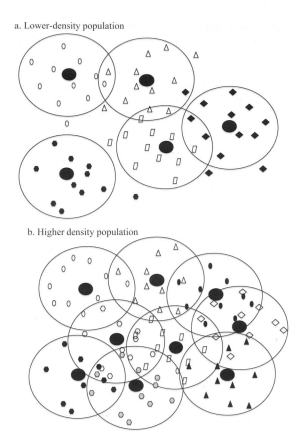

Figure 9.2   Effect of population density on spatial genetic structure in tree populations. Larger filled circles indicate tree locations, while open circles around them delineate seed shadows within which most seed will fall. Small symbols indicate seedlings from each maternal parent. The lower-density population (a) has less overlap in seed shadows and therefore a higher probability of neighboring seedlings inheriting identical alleles from the same mother tree.

(a) Kodiak Island, Alaska

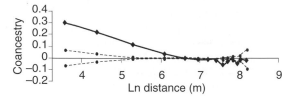

(b) Port McNeill, British Columbia

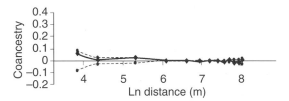

Figure 9.3   Spatial correlograms showing spatial genetic structure within populations of Sitka spruce. The solid line is the coancestry coefficient, and the dashed lines show the 95% confidence interval under the null hypothesis that genotypes are randomly distributed. The pairwise distance classes range from tens to thousands of meters (4 = 55 m, 6 = 403 m, and 8 = 2980 m). The positive coancestry coefficient in the Kodiak Island population (a) up to a distance of nearly 500 m indicates that individuals within 500 m of each other are more genetically similar to each other than are pairs of individuals taken at random from the population. From Gapare and Aitken (2005).

decrease. This simplified example considers only off-spring sharing a mother, but restricted pollen dispersal (e.g., via insect pollination) can also create SGS.

The extent of SGS is usually quantified using statistics that estimate spatial autocorrelation among genotypes at varying distances apart. The probability that two alleles are identical in state (Q) generally decreases with the spatial distance between them (r); thus SGS is characterized by the function $Q(r)$. Individuals sampled across a population are genotyped for a set of neutral markers, and the pairwise physical distance between all possible pairs of individuals is estimated. For two

individuals i and j, the probability that a random allele from i is identical to a random allele from j is the **coancestry** coefficient, F_{ij} (Vekemans and Hardy 2004):

$$F_{ij} = \frac{Q_{ij} - \bar{Q}}{1 - \bar{Q}} \qquad (9.8)$$

Pairs of individuals are pooled into pairwise distance classes, and the average coancestry for each distance class is plotted in a spatial correlogram (Figure 9.3). A full review of the statistical methods for assessing SGS is beyond the scope of this section. Vekemans and Hardy (2004) reviewed several statistical approaches for estimating patterns of relatedness for quantifying SGS.

Population density and SGS can vary substantially among populations if dispersal distance stays relatively constant. Gapare and Aitken (2005) found strong SGS and higher inbreeding levels in lower-density Sitka spruce populations at the northern and southern range margins, and low SGS in high-density central populations. Figure 9.3 illustrates the coancestry of individuals in different pairwise distance categories in one peripheral, geographically isolated population (Kodiak Island, Alaska), and in one population in the center of the species range (Port McNeill, British Columbia).

The positive coancestry coefficient in the Kodiak Island population up to a distance of nearly 500 m indicates that individuals within 500 m of each other are more genetically similar to each other than are pairs of individuals taken at random from the population. While these populations now have a similar density of mature trees, the Kodiak Island population was founded only ~500 years ago, and the majority of trees appear to have descended from relatively few founders that persist as very large, old trees in the population. As densities were low for the initial colonizing generations, the spatial genetic structure persists and is strong (Gapare and Aitken 2005). A similar SGS exists in populations at the southern periphery of the species range, not due to colonization dynamics as that population has likely reached a quasi-equilibrium between density and dispersal, but more likely due to low population density.

9.2.2 Effects of spatial distribution of relatives on inbreeding probability

The average relatedness of neighboring individuals and the population density impact the mating system and probability of inbreeding in a population. For example, geographically isolated Sitka spruce populations at the northern and southern range peripheries receive pollen from just two effective donors, on average, while those in the high-density populations in the center of the species range have tens of effective pollen donors in the pollen cloud they sample (Mimura and Aitken 2007). The peripheral, isolated populations have an effective self-pollination rate of 15 to 35% (see Example 9.1 for how the selfing rate is calculated). In contrast, high-density populations with no significant spatial genetic structure in the range center have an effective selfing rate of less than 5%. Note that this effective selfing can be due to either self-pollination or to biparental inbreeding (mating of relatives); in

the case of Sitka spruce, it appears to be largely due to the latter as a function of the proximity of related individuals (Figure 9.3).

Spatial genetic structure can also provide information on the size and distribution of individual clones in plants that reproduce vegetatively, usually through horizontal above- or below-ground spreading of individual genotypes. Estimating population sizes is difficult if the extent of clonality is not known, and it is often not easy to phenotypically distinguish multiple stems of a clone from multiple individuals. Travis *et al.* (2004) used analysis of SGS of the salt marsh plant smooth cordgrass to determine the extent of clonal structure in populations of different ages. They found that populations of younger plants had less clonal diversity than older populations, and also had relatively high rates of self-pollination, but concluded that inbreeding depression was likely high, as there were fewer inbred individuals in older cohorts.

9.3 GENETIC DIVERGENCE AMONG POPULATIONS AND GENE FLOW

The distribution of genetic variation among populations results from the interaction of the opposing effects of genetic drift, which causes subpopulations to diverge, and the cohesive effects of gene flow, which acts to make subpopulations more similar to each other. All neutral loci in the genome are expected to show similar patterns of divergence (i.e., F_{ST}). Nevertheless, we will see that the effects of natural selection on individual loci can cause them to display very different patterns of divergence among subpopulations.

9.3.1 Complete isolation

Let us initially consider a large random mating population that is subdivided into many completely isolated demes. We will consider the effect of this subdivision on a single locus with two alleles. Assume all HW conditions are valid except for small population size within each individual isolated subpopulation. Genetic drift will occur in each of the isolated demes; eventually, each deme will become fixed for one allele or the other.

What is the effect of this subdivision on our two measures of the genetic characteristics of populations: allele frequencies and genotype frequencies? If the initial allele frequency of the A allele in the large, random mating population was p, the allele frequency in our large, subdivided population will still be p because we expect p of the isolates to become fixed for the A allele, and $(1-p)$ of the isolates to become fixed for the a allele. Thus, subdivision (nonrandom mating) itself has no effect on overall allele frequencies.

We can see this effect in the guppy example from Table 6.3 where 16 subpopulations were founded by a single male and female from a large population. Genetic drift within each subpopulation acted to change allele frequencies at the two loci (AAT-1 and PGM-1) for four generations. However, the average allele frequencies over the 16 subpopulations at both loci are very close to the frequencies in the large founding population. Therefore, allele frequencies in the populations as a whole were not affected by subdivision.

However, the subdivision into 16 separate subpopulations did affect the genotypic frequencies of this population. We can use the F-statistics approach developed in the previous section to describe this effect at the AAT-1 locus. In this case, H_S is the mean expected heterozygosity averaged over the 16 subpopulations (0.344), and H_T is the HW heterozygosity (0.492) using the allele frequency averaged over all subpopulations (0.562). Therefore:

$$F_{ST} = 1 - (H_S/H_T) = 1 - (0.344/0.492) = 0.301$$

In words, the average heterozygosity of individual guppies in this population has been reduced by 31% because of the subdivision and subsequent genetic drift within the subpopulations.

We know from Section 6.3 that heterozygosity will be lost by genetic drift at a rate of $1/2N_e$ per generation. Therefore, we expect F_{ST} among completely isolated populations to increase as follows (modified from equation 6.3):

$$F_{ST} = 1 - \left(1 - \frac{1}{2N_e}\right)^t \qquad (9.9)$$

where N_e is the effective population size of each subpopulation and t is the number of generations (Figure 9.4). The application of this equation to real populations is limited because it assumes a large number of equal size subpopulations and constant population size.

9.3.2 Gene flow

In most cases, there will be some genetic exchange (**gene flow**) among demes within a species. We must

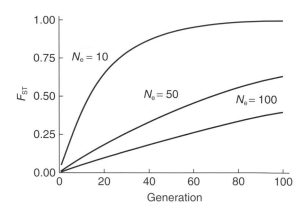

Figure 9.4   Expected increase in F_{ST} over time (generations) among completely isolated populations of different population sizes using equation 9.9.

therefore consider the effects of such partial isolation on the genetic structure of species. Let us first consider the simple case of two demes (A and B) of equal size that are exchanging individuals in both directions at a rate m. Therefore, m is the proportion of individuals reproducing in one deme that were born in the other deme. In this case:

$$q'_A = (1-m)q_A + mq_B$$
and
$$q'_B = mq_A + (1-m)q_B$$
$$(9.10)$$

For example, consider two previously isolated populations that begin to exchange migrants at a rate of $m = 0.10$ (10% exchange per generation). Assume that the allele frequency in population A is 1.0, and in population B it is 0.0. The above model can be used to predict the effects of gene flow between these two populations as shown in Figure 9.5.

Equilibrium will be reached when $q_A = q_B$, and q^* will be the average of the initial allele frequencies in the two demes; in this case; $q^* = 0.5$. In general, there are two primary effects of gene flow:

1 Gene flow reduces genetic differences between populations.

2 Gene flow increases genetic variation within populations.

Gene flow among populations is the cohesive force that holds together geographically separated populations into a single evolutionary unit – the species. In the rest of this chapter we will consider the interaction

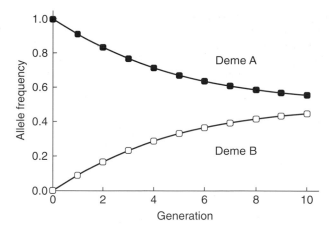

Figure 9.5 Expected changes in allele frequencies in two demes that are exchanging 10% of their individuals each generation ($m = 0.10$) using equation 9.9.

between the homogenizing effects of gene flow and the actions of genetic drift and natural selection that cause populations to diverge.

9.4 GENE FLOW AND GENETIC DRIFT

In the absence of other evolutionary forces, any gene flow between populations will bring about genetic homogeneity. With lower amounts of gene flow it will take longer, but eventually all populations will become genetically identical. However, we saw in Section 6.1 that genetic drift causes isolated subpopulations to become genetically distinct. Thus, the actual amount of divergence between subpopulations is a balance between the homogenizing effects of gene flow, making subpopulations more similar, and the disruptive effects of drift, causing divergence among subpopulations. We examine this using a series of models for different patterns of gene flow. All of these models will necessarily be much simpler than the actual patterns of gene flow in natural populations.

9.4.1 Island model

We will begin with the simplest model that combines the effects of gene flow and genetic drift. Assume that a population is subdivided into a series of ideal populations, each of local effective size N, which exchange individuals at a rate of m. Specifically, each subpopula-

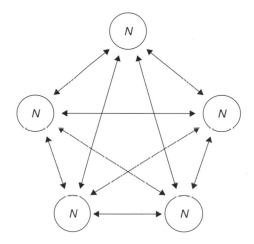

Figure 9.6 Pattern of exchange among five subpopulations under the island model of migration. Each subpopulation of local effective size N exchanges migrants with the other subpopulations with equal probability. More specifically, each subpopulation contributes an equal number of genes to a migrant pool. The total proportion of individuals in a subpopulation from the migrant pool is m per generation, and the rest $(1 - m)$ are drawn from the local population.

tion contributes an equal number of genes to a migrant pool. The total proportion of individuals in a subpopulation from the migrant pool is m per generation, and the rest $(1 - m)$ are drawn from the local population. This model is called the **island model of migration** (Figure 9.6).

As before, we will measure divergence among subpopulations (demes) using F_{ST}. Genetic drift within each deme will act to increase divergence among demes, that is, to increase F_{ST}. However, migration between demes will act to decrease F_{ST}. As long as $m > 0$, there will be some steady-state (equilibrium) value of F_{ST} at which the effects of drift and gene flow will be balanced.

Sewall Wright (1969) has shown that at equilibrium under the island model of migration with an infinite number of demes:

$$F_{ST} = \frac{(1-m)^2}{[2N-(2N-1)(1-m)^2]} \tag{9.11}$$

Fortunately, if m is small this approaches the much simpler:

$$F_{ST} \approx \frac{1}{(4mN+1)} \tag{9.12}$$

This approximation provides an accurate expectation of the amount of divergence under the island model (Figure 9.7). For example, the exact expected equilibrium value of F_{ST} with one migrant per generation ($mN = 1$) using equation 9.11 is 0.199. The approximate value using equation 9.12 is 0.200; the value resulting from the simulation shown in Figure 9.7 with 20 subpopulations ($F_{ST} = 0.215$) is very close to this expected value. One important result of this analysis is that very little gene flow is necessary for populations to be genetically connected.

Equations 9.11 and 9.12 assume an infinite number of subpopulations. The expected value of F_{ST} at equilibrium can be corrected as shown below to take into account a finite number (n) of subpopulations (Slatkin 1995):

$$F_{ST} \approx \frac{1}{(4mNa+1)} \tag{9.13}$$

where $a = \left(\dfrac{n}{n-1}\right)^2$

This effect is small unless there are very few subpopulations. For example, with 20 subpopulations as above, the expected value of F_{ST} with equation 9.13 is 0.184, rather than 0.200 with equation 9.12.

Equation 9.12 also provides a surprisingly simple result; the amount of divergence among demes depends only on the number of migrant individuals (mN), and

not the proportion of exchange among demes (m). The dependence of divergence only on the number of migrants irrespective of population size may seem counterintuitive. Remember, however, that the amount of divergence results from the opposing forces of drift and migration. The larger the demes are, the slower they are diverging through drift; thus, proportionally fewer migrants are needed to counteract the effects of drift. Small demes diverge rapidly through drift, and thus proportionally more migrants are needed to counteract drift.

This result has important implications for our interpretation of observed patterns of genetic divergence among populations (e.g., Figure 9.1). We expect to find approximately the same amount of divergence among demes of size 200 with $m = 0.025$, as we do with demes of size 50 with $m = 0.1$ ($0.025 \times 200 = 0.1 \times 50 = 5$ migrants per generation). One important outcome of this result is that species with larger local populations (N) are expected to have less genetic divergence among populations than species with the same amount of genetic exchange (m) with smaller local populations (Lowe and Allendorf 2010). We will see in Section 15.4.1 that this effect has important implications for understanding the relationship between genetic divergence and demographic connectivity among populations.

9.4.2 Stepping-stone model

In natural populations, migration is often greater between subpopulations that are near each other (Slatkin 1987). This violates the assumption of equal probability of exchange among all pairs of subpopulations with the island model of migration. The stepping-stone model of migration was introduced (Kimura and Weiss 1964) to take into account both short-range migration (which occurs only between adjacent subpopulations) and long-range migration (which occurs at random between subpopulations). Linear stepping-stone models (Figure 9.8) are useful for modeling populations with a one-dimensional linear structure, as occurs along a river, river valley, valley, or a mountain ridge, for example. Two-dimensional stepping-stone models are useful for modeling populations with a grid structure (or 2D checkerboard pattern) across the landscape.

The mathematical treatment of the stepping-stone model is much more complex than the island model. In general, migration in the stepping-stone model is

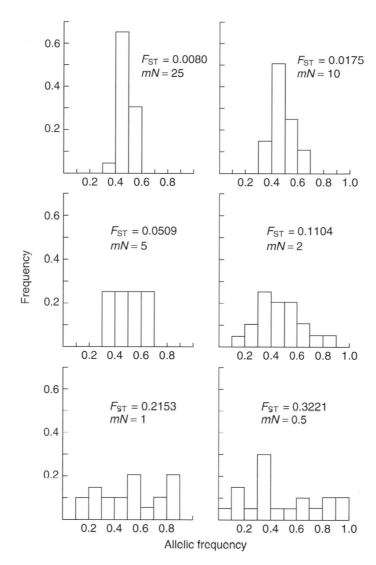

Figure 9.7 Simulated patterns of allelic frequency divergence with the island model of migration showing the effect of different amounts of migration. Computer simulations were carried out with 20 subpopulations ($N = 200$) and different amounts of migration (mN). Redrawn from Allendorf and Phelps (1981).

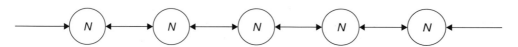

Figure 9.8 Pattern of exchange among subpopulations under the single-dimension stepping-stone model of migration. Each subpopulation of size N exchanges $m/2$ individuals with each adjacent subpopulation.

less effective at reducing differentiation caused by drift because subpopulations exchanging genes tend to be genetically similar to each other. Therefore, there will be greater differentiation (i.e., greater F_{ST}) among subpopulations with the stepping-stone than the island model for the same amount of genetic exchange (m). In addition, in the stepping-stone model, adjacent subpopulations should be more similar to each other than geographically distant populations (see Figure 9.1). With the island model of migration, genetic divergence will be independent of geographic distance (Figure 9.9).

9.5 CONTINUOUSLY DISTRIBUTED POPULATIONS

In some species, individuals are distributed continuously across large landscapes (e.g., coniferous tree species across boreal forests) and are not subdivided into discrete subpopulations by barriers to gene flow (Figure 9.10). Nonetheless, gene flow can be limited to relatively short distances, leading to increasing genetic differentiation as the geographic distance between individuals becomes greater (see Section 9.2.1). This

effect was originally described as **isolation by distance** by Wright (1943). Today, many papers use the term isolation by distance in a more general sense to refer to any pattern where genetic divergence increase with geographic distance (e.g., Figure 9.1).

The mathematics of the distribution of genetic variation in continuously distributed populations is complex (Felsenstein 1975, Epperson 2007). It is impossible to identify and sample discrete population units because no sharp boundaries exist. In this case, the **neighborhood** has been defined as the area from which individuals can be considered to be drawn at random from a random mating population (Wright 1943, 1946). This model assumes that dispersal distances are normally distributed about a mean of zero. In this case, Wright's neighborhood size (NS = the effective number of parents in a neighborhood) is:

$$NS = 4\pi\sigma^2 D \tag{9.14}$$

where D is density (number of individuals per unit area), π is pi, and σ^2 is the mean squared parent-offspring axial distance (that is, along one axis). NS can be thought of as the number of reproducing individu-

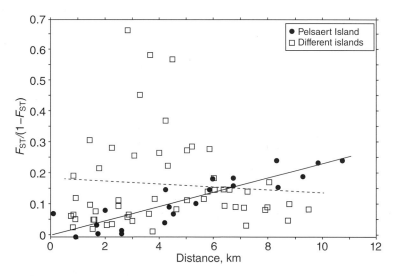

Figure 9.9   Observed relationship between genetic divergence $F_{ST}/(1 - F_{ST})$ and geographic distance in populations of the intertidal snail *Austrocochlea constricta* off the coast of Western Australia at 17 polymorphic allozyme loci. The open boxes are comparisons between pairs of populations on different islands. There is no relationship here between genetic divergence and geographic distance, as expected with the island model of migration. The closed circles show comparisons between local populations found on a single large island (Pelsaert Island). These show the 'isolation by distance' pattern expected with either the stepping-stone model of migration or isolation by distance in a continuously distributed population. From Johnson and Black (2006).

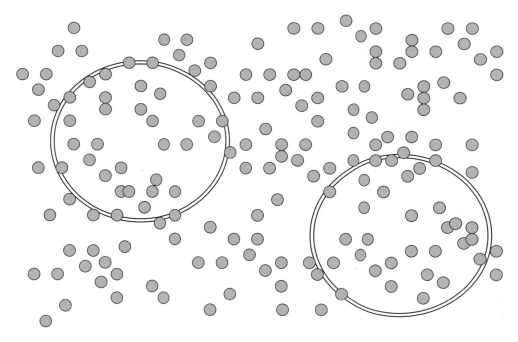

Figure 9.10 Continuous distribution of individuals where no sharp boundaries separate individuals (gray dots) into discrete groups. Nonetheless, genetic isolation arises over geographic distance because nearby individuals are more likely to mate with each other than with individuals that are farther away (isolation by distance). We can place a circle of the appropriate neighborhood size anywhere (see two circles) and individuals inside will represent a panmictic group in HW proportions (i.e., a genetic neighborhood: Wright 1946).

als in a circle of radius 2σ, and a circle of this size would include about 87% of the parents of individuals at the center (Wright 1946).

Chambers (1995) provides a helpful overview of the conservation implications of the role of subdivision in continuously distributed populations maintaining genetic variation. It is very difficult to come up with simple applications of this model to conservation. However, it does allow us to estimate the geographic distance at which individuals will become genetically differentiated due to limited gene flow. For example, if the mean gene flow distance is 1 km, then we would expect substantial genetic differentiation between individuals separated by, say, 5–10 km (Manel *et al.* 2003).

9.6 CYTOPLASMIC GENES AND SEX-LINKED MARKERS

Maternally inherited cytoplasmic genes and sex-linked markers generally show different amounts of differen-

tiation among populations than autosomal loci for several reasons. First, they usually have a smaller effective population size than autosomal loci and therefore show greater divergence due to genetic drift. In addition, differences in migration rates between males and females can cause large differences between cytoplasmic genes and sex-linked markers compared with autosomal loci.

9.6.1 Cytoplasmic genes

The expected amount of allele frequency differentiation with a given amount of gene flow is different for mitochondrial and chloroplast genes versus nuclear genes because of haploidy and uniparental inheritance. We can estimate F_{ST} values to compare the amount of allelic differentiation for nuclear and mitochondrial genes. However, since mtDNA is haploid, individuals are hemizygous rather than homozygous or heterozygous.

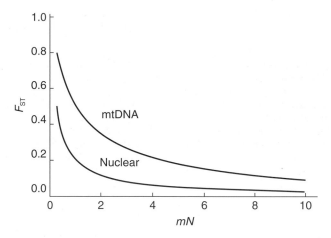

Figure 9.11 Expected values of F_{ST} with the island model of migration for a nuclear locus (equation 9.12) and mtDNA (and cpDNA; equation 9.15) assuming equal migration rates of males and females.

We generally expect more differentiation at mtDNA and cpDNA genes than for nuclear genes because of their smaller effective size. That is, the greater genetic drift with smaller effective population size will bring about greater differentiation in populations that are connected by the same amount of gene flow. If migration rates are equal in males and females, then we expect the following differentiation for mtDNA with the island model of migration (Birky *et al.* 1983):

$$F_{ST} \approx \frac{1}{(mN+1)} \qquad (9.15)$$

This expression is sometimes written to consider only females:

$$F_{ST} \approx \frac{1}{(2mN_f+1)} \qquad (9.16)$$

where N_f is the number of females in the population. Equations 9.15 and 9.16 are identical if there are an equal number of males and females in the population ($N_f = N_f$) because $(2N_f) = N$.

Thus, with equal migration rates in males and females, we expect approximately two to four times as much allele frequency differentiation at mitochondrial genes than at nuclear genes (Figure 9.11). We can see

this effect in the study of sockeye salmon shown in Figure 9.13 the F_{ST} at mtDNA was greater than the F_{ST} at all but one of the 20 nuclear loci examined.

This difference in F_{ST} for a nuclear locus and mtDNA is expected to be greater for species in which migration rates of males are greater than those of females (see Example 9.3). Similarly, we often see much greater F_{ST} values for cpDNA than nuclear genes in plants with maternal inheritance of cpDNA because most of the gene flow is via pollen (Petit *et al.* 2005) (Example 9.4). Larsson *et al.* (2009) present a valuable summary of the statistical power for detecting genetic divergence using nuclear and cytoplasmic markers.

9.6.2 Sex-linked loci

Genes on the Y-chromosome of mammals present a parallel situation to mitochondrial genes. The Y-chromosome is haploid and is only transmitted through the father. Thus, the expectations that we just developed for cytoplasmic genes also apply to Y-linked genes, except we must substitute the number of males for females. Comparison of the patterns of differentiation at autosomal, mitochondrial, and Y-linked genes can provide valuable insight into the evolutionary history of species and current patterns of gene flow (Example 9.5).

Example 9.3 Sex-biased dispersal of great white sharks

The great white shark is globally distributed in temperate waters off continental shelves (Pardini *et al.* 2001). Relatively little is known about the ecology and demography of this species because of its rarity and large size. Pardini *et al.* (2001) examined both mtDNA and microsatellite genotypes in great white sharks collected off the coasts of South Africa, Australia, and New Zealand.

Comparison of the control region of mtDNA revealed two major haplogroups (A and B) that have approximately 4% sequence divergence. A haplogroup is a group of similar haplotypes that share a common ancestor. For example, all of the mtDNA haplotypes from Chile in Figure 9.16 comprise a single haplogroup. No differences in haplogroup frequencies were found between sharks from Australia or New Zealand. However, sharks from South Africa were extremely divergent from sharks in Australia and New Zealand ($F_{ST} = 0.85$):

Population	Type A	Type B
Australia & NZ	48	1
South Africa	0	39

In striking contrast to this result, no allele frequency differences were found at five nuclear microsatellite loci among these regions.

Pardini *et al.* (2001) concluded that female great white sharks are philopatric and that males undertake long transoceanic movements. However, study of transoceanic movement with electronic tags and photographic identification indicate that females, as well as males, make transoceanic movements between these areas (Bonfil *et al.* 2005). Therefore, the difference between divergence of mtDNA and nuclear markers is apparently not based on differences in transoceanic migrations of males and females, but result from whether or not these migrants become reproductively integrated into the recipient population.

Example 9.4 Divergence at nuclear loci and cpDNA in the white campion

McCauley (1994) compared the distribution of genetic variation at seven allozyme loci and chloroplast DNA in white campion (Table 9.1). As expected, a much greater proportion of the variation was distributed among populations for the cpDNA marker ($F_{ST} = 0.674$) compared to the nuclear loci ($F_{ST} = 0.134$). However, this difference is even greater than expected with the island model of migration at equilibrium (see Figure 9.11). $F_{ST} = 0.134$ is expected to result from 1.6 migrants per generation with the island model (Figure 9.11 and equation 9.11). This amount of gene flow should result in an F_{ST} for cpDNA of 0.385 (equation 9.15). This value is outside of the 95% confidence interval for the estimated F_{ST} for cpDNA.

The simplest explanation for this discordance between nuclear loci and cpDNA is that most of the gene flow is from pollen rather than seeds. Therefore, migration rates will be greater for nuclear genes than for maternally inherited genes such as cpDNA. This same effect has been seen in many plants, especially ones that are wind pollinated (Ouborg *et al.* 1999).

Table 9.1 Estimates of F_{ST} from ten populations of the white campion at seven allozyme loci and cpDNA. The 95% confidence limits are provided in the bottom two rows (McCauley 1994).

Locus	F_{ST}
GPI	0.125
IDH	0.083
LAP	0.172
MDH	0.230
PGM	0.145
6-PGD	0.042
SKDH	0.083
Allozymes	0.134 (0.073–0.195)
cpDNA	0.674 (0.407–0.941)

Example 9.5   Y-chromosome isolation in a shrew hybrid zone

A Y-chromosome microsatellite locus and ten auto-somal microsatellite loci were typed across a hybrid zone between the Cordon and Valais races of the common shrew in western France (Balloux *et al.* 2000). There is a contact zone where the two races occur on either side of a stream. Gene flow is somewhat limited, but the two races show relatively little divergence at the autosomal microsatellite loci (F_{ST} = 0.02; Balloux *et al.* 2000, Brünner and Hausser 1996).

   Almost all gene flow in these shrews appears female-mediated, and male hybrids are generally unviable. No alleles were shared across the hybrid zone at the Y-linked locus (Figure 9.12). However, the

F_{ST} value between races at the Y-linked microsatellite loci is just 0.19; this low value does not reflect the absence of alleles shared between races because of the high within-race heterozygosity at this locus (this effect is discussed in Section 9.7). R_{ST} is an analogue of F_{ST} that takes the relative size of microsatellite alleles (i.e., allelic state) into consideration assuming a stepwise mutation model. The strong divergence is reflected in the value of R_{ST} (0.98). It is important to incorporate allele length (mutational) information when H_S is high, because mutations likely contribute to population differentiation when populations are long-isolated, as in this example (see Section 9.8).

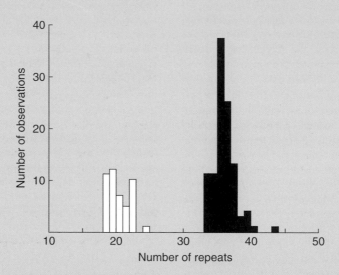

Figure 9.12   Allele frequency distribution of a Y-linked microsatellite locus (*L8Y*) in the European common shrew. Males of the Cordon race are represented by black bars, and Valais males by white bars. From Balloux *et al.* (2000).

9.7 GENE FLOW AND NATURAL SELECTION

We will now examine the effects of natural selection on the amount of genetic divergence expected among subpopulations in an island model of migration (Allendorf 1983). We previously concluded that the amount of divergence, as measured by F_{ST}, is dependent only upon the product of migration rate and deme size (mN). Does this simple principle hold when we combine

the effects of natural selection with the island model of migration? As we will see shortly, the answer is no.

9.7.1 Heterozygous advantage

Assume the following fitnesses hold within each deme:

	AA	Aa	aa
Fitness	$1 - s$	1.0	$1 - s$

Table 9.2 Simulation results (except top row) of steady-state F_{ST} values for 20 demes with selective neutrality ($s = 0.00$) or **heterozygous advantage** in which both homozygous phenotypes have a reduction in fitness of s in all demes. Each value is the mean of 20 repeats. Expected F_{ST} values from equation 9.12. From Allendorf (1983).

	mN						
	0.5	**1**	**2**	**5**	**10**	**25**	**N**
Expected	0.333	0.200	0.111	0.048	0.024	0.010	
$s = 0.00$	0.307	0.204	0.124	0.042	0.020	–	25
	0.335	0.183	0.108	0.048	0.026	0.012	50
	0.322	0.188	0.106	0.044	0.025	0.010	100
$s = 0.01$	0.283	0.164	0.067	0.050	0.022	–	25
	0.243	0.153	0.082	0.041	0.023	0.012	50
	0.178	0.124	0.093	0.038	0.036	0.011	100
$s = 0.05$	0.193	0.126	0.071	0.044	0.024	–	25
	0.133	0.107	0.062	0.034	0.024	0.009	50
	0.083	0.107	0.043	0.024	0.018	0.011	100
$s = 0.10$	0.122	0.104	0.053	0.043	0.022	–	25
	0.094	0.076	0.050	0.031	0.021	0.009	50
	0.041	0.029	0.032	0.022	0.010	0.007	100

Because of its complexity, one of the best ways to explore this system is using computer simulations. We will examine the results of simulations combining natural selection with the island model of migration in which there are 20 subpopulations. Natural selection will act to maintain a stable equilibrium of $p^* = 0.5$ in each deme. Thus, this model of selection will reduce the amount of divergence among demes (Table 9.2). The greater the value of s, the greater will be the reduction in F_{ST} (Table 9.2). Even relatively weak selection can have a marked effect on F_{ST}; for example, see $s = 0.01$ and $mN = 0.5$ in Table 9.2. It is also apparent that genetic divergence among demes is no longer only a function of mN. For a given value of mN, natural selection becomes more effective, and thus F_{ST} is reduced, as population size increases.

9.7.2 Divergent directional selection

Assume the following relative fitnesses in a population consisting of 20 demes:

	AA	Aa	aa
Demes 1–10	1	$1 - (t/2)$	$1 - t$
Demes 11–20	$1 - t$	$1 - (t/2)$	1

This pattern of divergent directional selection will act to maintain allele frequency differences among demes so that large differences can be maintained even with extensive genetic exchange. Again, selection is more effective with larger demes (Table 9.3).

9.7.3 Comparisons among loci

Gene flow and genetic drift are expected to affect all loci uniformly throughout the genome. However, the effects of natural selection will affect loci differently depending upon the intensity and pattern of selection and effective population size. As we have noted above, even fairly weak natural selection can have a substantial effect on divergence among large subpopulations. Therefore, surveys of genetic differentiation at many loci throughout the genome can be used to detect outlier loci that are candidates for the effects of natural selection.

Detecting locus-specific effects is crucial because only genome-wide effects inform us reliably about population demography and phylogenetic history, whereas locus-specific effects can help identify those genes important for fitness and adaptation. An example of a locus-specific effect is differential directional selection, whereby one allele is selected for in one environment, but the allele is disadvantageous in a different

Table 9.3 Simulation results (except top row) of steady-state F_{ST} values for 20 demes with divergent **directional selection**. Each value is the mean of 20 repeats. One homozygous genotype has a reduction in fitness of t in 10 demes; the other homozygous genotype has the same reduction in fitness in the other 10 demes. Heterozygotes have a reduction in fitness of one-half t in all demes. Expected F_{ST} values from equation 9.12. From Allendorf (1983).

| | *mN* | | | | | | |
	0.5	1	2	5	10	25	*N*
Expected	0.333	0.200	0.111	0.048	0.024	0.010	
$t = 0.00$	0.307	0.204	0.124	0.042	0.020	–	25
	0.335	0.183	0.108	0.048	0.026	0.012	50
	0.322	0.188	0.106	0.044	0.025	0.009	100
$t = 0.01$	0.334	0.170	0.107	0.056	0.022	–	25
	0.298	0.119	0.100	0.038	0.026	0.010	50
	0.300	0.185	0.115	0.035	0.023	0.010	100
$t = 0.05$	0.356	0.186	0.120	0.050	0.022	–	25
	0.462	0.268	0.149	0.055	0.026	0.011	50
	0.595	0.423	0.198	0.063	0.021	0.012	100
$t = 0.10$	0.470	0.245	0.163	0.047	0.029	–	25
	0.624	0.365	0.261	0.077	0.036	0.013	50
	0.805	0.657	0.443	0.159	0.063	0.019	100

environment. This selection would generate a large allele frequency difference (high F_{ST}) only at this locus relative to neutral loci throughout the genome (see Example 9.6). For example, just a 10% selection coefficient favoring different alleles in two environments can generate large differences with this pattern of selection between the selected locus ($F_{ST} = 0.657$) and neutral loci ($F_{ST} = 0.188$), as shown in Table 9.3 with local population sizes of $N = 100$ and $mN = 1$.

It is crucial to identify outlier loci not only because such loci might be under selection and help us to understand adaptive differentiation, but also because outlier loci can severely bias estimates of population parameters (e.g., F_{ST} or the number of migrants). Most estimates of population parameters assume that loci are neutral. For example, Allendorf and Seeb (2000) found with sockeye salmon that a single outlier locus with extremely high F_{ST} could bias high estimates of the mean F_{ST} from 0.09 to 0.20 (Figure 9.13): more than double!

In another example, Wilding et al. (2001) genotyped 306 AFLP loci in an intertidal snail (the rough periwinkle) collected along rocky ocean shorelines. Fifteen of the 306 loci had an F_{ST} that was substantially higher than expected for neutral loci in a comparison of two morphological forms (H and M) that were collected along the same shoreline (Figure 9.14). Interestingly,

these same 15 loci were also found to be outliers at other shoreline locations, supporting the hypothesis that these 15 loci are under selection. Furthermore, when these outlier loci are used to construct a similarity cluster diagram, the samples cluster together primarily on the basis of morphology and habitat type, rather than geographic distance between the populations; this provides further support for the hypothesis of selection at these 15 loci. This illustrates the importance of removing outlier loci when inferring historical relationships between populations (Section 16.6). Similarly, estimates of the time since populations diverged should be based only on neutral markers, as should estimates of migration rates (Luikart et al. 2003).

9.8 LIMITATIONS OF F_{ST} AND OTHER MEASURES OF SUBDIVISION

F_{ST} was developed by Sewall Wright long before the use of molecular markers to describe genetic variation in natural populations, and he assumed that loci had just two alleles. Nei (1977) developed G_{ST} as an analogue to F_{ST} when allozyme loci with more than two alleles were first used to describe population genetic structure. Here we consider some of the limitations of F_{ST} as a measure of genetic divergence among populations.

Example 9.6 Use of adaptive loci to detect genetic subdivision in marine fishes

Knowledge of the breeding structure of fish stocks is crucial for developing and implementing effective management strategies that are urgently needed to maintain sustainable fisheries (Kritzer and Sale 2004, Guest Box 18). However, population genetic studies of marine fishes generally have failed to detect genetic differences even between apparently geographically isolated subpopulations for which there is evidence of some reproductive isolation (Waples 1998). This failure results from the large population sizes and high gene flow among stocks of many marine fishes. Even very low exchange rates among stocks with large population sizes will be sufficient to eliminate genetic evidence of population differentiation at neutral loci.

For example, almost no genetic differentiation ($F_{ST} = 0.003$) was found at nine neutral microsatellite loci in Atlantic cod, but substantial differentiation ($F_{ST} = 0.261$) was found at the *Pan* I locus (Pampoulie *et al.* 2006), which previous studies has shown to be under natural selection (Pogson and Fevolden 2003). Recent analysis has suggested that *Pan* I allele frequencies are influenced by temperature, salinity, and depth (Case *et al.* 2005). The utility of divergent markers such as this for stock analysis would be greatly reduced if such differences were not stable and changed over a few generations. Fortunately, comparison of current patterns of genetic differentiation using otoliths going back up to 70 years demonstrated that these allele frequency differences have been stable, and therefore can be used as a reliable marker for stock identification (Nielsen *et al.* 2007).

Other loci in Atlantic cod have been found to be useful in describing genetic differentiation among populations. Nielsen *et al.* (2009) screened 98 gene-associated SNP polymorphisms in Atlantic cod. Eight of these SNPs demonstrated exceptionally high F_{ST} values and were considered to be subject to directional selection in local demes, or closely linked to loci under selection. Even on a limited geographic scale between the nearby North Sea and Baltic Sea populations, four loci displayed evidence of adaptive divergence. Analysis of archived otoliths from one of these populations indicated that these allele frequencies were stable over 24 years.

A similar result has been found with European flounder. Little genetic differentiation was found among subpopulations at nine microsatellite loci ($F_{ST} = 0.02$). However, substantial differentiation ($F_{ST} = 0.45$) was present at a heat-shock locus (*Hsc70*), which was selected as a candidate gene because of its known function (Hemmer-Hansen *et al.* 2007). Population genomic approaches allow us to identify genes involved in adaptive traits without prior information about which traits are important in the species in question. These adaptive genes can then be employed to describe spatial genetic structure for species in which neutral genetic markers have not been informative.

9.8.1 Genealogical information

One important limitation of F_{ST} (and related measures like G_{ST}) is that they do not consider the identity of alleles (i.e., genealogical degree of relatedness). For example, in the common shrew in Example 9.5, the F_{ST} for a Y-linked microsatellite is only 0.19 across a hybrid zone between races, even though the two races share no alleles at this locus. An examination of Figure 9.12 clearly shows that all of the alleles on either side of the hybrid zone are more similar to each other than to any of the alleles on the other side of the hybrid zone. A measure related to F_{ST}, called R_{ST}, uses information on the length of alleles at microsatellite loci (Slatkin 1995), and is much higher in this case ($R_{ST} = 0.98$).

Another measure of differentiation that uses information on allele genealogical relationships is Φ_{ST} (Excoffier *et al.* 1992, Michalakis and Excoffier 1996). Measures using genealogical information (like Φ_{ST} and R_{ST}) use the degree of differentiation between alleles as a weighting factor that increases the metric (e.g., F_{ST}) proportionally to the number of mutational differences between alleles. R_{ST} is analogous to F_{ST}, and is defined as the proportion of variation in allele length that is due to differences among populations. R_{ST} assumes that each mutation changes an allele's length by only one repeat unit, for example, a mutation adds or removes one dinucleotide 'CA' unit (see Example 4.1 and stepwise mutation in Section 12.1.2). This is important because if mutations cause only a one-step change, then any populations with alleles differing by a few steps will have experienced substantial recent gene flow, whereas populations with alleles differing by many steps will have had little or no gene flow (such that isolation has allowed accumulation of many

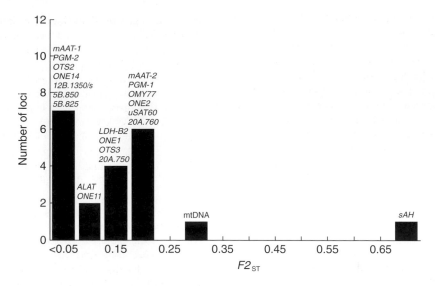

Figure 9.13 Genome-wide versus locus-specific effects, and the identification of outlier loci that are candidates for being under selection in sockeye salmon. Gene flow and genetic drift lead to similar genome-wide allele frequency differentiation (F_{ST}) among four populations for 19 nuclear loci with an F_{ST} less than 0.20. The $F2_{ST}$ values are based on two alleles at each locus so that loci with different heterozygosities can be compared (see Section 9.8.2). One nuclear locus (sAH) has a much greater $F2_{ST}$ and is a candidate for being under natural selection. From Allendorf and Seeb (2000).

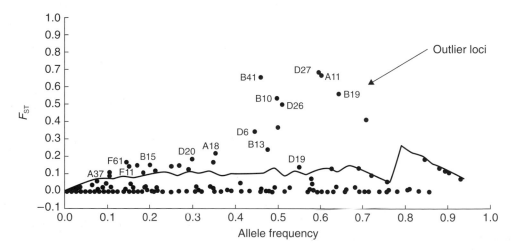

Figure 9.14 Locus-specific F_{ST} estimates between two morphological forms (H and M) of the rough periwinkle intertidal snail collected along the same shoreline show outlier loci from the neutral expectation (Thornwick Bay). Fifteen AFLP loci (dots above solid line) had exceptionally high F_{ST} values (>0.20) compared with the mean observed F_{ST} (<0.04) and the null distribution of 'neutral' F_{ST} values (~0.0–0.2). The solid line is the upper 99th percentile of the null distribution of F_{ST} for the simulated neutral loci. Very few outliers were found when comparing same-morphology populations (i.e., H vs. H, or M vs. M) from different geographic areas. From Wilding *et al.* (2001).

mutational steps between populations). This is the pattern that we see in Figure 9.12.

Gaggiotti and Foll (2010) have presented a method to estimate population-specific F_{ST} values rather than global or pairwise F_{ST} values. They define F'_{ST} as the probability that two genes chosen at random from the population share a common ancestor within than population. This allows for differences in local population sizes and migration rate, unlike the standard island model. Their approach has the potential to be extremely valuable in interpreting patterns of genetic population structure from a conservation perspective. For example, Cosentino *et al.* (2012) have used F'_{ST} to interpret the metapopulation structure of tiger salamanders (see Section 15.4).

9.8.2 High heterozygosity within subpopulations

F_{ST} (and its analogue G_{ST}) has limitations when using loci with high mutation rates and high heterozygosities, such as microsatellites. F_{ST} is biased downwards when variation *within* subpopulations (H_S) is high. The source of this bias is obvious: when variation within populations is high, the proportion of the total variation distributed between populations can never be very high (Hedrick 1999, Meirmans and Hedrick 2011). For example, if $H_S = 0.90$, F_{ST} cannot be higher than 0.10 ($1 - 0.90 = 0.10$; see equation 9.3).

Allendorf and Seeb (2000) used $F2_{ST}$ to compare F_{ST} between different marker types, to test whether they showed different patterns of genetic divergence among subpopulations because of natural selection (see Figure 9.13). $F2_{ST}$ is estimated by using the frequency of the most common allele at each locus, and combing (or binning) all other alleles into a single allele frequency, as recommended by McDonald (1994). The advantage of $F2_{ST}$ is that it allows valid comparison of divergence at different loci because it is based on two alleles at all loci. The disadvantage of $F2_{ST}$ is that much information is lost because of the binning together of alleles. It is possible, however, to estimate an $F2_{ST}$ for each allele by binning all other alleles (Chakraborty and Leimar 1987, Bowcock *et al.* 1991).

Hedrick (2005) introduced G'_{ST}, which is G_{ST} divided by its maximum possible values with the same overall allele frequencies. Thus, G'_{ST} has a range from 0 to 1, and was designed to be independent of H_S, (although see Ryman and Leimar 2008). If H_S is high, then G'_{ST}

can be much greater than G_{ST}. G'_{ST} is designed to be a standardized measure of G_{ST}, which accounts for different levels of total genetic variation at different loci (Meirmans and Hedrick 2011).

Jost (2008) was quite critical of the use of F_{ST} and G_{ST} as measures of differentiation, and he introduced D, which is a similar measure to G'_{ST}. D differs from G_{ST} in that G_{ST} measures deviations from panmixia, while D measures deviations from complete differentiation (Whitlock 2011). D and G_{ST} behave quite differently (Heller and Siegismund 2009, Ryman and Leimar 2009). D is zero when all populations are identical, monotonically increases with increasing divergence between populations, and tends to 1 as different populations have no shared alleles.

Both G'_{ST} and D have been used increasingly in the literature since they were introduced. Nevertheless, they do not solve the problem of measuring divergence at loci with high within-subpopulation heterozygosity, which they were designed to address. In addition, they have serious problems themselves (Ryman and Leimar 2008, 2009, Heller and Siegismund 2009, Whitlock 2011). They are both insensitive to genetic drift and gene flow when the mutation rate is high relative to the migration rate (Ryman and Leimar 2008, 2009, Whitlock 2011). Furthermore, D is specific to the locus being measured even with selective neutrality, and so little can be inferred about the population demography from estimating D. Moreover, neither G'_{ST} nor D estimates quantities that can be interpreted in terms of population genetic theory. Jost (2009) showed that D is a useful measure of differentiation, but also warns that it ought not to be used for measuring migration. In contrast, F_{ST} measures a fundamental parameter in population genetics theory (Holsinger and Weir 2009). We may have imperfect statistical estimators of this quantity (such as G_{ST} for loci with high mutation rates), but the underlying quantity is inherently interesting, and these biases can be addressed with other techniques (e.g., R_{ST} in the case of microsatellite loci).

9.8.3 Other measures of divergence

Another widely used measure of population genetic differentiation is Nei's genetic distance (D; Nei 1972). This measure will increase linearly with time for completely isolated populations and the infinite allele model of mutation with selective neutrality. Nevertheless, D is often used and appears to perform relatively

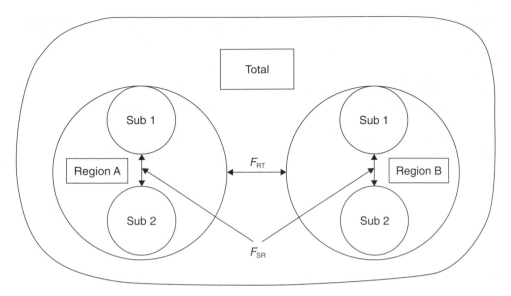

Figure 9.15 Organization of hierarchical population structure with two levels of subdivision: subpopulations within regions (F_{SR}) and regions within the total species (F_{RT}). Each region has two subpopulations. F_{SR} is the proportion of the total diversity due to differences between subpopulations within regions. F_{RT} is the proportion of the total diversity due to differences between regions.

well for non-isolated populations (Paetkau *et al.* 1997). Nei's (1978) unbiased *D* provides a correction for sample size. This correction is not so important for comparison between species, but can be for conservation cases where intraspecific populations are being compared. Without this correction, poorly sampled populations will on average appear to be the most divergent. Another reliable and widely used measure of genetic distance is Cavalli-Sforza and Edwards' chord distance (Cavalli-Sforza and Edwards 1967). There are numerous other genetic distance measures (e.g., see Paetkau *et al.* 1997) that are less widely used and beyond the scope of this book.

9.8.4 Hierarchical structure

Populations are often structured at multiple hierarchical levels, for example, locally and regionally. For example, several subpopulations (demes) might exist on each side of a barrier such as a river or mountain ridge. Here, two hierarchical levels are (1) the local deme level, and (2) the regional group of demes on

either side of the river (Figure 9.15). It is useful to identify such hierarchical structures and to quantify the magnitude of differentiation at each level to help guide conservation management (e.g., identification of management units and evolutionarily significant units; see Chapter 16). For example, if regional populations are highly differentiated but local demes within regions are not, managers should prioritize translocations between local demes and not between regional populations.

Hierarchical structure is often quantified using hierarchical *F*-statistics that partition the variation into local and regional components, such as the proportion of the total differentiation due to differences between subpopulations within regions (F_{SR}), and the proportion of differentiation due to differences between regions (F_{RT}). Hierarchical structure is also often quantified using AMOVA (analysis of molecular variance; Excoffier *et al.* 1992), which is analogous to the standard statistical approach ANOVA (analysis of variance). Sherwin (2010) has proposed using a hierarchical approach to describe genetic variation that is similar to Shannon's entropy-based diversity, which is the standard for ecological communities.

9.9 ESTIMATION OF GENE FLOW

Gene flow is important to measure in conservation biology because low or reduced gene flow can lead to local inbreeding and inbreeding depression, whereas high or increased gene flow can limit local adaptation and cause outbreeding depression. Measuring and monitoring gene flow can help to maintain viable populations (and metapopulations) in the face of changing environments and habitat fragmentation. Renewed gene flow (following isolation) can result in '**genetic rescue**', through heterosis in 'hybrid' offspring (Tallmon *et al.* 2004). Finally, rates of gene flow in animals are correlated with rates of dispersal; thus knowing rates of gene flow can help predict the likelihood of recolonization of vacant habitats following extirpation or over harvest (i.e., 'demographic rescue'). However, over 90% of the gene flow in plant species is due to pollen movement and not seed dispersal (Petit *et al.* 2005).

Rates of gene flow can be estimated in several ways using molecular markers. First, indirect estimates of average migration rates (mN) can be obtained from: (1) allele frequency differences (F_{ST}) among populations; (2) the proportion of private alleles in populations; or (3) a likelihood-based approach using information both on allele frequencies and private alleles (see below). The migration rate estimate is an average over the past tens to hundreds of generations (see below).

Second, direct estimates of current dispersal rates can be obtained using genetic tagging and a mark-recapture approach that directly identifies individual immigrants by identifying their 'foreign' genotypes (i.e., genotypes unlikely to originate from the local gene pool). This approach can give estimates of migration rates in the current generation. We now discuss the indirect and direct (assignment test) approaches, in turn, below.

9.9.1 F_{ST} and indirect estimates of *mN*

We can estimate the average number of migrants per generation (mN) by using the island model of migration (Figure 9.6, Guest Box 9). For example, an F_{ST} of 0.20 yields an estimate of one migrant per generation ($mN = 1$), under the island model of migration (Figure 9.11, nuclear markers). Less differentiation ($F_{ST} = 0.10$) leads to a higher estimate of gene flow ($mN = 2$). Equation 9.12 can be rearranged to allow estimation of the average number of migrants (mN) from F_{ST}, under the island model, as follows:

$$\widehat{mN} \approx \frac{(1 - F_{ST})}{4 F_{ST}} \tag{9.17}$$

The following assumptions are required for interpreting estimates of mN from the simple island model:
1 An infinite number of populations of equal size (see also Section 9.4.1. above).
2 That N and m are the same and constant for all populations (thus migration is symmetrical).
3 Selective neutrality and no mutation.
4 That populations are at migration-drift equilibrium (a dynamic balance between migration and drift).
5 Demographic equality of migrants and residents (i.e., natives and migrants have an equal probability of survival reproduction).

The assumptions of this simple model are unlikely to hold in natural populations. This has led to criticism of the usefulness of mN estimates from the island model approach (Whitlock and McCauley 1999). Performance evaluations using both simulations and analytical theory suggest that the approach gives reasonable estimates of mN even when certain assumptions are violated (Slatkin and Barton 1989, Mills and Allendorf 1996).

A major limitation to estimating mN from F_{ST} (and from other methods below) is that F_{ST} must be moderate to large ($F_{ST} > 0.05$–0.10). This is because the variance in estimates of F_{ST} (and thus confidence intervals on mN estimates) is high at low F_{ST}. Confidence intervals on one mN estimate could range, for example, from less than 10 up to 1000 (depending on the number and variability of the loci used). This high variance is unfortunate because managers often need to know whether, for example, mN is 5 versus 50, because 50 would be high enough to allow recolonization and demographic rescue on an ecological timescale, whereas 5 might not. The high variance of mN estimates at low F_{ST}, along with the model assumptions, means that we often cannot interpret mN estimates literally; instead we often use mN to roughly assess the approximate magnitude of migration rates (e.g., 'high' versus 'low').

Another limitation of indirect approaches is that few natural populations are at equilibrium, primarily because many generations are required to reach equilibrium. For example, if a population becomes fragmented, but N remains large, drift will be weak. In this case, many generations are required for F_{ST} to increase to the equilibrium level. The approximate number of generations required to approach equilibrium is given by the following expression: $1/[2m + 1/(2N)]$. If N is large and m is small, the time to equilibrium is large. Thus, F_{ST} will increase slowly in large, recently isolated population fragments, and the effects of reduced gene flow will not be detectable by indirect methods until after many generations of isolation. In such a case, direct estimates of gene flow (below) are preferred, to complement the indirect estimates. For conservation genetic purposes, fragments with large N are relatively less crucial to detect because they are relatively less susceptible to rapid genetic change.

If N is small, then drift will be rapid and we might detect increased F_{ST} after only a few generations. Such a scenario of severe fragmentation is obviously the most important to detect for conservation biologists. It will also be the most likely to be detectable using an indirect (e.g., F_{ST}-based) genetic monitoring approach.

In summary, although mN estimates from F_{ST} must be interpreted with caution, they can provide useful information about gene flow and population differentiation. Nevertheless the use of different and complementary methods (several indirect plus direct methods) is recommended (Neigel 2002).

9.9.2 Private alleles and *mN*

Another indirect estimator of mN is the private alleles method (Slatkin 1985). A private allele is one found in only one population. Slatkin showed that a linear relationship exists between mN and the average frequency of private alleles. This method works because if gene flow (mN) is low, populations will have numerous private alleles that arise through mutation, for example. The time during which a new allele remains private depends only upon migration rates, such that the proportion of alleles that are private decreases as migration rate increases. If gene flow is high, private alleles will be uncommon.

This method could be less biased than the F_{ST} island model method (above), when using highly polymorphic markers, because it is apparently less sensitive to problems of homoplasy created by back mutations, than is the F_{ST} method (Allen *et al.* 1995). Homoplasy is most likely when using loci with high mutation rates and back mutation, like some microsatellites (e.g., evolving under the stepwise mutation model).

For example, Allen *et al.* (1995) studied grey seals and obtained estimates of mN of 41 using the F_{ST} method, 14 using the R_{ST} method, and 5.6 from the private allele method. The lowest mN estimate might arise from the private allele method because this method could be less sensitive to homoplasy, which causes underestimation of F_{ST} or R_{ST}, and thus overestimation of mN (Allen *et al.* 1995). The values of mN from this study must be interpreted with caution as the assumptions of the island model are probably not met, and mN values are fairly high and thus have a high variance. Furthermore, the reliability of the private alleles method has not been thoroughly investigated for loci with potential homoplasy (e.g., microsatellites).

In another study, mN estimates from allozyme markers were highly correlated with dispersal capability among ten species of ocean shore fish (Waples 1987). Three estimators of mN were compared: Nei and Chesser's F_{ST}-based method (F_{STn}), Weir and Cockerham's F_{ST}-based method (F_{STw}) and the private alleles method. The two F_{ST}-based estimators gave highly correlated estimates of mN, whereas the private alleles gave less correlated estimates. This lower correlation could result from a low incidence of private alleles in some species. These species were studied with up to 19 polymorphic allozymes with heterozygosities ranging from 0.009 to 0.087 (mean 0.031). Low polymorphism markers might be of little use with the private alleles method because very few private alleles might exist. More studies are needed comparing the performance of different mN estimators (e.g., likelihood-based methods below) and different marker types (microsatellites versus allozymes or SNPs).

9.9.3 Maximum likelihood and the coalescent

A maximum likelihood estimator of mN was published by Beerli and Felsenstein (2001). This method is promising because, unlike classical methods (above), it does not assume symmetric migration rates or identical population sizes. Furthermore, likelihood-based methods use all the data in their raw form (Section A5), rather than a single summary statistic, such as F_{ST}. The statistic F_{ST} does not use information such as the proportion of alleles that are rare. Thus, the likelihood method should give less biased and more precise estimates of mN than classic moments-based methods (Beerli and Felsenstein 2001). Indeed, a recent empirical study of garter snakes (Bittner and King 2003) suggested that coalescent methods are likely to give more reliable estimates of mN than F_{ST}-based methods, because the F_{ST}-based methods are more biased by lack of migration-drift equilibrium and changing population size.

Beerli and Felsenstein (2001) stated that "Maximum likelihood methods for estimating population parameters, as implemented in MIGRATE and GENETREE will make the classical F_{ST}-based estimators obsolete . . .". While this is likely true for some scenarios, new methods and software should be used cautiously (and in conjunction with the classical methods), at least until performance evaluations have thoroughly validated the new methods (e.g., see Section A9). One problem with evaluating the performance of the many likelihood-based methods is they are computationally slow. For example, it can take days or weeks of computing time to obtain a single mN estimate (e.g., using 10–20 loci per population). This makes the validation of methods difficult because validation requires hundreds of estimates for each of numerous simulated scenarios (i.e., different migration rates and patterns, population sizes, mutation dynamics, and sample sizes). The software program MIGRATE (Beerli 2006) for likelihood-based estimates of mN is freely available (see also GENETREE from Bahlo and Griffiths 2000, see also Hey 2010).

The **coalescent** modelling approach (a 'backward looking' strategy of simulating genealogies) is usually used in likelihood-based analysis in population genetics (see Section A10). The coalescent is useful because it provides a convenient and computationally efficient way to generate random genealogies for different gene flow patterns and rates. The efficiency of constructing coalescent trees is important, because likelihood (see Section A5) involves comparisons of enormous numbers of different genealogies in order to find those genealogies (and population models) that maximize the likelihood of the observed data. The coalescent also facilitates the extraction of genealogical information from data (e.g., divergence patterns between microsatellite alleles or DNA sequences), by easily incorporating both random drift and mutation into population models. Traditional estimators of gene flow sometimes do not use genealogical information, and are based on 'forward looking' models for which simulations are slow and probability computations are difficult.

9.9.4 Assignment tests and direct estimates of *mN*

Direct estimates of migration (mN) can be obtained by directly observing migrants moving between populations. Direct estimates of mN have traditionally been obtained by marking many individuals after birth and following them until they reproduce, or by tracking pollen dispersal by looking for the spread of rare alleles or morphological mutants in seeds or seedlings. The number of dispersers that breed in a new (non-natal) population then becomes the estimated mN.

An advantage of direct estimates is that they detect migration patterns of the current generation without the assumption of population equilibrium (migration–mutation–drift equilibrium). This allows up-to-date monitoring of movement and more reliable detection of population fragmentation (reduced dispersal) without waiting for populations to approach equilibrium (see above).

An important limitation of direct estimates is that they might not detect pulses of migrants that can occur only every 5–10 years, as in species where dispersal is driven by cyclical population demography or periodic weather conditions. Unlike direct estimates, indirect estimates of mN estimate the average gene flow over many generations and thus will incorporate effects of pulse migration. For example, 10 migrants every 10 generations will have the same impact on indirect mN estimates as one migrant per generation for 10 generations.

Another limitation of direct estimates is that they often cannot estimate rates of 'evolutionarily effective' gene flow. Direct estimates of mN only assume that an observed migrant will reproduce and pass on genes with the same probability as a local resident individual. However, migrants might have a reduced mating success if they cannot obtain a local territory, for example. Alternatively, migrants might have exceptionally high mating success if there is a 'rare male' or 'foreign individual' advantage. Furthermore, immigrants could produce offspring more fit than local individuals if heterosis occurs following crossbreeding between immigrants and residents. Heterosis can lead to more gene flow than expected from neutral theory, for any given number of migrants (see 'genetic rescue', Section 15.5). Because direct observation of migrants generally does not detect local mating success (effective gene flow), direct observations generally only estimate **dispersal** and not gene flow (i.e., migration), unless we assume that observed migrants reproduce.

Unfortunately, direct estimates of mN are difficult to obtain using traditional field methods of capture–mark–recapture. Following individuals from their birthplace until reproduction is extremely difficult or impossible for many species.

Assignment tests offer an attractive alternative to the traditional capture–mark–recapture approach to making direct estimates of mN (Wilson and Rannala

2003). For example, we can genotype many individuals in a single population sample, and then determine the proportion of 'immigrant' individuals (i.e., individuals with a foreign genotype) that is unlikely to have originated locally. For example, a study of the inanga revealed that one individual sampled in New Zealand had an extremely divergent mtDNA haplotype, which was very similar to the haplotypes found in Tasmania (Figure 9.16). It is likely that the individual (or one of its maternal ancestors) originated in Tasmania and

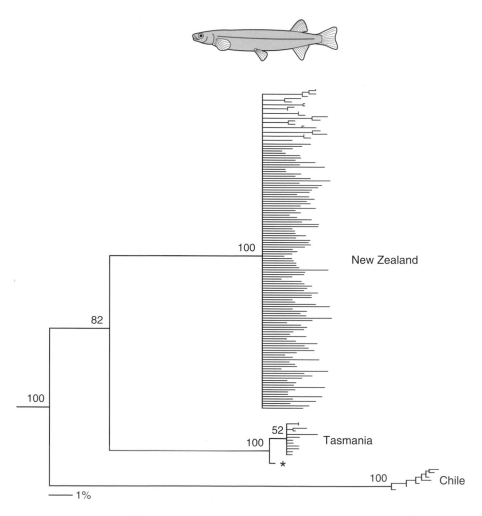

Figure 9.16 Detection of a migrant between populations of inanga using a phylogram derived from mtDNA control region sequences. One mtDNA type (marked with a star) sampled in New Zealand was very similar to the mtDNA types found in Tasmania. This suggests that a small amount of gene flow occurs between the New Zealand populations and Tasmania. From Waters *et al.* (2000).

migrated to New Zealand. The inanga spawns in fresh-water, but spends part of its life history in the ocean.

One problem with using only mtDNA is that we cannot estimate male-mediated migration rates (because mtDNA is maternally inherited). Further, the actual migrant could have been the mother or grand-mother of the individual sampled. We could test whether the migrant or its mother was the actual immigrant by genotyping many autosomal markers (e.g., microsatellites). For example, if a parent was the migrant then only half of the individual's genome (alleles) would have originated from another popula-tion. We can estimate the proportion of an individual's genome arising from each of two parental populations via admixture analysis (see Example 22.5).

Assignment tests based on multiple autosomal makers are useful for identifying immigrants. For example, for a candidate immigrant, we first remove the individual from the dataset and then compute the expected frequency of its genotype (p^2) in each candi-date population of origin by using the observed allele frequencies (p) from each population (Figure 9.17). If the likelihood for one population is far higher than the other, we 'assign' the individual to the most likely pop-ulation. The likelihood can be computed as the fre-quency of the genotype in the population (expected under HW proportions). Computing the multilocus assignment likelihood requires multiplication together of single-locus probabilities (multiplication rule), and thus requires the assumption of independence among loci (e.g., no gametic disequilibrium).

The power of assignment tests increases with the amount of differentiation among subpopulations. There-fore, outlier loci with high differentiation (Example 9.6) can be extremely valuable for individual assign-ment (Hansen *et al.* 2007, Ackerman *et al.* 2011) For example, Karlsson *et al.* (2011) were interested in developing genetic markers to detect potentially harm-ful introgression from farmed Atlantic salmon into wild populations in Norway. They found very low overall genetic divergence ($F_{ST} = 0.016$) at 4514 SNP loci. However, they identified a set of 200 SNP loci having much higher F_{ST} values (0.094), apparently due to domestication selection in the farmed fish. They then developed a panel of 60 SNPs that collectively are diagnostic in identifying individual salmon as being farmed or wild, regardless of their populations of origin.

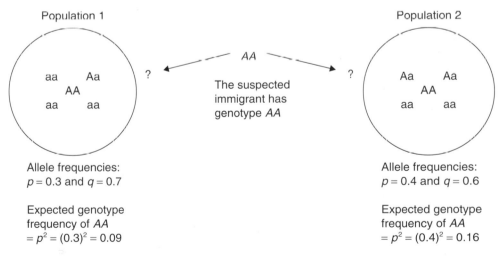

Figure 9.17 Simplified example of using an assignment test to identify an immigrant (*AA*). We first remove the individual in question from the dataset and then compute its expected genotype frequency (p^2) in each population using the observed allele frequencies for each population (p^1 and p^2, respectively), and assume Hardy-Weinberg proportions. If the individual with the genotype *AA* was captured in Population 1 but its expected genotype frequency is far higher in Population 2, then we could conclude the individual is an immigrant. The beauty of assignment tests is that they are relatively simple but potentially powerful if many loci (each with many alleles) are used. Note that obtaining the multilocus likelihoods generally requires multiplication of single locus probabilities (multiplication rule), and thus requires independent loci in gametic equilibrium.

9.10 POPULATION SUBDIVISION AND CONSERVATION

Understanding the genetic population structure of species is essential for conservation and management (see Guest Box 9). The techniques to study genetic variation and the genetic models that we have presented in this chapter allow us to understand rather quickly the genetic population structure of any species of interest. Understanding the amount of genetic differentiation among populations is crucial when developing a captive breeding program (see Chapter 19) or selecting individuals to be moved among populations for either demographic or genetic 'rescue' (see Chapter 15).

An understanding of spatial genetic structure (SGS) within populations can provide important information for the conservation and management of populations. Data on SGS can be used to estimate indirectly those dispersal rates and neighborhood sizes (e.g., Fenster *et al.* 2003) that are too laborious and time-consuming to estimate directly though tracking dispersal. For example, Solmsen *et al.* (2011) found strong SGS along 7 km of a dry riverbed in female African striped mice, but low SGS in males of the same species. From these data, they concluded that males disperse farther than females, and that males of lower fitness (lower body weight) disperse farther than larger males. Information about SGS can inform reserve design. If a population has strong SGS and if only a portion of a population is conserved, more genetic variation will be lost than for a population with no SGS, and variation distributed randomly across the population.

The application of genetic population structure of species, however, is often not straightforward and is sometimes controversial. For example, how 'distinct' does a population have to be to be considered a **distinct population segment**, in order to be listed under the US ESA (see Guest Box 9)? The application of genetic information to identify appropriate units for conservation and management is considered in detail in Chapter 16.

Population subdivision influences the evolutionary potential of a species, that is, the ability of a species to evolve and adapt to environmental change. To understand this, it is helpful to consider extremes of subdivision. For example, a species with no subdivision would have such high gene flow that local adaptation would not be possible (left side of Figure 9.18). Thus, the total range of types of multilocus genotypes would be limited. On the other hand, if subdivision is extreme then new beneficial mutations that arise will not readily spread across the species. Furthermore, subpopulations may be so small that genetic drift overwhelms natural selection. Thus local adaptation is limited and random change in allele frequencies dominates, so that harmful alleles may drift to high frequency or go to fixation. An intermediate amount of population subdivision will result in substantial genetic variation both within and between local populations; this population structure has the greatest evolutionary potential.

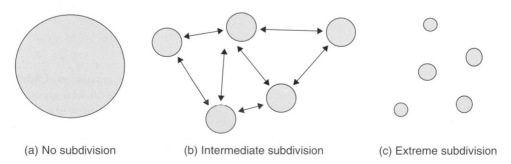

(a) No subdivision   (b) Intermediate subdivision   (c) Extreme subdivision

Figure 9.18   Range of possible degrees of population subdivision. Intermediate degrees of subdivision (b) generally yield the highest adaptive potential with possibilities for local adaption to local environments, yet with occasional gene flow and large enough local effective size to prevent rapid inbreeding and loss of variation.

Guest Box 9   Genetic population structure and conservation of fisher in Western North America
M.K. Schwartz and J.M. Tucker

The fisher is a medium-sized carnivore endemic to North America's northern forests. They are a dark brown mustelid with their head and shoulders appearing grizzled, and their chests often sporting white patches. It is this dense and warm coat, highly valued in the fur trade, along with the relative ease of trapping the species either directly or incidentally, that was partially responsible for their massive historic range contraction (Powell 1993). Fishers prefer forests that are diverse, with large diameter trees and ample structure. This includes stands with dead and decaying trees where females can establish their natal dens. Conversion of these forests though urbanization and forestry has been detrimental to many populations. Recently, due to habitat regeneration and improved trapping management, populations of fishers in the Eastern and Midwestern United States have recovered, but those in the Western US are still small and disjointed (Aubry and Lewis 2003, Vinkey *et al.* 2006). Here, we describe how population genetic substructure analyses have been used to understand the natural history and provide information for conservation of west coast fisher populations.

At the beginning of the 20th century, fishers were distributed throughout the mountainous region of the West Coast (Powell 1993). By the middle of the century, it was recognized that fishers in Oregon were gone, those in Washington were quickly disappearing or already absent, and the largest remaining population was in California (Aubry and Lewis 2003). This led to the reintroduction of fishers from British Columbia and the Midwestern US to Oregon in the 1950s, and the reintroduction of British Columbia fishers to the Olympic National Park, Washington, in 2008 (Aubry and Lewis 2003).

There is considerable genetic differentiation between West Coast fishers and other fisher populations in North America (Drew *et al.* 2003). The West Coast group shares only one of five mtDNA haplotypes at the control region with its nearest population in the Rocky Mountains. These two regions are also highly divergent at nuclear microsatellite loci (Schwartz, unpublished data).

Historical descriptions considered that fisher in California (CA) were a single connected population (Grinnell *et al.* 1937). Now, however, there are northern (CA-N) and southern (CA-S) populations that are separated by a large geographic gap (Figure 9.19). Genetic analyses have been used to test whether this isolation between CA-N and CA-S is recent, due to loss of habitat and population decline, or whether the populations were distinct before the effects of humans. Recent sequencing of the entire mtDNA genome found the CA-S population is fixed for a single haplotype that is distinct from the most similar haplotype in the CA-N population by nine substitutions (Knaus *et al.* 2011). These authors estimated that these populations have been isolated for thousands, rather than hundreds, of years, suggesting that there has been a break in the distribution of California fishers that preceded any land conversion associated with European settlement, which only occurred in the last hundreds of years.

Tucker *et al.* (in review) found similar results by comparing genetic divergence at nuclear loci between CA-N and CA-S in historic and contemporary samples. The F_{ST} between these contemporary populations was 0.374 at 10 microsatellite loci. Using a coalescent-based Bayesian analysis (see Section A9), Tucker *et al.* (in review) detected a 90% decline in fisher effective population size and dated the time of this decline at over a thousand years ago, consistent with the timing of the CA-N and CA-S divergence estimated from the whole mtDNA genome analysis.

Genetic data were a major part of the decision that found the West Coast population of fisher to be a distinct population segment (DPS) under the US Endangered Species Act (USFWS 2004; see Section 9.9 and Box 16.1). Genetic data were used to conclude both that the West Coast populations were (1) discrete and significant because "there is no naturally occurring genetic interchange" between West Coast and other fisher populations, and (2) "the

(Continued)

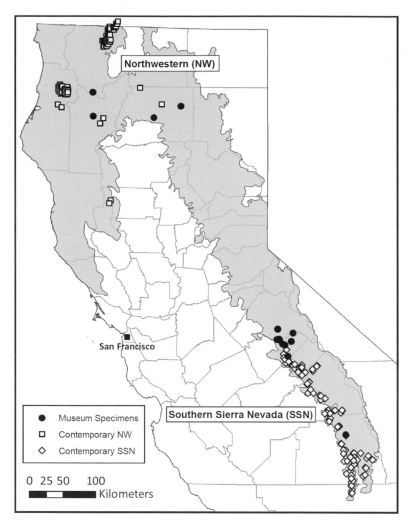

Figure 9.19 Locations of historic and contemporary samples of fisher from northern and southern California. The assumed historic range is shaded. From Tucker *et al.* (in review).

extinction of fishers in their West Coast range would also result in the loss of a significant genetic entity". In light of the recent genetic data presented here, the two California fisher populations potentially could be considered distinct population segments under the US ESA. These results also raise the issue of the possible effects of translocating individuals between these populations as part of a recovery plan. The high genetic divergence suggests that genetic exchange between these populations could reduce fitness and result in outbreeding depression (see Section 17.3).

CHAPTER 10

Multiple loci

Field cricket, Example 10.3

It is now generally understood that, as a consequence of selection, random genetic drift, co-ancestry, or gene flow, alleles at different loci may not be randomly associated with each other in a population. While this effect is generally regarded as a consequence of linkage, even genes on different chromosomes may be held temporarily or permanently out of random association by forces of selection, drift and nonrandom mating.

Richard C. Lewontin (1988)

Population geneticists recently have devoted much attention to the topic of gametic disequilibrium. The analysis of multiple-locus genotypic distributions can provide a sensitive measure of selection, genetic drift, and other factors that influence the genetic structure of populations.

David W. Foltz et al. (1982)

Chapter Contents

Conservation and the Genetics of Populations, Second Edition. Fred W. Allendorf, Gordon Luikart, and Sally N. Aitken.
© 2013 Fred W. Allendorf, Gordon Luikart and Sally N. Aitken. Published 2013 by Blackwell Publishing Ltd.

We have so far considered loci one at a time. Population genetic models become much more complex when two or more loci are considered simultaneously. Fortunately, many of our genetic concerns in conservation can be dealt with from the perspective of individual loci. Nevertheless, there are a variety of situations in which we must concern ourselves with the interactions between multiple loci. For example, genetic drift in small populations can cause nonrandom associations between loci to develop. Therefore, the consideration of multilocus genotypes can provide another method of detecting the effects of genetic drift in natural populations (Slatkin 2008). Recent papers describing genotypes at hundreds (Li and Merilä 2010) or thousands (Hohenlohe *et al.* 2010a) of loci in nonmodel organisms make it more important than ever to understand the interpretation of multilocus genotypes.

In addition, the genotype of individuals over many loci can be used to identify individuals genetically because the genotype of each individual (with the exception of identical twins or clones) is genetically unique if enough loci are considered. This genetic 'fingerprinting' capability has many potential applications in understanding populations, estimating population size (Chapter 14), and in applying genetics to problems in **forensics** (Chapter 22).

The nomenclature of multilocus genotypes is particularly messy and often inconsistent. It is difficult to find any two papers (even by the same author!) that use the same gene symbols and nomenclature for multilocus genotypes. Therefore, we have made a special effort to use the simplest possible nomenclature and symbols that are as consistent as possible with previous usage in the literature.

The term **linkage disequilibrium** is commonly used to describe the nonrandom association between alleles at two loci (Lewontin and Kojima 1960). However, this term is misleading because unlinked loci can be in so-called 'linkage disequilibrium'. Things are complicated enough without using misnomers that lead to additional confusion when considering multilocus models. The term **gametic disequilibrium** is a much more descriptive and appropriate term to use in this situation. We have chosen to use gametic disequilibrium in order to reduce confusion.

We will first examine general models describing associations between loci and their evolutionary dynamics from generation to generation. We will then explore the various evolutionary forces that cause nonrandom associations between loci to come about in natural populations (genetic drift, natural selection, population subdivision, and hybridization). Finally, we will compare various methods for estimating associations between loci in natural populations.

10.1 GAMETIC DISEQUILIBRIUM

We now focus our interest on the behavior of two autosomal loci considered simultaneously under all of our Hardy-Weinberg equilibrium assumptions. We know that each locus individually will reach a neutral equilibrium in one generation under Hardy-Weinberg conditions. Is this true for two loci considered jointly? We will see shortly that the answer is no.

Loci on different chromosomes will be unlinked so that heterozygotes at both loci (*AaBb*) will produce all four gametes (*AB*, *Ab*, *aB*, and *bb*) in equal frequencies ($r = 0.5$). Two loci that are close together on the same chromosome are generally linked so that the frequency of the parental gamete types (*AB* and *ab* in Figure 10.1) will be greater than the frequency of the nonparental gametes ($r < 0.5$). Some loci on the same chromosome can be far enough apart so that there is enough recombination to produce equal frequencies of all four gametes, such that they are unlinked ($r = 0.5$). Two loci that are on the same chromosome are **syntenic**, whether they are linked ($r < 0.5$) or unlinked ($r = 0.5$).

Allele frequencies are insufficient to describe genetic variation at multiple loci. Fortunately, however, we do not have to keep track of all possible genotypes. Rather we can use the gamete frequencies to describe nonrandom associations between alleles at different loci. For example, in the case of two loci that each has two alleles, there are just two allele frequencies, but there are nine different genotype frequencies. However, we can describe this system with just four gamete frequencies.

Let G_1, G_2, G_3, and G_4 be the frequencies of the four gametes *AB*, *Ab*, *aB*, and *ab* respectively, as shown below. If the alleles at these loci are associated randomly, then the expected frequency of any gamete type will be the product of the frequencies of its two alleles:

Gamete	Frequency	
AB	$G_1 = (p_1)(p_2)$	
Ab	$G_2 = (p_1)(q_2)$	
aB	$G_3 = (q_1)(p_2)$	(10.1)
ab	$G_4 = (q_1)(q_2)$	

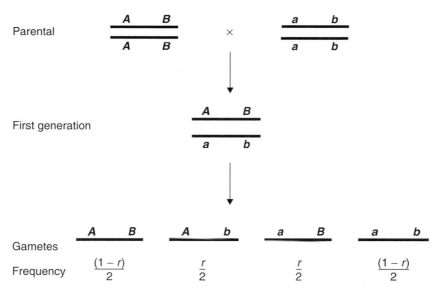

Figure 10.1 Outline of gamete formation in first-generation (F_1) hybrids between two parents homozygous for different alleles at two loci. The gametes produced by the F_1 hybrids are affected by the rate of recombination (r). These four gametes will be equally frequent (25% each) for unlinked loci ($r = 0.5$). There will be an excess of parental gametes (AB and ab in this case) if the loci are linked ($r < 0.5$).

Table 10.1 Genotypic array for two loci showing the expected genotypic frequencies in a random mating population.

	AA	**Aa**	**aa**
BB	G_1^2	$2G_1G_3$	G_3^2
Bb	$2G_1G_2$	$2G_1G_4 + 2G_2G_3$	$2G_3G_4$
bb	G_2^2	$2G_2G_4$	G_4^2

where (p_1; q_1) and (p_2; q_2) are frequencies of the alleles (A; a) and (B; b), at loci 1 and 2, respectively. The expected frequencies of two locus genotypes in a random mating population can then be found as shown in Table 10.1.

D is used as a measure of the deviation from random association between alleles at the two loci (Lewontin and Kojima 1960). D is known as the coefficient of gametic disequilibrium and is defined as:

$$D = (G_1G_4) - (G_2G_3) \qquad (10.2)$$

or

$$D = G_1 - p_1p_2 \qquad (10.3)$$

If alleles are associated at random in the gametes (as in expression 10.1), then the population is in gametic equilibrium and $D = 0$. If D is not equal to zero, the alleles at the two loci are not associated at random with respect to each other, and the population is said to be in gametic disequilibrium (Example 10.1). For example, if a population consists only of a 50:50 mixture of the gametes A_1B_1 and A_2B_2, then:

$G_1 = 0.5$

$G_2 = 0.0$

$G_3 = 0.0$

$G_4 = 0.5$

and

$$D = (0.5)(0.5) - (0.0)(0.0) = +0.25$$

The amount of gametic disequilibrium (i.e., the value of D) will decay from generation to generation as a function of the rate of recombination (r) between the two loci:

$$D' = D(1-r) \qquad (10.4)$$

So after t generations:

Example 10.1   Genotypic frequencies with and without gametic disequilibrium

Let us consider two loci at which allele frequencies are $p_1 = 0.4$ $(q_1 = 1 - p_1 = 0.6)$ and $p_2 = 0.7$ $(q_2 = 1 - p_2 = 0.3)$ in two populations. The two loci are randomly associated in one population, but show maximum nonrandom association in the other. The gametic frequency values below show the case of random association of alleles at the two loci (gametic equilibrium, $D = 0$) and the case of maximum positive disequilibrium ($D = +0.12$; see Section 10.1.1 for an explanation of the maximum value of D).

Gamete	$D = 0$	D(max)
$A\ B$	$(p_1)(p_2) = 0.28$	0.40
$A\ b$	$(p_1)(q_2) = 0.12$	0.00
$a\ B$	$(q_1)(p_2) = 0.42$	0.30
$a\ b$	$(q_1)(q_2) = 0.18$	0.30

In a random mating population, the following genotypic frequencies will result in each case as shown

below. The expected genotypic frequencies with $D = 0$ are shown without brackets, and the expected genotypic frequencies with maximum positive gametic disequilibrium are shown in square brackets:

	AA	**Aa**	**aa**	**Total**
BB	0.08	0.24	0.18	0.49
	[0.16]	[0.24]	[0.09]	[0.49]
Bb	0.07	0.20	0.15	0.42
	[0]	[0.24]	[0.18]	[0.42]
bb	0.01	0.04	0.03	0.09
	[0]	[0]	[0.09]	[0.09]
Total	0.16	0.48	0.36	
	[0.16]	[0.48]	[0.36]	

Notice that each locus is in Hardy-Weinberg proportions in the populations either with or without gametic disequilibrium.

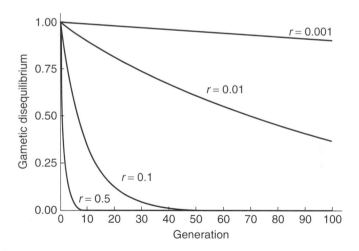

Figure 10.2   Decay of gametic disequilibrium (D_t/D_o) with time for various amounts of recombination (r) between the loci from equation 10.5.

$$D_{t'} = D_0(1-r)^t \tag{10.5}$$

If the two loci are not linked (i.e., $r = 0.5$), the value of D_t will be halved each generation until equilibrium is reached at $D = 0$. **Linkage** ($r < 0.5$) will delay the rate of decay of gametic disequilibrium. Nevertheless, D will

eventually be equal to zero, as long as there is some recombination ($r > 0.0$) between the loci. However, if the two loci are tightly linked, it will take many generations for them to reach gametic equilibrium (Figure 10.2).

We therefore expect that nonrandom associations of genotypes between loci (i.e., gametic disequilibrium) would be much more frequent between tightly linked loci. For example, Zapata and Alvarez (1992) summarized observed estimates of gametic disequilibrium between five allozyme loci in several natural populations of *Drosophila melanogaster* on the second chromosome. The effective frequency of recombination is the mean of recombination rates in females and males, assuming no recombination in males. Only pairs of loci with less than 15% recombination showed consistent evidence of gametic disequilibrium. In contrast, recent studies of species of conservation interest have found much greater gametic disequilibrium, even between loci on different chromosomes (see Section 10.7).

10.1.1 Other measures of gametic disequilibrium

D is a poor measure of the relative amount of disequilibrium at different pairs of loci because the possible values of D are constrained by allele frequencies at both loci. The largest possible positive value of D is either p_1q_2 or p_2q_1, whichever value is smaller; and the largest negative value of D is the lesser value of p_1p_2 or q_1q_2. We can see that the largest positive value of D occurs when G_1 is maximum. p_1 is equal to G_1 plus G_2, and p_2 is equal to G_1 plus G_3, therefore, the largest possible value of G_1 is the smaller of p_1 and p_2. We can see this in Example 10.1 in which the largest positive value of D occurs when G_1 is equal to p_1 which is less than p_2. Once the values of G_1, p_1, and p_2 are set, all of the other gamete frequencies must follow.

This allele frequency constraint of D reduces its value for comparing the amount of gametic equilibrium for the same loci in different populations, or for different pairs of loci in the same population. For example, consider two pairs of loci in complete gametic disequilibrium. In case 1 both loci are at allele frequencies of 0.5, while in case 2 both loci are at frequencies of 0.8. The following gamete frequencies result:

	Frequencies	
Gamete	Case 1	Case 2
AB	0.5	0.9
Ab	0.0	0.0
aB	0.0	0.0
ab	0.5	0.1

The value of D in case 1 will be +0.25, while it will be +0.09 in case 2.

Several other measures of gametic disequilibrium have been proposed that are useful for various purposes (Hedrick 1987). A useful measure of gametic disequilibrium should have the same range regardless of allele frequencies. This will allow comparing the amount of disequilibrium among pairs of loci with different allele frequencies.

Lewontin (1964) suggested using the parameter D' to circumvent the problem of the range of values being dependent upon the allele frequencies:

$$D' = \frac{D}{D_{max}} \qquad (10.6)$$

Thus, D' ranges from 0 to 1 for all allele frequencies. However, even D' is not independent of allele frequencies, and, therefore, is not an ideal measure of gametic disequilibrium (Lewontin 1988). Nevertheless, Zapata (2000) has concluded that the D' coefficient is a useful tool for the estimation and comparison of the extent of overall disequilibrium among many pairs of multiallelic loci.

The correlation coefficient (R) between alleles at the two loci has also been used to measure gametic disequilibrium:

$$R = \frac{D}{(p_1q_1p_2q_2)^{1/2}} \qquad (10.7)$$

R has a range of values between −1.0 and +1.0. However, this range is reduced somewhat if the two loci have different allele frequencies. Both D' and R will decay from generation to generation by a rate of $(1 - r)$, as does D, because they are both functions of D.

10.1.2 Associations between cytoplasmic and nuclear genes

Just as with multiple nuclear genes, nonrandom associations between nuclear loci and mtDNA genotypes may occur in populations.

Gamete	Frequency	
AM	G_1	
Am	G_2	(10.8)
aM	G_3	
am	G_4	

Again, D is a measure of the amount of gametic equilibrium and is defined as in expression 10.2. D between nuclear and cytoplasmic genes will decay at a rate of one-half per generation, just as for two unlinked nuclear genes. That is:

$$D' = D(0.5) \tag{10.9}$$

and, therefore,

$$D_t = D(0.5)^t \tag{10.10}$$

10.2 SMALL POPULATION SIZE

Nonrandom associations between loci will be generated by genetic drift in small populations. We can see this readily in the extreme case of a bottleneck of a single individual capable of reproducing by selfing because a maximum of two gamete types can occur within a single individual. Conceptually, we can imagine the four gamete frequencies to be analogous to four alleles at a single locus. Changes in gamete frequencies from generation to generation caused by drift will often result in nonrandom associations between alleles at different loci. The expected value of D due to drift is zero. Nevertheless, drift-generated gametic disequilibria may be very great and are equally likely to be positive or negative in sign. For example, genome-wide investigations in humans have found that large blocks of gametic disequilibrium occur throughout the genome in human populations. These blocks of disequilibrium are thought to have arisen during an extreme population bottleneck that occurred some 25,000 to 50,000 years ago (Reich *et al.* 2001).

Gametic disequilibrium produced by a single generation of drift may take many generations to decay. Therefore, we would expect substantially more drift-generated gametic disequilibrium between closely linked loci. In fact, the expected amount of disequilibrium for closely linked loci is:

$$E(R^2) \approx \frac{1}{1+4Nr} \tag{10.11}$$

where R^2 is the square of the correlation coefficient (R) between alleles at the two loci (equation 10.7) (Hill and Robertson 1968, Ohta and Kimura 1969). For unlinked loci, the following value of R^2 is expected (Weir and Hill 1980):

$$E(R^2) \approx \frac{1}{3N} \tag{10.12}$$

Guest Box 10 discusses the use of genotype frequencies at many loci and equation 10.12 to estimate effective size in natural populations.

10.3 NATURAL SELECTION

Let us examine the effects of natural selection with constant fitnesses at two loci each with two alleles. We will designate the fitness of a genotype to be w_{ij}, where i and j are the two gametes that join to form a particular genotype. There are two genotypes that are heterozygous at both loci (AB/ab and Ab/aB); we will assume that both double heterozygotes have the same fitness (i.e., $w_{23} = w_{14}$).

	AA	Aa	aa
BB	w_{11}	w_{13}	w_{33}
Bb	w_{12}	$w_{23} = w_{14}$	w_{34}
bb	w_{22}	w_{24}	w_{44}

The frequency of the AB gamete after one generation of selection will be:

$$G_{1'} = \frac{G_1(G_1 w_{11} + G_2 w_{12} + G_3 w_{13} + G_4 w_{14}) - r w_{14} D}{\bar{w}} \tag{10.13}$$

where $\bar{w}$ is the average fitness of the population. We can simplify this expression by defining $\bar{w}_i$ to be the average fitness of the ith gamete.

$$\bar{w}_i = \sum_{j=1}^{4} G_j w_{ij} \tag{10.14}$$

and then

$$\bar{w} = \sum_{i=1}^{4} G_i \bar{w}_i \tag{10.15}$$

and

$$G_{1'} = \frac{G_1 \bar{w}_1 - r w_{14} D}{\bar{w}} \tag{10.16}$$

We can derive similar recursion equations for the other gamete frequencies.

$$G_{2'} = \frac{G_2 \bar{w}_2 + r w_{14} D}{\bar{w}}$$

$$G_{3'} = \frac{G_3 \bar{w}_3 + r w_{14} D}{\bar{w}} \qquad (10.17)$$

$$G_{4'} = \frac{G_4 \bar{w}_4 - r w_{14} D}{\bar{w}}$$

	A_1A_1	A_1A_2	A_2A_2
B_1B_1	w_{11}	w_{11}	w_{11}
B_1B_2	w_{12}	w_{12}	w_{12}
B_2B_2	w_{22}	w_{22}	w_{22}

where $w_{11} < w_{12} < w_{22}$.

Imagine that the favored B_2 allele is a new mutation at the B locus. In this case, the selective advantage of the B_2 allele frequencies may carry along either the A_1 or A_2 allele, depending upon which allele is initially associated with the B_2 mutation. This is known as **genetic hitchhiking** and will result in a so-called **selective sweep**. The magnitude of this effect depends upon the selection differential, the amount of recombination (r), and the initial gametic array (Figure 10.3). A selective sweep will reduce the amount of variation at loci that are tightly linked to the locus under selection.

For example, low-activity alleles at the glucose-6-phosphate dehydrogenase locus in humans are thought to reduce risk from the parasite responsible for causing malaria (Tishkoff *et al.* 2001). The pattern of gametic disequilibrium between these alleles and closely linked microsatellite loci suggests that these alleles have increased rapidly in frequency by natural selection since the onset of agriculture in the past 10,000 years.

There are no general solutions for selection at two loci. That is, there is no simple formula for the equilibria and their stability. However, a number of specific models of selection have been analyzed. The simplest of these is the additive model where the fitness effects of the two loci are summed to yield the two-locus fitnesses. Another simple case is the multiplicative model where the two-locus fitnesses are determined by the product of the individual locus fitnesses. In both of these cases, heterozygous advantage at each locus is necessary and sufficient to insure stable polymorphisms at both loci.

In some cases, the multilocus fitness cannot be predicted by either the additive or multiplicative combination of fitnesses at individual loci (Phillips 2008). Such interaction between loci is referred to as **epistasis** (i.e., the interaction of different loci such that the multiple locus phenotype is different from that predicted by simply combining the effects of each individual locus). The study of epistasis, or interactions between genes, is fundamentally important to understanding the structure and function of genetic pathways and the evolutionary dynamics of complex genetic systems.

A detailed examination of the effects of natural selection at two loci, including epistasis, is beyond the scope of our consideration. Interested readers are directed to appropriate population genetics sources (e.g., Hartl and Clark 1997, Phillips 2008, Hedrick 2011). We will consider two situations of selection at multiple loci that are particularly relevant for conservation.

10.3.1 Genetic hitchhiking

Natural selection at one locus can affect closely linked loci in many ways. Let us first consider the case where directional selection occurs at one locus (B) and the second locus is selectively neutral (A). The following fitness set results:

10.3.2 Associative overdominance

Selection at one locus can also affect closely linked neutral loci when the genotypes at the selected locus are at an equilibrium allele frequency. Consider the case of heterozygous advantage where, using the previous fitness array, $w_{11} = w_{22} = 1.0$ and $w_{12} = (1 + s)$. The effective fitnesses at the A locus are affected by selection at the B locus (s) and D; the marginal fitnesses are the average fitness at the A locus considering the two-locus genotypes. These would be the estimated fitnesses at the A locus if only that locus were observed. If D is zero then all the genotypes at the A locus will have the same fitness. However, if there is gametic disequilibrium (i.e., D is not equal to zero), then heterozygotes at the A locus will experience a selective advantage because of selection at the B locus.

This effect has been called **associative overdominance** (Ohta 1971) or **pseudo-overdominance** (Carr and Dudash 2003). This pattern of selection has also been called marginal overdominance (Hastings 1981). However, **marginal overdominance** has more generally been used for the situation where genotypes experience multiple environments and different alleles are favored in different environments (Wallace 1968). This situation can lead to an overall greater fitness of heterozygotes, even though they do not have a greater fitness in any single environment.

Heterozygous advantage is not necessary for linked loci to experience associative overdominance. Heterozygous individuals at a selectively neutral locus will have higher average fitnesses than homozygotes if the locus is in gametic disequilibrium with a locus having deleterious recessive alleles (Ohta 1971).

We can see this with the genotypic arrays in Example 10.1. Let us assume that the b allele is a recessive lethal (i.e., fitness of the bb genotype is zero). In the case of gametic equilibrium ($D = 0$), exactly q_2^2 ($0.3 \times 0.3 = 0.09$) of genotypes at the A locus have a

fitness of zero. Thus, the mean or marginal fitness at the A locus is $1 - 0.09 = 0.91$. However, in the case of maximum positive disequilibrium, only the aa genotypes have reduced fitness because the AA and Aa genotypes do not occur in association with the bb genotype. Thus, the fitness of AA, Aa, and aa are 1, 1, and 0.75. There are many more aa than AA homozygotes in the population; therefore, Aa heterozygotes have greater fitness than the mean of the homozygotes.

Associative overdominance is one possible explanation for the pattern seen in many species in which individuals that are more heterozygous at many loci have greater fitness. Example 10.2 presents an example where associative overdominance is most likely responsible for **heterozygosity-fitness correlations** (HFCs) in great reed warblers.

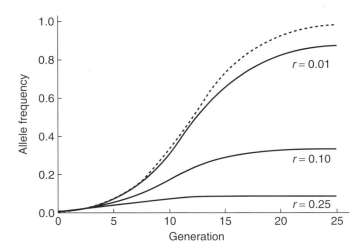

Figure 10.3 Effect of hitchhiking on a neutral locus that is initially in complete gametic disequilibrium with a linked locus that is undergoing directional selection ($w_{11} = 1.0$; $w_{12} = 0.75$; $w_{22} = 0.5$). r is the amount of recombination between the two loci. The dashed line shows the expected change at the selected locus.

Example 10.2   Associative overdominance explains heterozygosity-fitness correlations in great reed warblers

Individuals that are more heterozygous at many loci have been found to have greater fitness in many species (Hansson and Westerberg 2002, Szulkin et al. 2010). Such heterozygosity-fitness correlations (HFCs) have three possible primary explanations. First, the association may be a consequence of differences in inbreeding among individuals within a population. Inbred individuals will tend to be less heterozygous and experience inbreeding depression. Second, the

loci being scored may be in gametic disequilibrium with loci that affect the traits being studied, resulting in associative overdominance. Lastly, the associations may be due to heterozygous advantage at the loci being studied. This latter explanation seems unlikely for loci such as microsatellites that are generally assumed to be selectively neutral. There is some evidence that HFC at allozyme loci might be due to the loci themselves (Thelen and Allendorf 2001).

Hansson *et al.* (2004) distinguished between inbreeding and associative overdominance in great reed warblers by testing for HFC within pairs of siblings with the same pedigree. This comparison eliminated the reduced genome-wide heterozygosity of inbred individuals as an explanation, because full siblings have the same pedigree inbreeding coefficient (*F*). Fifty pairs of siblings were compared in which only one individual survived to adult age. Paired siblings were confirmed to have the same genetic parents (by molecular methods) and were matched for sex, size (length of the innermost primary feather), and body mass (when nine days old).

The surviving sib tended to have greater multilocus heterozygosity at 19 microsatellite loci (Figure 10.4; $P < 0.05$). In addition, the surviving sibs also had significantly greater d^2 values ($P < 0.01$). This measure is the squared difference in number of repeat units between the two alleles in a heterozygous individual, d^2 = (number of repeats at allele *A* − number of repeats at allele *B*)^2. The difference in repeat score between alleles carries information about the amount of time that has passed since they shared a common ancestral allele (see the coalescent in Section A10 of the Appendix). This assumes a single-step model of mutation. Heterozygotes with smaller values possess two alleles that are likely to have shared a common ancestral allele more recently than heterozygotes with larger d^2 values (see Figure 12.1). Therefore, heterozygotes with lower d^2 possess two alleles marking chromosomal segments that are more likely to carry the same deleterious recessive allele responsible for associative overdominance. The strong relationship between d^2 and recruitment suggests that associative overdominance is responsible for the observed HFC.

The studied population of great reed warblers was small and recently founded. Thirty-five of 162 pairwise tests for gametic disequilibrium were significant (uncorrected for multiple tests, $P < 0.05$), suggesting widespread gametic disequilibrium in this population because of its recent founding and small size (see Section 10.2). These authors conclude that associative overdominance is likely to be responsible for the HFC that they have observed. They also argue

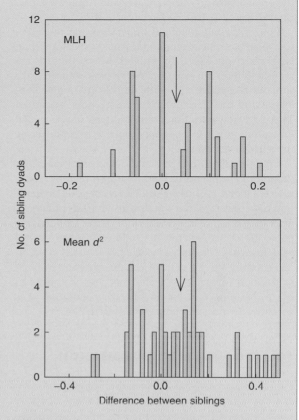

Figure 10.4 Difference between surviving and nonsurviving great reed warbler siblings (50 matched pairs) in multilocus heterozygosity (MLH; $P < 0.05$) and mean d^2 ($P < 0.01$). Arrows indicate mean difference. The greater MLH and d^2 of surviving birds apparently results from associative overdominance. From Hansson *et al.* (2004).

that gametic disequilibrium is likely to be responsible for many observations of HFC in other species, especially in cases of recently founded or small populations.

10.3.3 Genetic draft

We saw in Section 10.3.1 that directional selection at one locus can reduce the amount of genetic variation at closely linked loci following a selective sweep. This is a special case of a more general effect in which selection at one locus will reduce the effective population size of linked loci. This has been termed the **Hill-Robertson effect** (Hey 2000) because it was first discussed in a paper that considered the effect of linkage between two loci under selection (Hill and Robertson 1966). Observations with *Drosophila* have found that regions of the genome with less recombination tend to be less genetically variable, as would be expected with the Hill-Robertson effect (Begun and Aquadro 1992, Charlesworth 1996).

This effect has potential importance for conservation genetics. For example, we would expect a strong Hill-Robertson effect for mtDNA where there is no recombination. A selective sweep of a mutant with some fitness advantage could quickly fix a single haplotype and therefore greatly reduce genetic variation. Therefore, low variation at mtDNA may not be a good indicator of the effective population size experienced by the nuclear genome.

Gillespie (2001) has presented an interesting consideration of the effects of hitchhiking on regions near a selected locus. He has termed this effect **genetic draft** and has suggested that the stochastic effects of genetic draft may be more important than genetic drift in large populations. In general, it would reduce the central role thought to be played by effective population size in determining the amount of genetic variation in large populations. The potential effects of genetic draft seem not to be important for the effective population sizes usually of concern in conservation genetics.

10.4 POPULATION SUBDIVISION

Population subdivision will generate nonrandom associations (gametic disequilibrium) between alleles at multiple loci if the allele frequencies differ among subpopulations at both loci. This is an extension of the Wahlund principle, the excess of homozygotes caused by population subdivision at a single locus, to two loci (see Section 9.2) (Sinnock 1975). In general, for k equal-sized subpopulations:

$$D = \bar{D} + cov(p_1, p_2) \qquad (10.18)$$

where $\bar{D}$ is the average D value within the k subpopulations (Prout 1973, Nei and Li 1973).

This effect is important when two or more distinct subpopulations are collected in a single sample. For example, many populations of fish living in lakes consist of several genetically distinct subpopulations that reproduce in different tributary streams. Thus, a single random sample taken of the fish living in the lake will comprise several separate demes. Makela and Richardson (1977) have described the detection of multiple genetic subpopulations by an examination of gametic disequilibrium among many pairs of loci.

Cockerham and Weir (1977) have introduced a composite measure of gametic disequilibrium (D_C) that partitions gametic disequilibrium into two components: the usual measure of gametic disequilibrium, D, plus an added component that is due to the nonrandom union of gametes caused by population subdivision (D_B).

$$D_C = D + D_B \qquad (10.19)$$

In a random mating population, D and D_C will have the same value. We will see in the next section that the composite measure is of special value when estimating gametic disequilibrium from population samples. Campton (1987) has provided a helpful discussion of the derivation and use of the composite gametic disequilibrium measure.

10.5 HYBRIDIZATION

Hybridization between populations, subspecies, or species will result in gametic disequilibrium. Figure 10.1 can be viewed as the resulting genotypes and gametes in the first two generations of hybridization. The F_1 hybrid will be heterozygous for all loci at which the two taxa differ. The gametes produced by the F_1 hybrid will depend upon the linkage relationship of the two loci. If the two loci are unlinked, then all four gametes will be produced in equal frequencies because of recombination.

Table 10.2 shows the genotypes produced by hybridization between two taxa that are fixed for different alleles at two unlinked loci. This assumes that the two taxa are equally frequent and mate at random. We can see here that gametic disequilibrium (D) will be reduced by exactly one-half each generation. For unlinked loci, recombination will eliminate the association between loci in heterozygotes. However, only one-half of the population in a random mating population will be heterozygotes in the first generation. Recombination in the two homozygous genotypes will not have any effect. Therefore, gametic disequilibrium (D) will be reduced by exactly one-half each generation. A similar effect will occur in later generations even though more genotypes will be present. That is, recombination will only affect the frequency of gametes produced in individuals that are heterozygous at both loci ($AaBb$).

Gametic disequilibrium will decay at a rate slower than one-half per generation if the loci are linked.

Tight linkage will greatly delay the rate of decay of D. For example, it will take 69 generations for D to be reduced by one-half if there is 1% recombination between loci (see equation 10.5).

Gametic disequilibrium will also decay at a slower rate if the population does not mate at random because there is positive assortative mating of the parent types. This will reduce the frequency of double-heterozygotes in which recombination can act to reduce gametic disequilibrium. We can see this using equation 10.19. In this case, the D component of the composite measure (D_C)

will decline at the expected rate, but D_B will persist depending upon the amount of assortative mating. Random mating in a hybrid population can be detected by testing for Hardy-Weinberg proportions at individual loci.

These two alternative explanations of persisting gametic disequilibrium in a hybrid can be distinguished. Assortative mating will affect all pairs of loci (including cytoplasmic and nuclear associations), while the effect of linkage will differ between pairs depending upon their rate of recombination. Example 10.3 describes

Table 10.2 Expected genotype frequencies and coefficient of gametic disequilibrium (D) in a random mating hybrid swarm.

Genotypes	Genotype frequencies				
	Parental	First generation	Second generation	Third generation	Equilibrium
AABB	0.500	0.250	0.141	0.098	0.063
AABb			0.094	0.118	0.125
AAbb			0.016	0.035	0.063
AaBB			0.094	0.118	0.125
AaBb		0.500	0.312	0.267	0.250
Aabb			0.094	0.118	0.125
aaBB			0.016	0.035	0.063
aaBb			0.094	0.118	0.125
aabb	0.500	0.250	0.141	0.098	0.063
D	–	+0.250	+0.125	+0.063	0.000

Example 10.3   Cytonuclear disequilibrium in a hybrid zone of field crickets

Hybrid zones occur where two genetically distinct taxa are sympatric and hybridize to form at least partially fertile progeny. Observations of the distribution of multilocus genotypes within hybrid zones and the patterns of introgression across hybrid zones can provide insight into the patterns of mating and the fitnesses of hybrids that may contribute to barriers to gene exchange between taxa.

Harrison and Bogdanowicz (1997) described gametic disequilibrium in a hybrid zone between two species of field crickets, *Gryllus pennsylvanicus* and

G. firmus. These two species hybridize in a zone that extends from New England to Virginia in the US. Analyses of four anonymous nuclear loci, allozymes, mtDNA, and morphology at three sites in Connecticut indicate that nonrandom associations between nuclear markers, between nuclear and mtDNA (Figure 10.5), and between genotypes and morphology persist primarily because of more frequent matings between parental types. That is, the crickets at these three sites in this hybrid appear to be primarily parental with a few F1 individuals and even fewer later generation hybrids.

(Continued)

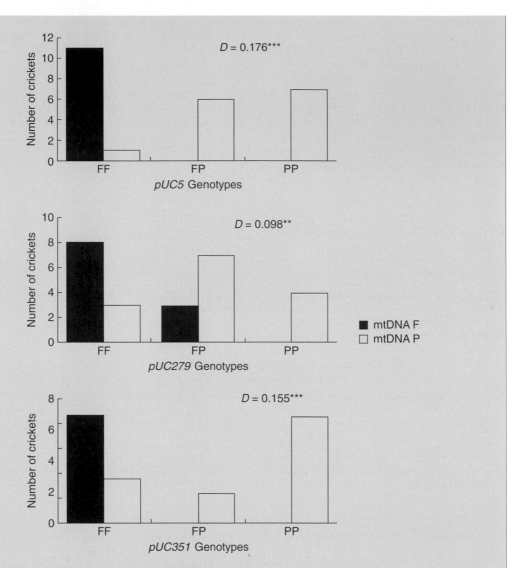

Figure 10.5 Gametic disequilibrium between mtDNA and three nuclear loci in a hybrid zone between two species of field crickets, *Gryllus pennsylvanicus* (P) and *G. firmus* (F). The mtDNA from *G. firmus* (F) is significantly more frequent for homozygotes (FF) for the *G. firmus* nuclear allele at all three loci. From Harrison and Bogdanowicz (1997). ***P < 0.001; **P < 0.01.

These two species of field crickets are genetically similar. There are no fixed diagnostic differences at allozyme loci, and more than 50 anonymous nuclear loci had to be screened to find four that were diagnostic. These two taxa meet the criteria for species according to some **species concepts** but not others. Regardless, as we will see in Chapter 16, the long-term persistence of parental types throughout an extensive hybrid zone indicates that these species are clearly distinct biological units.

the multilocus genotypes in a natural hybrid zone between two species of crickets. In this case, most genotypes are similar to the parental taxa and gametic disequilibrium persists over all loci because of assortative mating. Forbes and Allendorf (1991) have described a true **hybrid swarm** in which mating is at random (all loci are in Hardy-Weinberg proportions), but gametic disequilibrium persists at linked loci (Example 10.4).

We will examine hybridization and its genotypic effects again in Chapter 17 when we consider the effects of hybridization on conservation.

10.6 ESTIMATION OF GAMETIC DISEQUILIBRIUM

There is no simple way to estimate gametic equilibrium values from population data (Kalinowski and Hedrick 2001, Barton 2011). As described in the next section, even the simplest case of two alleles at a pair of loci is complicated. Estimation becomes more difficult when we consider that virtually all loci have more than two alleles, and we often have genotypes from many loci. There are a total of $n(n-1)/2$ pairwise combinations of loci if we examine n loci. So with 10 loci, each with just two alleles, there are a total of 45 combinations of two locus gametic equilibrium values to estimate.

10.6.1 Two loci with two alleles each

Let us consider the simplest case of two alleles at a pair of loci (see genotypic array in Table 10.1). The gamete types (e.g., AB or Ab) cannot be observed directly but must be inferred from the diploid genotypes. For example, $AABB$ individuals can only result from the union of two AB gametes, and $AABb$ individuals can only result from the union of an AB gamete and an Ab gamete. Similar inferences of gametic types can be made for all individuals that are homozygous at one or both loci. In contrast, gamete frequencies cannot be inferred from double heterozygotes ($AaBb$) because they may result from either union of AA and bb gametes or Ab and aB gametes. Consequently, gametic disequilibrium cannot be calculated directly from diploids.

Several methods are available to estimate gametic disequilibrium values in natural populations when the two gametic types of double heterozygotes cannot be

---

Example 10.4   Gametic disequilibrium in a hybrid swarm

Forbes and Allendorf (1991) studied gametic disequilibrium in a hybrid swarm of cutthroat trout. They observed the following genotypic distribution between two closely linked diagnostic allozyme loci. At both loci, the upper-case allele (A and B) designates the allele fixed in the Yellowstone cutthroat trout and the lower-case allele (a and b) is fixed in westslope cutthroat trout. There is a large excess of parental gamete types. The allele frequencies at the two loci are $p_1 = 0.589$ and $p_2 = 0.518$. The expected genotypes if $D = 0$ are presented in parentheses:

ME-4	LDH-A2			
	AA	**Aa**	**aa**	**Total**
BB	7	0	0	7
	(2.6)	(3.6)	(1.3)	
Bb	3	12	0	15
	(4.8)	(6.8)	(4.9)	
bb	0	1	5	6
	(1.2)	(2.2)	(1.0)	
Total	10	13	5	

The estimated value of D in this case is 0.213 and $D' = 1.000$ using the EM method described in Section 10.6.1. The estimated gamete frequencies are presented below:

Gamete	$D = 0$	$D = 0.213$
A B	$(p_1)(p_2) = 0.305$	0.518
A b	$(p_1)(q_2) = 0.284$	0.071
a B	$(q_1)(p_2) = 0.213$	0.000
a b	$(q_1)(q_2) = 0.198$	0.411

---

distinguished. The simplest way is to ignore them, and simply estimate D from the remaining eight genotypic classes. The problem with this method is that double heterozygous individuals may represent a large proportion of the sample (Example 10.4), and their exclusion from the estimate will result in a substantial loss of information.

The best alternative is the **expectation maximization** (EM) algorithm, which provides a maximum likelihood estimate of gamete frequencies assuming random mating (Hill 1974). We previously used the EM approach in the case of a null allele where not all genotypes could be distinguished at a single locus (Section 5.4.2). This approach uses an iteration procedure along with the maximum likelihood estimate of the gamete frequencies:

$$\hat{G}_1 = \left(\frac{1}{2N}\right) \left(2N_{11} + N_{12} + N_{21} + \frac{N_{22}\hat{G}_1(1-\hat{p}_1-\hat{p}_2+\hat{G}_1)}{\hat{G}_1(1-\hat{p}_1-\hat{p}_2-\hat{G}_1)} + (\hat{p}_1-\hat{G}_1)(\hat{p}_2-\hat{G}_1) \right)$$

$$(10.20)$$

where N is the sample size, N_{11} is the number of $AABB$ genotypes observed, N_{12} is the number of $AaBB$ genotypes observed, and N_{21} is the number of $AABb$ genotypes observed. This expression is not as opaque as it first appears. The first three sums in the right-hand parentheses are the observed numbers of the G_1 gametes in genotypes that are homozygous for at least one locus. The fourth value is the expected number of copies of the G_1 gamete in the double heterozygotes.

We need to make an initial estimate of gamete frequencies and then iterate using this expression. Our initial estimate can either be the estimate of gamete frequencies with $D = 0$, or we can use the procedure described in the previous paragraph to initially estimate D from the remaining eight genotypic classes. The other three gamete frequencies can be solved directly once we estimate G_1 and the single locus allele frequencies. Iterations can sometimes converge on different gamete values, depending upon the initial gamete frequencies (Excoffier and Slatkin 1995). Kalinowski and Hedrick (2001) presented a detailed consideration of the implications of this problem when analyzing datasets with multiple loci.

It is crucial to remember that the EM algorithm assumes random mating and Hardy-Weinberg proportions. The greater the deviation from expected Hardy-Weinberg proportions, the greater the probability that this iteration will not converge on the maximum likelihood estimate. Stephens et al. (2001) have provided an algorithm to estimate gamete frequencies that assumes that the gametes in the double heterozygotes are likely to be similar to the other gametes in the sample. This method is likely to be less sensitive to nonrandom mating in the population being sampled.

10.6.2 More than two alleles per locus

The numbers of possible multilocus genotypes expands rapidly when we consider more than two alleles per locus. For example, there are six genotypes and nine gamete types at a single locus with three alleles. Therefore, there are $6 \times 6 = 36$ diploid genotypes and $9 \times 9 = 81$ possible combinations of gametes at two loci each with three alleles. D values for each pair of alleles at two loci can be estimated and tested statistically (Kalinowksi and Hedrick 2001). The EM iteration procedure is more likely to converge to a value other than the maximum likelihood solution as the number of alleles per locus increases. Therefore, it is important to start the iteration from many different starting points with highly polymorphic samples.

So far we have considered genotypes at a pair of loci. There is much more information available if we consider the distribution of genotypes over many loci simultaneously. This is not a simple problem (Waits et al. 2001)! And, there are much larger datasets that are being used. As we saw in Guest Box 9, over 45,000 SNP loci have been examined in the stickleback.

10.7 MULTIPLE LOCI AND CONSERVATION

The interpretation of multilocus genotypes is becoming increasingly important in conservation because of advances in techniques to screen many loci, and advances in data analysis. The more loci examined, the more pairs of loci we will happen to sample that are on the same chromosome, for which we are much more likely to detect gametic disequilibrium (Figure 10.1).

Moreover, recent studies have found gametic disequilibrium even between unlinked pairs of loci on different chromosomes (Example 10.5, Bensch et al. 2006, Slate and Pemberton 2007). This is perhaps not unexpected. We saw in Section 10.2 that small population size in itself can produce substantial amounts of gametic disequilibrium. In addition, the rate of hybridization between subpopulations in many species has also increased because of human activities (Slate and Pemberton 2007, see Chapter 17). Thus, many of these populations that are of conservation interest might have substantial gametic disequilibrium because of hybridization, population subdivision, or small population size.

Example 10.5 Extensive gametic disequilibrium at microsatellite loci in the Siberian jay

Li and Merilä (2010) estimated gametic disequilibrium between 103 microsatellite loci in a semi-isolated population of Siberian jay from western Finland. This subpopulation has been the subject of a long-term field study for over 30 years.

A linkage map for this population was constructed from pedigrees through direct field observations in combination with verification of parentage using microsatellite genotypes (Jaari et al. 2009). Recombination rates were estimated by the examination of 311 progeny fathered by 85 males and mothered by 95 females. A

total of 107 microsatellite loci were assigned to nine autosomal and one Z-chromosome specific linkage groups. Ten loci could not be assigned to any linkage group. Six of the 103 loci were found to be sex-linked; three of these were in a pseudoautosomal region found on both the Z and W chromosomes, and three were Z-chromosome specific. As has been found in many species, there was less recombination in males than in females. On average, there was 28% greater recombination in females than in males. Figure 10.6 shows the comparative linkage map for one of the autosomal linkage groups.

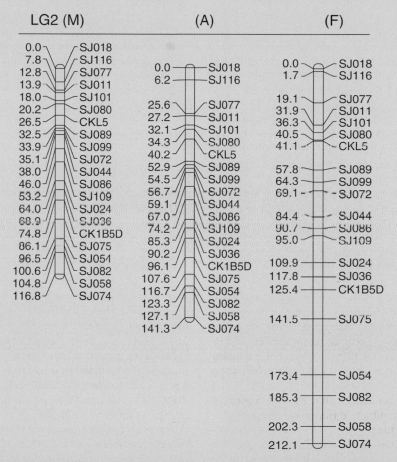

Figure 10.6 Linkage group 2 of the Siberian jay for males (M), females (F), and the average recombination rates of males and females (A). The names of the loci are on the right, and the total map distances on the left in centimorgans (cM). One cM corresponds to 1% recombination. There is greater recombination for this linkage group in females than in males, as indicated by the greater distances between loci on the female. From Jaari et al. (2009).

(Continued)

A total of 97 autosomal and 6 sex-linked loci were genotyped to estimate gametic disequilibrium in the wild population (Li and Merilä 2010). As expected, the amount of gametic disequilibrium between pairs of linked loci declined as the rate of recombination increased (Figure 10.7). Unlike the data from *Drosophila* described in Section 10.1, substantial gametic disequilibrium was found between pairs of loci separated by much more than 15% recombination. Significant ($P < 0.05$) gametic disequilibrium was even found in 83% of unlinked marker pairs on different chromosomes. As expected, the amount of gametic disequilibrium between pairs of loci on different chromosomes and unlinked pairs of loci on the same chromosome was quite similar: $D' = 0.356$ versus $D' = 0.354$.

The overall amount of gametic disequilibrium in the population is surprisingly high. This gametic disequilibrium probably results at least partially from the small effective population size ($N_e = 170$, Fabritius 2010). In addition, pedigree analysis over many generations revealed five different extended family groups in this population. Such subdivision is expected to increase gametic disequilibrium, as we saw in Section 10.4.

The substantial gametic disequilibrium in this population has important implications, regardless of its cause. These observations also emphasize how misleading it is to use the term 'linkage disequilibrium' to refer to nonrandom associations between loci, as we discussed at the beginning of this chapter. In this case, most pairs of loci found to be in 'linkage' disequilibrium are actually unlinked.

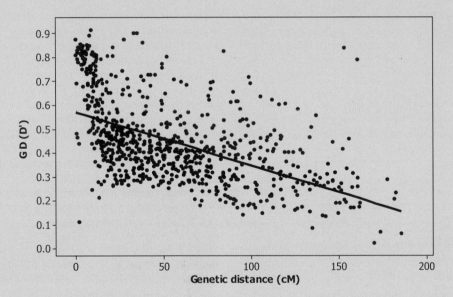

Figure 10.7   Gametic disequilibrium, as measured by D' (equation 10.6), for pairs of loci on the same chromosomes separated by different amounts of recombination in the Siberian jay. Data from Li and Merilä (2010).

Guest Box 10   Estimation of effective population size using gametic disequilibrium
Robin S. Waples

A variety of methods are now widely used to estimate N_e in natural populations, but this is a relatively recent development. Early applications (e.g., Pollak 1983) focused on large populations, which can be difficult to enumerate in the wild. However, the signal of genetic drift that these methods respond to is proportional to $1/N_e$ (see equation 6.2), which means that they are most effective for studying small populations. In the last two decades, this fact has been widely exploited by those interested in conservation or studying evolutionary processes in nature. Most applications have used the temporal method (Krimbas and Tsakas 1971, Nei and Tajima 1981), which measures allele frequency change between samples taken at different times (see Section 7.6 and Guest Box 7). However, any temporal comparison requires at least two samples, each of which could be used to generate an estimate of N_e using gametic disequilibrium. This approach is commonly called the 'linkage disequilibrium' (LD) method, even though it is generally assumed that the loci being used to estimate N_e are not linked. The LD method to estimate N_e takes advantage of the generation of gametic disequilibrium in small populations (see Section 10.2 and equation 10.12).

Single-sample LD estimates are easy to calculate and their performance has been extensively evaluated with simulated data (England *et al.* 2010, Tallmon *et al.* 2010, Waples and Do 2010). Precision can limit practical usefulness of genetic methods for estimating N_e, and an advantage the LD method has in this respect is that the number of useful data points increases with the square of the numbers of loci and alleles, rather than linearly as in the temporal method. If L loci are used, the number of pairwise comparisons of loci is $[L(L-1)]/2$, and many more comparisons of different pairs of alleles are possible.

Most applications of the LD method assume the loci are not linked (e.g., Park 2011), and the disequilibria are due to genetic drift from a finite number of parents. In theory, power is actually higher for linked markers, but it is necessary to know the recombination fraction (Hill 1981), which is rare for non-model species. Moreover, it is now becoming apparent that the amount of recombination between loci can differ markedly among individuals and populations (Dumont and Payseur 2011).

Disequilibria at unlinked loci decay quickly (see Figure 10.2), so the LD method generally provides an estimate of N_e in the parental generation, although recent bottlenecks can downwardly bias estimates for a few generations (Waples 2005a). If the population is relatively small ($N_e < 100$), reasonably precise estimates can be obtained using samples of 25–50 individuals with 5–10 moderately variable loci. Considerably more data are needed to achieve comparable precision if the population is relatively large ($N_e \sim 500$–1000 or higher). As in most population genetics models, underlying theory for the LD method assumes discrete generations. For age-structured populations, samples from a single age cohort has the simplest interpretation: the estimate is the effective number of parents that produced the cohort. More work needs to be done to understand how estimates based on mixed-age samples relate to effective size per generation when generations overlap.

Use of gametic disequilibrium to estimate the effective population size of introduced red foxes on Phillip Island, Australia, illustrates many of these points (Figure 10.8, Berry and Kirkwood 2010). The foxes on this island have had a major effect on livestock and breeding colonies of little penguins. Samples of all foxes killed each year were taken from 1994 through 2008. Sample sizes were small (<50 each year; median = 25), but 34 microsatellite loci were used, and these provided $[(34 * 33)/2] = 561$ different locus-by-locus comparisons and over 1000 total pairwise allelic comparisons for most of the yearly estimates. Furthermore, the population is small (all yearly effective size estimates were in the range 10–31) so precision was high and confidence intervals were narrow. Because the samples were of mixed-age individuals, it is not possible, as noted above, to determine exactly how the effective size estimates relate to N_e per generation.

(Continued)

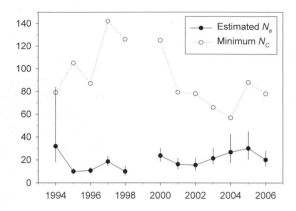

Figure 10.8 Annual estimates of effective population size and minimum census size in an introduced population of red fox on Phillip Island, Australia. Estimates of N_e were based on the amount of gametic disequilibrium using the program LDNE (Waples and Do 2008). Lines indicate the 95% confidence intervals. Minimum census size (N_C) was based on the number of unique genotypes observed. An aggressive program to control red foxes was implemented in 2000. Data from Berry and Kirkwood (2010).

Nevertheless, the consistently low estimates compared with the census size (N_C) suggest that only a fraction of adults successfully reproduce. Minimum census size was based on the number of unique genotypes observed (see Section 14.1).

A surprising observation of this genetic monitoring effort was that effective size did not decline following implementation of an aggressive fox control program in 2000, even though census size did decline (Figure 10.8). The authors concluded that the population experienced a density-dependent release from reproductive suppression; when a fraction of the adults were removed, others that had not had a chance to reproduce now became successful breeders. This combination of genetic and demographic information has caused managers to rethink their strategy for fox control. These results also show that the N_e/N_C ratio can change over time within a population as a result of density dependent effects, and this argues for some caution in the use of N_e as a metric for monitoring changes in abundance (Ardren and Kapuscinski 2003, Tallmon *et al.* 2010).

Applications of the LD method, and other single-sample estimators, are likely to continue to increase in the future, spurred both by growing concern for the conservation status of small populations in fragmented landscapes and the ready availability of numerous genetic markers (Luikart *et al.* 2010). As large numbers of SNP markers become widely used for non-model organisms, it will be important to evaluate the assumption that pairs of loci are unlinked.

CHAPTER 11

Quantitative genetics

Bighorn sheep, Guest Box 11

An overview of theoretical and empirical results in quantitative genetics provides some insight into the critical population sizes below which species begin to experience genetic problems that exacerbate the risk of extinction.

Michael Lynch (1996)

Most of the major genetic concerns in conservation biology, including inbreeding depression, loss of evolutionary potential, genetic adaptation to captivity, and outbreeding depression, involve quantitative genetics.

Richard Frankham (1999)

Chapter Contents

Conservation and the Genetics of Populations, Second Edition. Fred W. Allendorf, Gordon Luikart, and Sally N. Aitken.
© 2013 Fred W. Allendorf, Gordon Luikart and Sally N. Aitken. Published 2013 by Blackwell Publishing Ltd.

Most phenotypic differences among individuals within natural populations are quantitative rather than qualitative. Some individuals are larger, stronger, or can run faster than others. Such phenotypic differences cannot be classified by certain characteristics, such as wrinkled or smooth peas, or bands on a gel. The inheritance of quantitative traits is usually complex, and many genes are involved (i.e., they are polygenic). In addition to genetics, the environment to which individuals are exposed will also affect their phenotype (see Figure 2.2). Understanding genetic variation in quantitative traits is important for conservation, as natural selection acts directly on phenotypes, not on genotypes, and these traits therefore reflect the way a species is adapted to its environment. The single locus genetic models that we have been using until this point are inadequate for understanding this variation. Instead of considering only the effects of one or two genes at a time, we will expand our examination to inheritance of polygenic traits, and partition the genetic basis of such phenotypic variation into various sources using statistical procedures.

The study of quantitative genetics began shortly following the rediscovery of Mendel's principles to resolve the controversy of whether discrete Mendelian factors (genes) could explain the genetic basis of continuously varying characters (Lynch and Walsh 1998). The theoretical basis of quantitative genetics was developed primarily by R.A. Fisher (1918) and Sewall Wright (1921). The empirical aspects of quantitative genetics were developed primarily from applications to improve domesticated animals and agricultural crops (Lush 1937, Falconer and Mackay 1996).

The models of quantitative genetics have been applied to understanding genetic variation in natural populations only in the last 30 or so years (Roff 2007), and the past decade has seen the emergence of several new genetic and analytical tools for this purpose. The abundance of molecular markers now available makes it possible to identify **quantitative trait loci** (QTLs) – the specific chromosomal regions that influence variation in continuous traits (Barton and Keightley 2002), and even identify single nucleotide polymorphisms (SNPs) contributing to quantitative trait variation and adaptation (Stinchcombe and Hoekstra 2008, Stapley *et al.* 2010). Understanding the evolutionary effects of QTLs will allow us to improve our understanding of how genes influence phenotypic variation and improve our understanding of the genetic architecture of phenotypic variation (e.g., the

number of genes involved and their patterns and levels of variation).

The principles of quantitative genetics can also be applied to a variety of problems in conservation (reviews by Barker 1994, Lynch 1996, Storfer 1996, Lande 1996, Frankham 1999, Kruuk 2004, Kramer and Havens 2009). We saw in Chapter 2 that pink salmon on the west coast of North America have become smaller at sexual maturity over a period of 25 years (see Figure 2.5). This apparently resulted from the effects of a size-selective fishery in which larger individuals had a higher probability of being caught (Ricker 1981). Understanding the quantitative genetic basis of traits is essential for predicting genetic changes that are likely to occur in captive propagation programs as populations become 'adapted' to captivity (see Chapter 19), or to determine whether adequate adaptive variation exists for a population to adapt to new environmental conditions or threats (Sgró *et al.* 2011). Quantitative genetic studies of population differentiation can be used to select source populations for ecological restoration or reintroductions. As quantitative genetic studies provide information about local adaptation, and genetic markers provide information about gene flow, genetic drift, and population history, these types of data are complementary, rather than redundant, for developing conservation strategies.

This chapter provides a conceptual overview of the application of quantitative genetics to problems in conservation. Our emphasis is on the interpretation of results of quantitative genetic experiments with model species in laboratory and crop species, and on more recent studies of quantitative genetic variation in natural populations. Detailed consideration of quantitative genetic principles can be found in Falconer and Mackay (1996), as well as Lynch and Walsh (1998).

11.1 HERITABILITY

There are three major types of quantitative characters that are affected by a combination of polygenic inheritance and the environment:

1 **Continuous characters.** These characters are continuously distributed in populations (e.g., weight, height, and body temperature), and often have an approximately normal frequency distribution. For example, Figure 11.1 shows the body length of pink salmon at sexual maturity in an experimental population from Alaska (Funk *et al.* 2005).

2 Meristic characters. The values of these characters are restricted to integers; that is, they are countable (e.g., number of vertebrae, number of fingerprint ridges, or number of seed per fruit). For example, Figure 11.2 shows the distribution of the total number (left plus right) of gill rakers on the upper gill arch in the same population of pink salmon as Figure 11.1.

3 Threshold characters. These are characters in which individuals fall into a few discrete states, but there is an underlying continuously distributed

genetic basis to the trait in question (e.g., alive or dead, diseased or healthy).

As we discussed in Chapter 2, phenotypes are the joint products of genotypes and environments. The total amount of phenotypic variation for a quantitative trait within a population can be thought of as arising from two major sources: environmental differences between individuals and genetic differences between individuals. Writing this statement in the form of a simple mathematical model, we have:

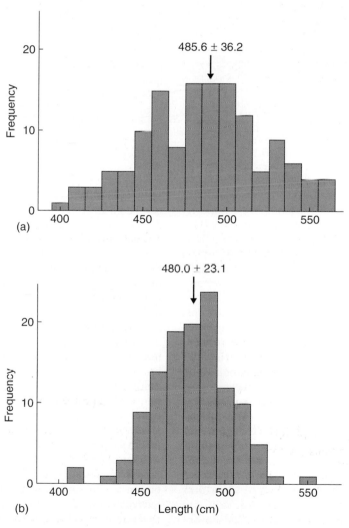

(a)

(b)

Length (cm)

Figure 11.1 Body length of (a) male and (b) female pink salmon at sexual maturity in a population from Alaska. Arrows indicate mean (± standard deviation). There is no significant difference in the mean length of males and females in this population. However, males have a significantly greater variance of body length. Data from Funk *et al.* (2005).

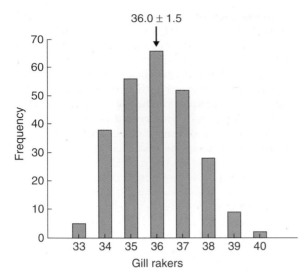

Figure 11.2   Distribution of the total (left plus right) of gill rakers on the upper gill arch in the same population of pink salmon as Figure 11.1. Arrow indicates mean (± standard deviation). There are no differences between males and females for this trait. Data from Funk *et al.* (2005).

$$V_P = V_E + V_G \qquad (11.1)$$

where V is variance, a statistical measure of variation equal to the standard deviation squared, and V_P, V_E and V_G are the phenotypic, environmental, and genetic variances of a trait. The genetic differences between individuals (V_G) can be attributed to three different sources of variance:

V_A = additive effects

(effects of allele substitution)

V_D = dominance effects

(effects of interactions between alleles)

V_I = epistatic effects

(effects of interactions between loci)

Therefore,

$$\begin{aligned} V_P &= V_E + V_G \\ &= V_E + V_A + V_D + V_I \end{aligned} \qquad (11.2)$$

11.1.1 Broad-sense heritability

Heritability is a measure of the relative influence of genetics versus environmental factors in determining phenotypic differences among individuals for a trait within a population. To determine the heritability of a trait, you need to know the relationships among individuals that have been measured. Heritability in the broad sense (H_B) is the proportion of the phenotypic variability that results from genetic differences between individuals. That is:

$$H_B = \frac{V_G}{V_P} = \frac{V_A + V_D + V_I}{V_P} \qquad (11.3)$$

For example, Sewall Wright removed virtually all of the genetic differences between guinea pigs within lines by continued sister–brother matings for many generations. The total phenotypic variance (V_P) in the population as a whole (consisting of many separate inbred lines) for the amount of white spotting was 573. The average variance within the inbred lines was 340; this must be equal to V_E because genetic differences among individuals within the lines were removed through inbreeding. Thus:

$$V_P = V_E + V_G$$
$$573 = 340 + V_G$$
$$V_G = 573 - 340 = 233$$

and

$$H_B = 233/573 = 0.409$$

11.1.2 Narrow-sense heritability

Conservation biologists are often interested in predicting how a population will respond to selection when individuals differ in survival and reproductive success (Stockwell and Ashley 2004). Similarly, animal and plant breeders are often interested in improving the performance of agricultural species for specific traits of interest (e.g., growth rate or egg production). However, broad-sense heritability may not provide a good prediction of the response. We will see shortly that a trait may not respond to selective differences, even though variation for the trait is largely based upon genetic differences between individuals, if that variation is due to dominance or epistasis.

Another definition of heritability is commonly used because it provides a measure of the genetic resemblance between parents and offspring or between other related individuals, and therefore predicts the response of a trait to selection. Heritability in the narrow sense (H_N) is the proportion of the total phenotypic variation that is due only to additive genetic differences (V_A) among individuals. It provides a useful measure of the evolvability of a phenotypic trait in a population.

$$H_N = \frac{V_A}{V_P} \qquad (11.4)$$

If all genetic variation for a trait is additive, then H_B and H_N will be equal. If not specified, heritability generally (but not always) refers to narrow-sense heritability, both in this chapter and in the published literature.

We can see the need for the distinction between broad- and narrow-sense heritability in the following hypothetical example. Assume that differences in length at sexual maturity are determined primarily by a single locus with two alleles in a fish species, there are no environmental effects, and that all genetic variation is due to dominance, with individuals heterozygous at this locus longer than either of the two homozygotes. In this example, the broad sense heritability for this trait is 1.00 since all of the variation is due to genetic differences.

Let us do a thought selection experiment in a random mating population in which these two alleles are equally frequent:

Genotype	Length	Frequency
A_1A_1	10 cm	$p^2 = 0.25$
A_1A_2	12 cm	$2pq = 0.50$
A_2A_2	10 cm	$q^2 = 0.25$

Thus, half of the fish in this population are approximately 12 cm long and the other half are 10 cm long. What would happen if we selected only the longer 12 cm fish for breeding in an attempt to produce larger fish? All of the fish selected for breeding will be heterozygotes. The progeny from $A_1A_2 \times A_1A_2$ matings will segregate in Mendelian proportions of 25% A_1A_1: 50% A_1A_2: 25% A_2A_2. Therefore, the progeny generation is expected to have the same genotype and phenotype frequencies as the parental generation. That is, there will be no response to selection even though all of the phenotypic differences have a genetic basis ($H_B = 1.00$).

In this example at $p = 0.5$, all of the phenotypic differences are due to dominance effects (V_D) resulting from the interaction between the A_1 and A_2 alleles in the heterozygotes. Thus, there is no response to selection, and the narrow-sense heritability is zero ($H_N = 0$). This hypothetical population would have responded to this selection if the two alleles did not have equal frequency, although it would only take one generation of selecting only those fish with 12 cm length to reach

$p = q = 0.5$. We will see in Section 11.2.1 that heritability will be different at different allele frequencies.

Heritability is another area where the nomenclature and the symbols used in publications can cause confusion to the reader. Narrow-sense heritability is often represented by h^2 and broad-sense heritability sometimes by H^2. The square in these symbols is in recognition of Wright's (1921) original description of the resemblance between parents and offspring using his method of path analysis in which, under the additive model of gene action, an individual's phenotype is determined by $h^2 + e^2$, where e^2 represents environmental effects and h^2 is the proportion of the phenotypic variance due to the genotypic value (see Lynch and Walsh 1998, appendix 2). We have chosen not to use these symbols in the hope of reducing possible confusion.

11.1.3 Estimation of heritability

Heritability can be estimated by several different methods that all depend upon comparing the relative phenotypic similarity of individuals with known or inferred genetic relationships.

11.1.3.1 Parent–offspring regression

One of the most direct ways to estimate heritability is by regressing the progeny phenotypic values on the parental phenotypic values for a trait. The narrow-sense heritability can be estimated by the slope of the regression of the offspring phenotypic value on the mean of the two parental values (called the mid-parent value).

Alatalo and Lundberg (1986) estimated the heritability of tarsus length in a natural population of the pied flycatcher using data from 338 nest boxes in Sweden. The narrow-sense heritability was estimated by twice the slope of the regression line of the progeny value on the maternal value. The slope is doubled in this case because only the influence of the maternal parent was considered. The male parents were not included in this analysis because previous results had shown that nearly 25% of progeny were the result of extra-pair copulations rather than mating with the social father. Heritability was estimated to be 0.496 (2×0.248) in an examination of nests in which nestlings were reared by their maternal mother (Figure 11.3a). In addition, 54 clutches were exchanged as eggs between parents to separate genetic and post-hatching environmental effects. There was no detectable resemblance at all

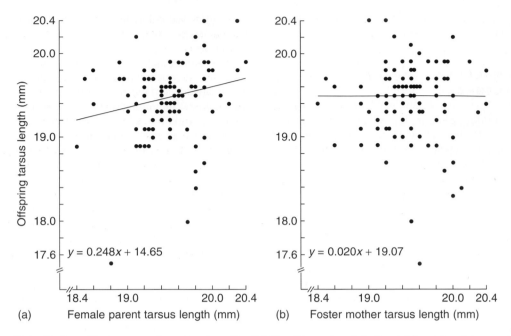

Figure 11.3 Mother–offspring regression estimation of heritability of tarsus length in the pied flycatcher: (a) regression with mother so the narrow-sense heritability is twice the slope ($H_{N} = 0.496$); (b) regression with foster mother. Each point represents the mean tarsus length of progeny from one nest. From Alatalo and Lundberg (1986).

between foster mothers and their progeny, suggesting **maternal effects** during incubation and juvenile growth are small (Figure 11.3b).

The relatively high heritability in this example is somewhat typical of estimates for morphological traits in natural populations (Roff 1997). Our reanalysis in Chapter 2 of Punnett's (1904) data on vertebrae number in velvet belly sharks results in a heritability estimate of 0.63 ($P < 0.01$; see Figure 2.6). Similarly high heritabilities for a variety of meristic traits (vertebrae number, fin rays, etc.) have been reported in many fish species (reviewed in Kirpichnikov 1981). Many morphological characters in bird species also have high narrow-sense heritabilities (e.g., Table 11.1).

11.1.3.2 Progeny testing

Another approach to estimating heritability is to use phenotypic similarity among half- or full-siblings to evaluate quantitative genetic variation and **breeding value** of parents in an approach called 'progeny testing'. In animal breeding, such an approach is sometimes called the 'sire model', where the genetic

Table 11.1 Heritability estimates from parent–offspring regression for morphological traits for three species of Darwin's finches in the wild (Grant 1986). Heritability ranges between zero and one. However, estimates of heritability can be greater than 1.0; for example, the slope of the regression of progeny on mid-parent values for weight and bill length was greater than 1.0 in *G. conirostris*.

Character	*G. fortis*	*G. scandens*	*G. conirostris*
Weight	0.91***	0.58	1.09***
Wing cord length	0.84***	0.26	0.69*
Tarsus length	0.71***	0.92***	0.78**
Bill length	0.65***	0.58*	1.08***
Bill depth	0.79***	0.80*	0.69***
Bill width	0.90**	0.56*	0.77**

*$P < 0.05$, **$P < 0.01$, ***$P < 0.001$.

value of a sire is determined from the performance of his offspring.

To separate genetic and environmental variances for a phenotypic trait, particularly when genetic variances are smaller than environmental variances in wild populations, individuals are ideally raised and phenotyped in a common environment with replication. **Common garden experiments**, a term used even for animals, have designs that facilitate the statistical separation of genetic and environmental effects (see Section 2.5). They are common for plant species, both wild and domesticated, as well as for domesticated and model animal species. However, these experiments are difficult or impossible to conduct for most wild animal species, or for highly endangered plants.

Progeny testing methods to estimate heritability from common gardens utilize the extent to which related individuals carry genes that are identical by descent. For example, if you collect seed from individual plants of an outcrossing species, plants grown from those seeds will all have the same mother but can have different fathers. If you assume that the seeds are half-siblings, then on average they will share one quarter of their genes from their mother by descent, so the variance among half-sib families (V_F) reflects one quarter of the additive genetic variation (V_A):

$$V_A = 4V_F \tag{11.5}$$

and

$$H_N = \frac{4V_F}{V_P} \tag{11.6}$$

This approach can bias heritability estimates upwards if maternal effects exist, for example, non-genetic differences among mother plants in seed weight that influence phenotypic traits, or if some individuals from the same mother are full- rather than half-siblings. If known full-sibling families are used, then additive and non-additive genetic variation cannot be separated.

11.1.3.3 Animal model

A big step forward in the estimation of genetic variance components and heritabilities for populations in the wild has been the development of a more analytically complex method that uses a maximum likelihood mixed model approach called the '**animal model**' (Kruuk 2004). The name refers to the estimation of additive genetic value of individual animals rather than groups of related individuals. It is now widely used in both plant and animal breeding programs. This model uses multigenerational pedigree information to partition phenotypic variation into additive genetic and environmental components. Best linear unbiased prediction models are used to estimate variance components (see Kruuk 2004 for details). With the animal model, environmental variation in the wild can be further partitioned into various sources. This approach can also be used to estimate effects of common maternal environments.

The complexity of genetic and environmental effects on phenotypic variance and response to selection has been well illustrated for an island population of Soay sheep in Scotland using the animal model. Complete records of animal births, deaths, and phenotypes have been kept since 1985. Wilson *et al.* (2006) used the animal model to study natural selection in this population under varying environmental conditions. They found that when environmental conditions are harsh, selection for increased birthweight is strong but the response is constrained by low genetic variance. When conditions were favorable, genetic variance was higher; however, selection was weak. Fluctuating selection pressures and differences among environments in phenotypic expression of genetic variation may limit rates of evolution but maintain genetic variation.

The animal model can be applied to populations in the wild with known pedigrees, typically populations of mammals or birds that are the focus of long-term studies that tag individual animals and record their parentage. Pedigrees can also be inferred from genetic markers, especially if individuals are genotyped for a large number of loci (Wang 2002, Jones and Wang 2010). Coltman *et al.* (2003) combined these approaches, using observed mother–offspring relationships and marker-inferred father–offspring relationships to study selection responses due to hunting in a bighorn sheep population (Guest Box 11). DiBattista *et al.* (2009) inferred the pedigrees of a large population of lemon sharks in the Bahamas with several hundred parents and over a thousand offspring using microsatellite markers. They were then able to estimate the heritability of size and morphological traits using the animal model. Methods for estimating heritability have also been developed that assess pairwise relatedness and phenotypic similarity of individuals in the wild (Ritland 2000), but this approach does not perform as well as those based on reconstructing pedigrees, and resulting estimates of heritability tend to be down-wardly biased (Coltman 2005, Jones and Wang 2010).

11.1.4 Genotype-by-environment interactions

We saw in Figure 2.12 that the environment can have a profound effect on the phenotypes of yarrow plants resulting from a particular genotype. We also saw evidence for local adaptation with reciprocal transplants in *Gilia capitata* (Figure 8.1) in which plants had greater fitness in their native habitat. In a statistical sense, these are examples of **interactions** between genotypes and environments. We can expand our basic model to include these important interactions as follows:

$$V_P = V_E + V_G + V_{G \times E} \tag{11.7}$$

Genotype-by-environment (G × E) interactions are of major concern in conservation biology when translo-cating individuals to alleviate inbreeding depression (genetic rescue, Chapter 15), when source populations are being chosen for ecological restoration, and when reintroducing captive populations into the wild (Chapter 19). Species that show strong genotype-by-environment interaction will likely need more popula-tions conserved to capture the adaptive genetic diversity present in the species as a whole, and will need more local populations used as sources of individuals for res-toration to avoid maladaptation. We will see in Chapter 19 that many traits that are advantageous in captivity may greatly reduce the fitness of individuals in the wild.

11.2 SELECTION ON QUANTITATIVE TRAITS

Evolutionary change by directional natural selection can be thought of as a two-step process. First, there must be phenotypic variation for the trait that results in differential survival or reproductive success (i.e., fitness). Second, there must be additive genetic varia-tion for the trait ($H_N > 0$; Fisher 1930). Heritability in the narrow sense can also be estimated by the response to selection. This is usually called the 'realized' herita-bility (Figure 11.4). If a trait does not respond at all to directional selection, then there is no additive genetic variation and $H_N = 0$. If the mean of the selected progeny is equal to the mean of the selected parents, then $H_N = 1$. Generally, the mean of the selected progeny will be somewhere in between these two extremes ($0 < H_N < 1.0$). Francis Galton, a cousin of Charles Darwin (Provine 2001), coined the expression "regression" to describe the general tendency for

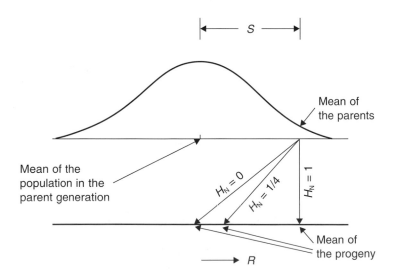

Figure 11.4 Illustration of the meaning of narrow-sense heritability based upon a selection experiment and equation 11.8. The *x*-axis represents phenotype, and the *y*-axis represents frequency of individuals within a population. If there is no response to selection, then $H_N = 0$; if the mean of the progeny from selected parents is equal to the mean of the selected parents, then $H_N = 1$. Modified from Crow (1986).

progeny of selected parents to "regress" towards the mean of the unselected population.

In this case, heritability is the response to selection divided by the total selection differential. Thus:

$$H_N = \frac{R}{S} \qquad (11.8)$$

where S is the selection differential, defined as the difference in the means between the selected parents and the whole population, and R is the response to selection, which is the difference between the mean of the

progeny of the selected parents and mean of the whole population in the previous generation. Equation 11.8 is often written in the form of the breeders' equation to allow prediction of the expected genetic gain (equivalent to response to selection R) from artificial selection:

$$R = H_N S \qquad (11.9)$$

Figure 11.5 illustrates two generations of artificial selection for a trait with a heritability of 0.33. The breeder selects by truncating the population and uses only individuals above a certain threshold as breeders.

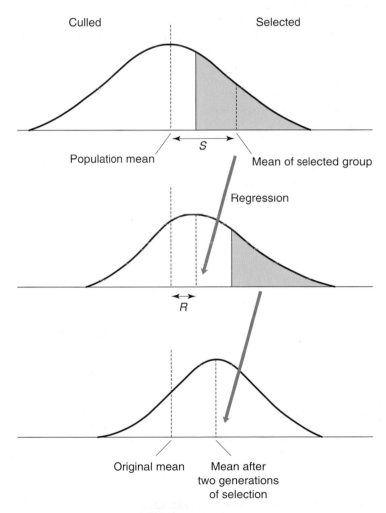

Figure 11.5 Two generations of selection for a trait with a heritability of 0.33. The progeny mean moves one-third (0.33) of the distance from the population mean towards the mean of the selected parents (shaded) in each generation. From Crow (1986).

The mean of the progeny from these selected parents will regress two-thirds $(1 - 0.33)$ of the way towards the original population mean.

11.2.1 Heritabilities and allele frequencies

Heritability for a particular trait is not constant. It will generally vary among populations that are genetically divergent. However, it may also vary within a single population under different environmental conditions or over time. In addition, heritability within a population will not be constant even within the same environment over generations, because it is influenced by allele frequencies; however, if many genes are involved with predominantly additive effects, it will stay more similar over time than if few genes and non-additive effects are involved.

Let us examine this effect with a simple single-locus model in which all phenotypic variation has a genetic basis, with notation for fitness (w) following Chapter 8, and w_{A1} and w_{A2} indicating the additive effects of alleles A_1 and A_2 on fitness. In a single-locus case such as this, heritability can be calculated directly by calculating the appropriate variances:

$$V_G = p^2(w_{11} - \overline{w})^2 + 2pq(w_{12} - \overline{w})^2 + q^2(w_{22} - \overline{w})^2 \tag{11.10}$$

$$V_A = 2[p(w_{A1} - \overline{w})^2 + q(w_{A2} - \overline{w})^2] \tag{11.11}$$

where

$$w_{A1} = \frac{p^2(w_{11}) + pq(w_{12})}{p} \tag{11.12}$$

and

$$w_{A2} = \frac{pq(w_{12}) + q^2(w_{22})}{q} \tag{11.13}$$

V_A is very closely related to average heterozygosity for the gene determining this trait, and it will also be maximum at $p = q = 0.5$. Since we have assumed that V_E is zero, the narrow-sense heritability is the additive genetic variance divided by the total genetic variance.

$$H_N = \frac{V_A}{V_G} \tag{11.14}$$

We can use this approach to estimate heritability in the example of height in a plant determined by a single locus with two alleles, A_1 and A_2. We can use equations

11.10–11.14 to estimate heritability over all possible allele frequencies (Figure 11.6). Here we consider three cases: (A) purely additive alleles, where the heterozygotes are intermediate to the homozygotes; (B) complete dominance of allele A_2, where the heterozygotes have the same phenotype as the taller homozygotes (A_2A_2); and (C) overdominance, where the heterozygotes are taller than either homozygotes.

Genotype	Additive (A)	Dominant (B)	Overdominant (C)
A_1A_1	20 cm	20 cm	20 cm
A_1A_2	22 cm	24 cm	24 cm
A_2A_2	24 cm	24 cm	20 cm
H_N	Always 1.0	High when q is low	0 when $p = 0.5$ >0 elsewhere

There are several important features of the relationship between allele frequency and heritability in this model (Figure 11.6). If all of the genetic variance is additive (case A), heritability is always 1.0. In the case of dominance (case B), heritability is high when the dominant allele (A_2) is rare but low when this allele is common. As we saw in Chapter 8, selection for a high-frequency dominant allele is not effective, so heritability will be low. Finally in the case of overdominance (case C), heritability is high when either of the alleles is rare because increasing the frequency of a rare allele will increase the frequency of heterozygotes and thus the population will be taller in the next generation. That is, the population will respond to selection when either allele is rare, and therefore heritability will be high. However, the frequency of heterozygotes is greatest when the two alleles are equally frequent, and heritability will be zero.

While these single-locus examples are useful for illustrating the principles of quantitative genetics, heritability, and the effects of additive and non-additive genetic variation, the majority of phenotypic traits with a continuous distribution are determined by several to many genes as well as affected by environmental variation. Since alleles at different loci combine to generate a phenotype, different genotypes may result in the same phenotype. A classic study of DDT resistance in *Drosophila* provides an example of this (Crow 1957). A DDT-resistant strain was produced by raising flies in a large experimental cage with the inner walls painted

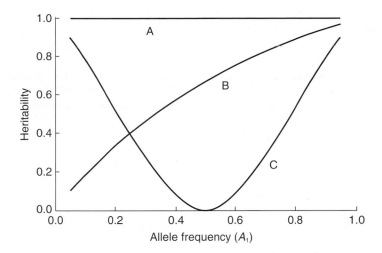

Figure 11.6 Hypothetical example of heritability (V_A/V_G) for plant height determined by a single locus with two alleles: (A) the additive case where heterozygotes are intermediate to both homozygotes; (B) complete dominance so that the heterozygotes have the same phenotype as the taller homozygotes (A_2A_2); and (C) the case of overdominance where the heterozygotes are taller than either of the homozygotes. It is assumed that all of the variability in height is genetically determined (i.e., $V_E = 0$).

with DDT. The concentration of DDT was increased over successive generations as the flies became more and more resistant, until over 60% of the flies survived doses of DDT that initially killed over 99% of all flies. Flies from the resistant strain were then mated to flies from laboratory strains that had not been selected for DDT resistance. The F_1 flies were mated to produce an F_2 generation that had all possible combinations of the three major chromosomal pairs, as identified by marker loci on each chromosome (Figure 11.7).

The results of this elegant experiment demonstrate that genes affecting DDT resistance occur on all three chromosomes. The addition of any one of the three chromosomes (Figure 11.7) increases DDT resistance. Furthermore, the effects of different chromosomes are cumulative, contributing to additive genetic variation: that is, the more copies of chromosomes from the resistant strain, the more resistant the F_2 flies were to DDT.

Knowing whether a particular trait is affected primarily by a single gene or by many genes with small effects can influence our recommendations based upon genetic considerations. For example, captive golden lion tamarins suffer from a diaphragmatic defect that seems to be hereditary (Bush *et al.* 1996). The presence or absence of this condition is one criterion used in the selection of individuals for release into the wild (Example 11.1).

11.2.2 Genetic correlations

Genetic correlations are a measure of the genetic association of different traits. Such correlations can result from two underlying causes. First, many genes affect more than a single trait and so they can cause simultaneous effects on different aspects of the phenotype; this is known as **pleiotropy**. For example, a gene that increases growth rate is likely to affect both stature and weight, and this pleiotropy will cause genetic correlations between the two traits. **Gametic disequilibrium**, the non-random association of alleles at different loci, can also result in genetic correlations between traits. It can result from either close physical linkage of genes on a chromosome, or from population structure. Selection (either natural or artificial) for a particular trait will often result in a correlated response in the mean value of another trait because of genetic correlations.

Figure 11.8 demonstrates a genetic correlation between two meristic traits in the experimental population of pink salmon from Figure 11.1. The parental values for one trait, pelvic rays, were a fairly good predictor of the progeny phenotypes for another trait, pectoral rays ($P < 0.05$). The reciprocal regression (progeny pelvic rays on mid-parent pectoral rays) is also significant ($P < 0.01$). The actual point estimate of the genetic correlation between these traits (0.64)

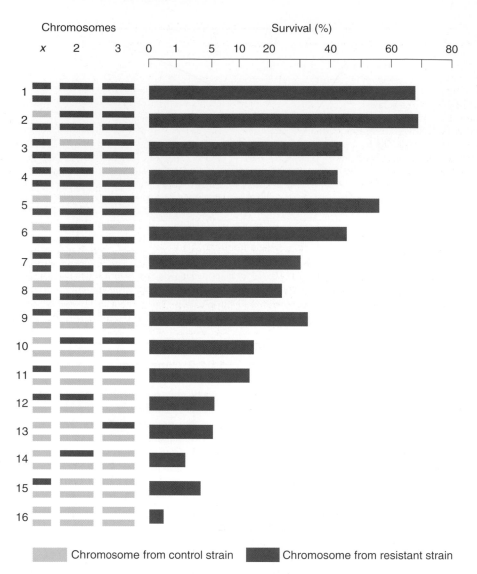

Figure 11.7   Results of an experiment demonstrating that numerous genes with largely additive effects determine DDT resistance in *Drosophila* (Crow 1957). The addition of any one of the chromosomes from the strain selected for DDT resistance increases DDT resistance. Redrawn from Crow (1986).

takes into account both of these parental–progeny relationships (Funk *et al.* 2005):

$$r_A = \frac{\text{cov}_{XY}}{\sqrt{(\text{cov}_{XX}\,\text{cov}_{YY})}} \qquad (11.15)$$

where cov_{XY} is the 'cross-variance' that is obtained from the product of the value of trait X in parents and the value of trait Y in progeny, and cov_{XX} and cov_{YY} are

the progeny–parent covariance for traits X and Y separately. We can see from Figure 11.8 that if we selected by using parents with many pelvic rays, the number of pectoral rays in the progeny would increase. Genes that decrease developmental rate tend to increase counts for a suite of meristic traits in the closely related rainbow trout (Leary *et al.* 1984). Additive genetic correlations can also be calculated from progeny experiments using the additive genetic covariance cov_A

Example 11.1 Should golden lion tamarins with diaphragmatic defects be released into the wild?

Golden lion tamarins held in captivity have a relatively high frequency of diaphragmatic defects detected by radiography (Bush *et al.* 1996). No true hernias were observed, but 35% of captive animals had marked defects in the muscular diaphragm that provided a potential site for liver or gastrointestinal herniation. Only 2% of wild living animals had this defect. This difference between captive and wild animals could be the result of relaxation of selection against the presence of such defects in captivity, or genetic drift in captivity due to small population size (Chapter 19).

All captive-born animals that are candidates for release in the wild are now screened for this defect before release. Individuals with a relatively severe defect are disqualified for reintroduction. Approximately 10% of all captive animals are expected to be disqualified using this criterion. This procedure is designed to protect the wild population against an increase in a potentially harmful defect that may have a genetic basis.

This screening procedure seems appropriate. However, its effectiveness will depend upon the genetic basis of this defect. If it is caused by a single gene with a major effect, then this selection is expected to be effective in limiting the increase of the genetic basis for this defect in the wild. If, however, this defect is caused by many genes with small effect, this selective removal of only those animals with severe defects will have little effect.

In addition, there is a potential concern with such selection of animals for reintroduction. As we have seen, selection can reduce the amount of genetic variation. Disqualifying, say, 25% or 50% of candidates for reintroduction could potentially reduce genetic variation in the reintroduced animals. For example, Ralls *et al.* (2000) concluded that selective removal of an allele responsible for chondrodystrophy in California condors would not be advisable because of the potential reduction of genetic variation in the captive population. Thus, there is no simple answer to the question of whether or not animals with defects should be released into the wild.

---

between traits X and Y, and V_A for trait X and trait Y in the denominator (Falconer and Mackay 1996).

When two traits are genetically correlated in a direction that means selection increasing fitness for one trait will also increase fitness for the other, the correlation is reinforcing and will enhance evolutionary response (Etterson and Shaw 2001). In contrast, if two traits are correlated such that an increase in fitness in one results in a decrease in fitness in the other, the correlation is antagonistic, and evolutionary responses will be inhibited. Strong antagonistic correlations often reflect biological tradeoffs among traits, and will slow or prevent responses to selection despite the presence of additive genetic variation for each trait involved.

11.3 FINDING GENES UNDERLYING QUANTITATIVE TRAITS

The field of quantitative genetics developed to allow the genetic analysis of traits affected by multiple loci for which it was impossible to identify individual genes having major phenotypic effects. Formal genetic analysis could only be performed for traits in which discrete (qualitative) phenotypes (round or wrinkled seeds, brown or albino, etc.) could be identified to test their

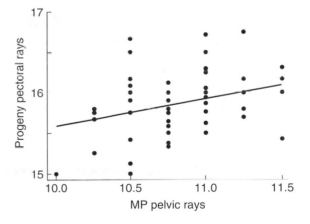

Figure 11.8 Genetic correlation between meristic traits in pink salmon. Regression of the number of rays in the pectoral fins of progeny on the mid-parent (MP) values of number of rays in the pelvic fins (slope = 0.342; $P < 0.023$). Data from Funk *et al.* (2005).

mode of inheritance. Classic quantitative genetics treated the genome as a black box and employed estimates of a variety of statistical parameters (e.g., heritabilities, genetic correlations, and the response to selection) to describe the genetic basis of continuous (quantitative) traits. However, the actual genes affecting these traits could not be identified with this approach. Alan Robertson (1967) described this phenomenon as the "*fog of quantitative variation*".

The fog is now lifting. The explosion in high-throughput sequencing has made it possible to identify the chromosomal regions, genes, and even single nucleotide polymorphisms (SNPs) involved in quantitative traits, even for genes with a relatively small effect on genotypes. High-throughput sequencing has facilitated this in four ways, by: (1) reducing the cost and time of genotyping many individuals for many markers; (2) allowing for the precise mapping of those genetic markers within the genome; (3) making feasible the testing of associations between particular markers and phenotypes within either mapping populations or in natural populations; and (4) allowing for the precise estimation of pedigrees and relatedness in natural populations (Stapley *et al.* 2010).

Until recently it was only possible to locate genes of major effect to within a relatively large chromosomal region spanning many genes, but it is now possible to identify specific regions of the genome called quantitative trait loci (QTLs), specific genes within those regions, and in some cases even the causal SNPs that are responsible for variation in continuous traits. Understanding the effects of these potentially adaptive loci allows us to determine how specific genes influence phenotypic variation, directly assess adaptive genetic variation, and determine how many genes affect phenotypic traits of particular interest.

11.3.1 QTL mapping

QTL mapping involves searching for polymorphic genetic markers that segregate with phenotypic variation for a trait. A high-density genetic linkage map is developed through genotyping large numbers of F_1 or F_2 progeny of a controlled cross between phenotypically different parents for a large number of genetic markers, often amplified fragment length polymorphisms (AFLPs), that are distributed across the genome. Those markers or pairs of adjacent markers that segregate with variation for the phenotypic trait of interest must be located close to one or more causal genes affecting that trait. QTL mapping is not usually seeking the causal gene or sequence variation within the gene that results in phenotypic variation, but is rather seeking a marker physically linked on a chromosome

to that gene. Moving from a QTL to a causal gene within the region involved has proved to be quite difficult (Mackay 2001, Stinchcombe and Hoekstra 2008).

Table 11.2 illustrates this approach schematically for a chromosome that contains a QTL with two alleles, where the Q^+ allele causes a greater phenotypic value than the Q^- allele. If the marker locus is close to the QTL, then genotypes at the marker locus (B in this case) will have different mean phenotypes because of linkage. Some loci farther away on the same chromosome (A in this case) will show independent segregation because of recombination, and therefore all three genotypes at the marker will have the same phenotypic mean. Because of the reliance on gametic disequilibrium that may exist in one population but not another, and because markers that segregate in one population may be fixed in another, QTLs are often population-specific and even pedigree-specific (Mackay 2001).

The size of the difference in mean phenotypic effect associated with a QTL marker will depend upon the magnitude of effect of the QTL and the amount of recombination between the marker and the QTL. The tighter the linkage, the greater will be the difference in mean phenotypes between genotypes at the marker locus. Figure 11.9 shows the expected relationship between the mean phenotype of the marker locus and the map distance between the marker locus and the QTL. The genetic map distance is measured in centiMorgans (cM): 1 cM equals 1% recombination. Therefore, loci that are at least 50 cM apart will segregate independently and are said to be unlinked. Loci on the same chromosome are syntenic. In this example, the A locus and the QTL are syntenic but unlinked.

This approach has been used to determine the genetic basis of the derived lifecycle mode of paedomorphosis in the Mexican axolotl (Voss 1995, Voss and Shaffer 1997). Mexican axolotls do not undergo metamorphosis and have a completely aquatic lifecycle. In contrast, closely related tiger salamanders undergo metamorphosis in which their external gills are absorbed and other changes occur before they become terrestrial. Voss (1995) crossed these two species and found that their F_1 hybrids underwent metamorphosis. He then backcrossed the F_1 hybrids to a tiger salamander and found evidence for a major single gene controlling most of the variation in this life history difference, with additional loci having smaller effects.

Voss and Shaffer (1997) used a similar backcross approach in two crosses to search for QTLs affecting metamorphosis. They scored 262 AFLP marker loci in

Table 11.2 Relationship between marker genotypes at a linked marker locus (*B*) and an unlinked locus (*A*) with a QTL. The Q^+ allele at the QTL causes an increase in the phenotype. The strong association between the phenotype of interest and genotypes at the *B* locus suggests that a QTL affecting the trait is linked to the *B* locus (see Figure 11.9). Modified from Kearsey (1998).

	F_1		
A_1		B_1	Q^+
A_2		B_2	Q^-

	Frequencies of F_2 genotypes			
Marker genotype	Q^+Q^+	Q^+Q^-	Q^-Q^-	**Mean phenotype**
A_1A_1	0.25	0.50	0.25	Intermediate
A_1A_2	0.25	0.50	0.25	Intermediate
A_2A_2	0.25	0.50	0.25	Intermediate
B_1B_1	Most	Few	Rare	High
B_1B_2	Few	Most	Few	Intermediate
B_2B_2	Rare	Few	Most	Low

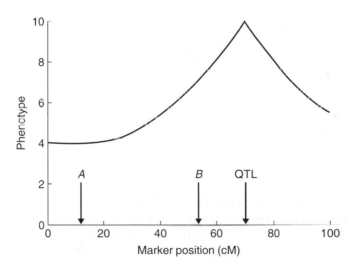

Figure 11.9 Hypothetical relationship between the mean phenotype of marker loci in the F_2 generation on a chromosome containing a QTL at map position 70 (see Table 11.2).

the backcross progeny. Only one region of the genome, which contained three AFLP markers, was associated with this lifecycle difference. They hypothesized that a dominant allele (*MET*) at a QTL in this region was associated with the metamorphic phenotype in the tiger salamander, and the recessive *met* allele from the Mexican axolotl was associated with paedomorphosis.

This result has been confirmed by additional analysis of nearly 1000 segregating marker loci (Smith *et al.* 2005b). No other regions were found to contribute to this life history difference. It is interesting to note that this same QTL also contributes to continuous variation for the timing of metamorphosis in the tiger salamander (Voss and Smith 2005).

The increased availability of markers (e.g., AFLPs or SNPs) and decreased cost of genotyping has facilitated QTL mapping studies in model and domesticated species, but the need for controlled crosses limits this approach in natural populations. With the advent of high-throughput sequencing, the search for the genetic basis of adaptive-trait variation has shifted from QTL mapping to candidate loci and whole genome scan approaches. If a city is used as a metaphor for a genome, QTL mapping searches for the a neighborhood or street containing a gene or genes affecting a phenotypic trait, while candidate genes approaches or whole genome scans are seeking the specific house addresses of the genes involved.

11.3.2 Candidate gene approaches

Candidate genes are genes of known function that are suspected of having a substantial influence on a phenotypic trait of interest. While the QTL approach is top-down, in that it begins with the phenotype in order to search for responsible genes, the candidate gene approach begins with the genes implicated in phenotypic trait variation in other species and searches for phenotypic effects of polymorphisms in those genes in target species (see Figure 2.2).

The DNA sequences, functions, and effects on phenotypes of many genes are now known for fully sequenced model organisms. To search for the genetic basis of quantitative trait variation in natural populations of non-model species, candidate genes are first sequenced in a small number of individuals to identify SNPs. SNPs are then genotyped or candidate genes are resequenced in many individuals. The SNPs can then be tested for significant associations with phenotypic traits, or for significant associations with environmental factors, after controlling for relatedness and population structure in an approach called **association mapping**.

Candidate gene approaches have been particularly successful in identifying polymorphisms associated with phenotypic variation in coat color in wild mammal populations. For example, Hoekstra *et al.* (2006) identified a single amino-acid change in the candidate gene *MC1R* in beach mice that explained 10–36% of the phenotypic variation in coat color (Figure 11.10). See Stinchcombe and Hoekstra (2008) and Stapley *et al.* (2010) for more information on applications of candidate genes and genome scans to understanding adaptation in wild populations.

11.3.3 Genome-wide association mapping

The third approach to identifying genetic polymorphisms underlying phenotypic variation is to conduct whole genome scans by genotyping many markers (usually SNPs, AFLPs, or microsatellites) distributed across the genome or across the transcriptome, and to test all markers for associations with phenotypes (association mapping). This approach requires no *a priori* knowledge about individual gene function, but requires large genomic resources and computational power. Markers can be selected at random, or they can be based on expressed sequence tags (ESTs) from transcriptome sequencing, an approach that increases the chance of markers being linked to loci under selection. Like candidate gene approaches, association mapping of markers in whole genome scans requires controlling for neutral population structure (Anderson *et al.* 2011). The resources required for whole genome scans remain beyond those available for most non-model species of conservation concern.

Genome-wide association studies with model species can potentially identify candidate genes affecting important phenotypic traits for non-model species. Fournier-Level *et al.* (2011) used a genome-wide association study of ~213,000 SNPs to identify loci associated with local adaptation to climate in *Arabidopsis* plants originating from over 900 locations across Europe and growing in four common gardens in different countries. Interestingly, they found that SNPs conferring higher fitness in one environment often had neutral effects on fitness in other environments. Genes containing SNPs associated with adaptation to climate in this study may be a valuable resource for candidate genes approaches in non-model plant species.

11.4 LOSS OF QUANTITATIVE GENETIC VARIATION

The loss of genetic variation (heterozygosity and allelic diversity) via genetic drift will affect quantitative variation as well as neutral molecular variation. As a result, the ability of small populations to adapt and evolve is expected to be lower than that of large populations, as genetic drift will erode the amount of additive genetic variation available for selection, and new beneficial mutations will be accumulated more slowly (Willi *et al.* 2006). However, the rate at which genetic variation will be lost will depend upon a large number

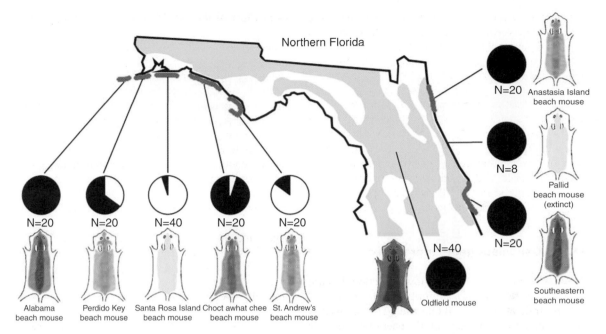

Figure 11.10 Frequency of alleles for the *melanocortin-1 receptor* (*MC1R*) gene and association with average coat color phenotype in one mainland and eight beach mouse subspecies in northern Florida. The *MC1R* locus explains 10–36% of the phenotypic variation in coat color in beach mice. The light grey area represents the distribution of the mainland subspecies, and the dark grey areas indicate the distributions of the beach mouse subspecies. Circles indicate the relative allele frequencies of the light (white) and dark (black) alleles identified. From Hoekstra *et al.* (2006).

of factors, including the number of loci affecting a trait, the amount of dominance or epistasis, and the strength and type of selection.

11.4.1 Effects of genetic drift and bottlenecks

We saw in Chapter 6 that heterozygosity at neutral loci will be lost at a rate of $1/(2N_e)$ per generation. We can relate the effects of genetic drift on allele frequencies at a locus with two alleles to additive quantitative variation as follows (Falconer and Mackay 1996):

$$V_A = \sum 2pq[a + d(q - p)]^2 \qquad (11.16)$$

where p and q are allele frequencies so that $2pq$ is the expected frequency of heterozygotes, a is half the phenotypic difference between the two homozygotes, and d is the dominance deviation.

This model results in the simple prediction that V_A, and therefore H_N, will be lost at the same rate as neutral

heterozygosity in the case where all of the variation is additive ($d = 0$).

Can heterozygosity for neutral markers predict within-population quantitative variation and evolvability in small populations? In a comprehensive review of limits to adaptive capacity in small populations, Willi *et al.* (2006) summarized laboratory studies that maintained either high levels of inbreeding or small effective population sizes over ten generations, and estimated H_N (Figure 11.11). At low levels of inbreeding or larger effective population sizes, heritabilities were often higher than in outbred control lines, suggesting that variation due to dominance or epistasis was being expressed. This increased phenotypic variation and heritability, however, is unlikely to enhance adaptive responses to selection as it is based on the expression of recessive deleterious alleles (Willi *et al.* 2006). At higher levels of inbreeding or smaller effective population sizes, however, heritability estimates decreased below that of outbred controls and tracked expected values based on an additive genetic model with no selection, paralleling predictions for heterozygosity.

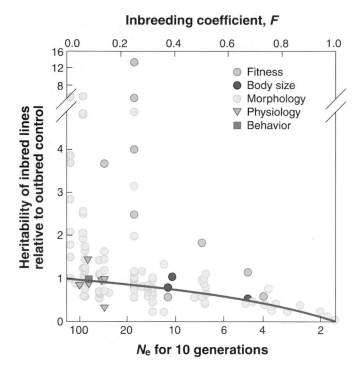

Figure 11.11 The effects of inbreeding coefficient (upper axis) or the effective population size that would result in the same inbreeding coefficient (lower axis) on heritability estimates in laboratory studies of experimental populations. Study organisms include *Drosophila* (ten studies), house flies (five studies), *Tribolium* beetles (two studies), two plant species, laboratory mice and a butterfly. Heritability estimates are expressed as a ratio relative to outbred control populations. The solid line indicates the expected decline for purely additive genetic variation. From Willi *et al.* (2006). See Color Plate 5.

This suggests that the effects of population bottlenecks on additive genetic variation in the short term are unpredictable and trait-specific, with heritability often increasing. However, in the longer term, the genetic variation available for adaptation is expected to decline, reducing the capacity for populations to adapt to new conditions.

Field studies of the relationship between population size and population V_A have produced mixed results (Willi *et al.* 2006). While some studies have found that smaller populations have lower H_N than larger populations, others have found no relationship. A meta-analysis by Willi *et al.* found the correlation between V_A and H_N and population census size was only 0.0044. They summarized the possible reasons why population size and V_A are not correlated in some field studies as: (1) the current population size may have decreased recently and the genetic effect of that decline has not yet manifested; (2) if stabilizing rather than directional selection is acting, then V_A will not decrease until popu-

lation size is very low; (3) gene flow among populations may be increasing effective population size; and (4) the smallest, most vulnerable populations may have already been extirpated prior to being sampled.

11.4.2 Effects of selection

In general, natural selection will use up or remove additive genetic variation for a trait, so alleles that increase fitness should increase in frequency until they reach fixation, while alleles associated with reduced fitness will be reduced in frequency and eventually be lost from the population. Therefore, we expect that traits that are strongly associated with fitness should have lower heritabilities than traits that are under weak or no natural selection. A vast body of empirical information in many species supports this expectation (Mousseau and Roff 1987, Roff and Mousseau 1987). For example, Kruuk *et al.* (2000) found a strong negative association

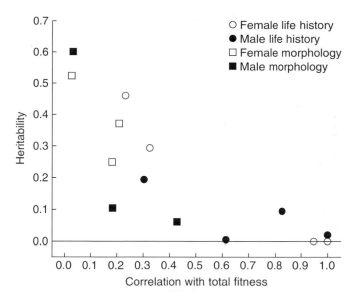

Figure 11.12 Relationship between the narrow-sense heritability of traits and their correlation with total fitness in a population of red deer. These results demonstrate that morphometric traits tend to have greater heritabilities than life history traits, and traits with a greater effect on fitness tend to have lower heritabilities. From Kruuk *et al.* (2000).

between the heritability of traits and their association with fitness in a wild population of red deer (Figure 11.12). However, Houle (1992) analyzed *Drosophila* heritability estimates from many studies, and concluded that life history traits have lower H_N, not due to lower V_A, but due to higher environmental variance, V_E, resulting in greater phenotypic variance, V_P. He suggested that the additive genetic coefficient of variation, estimated as V_A divided by the trait mean, is a better measure of adaptive variation.

We do not expect strong selection for a trait to remove all additive genetic variation for highly polygenic traits. A famous long-term selection experiment in maize in Illinois has been selecting separate lines for high and low oil and protein content for over 100 years (Dudley 2007). Oil and protein continue to increase, on average, in lines selected for increasing content (Figure 11.13). The average oil content of high-oil lines is now more than four-fold higher than the initial population, and protein has increased three-fold! Less surprising is that limits to selection on the low oil and low protein lines eventually plateaued, although they still contained genetic variation as evidenced by the responses to selection in the reverse low selection lines (see RLO line in Figure 11.13). QTL studies have shown that many loci are involved in determining both oil and protein content. Explanations for the long-term continued response to

selection include (1) low initial frequencies of alleles conferring higher oil or protein content at many of these loci; (2) release of epistatic variation over time; (3) mutations in QTLs underlying these traits adding phenotypic variation; and (4) a change in environments over time, for example, increased availability of nitrogen from fertilization may have allowed alleles for higher protein to be expressed and selected (Dudley 2007).

11.5 DIVERGENCE AMONG POPULATIONS

Quantitative genetics can also be used to understand genetic differentiation among local populations within species (Merilä and Crnokrak 2001). This approach has been especially interesting when used to understand the patterns of local adaptation caused by differential natural selection acting on heritable traits. We also expect local populations of species to differ for quantitative traits because of the effects of genetic drift. Understanding the relative importance of genetic drift and natural selection as determinants of population differentiation is an important goal when studying quantitative traits.

We saw in Chapter 9 that the amount of genetic differentiation among populations is often estimated by

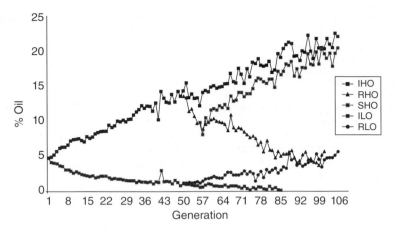

Figure 11.13 The effects of over a century of divergent selection for oil content in corn, showing the persistence of additive genetic variation. All selection lines started with the same open-pollinated corn cultivar in 1896. Each line indicates selection on a different strain of corn. IHO, Illinois High Oil, was selected continually for increased oil content for 106 generations. RHO, Reverse High Oil, was selected first for high oil for 48 generations, and then for low oil for the remaining generations. SHO, Switchback High Oil, was selected for high oil for 48 generations, low oil for seven generations, and then high oil for the remaining generations. ILO, Illinois Low Oil, was selected for decreased oil content for 106 generations. RLO, Reverse Low Oil was selected for low oil content for 48 generations and then for high oil for the remaining generations. From Dudley (2007).

the fixation index F_{ST}, which is the proportion of the total genetic variation that is due to genetic differentiation among local populations. An analogous measure of population differentiation for quantitative traits has been termed Q_{ST} (Spitze 1993):

$$Q_{ST} = \frac{V_{AB}}{2V_{AW} + V_{AB}} \qquad (11.17)$$

where V_{AB} is the additive genetic variation due to differences among populations, and V_{AW} is the mean additive genetic variation within populations. Q_{ST} is expected to have the same value as F_{ST} if it is estimated from the allele frequencies at the loci affecting the quantitative trait under investigation. This assumes that the local populations are in Hardy-Weinberg proportions ($F_{IS} = 0$) and the loci are in gametic equilibrium.

Q_{ST} is more difficult to estimate than F_{ST} because it requires common garden experiments containing multiple populations to distinguish genetic differences among populations from environmental influences on the trait. It also requires known genetic relationships among individuals within those populations to allow for partitioning within-population variation into genetic and environmental components similar to estimating heritability. These requirements make it difficult to estimate Q_{ST} except in those plant and animal species that can be readily raised in experimental conditions (e.g., *Drosophila* and *Daphnia*).

Some authors have used P_{ST}, the proportion of total phenotypic variation due to differences among populations, as a surrogate for Q_{ST} in species where it is difficult to estimate additive genetic variation within and among populations (Leinonen *et al.* 2008, Raeymaekers *et al.* 2007). Environmental effects can obscure the true amount of quantitative trait divergence if quantitative divergence is estimated from wild phenotypes. Population divergence can be overestimated in cases where phenotypic variation is primarily a plastic response to the environment, or underestimated in cases where the environmental effects produce reduced phenotypic variation, even if genetic divergence is high. A meta-analysis by Leinonen *et al.* (2008) concluded that studies using P_{ST} do not tend to yield higher estimates than common garden Q_{ST} studies that use estimated additive genetic variation within and among populations.

The comparison of F_{ST} and Q_{ST} can provide valuable insight into the effects of natural selection on quantitative traits. F_{ST} is relatively easy to measure at a wide variety of molecular markers that are generally assumed not to be strongly affected by natural selection. Thus, the value of F_{ST} depends only on local effective population sizes (genetic drift) and dispersal (gene

Table 11.3 Possible relationships in natural populations of divergence at neutral molecular markers (F_{ST}) and a quantitative trait (Q_{ST}).

Result	Interpretation
$Q_{ST} > F_{ST}$	Differential directional selection on the quantitative trait
$Q_{ST} = F_{ST}$	The amount of differentiation at the quantitative trait is similar to that expected by genetic drift alone
$Q_{ST} < F_{ST}$	Natural selection favoring the same phenotype in different populations

flow) among local populations. Consequently, differences between F_{ST} and Q_{ST} can be attributed to the effects of natural selection on the quantitative trait.

There are three possible relationships between F_{ST} and Q_{ST} (Table 11.3). First, if $Q_{ST} > F_{ST}$, then the degree of differentiation in the quantitative trait exceeds that expected by genetic drift alone, and consequently, directional natural selection favoring different phenotypes in different populations must have been involved to achieve this much differentiation. Second, if Q_{ST} and F_{ST} estimates are roughly equal, the observed degree of differentiation at the quantitative trait is the same as expected with genetic drift alone. Finally, if $Q_{ST} < F_{ST}$, the observed differentiation is less than that to be expected on the basis of genetic drift alone. This means that natural selection must be favoring the same mean phenotype in different populations. While these comparisons are simple qualitatively, making statistically rigorous comparisons is challenging. A single trait provides a single Q_{ST} estimate; yet this must be compared to the distribution of F_{ST} estimates across loci rather than only to the mean to determine whether the difference between the two is significant (Whitlock 2008, Whitlock and Guillaume 2009).

Laboratory experiments with house mice have supported the validity of this approach to understand the effects of natural selection on quantitative traits (Morgan *et al.* 2005). Comparison of Q_{ST} and F_{ST} in laboratory lines with known evolutionary history generally produced the correct evolutionary inference in the interpretation of comparisons within and between lines. In addition, Q_{ST} was relatively greater than F_{ST} for those traits for which strong directional selection was applied between lines.

11.6 QUANTITATIVE GENETICS AND CONSERVATION

How should we incorporate quantitative genetic information into conservation and management programs? Quantitative genetics has not played a major role in conservation genetics to date, a field that has been dominated by studies of molecular genetic variation at individual loci. Some have argued that quantitative genetic approaches may be more valuable than molecular genetics in conservation, since quantitative genetics allows us to study traits associated with fitness rather than just markers that are neutral or nearly neutral with respect to natural selection (Storfer 1996). This might be particularly important in light of anthropogenic climate change (Kramer and Havens 2009, Sgrò *et al.* 2011). Others suggest that quantitative genetic parameters are too difficult to estimate and have errors too large to be useful.

Quantitative genetics provides an invaluable conceptual basis for understanding genetic variation and evolutionary potential in populations. This approach is important for many crucial issues in conservation. For example, quantitative genetics is essential for understanding inbreeding depression (Chapter 13). If inbreeding depression is caused by a few loci with major effects, then the alleles responsible for inbreeding depression may be 'purged' from a population. However, inbreeding depression caused by many loci with small effects will be extremely difficult to purge (see discussion in Section 13.6). Similar considerations come into play with such important issues as the loss of evolutionary potential in small populations, and the minimum effective population size required to maintain adequate genetic variation to increase the probability of long-term survival of populations and species.

Small populations experiencing prolonged bottlenecks will eventually lose additive genetic variation. While genetic variation can increase in the short term following a bottleneck, this effect is not expected to increase the adaptive potential of populations in the longer term (Willi *et al.* 2006). Individuals in small, threatened populations often have reduced mean fitness due to factors including environmental stress and inbreeding, and the loss of adaptive variation will further contribute to their risk of population extirpation and species extinction.

The application of quantitative genetics to conservation has differed between plants and animals. With plant species, there is a history of using common garden experiments to guide conservation activities, including identifying suitable seed source populations for species and ecosystem restoration and quantifying how far seed can be moved from collection population to restoration sites or for genetic rescue without resulting in maladaptation, particularly for tree species but increasingly for other types of plants (Hufford and Mazer 2003). Quantitative genetic approaches and common garden experiments are being used to predict the capacity for adaptation and impacts of climate change on local populations (Aitken *et al.* 2008, Kramer and Havens 2009). Information on the additive and non-additive genetic variation, heritability and extent of resistance to an introduced pathogen has also been critical for the development of conservation strategies in some cases. For example, whitebark pine and limber pine populations are being inoculated with the introduced fungal pathogen causing white pine blister rust in common gardens to determine the quantitative genetic basis and frequency of resistance (Schoettle and Sniezko 2007). This information is being used to model the demographic viability of populations under various scenarios to inform conservation strategies (Schoettle *et al.* 2011).

For most animal species, common garden experiments are not feasible, and estimates of heritability or Q_{ST} in the wild are often problematic and may not be informative. However, in some cases genomic techniques are now allowing for estimation of the amount and distribution of quantitative trait variation using marker-based approaches. With the identification of candidate genes for adaptation in model species, it is rapidly becoming feasible to compare F_{ST} values for neutral markers with those loci that may be of adaptive significance to identify markers for local adaptation (Stapley *et al.* 2010). Heritability estimates in the wild are improving due to marker-based pedigree reconstruction and the animal model. The large standard errors associated with heritability estimates (Storfer 1996) and the statistical challenges of Q_{ST} analysis still make the use of neutral molecular estimates for population diversity and divergence more attractive for many species. Nevertheless in some cases, estimates of quantitative genetic parameters for specific adaptive traits will provide useful information about the capacity of populations to adapt to specific conditions, such as climatic changes, or to persist in the face of introduced diseases.

11.6.1 Response to selection in the wild

Quantitative genetics provides a framework for understanding and predicting the possible effects of selection

acting on wild populations (see Guest Box 8). For example, many commercial fisheries target particular age or size classes within a population (Conover and Munch 2002). In general, larger individuals are more likely to be caught than smaller individuals. The expected genetic effect (i.e., response, R) of such selectivity on a particular trait depends upon the selection differential (S), and the narrow-sense heritability (H_N), as we saw in equation 11.9. We saw in Figure 2.5 that the mean size of pink salmon caught off the coast of North America declined between 1950 and 1974 (12 generations), apparently because of size-selective harvest; approximately 80% of the returning adult pink salmon were harvested in this period (Ricker 1981). The mean reduction in body weight in 97 populations over these years was approximately 28%.

As we have seen, additive genetic variation is necessary for a response to natural selection. Empirical studies have found that virtually every trait that has been studied has some additive genetic variance (i.e., $H_N > 0$; Roff 2003). Thus, evolutionary change may be slowed by a loss of genetic variation for a trait, but it is not expected to be prevented in most cases. An exception to this has been found in a rainforest species of *Drosophila* in Australia (Hoffmann *et al.* 2003). These authors found no response to selection for resistance to desiccation after 30 generations of selection! A parent–offspring regression analysis estimated a narrow-sense heritability of zero with the upper 95% confidence value of 0.19. This result is especially puzzling since there are clinal differences between populations for this trait. This suggests there has been a history of selection and response for this trait. This is not the only report of lack of response to traits associated with the effects of global warming. Baer and Travis (2000) suggested that the lack of genetic variation was responsible for a lack of response to artificial selection for acute thermal stress tolerance in a live-bearing fish.

Genetic correlations among traits can also constrain the response to selection in the wild when there is substantial additive genetic variation for a trait (Section 11.2.2). For example, Etterson and Shaw (2001) studied the evolutionary potential of three populations of a native annual legume in tallgrass prairie fragments in North America to respond to the warmer and more arid climates predicted by global climate models. Despite substantial heritabilities for the traits under selection, between-trait genetic correlations that were antagonistic to the direction of selection will limit the adaptive evolution of these populations in response to warming. The predicted rates of evolutionary response, taking genetic correlations into account, were much slower than the predicted rate of change with heritabilities alone, unless genetic correlations are favorable between traits and thus reinforcing (i.e., the correlated effect from selection on one trait increases fitness in the other).

11.6.2 Can molecular genetic variation within populations estimate quantitative variation?

As we saw in Section 11.4, we expect an equivalent loss of molecular heterozygosity and additive genetic variation during a bottleneck. Therefore, comparison of heterozygosity may provide a good estimate of the loss of quantitative variation. Briscoe *et al.* (1992) studied

the loss of molecular and quantitative genetic variation in laboratory populations of *Drosophila*. They found that substantial molecular and quantitative genetic variation was lost during captivity, even though census sizes were of the order of 5000 individuals. More importantly for our question, they found that heterozygosity at nine allozyme loci provided an excellent estimate of the loss of quantitative genetic variation as measured by H_N for sternopleural bristle number. Some studies from natural populations have also found that molecular genetic variations support this result (Example 11.2). However, a meta-analysis by Reed and Frankham (2001) found no relationship between heterozygosity and heritability across 19 studies.

The relationship between molecular and quantitative variation is not so simple. Bottlenecks have been

---

Example 11.2 A tale of two pines: lack of molecular variation corresponds to a lack of quantitative variation in red pine but not stone pine

Conifers are generally genetically diverse species, with high levels of variation for molecular markers and quantitative traits due to large populations, low F_{ST} values due to high levels of gene flow via pollen, and relatively high Q_{ST} values for some quantitative traits indicating strong adaptation to local climates (Savolainen *et al.* 2007, see also Example 2.2 for Sitka spruce). Two pines, red pine and stone pine, do not fit this norm. The red pine is a common species with a broad range across eastern North America. The stone pine is a common circum-Mediterranean species well known for its edible seeds, which are a common source of 'pine nuts'.

Molecular genetic variation is extremely low in red pine. Allozyme studies have found little or no variation; three of the four rare alleles detected in red pine were null alleles that produced no detectable enzyme (Fowler and Morris 1977, Allendorf *et al.* 1982, Simon *et al.* 1986, Mosseler *et al.* 1991). The estimated expected heterozygosity at the species level was 0.001 (Allendorf *et al.* 1982). Mosseler *et al.* (1992) also found almost no variation for a number of randomly amplified polymorphic markers in red pine. Some genetic variation has been found in chloroplast microsatellites from red pine (Echt *et al.* 1998). For example, Walter and Epperson (2001) found genetic variation in ten chloroplast microsatellite loci in individuals collected throughout the range of red pine. Only six choloroplast haplotypes were found, and 78% of all trees had the same haplotype. Studies have

also detected very little genetic variation for quantitative characters in this species (Fowler and Lester 1970).

Stone pine is nearly as genetically depauperate as red pine for nuclear genetic markers. It has an expected heterozygosity for allozymes at the species level of just 0.015 (Fallour *et al.* 1997). It has even less chloroplast diversity than red pine, with just four haplotypes found using 12 cpDNA microsatellite markers across the species range (Vendramin *et al.* 2008). Of 34 populations sampled, 29 were fixed for a single haplotype. However, quantitative variation has been found among populations for a number of growth traits in a common garden experiment (Court-Picon *et al.* 2004), and broad-sense heritability (H_B) has been estimated as 0.17 and 0.20 for cone weight and seed production, respectively, in a grafted clone bank (Vendramin *et al.* 2008).

Both of these pines likely underwent a severe and prolonged bottleneck associated with glaciation within the last 20,000 years. There has not been enough time for these species to recover genetic variation for molecular genetic markers. However, stone pine either maintained some quantitative genetic variation through this bottleneck or has recovered some variation through mutations at loci underlying those traits since the bottleneck. This illustrates the complementary nature of molecular genetic markers and quantitative traits, and some of the uncertainties around predicting one from the other.

found to increase the heritability for some traits, but will reduce additive genetic variation for others and are always expected to reduce molecular genetic variation. In addition, different types of genetic variation will recover at different rates from a bottleneck because of different mutation rates (Section 12.5). The overall high effective mutation rate for quantitative traits because they are polygenic (i.e., a mutation at any one of the causal loci can result in variation for that trait) means that quantitative genetic variation may recover from a bottleneck more quickly than molecular quantitative variation (Lande 1996, Lynch 1996) (see Example 11.2). In addition, even populations with fairly low effective population sizes can maintain enough additive genetic variation for substantial adaptive evolution (Lande 1996). We also expect a weak correlation between molecular and quantitative genetic variation because of statistical sampling. Substantial variation in additive genetic variation among small populations is expected (Lynch 1996). There may also be large differences among quantitative traits because they will be under different selection pressures, with some under divergent selection resulting in greater differentiation among populations, and some under stabilizing selection with the same phenotype favored in all populations.

Therefore, the amount of molecular genetic variation within a population should be used carefully to make inferences about quantitative genetic variation. The closer the relationship between the populations being compared, the more informative the comparison will be. The strongest case is the comparison of a single population at different times. Reduced molecular genetic variation over time is likely to reflect loss of genetic variation at the genes responsible for additive genetic variation. However, comparisons between species will not be very informative for the reasons described in the preceding paragraph. Therefore, low molecular genetic variation within a species should not be taken to mean that adaptive evolution is not possible because of the absence of additive genetic variation.

11.6.3 Does population divergence for molecular markers estimate divergence for quantitative traits?

Is F_{ST} a good predictor of adaptive divergence, Q_{ST}? Leinonen et al. (2008) conducted a meta-analysis of published F_{ST} and Q_{ST} estimates. The correlation between the two was positive but weak. Adaptive divergence of populations was better predicted by F_{ST} for species with greater population differentiation; this relationship broke down for low-F_{ST} species, suggesting that neutral markers predict adaptive divergence poorly for widespread species with large populations and high levels of gene flow (e.g., wind-pollinated tree species and marine fish). In addition, differentiation at quantitative traits (Q_{ST}) typically exceeds differentiation at molecular markers (F_{ST}). This suggests a prominent role for natural selection in determining patterns of differentiation at quantitative trait loci.

A comparison of quantitative and molecular markers is a useful approach for understanding the role of natural selection and drift in determining patterns of differentiation in natural populations. Nevertheless, caution is needed in making these comparisons and in generalizing their results. Rigorous statistical comparisons are challenging, as previously discussed (Whitlock and Guillaume 2009). It is also somewhat surprising that Q_{ST} almost always exceeds F_{ST} (Leinonen et al. 2008) given that so many quantitative traits seem to be under stabilizing selection. This result suggests that different local populations almost always have different optimum phenotypic values. Another interpretation is that the optimum mean value is similar in different populations but that environmental differences result in different combinations of genotypes producing a similar phenotype (see countergradient selection in Section 2.5). This may also result from a bias in the traits selected for study in a given organism. Researchers are more likely to phenotype populations for a trait they suspect is related to local adaptation.

Once again, in the absence of quantitative genetic information, the amount of molecular genetic variation between populations should be used carefully to make inferences about quantitative genetic variation. Substantial molecular genetic divergence between populations suggests some isolation between these populations, and therefore provides strong evidence for the *opportunity* for adaptive divergence. And it is fair to say that some adaptive differences are *likely* to occur between populations that have been isolated long enough to accumulate substantial molecular genetic divergence. However, the reverse is not true. Lack of molecular genetic divergence should not be taken to suggest that adaptive differences do not exist. As we saw in Section 9.7, even fairly weak natural selection can have a profound effect on the amount of genetic divergence among populations.

Guest Box 11   Response to trophy hunting in bighorn sheep
David W. Coltman

Bighorn sheep populations are often managed to provide a source of large-horned rams for trophy hunting. In many places, strict quotas of the number of rams that may be harvested each year are enforced through the use of a lottery system for hunting permits. However, in other parts of their endemic range, any ram that reaches a minimum legal horn size can be taken during the annual autumn hunting season. In one population of bighorn sheep at Ram Mountain, Alberta, Canada, a total of 57 rams were harvested under such an unrestricted management regime over a 30-year period. This corresponded to an average harvest rate of about 40% of the legal sized rams in a given year, with the average age of a ram at harvest of six years. Since rams in this population do not generally reach their peak reproductive years until eight years of age (Coltman *et al.* 2002), hunters imposed an artificial selection pressure on horn size that had the potential to elicit an evolutionary response, provided that the horn size was heritable.

The heritability of horn size, or any other quantitative trait, can be estimated using pedigree infor-

mation. Mother–offspring relationships in the Ram Mountain population were known through observation, and father–offspring relationships were determined using microsatellites for paternity (Marshall *et al.* 1998) and sibship analyses (Goodnight and Queller 1999). An 'animal model' analysis (see Section 11.1.3.3) was conducted, which uses relatedness across the entire pedigree to estimate narrow-sense heritability (H_N) using maximum likelihood. Horn length was found to be highly heritable, with $H_N = 0.69$ (Coltman *et al.* 2003).

Examination of individual 'breeding values' (which is twice the expected deviation of each individual's offspring from the long-term population mean) revealed that rams with the highest breeding values were harvested earliest (Figure 11.14a) and therefore had lower fitness than rams of lower breeding value. As a consequence the average horn length observed in the population has declined steadily over time (Figure 11.14b). Unrestricted harvesting has therefore contributed to a decline in the trait that determines trophy quality by selectively targeting rams of high genetic quality before their reproductive peak.

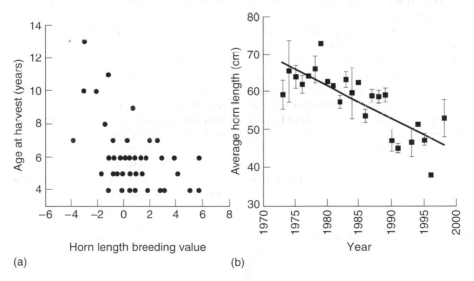

(a)

(b)

Figure 11.14   (a) Relationship between the age at harvest for trophy-harvested rams and their breeding value. (b) Relationship between mean (±SE) horn length (cm) of 4-year-old rams and year (*n* = 119 rams).

CHAPTER 12

Mutation

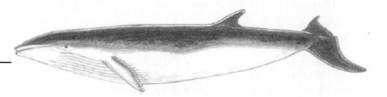

Minke whale, Example 12.2

Mutation is the ultimate source of all the genetic variation necessary for evolution by natural selection; without mutation evolution would soon cease.

Michael C. Whitlock and Sarah P. Otto (1999)

Mutations can critically affect the viability of small populations by causing inbreeding depression, by maintaining potentially adaptive genetic variation in quantitative characters, and through the erosion of fitness by accumulation of mildly deleterious mutations.

Russell Lande (1995)

Chapter Contents

Conservation and the Genetics of Populations, Second Edition. Fred W. Allendorf, Gordon Luikart, and Sally N. Aitken.
© 2013 Fred W. Allendorf, Gordon Luikart and Sally N. Aitken. Published 2013 by Blackwell Publishing Ltd.

Mutations are errors in the transmission of genetic information from parents to progeny. The process of mutation is the ultimate source of all genetic variation in natural populations. Nevertheless, this variation comes at a cost because most mutations that have phenotypic effect are harmful (deleterious). Mutations occur both at the chromosomal level and the molecular level. As we will see, mutations may or may not have a detectable effect on the phenotype of individuals.

An understanding of the process of mutation is important for conservation for several reasons. The amount of standing genetic variation within populations is largely a balance between the gain of genetic variation from mutations and the loss of genetic variation from genetic drift. Thus, an understanding of mutation is needed to interpret patterns of genetic variation observed in natural populations.

Moreover, the increased homozygosity of deleterious mutations is the primary source of inbreeding depression (see Chapter 13). The frequency of deleterious mutations results from a balance between mutation and natural selection (see Section 12.3). We have seen that natural selection is less effective in small populations (see Section 8.5). Therefore, deleterious mutations will tend to accumulate more rapidly in small populations; this effect can further threaten the persistence of small populations (see Section 14.7). On the other hand, the rate of adaptive response to environmental change is proportional to the amount of standing genetic variation for fitness within populations. Thus, long-term persistence of populations may require large population sizes in order to maintain important adaptive genetic variation.

Unfortunately, there are few empirical data available concerning the process of mutation because mutation rates are so low. The data that are available generally come from a few model organisms (e.g., mice, *Drosophila*, or *Arabidopsis*) that are selected because of their short generation time and suitability for raising a large number of individuals in the laboratory. However, we must be careful in generalizing results from such model species; the very characteristics that make these organisms suitable for these experiments may make them less suitable for generalizing to other species. For example, the per-generation mutation rate tends to be greater for species with longer generation length (Drake *et al.* 1998).

How common are mutations? On a per-locus or per-nucleotide level they are rare. For example, the rate of mutation for a single nucleotide is of the order of a few per billion gametes per generation. However, from a genome-wide perspective, mutations are actually very common. The genome of most species consists of billions of base pairs. Therefore, it has been estimated (Lynch *et al.* 1999) that each individual may possess hundreds of new mutations! Fortunately, almost all of these mutations are in nonessential regions of the genome and have no phenotypic effect.

We will consider the processes resulting in mutations and examine the expected relationships between mutation rates and the amount of genetic variation within populations. We will examine evidence for both harmful and advantageous mutations in populations. Finally, we will examine the effects of mutation rates on the rate of recovery of genetic variation following a population bottleneck.

12.1 PROCESS OF MUTATION

Chromosomes and DNA sequences are normally copied exactly during the process of replication and are transmitted to progeny. Sometimes, however, errors occur that produce new chromosomes or new DNA sequences. Empirical information on the rates of mutation is hard to come by because mutations are so rare. Thousands of progeny must be examined to detect mutational events. Thus, estimating the rates of mutations or describing the types of changes brought about by mutation is generally incredibly difficult. Most of our direct information about the process of mutation comes from model organisms (see Example 12.1).

Mutations can occur in germline cells, and these gametic mutations occur primarily during meiosis. In animals, only gametic mutations result in genetic variation passed on to progeny. However, in plants, mutations during mitosis in somatic cells can result in variation that can be passed when reproductive structures are produced by lineages of vegetative cells, e.g., flowers produced at the top of the plant or ends of branches. In long-lived or large organisms such as trees, somatic mutations may be a substantive source of variation. Ally *et al.* (2010) found evidence that somatic mutations in long-lived aspen clones reduce fertility and may eventually lead to senescence.

Most mutations with phenotypic effects tend to reduce fitness (Figure 12.1). Thus, as we will see in Chapter 14, the accumulation of mutations can decrease the probability of survival of small populations. Nevertheless, rare beneficial mutations are

Example 12.1   Coat color mutation rate in mice

Schlager and Dickie (1971) presented the results of a direct and massive experiment to estimate the rate of mutation to five recessive coat color alleles in mice: albino, brown, non-agouti, dilute, and leaden. They examined more than seven million mice in 28 inbred strains. Overall, they detected 25 mutations in over two million gene transmissions for an average mutation rate of 1.1×10^{-5} per gene transmission. As expected, the reverse mutation rate (from the recessive to the dominant allele) was much lower, approxi-

mately 2.5×10^{-6} per gene transmission. The reverse mutation rate is expected to be lower because there are more ways to eliminate the function of a gene than to reverse a defect. This assumes that the recessive coat color mutations are cause by mutations that cause loss of function (similar to the null alleles for allozyme loci).

Guest Box 12 presents a modern molecular study of the mutation process in pocket mice that produces a melanistic or dark morph.

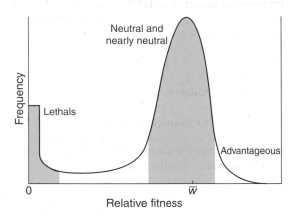

Figure 12.1   Hypothetical frequency of the fitness of new mutations relative to the mean fitness of the population ($\bar{w}$). Redrawn from Hedrick (2011).

important in adaptive evolutionary change (Elena *et al.* 1996). In addition, some recent experimental work with the plant *Arabidopsis thaliana* has suggested that nearly half of new spontaneous mutations increase fitness (Shaw *et al.* 2002); however, this result has been questioned in view of data from other experiments (Bataillon 2003, Keightley and Lynch 2003).

Mutations occur randomly, but there is some evidence that some aspects of the process of mutation may be an adaptive response to environmental conditions. There has been an ongoing controversy that mutations in prokaryotes may be directed toward particular environmental conditions (Lenski and

Sniegowski 1995). In addition, there is some evidence that the rate of mutations in eukaryotes may increase under stressful conditions, and thus create new genetic variability that may be important in adaptation to changing environmental conditions (Capy *et al.* 2000). However, there is little evidence for an influence of the environment on the effects of new mutations (Halligan and Keightley 2009).

12.1.1 Chromosomal mutations

We saw in Chapter 2 that rates of chromosomal evolution vary tremendously among different taxonomic groups. There are two primary factors that may be responsible for differences in the rate of chromosomal change: (1) the rate of chromosomal mutation; and (2) the rate of incorporation of such mutations into populations (Rieseberg 2001). Differences between taxa in rates of chromosomal change may result from differences in either of these two effects.

White (1978) estimated a general mutation rate for chromosomal rearrangements on the order of one per 1000 gametes in a wide variety of species from lilies to grasshoppers to humans. Lande (1979) considered different forms of chromosomal rearrangements in animals and produced a range of estimates between 10^{-4} and 10^{-3} per gamete per generation. There is evidence in some groups that chromosomal mutation rates may be substantially higher than this. For example, Porter and Sites (1987) detected spontaneous chromosomal mutations in five of 31 males that were examined.

The apparently tremendous variation in chromosomal mutation rates suggests that some of the differences between taxa could result from differences in mutation rates. In addition, there is some evidence that chromosomal polymorphisms may contribute to increased chromosomal mutation rates. That is, chromosomal mutation rates may be greater in chromosomal heterozygotes than homozygotes (King 1993). We will see in Section 12.1.4 that genomes with more transposable elements may have higher chromosomal mutation.rates.

12.1.2 Molecular mutations

There are several types of molecular mutations in DNA sequences: (1) substitutions, the replacement of one nucleotide with another; (2) recombinations, the exchange of a sequence from one homologous chromosome to the other; (3) deletions, the loss of one or more nucleotides; (4) insertions, the addition of one or more nucleotides; and (5) inversions, the rotation by 180° of a double-stranded DNA segment of two or more base pairs (see Graur and Li 2000).

The rate of spontaneous mutation is very difficult to estimate directly because of their rarity. Mutation rates are sometimes estimated indirectly by an examination of rates of substitutions over evolutionary time in regions of the genome that are not affected by natural selection. The expected rate of substitution per generation will be equal to the mutation rate for selectively neutral mutations (Kimura 1983). The average rate of substitution in mammalian nuclear DNA has been estimated to be $3–5 \times 10^{-9}$ nucleotide substitutions per nucleotide site per year (Graur and Li 2000). Thus, mutation rates for SNPs are on the order of 10^{-8} mutants per gene transmission (Nachman and Crowell 2000). The mutation rate, however, varies enormously between different regions of the nuclear genome. The rate of mutation in mammalian mitochondrial DNA has been estimated to be at least ten times higher than the average nuclear rate.

The mutation rate for protein coding loci (e.g., allozymes) is very low. In addition, not all DNA mutations will result in a change in the amino-acid sequence because of the inherent redundancy of the genetic code. Nei (1987, p. 30) reviewed the literature on direct and indirect estimates of mutation rates for allozyme loci. Most direct estimates of mutation rates in allozymes have failed to detect any mutant alleles; for example, Kahler et al. (1984) examined a total of 841,260 gene transmissions from parents to progeny at five loci and failed to detect any mutant alleles. General estimates of mutation rates for allozyme loci are of the order of 10^{-6} to 10^{-7} mutants per gene transmission (Nei 1987).

The rate of mutation at microsatellite loci is much greater than other regions of the genome because of the presence of simple sequence repeats (Li et al. 2002). Two mechanisms are thought to be responsible for mutations at microsatellite loci: (1) mispairing of DNA strands during replication; and (2) recombination. Estimates of mutation rates at microsatellite loci have generally been approximately 10^{-3} mutants per gene transmission (Ellegren 2000a) (Table 12.1). Microsatellite mutations appear largely to follow the **stepwise mutation model** (SMM) where single repeat units are added or deleted with near-equal frequency (Valdes

Table 12.1 Mutations at the *OGO1c* microsatellite locus in pink salmon (Steinberg et al. 2002). Approximately 1300 parent–progeny transmissions were observed in 50 experimental matings. Mutations were found only in the four families shown. The mutant allele is indicated by bold-faced type and the most likely progenitor of the mutant allele is underlined. All of the putative mutations differ by one repeat unit from their most likely progenitor. The overall mutation rate estimated from these data is 3.9×10^{-3} (5/1300).

Dam	Sire	Progeny genotypes				
a/b	*c/d*	*a/c*	*a/d*	*b/c*	*b/d*	**Mutant genotypes**
342/350	408/<u>474</u>	1	1	3	3	342/**478**
295/366	303/<u>362</u>	1	2	4	2	295/**366**
269/420	346/<u>450</u>	8	16	10	8	420/**446** (2)
348/348	309/<u>448</u>	5	4	0	0	348/**444**

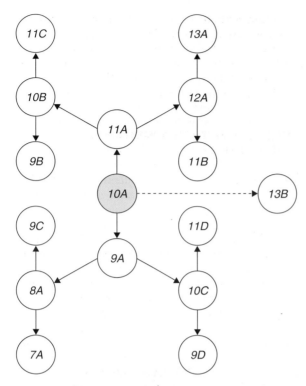

Figure 12.2 Pattern of mutation for microsatellites beginning with a single ancestral allele (shaded circle in the middle) with 10 repeats. Most mutations are a gain or loss of a single repeat (stepwise mutation model, SMM). The dashed arrow shows a multiple-step mutation from 9 to 13 repeats. Alleles are designated by the number of repeats and a letter which distinguishes homoplasic alleles, which are alike in state (number of repeats) but differ in origin.

et al. 1993) (Figure 12.2). However, the actual mechanisms of microsatellite mutation are much more complicated than this simple model (Estoup and Angers 1998, Li *et al.* 2002, Anmarkrud *et al.* 2008).

New mutations sometimes occur in clusters because they occur early during gametogenesis (Woodruff *et al.* 1996). Woodruff and Thompson (1992) found as many as 20% of new mutations in *Drosophila* represented clusters of identical mutant alleles sharing a common pre-meiotic origin. Cluster mutations at microsatellite loci have been found in several other species (Jones *et al.* 1999, Steinberg *et al.* 2002). The occurrence of clustered mutations results in non-uniform distributions of novel alleles in a population, which could influence interpretations of mutation rates and

patterns as well as estimates of genetic population structure. For example, Woodruff *et al.* (1996) have shown that mutant alleles that are part of clusters are more likely to persist and be fixed in a population than mutant alleles entering the population independently.

12.1.3 Quantitative characters

As we saw in Chapter 11, the amount of genetic variation in quantitative characters for morphology, physiology, and behavior that can respond to natural selection is measured by the additive genetic variance (V_A). The rate of loss of additive genetic variance due to genetic drift in the absence of selection is the same for the loss of heterozygosity (i.e., $1/2N_e$). The effective mutation rate for quantitative traits is much higher than the rate for single gene traits because mutations at many possible loci can affect a quantitative trait. The input of additive genetic variance per generation by mutation is V_m. The expected genetic variance at equilibrium between these two factors is $V_A = 2N_eV_m$ (Lande 1995, 1996).

Estimates of mutation rates for quantitative characters are very rare and somewhat unreliable. It is thought that V_m is roughly of the order of 10^{-3} V_A (Lande 1995). However, some experiments suggest that the great majority of these mutations are highly detrimental and therefore are not likely to contribute to the amount of standing genetic variation within a population. Thus, the effective V_m responsible for much of the standing variation in quantitative traits in natural populations may be an order of magnitude lower, 10^{-4} V_A (Lande 1996, Barton and Keightley 2002, Mackay 2010).

12.1.4 Transposable elements, mutation rates, and stress

Much of the genome of eukaryotes consists of sequences associated with transposable elements that possess an intrinsic capability to make multiple copies and insert themselves throughout the genome (Graur and Li 2000). For example, approximately half of the human genome consists of DNA sequences associated with transposable elements (Lynch 2001). This activity is analogous to the 'cut and paste' mechanism of a word processor. Transposable elements are potent agents of mutagenesis (Kidwell 2002). For example,

Clegg and Durbin (2000) found that nine out of ten mutations affecting flower color in the morning glory were the result of the insertion of transposable elements into genes. A consideration of the molecular basis of transposable elements is beyond our consideration (see chapter 7 of Graur and Li 2000). Nevertheless, the mutagenic activity of these elements is of potential significance for conservation.

Transposable elements can cause a wide variety of mutations. They can induce chromosomal rearrangements such as deletions, duplications, inversions, and reciprocal translocations. Kidwell (2002) has suggested that "transposable elements are undoubtedly responsible for a significant proportion of the observed karyotypic variation among many groups". In addition, transposable elements are responsible for a wide variety of substitutions in DNA sequences, ranging from insertion of the transposable element sequence to substitutions, deletions, and insertions of a single nucleotide (Kidwell 2002).

Stress has been defined as any environmental change that drastically reduces the fitness of an organism (Hoffmann and Parson 1997). McClintock (1984) first suggested that transposable element activity could be induced by stress. A number of transposable elements in plants have been shown to be activated by stress (Grandbastien 1998, Capy *et al.* 2000). Some transposable elements in *Drosophila* have been shown to be activated by heat stress, but other studies have not found an effect of heat shock (Capy *et al.* 2000). In addition, hybridization has also been found to activate transposable elements and cause mutations (Kidwell and Lisch 1998).

12.2 SELECTIVELY NEUTRAL MUTATIONS

Many mutations in DNA sequences have no phenotypic effect so that they are neutral with regard to natural selection (e.g., mutations in noncoding regions). In this case, the amount of genetic variation within a population will be a balance between the gain of variation by mutation and the loss by genetic drift (Figure 12.3). The distribution of neutral genetic variation among populations is primarily a balance between these two forces. Gene flow among subpopulations retards the process of differentiation until eventually a steady state may be reached between the opposing effects of gene flow and genetic drift. However, the process of mutation may also contribute to allele frequency divergence among populations in cases where the mutation rate approaches the migration gene flow rate.

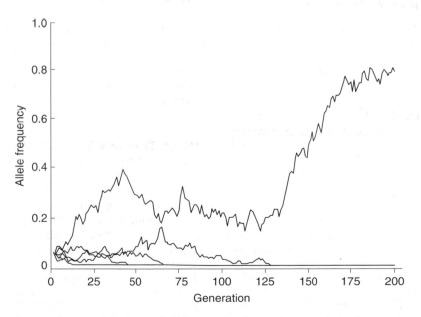

Figure 12.3 Simulations of genetic drift of neutral alleles introduced into a large population by mutation.

12.2.1 Genetic variation within populations

The amount of genetic variation within a population at equilibrium will be a balance between the gain of variation as a function of the neutral mutation rate (μ) and the loss of genetic variation by genetic drift as a function of effective population size (N_e).

We will first consider the so-called infinite allele model (IAM) in which we assume that every mutation creates a new allele that has never been present in the population. This model is appropriate if we consider variation in DNA sequences. A gene consists of a large number of nucleotide sites, each of which may be occupied by one of four bases (A, T, C, or G). Therefore, the total number of possible allelic states possible is truly a very large number! For example, there are over one million possible alleles if we consider just 10 base pairs ($4^{10} = 1,048,576$). In this case, the average expected heterozygosity (H) at a locus (or over many loci with the same mutation rate) is:

$$H = \frac{4N_e\mu}{(4N_e\mu+1)} = \frac{\theta}{\theta+1} \tag{12.1}$$

where μ is the neutral mutation rate and $\theta = 4N_e\mu$ (Kimura 1983). This relationship can be used to estimate effective population size if we know the mutation rate (see Example 12.2).

The much greater variation at microsatellite loci compared with allozymes results from the differences in mutation rates that we discussed in Section 12.1.2. Figure 12.4 shows the equilibrium heterozygosity for microsatellites, allozymes, and SNPs using equation 12.1. With these mutation rates, we expect a heterozygosity of 0.001 at SNPs, 0.038 at allozyme loci, and 0.80 at microsatellite loci with an effective population size of 10,000. However, we also expect a substantial amount of variation in heterozygosity between loci, especially loci with lower mutation rates (Figure 12.5).

The heterozygosity values for microsatellite loci in Figures 12.4 and 12.5 are likely to be overestimates because of several important assumptions. Microsatellite mutations tend to occur in steps of the number of repeat units. Therefore, each mutation will not be unique, but rather will be to an allelic state (say 11 copies of a repeat) that already occurs in the population. This is called **homoplasy**, in which two alleles that are identical in state have different origins (e.g., alleles *11C* and *11D* in Figure 12.2). Therefore the actual expected heterozygosity is less than predicted by equation 12.1. Allozymes also tend to follow a stepwise model of mutation (Ohta and Kimura 1973), but this will have a smaller effect because of the smaller number of alleles present in a population with a lower mutation rate.

It is also important to remember that the mutation rate used here (μ) is the neutral mutation rate. Mutations in DNA sequence within some regions of the genome are likely not to be selectively neutral. Therefore, different regions of the genome will have different effective neutral mutation rates, even though the

---

Example 12.2   How many whales are there in the ocean?

Equation 12.1 can also be used to estimate the effective population size of natural populations if we know the mutation rate (μ). For example, Roman and Palumbi (2003) estimated the historic (prewhaling) number of humpback, fin, and minke whales in the North Atlantic Ocean by estimating θ for the control region of mtDNA. In the case of mtDNA, $\theta = 2N_{e(f)}\mu$ because of maternal inheritance and haploidy. Roman and Palumbi (2003) used a range of mutation rates based on observed rates of divergence between mtDNA of different whale species.

Their genetic estimates of historic population sizes for humpback, fin, and minke whales are far greater than those previously calculated and are 6 to 20 times higher than the current population estimates for these species. This discrepancy is crucial for conservation because the International Whaling Commission management plan uses the estimated historic population sizes as guidelines for setting allowable harvest rates. We should be careful using estimates of N_e with this approach because there are a host of pitfalls (e.g., how reliable are our estimates of mutation rate?). Roman and Palumbi (2003) provide a useful discussion of the limitations of this method for estimating N_e. Palsbøll et al. (in press) have provided an insightful critique of using this method to estimate historic population sizes.

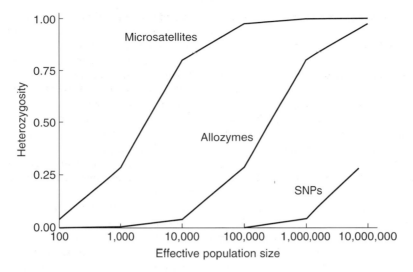

Figure 12.4 Expected heterozygosity in populations of different size using equation 12.1 for microsatellites ($\mu = 10^{-4}$), allozymes ($\mu = 10^{-6}$), and SNPs ($\mu = 10^{-8}$).

actual rate of molecular mutations is the same. For example, mutations in protein coding regions may affect the amino-acid sequence of an essential protein and thereby reduce fitness. Such mutations will not be neutral and therefore will not contribute to the amount of variation maintained by our model of drift–mutation equilibrium considered here. In these regions, so-called **purging** selection stops these mutations from reaching high frequencies in a population. In contrast, mutations in DNA sequence in regions of the genome that are not functional are much more likely to be neutral. This expectation is supported by empirical results; exons, which are the coding regions of protein loci, are much less variable than the introns which do not encode amino acids (Graur and Li 2000).

12.2.2 Population subdivision

The process of mutation may also contribute to allele frequency divergence among populations (see Section 9.8.1) (Ryman and Leimar 2008). The relative importance of mutation for divergence (e.g., F_{ST}) depends primarily upon the relative magnitude of the rates of migration and of mutation. Under the IAM of mutation with the island model of migration (Crow and Aoki 1984), the expected value of F_{ST} is approximately:

$$F_{ST} = \frac{1}{(1 + 4Nm + 4N\mu)} \tag{12.2}$$

Greater mutation rates will increase F_{ST} when new mutations are not dispersed at sufficient rates to attain equilibrium between genetic drift and gene flow. Under these conditions, new mutations may drift to substantial frequencies in the population in which they occur before they are distributed among other populations via gene flow (Neigel and Avise 1993). Mutation rates for allozymes and SNPs are generally less than 10^{-6} so that divergence at these loci is unlikely to be affected by different mutation rates unless the subpopulations are completely isolated. Mutation rates for some microsatellite loci may be as high as 10^{-2} (Anmarkrud et al. 2008).

Mutation under the IAM model may accelerate the rate of divergence at microsatellite loci among subpopulations that are very large and are connected by little gene flow. The actual expected effect of mutation is much more complicated than this, and it depends upon the mechanisms of mutation. For example, constraints on allele size at microsatellite loci under the stepwise mutation model (SMM) may reverse the direction of this effect (i.e., decrease the rate of divergence) under some conditions (Nauta and Weissing 1996).

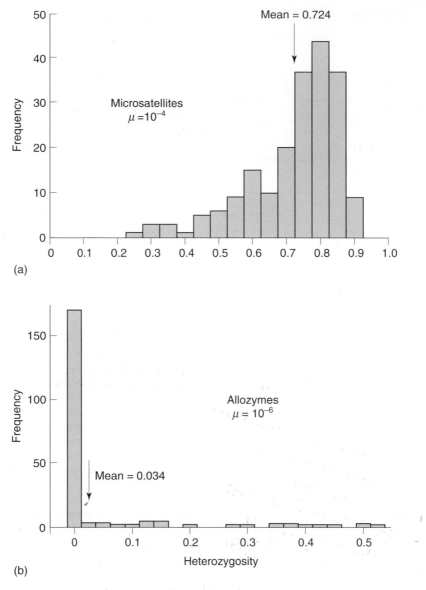

(a)

(b)

Figure 12.5 Simulated heterozygosities at 200 loci in a population with N_e = 10,000 and the infinite allele model of mutation produced with the program EASYPOP (Balloux 2001). (a) Microsatellite loci with $\mu = 10^{-4}$. (b) Allozyme loci with $\mu = 0^{-6}$. The expected heterozygosities are 0.800 and 0.038 (equation 12.1). The expected heterozygosity for SNPs with $\mu = 10^{-8}$ is <0.001; all loci were monomorphic in most simulations of SNPs. Thus, many basepairs must be screened in order to detect SNPs with high heterozygosity (>0.2).

In general, mutations will have an important effect on population divergence only when the migration rates are very low (say 10^{-3} or less) and the mutation rates are high (10^{-3} or greater; Nichols and Freeman 2004, Epperson 2005). However, as we saw in Section 9.8, F_{ST} will underestimate genetic divergence at loci with very high within-deme heterozygosities (H_S; Hedrick 1999). Large differences in H_S caused by differences in mutation rates among loci (e.g., Steinberg et al. 2002) can result in discordant estimates of F_{ST} among

Table 12.2   Estimates of effective population size (N_e) with computer simulations using equation 12.1 for a series of 20 subpopulations (local $N_e = 200$) that are connected by different amounts of gene flow with an island model of migration (EASYPOP, Balloux 2001). A mutation rate of 10^{-4} was used to simulate the expected heterozygosities at 100 microsatellite loci. The simulations began with no genetic variation in the first generation and ran for 10,000 generations. F_{ST}^* is the expected F_{ST} with this amount of gene flow corrected for a finite number of populations (Mills and Allendorf 1996). $\hat{N}_e$ is the estimated effective population size based upon the mean expected local heterozygosity (H_S) using equation 12.1.

mN	H_T	H_S	F_{ST}	F_{ST}^*	$\hat{N}_e$
0	0.814	0.076	0.907	1.000	205
0.5	0.665	0.477	0.283	0.311	2274
1.0	0.635	0.516	0.187	0.184	2667
2.0	0.621	0.558	0.100	0.101	3156
5.0	0.618	0.592	0.041	0.043	3630
10.0	0.606	0.594	0.020	0.022	3665

microsatellite loci. This may result in an underestimation of both the degree of genetic divergence among populations if all loci are pooled for analysis, and the estimation of F_{ST} (see Olsen *et al.* 2004b).

We saw in the previous section that long-term N_e can be estimated using the amount of heterozygosity in a population if we know the mutation rate. However, we also know from Chapter 9 that the amount of gene flow affects the amount of genetic variation in a population. Therefore, estimates of N_e using equation 12.1 may be overestimates because they reflect the total N_e of a series of populations connected by gene flow rather than the N_e of the local population. Consider two extremes. In the first, a population on an island is completely isolated from the rest of the members of its species ($mN = 0$). In this case, estimates of N_e using equation 12.1 will reflect the local N_e. In the other extreme, a species consists of a number of local populations that are connected by substantial gene flow (say $mN = 100$); in this case the estimates of N_e using equation 12.1 will reflect the combined N_e of all populations (Table 12.2).

12.3 HARMFUL MUTATIONS

Most mutations that affect fitness have a detrimental effect (Figure 12.1). Natural selection acts to keep these mutations from increasing in frequency. Consider the joint effects of mutation and selection at a single locus with a normal allele (A_1) and a mutant allele (A_2) that reduces fitness as shown below:

$A_1 A_1$	$A_1 A_2$	$A_2 A_2$
1	$1 - hs$	$1 - s$

where s is the reduction in fitness of the homozygous mutant genotype and h is the degree of dominance of the A_2 allele. A_2 is recessive when $h = 0$, dominant when $h = 1$, and partially dominant when h is between 0 and 1.

If the mutation is recessive ($h = 0$), then at equilibrium:

$$q^* = \sqrt{\frac{\mu}{s}} \tag{12.3}$$

When A_2 is partially dominant, q will generally be very small and the following approximation holds (Figure 12.6):

$$q^* \approx \frac{\mu}{hs} \tag{12.4}$$

See Lynch *et al.* (1999) for a consideration of the importance of mildly deleterious mutations in evolution and conservation.

12.4 ADVANTAGEOUS MUTATIONS

Genetic drift plays a major role in the survival of advantageous mutations even in extremely large populations. That is, most advantageous mutations will

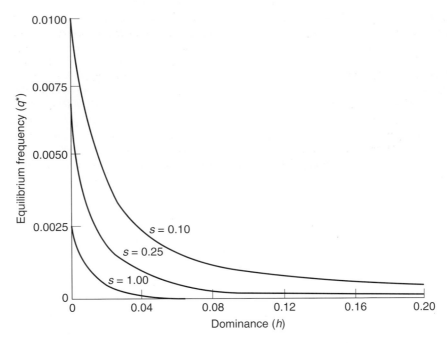

Figure 12.6 The expected equilibrium frequency of a deleterious allele (q^*) with mutation–selection balance ($\mu = 10^{-5}$) for different degrees of dominance (h) and intensity of selection (s). From Hedrick (2011).

be lost during the first few generations because new mutations will always be rare. The initial frequency of a mutation will be one over the total number of gene copies at a locus (i.e., $q = 1/2N$). Even a greatly advantageous allele that is recessive will have the same probability of initial persistence in a population because the advantageous homozygotes will not occur until the allele happens to drift to a relatively high frequency. For example, a new mutation will have to drift to a frequency over 0.20 before even 5% of the population will be homozygotes with the selective advantage. Therefore, the great majority of advantageous mutations that are recessive will be lost.

Dominant advantageous mutations have a much greater chance of surviving the initial period because their fitness advantage will be effective in heterozygotes that carry the new mutation. However, even most dominant advantageous mutations will be lost within the first few generations because of genetic drift. For example, over 80% of dominant advantageous mutations with a selective advantage of 10% will be lost within the first 20 generations (Crow and Kimura 1970, p. 423). This effect can be seen in a simple

example. Consider a new mutation that arises which increases the fitness of the individual that carries it by 50%. However, even if the individual that carries this mutation contributes three progeny to the new generation, there is a 0.125 probability that none of the progeny carry the mutation because of the vagaries of Mendelian segregation ($0.5 \times 0.5 \times 0.5 = 0.125$).

Gene flow and spread of advantageous mutations may be an important cohesive force in evolution (Rieseberg and Burke 2001). Ehrlich and Raven (1969) argued in a classic paper that the amounts of gene flow in many species are too low to prevent substantial differentiation among subpopulations by genetic drift or local adaptation so that local populations are essentially independently evolving units in many species. We saw in Chapter 9 that even one migrant per generation among subpopulations can cause all alleles to be present in all subpopulations. However, even much lower amounts of gene flow can be sufficient to cause the spread of an advantageous allele (say $s > 0.05$) throughout the range of a species (Rieseberg and Burke 2001). The rapid spread of such advantageous alleles may play an important role in maintaining the genetic integration of subpopulations connected by very small amounts of genetic exchange.

12.5 RECOVERY FROM A BOTTLENECK

The rate of recovery of genetic variation from the effects of a bottleneck will depend primarily upon the mutation rate (Lynch 1996). The equilibrium amount of neutral heterozygosity in natural populations (see equation 12.1) will be approached in a timescale equal to the shorter of $2N_e$ or $1/(2\mu)$ generations (Kimura and Crow 1964).

We can see this expectation in Figure 12.7 for microsatellites and allozymes. In these simulations of 100 typical microsatellite and allozyme loci, the expected heterozygosity at microsatellite loci returned to 50% of that expected at equilibrium after 2000 generations in populations of 5000 individuals. It took approximately three times as long at the loci with mutation rates typical of allozymes. In this case, $1/(2\mu)$ is 5000 gen-

erations for microsatellites and 100 times that for allozymes. However, $2N_e$ is 10,000 for both types of markers.

As we saw in Section 12.1.3, the estimated mutation rates (V_m) for phenotypic characters affected by many loci (quantitative characters) are similar to the rates of mutations at microsatellite loci. Therefore, we would expect quantitative genetic variance for phenotypic characters to be restored at rates comparable to those of microsatellites (Lande 1996). Thus, recovery of microsatellite variation following a severe bottleneck may be a good measure of the recovery of polygenic variation for fitness traits.

Figure 12.8 provides a simplistic representation of the effects of a severe population bottleneck on different sources of genetic variation. Microsatellites, allozymes, and quantitative traits are all expected to lose genetic variation at approximately the same rates. However, mtDNA will lose genetic variation more rapidly because of its smaller N_e. The rates of recovery of variation will depend upon the mutation rates for these different sources of genetic variation.

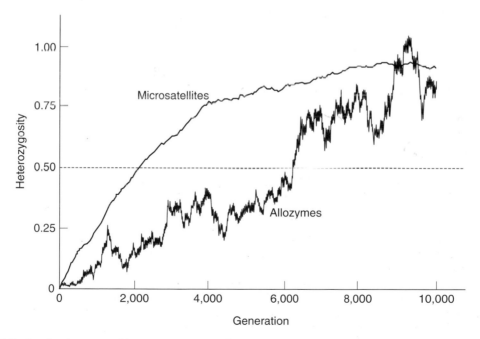

Figure 12.7   Simulated recovery of heterozygosity at 100 loci in a population of 5000 individuals following an extreme bottleneck using EASYPOP (Balloux 2001). The initial heterozygosity was zero. The mutation rates are 10^{-4} for microsatellites and 10^{-6} for allozymes. Heterozygosity is standardized as the mean heterozygosity over all 100 loci divided by the expected equilibrium heterozygosity using equation 12.1 (0.670 and 0.020, respectively).

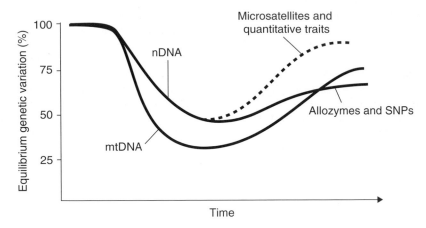

Figure 12.8 Diagram of relative expected effects of a severe population bottleneck on different types of genetic variation. The smaller N_e for mtDNA causes more genetic variation to be lost during a bottleneck. The rate of recovery following a bottleneck is largely determined by the mutation rate (see Section 12.1).

Guest Box 12   Color evolution via different mutations in pocket mice
Michael W. Nachman

Mutation is the ultimate source of genetic variation, yet the specific mutations responsible for evolutionary change have rarely been identified. We have been studying the genetic basis of color variation in pocket mice from the Sonoran Desert to try to find the mutations responsible for adaptive melanism. This research seeks to answer questions such as: Does adaptation result from a few mutations of major effect or many mutations of small effect? What kinds of genes and mutations underlie adaptation? Do these mutations change gene structure or gene regulation? Do similar phenotypes in different populations arise independently, and if so, do they arise from mutations at the same gene or from mutations at different genes?

Rock pocket mice are granivorous rodents well adapted to life in the desert. They remain hidden in burrows during the day, are active only at night, feed primarily on seeds, and do not drink water. In most places, these mice live on light-colored rocks and are correspondingly light in color. In several different places in the Sonoran Desert, these mice live on dark basalt of recent lava flows, and the mice in these populations are dark in color (see Figure 8.13). The close match between the color of the

mice and the color of the rocks is presumed to be an adaptation to avoid predation. Owls are among the primary predators, and even though owls hunt at night, they are able to discriminate between mice that match and do not match their background under conditions of very low light.

The genetic basis of melanism is amenable to analysis because of the wealth of background information on the genetics of pigmentation in laboratory mice and other animals (Bennett and Lamoreux 2003). We developed markers in several candidate genes and we then looked for nonrandom associations between genotypes at these genes and color phenotypes in populations of pocket mice near the edge of lava flows, where both light and dark mice are found together.

This search revealed that mutations in the gene encoding the melanocortin-1 receptor (*MC1R*) are responsible for color variation in one population in Arizona (Nachman *et al.* 2003). This receptor is part of a signaling pathway in melanocytes, the specialized cells in which pigment is produced. This work shows that an important adaptation – melanism – is caused by one gene of major effect in this case. Moreover, only four amino-acid changes distinguish light and dark animals, demonstrating that relatively few mutations are involved. These mutations produce a hyperactive receptor that results in the production of more melanin in melanocytes and therefore in darker mice.

Surprisingly, nearly phenotypically identical dark mice have arisen independently in this species in several populations in New Mexico (Hoekstra and Nachman 2003). In these mice from New Mexico, however, *MC1R* is not responsible for the differences in color. While the specific genes responsible for melanism in New Mexico have not yet been found, it is clear that the genetic basis of melanism in these populations is different from the genetic basis of melanism in the population from Arizona. In contrast to our results, Gross *et al.* (2009) found that two populations of the cavefish *Astyanax mexicanus* independently evolved reduced pigmentation as an adaptation to the cave environment by different mutations at *MC1R*.

These results demonstrate that there may be different genetic solutions to a common evolutionary problem. Thus, the genetic basis of local adaptation may be different for isolated populations subjected to the same selective regime. Similarly, Cohan and Hoffmann (1986) found that isolated laboratory populations of *Drosophila* became adapted to high ethanol concentrations by different physiological mechanisms involving changes at different loci. From a conservation perspective, it is important to recognize that phenotypically similar populations may be quite distinct genetically. This has important implications, for example, for planning translocations of individuals between populations (e.g., to supplement declining populations). Translocations between phenotypically similar but isolated populations may result in reduced fitness when the translocated individuals mate with native individuals.

PART III

GENETICS AND CONSERVATION

CHAPTER 13

Inbreeding depression

Monkeyflower, Example 13.1

That any evil directly follows from the closest interbreeding has been denied by many persons; but rarely by any practical breeder; and never, as far as I know, by one who has largely bred animals which propagate their kind quickly. Many physiologists attribute the evil exclusively to the combination and consequent increase of morbid tendencies common to both parents: and that this is an active source of mischief there can be no doubt.

Charles Darwin (1896, p. 94)

Probably the oldest observation about population genetics is that individuals produced by matings between close relatives are often less healthy than those produced by mating between more distant relatives.

Anthony R. Ives and Michael C. Whitlock (2002)

Chapter Contents

Conservation and the Genetics of Populations, Second Edition. Fred W. Allendorf, Gordon Luikart, and Sally N. Aitken.
© 2013 Fred W. Allendorf, Gordon Luikart and Sally N. Aitken. Published 2013 by Blackwell Publishing Ltd.

In 1965, the prominent population geneticist Richard C. Lewontin reviewed a book entitled *The Theory of Inbreeding* by Sir R.A. Fisher. In his review, Lewontin emphasized the central importance of inbreeding for understanding population genetics, as well as the difficulty of understanding 'inbreeding':

> *Notions of inbreeding lie at the very heart of genetics of sexual organisms, and every discovery in classical and population genetics has depended on some sort of inbreeding experiment. But a full understanding of the theory and ramifications of inbreeding always seems to evade us, just.*

Inbreeding is one of those topics that appears relatively straightforward, but becomes more complex the deeper we examine it.

The term 'inbreeding' is used to mean many different things in population genetics. Jacquard (1975) described five different effects of nonrandom mating that are measured by inbreeding coefficients. The multiple uses of 'inbreeding' can sometimes lead to confusion, so it is important to be precise when using this term. Templeton and Read (1994) have described three different phenomena of special importance for conservation that are all measured by 'inbreeding coefficients':

1 Genetic drift (F_{ST}, see Section 9.1);
2 Nonrandom mating within local populations (F_{IS}; see Section 9.1); and
3 The increase in genome-wide homozygosity (pedigree F) caused by matings between related individuals (e.g., father–daughter mating in ungulates, or matings between cousins in birds).

Keller and Waller (2002) provide an exceptionally clear presentation of these three different uses of inbreeding (see their Box 1). We will focus on this last meaning in this chapter.

Inbreeding (mating between related individuals) will occur in both large and small populations. In large populations, inbreeding may occur by self-fertilization or by nonrandom mating because of a tendency for related individuals to mate with each other. For example, in many tree species, nearby individuals are more likely to be related than trees farther apart, and have a higher probability of mating with each other because of geographic proximity (see Section 9.2, Hall *et al.* 1994). However, substantial inbreeding will occur in small populations even with random mating, because all or most individuals within a small population will be related. In an extreme example of a population of size two, after one generation, only brother–sister matings are possible. In a slightly larger population with 10 breeders, even the most distantly related individuals will be cousins after only a few generations. This has been called the "inbreeding effect" of small populations (Crow and Kimura 1970, p. 101).

Inbred individuals generally have reduced fitness in comparison with non-inbred individuals from the same population because of their increased homozygosity. **Inbreeding depression** is the reduction in fitness (or phenotype value) of progeny from matings between related individuals, compared with the fitness of progeny between unrelated individuals (Example 13.1). Inbreeding depression in natural populations will contribute to the extirpation of populations under some circumstances (see Chapter 14, Keller and Waller 2002, Allendorf and Ryman 2002).

The importance of inbreeding depression has been debated since the time of Darwin, as we can see from the quote above. Even today, the existence and importance of inbreeding depression is denied by some conservationists and policymakers (Räikkönen *et al.* 2009, Sarre and Georges 2009). Nevertheless, there is now overwhelming evidence that inbreeding depression is an important consideration in the persistence of populations (Frankham 2010, see Chapter 14). In this chapter, we will review the evidence for inbreeding depression and consider how it can be detected and measured in order to predict its effects in natural populations.

13.1 PEDIGREE ANALYSIS

An individual is 'inbred' if its mother and father share a common ancestor. This definition must be put into perspective because any two individuals in a population are related if we trace their ancestries back far enough. We must, therefore, define inbreeding relative to some 'base' population in which we assume all individuals are unrelated to one another. We usually define the base population operationally as those individuals in a pedigree beyond which no further information is available (Ballou 1983).

Inbred individuals will have increased homozygosity and decreased heterozygosity over their entire genome. The pedigree inbreeding coefficient (F) is the expected increase in homozygosity for inbred individuals relative to the base (non-inbred) population; it is also the expected decrease in heterozygosity throughout the

Example 13.1 Inbreeding depression in the monkeyflower

The monkeyflower is a self-compatible wild flower that occurs throughout western North America, from Alaska to Mexico. Willis (1993) studied two annual populations of this species on adjacent mountains about 2 km apart in the Cascade Mountains of Oregon. Seeds were collected from both populations and germinated in a greenhouse. Hand pollinations produced self-pollinations and pollinations from another randomly chosen plant from the same population. Seeds resulting from these pollinations were germinated in the greenhouse, and randomly chosen seedlings were transplanted back into their original population. The transplanted seedlings were marked and followed throughout the course of their life. Cumulative inbreeding depression through several life history stages was estimated by the proportional reduction in fitness in selfed versus outcrossed progeny $(1 - w_s/w_o)$. Substantial inbreeding depression was detected in both populations (Figure 13.1). A similar set of seedlings was maintained in the greenhouse. The amount of inbreeding depression in the greenhouse was similar to that found in the wild.

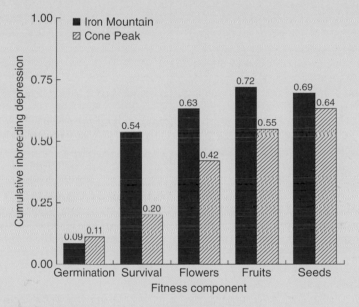

Figure 13.1 Cumulative inbreeding depression in two wild populations of the monkeyflower. Inbreeding depression was measured as the proportional reduction in fitness in selfed versus outcrossed progeny $(1 - w_s/w_o)$. From Willis (1993).

genome. F ranges from zero (for non-inbred individuals) to one (for totally inbred individuals).

An inbred individual may receive two copies of the same allele that was present in a common ancestor of its parents. Such an individual is **identical by descent** at that locus (i.e., **autozygous**). The probability of an individual being autozygous is its pedigree inbreeding coefficient, F. All autozygous individuals will be homozygous unless a mutation has occurred in one of the two copies descended from the ancestral allele in the base population. The alternative to being autozygous is **allozygous**. Allozygous individuals possess two alleles descended from the different ancestral alleles in the base population. Figure 13.2 illustrates the relationship of the concepts of autozygosity, allozygosity, homozygosity, and heterozygosity.

Is it really necessary to introduce these two new terms? Yes. Autozygosity and allozygosity are related to

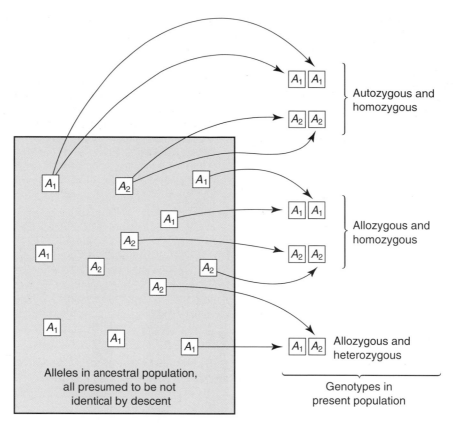

Figure 13.2   Patterns and definitions of genotypic relationships with pedigree inbreeding. Autozygous individuals in the present population contain two alleles that are identical by descent from a single gene in the ancestral population. In contrast, allozygous individuals contain two alleles derived from different genes in the ancestral population. Redrawn from Hartl and Clark (1997).

homozygosity and heterozygosity, but they refer to the descent of alleles through Mendelian inheritance rather than the molecular state of the allele in question. This distinction is important when considering the effects of pedigree inbreeding on individuals. We assume that the founding individuals in a pedigree are allozygous for two unique alleles. However, these allozygotes may either be homozygous or heterozygous at a particular locus, depending upon whether the two alleles are identical in state or not. For example, an allozygote would be homozygous if it had two alleles that are identical in DNA sequence.

We can see this using Figure 12.2. An individual with one copy of the *10A* allele and one copy of the *10C* allele (i.e., *10A/10C*) would be homozygous in

state for 10 repeats, but would be allozygous. In contrast, an individual with two copies of the *10B* allele (*10B/10B*) would be homozygous and autozygous.

13.1.1 Estimation of the pedigree inbreeding coefficient

Several methods are available for calculating the pedigree inbreeding coefficient. We will use the method of path analysis developed by Sewall Wright (1922). Figure 13.3 shows the pedigree of an inbred individual X. By convention, females are represented by circles, and males are represented by squares in pedigrees. Diamonds are used either to represent individuals whose

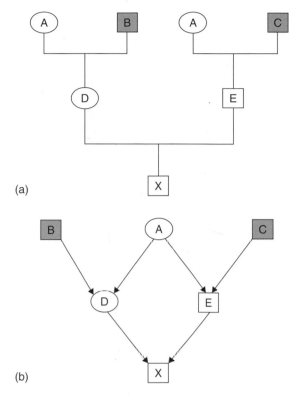

(a)

(b)

Figure 13.3 Calculation of the inbreeding coefficient for individual X using path analysis. (a) Conventional representation of pedigree for an individual whose mother and father had the same mother (individual A). (b) Path diagram to represent this pedigree to calculate the inbreeding coefficient. Shaded individuals in (a) need not be included in (b) because they are not part of the path through the common ancestor (A) and therefore do not contribute to the inbreeding of individual X.

sex is unknown or to represent individuals whose sex is not of concern.

What is the inbreeding coefficient of individual X in Figure 13.3a? The first step is to draw the pedigree as shown in Figure 13.3b so that each individual appears only once. Next we examine the pedigree for individuals who are ancestors of both the mother and father of X. If there are no such common ancestors, then X is not inbred, and $F_X = 0$. In this case, there is one common ancestor, individual A. Next, we trace all of the paths that lead from one of X's parents, through the common ancestor, and then back again to the other

parent of X. There is only one such path in Figure 13.3 (D<u>A</u>E); it is helpful to keep track of the common ancestor by underlining.

The inbreeding coefficient of an individual can be calculated by determining N, the number of individuals in the loop (not including the individual of concern) containing the common ancestor of the parents of an inbred individual. If there is a single loop then:

$$F = (1/2)^N (1 + F_{CA}) \tag{13.1}$$

where F_{CA} is the inbreeding coefficient of the common ancestor. The term $(1 + F_{CA})$ is included because the probability of a common ancestor passing on the same allele to two offspring is increased if the common ancestor is inbred. For example, if the inbreeding coefficient of an individual is 1.0, then it will always pass on the same allele to two progeny. If there is more than one loop, then the inbreeding coefficient is the sum of the F values from the separate loops.

$$F = \sum [(1/2)^N (1 + F_{CA})] \tag{13.2}$$

In the present case (Figure 13.3), there is only one loop with $N = 3$ and the common ancestor (A) is not inbred, therefore:

$$F_X = (1/2)^3 (1 + 0) = 0.125$$

This means that individual X is expected to be identical by descent (IBD) at 12.5% of his loci. Or, stated another way, the expected heterozygosity of individual X is expected to be reduced by 12.5%, compared with individuals in the base population. See Example 13.2 for calculating F when the common ancestor is inbred ($F_{CA} > 0$).

Figure 13.5 shows a complicated pedigree obtained from a long-term population study of the great tit in the Netherlands (van Noordwijk and Scharloo 1981). They have shown that the hatching of eggs is reduced by approximately 7.5% for every 10% increase in F. Ten different loops contribute to the inbreeding of the individual under investigation (Table 13.1). The total inbreeding coefficient of this individual is 0.1445.

On a curious historical note, both Sewall Wright and Charles Darwin displayed inbreeding within their immediate families. Sewall Wright's parents were first cousins (Provine 1986, p. 1); Wright calculated his own F value as 0.0625. Charles Darwin and his wife Emma Wedgwood were first cousins (Berra et al. 2010). In

Example 13.2 Calculating pedigree *F*

Figure 13.4 shows a pedigree in which a common ancestor of an inbred individual is inbred. What is the inbreeding coefficient of individual K in this figure?

There is one loop that contains a common ancestor of both parents of K (I<u>G</u>J). Therefore, using equation 13.2, $F_K = (1/2)^3(1 + F_G)$. The common ancestor in this

loop, G, is also inbred; there is one loop with three individuals through a common ancestor for individual G (D<u>B</u>E). Therefore, $F_G = (1/2)^3(1 + F_B) = 0.125$. Individual B is not inbred ($F_B = 0$) since she is a founder in this pedigree. Therefore, $F_G = 0.125$, and $F_K = (1/2)^3(1.125) = 0.141$.

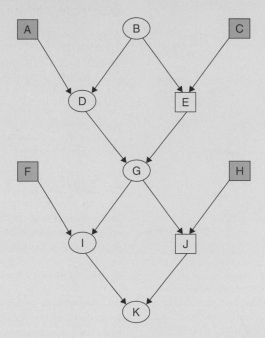

Figure 13.4 Hypothetical pedigree in which the common ancestor (G) of an inbred individual's (K) parents is also inbred. Shaded individuals are not part of the path through either of the common ancestors (individuals B and G) and therefore do not contribute to the inbreeding of individuals G and K.

addition, Charles and Emma also shared a set of distant grandparents going back several generations (Figure 13.6). An analysis of the Darwin/Wedgwood family tree has concluded that inbred children had significantly reduced probability of surviving through childhood compared to non-inbred children (Berra *et al.* 2010).

13.2 GENE DROP ANALYSIS

The pedigree analysis in the previous section provides an estimate of the increase in homozygosity and reduc-

tion in heterozygosity due to inbreeding. However, as we have seen in previous chapters, we are also interested in the loss of allelic diversity, as well as heterozygosity. A simple computer simulation procedure called **gene drop** analysis has been developed for more detailed pedigree analysis (MacCluer *et al.* 1986).

In this procedure, two unique alleles are assigned to each individual in the base population. Monte Carlo simulation methods are used to assign a genotype to each progeny based upon its parents' genotypes and Mendelian inheritance (Figure 13.7). This procedure is followed throughout the pedigree until each individual

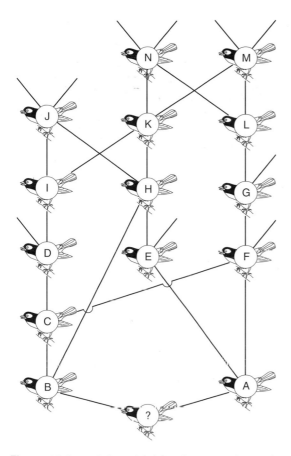

Figure 13.5 Complicated pedigree from a population of great tits in the Netherlands. The inbreeding coefficient of the bottom individual (?) is 0.1445 (see Table 13.1). From van Noordwijk and Scharloo (1981).

is assigned a genotype. This simulation is then repeated many times (say 10,000). Analysis of the genotypes in individuals of interest can provide information about the expected inbreeding coefficient, decline in heterozygosity, expected loss of allelic diversity, and many other characteristics that may be of interest (Example 13.3).

13.3 ESTIMATION OF F WITH MOLECULAR MARKERS

To understand the effect of inbreeding on fitness in natural populations, it is necessary to know the inbreeding coefficient of individuals. Unfortunately, it is difficult to estimate pedigree-based inbreeding coefficients in the field because relationships among individuals are not usually available. Pedigrees of wild populations are likely to be only a few generations deep, have gaps, and be inaccurate (Pemberton 2008). However, an individual's inbreeding coefficient can be estimated from the degree of homozygosity at molecular markers of its genome relative to the genomes of other individuals within the same population.

We saw in Section 13.1 that the pedigree inbreeding coefficient, F, is the expected increase in homozygosity due to identity by descent. For example, the offspring of a full-sib mating will have only 75% of the heterozygosity in their parents (Table 13.2). Offspring produced by half-sib matings will have only 87.5% of the heterozygosity observed in their parents. Therefore, we expect individual heterozygosity (H) at many loci to be reduced by a value of F. Pedigree F can be estimated by comparison of multilocus heterozygosity of

Table 13.1 Calculation of inbreeding coefficient of the individual (?) at the bottom of the pedigree shown in Figure 13.5.

Path	Length (N)	Common ancestor	F_{CA}	$(1/2)^N(1 + F_{CA})$
BC**F**A	4	F	0	0.0625
B**H**EA	4	H	0	0.0625
BCDI**K**HEA	8	K	0	0.0039
BCDI**J**HEA	8	J	0	0.0039
BHK**M**LGFA	8	M	0	0.0039
BHK**N**LGFA	8	N	0	0.0039
BCDIK**N**LGFA	10	N	0	0.0010
BCDIK**M**LGFA	10	M	0	0.0010
BCFGL**M**KHEA	10	M	0	0.0010
BCFGL**N**KHEA	10	N	0	0.0010
Total	–	–	–	0.1445

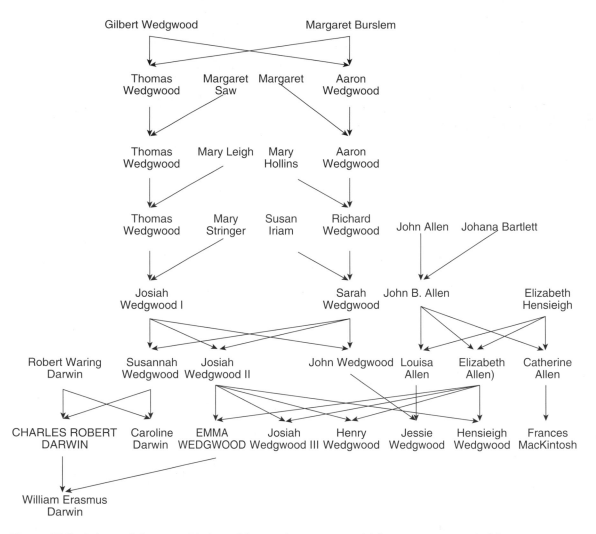

Figure 13.6 Pedigree of the Darwin/Wedgwood dynasty showing just one (William Erasmus Darwin) of first cousins Charles Darwin and Emma Wedgewood's ten children. Modified from Berra *et al.* (2010).

individuals over many loci, and the availability of large numbers of genetic markers for some taxa (e.g., SNPs) is improving the accuracy of such estimates (Jones and Wang 2010).

For example, individual inbreeding coefficients have been estimated using molecular markers in a wolf population from Scandinavia (Hedrick *et al.* 2001, Ellegren 1999). Twenty-nine microsatellite loci were examined in captive gray wolves for which the complete pedigree is known (Figure 13.9). The distribution of individual heterozygosity ranged from about 0.20 to 0.80. The pedigree inbreeding coefficient was significantly correlated with heterozygosity (Figure 13.10; $r^2 = 0.52$, $P < 0.001$). Thus, the 29 microsatellite loci in this case provide an accurate indicator of individual inbreeding level.

Precise estimation of inbreeding coefficients using heterozygosity requires many loci because the variance in heterozygosity estimates is large (i.e., confidence intervals are wide; Pemberton 2004). The reliability of using multiple locus heterozygosity to estimate the inbreeding coefficient of individuals depends primarily

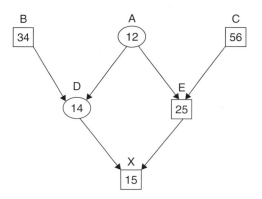

Figure 13.7 Pedigree from Figure 13.3 showing one possible outcome of gene dropping in which genotypes are assigned to descendants by Monte Carlo simulation of Mendelian segregation, beginning with two unique alleles in each founder.

Table 13.2 Expected decline of genome-wide heterozygosity ($1-F$) with different modes of inbreeding (modified from Dudash and Fenster 2000). The rate of loss per generation increases with the relatedness of parents. With selfing, 50% of the heterozygosity is lost per generation because one-half of the offspring from a heterozygote (Aa) will be homozygotes (following the 1:2:1 Mendelian ratios of 1 AA: 2 Aa: 1 aa).

	Heterozygosity remaining		
Generation	Half-sib	Full-sib	Selfing
0	1.000	1.000	1.000
1	0.875	0.750	0.500
2	0.781	0.625	0.250
3	0.695	0.500	0.125
4	0.619	0.406	0.062
5	0.552	0.328	0.031
10	0.308	0.114	0.008

Example 13.3 Gene drop analysis

Haig *et al.* (1990) have presented an example of gene drop analysis in their consideration of the effect of different captive breeding options on genetic variation in Guam rails (Figure 13.8). The upper part of the figure shows a sample pedigree in which three founders (A, B, C) are given six unique alleles. The lower part of the figure shows one result of a gene drop simulation. Two of the birds (G and I) are heterozygous; the heterozygosity has thus declined 50%. Only three of the initial six alleles are present in the living population (1, 4, and 6); thus 50% of the alleles have been lost. The proportional representation of each of the founders can also be calculated from this result. Four of the eight genes are descended from A (50%); one of the eight from B (12.5%); and three of the eight genes from C (37.5%). This simulation would be repeated 10,000 times to get statistical estimates of these parameters.

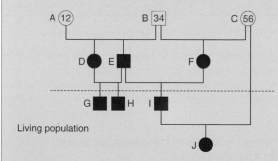

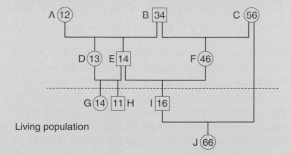

Figure 13.8 Example of a gene drop analysis in a hypothetical captive population with three founders. From Haig *et al.* (1990).

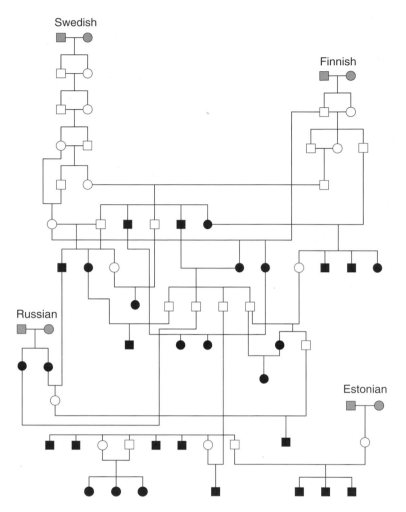

Figure 13.9 Pedigree of a captive gray wolf population. The black symbols are those animals included in the study evaluating the use of 29 microsatellite loci to estimate inbreeding coefficient. The gray individuals are the four founder pairs (assumed $F = 0$) from four countries. From Hedrick *et al.* (2001).

upon the number of loci used and the variance of the inbreeding coefficient among individuals (Slate *et al.* 2004, Balloux *et al.* 2004). Slate *et al.* (2004) presented the predicted correlation coefficient between multiple locus heterozygosity and F as a function of the variance of F among individuals and the number of loci used. Pemberton (2008) has argued that using molecular markers to reconstruct pedigrees within wild populations provides the most powerful method to test for effects of inbreeding in wild populations. However, this will depend upon the depth of the pedigree. Relative overall genome heterozygosity might provide a

better metric for detecting inbreeding depression in wild populations when only a shallow pedigree (say two or three generations) can be reconstructed (Balloux *et al.* 2004, Pemberton 2008).

13.4 CAUSES OF INBREEDING DEPRESSION

Inbreeding depression can result from either increased homozygosity or reduced heterozygosity (Crow 1948). This may sound like double-talk, but there is an impor-

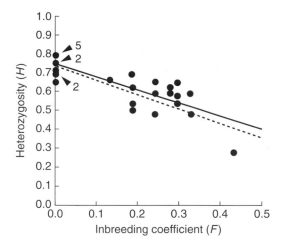

Figure 13.10 Relationship between individual heterozygosity (H) at 29 microsatellite loci and inbreeding coefficient (F) in a captive wolf population. The solid line represents the regression of H on F, and the dashed line is the expected relationship between H and F, assuming an H of 0.75 in non-inbred individuals. From Ellegren (1999).

tant distinction to be made here. Increased homozygosity leads to expression of a greater number of deleterious recessive alleles in inbred individuals, thereby lowering their fitness. Reduced heterozygosity reduces fitness of inbred individuals at loci at where the heterozygotes have a selective advantage over homozygote types (heterozygous advantage or overdominance; Section 8.2). Thus, the reduction in fitness caused by inbreeding (inbreeding load) has two primary sources (see Box 13.1).

The primary cause of inbreeding depression is an increase of homozygosity for deleterious recessive alleles (Charlesworth and Willis 2009). The probability of being homozygous for rare alleles increases surprisingly rapidly with inbreeding. Consider the effect of an average inbreeding coefficient of $F = 0.10$ on expression of a recessive lethal allele at a frequency of $q = 0.10$. The proportion of heterozygotes will be reduced by 10% compared to $F = 0$ in one generation, and each of the homozygotes will be increased by half of that amount (see Section 9.1):

Box 13.1 What is genetic load?

Genetic load is the relative difference in fitness between the theoretically fittest genotype within a population and the average genotype in that population (Crow 1970, Wallace 1991). There are many underlying mechanisms for genetic load in populations. The term "load" was first used by Muller (1950) in his consideration of the possible effects of increased mutation rates because of nuclear weapons. In this case, the **mutation load** is the reduction in fitness caused by the presence of deleterious alleles introduced by mutation. We saw in Section 8.2 that natural selection is not very effective at removing deleterious recessive alleles introduced by mutation because most of the deleterious alleles are hidden in heterozygotes.

There are four primary sources of genetic load that we need to consider in conservation:

1 **Mutation load:** The decrease in fitness caused by the accumulation of deleterious mutations (see Section 12.3).

2 **Segregation load:** The decrease in fitness caused by heterozygous advantage. When the fittest genotype is heterozygous, homozygotes with lower fitness will be produced by Mendelian segregation (see Section 8.2).

3 **Drift load:** The decrease in fitness caused by the increase in frequency of deleterious alleles resulting

from genetic drift (see Section 13.6). Drift load is caused by the increase in frequency of deleterious alleles that are maintained in a population at equilibrium between mutation and selection (see Section 12.3). In the extreme, alleles that contribute to **drift load** can become fixed in small populations. This fixed genetic load will cause the reduction in fitness of all individuals in small populations (see Figure 13.19).

4 **Migration load:** The reduction in fitness caused by the migration into a population of individuals that are less adapted to the local environment than native individuals (see Section 17.2).

The **inbreeding load** is the reduction in fitness of inbred individuals (see Figure 13.13). As we considered in Section 13.4, this reduction is caused by both the increase in homozygosity of deleterious recessive alleles and the reduction in heterozygosity at loci with heterozygous advantage. Thus, the inbreeding load results from a combination of mutation and segregation load (Charlesworth and Willis 2009). See Wallace (1991) for a clear discussion of several other types of genetic load.

We will consider **migration load** in more detail in Chapter 17.

	AA	**Aa**	**aa**
Expected	$p^2 + pqF$	$2pq - 2pqF$	$q^2 + pqF$
F = 0	0.810	0.180	0.010
F = 0.10	0.819	0.162	0.019

Thus, the expected proportion of individuals to be affected by this deleterious allele (*a*) will nearly double (from 0.010 to 0.019) with just a 10% increase in inbreeding. The increase in the number of affected individuals is even greater for less frequent alleles (Crow and Kimura 1970, p. 74). For example, we expect over ten times the number of homozygous individuals for a deleterious recessive allele at a frequency of $q = 0.01$ when $F = 0.10$ compared with $F = 0$.

It is crucial to know the mechanisms causing inbreeding depression because it affects the ability of a population to 'adapt' to inbreeding. A population could adapt to inbreeding if inbreeding depression is caused by deleterious recessive alleles that potentially could be removed (purged by selection). However, inbreeding depression caused by heterozygous advantage cannot be purged because overdominant loci will always suffer reduced fitness as homozygosity increases due to increased inbreeding.

Both increased homozygosity and decreased heterozygosity are likely to contribute to inbreeding depression, but it is thought that increased expression of deleterious recessive alleles is the more important mechanism (Charlesworth and Charlesworth 1987, Ritland 1996, Carr and Dudash 2003, Charlesworth and Willis 2009). For example, Remington and O'Malley (2000) performed a genome-wide evaluation of inbreeding depression caused by selfing during embryonic viability in loblolly pines. Nineteen loci were found that contributed to inbreeding depression; 16 loci showed predominantly recessive action. Evidence for heterozygous advantage was found at three loci.

13.5 MEASUREMENT OF INBREEDING DEPRESSION

Some inbreeding depression is expected in all species (Hedrick and Kalinowski 2000, see Guest Box 19). Deleterious recessive alleles are present in the genome of all species because they are continually introduced by mutation, and natural selection is inefficient in removing them because most copies are 'hidden' phenotypically in heterozygotes that do not have reduced fitness (see Chapter 12). Therefore we expect all species to show some inbreeding depression due to the increase in homozygosity of recessive deleterious alleles. For example, Figure 13.11 shows inbreeding depression

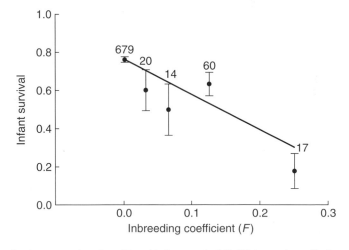

Figure 13.11 Relationship between inbreeding (*F*) and infant survival (± SE) in captive callimico monkeys. Callimico show a 33% reduction in survival resulting from each 10% increase in inbreeding ($P < 0.001$). Data are from 790 captive-born callimico, 111 of which are inbred. The numbers above the bars are the number of individuals studied at each inbreeding level. From Lacy *et al.* (1993).

for infant survival in a captive population of callimico monkeys.

13.5.1 Lethal equivalents

The effects of inbreeding depression on survival are often measured by the mean number of **lethal equivalents** (LEs) per diploid genome. A lethal equivalent is a set of deleterious alleles that would cause death if homozygous. Thus, one lethal equivalent may either be a single allele that is lethal when homozygous, two alleles each with a probability of 0.5 of causing death when homozygous, or 10 alleles each with a probability of 0.10 of causing death when homozygous.

We can see the effect of 1 LE in the example of a mating between full sibs that will produce a progeny with an F of 0.25 (Figure 13.12). Individuals A and B each carry one lethal allele (a and b, respectively). The probability of individual E being homozygous for the a allele is $(1/2)^4 = 1/16$; similarly, there is a 1/16 probability that individual E will be homozygous bb. Thus, the probability of E *not* being homozygous for a recessive allele at either of these two loci is $(15/16)(15/16) = 0.879$. Thus, 1 LE per diploid genome will result in approximately a 12% reduction $(1 - 0.879)$ in survival of individuals with an F of 0.25.

The number of LEs present in a species or population is generally estimated by regressing survival on the inbreeding coefficient (Figure 13.13). The effects of inbreeding on the probability of survival, S, can be expressed as a function of F (Morton *et al.* 1956):

$$S = e^{-(A+BF)}$$
$$\ln S = -A - BF \tag{13.3}$$

where e^{-A} is survival in an outbred population and B is the rate at which fitness declines with inbreeding (Hedrick and Miller 1992). B is the reduction in survival expected in a completely homozygous individual. Therefore, B is the number of LEs per gamete, and $2B$ is the number of LEs per diploid individual. B is estimated

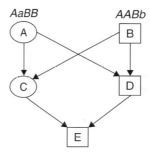

Figure 13.12   Effect of a single lethal equivalent (LE) on survival of inbred progeny produced by a mating between full-sibs ($F = 0.25$). Individuals A and B each carry one lethal allele (a and b, respectively). The probability of E being homozygous for the a allele is $(1/2)^4 = 1/16$; similarly, there is a 1/16 probability that individual E will be homozygous bb.

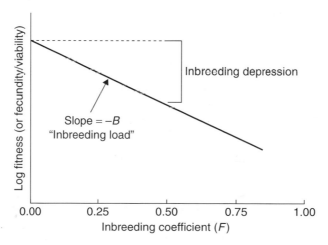

Figure 13.13   Relationship between inbreeding coefficient (F) and reduction in fitness. Inbreeding depression is the reduction in fitness of inbred individuals and measured by the number of lethal equivalents per gamete (B). Redrawn from Keller and Waller (2002).

by the slope of the weighted regression of the natural log of survival on F. The callimico monkeys shown in Figure 13.11 have an estimated 7.90 LEs per individual ($B = 3.95$; Lacy *et al.* 1993).

13.5.2 Estimates of inbreeding depression

The range of LEs per individual estimated for mammal populations ranges from about 0 to 30 in captivity (Ralls *et al.* 1988). The median number of LEs per diploid individual for captive mammals was estimated to be 3.14 (Figure 13.14). This corresponds to about a 33% reduction of juvenile survival, on average, for offspring with an inbreeding coefficient of 0.25. However, this is an underestimate of the magnitude of inbreeding depression in populations because it considers the reduction of fitness for only one life-history stage (juvenile survival), and ignores all others (e.g., adult survival, embryonic survival, fertility, etc.). As we saw in Example 13.1, inbreeding depression becomes greater as we consider more life-history stages.

In addition, captive environments are less stressful than natural environments, and stress typically increases inbreeding depression (see below). A meta-analysis of more recent published estimates found an overall average of 12 diploid LEs in wild mammal and bird populations over the life-history of species

(O'Grady *et al.* 2006). As we will see in Chapter 14, this amount of inbreeding depression can have a substantial effect on the viability of populations.

There are fewer estimates of the number of LEs using pedigree analysis in plant species. Most studies of inbreeding depression in plants compare selfed and outcrossed progeny from the same plants (see Example 13.1). In this situation, inbreeding depression is usually measured as the proportional reduction in fitness in selfed versus outcrossed progeny ($\delta = 1 - w_s/w_o$). These can be converted by:

$$\delta = 1 - \frac{w_s}{w_o} = 1 - e^{-B/2} \qquad (13.4)$$

and

$$B = -2\ln(1 - \delta) \qquad (13.5)$$

Some plant species show a tremendous amount of inbreeding depression. For example, Figure 13.15 shows estimates of inbreeding depression for embryonic survival in 35 individual Douglas-fir trees based upon comparison of the production of sound seed by selfing and crossing with pollen from unrelated trees. On average, each tree contained approximately 10 LEs. This is equivalent to over a 90% reduction in embryonic survival of progeny produced by selfing! Perhaps

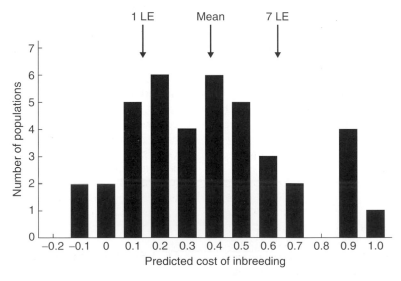

Figure 13.14   Distribution of the estimated cost of inbreeding in progeny with an inbreeding coefficient (F) of 0.25 in 40 captive mammal populations. Cost is the proportional reduction in juvenile survival. Mean LE = 3.14. From Ralls *et al.* (1988).

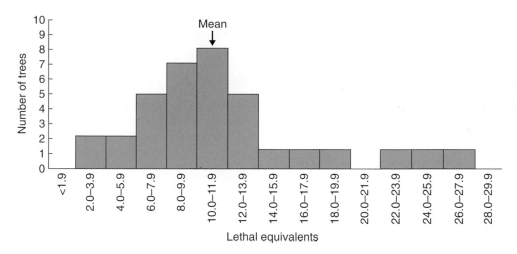

Figure 13.15 Inbreeding depression measured by the observed number of LEs for embryonic survival of seeds from 35 individual Douglas-fir parent trees, based upon comparison of the number of sound seed resulting from self-pollination compared with pollination by unrelated trees. Redrawn from Sorensen (1969).

more interesting, however, is the wide range of LEs in different trees (Figure 13.15). Conifers in general seem to have high inbreeding depression, perhaps because they are typically outcrossing species with large effective population sizes, but inbreeding depression is especially great in Douglas-fir (Sorensen 1999).

Most studies of inbreeding depression have been made in captivity or under controlled conditions, but experiments with both plants (e.g., Dudash 1990) and animals (e.g., Jiménez *et al.* 1994) have found that inbreeding depression is more severe in natural environments (see Example 13.4). Estimates of inbreeding depression in captivity are likely to be severe underestimates of the true effect of inbreeding in the wild (but not always; see Armbruster *et al.* 2000). Example 13.4 presents an interesting experiment in which inbreeding depression was measured in the same population under captive and wild conditions in different stages of the life-cycle.

Crnokrak and Roff (1999) reviewed the empirical literature on inbreeding depression for wild species. They tested this for mammals by comparing traits directly related to survival in wild mammals to the findings of Ralls *et al.* (1988) for captive species. They concluded that in general "the cost of inbreeding under natural conditions is much higher than under captive conditions". The cost of inbreeding in terms of survival was much higher in wild than in captive mammals.

In addition, there is evidence that inbreeding depression is more severe under environmental stress and challenge events (e.g., extreme weather, pollution, or disease; Armbruster and Reed 2005). Bijlsma *et al.* (1997) found a synergistic interaction between stress and inbreeding with laboratory *Drosophila* so that the effect of environmental stress is greatly enhanced with greater inbreeding. These conditions may occur only occasionally, and thus it will be difficult to measure their effect on inbreeding depression. For example, Coltman *et al.* (1999) found that individual Soay sheep that were more heterozygous at 14 microsatellite loci had greater overwinter survival rates in harsh winters, apparently due to greater resistance to nematode parasites. In addition, this effect disappeared when the sheep were treated with antihelminthics (Figure 13.17). A recent meta-analysis of the effect of the environment on inbreeding depression found that inbreeding depression is a linear function of stress, and that a population experiences one additional lethal equivalent for each 30% reduction in fitness caused by the stressful environment (Fox and Reed 2011).

13.5.3 Founder-specific inbreeding effects

Recent work indicates that the intensity of inbreeding depression can differ greatly depending on which

Example 13.4   Inbreeding depression for marine survival in anadromous rainbow trout (steelhead)

Thrower and Hard (2009) performed an elegant experiment to estimate the amount of inbreeding depression for a salmonid fish under captive and wild conditions. They were motivated by the absence of studies on inbreeding depression in salmonid fish except under hatchery conditions. Twenty wild steelhead (15 females and 5 males) were captured in 1996 from Sashin Creek in southeast Alaska, and mated in a hatchery. The progeny of these fish were then used in an experiment to measure inbreeding depression.

The F_1 progeny of the initial wild steelhead were used to produce an F_2 generation over five different years. Inbreeding depression was estimated by comparing the performance of F_2 progeny that were outbred ($F = 0$) and inbred fish produced by full-sib mating ($F = 0.25$). These F_2 progeny were raised in the hatchery during their freshwater life-history phase, and then released into the ocean. Surviving progeny were captured when they returned to spawn after spending two or three years in the ocean.

In captivity, no consistent difference was found in the growth or survival of outbred versus inbred progeny over the five-year course of the experiment.

In great contrast, consistent differences were found in the marine survival of non-inbred and inbred progeny (Figure 13.16). The survival of non-inbred progeny was significantly greater ($P < 0.001$) in all five years. On average, the marine survival of inbred progeny was reduced by 71%. This is equivalent to 10.8 diploid LE per individual just for this single phase of the life-history of these fish.

These results emphasize that measuring inbreeding depression in captivity can be a poor indicator of inbreeding depression in the wild environment. Specifically, these results indicate that inbreeding depression can be a major hazard to the persistence of small wild populations of salmonid fishes.

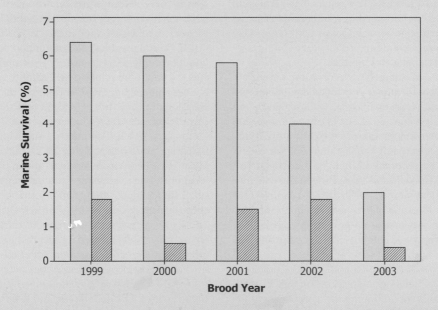

Figure 13.16   Marine survival rates for anadromous rainbow trout (steelhead) released from a hatchery in Alaska. Non-inbred fish (shaded bars) had significantly greater marine survival than inbred fish (hatched bars) in all five years of the experiment. Inbred fish ($F = 0.25$) were produced by mating full-sibs. The overall relative marine survival of inbred fish was only 29% of the non-inbred fish. Redrawn from Thrower and Hard (2009).

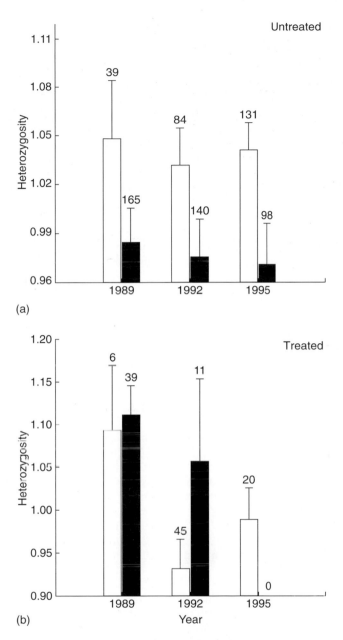

Figure 13.17 Relative observed individual heterozygosity (+ SE) of Soay sheep during three severe winters with high mortality. (a) Sheep that died (black-filled bars) had far lower heterozygosity than sheep that lived (open bars), when not treated with antihelminthics. (b) However, when treated with antihelminthics (to remove intestinal parasites), no difference in survival was detected between inbred (filled) and outbred individuals. Parasite load was higher in inbred individuals (with low heterozygosity), leading to their increased mortality during the stress of severe winters. A standardized relative heterozygosity was used because not all individuals were genotyped at all loci such that H = (proportion heterozygous typed loci/mean heterozygosity at typed loci). Numbers above bars indicate sample size. From Coltman *et al.* (1999).

individuals happen to be the founder that is a common ancestor of the parents of inbred individuals (see Guest Box 19, Lacy *et al.* 1996, Lacy and Ballou 1998, Casellas *et al.* 2008). This suggests that the genetic load is unevenly spread among founder genomes and supports the notion that inbreeding depression sometimes results from major effects at a few loci. We can see this clearly in Figure 13.15, which shows that the number of LEs for embryonic survival in individual Douglas-firs varied from less than four to more than 25. The effects of inbreeding depression will be much greater if the common ancestor of both parents carried 25 rather than 4 LEs.

This effect has also been detected by using the founder-specific partial F coefficient (Lacy *et al.* 1996, Gulisija *et al.* 2006). This allows the increase in homozygosity due to inbreeding to be attributed to a particular founder. For example, the overall inbreeding coefficient of the individual at the bottom of the pedigree shown in Figure 13.5 is 0.1445. This can be partitioned into the following founder-specific partial F coefficients for the six common ancestors: $F_F = 0.0625$, $F_H = 0.0625$, $F_K = 0.0039$, $F_J = 0.0039$, $F_M = 0.0059$, $F_N = 0.0059$.

A study with Ripollesa domestic sheep found that most of the inbreeding depression resulted from individuals being identical by descent for genes from just two of nine founders (Casellas *et al.* 2009). Managing founder-specific inbreeding depression using partial inbreeding coefficients could be extremely effective in cases where inbreeding depression results primarily from a few loci with major effects; such partial inbreeding coefficients could be useful when selecting potential matings in a captive population or when choosing individuals for release into wild populations.

13.5.4 Are there species without inbreeding depression?

Some have suggested that some species or populations are unaffected by inbreeding (e.g., Shields 1993). However, deleterious recessive alleles will be present in every population because of the process of mutation (see Section 12.3). Therefore, every population will have some inbreeding depression. In addition, lack of statistical evidence for inbreeding depression does not demonstrate the absence of inbreeding depression. This is especially true because of the low power to detect even a substantial effect of inbreeding in many studies, because of small sample sizes and confounding factors (Kalinowski and Hedrick 1999, see Guest Box 13). In a comprehensive review of the evidence for inbreeding depression in mammals, Lacy (1997) was unable to find "statistically defensible evidence showing that any mammal species is unaffected by inbreeding". Inbreeding depression for disease resistance has even been found in the close inbreeding naked mole-rat, which for many years was thought to be impervious to inbreeding depression (Example 13.5).

Measuring inbreeding depression in the wild is extremely difficult for several reasons. First, pedigrees are generally not available and the alternative molecular-based estimates might not be precise (see Section 13.3). Thus, the power to detect inbreeding depression is low in most studies. Second, fitness is difficult to measure in the wild. Inbreeding depression might occur only in certain life-history stages (e.g., zygotic survival or lifetime reproductive success) or under certain stressful conditions (e.g., severe winters or high predator density). Most studies in wild populations do not span enough years or life-history stages to measure inbreeding depression reliably. Finally, the detection of inbreeding depression requires the comparison of inbred and non-inbred individuals.

13.6 GENETIC LOAD AND PURGING

Small populations can be **purged** of deleterious recessive alleles by natural selection (Templeton and Read 1984, Crnokrak and Barrett 2002). Deleterious recessive alleles may reach substantial frequencies in large random mating populations because most copies are present in heterozygotes and are therefore not affected by natural selection. For example, over 5% of the individuals in a population in Hardy-Weinberg proportions will be heterozygous for an allele that is homozygous in only one out of 1000 individuals. Such alleles will be exposed to natural selection in inbred or small populations, and thereby be reduced in frequency or eliminated. Thus, populations with a history of inbreeding because of nonrandom mating (e.g., self-pollinating plants) or small N_e (e.g., a population bottleneck) may be less affected by inbreeding depression because of the purging of deleterious recessive alleles.

Reduced differences in fitness between inbred and non-inbred individuals within a population after it has gone through a bottleneck is not evidence for purging (Figure 13.19). Many nonlethal deleterious alleles might become fixed in such populations by genetic

Example 13.5   Inbreeding depression revealed by a disease challenge in the habitually inbreeding naked mole-rat

The naked mole-rat is a burrowing rodent native in east Africa that lives in eusocial colonies of 75 to 80 individuals in complex systems of burrows in arid deserts (Maree and Faulkes 2008). The tunnel systems built by naked mole-rats can stretch up to two or three miles in cumulative length.

The eusocial naked mole-rat has been considered a classic example of a habitual inbreeder that is not affected by inbreeding depression because of purging: "The only mammal species that has been shown to undergo continuous close inbreeding with no obvious effects of inbreeding depression" (Bromham and Harvey 1996). A microsatellite study found that over 80% of all mating occurs between first-degree relatives in the wild (Reeve *et al.* 1990). No evidence of inbreeding depression was found in 25 years of captive breeding (Ross-Gillespie *et al.* 2007).

A virulent enteric coronavirus swept unchecked through a captive naked mole-rat study population,

causing acute diarrhea, dehydration, and severe enteric hemorrhaging (Ross-Gillespie *et al.* 2007). No attempts were made to medicate infected animals. Mortality was monitored daily and dead animals were removed immediately for identification and confirmation of the presence of disease symptoms.

The severe symptoms associated with the coronavirus killed 161 of 365 animals (44%) in just eight weeks. Survival was significantly lower among more inbred animals (Figure 13.18). Offspring produced by half-sibling ($F = 0.125$) and full-sibling ($F = 0.250$) parent pairs were two to three times more likely to die than the offspring of unrelated parents ($F = 0$). Inbreeding depression in survival was measured as 2.3 diploid LEs per individual. This study demonstrates how loss of genetic heterozygosity through inbreeding may render populations vulnerable to local extinction from emerging infectious diseases, even when other indications of inbreeding depression are absent.

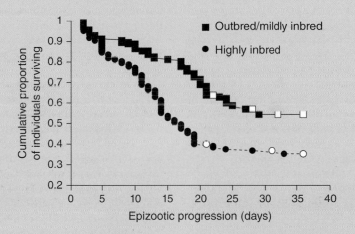

Figure 13.18   Proportional survival of highly inbred ($F \geq 0.25$) versus outbred and mildly inbred naked mole-rats ($0 \leq F \leq 0.125$) through the course of the coronavirus outbreak. Open markers denote values for which the likelihood function was adjusted to include still-living individuals. From Ross-Gillespie *et al.* (2007).

drift. The fixation of these alleles will cause a reduction in fitness of all individuals following the bottleneck, relative to the individuals in the population before the bottleneck (**drift load**, Box 13.1). However, inbreeding depression will appear to be reduced following fixation of deleterious alleles because of depressed fitness of

outbred individuals ($F = 0$) rather than by increased inbred fitness (**fitness rebound**) (Byers and Waller 1999). To test for purging, the fitness of individuals in the post-bottleneck population must be compared with the fitness of individuals in the pre-bottleneck population. Alternatively, we might test for purging by

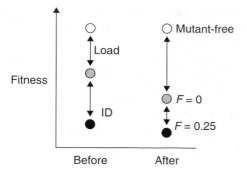

Figure 13.19 Diagram showing decreased inbreeding depression in a population before and after a bottleneck in which all of the inbreeding depression is due to increased homozygosity for deleterious recessive mutant alleles. The open circles show the average fitness of hypothetical 'mutant-free' individuals that have no deleterious alleles. The shaded circles show the average fitness of individuals of individuals produced by random mating. The dark circles show the average fitness of individuals with an F of 0.25. This illustrates that reduced inbreeding depression following a bottleneck can be caused by an increase in the fixed genetic load rather than by purging.

comparing the fitness of offspring from resident (relatively inbred) individuals versus the outbred offspring of crosses between residents and individuals from the pre-bottleneck population.

Reviews have found little evidence for purging in plant and animal populations (Gulisija and Crow 2007, Leberg and Firmin 2008). Byers and Waller (1999) found that only 38% of the 52 studies in plant populations found evidence of purging. When purging was found, it removed only a small proportion of the total inbreeding depression (roughly 10%). These authors concluded that "purging appears neither consistent nor effective enough to reliably reduce inbreeding depression in small and inbred populations" (Byers and Waller 1999).

Reviews of evidence for purging in animals have come to similar conclusions. Ballou (1997) found evidence for a slight decline in inbreeding depression in neonatal survival among descendants of inbred animals in a comparison of 17 captive mammal species. However, he found no indication for purging in weaning survival or litter size in these species. Boakes et al. (2007) reviewed the effects of inbreeding in 119 captive populations of mammals, birds, reptiles, and amphibians. Inbreeding depression for neonatal sur-

vival was significant across the 119 populations, although the severity of inbreeding depression varied greatly among taxa. Purging was found to be significant in 14 of the populations, and purging had a significant effect when the entire data-set was analyzed. However, the change in inbreeding depression due to purging averaged across the 119 populations was just 1%. Both Ballou (1997) and Boakes et al. (2007) have concluded that purging is not likely to be strong enough to be of practical use in eliminating inbreeding depression in populations of conservation interest.

Speke's gazelle has been cited as an example of the effectiveness of purging in reducing inbreeding depression in captivity (Templeton and Read 1984). Willis and Wiese (1997), however, have concluded that this apparent purging may have been due to the data analysis rather than purging itself; this interpretation has been disputed by Templeton and Read (1998). Ballou (1997), in his reanalysis of the Speke's gazelle data, found that purging effects were minimal and nonsignificant. Perhaps most importantly, he found that the inbreeding effects in Speke's gazelle were the greatest in any of the 17 mammal species he examined. Kalinowski et al. (2000) have recently concluded that the apparent purging in Speke's gazelle is the result of a temporal change in fitness and not a reduction in inbreeding depression (see response by Templeton 2002).

We would expect purging to be very effective in haplodiploid species (see Section 14.4) in which males are haploid, so deleterious recessive alleles are exposed to natural selection in males every generation and therefore should be purged relatively efficiently. Nevertheless, inbreeding depression has been found to be substantial in a variety of haplodiploid taxa: the insect order Hymenoptera (Antolin 1999), mites (Saito et al. 2000), and rotifers (Tortajada et al. 2009).

Some recent papers have presented evidence for a reduction in the intensity of inbreeding depression in the first few generations of inbreeding (e.g., Fox et al. 2008, Larsen et al. 2011). However, closer examination of these results indicates that the decrease in inbreeding depression is not due to 'purging' as defined above, as the selective removal of deleterious alleles contributing to inbreeding depression. Rather, the reduction in inbreeding depression observed in these experiments resulted primarily from the extinction of some experimental lines with greater inbreeding depression. For example, Lynch and Walsh (1998, pp. 274–276) summarized two sets of inbreeding experiments with the mouse in which full-sib mating resulted in a rapid reduction in litter size averaged over all inbred lines. However, only 10% of the original lines survived over this period, and the mean of all lines returned to pre-inbreeding levels after some ten generations.

The empirical evidence is now clear: purging is not an effective mechanism to reduce inbreeding depression in the conservation of plants and animals (Leberg and Firmin 2008). Nevertheless, purging of alleles can become important with long-term continuous inbreeding (Gulisija and Crow 2007). For example, Latter *et al.* (1995) found that *Drosophila* strains that experienced slow inbreeding over 200 generations had considerably less inbreeding depression. This might explain why populations of species that have had small effective population sizes have persisted over long periods of time. For example, the population of tuatara on North Brother Island (Example 16.3) has persisted as an isolated small population for approximately 10,000 years or approximately 200 generations (Table 7.6).

13.6.1 Why is purging not more effective?

Failure of purging to decrease inbreeding depression can be explained by several mechanisms. First, purging is expected to be most effective in the case of lethal or semi-lethal recessive alleles (Lande and Schemske 1985, Hedrick 1994); however, when inbreeding depression is very high (more than 10 LE), even lethals may not be purged except under very close inbreeding (Lande 1994). Second, the lack of evidence for purging is consistent with the hypothesis that a substantial proportion of inbreeding depression is caused by many recessive alleles with minor deleterious effects. Alleles with minor effects are unlikely to be purged by selection, because selection cannot efficiently target harmful alleles when they are spread across many different loci and different individuals. Third, it is not possible to purge the genetic load (segregation load) at overdominant loci as mentioned above. On the contrary, the loss of alleles at overdominant loci generally reduces heterozygosity and thus fitness.

Husband and Schemske (1996) found that inbreeding depression for survival after early development and reproduction and growth was similar in selfing and non-selfing plant species (also see Ritland 1996). They suggested that this inbreeding depression is due primarily to mildly deleterious mutations that are not purged, even over long periods of time. Willis (1999)

found that most inbreeding depression in the monkey-flower (see Example 13.1) is due to alleles with small effect, and not to lethal or sterile alleles. Bijlsma *et al.* (1999) found that purging in experimental populations of *Drosophila* is effective only in the environment in which the purging occurred, because additional deleterious alleles were expressed when environmental conditions changed.

Ballou (1997) suggested that **associative overdominance** may also be instrumental in maintaining inbreeding depression. Associative overdominance occurs when heterozygous advantage or deleterious recessive alleles at a selected locus results in apparent heterozygous advantage at linked loci (Section 10.3.2, Pamilo and Pálsson 1998). Kärkkäinen *et al.* (1999) have provided evidence that most of the inbreeding depression in the self-incompatible perennial herb *Arabis petraea* is due to overdominance or associative overdominance.

Inbreeding depression due to heterozygous advantage cannot be purged. However, it is unlikely that heterozygous advantage is a major mechanism for inbreeding depression (Charlesworth and Charlesworth 1987). Nevertheless, there is recent strong evidence for heterozygous advantage at the major histocompatibility complex (MHC) in humans (Thoss *et al.* 2011). Black and Hedrick (1997) found evidence for strong heterozygous advantage (nearly 50%) at both *HLA-A* and *HLA-B* in South Amerindians. Carrington *et al.* (1999) found that heterozygosity at *HLA A, B,* and *C* loci was associated with extended survival of patients infected with the human immunodeficiency virus (HIV). Strong evidence for the selective maintenance of MHC diversity in vertebrate species comes from other approaches as well (see references in Carrington *et al.* 1999).

13.7 INBREEDING AND CONSERVATION

The effects of inbreeding depression resulting from bottlenecks are difficult to predict in any given population for a variety of reasons (Bouzat 2010). Even if a population's history of inbreeding is known, it can be difficult to predict the cost of inbreeding on fitness (Guest Box 19). And, as discussed in Section 13.5.3, the reduction in fitness caused by inbreeding can vary greatly depending upon which individual is the common ancestor of an inbred individual's parents.

The inability to predict accurately the magnitude of inbreeding depression makes it difficult to incorporate inbreeding depression into models of population viability, even when a population's history and biology is well known. Nevertheless, managers must recognize that substantial inbreeding depression is likely to occur in any small population, especially under changing environments or stressful conditions.

Skepticism about the importance of inbreeding depression was justified in the early days of conservation biology. For example, Graeme Caughley questioned a central role for genetics in conservation biology in his 1994 review of conservation biology. He argued that there was no evidence for "genetic malfunction" leading to the extinction of a population or species. Many used Caughley's review as a basis to dismiss genetics as of minor relevance to conservation biology. In the next chapter, we will consider the evidence that has accumulated over the nearly 20 years since Caughley's review that unambiguously demonstrates the importance of inbreeding in the viability of populations (Frankham 2010).

Guest Box 13 Inbreeding depression in song sparrows
Lukas F. Keller

Given the long-standing interest in inbreeding and its central role in population genetics, one might think that all that can be known about inbreeding depression has been known for a long time. Yet, in the early 1990s, over a hundred years after Charles Darwin published a seminal book on the effects of inbreeding in plants (Darwin 1876), a debate was in full swing as to whether inbreeding had any significant effects on animals in the wild (e.g., Caro and Laurenson 1994, Caughley 1994).

Just at that time, a study of song sparrows on Mandarte Island, an islet off the west coast of Canada, had reached a state where analyses of inbreeding depression in fitness and other complex traits became possible (Smith *et al.* 2006). Armed with estimates of individual inbreeding coefficients and life-history data that covered not only an individual's complete life-cycle but also that of its offspring and grand-offspring, we tested whether inbreeding had any detectable effects on free-living animals.

What we found mirrored the results of decades of research in agricultural and laboratory settings: inbreeding depression was evident and sometimes severe in many different traits, including seasonal and lifetime reproductive success (Keller 1998, Keller *et al.* 2006), survival (Keller *et al.* 1994), immune response (Reid *et al.* 2003, 2007), a male's song repertoire size (Reid *et al.* 2005), and even such complex traits as senescence (Keller *et al.* 2008). Our results also provided a sobering illustration of the fact that many studies might lack the statistical power necessary to confirm even substantial inbreeding depression. Our analysis in 1998 using 21 years of data revealed that lifetime reproductive success of males declined with inbreeding, but the effect was not statistically significant (Keller 1998). Statistical significance was only revealed when we included 28 years of data in a later reanalysis, which showed that a 10% increase in inbreeding decreased male lifetime reproductive success by 25% (8 LEs; Figure 13.20, see Keller *et al.* 2006). This supports the notion expressed in this chapter that it may be best to assume that endangered species will exhibit inbreeding depression, whether or not it has actually been demonstrated in a particular case.

Inbreeding depression is expected to vary a great deal among species, populations, traits, environments, and even among founder lineages (see Section 13.5.3). This is also what we found in Mandarte's song sparrows. Some traits showed absolutely no inbreeding depression (e.g. the sex-ratio of nestlings; Postma *et al.* 2011) while others were very sensitive to inbreeding (>20 LEs for male song repertoire size). Inbreeding depression varied not only among traits but also within traits among years. Some of this variation was due to variation in environmental conditions such as spring temperature and rainfall (Marr *et al.* 2006). In some traits we found the commonly observed pattern (e.g., Armbruster and Reed 2005) of increasing

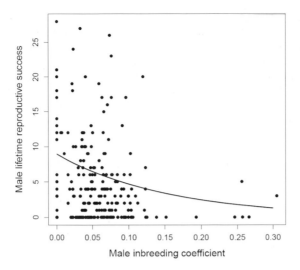

Figure 13.20 Inbreeding depression in male lifetime reproductive success of song sparrows on Mandarte Island, Canada. Lifetime reproductive success was measured as the total number of offspring that survived to independence from parental care over the lifetime of a male. The line is the fit of a negative binomial generalized linear model.

inbreeding depression with increasing environmental stress. However, contrary to expectation, inbreeding depression sometimes was less pronounced when environmental conditions were stressful, suggesting that the underlying causes of environment-dependent inbreeding depression are far from resolved (Waller *et al.* 2008, Cheptou and Donohue 2011).

Environment-dependent inbreeding depression is of great interest to conservation biologists because it can increase the risk of population extinction (e.g., Liao and Reed 2009). However, environment-dependent inbreeding depression also has important ecological and evolutionary consequences that are still poorly understood, both empirically and theoretically (Cheptou and Donohue 2011). As Lewontin (1965) put it so eloquently in the quotation at the beginning of this chapter: "... a full understanding of the theory and ramifications of inbreeding always seems to evade us, just."

CHAPTER 14

Demography and extinction

Clarkia pulchella, *Section 14.2*

As some of our British parks are ancient, it occurred to me that there must have been long-continued close interbreeding with the fallow-deer (Cervus dama) kept in them; but on inquiry I find that it is a common practice to infuse new blood by procuring bucks from other parks.

Charles Darwin (1896, p. 99)

What are the minimum conditions for the long-term persistence and adaptation of a species or a population in a given place? This is one of the most difficult and challenging intellectual problems in conservation biology. Arguably, it is the quintessential issue in population biology, because it requires a prediction based on a synthesis of all the biotic and abiotic factors in the spatial-temporal continuum.

Michael E. Soulé (1987b)

Chapter Contents

Conservation and the Genetics of Populations, Second Edition. Fred W. Allendorf, Gordon Luikart, and Sally N. Aitken.
© 2013 Fred W. Allendorf, Gordon Luikart and Sally N. Aitken. Published 2013 by Blackwell Publishing Ltd.

The quote from Darwin above shows that both evolutionary biologists and wildlife managers have recognized for over 100 years that the persistence of small isolated populations may be threatened by inbreeding. Nevertheless, the potential harmful effects of inbreeding and the importance of genetics in the persistence of populations have been somewhat controversial and remain so to this day (Gaggiotti 2003, Frankham 2010).

There are a variety of reasons for this controversy. Some have suggested that inbreeding is unlikely to have significant harmful effects on individual fitness in wild populations. Others have suggested that inbreeding may affect individual fitness, but is not likely to affect **population viability** (Caro and Laurenson 1994). Still others have argued that genetic concerns can be ignored when estimating the viability of small populations because they are in much greater danger of extinction due to stochastic demographic effects (Lande 1988, Pimm *et al.* 1988). Finally, some have suggested that it may be best not to incorporate genetics into demographic models because genetic and demographic 'currencies' are difficult to combine, and we have insufficient information about the effects of inbreeding in most wild populations (Beissinger and Westphal 1998).

The disagreement over whether or not genetics should be considered in demographic predictions of population persistence has been unfortunate and misleading (Guest Box 1). It is extremely difficult to separate genetic and demographic factors when assessing the causes of population extinction. This is because inbreeding depression initially usually causes subtle reductions in birth and death rates that interact with other factors to increase extinction probability (Mills *et al.* 1996). Obvious indications of inbreeding depression (severe congenital birth defects, monstrous abnormalities or otherwise easily visible fitness deficiencies) are not likely to be detectable until after severe inbreeding depression has accumulated in a population.

Extinction is a demographic process that will be influenced by genetic effects under some circumstances. The key issue is to determine under what conditions genetic concerns are likely to influence population persistence (Nunney and Campbell 1993). There have been important recent advances in our understanding of the interaction between demography and genetics in order to improve the effectiveness of our attempts to conserve endangered species (e.g.,

Landweber and Dobson 1999, Oostermeijer *et al.* 2003).

Perhaps most importantly, we need to recognize when management recommendations based upon strict demographics or genetics may actually be in conflict with each other. For example, Ryman and Laikre (1991) have considered supportive breeding in which a portion of wild parents are brought into captivity for reproduction and their offspring are released back into the natural habitat where they mix with wild individuals. Programs like this are carried out in a number of species to increase population size and thereby temper stochastic demographic fluctuations. Under some circumstances, however, supportive breeding may reduce effective population size and cause a reduction in heterozygosity that may have harmful effects on the population (Ryman and Laikre 1991). This example demonstrates a conflict in that supplemental breeding can provide demographic benefits yet be genetically detrimental.

The primary causes of species extinction today are deterministic and result from human-caused habitat loss, habitat modification, and overexploitation (Caughley 1994, Lande 1999). Reduced genetic diversity in plants and animals is generally a symptom of endangerment, rather than its cause (Holsinger *et al.* 1999). Nevertheless, genetic effects of small populations have an important role to play in the management of many threatened species. For example, Ellstrand and Elam (1993) examined the population sizes of 743 sensitive plant taxa in California. Over 50% of the occurrences contained fewer than 100 individuals. In general, those populations that are the object of management schemes are often small and therefore are likely to be susceptible to the genetic effects of small populations. Many parks and nature reserves around the world are small and becoming so isolated that they are more like 'megazoos' than healthy functioning ecosystems. Consequently many populations will require management (including genetic management) to insure their persistence (Ballou *et al.* 1994).

In this chapter, we will consider the effects of inbreeding depression, along with several other genetic factors that can reduce the probability of persistence of small populations. It is important to be aware that genetic problems associated with small populations go beyond inbreeding depression and the loss of heterozygosity. Finally, we synthesize the current debate about the population size needed to increase the probability of long-term persistence of populations.

14.1 ESTIMATION OF CENSUS POPULATION SIZE

The number of individuals in a population is perhaps the most fundamental demographic characteristic of a population. Accurate estimates of abundance or **census population size** (N_C) are essential for effective conservation and management (Sutherland 1996). Moreover, the rate of loss of genetic variation in an isolated population will be primarily affected by the number of breeding individuals in the population. And, it seems that it should be relatively easy to estimate population size compared with the obvious difficulties of estimating other demographic characteristics, such as the gender-specific age distribution of individuals in a population. However, estimating the number of individuals in a population is usually difficult even under what may appear to be straightforward situations (Luikart *et al.* 2010). For example, estimating the number of grizzly bears in the Yellowstone ecosystem has been an especially contentious issue (Eberhardt and Knight 1996). This is perhaps surprising for a large mammal that is fairly easy to observe and occurs within a relatively small geographic area.

Genetic analyses can provide help in estimating the number of individuals in a population (Luikart *et al.* 2010). A variety of creative methods have been applied to this problem (Schwartz *et al.* 1998). For example, we saw in Section 12.2.1 that the amount of variation within a population can be used to estimate effective population sizes, which can be modified to estimate historic census population sizes (Roman and Palumbi 2003). Here we consider two primary genetic methods for estimating population size. Bellemain *et al.* (2005) provided an excellent comparison of these methods.

14.1.1 One sample

The simplest method for estimating the minimum size of a population is from the number of unique genotypes observed. Kohn *et al.* (1999) used feces from coyotes to genotype three hypervariable microsatellite loci in a $15\,km^2$ area in California near the Santa Monica Mountains. They detected 30 unique multilocus genotypes in 115 feces samples. Thus, their estimate of the minimum N_C was 30.

The actual N_C of a population can be much greater than the number of genotypes detected, depending on what proportion of the population was sampled. For example, it is likely that not all coyotes in this population were sampled in the collection of 115 feces. However, the estimate of total population size can be modified to take into account the probability of not sampling individuals. The cumulative number of unique multilocus genotypes (y) can be expressed as a function of the number of feces sampled (x), and the asymptote of this curve (a) can be estimated by iterative nonlinear regression to provide an estimate of local population size:

$$y = \frac{(ax)}{b+x} \tag{14.1}$$

where b is the rate of decline in value of the slope (Kohn *et al.* 1999). In this case, the estimate was 38 individuals with a 95% confidence interval of 36–40 coyotes (Figure 14.1). Eggert *et al.* (2003) have provided an alternative estimator that behaves similarly to equation 14.1 (Bellemain *et al.* 2005).

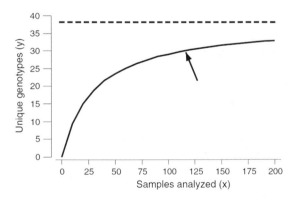

Figure 14.1 Use of rarefaction to estimate the number of coyotes from feces sampled in an area near Los Angeles. Plot of the average number of unique genotypes (y) discovered as a function of the number of samples (x) using the equation $y = (ax)/(b + x)$, where a is the population size asymptote, and b is a constant, which is the rate of decline in the value of the slope. Kohn *et al.* (1999) found 30 unique genotypes in 115 samples analyzed (see arrow), resulting in an estimate of 38 individuals in the population. Drawn using information in Kohn *et al.* (1999).

14.1.2 Two sample: capture–mark–recapture

A mark–recapture approach can also be used with genetic data to estimate population size (Bellemain *et al.* 2005) (Example 14.1). The multilocus genotypes of individuals can be considered as unique 'tags' that exist in all individuals and are permanent.

The simplest mark–recapture method to estimate population size is the Lincoln-Peterson index (Lincoln 1930):

$$N_c = \frac{(N1)(N2)}{R} \qquad (14.2)$$

where $N1$ is the number of individuals in the first sample, $N2$ is the total number of individuals in the second sample, and R is the number of individuals recaptured in the second sample. For example, suppose that ten ($N1$) animals were captured in the first sample, and one-half of the ten ($N2$) animals captured in the second sample were marked ($R = 5$). This would suggest that the ten animals in the first sample represented one-half of the population so that the total population size would be 20.

$$N_c = \frac{(10)(10)}{5} = 20$$

Genetic capture mark–recapture is potentially a very powerful method for estimating population size over large areas (Bellemain *et al.* 2005). It also can be non-invasive (that is, does not require handling or manipulating of animals). However, there are a variety of potential pitfalls (Taberlet *et al.* 1999, Luikart *et al.* 2010). Perhaps the greatest problem is getting random samples of the population. That is, heterogeneous capture probabilities because of geography or behavior can cause serious estimation problems (Boulanger *et al.* 2008). Another potential problem is the failure to distinguish individuals due to using too few or insufficiently variable loci. For example, a new capture might be erroneously recorded as a recapture if the genotype is not unique (due to low marker polymorphism). This has been termed the **shadow effect** (Mills *et al.* 2000). The shadow effect can result in an underestimation of population size and will also affect confidence intervals (Waits and Leberg 2000). Finally, genotyping errors can generate false unique genotypes and thereby cause an overestimate of population size (Waits and Leberg 2000).

It is also possible to estimate the census size of a population by identifying the number of parent–offspring pairs in samples of adults and juveniles (Skaug 2001, Nielsen *et al.* 2001, Bravington and Grewe 2007). This method is analogous to capture–mark–recapture methods, but it relies upon identifying the number of juveniles for which parents can be identified. In mark–recapture terms, each juvenile 'marks' two adults, which might subsequently be recaptured, allowing us to estimate the number of adults. This technique has been used with minke whales (Skaug

---

Example 14.1 Genetic 'tagging' of humpback whales

Palsbøll *et al.* (1997) used a genetic capture–mark–recapture approach to estimate the number of humpback whales in the North Atlantic Ocean. Six microsatellite loci were analyzed in samples collected on the breeding grounds by skin biopsy or from sloughed skin in 1992 and 1993. A total of 52 whales sampled in 1992 were 'recaptured' in 1993 as shown below:

	Females	Males	Total
1992	231	382	613
1993	265	408	673
Recaptures	21	31	52

Substitution into equation 14.1 provides estimates of 2915 female and 5028 male humpback whales in this population. Palsbøll *et al.* (1997) used a more complex estimator that has better statistical properties and estimated the North Atlantic humpback whale population to be 2804 females (95% confidence interval of 1776–4463) and 4894 males (95% confidence interval of 3374–7123). The total of 7698 whales was at the upper range of previous estimates based on photographic identification.

2001), male humpback whales (Nielsen *et al.* 2001), and bluefin tuna (Bravington and Grewe 2007).

14.2 INBREEDING DEPRESSION AND EXTINCTION

We saw in Chapter 13 that inbreeding depression is a universal phenomenon. In this section we will examine when inbreeding depression is likely to affect population viability. Three conditions must hold for inbreeding depression to reduce the viability of populations:

1 Inbreeding must occur.
2 Inbreeding depression must occur.
3 The traits affected by inbreeding depression must reduce population viability.

Conditions 1 and 2 will hold to some extent in all small populations. As discussed earlier and below, matings between relatives must occur in small populations, and some deleterious recessive alleles will be present in all populations. However, condition 3 is the crux of the controversy. There is little empirical evidence that tells us when inbreeding depression will affect population viability and how important that effect will be.

For inbreeding depression to affect population viability it must affect traits that influence population viability. For example, Leberg (1990) found that eastern mosquitofish populations founded by two siblings had a slower growth rate than populations founded by two unrelated founders. However, it has been difficult to isolate genetic effects in the web of interactions that affect viability in wild populations (Soulé and Mills 1998) (Figure 14.2). Laikre (1999) noted that many factors interact when a population is driven to extinction, and it is generally impossible to single out 'the' cause.

Some early authors asserted that there is no evidence for genetics affecting population viability (Caro and Laurenson 1994):

Although inbreeding results in demonstrable costs in captive and wild situations, it has yet to be shown that inbreeding depression has caused any wild population to

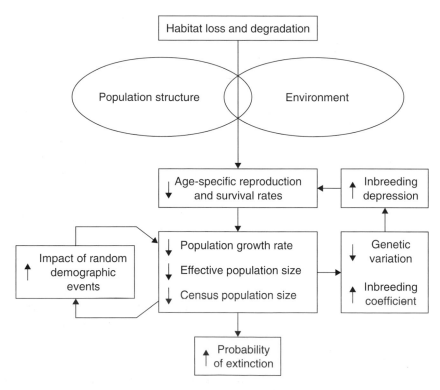

Figure 14.2 Simplified extinction vortex showing interactions between demographic and genetic effects of habitat loss and isolation that can cause increased probability of extinction. Redrawn from Soulé and Mills (1998).

decline. Similarly, although loss of heterozygosity has detrimental impact on individual fitness, no population has gone extinct as a result.

This observation prompted several papers reviewed below that tested for evidence of the importance of genetics in population declines and extinction. Now there is clear consensus that inbreeding can reduce the viability of populations in the wild (Frankham 2010).

14.2.1 Evidence that inbreeding depression affects population dynamics

Newman and Pilson (1997) founded a number of small populations of the annual *Clarkia pulchella* by planting individuals in a natural environment. All populations were founded by the same number of individuals (12); however, in some populations the founders were unrelated (high N_e treatment) and in some they were related (low N_e treatment). All populations were demographically equivalent (that is, the same N_C) but differed in the effective population size (N_e) of the founding population. A significantly greater proportion of the populations founded by unrelated individuals persisted throughout the course of the experiment (Figure 14.3).

Saccheri *et al.* (1998) found that the extinction risk for local populations of the Glanville fritillary butterfly increased significantly with decreasing heterozygosity at seven allozyme loci and one microsatellite locus, after accounting for the effects of environmental factors. Larval survival, adult longevity, and hatching rates of eggs were all reduced by inbreeding, and were thought to be the fitness components responsible for the relationship between heterozygosity and extinction.

Westemeier *et al.* (1998) monitored greater prairie chickens for 35 years and found that egg fertility and hatching rates of eggs declined in Illinois populations after these birds became isolated from adjacent populations during the 1970s. These same characteristics did not decline in adjacent populations that remained large and widespread. These results suggested that the decline of birds in Illinois was at least partially due to inbreeding depression. This conclusion was supported by the observation that fertility and hatching success recovered following translocations of birds from the large adjacent populations (Figure 14.4).

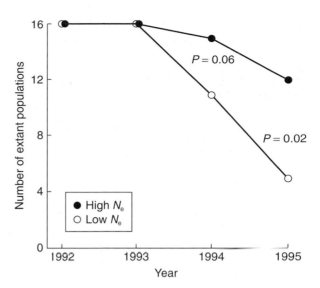

Figure 14.3 Population survival curves for populations of *Clarkia pulchella* founded by related (low N_e) and unrelated founders (high N_e). All populations were founded with 12 individuals; however, in some populations the founders were unrelated (high N_e treatment) and in some they were related (low N_e treatment). From Newman and Pilson (1997).

Madsen *et al.* (1999) studied an isolated population of adders in Sweden that declined dramatically some 35 years ago and has since suffered from severe inbreeding depression. The introduction of 20 males from a large and genetically variable population of adders resulted in a dramatic demographic recovery of this population. This recovery was brought about by increased survival rates, even though the number of litters produced by females per year actually declined during the initial phase of recovery.

The previous examples are from populations that were highly inbred and contained little heterozygosity. In an important, more general result, Reed *et al.* (2007) found that reduction in fitness caused by inbreeding depression affected the population dynamics in seven wild populations from two species of wolf spiders (*Rabidosa*). Differences in population growth rates were especially pronounced during stressful environmental conditions and in smaller populations (<500 individuals).

The relationship between genetic variation and rates of extinction are especially important when populations are faced with environmental stress (e.g., climate change, see Chapter 21). Agashe (2009) found that

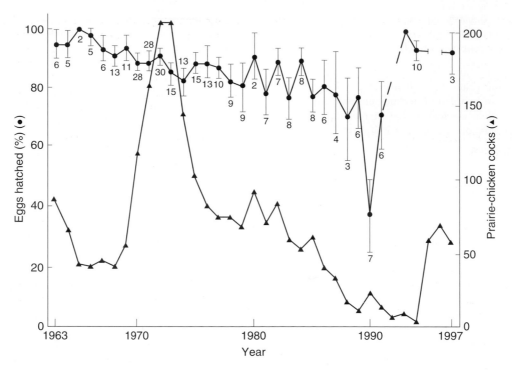

Figure 14.4   Annual means for success of greater prairie chicken eggs in 304 fully incubated clutches (circles) and counts of males (triangles) on booming grounds. Translocations of nonresident birds began in August 1992. From Westemeier *et al.* (1998).

laboratory populations of flour beetles were more likely to go extinct when challenged with a new food resource if the founding population contained less genetic variation.

14.2.2 Are small populations doomed?

The concepts and results presented here should not be taken to mean that populations that have lost substantial genetic variation because of a bottleneck are somehow 'doomed' or are not capable of recovery (Lesica and Allendorf 1992). An increase in frequency of some deleterious alleles and loss of genome-wide heterozygosity is inevitable following a bottleneck. However, the magnitude of these effects on fitness-related traits (survival, fertility, etc.) might not be large enough to constrain recovery. For example, the tule elk of the Central Valley of California has gone through a series of bottlenecks since the 1849 gold rush (McCullough *et al.* 1996). Simulation analysis was used to

estimate that tule elk have lost approximately 60% of their original heterozygosity (McCullough *et al.* 1996). Analyses of allozymes (Kucera 1991) and microsatellites (Williams *et al.* 2004) have confirmed relatively low genetic variation in tule elk. Nevertheless, the tule elk has shown a remarkable capacity for population growth, and today there are 22 herds totaling over 3000 animals (McCullough *et al.* 1996). Tule elk still may be affected by the genetic effects of the bottleneck in the future if they face some sort of stress (see Example 13.5). Red pine provides another example of a widespread species that has very little genetic variation (see Example 11.2).

Some have argued that the existence of species and populations that have survived bottlenecks is evidence that inbreeding is not necessarily harmful (Simberloff 1988, Caro and Laurenson 1994). However, we need to know how many similar populations went extinct following such bottlenecks to interpret the significance of such observations. For example, the creation of inbred lines of mice usually results in the loss of many of the lines (Lynch and Walsh 1998, pp. 274–276). This argument is similar to using the existence of 80-year-old smokers as evidence that cigarette smoking is not harmful. Only populations that have survived a bottleneck can be observed after the fact. Soulé (1987c) referred to this as this the "fallacy of the accident".

14.3 POPULATION VIABILITY ANALYSIS

Predictive demographic models are essential for determining whether or not populations are likely to persist in the future. Such risk assessment is essential for identifying species of concern, setting priorities for conservation action, and developing effective recovery plans. For example, one of the criteria for being included on the IUCN Red List (IUCN 2001) is the probability of extinction within a specified period of time. Some quantitative analysis is needed to estimate the extinction probability of a taxon based on known life history, habitat requirements, threats and any specified management options. This approach has come to play an important role in developing conservation policy (Shaffer *et al.* 2002).

Population viability analysis (PVA) is the general term for models that take into account a number of processes affecting population persistence to simulate the demography of populations in order to calculate the risk of extinction or some other measure of population viability (Ralls *et al.* 2002). The first use of this approach was by Craighead *et al.* (1973) who used a computer model of grizzly bears in Yellowstone National Park. They demonstrated that closing the park dumps to bears and the park's approach to problem bears was driving the population to extinction. McCullough (1978) developed an alternative model that came to different conclusions about the Yellowstone grizzly bear population. Both of these models were deterministic models in which the same outcome will always result with the same initial conditions and parameter values (e.g., stage-specific survival rates).

Mark Shaffer (1981) developed the first PVA model that incorporated chance events (stochasticity) into population persistence. Shaffer described four sources of uncertainty: **demographic stochasticity**, **environmental stochasticity**, **natural catastrophes**, and **genetic stochasticity** (also see Shaffer 1987).

Incorporation of stochasticity into PVA was a crucial step in attempts to understand and predict the population dynamics of populations. Many aspects of population dynamics are processes of sampling rather than completely deterministic events (e.g., stage-specific survival, sex determination, and transmission of alleles in heterozygotes, etc.). The predictability of an outcome decreases in a sampling process as the sample size is reduced. For example, in a large population the sex ratio will be near 50:50. However, this might not be true in a small population in which a large excess of males may significantly reduce the population growth rate (Leberg 1998).

In addition, there will be synergistic interactions between demographic processes and genetic effects (Figure 14.2). Fluctuations in population size may result in genetic bottlenecks during which inbreeding may occur and substantial genetic variation may be lost. Even if the population grows and recovers from the bottleneck, it will carry the legacy of this event in its genes. The loss of genetic variation during a bottleneck may have a variety of effects on demographic parameters (survival, reproductive rate, etc.). This may lead to large fluctuations in population size, increasing the probability of extinction. These interactions have been called "**extinction vortices**" (Gilpin and Soulé 1986), and consideration of these interactions is a central part of PVA (Lacy 2000b).

14.3.1 The VORTEX simulation model

The complexity of the factors affecting population persistence means the useful PVA models must also be complex. It is possible for individuals to develop their own computer program to model population viability. The development of such a model, however, requires a lot of time, and there is always a good probability that such a model will contain some programming errors. The usual alternative is to use an available software package for PVA that serves the same role as commercially available statistical packages. A number of such packages are available (Brook *et al.* 2000). We have chosen to present results using VORTEX because of its power, user friendliness, and widespread use (Lacy *et al.* 2009, Miller and Lacy 2005).

PVA requires information on birth and survival rates, reproductive rates, habitat capacity, and many other factors. It is important to understand the basic structure of the model being used in order to interpret the results. Figure 14.5 shows the relationships among

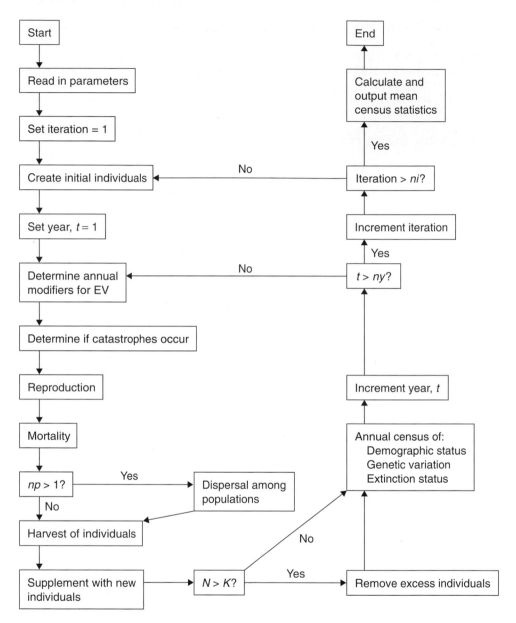

Figure 14.5 Flow chart of the primary components that occur within each subpopulation with the VORTEX simulation model. From Lacy (2000b). EV, environmental variation; K, carrying capacity; N, subpopulation size; ni, number of iterations; np, number of subpopulations; ny, years simulated; t, year.

the primary life history, environmental, and habitat components used by VORTEX.

As we saw in Section 6.5, a population growing exponentially increases according to the equation:

$$N_t = N_0 e^r \tag{14.3}$$

where N_0 is the initial population size ($t = 0$), N_t is the number of individuals in the population after t units of time (years in VORTEX), r is the exponential growth rate, and the constant e is the base of the natural logarithm (approximately 2.72). A population is growing if $r > 0$ and is declining if $r < 0$. Population size is stable if $r = 0$. Lambda (λ) is the factor by which the population increases during each time unit; that is,

$$N_{t+1} = N_t \lambda \tag{14.4}$$

Let us use VORTEX to consider PVA of grizzly bears from the Rocky Mountains of the US. Figure 14.6 shows a summary of the VORTEX input values used. The actual values are taken from Harris and Allendorf (1989), but have been modified for use here.

These life-history values result in a deterministic intrinsic growth rate (r) of 0.005 ($\lambda = 1.005$). Therefore, our simulated grizzly bear population is expected to increase by a factor of 1.005 each year (Figure 14.7). That is, if there are 1000 bears in year $t = 0$, there will be 1005 bears in year $t = 1$ (1000 * 1.005) and 1010 bears in year $t = 2$. The generation interval for grizzly bears is approximately 10 years. Therefore, this growth rate will result in just over a 5% increase in population size after one generation. It is important to look at the deterministic projections of population growth in any analysis with VORTEX. If r is negative, then λ will be less than 1, and the population is in deterministic decline (the number of deaths outpaces the number of births) and will become extinct even in the absence of any stochastic fluctuations.

We can use VORTEX to examine how much stochastic variability in population growth we may expect (Figure 14.7). On average, equation 14.3 does a good job of predicting growth rate. However, there is a wide range of results from each simulation, even though the same input values were used (Figure 14.8). The differences among runs results from VORTEX using random numbers to mimic the life history of each individual.

What will happen if we incorporate genetic effects (inbreeding depression) into this model? Genetic effects due to inbreeding and loss of variation will come into play when the population size becomes small. For example, one of the runs reached an N of 34 after 100 years. This population then proceeded to grow very quickly and exceeded 500 bears 90 years later. However, what would have happened if we had kept track of pedigrees within this population and then reduced juvenile survival as a function of the inbreeding coefficient (F)? Remember that the effective population size of grizzly bears is approximately one-quarter of the population size (Example 7.1). Therefore, the N_e was much smaller than 34 during this period. The increased juvenile mortality of progeny produced by the mating of related individuals would have hindered this population's recovery.

We can incorporate inbreeding depression with VORTEX by assigning a number of lethal equivalents (LEs) associated with decreased survival during the first year of life. Figure 14.9 shows the effects of inbreeding depression with these life history values on population persistence in 1000 simulation runs for 0, 3, and 6 LEs per diploid genome. In the absence of any inbreeding depression (zero LEs), the persistence probability is similar in the first and second hundred years of the simulations. However, even moderate inbreeding depression (3 LEs) reduces the probability of population persistence by approximately 25% in the second hundred years.

Note that even strong inbreeding depression has no effect on population persistence until after 100 years (approximately 10 generations) because it will take many generations for inbreeding relationships to develop within a population. This is an important point to recognize when considering management of real populations. As we saw in Guest Box 7, Yellowstone grizzly bears have now been completely isolated for nearly 100 years. Some have argued that there is no reason to be concerned about possible harmful genetic effects of this isolation because the population has persisted and seems to be doing well. However, it would be very difficult to detect inbreeding depression in a wild population of grizzly bears because we do not have good estimates of vital rates. In addition, Figure 14.8 shows that some populations that have accumulated substantial inbreeding depression may show positive growth rates. The effects of inbreeding depression are expected to be seen more quickly in species with shorter generation intervals.

Simulations by Liao and Reed (2009) found that extinction times decreased some 23% when interactions between inbreeding and stress interaction were

```
VORTEX 9.42 -- simulation of population dynamics

  1 population(s) simulated for 200 years, 1 iterations
  Extinction is defined as no animals of one or both sexes.
  No inbreeding depression
  EV in reproduction and mortality will be concordant.

  First age of reproduction for females: 5    for males: 5
  Maximum breeding age (senescence): 30
  Sex ratio at birth (percent males): 50

Population 1: Population 1

  Polygynous mating;
     % of adult males in the breeding pool = 100

  % adult females breeding = 33
    EV in % adult females breeding: SD = 9

    Of those females producing progeny, ...
      28.00 percent of females produce 1 progeny in an average year
      44.00 percent of females produce 2 progeny in an average year
      28.00 percent of females produce 3 progeny in an average year

    % mortality of females between ages 0 and 1 = 20
      EV in % mortality: SD = 4
    % mortality of females between ages 1 and 2 = 18
      EV in % mortality: SD = 4
    % mortality of females between ages 2 and 3 = 15
      EV in % mortality: SD = 4
    % mortality of females between ages 3 and 4 = 15
      EV in % mortality: SD = 4
    % mortality of females between ages 4 and 5 = 15
      EV in % mortality: SD = 4
    % mortality of adult females (5<=age<=30) = 12
      EV in % mortality: SD = 4

    (Same mortality values for males)

    Initial size of Population 1: 100
      (set to reflect stable age distribution)
    Carrying capacity = 1000
      EV in Carrying capacity = 0

    Animals harvested from Population 1, year 1 to year 1 at 1 year
  intervals: 0

    Animals added to Population 1, year 1 through year 1 at 1 year
  intervals: 0
```

Figure 14.6 VORTEX input summary for population viability analysis of grizzly bears. This output has been slightly modified from that produced by the program. EV is the environmental variation for the parameter. Values used are modified from Harris and Allendorf (1989).

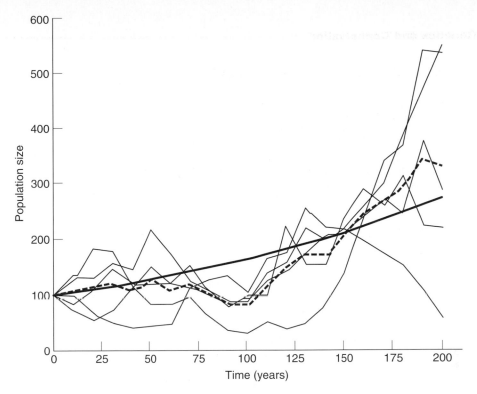

Figure 14.7 Stochastic variability in the growth of a grizzly bear population in five VORTEX simulations using input from Figure 14.6. The dark solid line is the expected growth rate with $r = 0.005$ ($\lambda = 1.005$) and an initial population size (N_0) of 100, using equation 14.4. The dark dashed line is the mean of the five simulated populations.

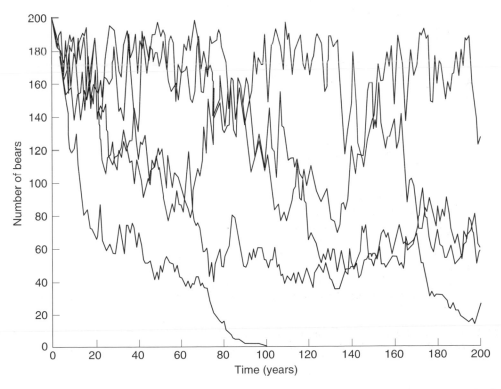

Figure 14.8 Results of five different VORTEX simulations of grizzly bears using input from Figure 14.6, except for the carrying capacity and initial population size. Each simulated population began with 200 bears and had a carrying capacity of 200 individuals. Inbreeding depression was incorporated as 3 LEs that increased mortality in the first year of life.

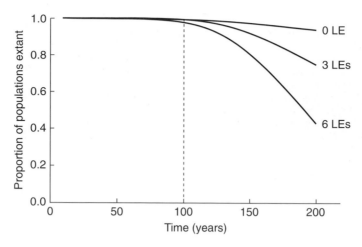

Figure 14.9 Effects of inbreeding depression on the persistence of a grizzly bear population based on VORTEX simulations using the values in Figure 14.6, except for the carrying capacity and initial population size. Each point represents the proportion of 1000 simulated populations that did not go extinct during the specified time period. Simulated populations began with 200 bears and had a carrying capacity of 200. Inbreeding depression was incorporated as a different number of lethal equivalents (0, 3, and 6) that increased mortality in the first year of life.

included. Fox and Reed (2011) have suggested that Liao and Reed (2009) significantly underestimated the effect of inbreeding–stress interactions because they held the interaction constant rather than increasing with the increasing stress. Inclusion of the inbreeding–stress interaction in viability modeling is crucial. This is especially important when populations are of intermediate size and are considered relatively safe from environmental and genetic stresses acting independently (Liao and Reed 2009).

14.3.2 What is a viable population?

An evaluation of the viability of a population requires identifying the time horizon of concern and the required probability of persistence or remaining above some minimum population size. There is no generally accepted time horizon or level of risk with regard to species extinctions (Shaffer *et al.* 2002). The World Conservation Union (IUCN) have offered standard criteria for placing taxa into categories of risk (Table 14.1). These criteria include predictions of the probability of extinction, as well as a variety of other alternative criteria (e.g., reduction in population size, current geographic range, or current population size). For example,

a species may be considered to be facing an extremely high risk of extinction in the wild (i.e., Critically Endangered) if its current population size is less than 50 mature individuals, without performing a PVA.

Early applications of PVA often set out to determine the minimum population size at which a population was likely to persist over some timeframe. The minimum viable population (MVP) concept was used to identify a goal or target for recovery actions. For example, one of the early grizzly bear recovery plans used the results of Shaffer's early work to set recovery targets for four of the six populations between 70 and 90 bears (Allendorf and Servheen 1986). This recommendation was based largely on simulation results of Shaffer and Samson (1985) who reported that only 2% of grizzly bear populations beginning with 50 adults became extinct after 100 years. However, 56% of these populations were extinct after 115 years!

The term MVP has fallen out of favor for a variety of reasons. Many feel that the goal of conservation should not be to set a minimum number of individuals or minimal distribution of a species. However, the concept of MVP is reasonable if we build in an appropriate margin of safety. Nevertheless, the term is not needed as long as we define the timeframe and probability of persistence that we are willing to accept.

Table 14.1 Examples of demographic criteria for evaluating the results of population viability analyses.

Source	Status	Probability of extinction	Timeframe
Shaffer (1978)	Minimum viable population (MVP)	<5%	100 years
Shaffer (1981)	MVP	<1%	1000 years
Thompson (1991)	Endangered	>5%	100 years
Rieman et al. (1993)	Low threat	<5%	100–200 years
	High threat	>50%	100–200 years
AEPBCA*	Vulnerable	>10%	Medium-term future
	Endangered	>20%	Near future
	Critically endangered	>50%	Immediate future
IUCN	Vulnerable	>10%	100 years
	Endangered	>20%	20 years or 5 generations
	Critically endangered	>50%	10 years or 3 generations

*Australian Environment Protection and Biodiversity Conservation Act 1999

14.3.2.1 Demographic criteria

Table 14.1 lists a variety of demographic criteria that have been used or suggested in the literature. There is no correct set of universal criteria to be used. Setting the timeframe and minimum probability of persistence are policy decisions that need to be specific for the situation at hand. Nevertheless, biological considerations should be used to set these criteria. Shorter periods have been recommended because errors are propagated each time step in longer time periods (Beissinger and Westphal 1998). However, we should also be concerned with more than just the immediate future with which we can provide reliable predictions of persistence. The analogy of the distance we can see into the 'future' using headlights while driving at night is appropriate here. We can only see as far as our headlights reach, but we need to be concerned about what lies beyond them (Shaffer et al. 2002). Population viability should be predicted on both short (say 10 generations) and long (more than 20 generations) timeframes.

There is a wide range of values presented in Table 14.1. The most stringent is the 99% probability of persistence for 1000 years used by Shaffer in 1981. The IUCN values are the closest thing to generally accepted standards and are fundamentally sound. They incorporate the concept of both short-term urgency (10 or 20 years) and long-term concerns. They also take into account that the appropriate timeframe will differ depending on the generation interval of the species

under concern. For example, tuatara (see Example 1.1) do not become sexually mature until after 20 years, and their generation interval is approximately 50 years. Therefore, 10 years is just one-fifth of a tuatara generation, but it would represent 10 generations for an annual plant.

14.3.2.2 Genetic criteria

Persistence over a defined time period is not enough. We are also concerned that the loss of genetic variation over the time period does not threaten the long-term persistence of the population or species under consideration. A variety of authors have suggested genetic criteria to be used in evaluating the viability of populations. Soulé et al. (1986) suggested that the goal of captive breeding programs should be to retain 90% of the heterozygosity in a population for 200 years. By necessity, these kinds of guidelines are somewhat arbitrary. Nevertheless, the genetic goal of retaining at least 90–95% of heterozygosity over 100–200 years seems reasonable for a PVA (Allendorf and Ryman 2002). A loss of heterozygosity of 10% is equivalent to a mean inbreeding coefficient of 0.10 in the population.

14.3.2.3 Fuzzy criteria

Some authors have proposed fuzzy logic to deal with the inherent uncertainty in classifying vulnerability, such as in the IUCN criteria (Todd and Burgman 1998, Akçakaya *et al.* 2000, Regan *et al.* 2000). Fuzzy logic considers degrees of truth rather than on a simple either/or classification (e.g., endangered/not endangered). Cheung *et al.* (2005) have proposed a fuzzy logic expert system to estimate intrinsic extinction vulnerabilities of marine fishes to fishing. Goodman (2002) has argued that the inherent subjectivity of fuzzy set theory, in contrast to a Bayesian approach, makes it poorly suited for scientific treatment of evidence in public decision-making.

14.3.3 Are plants different?

PVA has been used primarily with vertebrates. The VORTEX model itself was designed to model the life history of mammals and birds (Lacy 2000b). In his review of plant PVAs, Menges (2000) pointed out that previous reviews of PVAs included less than 1% plant species. Plants provide special challenges for PVA (e.g., seed banks, clonal growth, and periodic recruitment). Nevertheless, plant PVAs have proven useful in guiding conservation and management (Menges 2000).

In addition, most of the guidelines used to determine extinction risk and conservation status have been based on vertebrates, in which characteristics such as body size, fecundity, and geographic range have been found to be important (Knapp 2011). Davies *et al.* (2011) tested the effectiveness of using these largely vertebrate-based methods to estimate extinction risk with plants and concluded that they do not provide good estimates of extinction risk for the flora of the Cape of South Africa. They concluded that extinction risk is greater for young and fast-evolving lineages and cannot be predicted by comparison of life history traits.

14.3.4 Beyond viability

Population viability analyses have great value beyond simply predicting the probability of extinction. Perhaps more importantly, PVA can be used to identify threats facing populations and identify management actions to increase the probability of persistence. This can be done by **sensitivity testing**, in which a range of possible values for uncertain parameters are tested to determine what effects those uncertainties might have on the results. In addition, such sensitivity testing reveals which components of the data, model, and interpretation have the largest effect on population projections. This will indicate which aspects of the biology of the population and its situation contribute most to its vulnerability and, therefore, which aspects might be most effectively targeted for management. In addition, uncertain parameters that have a strong impact on results are those which might be the focus for future research efforts, to better specify the dynamics of the population. Close monitoring of such parameters might also be important for testing the assumptions behind the selected management options and for assessing the success of conservation efforts (Example 14.2).

Example 14.2   PVA of the Sonoran pronghorn

The pronghorn is endemic to western North America, and it has received high conservation priority because it is the only species in the family Antilocapridae. The Sonoran pronghorn is one of five subspecies and was listed as endangered under the US ESA in 1967 (Hosack *et al.* 2002). The pronghorn resembles an antelope in superficial physical characteristics, but it has a variety of unusual morphological, physiological, and behavioral traits (Byers 1997).

The Sonoran subspecies is restricted to approximately 44,000 hectares in southwestern Arizona. There were approximately 200 individuals in this population based on census estimates in the 1990s. A group of 22

biologists from a variety of federal, state, tribal, university, and environmental organizations convened a PVA workshop in September 1996. Nine primary questions and issues were identified as key to pronghorn recovery. All of these questions were explored with PVA simulation modeling during the workshop. The final three of those questions are presented below:

7   Can we identify a population size below which the population is vulnerable, but above which it could be considered for downlisting to a less threatened category?
8   Which factors have the greatest influence on the projected population performance?

9 How would the population respond (in numbers and in probability of persistence) to the following possible management actions: increase in available habitat; cessation of any research that subjects animals to the dangers of handling; exchange of some pronghorn with populations in Mexico; and supplementation of the wild population from a captive population?

Estimates of the life history parameters used by VORTEX were provided by participants of the workshop. Some of the values were available from field data, but there were no quantitative data available for many parameters. For these parameters, the field biologists provided 'best guesses'. The participants performed a sensitivity analysis to evaluate the response of the simulated populations to uncertainty by varying eight parameters: inbreeding depression, fecundity, fawn survival, adult survival, effects of catastrophes, harvest for research purposes, carrying capacity, and size and sex/age structure of the initial population.

Results indicated that the Sonoran pronghorn population had a 23% probability of extinction within 100 years using the best parameter estimates. This probability increased markedly if the population fell below some 100 individuals. Sensitivity analysis indicated that fawn survival rates had the greatest effects on population persistence (Figure 14.10). Sensitivity analysis also indicated that short-term emergency provisioning of water and food during droughts would substantially increase the probability of population persistence. The workshop concluded that this population is at serious risk of extinction, but that a few key management actions could greatly increase the probability of the population persisting for 100 years.

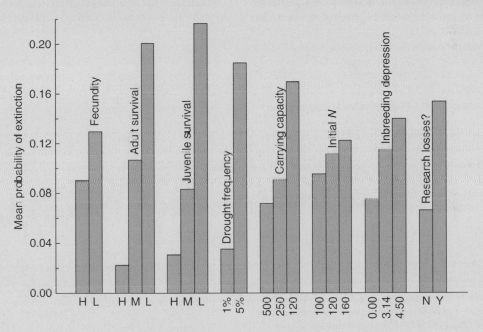

Figure 14.10   Results of PVA on the Sonoran pronghorn. The bars indicate the probability of extinction within 100 years for various values of eight parameters that were varied during sensitivity testing (H, high; M, medium; L, low; N, no; Y, yes). Extinction probabilities were fairly insensitive to some parameters (e.g., initial population size), but were greatly affected by others (e.g., adult and juvenile survival rates). From Hosack *et al.* (2002).

14.4 LOSS OF PHENOTYPIC VARIATION

Inbreeding depression is not necessary for the loss of genetic variation to affect population viability. Reduction in variability itself, even without a reduction in individual fitness because of inbreeding, can reduce population viability (Conner and White 1999, Fox 2005). For example, we saw in Section 6.4 that there can be extensive loss of allelic diversity caused by bottlenecks that are too large to affect the loss of heterozygosity and inbreeding. Loci associated with disease resistance (e.g., MHC) often have many alleles (Section 6.7.2). Thus, loss of allelic diversity at loci associated with disease resistance is expected to increase the vulnerability of populations to extinction.

Honey bees present a fascinating example of the potential importance of genetic variation itself (Jones *et al.* 2004). Honey bee colonies have different amounts of genetic variation depending on how many males the queen mates with. Brood nest temperatures tend to be more stable in colonies in which the queen has mated with multiple males (Figure 14.11). Honey bee workers regulate temperature by their behavior; they fan out hot air when the temperature is perceived as being too hot, and cluster together and generate metabolic heat when the temperature is perceived as being too low. Increased genetic variation for response thresholds produces a more graded response to temperature and results in greater temperature regulation within the hive.

14.4.1 Life history variation

Individual differences in life history (age at first sexual maturity, clutch size, etc.) that have at least a partial genetic basis occur in virtually all populations of

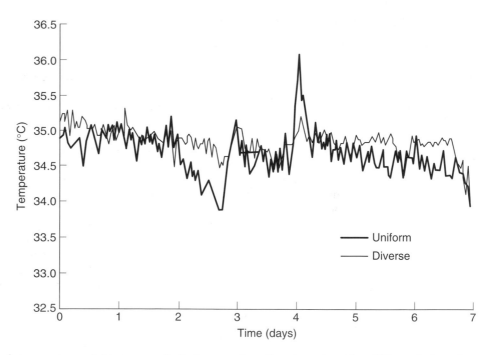

Figure 14.11   Hourly temperature variation in a genetically diverse and a uniform honey bee colony. This graph shows the average hourly temperature for a representative pair of experimental colonies that differed only in the number of males with which the queen mated. The uniform colony queen mated with a single male; the diverse colony queen mated with multiple males. From Jones *et al.* (2004).

plants and animals. Many of these differences may have little effect on individual fitness because of a balance or tradeoff between advantages and disadvantages. Nevertheless, the loss of this life history variability among individuals may reduce the likelihood of persistence of a population (Conner and White 1999, Fox 2005).

Agashe (2009) tested the effect of life history variation on population persistence experimentally by founding four populations of flour beetles in the laboratory with different amounts of ecologically relevant heritable variation. He followed the population dynamics over eight generations. He found that population stability and persistence increased with greater variation in all three different 'habitats' that he used.

Pacific salmon return to freshwater from the ocean to spawn and then die. In most species, there are individual differences in age at reproduction that often have a substantial genetic basis (Hankin *et al.* 1993). For example, Chinook salmon usually become sexually mature at age 3, 4, or 5 years. The greater fecundity of older females (because of their greater body size) is balanced by their lower probability of survival to maturity. These different life history types have similar fitnesses. Pink salmon are exceptional in that all individuals become sexually mature and return from the ocean to spawn in fresh water at two years of age (Heard 1991). Therefore, pink salmon within a particular stream comprise separate odd- and even-year populations that are reproductively isolated (Aspinwall 1974).

Consider a hypothetical comparison of two streams for purposes of illustration. The first stream has separate odd- and even-year populations, as is typical for pink salmon. In the second stream, there is phenotypic (and genetic) variation for the time of sexual maturity so that approximately 25% of the fish become sexually mature at age 1 and 25% of the fish become sexually mature at age 3; the remaining 50% of the population becomes mature at age 2.

All else being equal, we would expect the population with variability in age of return to persist longer than the two reproductively isolated populations. The effective population size (N_e) of the odd- and even-populations would be one-half the N_e of the single reproductive population with life history variability (Waples 1990). Thus, inbreeding depression would accumulate twice as rapidly in the two reproductively isolated populations than in the single variable population. The two smaller populations also would each be

more susceptible to extinction from demographic, environmental, and catastrophic stochasticity. For example, a catastrophe that resulted in complete reproductive failure for one year would cause the extinction of one of the populations without variability.

Greene *et al.* (2010) has provided empirical evidence for this effect. They compared population growth rates (as measured by recruits per spawner) and life history variation (length of freshwater and ocean residence) in nine populations of sockeye salmon from Bristol Bay, Alaska. There was an increasingly positive correlation between population growth rate and life history variation over time. The correlation was negative in the short term (less than 5 years), but increasingly positive from 5 to 20 years. These results suggest that in the short term, certain life-history types are favored by natural selection each year, but the types that are favored change among years. Thus, populations with greater life history diversity are more stable over long periods of time. The authors suggested this 'portfolio effect' of diversity is analogous to the financial stability expected from a diversified investment strategy.

14.4.2 Mating types and sex determination

The occurrence of separate genders or mating types is another case where the loss of phenotypic variation can cause a reduction in population viability without a reduction in the fitness of inbred individuals. Approximately 50% of flowering plant species have genetic incompatibility mechanisms (see Guest Box 14, de Nettancourt 1977). In one of these self-incompatibility systems, an individual's mating type is determined by its genotype at the self-incompatibility (S) locus (Richards 1986). Pollen grains can only fertilize plants that do not have the same S-allele as carried by the pollen. Homozygotes cannot be produced at this locus, and the minimum number of alleles at this locus in a sexually reproducing population is three. Smaller populations are expected to maintain many fewer S-alleles than larger populations at equilibrium (see Section 8.4.3, Wright 1960).

Les *et al.* (1991) considered the demographic importance of maintaining a large number of S-alleles in plant populations. A reduction in the number of S-alleles because of a population bottleneck will reduce the frequency of compatible matings and may result in reduced levels of seed set. Demauro (1993) reported that the last Illinois population of the lakeside daisy was effectively extinct even though it consisted of approximately 30 individuals, because all plants apparently belonged to the same mating type. Reinartz and Les (1994) concluded that some one-third of the remaining 14 natural populations of *Aster furactus* in Wisconsin had reduced seed sets because of a diminished number of *S*-alleles.

A similar effect can occur in the nearly 15% of animal species that are haplodiploid, in which sex is determined by genotypes at one or more hypervariable loci (ants, bees, wasps, thrips, whitefly, certain beetles, etc.; Crozier 1971). Heterozygotes at the sex-determining locus or loci are female, and the hemizygous haploids or homozygous diploid individuals are male (Packer and Owen 2001). Diploid males have been detected in over 30 species of Hymenoptera, and evidence suggests that single-locus sex determination is common. Most natural populations have been found to have 10–20 alleles at this locus. Therefore, loss of allelic variation caused by a population bottleneck will increase the number of diploid males produced by increasing homozygosity at the sex-determining locus or loci.

Diploid males are often inviable, infertile, or give rise to triploid female offspring (Packer and Owen 2001). Thus, diploid males are effectively sterile, and will reduce a population's long-term probability of persisting both demographically and genetically (Zayed and Packer 2005, Hedrick *et al.* 2006). The decreased numbers of females will reduce the foraging productivity of the nest in social species or reduce the population size of other species. In addition, the skewed sex ratio will reduce effective population size and lead to further loss of genetic variation throughout the genome because of genetic drift.

A much weaker gender effect may occur in animal species in which sex is determined by three or more genetic factors. Leberg (1998) found that species with multiple-factor sex determination (MSD) can experience large decreases in viability relative to species with simple sex determination systems in the case of very small bottlenecks. This effect results from increased demographic stochasticity because of greater deviations from a 1:1 sex ratio, not because of any reduction in fitness. MSD is rare, but it has been described in fish, insects, and rodents. For example, a recent genomics study of zebrafish found regions on two chromosomes that had a major effect on sex determination (Bradley *et al.* 2011).

14.4.3 Phenotypic plasticity

Phenotypic plasticity has the potential to affect population viability when the environment is changing stochastically. Temporal variation in the climate can provide a challenge to the persistence of populations. Phenotypic plasticity is the ability of a genotype to produce different phenotypes under different environmental conditions. Reed *et al.* (2010) developed an individual-based model in which phenotypes could respond to a temporally fluctuating environment, and fitness depended on the match between the phenotype and a randomly fluctuating trait optimum. They found that when cue and optimum were tightly correlated, plasticity buffered absolute fitness from environmental variability, and population size remained high and relatively invariant. In contrast, when this correlation weakened and environmental variability was high, strong plasticity reduced population size, and populations with excessively strong plasticity had substantially greater extinction probability. They suggested that population viability analyses should include more explicit consideration of how phenotypic plasticity influences population responses to environmental change.

Reed *et al.* (2011) considered how natural selection, phenotypic plasticity, and demography will affect population persistence in the context of climate change. They pointed out that the limits to plasticity and evolutionary potential across traits, populations, and species, and feedbacks between adaptive and demographic responses, are poorly understood. They concluded that understanding the extent and type of phenotypic plasticity is crucial to understanding the resilience and probabilities of persistence of populations and species facing climate change. (See Chapter 21 for more about phenotypic plasticity and climate change.)

14.5 LOSS OF EVOLUTIONARY POTENTIAL

The loss in genetic variation caused by a population bottleneck can cause a reduction in a population's ability to respond adaptively to future environmental changes through natural selection. Bürger and Lynch (1995) predicted, on the basis of theoretical considerations, that small populations (N_e less than 1000) are more likely to go extinct due to environmental change because they are less able to adapt than are large populations.

The ability of a population to evolve is affected both by heterozygosity and the number of alleles present. Heterozygosity is relatively insensitive to bottlenecks in comparison with allelic diversity (Allendorf 1986). Heterozygosity is proportional to the amount of genetic variance at loci affecting quantitative variation (James 1971). Thus, heterozygosity is a good predictor of the potential of a population to evolve immediately following a bottleneck. Nevertheless, the long-term response of a population to selection is determined by the allelic diversity remaining following the bottleneck or introduced by new mutations (Robertson 1960, James 1971).

The effect of small population size on allelic diversity is especially important at loci associated with disease resistance. Small populations are vulnerable to extinction by epidemics, and loci associated with disease resistance often have an exceptionally large number of alleles. For example, Gibbs et al. (1991) described 37 alleles at the major histocompatibility complex (MHC) in a sample of 77 adult blackbirds. Allelic variability at MHC is thought to be especially important for disease resistance (Edwards and Potts 1996, Black and Hedrick 1997). For example, Paterson et al. (1998) found that some microsatellite alleles within the MHC of Soay sheep are associated with parasite resistance and greater survival.

The effect of loss of variation due to inbreeding on the response to natural selection has been demonstrated in laboratory populations of *Drosophila* by Frankham et al. (1999). They subjected several different lines of *Drosophila* to increasing environmental stress by increasing the salt (NaCl) content of the rearing medium until the line went extinct. Outbred lines performed the best; they did not go extinct until the NaCl concentration reached an average of 5.5% (Figure 14.12). Highly inbred lines went extinct at a NaCl concentration of 3.5%. Lines that experienced an expected 50–75% loss of heterozygosity due to inbreeding went extinct at a mean of roughly 5% NaCl. Thus, loss of genetic variation due to inbreeding made these lines less able to adapt to continuing environmental change.

14.6 MITOCHONDRIAL DNA

Recent results have suggested that mutations in mitochondrial DNA (mtDNA) might decrease the viability of small populations (Gemmell et al. 2004). Mitochondria are generally transmitted maternally so that deleterious mutations that affect only males will not be subject to natural selection (Dowling et al. 2008), and recent empirical evidence has supported this expectation (Innocenti et al. 2011). Sperm are powered by a group of mitochondria at the base of the flagellum, and even a modest reduction in power output may reduce male fertility yet have little effect on females. A study of human fertility has found that mtDNA haplogroups are associated with sperm function and male fertility (Ruiz-Pesini et al. 2000). In addition, the mitochondrial genome has been found to be responsible for cytoplasmic male sterility, which is widespread in plants (Schnable and Wise 1998).

The viability of small populations may be reduced by an increase in the frequency of mtDNA genotypes that lower the fitness of males. Since females and males are haploid for mtDNA, it has not been recognized that mtDNA may contribute to the increased genetic load of small populations. The effective population size of the mitochondrial genome is generally only one-quarter that of the nuclear genome, so that mtDNA mutations are much more sensitive to genetic drift and population bottlenecks than nuclear loci (see Section 7.7).

Whether or not an increase in mtDNA haplotypes that reduce male fertility will affect population viability will depend on the mating system and reproductive biology of the particular population. However, it seems likely that reduced male fertility may decrease the number of progeny produced under a wide array of circumstances. At a minimum, the presence of mtDNA genotypes that reduce the fertility of some males would increase the variability in male reproductive success and thereby decrease effective population size. This would increase the rate of loss of heterozygosity and other effects of inbreeding depression that can reduce population viability.

14.7 MUTATIONAL MELTDOWN

Wright (1931, p. 157) first suggested that small populations would continue to decline in vigor slowly over time because of the accumulation of deleterious mutations that natural selection would not be effective in

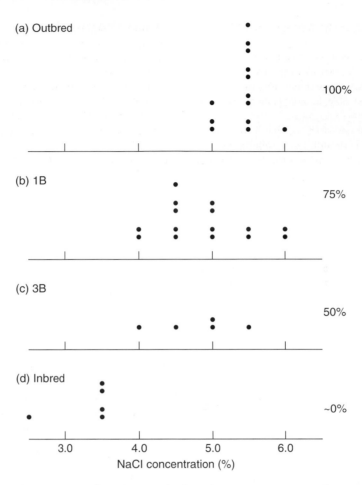

Figure 14.12 Results of an experiment demonstrating that loss of genetic variation can reduce a population's ability to respond by natural selection to environmental change. Lines of *Drosophila* with different relative amounts of expected heterozygosity (as indicated on the right of the figure) were exposed to increasing NaCl concentrations. Each dot represents the NaCl concentration at which lines went extinct. Modified from Frankham *et al.* (1999).

removing because of the overpowering effects of genetic drift (recall Section 8.5). More recent papers have considered the expected rate and importance of this effect for population persistence (Lynch and Gabriel 1990, Gabriel and Bürger 1994, Lande 1995). As deleterious mutations accumulate, population size may decrease further and thereby accelerate the rate of accumulation of deleterious mutations. This feedback process has been termed 'mutational meltdown'.

Lande (1994) concluded that the risk of extinction through this process "may be comparable in importance to environmental stochasticity and could substantially decrease the long-term viability of populations with effective sizes as a large as a few thousand". The expected timeframe of this process is hundreds or thousands of generations. Experiments designed to detect empirical evidence for this effect have had mixed results (e.g., Lynch *et al.* 1999).

14.8 LONG-TERM PERSISTENCE

When considering longer periods than those of a typical PVA, avoiding the loss of genetic variation is not enough for persistence. Environmental conditions are likely to change over time, and a viable population must be large enough to maintain sufficient genetic variation for adaptation to such changes. Evolutionary response to natural selection is generally thought to involve a gradual change of quantitative characters through allele frequency changes at the underlying loci, and discussions on the population sizes necessary to uphold 'evolutionary potential' have focused on retention of additive genetic variation of such traits.

14.8.1 How large do populations need to be to maintain sufficient genetic variation?

There is some disagreement among geneticists regarding how large a population must be to maintain 'normal' amounts of additive genetic variation for quantitative traits (Franklin and Frankham 1998, Lynch and Lande 1998). The suggestions for the effective sizes needed to retain evolutionary potential range from 500 to 5000. The logic underlying these contrasting recommendations is somewhat arcane and confusing. We, therefore, review some of the mathematical arguments used to support the conflicting views.

Franklin (1980) was the first to make a serious attempt to provide a direct estimate of the effective size necessary for retention of additive genetic variation (V_A) of a quantitative character. He argued that for evolutionary potential to be maintained in a small population, the loss of V_A per generation must be balanced by new variation due to mutations (V_m). V_A will be lost at the same rate as heterozygosity ($1/2N_e$) at selectively neutral loci, so the expected loss of additive genetic variation per generation is $V_A/2N_e$ (see also Lande and Barrowclough 1987, Franklin and Frankham 1998). Therefore:

$$\Delta V_A = V_m - \frac{V_A}{2N_e} \tag{14.5}$$

At equilibrium between loss and gain, ΔV_A is zero, and:

$$N_e = \frac{V_A}{2V_m}. \tag{14.6}$$

Using abdominal bristle number in *Drosophila* as an example, Franklin (1980) also noted that $V_m \approx 10^{-3}V_E$, where V_E is the environmental variance (i.e., the variation in bristle number contributed from environmental factors). Furthermore, assuming that V_A and V_E are the only major sources of variation, the heritability (H_N, the proportion of the total phenotypic variation that is due to additive genetic effects; Chapter 11) of this trait is $H_N = V_A/(V_A + V_E)$, and $V_E/(V_A + V_E) = 1 - H_N$. Thus, equation 14.6 becomes (cf. Franklin and Frankham 1998):

$$N_e = \frac{V_A}{2x10^{-3}V_E} = 500\frac{V_A}{V_E} = 500\frac{H_N}{1-H_N} \tag{14.7}$$

The heritability of abdominal bristle number in *Drosophila* is about 0.5. Therefore, the approximate effective size at which loss and gain of V_A are balanced (i.e., where evolutionary potential is retained) would be 500.

Lande (1995) reviewed the recent literature on spontaneous mutation and its role in population viability. He concluded that the approximate relation between mutational input and environmental variance observed for bristle count in *Drosophila* ($V_m \approx 10^{-3}V_E$) appears to hold for a variety of quantitative traits in several animal and plant species. He also noted, however, that a large number of new mutations seem to be detrimental, and that only about 10% are likely to be selectively neutral (or nearly neutral), contributing to the potentially adaptive additive variation of quantitative traits. Consequently, he suggested that a more appropriate value of V_m is $\approx 10^{-4}V_E$, and that Franklin's (1980) estimated minimum N_e of 500 necessary for retention of evolutionary potential should be raised to 5000.

In response, Franklin and Frankham (1998) suggested that Lande (1995) overemphasized the effects of deleterious mutations and that the original estimate of $V_m \approx 10^{-3}V_E$ is more appropriate. They argued that empirical estimates of V_m typically have been obtained from long-term experiments where a large fraction of the harmful mutations have had the opportunity of being eliminated, such that a sizeable portion of those mutations have already been accounted for. They also pointed out that in most organisms, heritabilities of quantitative traits are typically smaller than 0.5, and that this is particularly true for fitness-related characters. As a result, the quotient $H_N/(1 - H_N)$ in equation

14.7 is typically expected to be considerably smaller than unity, which reduces the necessary effective size. Franklin and Frankham (1998) concluded that an N_e of the order 500 to 1000 should be generally appropriate.

Lynch and Lande (1998) criticized the conclusions of Franklin and Frankham (1998) and argued that much larger effective sizes are justified for the maintenance of long-term genetic security. They maintain that the problems with harmful mutations must be taken seriously. An important point is that a considerable fraction of new mutations are expected to be only mildly deleterious with a selective disadvantage of less than 1%. Such mildly deleterious mutations behave largely as selectively neutral ones and are not expected to be 'cleansed' from the population by selective forces even at effective sizes of several hundred individuals. In the long run, the continued fixation of mildly deleterious alleles may reduce population fitness to the extent that it enters an extinction vortex (i.e., mutational meltdown, Lynch et al. 1995).

According to Lynch and Lande (1998) there are several reasons why the minimum N_e for long-term conservation should be at least 1000. At this size, at least the *expected* (average) amount of additive genetic variation of quantitative traits is of the same magnitude as for an infinitely large population, although genetic drift may result in considerably lower levels over extended periods of time. Furthermore, Lynch and Lande (1998) considered populations with $N_e > 1000$ highly unlikely to succumb to the accumulation of unconditionally deleterious alleles (i.e., alleles that are harmful under all environmental conditions) except on extremely long timescales. They also stressed, however, that many single locus traits, such as disease resistance, require much larger populations for the maintenance of adequate allele frequencies (Lande and Barrowclough 1987), and suggest that effective target sizes for conservation should be of the order of 1000–5000.

14.8.2 Return of the MVP

There has been a recent revival of the use of the minimum viable population concept (MVP) in conservation. Traill et al. (2010a) have argued on the basis of both demography and genetics that at least 5000 individuals are required for populations to have an acceptable probability of long-term persistence. There is nothing new in this number itself, as discussed in the

previous section, if we consider that the N_e is much smaller than census size (see Section 7.10). However, these authors also suggested that conservation funding should be prioritized on the basis of a linear function of how far a species falls below the 5000 threshold.

Clements et al. (2011) elaborated on this proposal and used the 5000 threshold as the basis for their proposed SAFE index (Species Ability to Forestall Extinction). Flather et al. (2011) have disagreed and argued that the use of a single "magic" number of 5000 individuals is overly simplistic and not useful. They also conclude that the proposed 5000 threshold is not supported by either theory or empirical data. For example, the genetic justification for the 5000 threshold is based on the need for N_e to be at least 500, assuming an N_e/N_C ratio of 0.10 (Traill et al. 2010a). As we saw in Chapter 7, there is a great deal of variability between species in the N_e/N_C ratio, so using 5000 as a general guideline is not justified.

14.9 THE 50/500 RULE

The 50/500 rule was introduced by Franklin (1980). He suggested that as a general rule-of-thumb, in the short term the effective population size should not be less than 50, and in the long term the effective population size should not be less than 500. The short-term rule was based upon the experience of animal breeders who have observed that natural selection for performance and fertility can balance inbreeding depression if ΔF is less than 1%; this corresponds to an effective population size using $\Delta F = 1/2N_e$ (equation 6.2). There is experimental evidence with house flies, however, that the N_e might have to be greater than 50 to escape extinction even in the short term (Reed and Bryant 2000). The basis of the long-term rule was discussed in detail in the previous section.

There are many problems with the use of simple rules such as this in a complicated world. There are no real thresholds (such as 50 or 500) in this process; the loss of genetic variation is a continuous process. The theoretical and empirical basis for this rule is not strong and has been questioned repeatedly in the literature. In addition, such simple rules can and have been misapplied. We once heard a biologist for a management agency use this rule to argue that genetics need not be considered in developing a habitat management plan that affected many species. After all, if N_e is less than 50, then the population is doomed so we do not need to be concerned with genetics, and if N_e is greater than 50, then the population is safe so we do not need to be concerned with genetics.

Nevertheless, we believe that the 50/500 rule is a useful guideline for the management of populations (Jamieson and Allendorf in press). Its function is analogous to a warning light on the dashboard of a car. If the N_e of an isolated population is less than 50, we should be concerned about possible increased probability of extinction because of genetic effects. These numbers should not, however, be used as targets. When the low fuel light comes on in your car, you do not stop filling the fuel tank once the light goes off. It is also important to remember that 50/500 is based only on genetic considerations. Some populations may face substantial risk of extinction because of demographic stochasticity before they are likely to be threatened by genetic concerns (Lande 1988, Pimm *et al.* 1988).

Discussion of guidelines for population sizes adequate for long-term persistence of populations from a genetics perspective will continue. Regardless of the precise value of this figure, there is agreement that the long-term goal for *actual* population sizes to insure viability should be thousands of individuals, rather than hundreds. Nevertheless, the rigid application of these guidelines to specific cases is problematic. For example, one of the authors (C.J.A. Bradshaw) of the SAFE index paper (Clements *et al.* 2011) has argued in the public press that the kakapo (see Example 19.1) is doomed to extinction because of its small population size (less than 100 individuals), no matter how many resources we invest (Jamieson and Allendorf, in press). Conservation priorities also need to consider what will be lost. The MVP approach alone ignores the value of the kakapo as the world's only flightless, lek-building parrot and the sole representative of a monotypic genus and family (see Section 1.2).

Guest Box 14 Management implications of loss of genetic diversity at the self-incompatibility locus for the button wrinklewort
A. G. Young, M. Pickup, and B. G. Murray

The perennial grassland herb button wrinklewort occurs in the temperate grasslands of Australia's southeast. This ecosystem has been substantially reduced in extent and condition over the last 150 years due to pasture improvement for sheep grazing. The fate of button wrinklewort populations has paralleled that of its habitat, with the species now persisting in only 27 populations ranging in size from as few as seven to approximately 90,000 plants, but over half of the populations have fewer than 200 reproductive individuals.

Many smaller populations are declining (Morgan 1999, Young *et al.* 2000a), especially in the southern part of the species range, and demographic monitoring has shown strong relationships between population size and seed set, with small populations of <200 individuals setting less than a third of the seed of those with more than 1000 flowering plants (Young *et al.* 2000a). Population simulation modeling reveals that these small populations exhibit significantly reduced population viability (Young *et al.* 2000a).

Like many of the Asteraceae, the button wrinklewort has a genetically controlled sporophytic self-incompatibility system (Young *et al.* 2000b) (see Section 14.4.2). Simultaneous analysis of pollinator limitation, inbreeding levels, and genetic diversity at the self-incompatibility locus shows that reproductive failure in small populations is primarily due to low S-allele richness, leading to genetic mate limitation which reduces fertilization success (see Figure 14.13) (Young *et al.* 2000a, Young and Pickup 2010).

Interpopulation crossing studies show that mate limitation can be eliminated and seed set restored in small populations by introducing new S-alleles (Pickup and Young 2008). This suggests that genetic rescue of small populations could be achieved by increasing S-allele richness through translocation of plants among populations – especially from large,

(Continued)

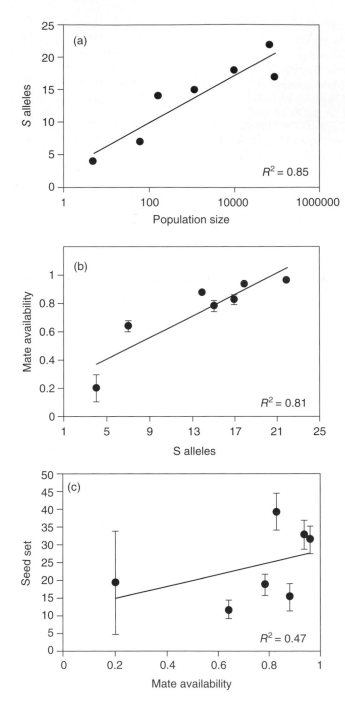

Figure 14.13 Relationships between population parameters and reproductive output in the self-incompatible button wrinklewort. (a) Effect of population size on *S*-allele richness; (b) effect of *S*-allele richness on genetic mate availability as a proportion of the overall reproductive population (± SE); (c) effect of mate availability on reproductive success measured by seed set (± SE). Modified from Young and Pickup (2010).

genetically diverse and demographically viable, northern populations to small declining southern ones.

However, cytogenetic analysis indicates that the situation is complicated by considerable chromosomal variation (Murray and Young 2001), with northern populations being primarily diploid $2n = 22$, while in the south the majority of the populations are $2n = 44$ autotetraploids. Despite maintaining higher allelic richness on average than equivalent-sized diploid populations (Brown and Young 2000), polyploid populations are more mate-limited owing to the greater likelihood of matching S-alleles among tetraploid genotypes (Young et al. 2000b). While this makes the inclusion of novel genetic material even more of an imperative for small southern populations, importing S-alleles

from northern diploid populations presents a range of genetic problems: (1) diploid × tetraploid crosses produce substantially fewer fruits than crosses within ploidy level; (2) the triploid progeny of diploid × tetraploid crosses have reduced pollen fertility due to production of unbalanced gametes during meiosis; (3) backcrossing of triploids to either diploids or tetraploids produces a range of aneuploids with low fertility (Young and Murray 2000).

Taken together, these data argue for the introduction of new genetic material into small button wrinklewort populations that currently have fewer than 200 reproductive plants, as this is likely to increase reproductive success and enhance population viability. However such genetic augmentation activities should only be undertaken between populations of the same chromosomal race.

CHAPTER 15

Metapopulations and fragmentation

White campion, Section 15.5

An important case arises where local populations are liable to frequent extinction, with restoration from the progeny of a few stray immigrants. In such regions the line of continuity of large populations may have passed repeatedly through extremely small numbers even though the species has at all times included countless millions of individuals in its range as a whole.

Sewall Wright (1940)

Theoretical results have shown that a pattern of local extinction and recolonization can have significant consequences for the genetic structure of subdivided populations; consequences that are relevant to issues in both evolutionary and conservation biology.

David E. McCauley (1991)

Chapter Contents

Conservation and the Genetics of Populations, Second Edition. Fred W. Allendorf, Gordon Luikart, and Sally N. Aitken.
© 2013 Fred W. Allendorf, Gordon Luikart and Sally N. Aitken. Published 2013 by Blackwell Publishing Ltd.

The models of genetic population structure that we have examined to this point have assumed a connected series of equal-sized populations in which the population size is constant. However, the real world is much more complicated than this. Local populations differ in size, and local populations of some species may go through local extinction events and then be recolonized by migrants from other populations. These events will have complex, and sometimes surprising, effects on the genetic population structure and evolution of species.

Understanding the genetic effects of habitat fragmentation is becoming increasingly important because of ongoing loss of habitat. Many species that historically were nearly continuously distributed across broad geographic areas are now restricted to increasingly smaller and more isolated patches of habitat. In this chapter, we will combine genetic and demographic models to understand the distribution of genetic variation in species. We will also consider how these processes affect the viability of populations.

15.1 THE METAPOPULATION CONCEPT

Sewall Wright (1940) was the first to consider the effects of extinction of local populations on the genetic structure and evolution of species. He considered the case where local populations are liable to frequent extinction and are restored with the "progeny of a few stray immigrants" (Figure 15.1).

Wright pointed out that such local extinctions and recolonization events would act as bottlenecks that would make the effective population size of a group of local subpopulations much smaller than expected based on the number of individuals present within the subpopulations. Therefore, many of the subpopulations would be derived from a few local subpopulations that persist for long time periods. In modern terms, the genes in many of the subpopulations would 'coalesce' to a single gene that was present in a 'source' subpopulation in the relatively recent past.

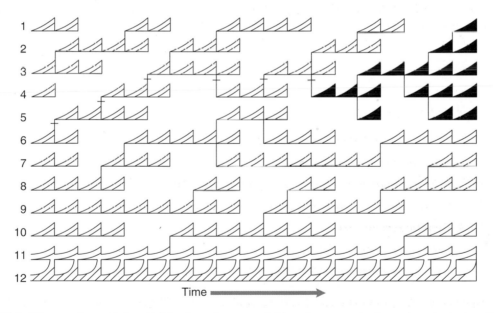

Figure 15.1 Diagram of a species in which local populations are liable to frequent extinction and recolonization. Time proceeds from left to right. Twelve different local patches are represented by a horizontal row (numbered 1 through 12). Note that the bottom two local populations never go extinct, whereas all others go extinct every 2–9 time steps. For example, the subpopulation in patch 7 went extinct at the end of time steps 2, 6, and 15. The darkly shaded subpopulations in the upper-right have passed through small groups of migrants six times. Modified from Wright (1940).

He also considered that genetic drift during such periodic bottlenecks would provide a mechanism for fixation of chromosomal rearrangements that are favorable when homozygous. Such arrangements are selected against in heterozygotes (see Section 8.2) and therefore will be selectively removed in large populations where natural selection is effective. Wright felt that such "nonadaptive inbreeding effects" in local populations might create a greater diversity of multi-locus genotypes and thus make natural selection more effective within the species as a whole. He later suggested that differential rates of extinction and recolonization among local populations could result in intergroup selection that could lead to the increase in frequency of traits that were "socially advantageous" but individually disadvantageous (Wright 1945).

The term **metapopulation** was introduced by Richard Levins (1970) to describe a "population of populations". In Levins' model, a metapopulation is a group of small populations that occupy a series of similar habitat patches isolated by unsuitable habitat. The small local populations have some probability of extinction (*e*) during a particular time interval. Empty habitat patches are subject to recolonization with probability (*c*) by individuals from other patches that are occupied. Metapopulation dynamics are a balance between extinction and recolonization so that at any particular time some proportion of patches are occupied (*p*) and some are extinct (1–*p*). At equilibrium:

$$p^* = \frac{c}{c+e} \tag{15.1}$$

The concept of metapopulations has become a valuable framework for understanding the conservation of populations and species (Hanski and Gilpin 1997, Dobson 2003). It is ironic that Levins (1969) originally developed this model in order to determine better strategies for controlling agricultural insect pests.

The general definition of a metapopulation is a group of local populations that are connected by dispersing individuals (Hanski and Gilpin 1991). More realistic models have incorporated differences in local population size and differential rates of exchange among populations, as well as differential rates of extinction and colonization (Figure 15.2). In general, larger patches are less likely to go extinct because they will support larger populations. Patches that are near other occupied patches are more likely to be recolonized. In addition, immigration into a patch that

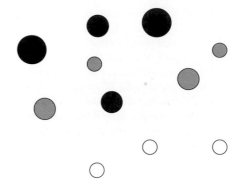

Figure 15.2 The pattern of occupancy of habitat patches of different sizes and isolation in a metapopulation. Darker shading indicates higher probability that a patch will be occupied. Large patches that are close to other populations have the greatest probability of being occupied.

decreases the extinction rate for either demographic or genetic reasons has been called the **rescue effect** (Brown and Kodric-Brown 1977, Ingvarsson 2001).

15.2 GENETIC VARIATION IN METAPOPULATIONS

It is important to consider both spatial and temporal scales in considering the genetic size of metapopulations. Slatkin (1977) described the first metapopulation genetic models. Hanski and Gilpin (1991) described three spatial scales for consideration (Figure 15.3):

1 The **local scale** is the scale at which individuals move and interact with one another in their course of routine feeding and breeding activities.
2 The **metapopulation scale** is the scale at which individuals infrequently move from one local population to another, typically across habitat that is unsuitable for their feeding and breeding activities.
3 The **species scale** is the entire geographic range of a species; individuals typically have no possibility of moving to most parts of the range. Metapopulations on opposite ends of the range of a species do not exchange individuals, but they remain part of the same genetic species because of movement among intermediate metapopulations.

The effect of metapopulation structure on the pattern of genetic variation within a species depends upon the spatial and temporal scale under considera-

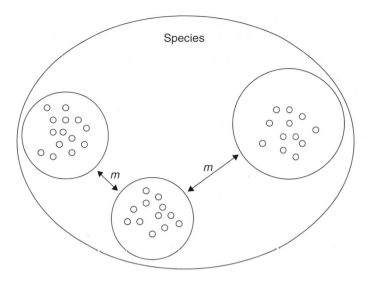

Figure 15.3 Hierarchical spatial organization of a species consisting of three metapopulations, each consisting of a cluster of a local populations that each exchange individuals. A small amount of gene flow between the three metapopulations (*m*) maintains the genetic integrity of the entire species. The two metapopulations on opposite ends of the range do not exchange individuals, but they remain part of the same genetic species because of movement between the intermediate metapopulation.

tion. We can use our models of genetic subdivision introduced in Section 9.1 to see this relationship (Waples 2002). Effective population size is a measure of the rate of loss of heterozygosity over time. The short-term effective population size is related to the decline of the expected average heterozygosity within subpopulations (H_S). The long-term effective population size is related to the decline of the expected heterozygosity if the entire metapopulation were panmictic (H_T).

Consider a metapopulation consisting of six subpopulations of 25 individuals each that are 'ideal' as defined in Section 7.1, so that $N_e = N = 25$. The total population size of this metapopulation is $6 \times 25 = 150 = N_T$. The subpopulations are connected by migration under the island model of population structure, so that each subpopulation contributes a proportion *m* of its individuals to a global migrant pool every generation, and each subpopulation receives the same proportion of migrants drawn randomly from this migrant pool (Section 9.4). The rate of decline of both H_S and H_T will depend upon the amount of migration among subpopulations (Figure 15.4).

In the case of complete isolation, the local effective population size is $N = 25$ and heterozygosity within each subpopulation declines at a rate of $1/2N = 1/50 = 2\%$

per generation. However, different alleles will be fixed by chance in different subpopulations. Therefore, heterozygosity in the global metapopulation (H_T) will become 'frozen' and will not decline. This can be seen in Figure 15.4a. Five of the six isolated subpopulations went to fixation within the first 100 generations.

In the other extreme of near panmixia among the subpopulations, the local effective population size will be N_T so that heterozygosity within each subpopulation declines at a rate of $1/2N_T = 1/300 = 0.3\%$ per generation. In this case, the heterozygosity in the global metapopulation (H_T) will decline at the same rate as the local subpopulations. Eventually all subpopulations will go to fixation for the same allele so that H_T will become zero (Figure 15.4c).

Thus, complete isolation will result in a small short-term effective population size, but greater long-term effective population size. The case of effective panmixia has the extreme opposite effect, that is, greater short-term effective population size, but smaller long-term effective population size (Figure 15.5).

The case of an intermediate amount of gene flow has the best of both worlds. The introduction of new genes by migration will maintain greater heterozygosities within local populations than the case of complete isolation. However, a small amount of migration will

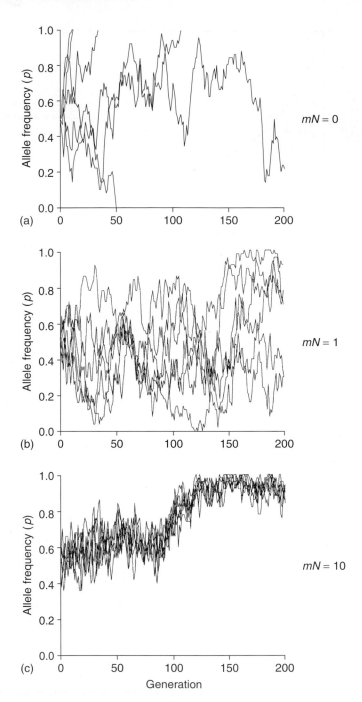

Figure 15.4   Changes in allele frequency in six subpopulations each of $N = 25$ connected by varying amounts of migration under the island model. The top graph shows the case of complete isolation ($m = 0$). The middle graph shows the case of one migrant per generation ($mN = 1$; $m = 0.04$). The bottom graph shows the case near panmixia among subpopulations ($mN = 10$; $m = 0.4$). The graphs were drawn using the *Populus* simulation program (Alstad 2001).

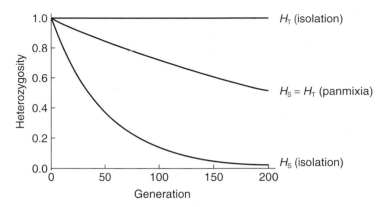

Figure 15.5 Expected decline in local (H_S) and total (H_T) heterozygosity in a population with six subpopulations of $N = 25$ each. In the case of effective panmixia ($mN = 10$) the decline in both local and global heterozygosities are equivalent and are equal to ($1/2N_T = 1/300$ per generation). In the case of complete isolation, local heterozygosity declines at an rate of $1/2N = 0.02$ per generation, but global heterozygosity is constant because random drift within local subpopulations causes the fixation of different alleles.

not be enough to restrain the subpopulations from drifting to near-fixation of different alleles. Therefore, a small amount of gene flow will maintain nearly the same amount of heterozygosity within local subpopulations as the case of effective panmixia and will maintain long-term heterozygosity at nearly the same rate as the case of complete isolation (Figure 15.5).

15.3 EFFECTIVE POPULATION SIZE OF METAPOPULATIONS

The effective population size of a metapopulation is extremely complex. Wright (1943) has shown that in the simplest case, when a metapopulation of size N_T is divided into many identical partially isolated islands that each contribute equally to the migrant pool migration, then:

$$N_{eT} \approx \frac{N_T}{1 - F_{ST}} \qquad (15.2)$$

where N_{eT} is the long-term effective population size of the metapopulation (Nunney 2000, Waples 2002). Thus, increasing population subdivision (as measured by F_{ST}) will increase the long-term effective population size of the metapopulation. Equation 15.2 also indicates that the effective size of the metapopulation will be greater than the sum of the N_es of the subpopulations when there is divergence among the subpopula-

tions. In fact N_{eT} approaches infinity as m approaches zero (see Figure 15.4a).

The validity of equation 15.2 and our conclusions for natural populations depend upon the validity of our assumptions of no local extinction ($e = 0$) and N within subpopulations being constant and equal. However, in the classic metapopulation of Levins (1970), extinction and recolonization of patches (subpopulations) is common. Wright (1940) pointed out that in the case of frequent local extinctions, the long-term N_e can be much smaller than the short-term N_e because of the effects of bottlenecks associated with recolonization:

$$N_e(\text{long}) \ll N_e(\text{short})$$

For example, the entire ancestry of the darkly shaded group of related populations in Figure 15.1 have "passed through small groups of migrants six times in the period shown" (Wright 1940). Thus, these populations are expected to have low amounts of genetic variation, even thought their current size may be very large.

This effect can be seen in Figure 15.6, which shows a metapopulation consisting of three habitat patches from Hedrick and Gilpin (1997). The local populations in all three patches initially have high heterozygosity. The population in patch 1 goes extinct and is recolonized by a few individuals from patch 2 in generation 20, resulting in low heterozygosity. The population in patch 1 goes extinct and is recolonized again from

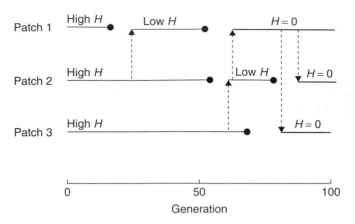

Figure 15.6   Effect of local extinction and recolonization by a few founders on the heterozygosity (*H*) in a metapopulation consisting of three habitat patches. Redrawn from Hedrick and Gilpin (1997).

patch 2. However, the few colonists from patch 2 have low heterozygosity because of an earlier extinction and recolonization in patch 2; this results in near zero heterozygosity in patch 1. Patches 2 and 3 are later recolonized by migrants from patch 1, so heterozygosity is zero in the entire metapopulation.

Hedrick and Gilpin (1997) explored a variety of conditions with computer simulations to estimate long-term N_{eT} as a function of decline in H_T. They found that the rate of patch extinction (*e*) and the characteristics of the founders were particularly important. Slatkin (1977) described two extreme possibilities regarding founders. In the "**propagule** pool" model, all founders come from the same founding local population. In the "migrant pool" model, founders are chosen at random from the entire metapopulation. As expected, high rates of patch extinction greatly reduce N_{eT}. In addition, if vacant patches were colonized by a few founders, or if the founders came from the same subpopulation, rather than the entire metapopulation, H_T and N_{eT} were greatly reduced.

Relaxing the assumption that all subpopulations contribute an equal number of migrants also affects long-term N_{eT} as a function of decline in H_T. Nunney (1997) considered the case where differential productivity of the subpopulations brings about differential contributions to the migrant pool due to the accumulation of random differences among individuals in reproductive success. In this case, the effective size of a metapopulation (N_{eT}) is reduced due to increasing F_{ST}

by what Nunney (1997) has called "interdemic genetic drift".

We expect metapopulation dynamics to have a greater effect on variation at mtDNA than nuclear DNA because of the smaller effective population size of mtDNA (see Chapter 7). Grant and Leslie (1993) found that a variety of vertebrate species (mammals, birds, and fish) in southern Africa show greatly reduced amounts of variation at mtDNA compared with nuclear variation relative to vertebrate species in the northern hemisphere. For example, a cichlid fish (*Pseudocrenilabrus philander*) had unusually high amounts of genetic variation both within ($H_S = 6.2\%$) and between populations ($F_{ST} = 0.30$) at allozyme loci. However, this species has virtually no genetic variation at mtDNA. Grant and Leslie (1993) suggested that the absence of genetic variation at mtDNA results from cycles of drought and rainfall in the semiarid regions of Africa that have caused relatively frequent local extinctions and recolonizations, but which have not been severe enough to cause the loss of much nuclear variability.

It is clear that the effect of metapopulation structure on the effective population size of natural populations is complex. The long-term effective population size may either be greater or less than the sum of the local N_es, depending on a variety of circumstances: rates of extinction and recolonization, patterns of migration, and the variability in size and productivity of subpopulations. It is especially important to distinguish between local and global effective population size because these

Plate 1 (Figure 2.15) Photograph of sockeye salmon (large) and kokanee (small) on the breeding grounds in Dust Creek a tributary of Takla Lake, British Columbia. Both kokanee and sockeye salmon have bright red bodies and olive green heads. Photo by Chris Foote.

Conservation and the Genetics of Populations, Second Edition. Fred W. Allendorf, Gordon Luikart, and Sally N. Aitken.
© 2013 Fred W. Allendorf, Gordon Luikart and Sally N. Aitken. Published 2013 by Blackwell Publishing Ltd.

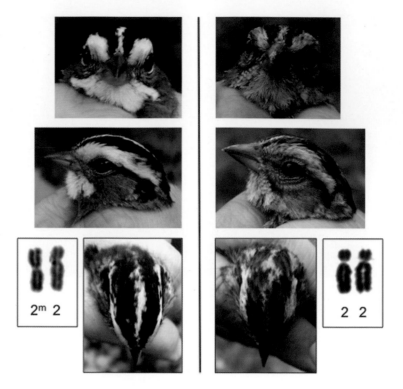

Plate 2 (Figure 3.11)   White and tan plumage morphs of the white-throated sparrow. Morph of an individual is absolutely associated with the presence (ZAL2^m/ZAL2 = white) or absence (ZAL2/ZAL2 = tan) of a chromosomal rearrangement. Photo by Elaina Tuttle.

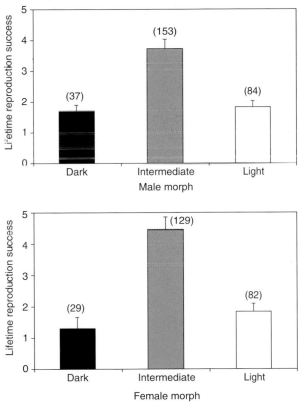

Plate 3 (Figure 8.4) Mean lifetime reproductive success (+ standard error) in three color morphs (above) of male and female European common buzzards. From Boerner and Krüger (2009). Numbers above error bars are the sample sizes. Photo by Oliver Krüger.

Plate 4 (Figure 8.13) Light and dark phenotype of rock pocket mice on light-colored rocks and dark lava. From Nachman *et al.* (2003).

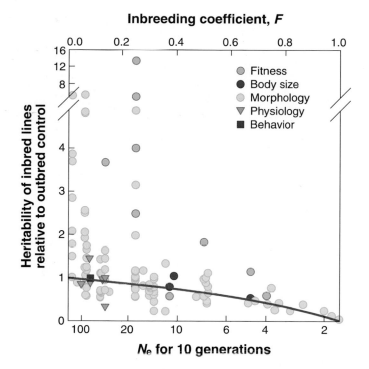

Plate 5 (Figure 11.11) The effects of inbreeding coefficient (upper axis) or the effective population size that would result in the same inbreeding coefficient (lower axis) on heritability estimates in laboratory studies of experimental populations. Study organisms include *Drosophila* (ten studies), house flies (five studies), *Tribolium* beetles (two studies), two plant species, laboratory mice and a butterfly. Heritability estimates are expressed as a ratio relative to outbred control populations. The solid line indicates the expected decline for purely additive genetic variation. From Willi *et al.* (2006).

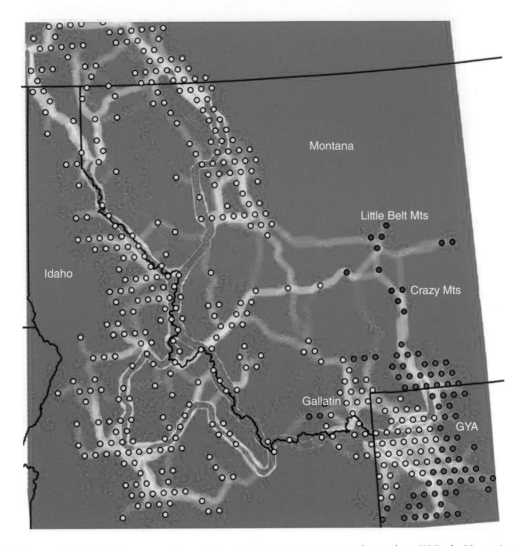

Plate 6 (Figure 15.11) Landscape genetics analysis of wolverine connectivity in the northern US Rocky Mountains, showing cumulative least-cost paths between systematically placed locations (circles) in spring snow cover cells. Paths in orange are predicted to be used more often than those in cooler colors. The color of the circle corresponds to the average cost distance between that location and all other locations, based on our models. The graph was divided into four modes (three within the northern US Rockies, and one between the Greater Yellowstone Area and Colorado). The yellow mode has the lowest average cost distances. Rockies), the blue bars the next lowest, the pink bars (Crazy and Little Belt Mountains) have the greatest average cost distances in the northern US Rocky Mountains, and the green bars show the distances between all points from Colorado to Greater Yellowstone. Modified from Schwartz, M.K. *et al.* (2009).

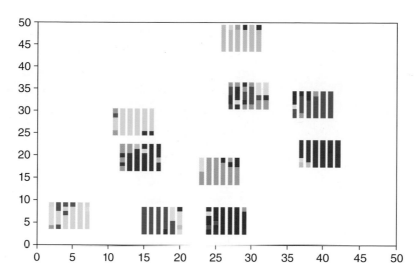

Plate 7 (Figure 16.13) Potential problems when using individual-based methods to identify discrete subpopulations in the case of isolation by distance. Assignments using STRUCTURE (Pritchard *et al.* 2000) when 360 individuals are sampled in ten clusters of individuals from simulations of a continuously distributed population. Each of the ten sample locations consists of six vertical bars representing six individuals within each bar (36 individuals per sampling location). Each square represents an individual. Each square's shading indicates the cluster to which STRUCTURE assigned the individual, based on maximum Q. The values on the *x* and *y* axis are cell references in the 50×50 grid containing 250 individuals total. Modified from Schwartz and McKelvey (2009).

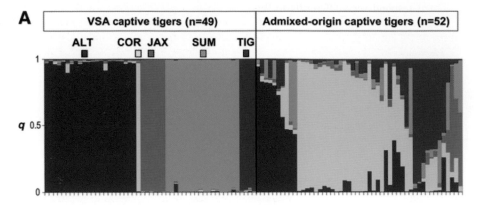

Plate 8 (Figure 22.3) STRUCTURE population cluster analysis of 101 captive tigers. The left panel shows 49 tigers assigned to a single subspecies based on 134 reference tigers with verified subspecies ancestry (VSA). The right panel shows 52 tigers with apparent admixed origins. Each individual is represented by a thin vertical bar (defined by tick marks under colored regions on the *x* axis) partitioned into five colored segments representing individual ancestry (*q*) to the five indicated tiger subspecies. ALT, Amur; COR, Indochinese; JAX, Malayan; SUM, Sumatran; TIG, Bengal. From Luo *et al.* (2008).

two parameters often respond very differently to the same conditions. All of these factors should be considered in evaluating conservation programs for endangered species (Waples 2002).

15.4 POPULATION DIVERGENCE AND CONNECTIVITY

The effects of extinction and recolonization on the amount of divergence among populations (F_{ST}) is extremely complex (McCauley 1991). Gilpin (1991) has considered the effects of the relative rates of extinctions and recolonization on genetic divergence (Figure 15.7). If $e > c$, then the metapopulation is not viable. If both extinction and recolonization occur regularly and $c > e$, then patch coalescence will occur in which all patches descend from a single patch. For example, patches 1–5 (filled black) in Figure 15.1 coalesce to a single ancestral patch in seven steps back in time. If the rate of colonization is much greater than local extinctions ($c \gg e$) then all patches will have similar allele frequencies (panmixia). Allele frequency divergence ($F_{ST} > 0$) among local populations is expected only in a fairly narrow range of rates of colonization and extinction.

As we have seen, most species have a large amount of heterozygosity at allozyme and nuclear DNA loci. On this basis, Hedrick and Gilpin (1997) concluded that most species have not functioned as a classic Levins-type metapopulation during their evolutionary history (see Figure 15.7).

Perhaps most importantly, patterns of extinction and recolonization in nature may invalidate many inferences resulting from models that assume equilibrium (e.g., $F_{ST}^* = 1/(4mN+1)$ with the island model of migration, equation 9.12). The effects of metapopulation dynamics (local extinctions and recolonizations) depend largely on the number and origin of founders that recolonize patches. We began this chapter with a model by Wright (1940) in which the genetic differentiation of local populations was enhanced because patches were founded by only a few individuals, so that genetic differentiation was enhanced by bottlenecks. At the other extreme, extinctions and recolonizations may act as a form of gene flow and reduce genetic differentiation if patches are founded by individuals drawn from different patches (Slatkin 1977):

Few founders from one patch ⟷ Many founders from several patches

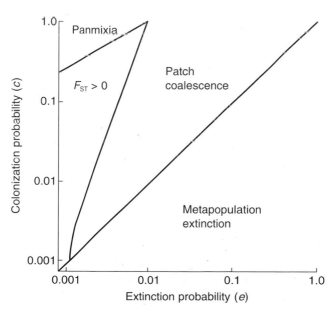

Figure 15.7 Effects of metapopulation dynamics on genetic divergence among local populations. From Gilpin (1991).

15.4.1 Genetic versus demographic connectivity

Dispersal can contribute significantly to population growth rates, gene flow, and, ultimately, species persistence. It also plays a major role in determining the rate at which populations or species can shift their range in response to changing environmental conditions (Chapter 21). Therefore, assessing the effects of dispersal is crucial to understanding population biology and evolution in natural systems (Wright 1951, Hanski and Gilpin 1997, Clobert *et al.* 2001). Likewise, effective protection of endangered species and management of economically important species often rely on estimates of 'population connectivity' (Mills and Allendorf 1996, Drechsler *et al.* 2003), a concept based on the dispersal of individuals among discrete populations, but which can have very different meanings and implications depending on how it is measured.

Demographically connected populations are those in which population growth rates (λ, r) or specific vital rates (survival and birth rates) are affected by immigration or emigration (Lowe and Allendorf 2010). Demographic connectivity is generally thought to promote population stability (e.g., $\lambda \geq 1.0$), and this stabilizing effect can occur at two different scales. In individual populations, demographic connectivity can promote stability by providing an immigrant subsidy that compensates for low survival or birth rates of residents (i.e., low local recruitment). Demographic connectivity can also promote the stability of metapopulations by increasing colonization of unoccupied patches (i.e., discrete subpopulations), even when the extinction rate of occupied patches is high (Levins 1970, Hanski 1998).

Genetic data are often used to assess "population connectivity" because it is difficult to measure dispersal directly at large spatial scales. As we saw in Chapter 9, however, genetic connectivity depends primarily upon the absolute number (mN) of dispersers among populations (see equation 9.12). In contrast, demographic connectivity depends upon the relative contributions to population growth rates of dispersal vs. local recruitment (i.e., survival and reproduction of residents) (Hastings 1993). Therefore, estimates of genetic divergence alone provide little information on demographic connectivity (Figure 15.8, Lowe and Allendorf 2010).

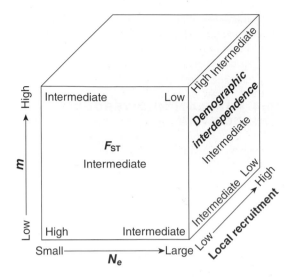

Figure 15.8 Combinations of migration rate (m), effective population size (N_e), and local recruitment resulting in different expected values of genetic divergence (F_{ST}) and demographic independence. Local recruitment is determined by births and deaths of resident individuals. Genetic divergence is primarily a function of mN, while demographic connectivity is primarily a function of m relative to local recruitment. From Lowe and Allendorf (2010).

15.5 GENETIC RESCUE

The term 'genetic rescue' was coined to describe the increase in viability of small populations following immigration resulting from the reduction of genetic load caused by inbreeding depression (Thrall *et al.* 1998). Genetic rescue is generally considered to occur when population growth rate or viability increases by more than can be attributed to just the demographic contribution of migrant individuals (Ingvarsson 2001). Genetic rescue can play a crucial role in the persistence of small natural populations and is an effective conservation tool under some circumstances (Tallmon *et al.* 2004, Hedrick and Fredrickson 2010, see Guest Box 15). Increasingly widespread evidence that genes from a pulse of immigrants into a local population often result in heterosis that increases the population growth rate also has important implications for the study of evolution and metapopulation dynamics. However, the occurrence of outbreeding depression following heterosis in the first generation in some cases

---

Example 15.1 Genetic rescue of an isolated population of bighorn sheep

Hogg *et al.* (2006) documented genetic rescue in a natural population of bighorn sheep on the National Bison Range, an isolated wildlife refuge, in Montana. The population was founded in 1922 with 12 individuals from Alberta, Canada. The mean population size from 1922 to 1985 was approximately 40 individuals. Starting in 1985, 15 individuals (mostly from Alberta, Canada) were introduced over a 10-year period.

The restored gene flow caused an increase in the expected heterozygosity at 8 microsatellite loci from 0.44 to more than 0.60. The gene flow also erased the genetic bottleneck signature consisting of a severe deficit of rare alleles (Hogg *et al.* 2006).

Survival and reproductive success were remarkably higher in the outbred than in inbred individuals. For example, the average annual reproductive success (number of lambs weaned) for females was 2.2 fold higher in outbred individuals compared with inbred resident animals. Average male annual reproductive success (number of lambs fathered) was 2.6 fold higher in outbred individuals. Survival for both females and males was higher in outbred individuals; average

lifespan was 2 years longer in outbred animals compared with inbred ones with only resident genes.

This study was exceptional in that individual-based measures of fitness were available through a long-term (25 year) study. Most studies of genetic rescue report only a correlative increase in population size with genetic variation (e.g., mean heterozygosity), with no direct evidence that the increased genetic variation causes the increase in population size. In correlative studies, environmental factors could be the cause of increase in population size. In this study the inbred and outbred individuals coexisted in the same environment. This allowed for control of, or removal of, environmental effects. Finally, individual-based measures of fitness rescue are generally better than population-based measures (e.g., increased population growth rate) because it is possible to recover individual fitness without actually increasing population size, for example, if an environmental challenge or disease outbreak prevents a population size increase even following increased individual fitness.

---

indicates that care is needed in considering the source of populations for rescue (see Chapter 17).

Recent studies report positive fitness responses to low levels of migration (gene flow) into populations that have suffered recent demographic declines and suggest that natural selection can favor the offspring of immigrants (Example 15.1). Madsen *et al.* (1999) studied an isolated population of adders in Sweden that declined dramatically some 35 years ago and that has since suffered from severe inbreeding depression. The introduction of 20 males from a large and genetically variable population of adders resulted in a dramatic demographic recovery of this population. This recovery was brought about by increased survival rates, even though the number of litters produced by females per year actually declined during the initial phase of recovery. A genetic rescue effect has been uncovered in experimental populations of house flies, but only following many generations in which it was not detected (Bryant and Reed 1999).

Two recent studies provide evidence that genetic rescue may be an important phenomenon. In experimentally inbred populations of a mustard (*Brassica*

campestris), one immigrant per generation significantly increased the fitness of four out of six fitness traits in treatment populations, compared with (no immigrant) control populations (Newman and Tallmon 2001). Interestingly, there was no fitness difference between one-immigrant and 2.5-immigrant treatments after six generations, but there was greater phenotypic divergence among populations in the one-migrant treatment, which could facilitate local adaptation in spatially structured populations subject to divergent selection pressures. In small, inbred white campion populations, Richards (2000) found that gene flow increased germination success and that the success of immigrant pollen correlated positively with the amount of inbreeding in recipient populations (see also Guest Box 13).

Genetic rescue may be of crucial importance to entire metapopulations by reducing local inbreeding depression and increasing the probability of local population persistence; in turn, this maintains a broad geographic range that buffers overall metapopulation extinction and provides future immigrants for other populations. In long-established plant populations, it is conceivable that plants emerging from long-dormant seed banks could also provide an intergenerational genetic rescue. Genetic rescue might also contribute to the spread of invasive species along the leading edge of invasion by supplying established, small populations with adequate genetic variation to respond to selection and adapt to the new environment (see Chapter 20).

15.5.1 Beyond genetic rescue: genetic restoration

As we have just seen, genetic rescue refers to the alleviation of inbreeding depression. However, we emphasized in Chapter 14 that inbreeding depression is not the only genetic concern with small populations. Hedrick (2005) has suggested that a broader framework to alleviate genetic problems with small and isolated populations is needed. He has suggested that the concept of "**genetic restoration**" be used to include efforts beyond reducing genetic load caused by inbreeding depression.

Migrant individuals can sometimes be extremely successful because their progeny will not be affected by inbreeding depression. This can reduce the effective population size of the 'rescued population' so that inbreeding depression can return relatively rapidly. Molecular genetic analysis of a natural 'genetic rescue' demonstrates that this inbreeding effect can be extremely large (Example 15.1). Gray wolves on Isle Royale in Lake Superior, North America, are isolated from mainland wolves by a channel of water. This population was founded around 1950, and typically consists of 25 or so wolves. By the late 1990s, the estimated mean inbreeding coefficient in this population was near 0.80, and many of the wolves had morphological abnormalities associated with inbreeding depression (Räikkönen et al. 2009). A single male wolf crossed this channel on ice during the winter of 1997, and quickly reduced the average inbreeding coefficient of wolves on the island (Figure 15.9a). However, this immigrant wolf was so successful that in less than 10 years he was an ancestor of every wolf on the island, and 56% of the genes in the population originated in this wolf, leading to further inbreeding (Figure 15.9b).

15.6 LANDSCAPE GENETICS

Landscape genetics is a rapidly growing interdisciplinary field that aims to assess the influence of landscape features and environmental variables on gene flow, dispersal, and genetic variation (Manel et al. 2003, Holderegger and Wagner 2008). Landscape genetics combines approaches from molecular population genetics, landscape ecology, and spatial statistics. The approaches can involve novel individual-based statistical assessments that use genetic patterns such as spatial discontinuities to identify population boundaries or to group individuals into populations (Guillot et al. 2005). Individual-based assessments help to prevent incorrect delineation of populations, which can occur when populations are identified using the geographic location or physical characters of individuals sampled. A priori population identification is often subjective (Pritchard et al. 2000).

Individual-based landscape genetic approaches can provide finer-scale assessments of genetic structure than traditional population genetic approaches (Section 16.4.2). Such approaches are crucial for precise geographic localization of genetic discontinuities caused by **landscape resistance barriers** or **secondary contact zones**. Nevertheless population-based approaches are also useful in landscape genetics (e.g., Epps et al. 2007, Wang et al. 2009). Figure 15.10 presents three possible landscape genetic approaches to detect barriers to gene flow.

Continuously distributed populations (see Section 9.5) are better investigated with landscape genetic models than metapopulation genetic models, as the latter use an a priori grouping of individuals into local populations. This is important because most species are not distributed in discrete demes. Analyzing continuously distributed individuals as discrete groups can lead to erroneous inferences about genetic structure and connectivity (see Section 16.4.2 and Figure 16.13).

15.6.1 Landscape connectivity and complex models

Functional landscape connectivity (i.e., the dispersal, gene flow, or disease transmission across landscapes) can be inferred using landscape genetic approaches. For example, Blanchong et al. (2008) used landscape genetic modeling to test for landscape features influencing the distribution of chronic wasting disease (CWD) in white-tailed deer in Wisconsin, US. The features tested were the Wisconsin River running east–west through the northern third of the region, and US Highway 18/151 running east–west through the

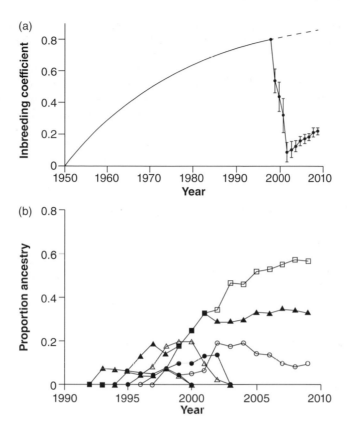

Figure 15.9 (a) The population inbreeding coefficient (F) averaged over each individual present in Isle Royale wolves from 1950 to 2009. The dashed line projects the expected increase in F if the single male wolf did not immigrate in 1997. (b) The proportion of ancestry of the immigrant wolf (open squares) and six native breeding wolves. The wolf with the second greatest contribution (closed triangles) was the first mate of the immigrant wolf. From Adams *et al.* (2011).

southern third of the region. Genetic differentiation between deer populations was greatest, and CWD prevalence lowest, in areas separated by a river, indicating that rivers reduced spread of disease from the geographic area of the disease origin. This result suggested that landscape genetics can help to predict populations at high risk of infection based on their genetic connectivity to infected host populations, and might also help target areas for disease surveillance and preventative measures such as increased harvest (e.g., along rivers to isolate or reduce contact rates between populations). To understand functional landscape connectivity, multiple sympatric species (e.g., hosts, parasites, predators, and prey) could be assessed using landscape genetic approaches.

Landscape genetics can be thought of as an extension of traditional metapopulation genetics models by allow-

ing more complex modeling of connectivity in heterogeneous landscapes. Recent modeling approaches allows testing among many realistic landscape genetic models (e.g., with a range of barrier configurations and strengths). This can yield a more detailed understanding of interactions between landscapes, environment, and connectivity. Such detailed understanding is important in conservation for achieving goals such as accurate and precise localization of corridors and barriers. For example, many traditional metapopulation models yield a single migration rate parameter (m) between demes or habitat patches, and consider only a simple homogeneous matrix of inhospitable habitat between demes. More recent landscape genetic models assume a complex matrix of multiple habitat types with different resistances to gene flow separating different pairs of demes or individuals.

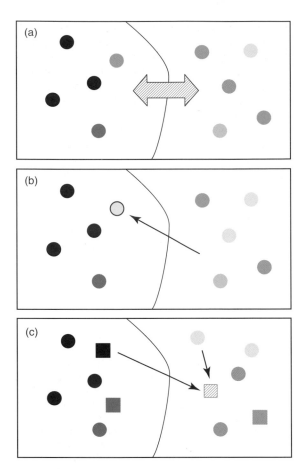

Figure 15.10 Three landscape genetic approaches to infer barriers to gene flow. The black line refers to a landscape feature, such as a river; the filled circles refer to the adult individuals; and different gray shadings of these circles refer to genetically similar genotypes. (a) Approach based on genetic distances between pairs of individuals. Here, the genetic distances between all pairs of individuals are determined and correlated with landscape structure. In the present case, the largest genetic distances occur across the landscape feature, that is, the hypothesized barrier (hatched doubleheaded arrow). (b) Recent gene flow assessed by assignment tests (see Section 9.9.4). The individuals are grouped *a priori* into two populations on either side of the landscape feature. An assignment test identifies one individual (circle with solid outer line) as a recent immigrant from the other side of the landscape feature (arrow). (c) Current gene flow assessed by parentage analysis of offspring (squares). The hatched offspring has one parent on the other side of the landscape feature (arrow). From Holderegger and Wagner (2008).

The use of relatively complex landscape models allows testing of multiple hypotheses in realistic settings to better understand the relative importance of different features such as roads and vegetation cover for gene flow and landscape connectivity. Complex habitat models allow quantification of the relative resistance of different pathways and the identification of multiple potential pathways for dispersal or gene flow across the landscape. Each potential path can have a different cumulative (total) resistance to movement. The path with the least cumulative resistance is referred to as the **least cost path** (Epps *et al.* 2007). The least cost path is often longer in geographic distance than the direct straight-line path because the direct path often crosses poor habitat with high ecological cost to movement. Researchers calculate the least cost paths under many different hypotheses (resistance models) and then compare each matrix of least cost paths to a matrix of genetic relatedness (or genetic distance) to assess the merit of each hypothesis. This approach is being widely used by ecologists and managers to understand habitat features that various wildlife and plants require for dispersal and gene flow.

15.6.2 Corridor mapping

Corridor mapping is facilitated by landscape genetic approaches in which individuals are sampled widely across large landscapes. For example, Epps *et al.* (2007) applied a population-based landscape genetic approach using least cost path modeling to identify corridors and assess gene flow between bighorn sheep populations. The corridors predicted from the genetic modeling were largely consistent with known intermontane movements of bighorn sheep. In addition, the authors determined that gene flow was highest in landscapes with more than 10% sloping terrain; thus, gene flow occurred over longer distances when steep escape terrain was available, consistent with bighorn sheep biology and the use of cliffs to avoid predators. Their work also linked reduced genetic variation to the construction of interstate highways and canals that have reduced connectivity and heterozygosity by approximately 15% in only 40 years. This diminished connectivity could reduce individual and population growth (Hogg *et al.* 2006) and threaten population persistence (Epps *et al.* 2005).

Understanding the effects of habitat and landscape features is crucial in predicting the effects of habitat loss and climate change on the future viability of many species. Example 15.2 presents a landscape genetic analysis of wolverines in the US Rocky Mountains to develop corridor maps in order to: (1) evaluate current connectivity among extant populations; (2) evaluate the potential of colonization from extant populations to areas of recent historical extirpation; and (3) predict the effects of projected climate change on wolverine populations in the future.

15.6.3 Landscape genomics

Most landscape genetic analyses have employed only 10 or 20 selectively neutral markers. **Landscape genomics**, on the other hand, is the simultaneous study of hundreds or thousands of markers, including markers in genes under selection and individuals sampled across an environmental gradient. While landscape genomics is, in one sense, simply landscape genetics with lots of data (thus reduced variance and increased precision), the qualitatively different nature of the markers (adaptive, nonindependent) and analytical approaches associated with these data are different enough to produce substantially different conceptual and computation approaches to hypothesis testing (Schwartz, M.K. *et al.* 2009, Allendorf *et al.* 2010).

Landscape genomics will help to identify evolutionarily significant units or distinct population segments by including both neutral and adaptive variation (see Chapter 16) (Funk *et al.* 2012, Gebremedhin *et al.* 2009). For example, neutral loci can help assess reproductive isolation, whereas adaptive gene markers can assess adaptive differentiation (Luikart *et al.* 2003). Landscape genomics will also help identify management units by improving precision and accuracy for localizing boundaries on the landscape that separate demographically independent populations (Palsbøll *et al.* 2007).

Example 15.2 Landscape genetics and connectivity of wolverines in the US Rocky Mountains

The wolverine is a stocky and muscular carnivore that superficially resembles a small bear more than it resembles its fellow mustelids. It has a reputation for ferocity and strength out of proportion to its size, with the documented ability to kill prey many times its size. The wolverine is found in the northern boreal forests of North America, Europe, and Asia. Wolverines have experienced a steady decline in numbers for over 100 years because of trapping and habitat fragmentation, but now there is some evidence of population growth and range expansion.

Landscape features that influence wolverine population substructure and gene flow have been largely unknown. Schwartz, M. K. *et al.* (2009) examined 210 wolverines at 16 microsatellite loci to infer connectivity and movement patterns among wolverine in the northern Rocky Mountains of the US in Montana, Idaho, and Wyoming. They constructed a pairwise genetic distance (proportion of shared alleles) matrix among all individuals. Previous work indicated that the distribution of persistent spring snow was the key environmental feature affecting wolverine presence and movement. They built hypothetical resistance surfaces using different values for resistance to movement of areas not having persistent spring snow. They then estimated the correlation between matrices of pairwise genetic distances and movement costs among all individuals.

Significant positive correlations between genetic distance and cost distance were detected for all models based on spring snow cover. Models simulating large preferences for dispersing within areas characterized by persistent spring snow explained the data better than a model based on straight-line geographic distance. In all cases, cost models based on snow cover had significantly stronger correlations with observed genetic distances than straight-line (Euclidian) distances. For all cost models based on snow cover, there was no relationship between genetic distance and Euclidean distance once the snow-based correlation was removed. The least-cost path analysis among all pairs of systematically gridded points suggested several important wolverine corridors connecting areas of spring snow cover (orange paths in Color Plate 6).

These results were used to derive empirically based least-cost corridor maps. These corridor maps were concordant with previously published population subdivision patterns based on mitochondrial DNA and indicate that natural colonization of the southern Rocky Mountains by wolverines will be difficult but not impossible (Figure 15.11). In 2009, a male wolverine with a GPS collar actually traveled from Wyoming to Colorado using a path nearly identical to the one predicted in this paper (M.K. Schwartz, personal communication).

(Continued)

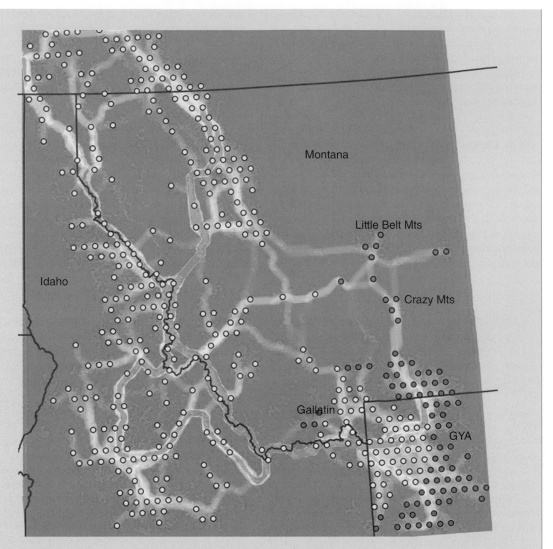

Figure 15.11 Landscape genetics analysis of wolverine connectivity in the northern US Rocky Mountains, showing cumulative least-cost paths between systematically placed locations (circles) in spring snow cover cells. Paths in orange are predicted to be used more often than those in cooler colors. The color of the circle corresponds to the average cost distance between that location and all other locations, based on our models. The graph was divided into four modes (three within the northern US Rockies, and one between the Greater Yellowstone Area and Colorado). The yellow mode has the lowest average cost distances. Rockies), the blue bars the next lowest, the pink bars (Crazy and Little Belt Mountains) have the greatest average cost distances in the northern US Rocky Mountains, and the green bars show the distances between all points from Colorado to Greater Yellowstone. Modified from Schwartz, M.K. *et al.* (2009). See Color Plate 6.

These genomic approaches might allow identification of adaptive genetic variation related to important traits, such as phenology, drought tolerance, disease resistance, or temperature tolerance, so that management may focus on maintaining adaptive genetic potential. In this context, a landscape genomic approach allows mapping of associations between adaptive genome regions and environmental gradients in space and time (Gaggiotti *et al.* 2009, Manel *et al.* 2010). For example, a recent study used a limited genome scan approach to test for adaptive genes under selection in populations of rainbow trout occurring in desert and montane streams from Idaho, US (Narum *et al.* 2010). Populations from the same environment (desert) had highly different allele frequencies from the montane populations at six of 96 genes, indicating local adaptation to differing climates and directional selection on these six ecologically relevant genes (Narum *et al.* 2010). Such studies could allow forecasting of the effects of environmental change on gene flow of adaptive alleles by predicting spatio-temporal landscape change and modeling of gene flow of adaptive alleles across future landscapes (see Chapter 21).

15.7 LONG-TERM POPULATION VIABILITY

There is sometimes confusion regarding when to address short- versus long-term genetic goals, and how they relate to the conservation of local populations versus entire species (Jamieson and Allendorf in press, see Section 14.9). Short-term goals are appropriate for the conservation of local populations. As indicated above, those goals are aimed at keeping the rate of inbreeding at a tolerable level. The effective population sizes at which this may be achieved (e.g., $N_e = 50$), however, are typically not large enough for new mutations to compensate for the loss of genetic variation through genetic drift. Some gene flow from neighboring populations is necessary to provide reasonable levels of genetic variation for quantitative traits to ensure long-term population persistence (see Section 14.8).

The long-term viability of a metapopulation, or species, is influenced by the number and complexity of the subpopulations. Metapopulation viability can be increased by the maintenance of a number of populations across multiple, diverse, and semi-independent environments, as illustrated in Example 15.3.

Example 15.3 Metapopulation structure and long-term productivity and persistence of sockeye salmon

Complex genetic population structure can play an important role in the long-term viability of populations and species. Sockeye salmon within major regions generally consist of hundreds of discrete or semi-isolated individual local demes (Hilborn *et al.* 2003). The amazing ability of sockeye salmon to return and spawn in their natal spawning sites results in substantial reproductive isolation among local demes. Local demes of sockeye salmon with major lake systems generally show pairwise F_{ST} values of 0.10 to 0.20 at allozyme and microsatellite loci, indicating relatively little gene flow.

These local demes occur in a variety of different habitats which, combined with the low amount of gene flow, results in a complex of locally adapted populations. Sockeye salmon spawning in tributaries to Bristol Bay, Alaska, display a wide variety of life history types associated with different breeding and rearing habitats. Bristol Bay sockeye salmon spawn in streams and rivers from 10 cm to several meters deep, in substrate ranging from small gravel to cobble.

Some streams have extremely clear water, while others spawn in sediment-laden streams just downstream from melting glaciers. Sockeye salmon also spawn on the beaches in lakes with substantial groundwater. Different demes spawn at different times of the year. The date of spawning is associated with the long-term average thermal regime experienced by incubating eggs, so that fry emerge in the spring in time to feed on zooplankton and aquatic insects. Fish from different demes have a variety of morphological, behavior, and life history differences associated with this habitat complexity.

Up to 40 million fish are caught each year in the Bristol Bay sockeye fishery, in several fishing areas associated with different major tributaries. There is a large year-to-year variability in overall productivity, but the range of the productivity of this fishery has been generally consistent for nearly 100 years (Figure 15.12). However, the productivity of different demes and major drainages has changed dramatically over

(Continued)

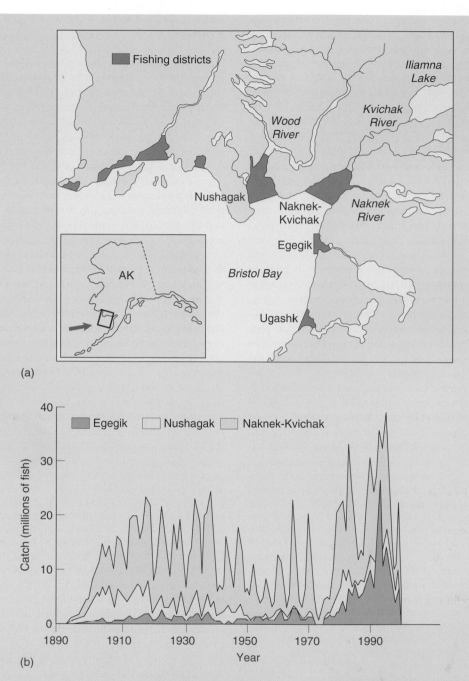

Figure 15.12 (a) Map of fishing districts (crosshatched areas) around Bristol Bay, Alaska. (b) Catch history of the three major sockeye salmon fishing areas within Bristol Bay. The overall productivity of the system has been generally stable, but the relative contributions of the three major areas have changed greatly. For example, the Egegik district (see map) generally contributed less than 5% to the fishery until 1975, but has been a major contributor since then. From Hilborn *et al.* (2003).

the years. The relative productivity of local demes has changed as the marine and freshwater climates change. Local reproductive units that are minor components of a mixed stock fishery during one climatic regime may dominate during others. Therefore, maintaining productivity over long timescales requires protecting against the loss of local populations during certain environmental regimes (Schindler *et al.* 2010).

The long-term stability of this complex system stands in stark contrast to the dramatic collapse and extirpation of a highly productive population of this species introduced into the Flathead River drainage of Montana (Spencer *et al.* 1991). The life-history form of this species that spends its entire life in freshwater is known as kokanee (Guest Box 2). Sockeye salmon were introduced into Flathead Lake in the early 20th century, and by the 1970s some 50,000 to 100,000 fish returned to spawn in one primary local population that supported a large recreational fishery. Opossum shrimp were introduced into Flathead Lake in 1983 and had a major effect on the food web in this ecosystem. A primary effect was the predation of opossum shrimp on the zooplankton which was the major food resource of the kokanee. This productive single deme of kokanee went from over 100,000 spawners in 1985 to extirpation just three years later.

This example illustrates the importance of maintaining multiple semi-isolated subpopulations with different life history to help insure long-term population and species persistence.

The accumulation of mildly deleterious mutations as considered in Section 14.7 may also affect the long-term viability of metapopulations (Higgins and Lynch 2001). Under some circumstances, metapopulation dynamics can reduce the effective population size so that even mutations with a selection coefficient as high as $s = 0.1$ can behave as nearly neutral and cause the erosion of metapopulation viability.

The long-term goal, where the loss of variation is balanced by new mutations, refers primarily to a global population, which may coincide with a species or subspecies that cannot rely on the input of novel genetic variation from neighboring populations. This global population may consist of one more or less panmictic unit, or it may be composed of multiple subpopulations that are connected by some gene flow, either naturally or through translocations (Mills and Allendorf 1996). It is the total assemblage of interconnected subpopulations forming a global population that must have an effective size meeting the criteria for long-term conservation (e.g., $N_e \geq 500$ to 1000). The actual size of this global population will vary considerably from species to species depending on the number and size of the constituent subpopulations and on the pattern of gene flow between them (Waples 2002).

Guest Box 15   Fitness loss and genetic rescue in stream-dwelling topminnows
Robert C. Vrijenhoek

The 'guppy-sized' livebearing topminnow *Poeciliopsis monacha* inhabits rocky arroyos in northwestern Mexico. The upstream portion of a small stream, the Arroyo de los Platanos, dried completely during a severe drought in 1976, but within two years fish recolonized this area from permanent springs that exist downstream. The founder population was homozygous at loci that were polymorphic in the source population, a loss of variation that corresponded with manifestations of inbreeding depression.

Fitness of the source and founder populations was compared with that of coexisting asexual forms of *Poeciliopsis* that experienced the same extinction/ recolonization event. The reproductive mode of cloning preserves heterozygosity and limits inbreeding depression in the asexual fish. Compared with local clones, the inbred founder population of *P. monacha* exhibited poor developmental stability (e.g., asymmetry in bilateral morphological traits, see Section 2.2) and an increased parasite load (Vrijenhoek and Lerman 1982, Lively *et al.* 1990). Genetic variation at allozyme loci in this species is associated with the ability to survive seasonal stresses in these desert streams – cold temperatures, extreme heat, and hypoxia (Vrijenhoek *et al.* 1992).

Prior to the extinction event in 1976, the sexual *P. monacha* constituted 76% of the fish population in the

(Continued)

upper Platanos, while 24% of the fish were asexual. After recolonization the sexual fish constituted no more than 10% for the next 5 years (10–15 generations). Corresponding frequency shifts did not occur downstream in permanent springs where levels of heterozygosity remained stable in *P. monacha*.

By 1983, *P. monacha* was nearly eliminated from several small pools in the upper Platanos while the clones flourished there. We rescued the founder population by transplanting 30 genetically variable females from a downstream location where *P. monacha* was genetically variable. By the following spring (2–3 generations), sexual *P. monacha* regained numerical dominance over the clones (Vrijenhoek 1989), and its parasite load dropped to levels that were typical of the permanent localities downstream (Lively *et al.* 1990). Restoration of genetic variability reversed inbreeding depression in *P. monacha* and restored its fitness relative to that of the competing fish clones (Figure 15.13).

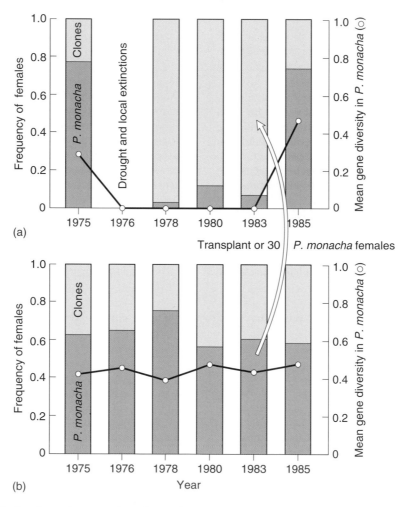

Figure 15.13 Population dynamics of *Poeciliopsis* topminnows. (A) The upper portion of the Arroyo de los Platanos. (B) The mainstream of the Arroyo de Jaguari. Histogram bars, arranged by year, record the frequencies of *P. monacha* females (dark gray) and the triploid clones (light gray) in each sample. The mean gene diversity in *P. monacha* across four polymorphic allozyme loci is traced by the black line. The dotted arrow indicates a single transplant of 30 *P. monacha* females into the upper Platanos population in 1983. From Vrijenhoek (1989).

Setting aside the special reproductive features of *Poeciliopsis*, it is easy to imagine similar interactions between a rare endangered species and its competitors and parasites. Furthermore, loss of heterozygosity in small populations and inbreeding depression can have manifold effects on fitness that might reduce a population's capacity to resist dis-placement by alien competitors and to combat novel diseases. The genotypic differences among individuals of a sexually reproducing species help to reduce intraspecific competition and provide the variability needed to persist in an evolutionary arms race with rapidly evolving parasites, pathogens and competitors (Van Valen 1973).

CHAPTER 16

Units of conservation

Flatwoods salamander, Section 16.7

The zoo directors, curators, geneticists and population biologists who attempt to pursue the elusive goal of preservation of adaptive genetic variation are now considering the question of which gene pools they should strive to preserve.

Oliver A. Ryder (1986)

It is widely recognised that status assessments and the conservation of biological diversity require that units below the species level be considered when appropriate. The Species at Risk Act includes 'subspecies, varieties or geographically or genetically distinct populations' in its definition of wildlife species.
Committee on the Status of Endangered Wildlife in Canada, COSEWIC (2010)

Chapter Contents

Conservation and the Genetics of Populations, Second Edition. Fred W. Allendorf, Gordon Luikart, and Sally N. Aitken.
© 2013 Fred W. Allendorf, Gordon Luikart and Sally N. Aitken. Published 2013 by Blackwell Publishing Ltd.

The identification of appropriate taxonomic and population units for protection and management is essential for the conservation of biological diversity. For species identification and classification, genetic principles and methods are relatively well developed; nevertheless, species identification can be controversial. Within species, the identification and protection of genetically distinct local populations should be a major focus in conservation because the conservation of many distinct populations is crucial for maximizing evolutionary potential and minimizing extinction risks (Hughes *et al.* 1997, Luck *et al.* 2003, Hilborn *et al.* 2003). Furthermore, the local population is often considered the functional unit in ecosystems (Luck *et al.* 2003). For example, the US Marine Mammal Protection Act (MMPA) seeks to maintain populations as functioning elements of their ecosystem (MMPA Regulations, 50 CFR 216).

Identification of population units is necessary so that management and monitoring programs can be efficiently targeted toward distinct or independent populations. Biologists and managers must be able to identify populations and geographic boundaries between populations in order to effectively plan harvesting quotas, avoid overharvesting in a population or area, and to devise translocations and reintroductions of individuals to prevent mixing of adaptively differentiated populations. In addition, it is sometimes necessary to prioritize among population units (or taxa) to conserve, because limited financial resources preclude conservation of all units (Ryder 1986).

Finally, many governments and agencies have established legislation and policies to protect intraspecific population units. This requires the identification of population units. For example, the US Endangered Species Act (ESA) allows listing and full protection of **distinct population segments** (DPSs) of vertebrate species (Box 16.1). Other countries also have laws that depend upon the identification of distinct taxa and populations for the protection of species and habits

---

Box 16.1   The US Endangered Species Act (ESA) and conservation units

The ESA of the United States is one of the most powerful pieces of conservation legislation ever enacted. It has been a major stimulus motivating biologists to develop criteria for identifying intraspecific population units for conservation. This is because the ESA provides legal protection for subspecies and "distinct population segments" (DPSs) of vertebrates, as if they were full species. According to the ESA:

The term 'species' includes any subspecies of fish or wildlife and plants, and any distinct population segment of any species of vertebrate fish or wildlife which interbreeds when mature.

However, the ESA does not provide criteria or guidelines for delineating DPSs. The identification of intraspecific units for conservation is controversial. This is not surprising given that the definition of a 'good species' is controversial (see Section 16.5). Biologists have vigorously debated the criteria for identifying DPSs and other conservation units ever since the US Congress extended full protection of the ESA to "distinct" populations, but did not provide guidelines.

The IUCN Red List allows for the separate assessment of geographically distinct populations. These subpopulations are defined as "geographically or otherwise distinct groups in the population between which there is little demographic or genetic exchange (typically one successful migrant individual or gamete per year or less) (IUCN 2001).

Legislation in other countries around the world has provisions that recognize and protect intraspecific units of conservation. For example, Canada passed the Species at Risk Act (SARA) in 2003. The SARA aims to "prevent wildlife species from becoming extinct, and to secure the necessary actions for their recovery". Under the SARA, "wildlife species" means a "species, subspecies, variety or geographically or genetically distinct population of animal, plant or other organism, other than a bacterium or virus, which is wild by nature".

In Australia, the Environment Protection and Biodiversity Conservation Act of 1999 (EPBC) also allows protection for species, subspecies, and distinct populations. But, like the ESA in the US, there are challenges with defining and identifying intraspecific units (Woinarski and Fisher 1999). Unlike the US ESA and SARA in Canada, the EPBC Act (1996) also recognizes and allows protection of ecological communities (an assemblage of native species that inhabits a particular area in nature).

(Box 16.1). Species and subspecies identification is based upon traditional, established taxonomic criteria as well as genetic criteria, although the criteria for species identification are sometimes controversial. The criteria for delineating intraspecific units for conservation has been highly controversial.

In this chapter, we examine the components of biodiversity and then consider methods to assess taxonomic and population relationships. We discuss the criteria, difficulties, and controversies in the identification of conservation units. We also consider the identification of appropriate population units for legal protection and for management actions such as supplemental translocations of individuals between geographic regions. Recall that in the previous chapter, we considered three spatial scales of genetic population structure for conservation: local population, metapopulation, and species.

16.1 WHAT SHOULD WE PROTECT?

Genes, species, and ecosystems are three primary levels of biodiversity (Figure 16.1) recognized by the Convention on Biological Diversity (CBD). There has been some controversy as to which level should receive pri-

ority for conservation efforts (e.g., Bowen 1999). However, it is clear that all three levels must be conserved for successful conservation of biodiversity. For example, it is as futile to conserve ecosystems without species, as it is to save species without large, healthy ecosystems. Nevertheless, a recent analysis found that genetic variation in wild animals and plants was generally not included in national plans to implement the CBD (Laikre *et al.* 2010a). In fact, fewer than 50% of the national plans that were reviewed included the goal of conserving genetic variation of wild populations.

An example of this kind of futility is that of the African rhinoceros, which is being protected, mainly in zoos and small nature reserves, but for which little habitat (free from poachers) is currently available. Without conserving vast habitats for future rhino populations, it seems pointless to protect rhinos in small nature reserves surrounded by armed guards and fences.

It is not too late for rhinos. Vast habitats do exist, and rhinos could be successful in these habitats if poaching is eliminated. In addition to conserving rhino species and their habitats, it is also important to conserve genetic variation within rhino species because variation is a prerequisite for long-term adaptive change and the avoidance of fitness decline through inbreeding depression. Clearly, it is important to recognize and conserve all levels of biodiversity: ecosystems, species, and genes.

The debate over whether to protect genes, species, or ecosystems is, in a way, a false trichotomy because each level is an important component of biodiversity as a whole. Nevertheless considering each level separately can help us appreciate the interacting components of biodiversity, and the different ways that genetics can facilitate conservation at different levels. Appreciation of each level can also promote understanding and multidisciplinary collaborations across research domains. Finally, a fourth level of biodiversity – that of genetically distinct local populations – is arguably the most important level for focusing conservation efforts (Figure 16.1). The conservation of multiple genetically distinct populations is necessary to ensure long-term species survival and the stable functioning of ecosystems (Luck *et al.* 2003, Example 15.3).

We can also debate which temporal component of biodiversity to prioritize for conservation: past, present, or future biodiversity. All three components are important, although future biodiversity often warrants special concern (Box 16.2).

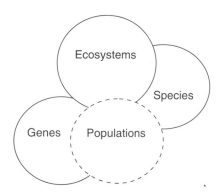

Read here papers

Figure 16.1 Primary levels of biodiversity recognized by the IUCN (solid circles), and a fourth level – populations – recognized as perhaps most crucial for long-term species persistence (Hughes *et al.* 1997, Luck *et al.* 2003). In reality, biodiversity exists across a continuum of many hierarchical levels of organization including genes, genomes (i.e., multilocus genotypes), local populations (dashed line), communities, ecosystems, and biomes. Additional levels of diversity include metapopulations, subspecies, genera, families, and so on.

Box 16.2   Temporal considerations in conservation: past, present, and future

What temporal components of biodiversity do we wish to preserve? Do we want to conserve ancient isolated lineages, current patterns of diversity (ecological and genetic), or the diversity required for future adaptation and for novel diversity to evolve? Most would agree 'all of the above'. All three temporal components are interrelated and complementary (Figure 16.2). For example, conserving current diversity helps insure future adaptive potential. Similarly, conserving and studying ancient lineages ('living fossils') can help us understand factors important for long-term persistence. Nevertheless one can argue that the most important temporal component to consider is future biodiversity: the ability of species and populations to adapt to future environments, for example, following global climate change. If populations do not adapt to future environments then biodiversity will decline, leading to loss of ecosystem functioning and services. Figure 16.2 illustrates how different temporal components of biodiversity (past, present, and future) can be related to different scientific disciplines (systematics, ecology, and evolutionary biology, respectively). These components are also often related to different hierarchical levels of biodiversity: species, ecosystems, and genes, respectively.

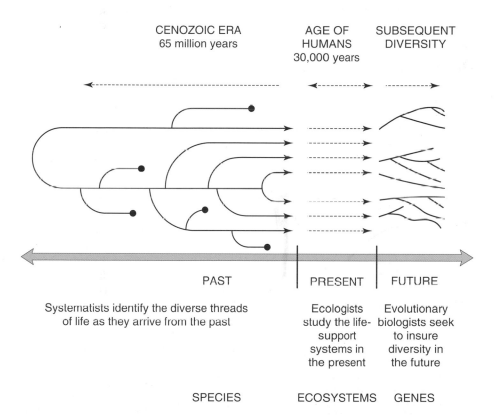

Figure 16.2   The temporal framework (past, present, and future), the corresponding disciplines (systematics, ecology, and evolutionary biology), and the levels of biodiversity (species, ecosystems, and genes) that are often considered when prioritizing biodiversity for conservation. Modified from Bowen (1999).

Another choice that is often debated is whether we should emphasize protecting the existing *patterns* of diversity or the *processes* that generate diversity (e.g., ecological and evolutionary processes themselves). Again the answer is, in general, both. It is clear that we should prioritize the preservation of the process of adaptation so that populations and species can continually adapt to future environmental changes. However, one important step toward preserving natural processes is to quantify, monitor, and maintain natural patterns of population subdivision and connectivity, for example, to identify intraspecific population units, boundaries, and corridors for dispersal in current and future environments. This would prevent extreme fragmentation and promote continued natural patterns of gene flow among populations.

How do we conserve the 'processes' of evolution, including adaptive evolutionary change? We must first maintain healthy habitats and large wild populations, because only in large populations can natural selection proceed efficiently (see Section 8.5). In small populations, genetic drift leads to random genetic change, which is generally nonadaptive. Drift can preclude selection from maintaining beneficial alleles and eliminating deleterious ones. To maintain evolutionary processes, we must also preserve multiple populations – ideally from different environments, so that selection pressures remain diverse and multilocus genotype diversity remains high. In this scenario, a wide range of local adaptations are preserved within species, as well as some possibility of adaptation to different future environmental challenges.

16.2 SYSTEMATICS AND TAXONOMY

The description and naming of distinct taxa is essential for most disciplines in biology. In conservation biology, the identification of taxa (taxonomy) and assessing their evolutionary relationships (systematics) is crucial for the design of efficient strategies for biodiversity management and conservation. For example, failing to recognize the existence of a distinct and threatened taxon can lead to insufficient protection and subsequent extinction. Identification of too many taxa (oversplitting) can waste limited conservation resources.

There are two fundamental aspects of evolution that we must consider: phenotypic change through time (**anagenesis**), and the branching pattern of reproductive relationships among taxa (**cladogenesis**). The two primary taxonomic approaches are based on these two aspects.

Historically, taxonomic classification was based primarily upon phenotypic similarity (**phenetics**), which reflects evolution via anagenesis; that is, groups of organisms that were phenotypically similar were grouped together. This classification is conducted using clustering algorithms (described below) that group organisms based exclusively on 'overall similarity' or outward appearance. For example, populations that share similar allele frequencies are grouped together into one species. In this example, the clustering by overall similarity of allele frequencies is phenetic. The resulting diagram (or tree) used to illustrate classification is called a **phenogram**, even if based upon genetic data such as allele frequencies.

A second approach is to classify organisms on the basis of their phylogenetic relationships (**cladistics**). Cladistic methods group together organisms that share **derived** traits (originating in a common ancestor), reflecting cladogenesis. Under cladistic classification, only **monophyletic** groups can be recognized, and only genealogical information is considered. The resulting diagram (or tree) used to illustrate relationships is called a **cladogram** (or sometimes, a **phylogeny**). Phylogenetics is discussed below in Section 16.3.

Our current system of taxonomy combines cladistics and phenetics, and it is sometimes referred to as evolutionary classification (Mayr 1981). Under evolutionary classification, taxonomic groups are usually classified on the basis of phylogeny. However, groups that are extremely phenotypically divergent are sometimes recognized as separate taxa even though they are phylogenetically related. A good example of this is birds (Figure 16.3). Birds were derived from a dinosaur ancestor, as evidenced from the fossil record showing reptiles with feathers (bird–reptile intermediates) (Prum 2003). Therefore, birds and dinosaurs are sister groups that should be classified together under a strictly cladistic classification scheme. However, birds underwent rapid evolutionary divergence associated with their development of flight. Therefore, birds are classified as a separate class while dinosaurs are classified as a reptile (class Reptilia). Sometimes in the literature dinosaurs are now referred to as 'nonavian dinosaurs', as a reminder of these relationships (Erickson *et al.* 2006).

There is a great deal of controversy associated with the 'correct' method of classification. We should use all kinds of information available (morphology,

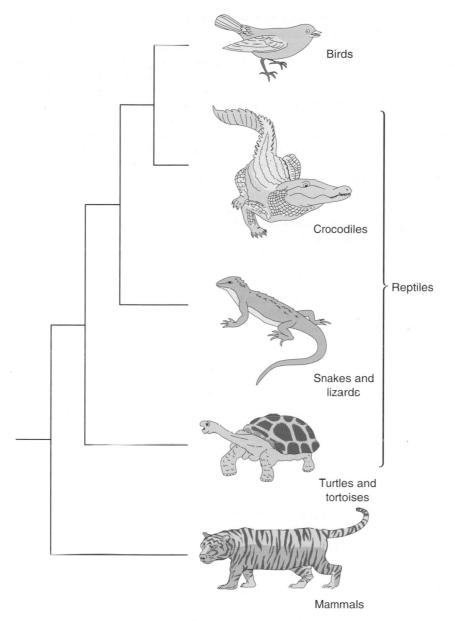

Figure 16.3 Phylogenetic relationships of birds, mammals, and reptiles. Note that crocodiles and birds are more closely related to each other than either is to other reptiles. That is, crocodiles share a more recent common ancestor with birds than they do with snakes, lizards, turtles, and tortoises. Therefore, the class Reptilia is not monophyletic.

physiology, behavior, life history, geography, parasite distributions, and genetics) and the strengths of different schools (phenetic and cladistic) when classifying organisms (Mayr 1981, see also Section 16.6).

16.3 PHYLOGENY RECONSTRUCTION

A phylogenetic tree is a pictorial summary that illustrates the pattern and timing of branching events in the evolutionary history of taxa (Figure 16.4). A phylogenetic tree consists of **nodes** for the taxa being considered, and branches that connect taxa and show their relationships. Nodes are at the tips of branches and at branching points representing extinct ancestral taxa (i.e., internal and ancestral nodes). A phylogenetic tree represents a hypothesis about relationships that is open to change as more taxa or characters are added. The same phylogeny can be drawn many different ways. Branches can be rotated at any internal node

without changing the relationship between the taxa, as illustrated in Figure 16.5.

Branch lengths are often proportional to the amount of genetic divergence between taxa. If the amount of divergence is proportional to time, a phylogeny can show time since divergence between taxa. Molecular divergence (through mutation and drift) will be proportional to time if mutation accumulation is stochastically constant (like radioactive decay). The idea that molecular divergence can be constant is called the **molecular clock** concept. In conservation biology, the molecular clock and divergence estimates can help identify distinct populations and prioritize them based on their distinctiveness or divergence times. One serious problem with estimating divergence times is that extreme genetic drift, such as founder events and bottlenecks, can greatly inflate estimates of divergence times, leading to long branch lengths and misleading estimates of phylogenetic distinctiveness (see Section 9.7).

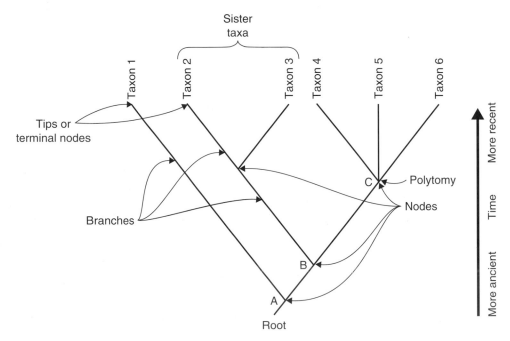

Figure 16.4 A phylogenetic tree (phylogeny). A **polytomy** (node 'C') occurs when more than two taxa are joined at the same node because data cannot resolve which shared a more recent common ancestor. A widely controversial polytomy 10–20 years ago was that of chimpanzees, gorillas, and humans. However, extensive genetic data now shows that chimps and humans are more closely related (i.e., sister taxa). From Freeman and Herron (1998).

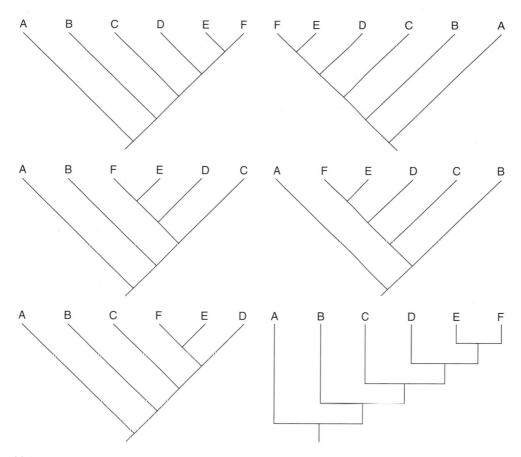

Figure 16.5 Six phylogenetic trees showing identical relationships among taxa. Branches can be rotated at nodes without changing the relationships represented on the trees. From Freeman and Herron (1998).

16.3.1 Methods

There are two basic steps in phylogeny reconstruction: (1) generate a matrix of character states (e.g., derived versus ancestral states); and (2) build a tree from the matrix. Cladistic methods use only shared derived traits, **synapomorphies**, to infer evolutionary relationships. Phenogram construction is based on overall similarity. Therefore, a phylogenetic tree may have a different topology from a phenogram using the same character state matrix (Example 16.1).

The actual construction of phylogenies is much more complicated than this simple example. It is sometimes difficult to determine the ancestral state of a character. Moreover, the number of possible evolutionary trees rises at an alarming rate with the number of taxa. For example, there are nearly 35 million possible rooted, bifurcating trees with just 10 taxa, and over 8×10^{21} possible trees with 20 taxa! In addition, there are a variety of other methods besides parsimony for inferring phylogenies (Hall 2004). The field of inferring phylogenies has been marked by more heated controversy than perhaps any other area of evolutionary biology (see Felsenstein 2004).

16.3.2 Gene trees and species trees

It is important to realize that different genes can result in different phylogenies, and that gene trees are often different from the true species phylogeny (Nichols 2001). Different gene phylogenies can arise due to four

Example 16.1 Phenogram and cladogram of birds, crocodiles, and lizards

As we have seen, birds and crocodiles are sister taxa based upon phylogenetic analysis, but crocodiles are taxonomically classified as reptiles because of their phenetic similarity with snakes, lizards, and turtles. These conclusions are based on a large number of traits. Here we will consider five traits in Table 16.1 to demonstrate how a different phenogram and cladogram can result from the matrix of character states.

Lizards and crocodiles are more phenotypically similar to each other than either is to birds because they share 3 out of 5 traits (0.60), while crocodile and birds share just 2 out of 5 traits (0.40). Thus, the following phenotypic similarity matrix results (Table 16.2). We can construct a phenogram based upon clustering together the most phenotypically similar groups (Figure 16.6a). The phenotypic similarity of lizards and crocodiles results from their sharing ancestral character states because of the rapid phenotypic changes that occurred in birds associated with adaptation to flight.

Parsimony methods were among the first to be used to infer phylogenies, and they are perhaps the

Table 16.1 Character states for five traits used to construct a phenogram and cladogram of lizards, crocodiles, and birds.

Taxon	Traits* 1	2	3	4	5
A Lizards	0	0	0	0	0
B Crocodiles	1	1	0	0	0
C Birds	1	1	1	1	1

*0, ancestral; 1, derived.
Traits: 1, heart (three- or four-chambered); 2, inner ear bones (present or absent); 3, feathers (present or absent); 4, wings (present or absent); and 5, hollow bones (present or absent).

Table 16.2 Phenotypic similarity matrix for of lizards, crocodiles, and birds based upon proportion of shared character states in Table 16.1.

	Lizards	Crocodiles	Birds
Lizards	1.0		
Crocodiles	0.6	1.0	
Birds	0.0	0.4	1.0

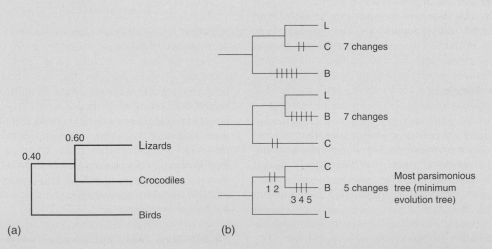

(a)   (b)

Figure 16.6 (a) Phenogram and (b) cladograms showing phenotypic and evolutionary relationships, respectively, among lizards, crocodile, and birds. Numbers in (a) are genetic distance estimates (e.g., 0.60 distance units between lizards and crocodiles). Vertical slashes in (b) on branches represent changes. Numbers below slashes on the bottom (most parsimonious) tree correspond to the traits (i.e., evolutionary change in traits) listed in Table 16.1.

easiest phylogenetic methods to explain and understand (Felsenstein 2004, p. 1). There are many possible phylogenies for any group of taxa. Parsimony is the principle that the phylogeny to be preferred is the one that requires the minimum amount of evolution. To use parsimony, we must search all possible phylogenies and identify the one or ones that minimize the number of evolutionary changes.

There are only three possible bifurcating phylogenies for lizards, crocodiles, and birds. Figure 16.6b shows these trees and the number of evolutionary changes from the ancestral to the derived trait to explain the character state matrix. The upper two phylogenies both require seven changes because certain evolutionary changes had to occur independently in the crocodile and bird branches. The bottom tree requires only five evolutionary changes to explain the character state matrix above. Thus, the bottom tree is the most parsimonious. Birds and crocodiles form a monophyletic group because they share two synapomorphies (traits 1 and 2).

main phenomena: lineage sorting and associated genome sampling error, sampling error of individuals or populations, natural selection, or introgression following hybridization. Thus, many independent genes or DNA sequences should be used when assessing phylogenetic relationships (Wiens *et al.* 2010).

16.3.2.1 Lineage sorting and sampling error

Ancestral **lineage sorting** occurs when different DNA sequences from a mother taxon are sorted into different daughter species such that lineage divergence times do not reflect population divergence times. For example, two divergent lineages can be sorted into two recently isolated populations, where less-divergent lineages might become fixed in different ancient daughter populations. Lineage sorting makes it important to study many independent DNA sequences, to avoid sampling error associated with sampling too few or an unrepresentative set of genetic characters (loci).

Sampling error of individuals occurs when too few individuals or nonrepresentative sets of individuals are sampled from a species, such that the inferred gene tree differs from the true species tree. For example, many early studies using mtDNA analysis included only a few individuals per geographic location, which could lead to erroneous phylogeny inference. Limited sampling is likely to detect only a subset of local lineages (i.e., alleles), especially when some lineages exist at low frequency.

We can use simple probability to estimate the sample size that we need to detect a rare lineage (haplotype) or allele. For example, how many individuals must we sample to have a greater than 95% chance of detecting an allele with frequency of 0.10 ($p = 0.10$)? Each time we examine one sample, we have a 0.90 chance ($1-p$) of not detecting the allele in question and a 0.10 chance (p) of detecting it. Using the product rule (see Box 5.1), the probability of not detecting an allele at $p = 0.1$ in a sample of size x is $(1-p)^x$. Therefore, the sample size required to have a 95% chance of sampling an allele with frequency of 0.10 is 29 haploid individuals or 15 diploids for nuclear markers: $(1-0.1)^{29} = 0.047$.

16.3.2.2 Natural selection

Directional selection can cause gene trees to differ from species trees if a rare allele increases rapidly to fixation because of natural selection (**selective sweep,** Section 10.3.1). For example, a highly divergent (ancient) lineage may be swept to fixation in a recently derived species. Here the ancient age of the lineages would not match the recent age of the newly derived species. In another example, balancing selection could maintain the same lineages in each of two long-isolated species, and lead to erroneous estimation of species divergence, as well as a phylogeny discordant with the actual species phylogeny and with neutral genes (e.g., Bollmer *et al.* 2007). To avoid selection-induced errors in phylogeny reconstruction, many loci should be used. Analysis of many loci can help to identify a locus with unusual phylogenetic patterns due to selection (as in Section 9.7). For example, selection might cause rapid divergence at one locus that is not representative of the rest of the genome or of the true species tree.

16.3.2.3 Introgression

Introgression also causes gene trees to differ from species trees. For example, hybridization and subsequent back-crossing can cause an allele from species X to introgress into species Y. This has happened between wolves and coyotes that hybridize in the northeastern US, where coyote mtDNA has introgressed into wolf populations. Here, female coyotes hybridize with male wolves, followed by the F_1 hybrids mating with wolves, such that coyote mtDNA introgresses into wolf populations (Roy *et al.* 1994). This kind of unidirectional introgression of maternally inherited mtDNA has been detected in deer, mice, fish, and many other species (Good *et al.* 2008). Introgression can also be unidirectional or asymmetric at nuclear loci, depending on demography and colonization history (Scascitelli *et al.* 2010).

16.3.2.4 MtDNA gene tree versus species tree

An example of a gene tree not being concordant with the species tree is illustrated in a study of mallard ducks and black ducks (Avise 1990). The black duck apparently recently originated (perhaps via rapid phenotypic evolution) from the more widely distributed mallard duck. This likely occurred when a peripheral mallard population became isolated, evolved into the black duck and became fixed for a single mtDNA lineage (e.g., via lineage sorting and/or selection). The mallard population is much larger and maintains several divergent mtDNA lineages, including the lineage fixed in the black duck (Figure 16.7). Thus, while the black duck is monophyletic, the mallard is paraphyletic relative to the black duck for mtDNA. Because the black duck mtDNA is common in the mallard, the black duck appears to be part of the mallard species, when considering only mtDNA data. However, the black duck has important phenotypic, adaptive, and behavioral differences meriting recognition as a separate species.

This duck example illustrates a problem that is likely to occur when identifying species from molecular data alone or from only one locus. It shows the importance

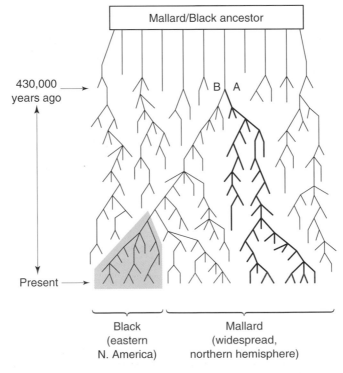

Figure 16.7 Simplified diagrammatic representation of the possible matriarchal ancestry of mallard and black ducks. The mtDNA lineage A is shown in dark lines, and the black duck portion of the phylogeny is shaded. From Avise (1990).

of considering non-molecular characteristics such as life history, morphology, and geography, along with molecular data from many loci (see Section 16.6).

For example, early work with mtDNA suggested that brown bears are **paraphyletic** with respect to polar bears (Talbot and Shields 1996). Recent phylogenetic analysis of the complete mtDNA genome, as well as geological and molecular age estimates of a 100,000-year-old **subfossil** bear specimen, indicated that polar bears adapted rapidly within approximately 20,000 years following their split from a brown bear precursor (see Figure 16.8a, Lindqvist *et al.* 2010). However, we must remember that this conclusion is based on the phylogeny of mtDNA, not these species. A recent analysis of over 9000 base pairs at 14 nuclear

loci indicated that polar bears are monophyletic and diverged from brown bears some 600,000 years ago (Figure 16.8b, Hailer *et al.* 2012). The authors suggested that polar bears carry brown bear mitochondrial DNA because of past hybridization and introgression. These results show the importance of examining nuclear as well as mtDNA, in determining relationships among populations and species.

16.4 GENETIC RELATIONSHIPS WITHIN SPECIES

Identifying populations and describing population relationships is often difficult but crucial for conservation

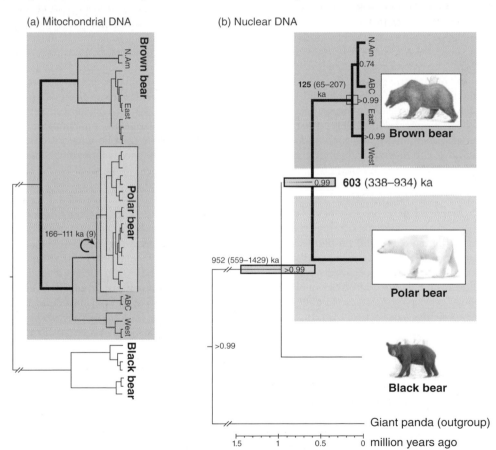

Figure 16.8 Phylogenetic trees of polar and brown bears based on (a) complete mtDNA genome sequences (Lindqvist *et al.* 2010), and (b) on 14 nuclear loci (Hailer *et al.* 2012). The circular arrow in (a) denotes mtDNA replacement in polar bears. Numbers next to nodes in (b) indicate statistical support, and gray bars are 95% highest credibility ranges for node ages (ka = thousands of years ago). See Hailer *et al.* (2012) for description of brown bear samples.

and management actions such as monitoring population status, measuring gene flow, and planning translocation strategies (Waples and Gaggiotti 2006). Population relationships are generally assessed using multilocus allele frequency data and statistical approaches for clustering individuals or populations with a dendrogram or tree in order to identify genetically similar groups.

Population trees and phylogenetic trees look similar to each other, but they display fundamentally different types of information (Kalinowski 2009). Phylogenies show the time since the most recent common ancestor (TMRCA) between taxa. Phylogenies represent relationships among taxa that have been reproductively isolated for many generations. A phylogeny identifies monophyletic groups – isolated groups that shared a common ancestor. Phylogenetic trees can be used both for species and for genes (e.g., mtDNA) (Nichols 2001). In the case of species, the branch points represent speciation events; in the case of genes, the branch points represent common ancestral genes.

Population trees, in contrast, generally identify groups that have similar allele frequencies because of ongoing genetic exchange (i.e., gene flow). The concept of TMRCA is not meaningful for populations with ongoing gene flow. Populations with high gene flow will have similar allele frequencies and cluster together in population trees.

The differences between population and phylogenetic trees, as described here, are somewhat oversimplified to help explain the differences. In reality there is a continuum in the degree of differentiation among populations in nature. Some populations within the same species may have been reproductively isolated for many generations. In this case, genealogical information and the phylogenetic approach can be used to infer population relationships (see Section 16.4.3).

The description of genetic population structure is the most common topic for a conservation genetics paper in the literature. Individuals from several different geographic locations are genotyped at a number of loci to determine the patterns and amounts of gene flow among populations. This population-based approach assumes that all individuals sampled from one area were born there, and represent a local breeding population. However, new powerful approaches have been developed that allow the description of population structure using an individual-based approach. That is, many individuals are sampled, generally over a wide geographic range, and then placed in population units on the basis of genotypic similarity.

16.4.1 Population-based approaches

A bewildering variety of approaches have been used to describe the genetic relationships among a series of populations. We will discuss several representative approaches.

The initial step in assessing population relationships, after genotyping many individuals, often is to conduct statistical tests for differences in allele frequencies between sampling locations (Waples and Gaggiotti 2006, Morin et al. 2009). For example, a chi-square test is used to test for allele frequency differences between samples (e.g., Roff and Bentzen 1989). If two samples are not significantly different, they are often pooled together to represent one population. Before pooling, it is important to understand the statistical power, given your sample size of loci and individuals, to detect structure if it exists and the difference between statistical and biological significance (Waples 1998, Taylor and Dizon 1999, Ryman et al. 2006). It can also be important to resample from the same geographic location in different years or seasons to test for sampling error and for stability of genetic composition through time. After distinct population samples have been identified, the genetic relationships (i.e., genetic similarity) among populations can be inferred.

16.4.1.1 Population dendrograms

Population relationships are often assessed by constructing a dendrogram based upon the genetic similarity of populations. The first step in dendrogram construction is to compute a genetic differentiation statistic (e.g., F_{ST} or Nei's D, Sections 9.1 and 9.8) between each pair of populations. A genetic distance can be computed using any kind of molecular marker (allozyme frequencies, DNA haplotypes) and a vast number of metrics (e.g., Cavalli-Sforza chord distance, Slatkin's R_{ST}, and Wright's F_{ST}, Section 9.8). This yields a **genetic distance matrix** (Table 16.3).

The second step is to use a clustering algorithm to group populations with similar allele frequencies (e.g., low F_{ST}). The most widely used cluster algorithms are UPGMA (unweighted pair group method with arithmetic averages) and neighbor-joining (Salemi and Van-Damme 2003). UPGMA clustering for dendrogram construction (Figure 16.9) is illustrated by a study assessing population relationships of a perennial lily from Florida (Example 16.2).

Neighbor-joining is one of the most widely used algorithms for constructing dendrograms from a

Example 16.2   Dendrogram construction via UPGMA clustering of lily populations

UPGMA (unweighted pair group method with arithmetic averages) clustering was used to assess relationships among five populations of a perennial lily (*Tofieldia racemosa*) from northern Florida (Godt *et al.* 1997). Allele frequencies from 15 polymorphic allozyme loci were used to construct a genetic distance matrix (Table 16.3) and subsequently a dendrogram using the UPGMA algorithm.

The UPGMA algorithm starts by finding the two populations with the smallest inter-population distance in the matrix. It then joins the two populations together at an internal node. In our lily example here, populations "FL1" and "FL2" are grouped together first because the distance (0.001) is the smallest (underlined in Table 16.3). Next, the mean distance from FL1 (and from FL2) to each other population is used to cluster taxa. The next shortest distance is the mean of FL3 to FL1 and FL3 to FL2 (i.e., the mean of 0.002 and 0.003; see asterisks in Table 16.3); thus FL3 is clustered as the sister group of FL1 and FL2. Next SC is clustered, followed by NC (Figure 16.9).

In this example, the genetic distance is correlated with the geographic distance, in that South Carolina is geographically and genetically closer to Florida populations than is North Carolina.

Table 16.3   Genetic distance (*D*; Nei 1972) matrix based upon allele frequencies at 15 allozyme loci for five populations of a perennial lily. Data from Godt *et al.* (1997). Asterisks and underlining are explained in Example 16.2.

	Population				
	FL1	**FL2**	**FL3**	**SC**	**NC**
FL1	–				
FL2	0.001	–			
FL3	0.003*	0.002*	–		
SC	0.029	0.032	0.030	–	
NC	0.059	0.055	0.060	0.062	–

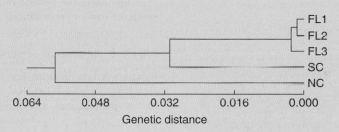

Figure 16.9   Dendrogram generated using the UPGMA clustering algorithm and the genetic distance matrix from Table 16.3. FL is Florida, SC and NC are South Carolina and North Carolina, respectively. From Godt *et al.* (1997).

distance matrix (Salemi and VanDamme 2003). Neighbor-joining is different from UPGMA in that the branch lengths for sister taxa (e.g., FL1 and FL2, Table 16.3) can be different, and thus can provide additional information on relationships between populations. For example, FL1 is more distant from FL3 than FL2 is from FL3 (Table 16.3). This is not evident in the UPGMA dendrogram (Figure 16.9), but would be in a neighbor-joining tree. It follows that neighbor-joining trees are especially useful when populations have substantial amounts of divergence. Other advantages include that

neighbor-joining is fast and thus useful for large datasets and for **bootstrap analysis** (see next paragraph), which involves the construction of hundreds of replicate trees from resampling the data. It also permits correction for multiple character changes when computing distances between taxa. Disadvantages include that it gives only one possible tree and it depends on the model of evolution used.

Bootstrap analysis is a widely used sampling technique for assessing the statistical error when the underlying sampling distribution is unknown. In dendrogram construction, we can bootstrap-resample across loci from the original data-set, meaning that we sample with replacement from our set of loci until we obtain a new set of loci, called a 'bootstrap replicate'. For example, if we have genotyped 12 loci, we randomly draw 12 numbers from 1 and 12 and these numbers (loci) become our bootstrap replicate data-set. We repeat this procedure 1000 times to obtain 1000 data-sets (and 1000 dendrograms). The proportion of the random dendrograms with the same cluster (i.e., branch group) will be the bootstrap support for the cluster. For example, the cluster containing FL1 with FL2 has a bootstrap support of 85 (i.e., 85% of 1000 bootstrap trees grouped FL1 with FL2).

16.4.1.2 Multidimensional representation of relationships among populations

Dendrograms cannot illustrate complex relationships among multiple populations because they consist of a one-dimensional branching diagram. Thus, dendrograms can oversimplify and obscure relationships among populations. Note that this one-dimensionality is not a limitation in using dendrograms to represent phylogenic relationships, as these can be represented by a one-dimensional branching diagram as long as there has not been secondary contact or hybridization following speciation. If secondary contact, hybridization, **horizontal gene transfer**, or multiple branching relationships exist among populations, this can potentially be represented by phylogenetic networks (Reeves and Richards 2007).

There are a variety of multivariate statistical techniques (e.g., principal component analysis, PCA) that summarize and can be used to visualize complex data-sets with multiple dimensions (e.g., many loci and alleles) so that most of the variability in allele frequencies can be extracted and visualized on a two-dimensional or three-dimensional plot (see Example 16.3). Related multivariate statistical techniques include PCoA (principal coordinates analysis), FCA (frequency correspondence analysis), and MDS (multidimensional scaling).

Example 16.3   How many species of tuatara are there?

We saw in Example 1.1 that tuatara on North Brother Island in Cook Strait, New Zealand, were described as a separate species primarily on the basis of variation at allozyme loci (Daugherty *et al.* 1990). A neighbor-joining dendrogram based on allele frequencies at 23 allozyme loci suggested that the North Brother tuatara population is highly distinct because it is separated on a long branch (Figure 16.10a).

More recent molecular genetic data, however, have raised some important questions about this conclusion. Analysis of mtDNA sequence data indicates that tuatara from North Brother and three other islands in Cook Strait are similar to each other, and they are all distinct from the Northern tuatara populations (Hay *et al.* 2003, 2010). Allele frequencies at microsatellite loci also support the grouping of tuatara from Cook Strait (Hay *et al.* 2010).

Principal component analysis (PCA) of the allozyme data supports the similarity of the Cook Strait tuatara populations (Figure 16.10b). Three major population groupings are apparent in the plot of the first two components of the PCA analysis (Figure 16.10b). The first component distinguishes between the Northern group and the Cook Strait populations; the North Brother population clusters closely with the Western Cook Strait populations on this axis. PC2 separates the North Brother population from the other populations. The North Brother population clusters with the other Cook Strait populations on PC1, which explains nearly 50% of the variation. The North Brother population is distinct only for the second main variance component (PC2), which explains 34% of the variation. These results suggest that the Cook Strait populations are much more genetically similar to each other than they are to the Northern populations.

North Brother Island is very small, and the tuatara on this island have substantially less genetic variation at microsatellite loci. Thus, the genetic distinctiveness of North Brother tuatara is likely due to small population size and rapid genetic drift, rather than long-term isolation that might warrant species status.

This example illustrates the limitations of one-dimensional tree diagrams and the possible loss (or oversimplification) of information when data are collapsed into one dimension.

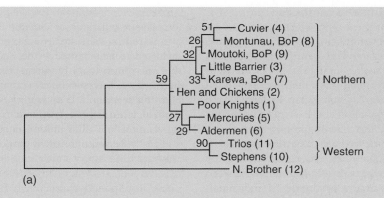

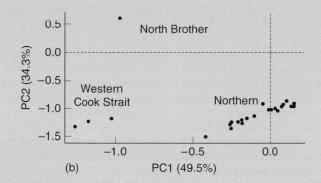

Figure 16.10 (a) Neighbor-joining dendrogram (the numbers on the branches are the bootstrap values), and (b) principle component analysis based on allele frequencies at 23 allozyme loci. (c) Locations of populations sampled. Open circles indicate where fossil remains have been found.

16.4.2 Individual-based methods

Individual-based approaches are used to assess population relationships through first identifying populations by delineating genetically similar clusters of individuals. Clusters of genetically similar individuals are often identified by building a dendrogram in which each branch tip is an individual. Second, we quantify genetic relationships among the clusters (putative demes).

Individual-based methods for assessing population relationships make no *a priori* assumptions about how many populations exist or where boundaries between populations occur on the landscape. If individual-based methods are not used, we risk wrongly grouping individuals into populations based on somewhat arbitrary traits (e.g., color) or an assumed geographic barrier (a river) identified by humans subjectively (see also Section 15.6).

One example of erroneous *a priori* grouping would be migratory birds that we sample on migration routes or on overwintering grounds. Here, we might wrongly group together individuals from different breeding populations, only because we sampled them together at the same geographic location. A similar error could be made in migratory butterflies, salmon, or whales, if we sample mixtures containing individuals from different breeding groups originating from different geographic origins.

An individual-based approach was used by Pritchard *et al.* (2000) to assess relationships among populations of the endangered Taita thrush in Africa. The authors built a tree of individuals based on pairwise genetic distance between individuals. Each individual was genotyped at seven microsatellite loci (Galbusera *et al.* 2000). The proportion of shared alleles between each pair of individuals was computed, and then a clustering algorithm (neighbor-joining) was used to group similar individuals together on branches. The geographic location of origin of individuals was also plotted on the branch tips to help identify population units. The analysis revealed three distinct populations represented by three discrete clusters of individuals (Figure 16.11).

This example illustrates a strength of the individual-based approach: the ability to identify migrants. Individuals (i.e., branches) labeled with 'N' and an asterisk '*' (bottom of tree, Figure 16.11) were sampled from the 'N' location (Ngangao) but cluster genetically with Mbololo (labeled 'M'). This suggests these individuals are migrants from Mbololo into the Ngangao population.

An individual-based and model-based approach that identifies populations as clusters of individuals was introduced by Pritchard *et al.* (2000). 'Model-based' refers to the use of a model with k populations (demes) that are assumed to be in Hardy-Weinberg and gametic equilibrium. This approach first tests whether our data fit a model with $k = 1$, 2, 3 or more populations. The method uses a computer algorithm to search for the set (k) of individuals that minimizes the amount of HW and gametic disequilibrium (Section 10.1) in the data. Many possible sets of individuals are tested. Once k is inferred (step 1), the algorithm estimates, for each individual, the (posterior) probability (Q) of the individual's genotype originating from each population (step 2). If an individual is equally likely to have originated from population X and Y, then Q will be 0.50 for each population.

For example, Bergl and Vigilant (2007) used 11 microsatellite loci to identify migrants between forest fragments containing the critically endangered Cross River gorilla in Nigeria and the Cameroon. Genotypes were obtained from between one-quarter and one-third of the estimated total population through the use of noninvasively collected DNA samples. The population structure inferred from microsatellites was consistent with geography and habitat fragmentation. Previous field surveys suggested that gorillas in different fragments were isolated from one another. However, the genetic data suggested some gene flow and identified four migrants between habitat fragments, along with individuals of admixed ancestry, suggesting that there has been relatively recent connectivity between many of the localities (Figure 16.12).

These results are encouraging for the conservation of the Cross River gorilla population. The data suggest that gorillas are able to move between localities and reproduce in the new location. Such dispersal and gene flow are crucial for the maintenance of long-term fitness and persistence of small, fragmented populations.

Individual-based analyses can also be conducted with many multivariate statistical methods (e.g., PCA) if individuals are used as the operational unit (instead of populations). These multivariate approaches make no prior assumptions about the population structure model (e.g., Hardy-Weinberg and gametic equilibrium are not assumed).

Individual-based methods are useful to identify cryptic subpopulations and localize population boundaries on the landscape. Once genetic boundaries (dis-

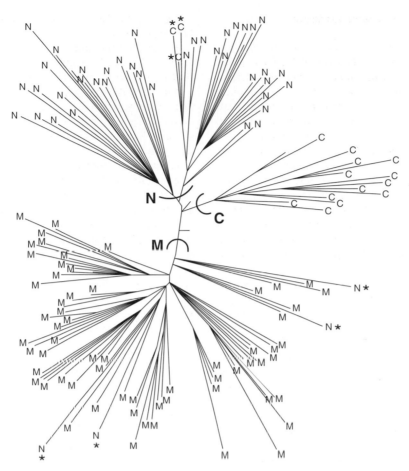

Figure 16.11 Neighbor-joining tree of individual Taita thrush based on the proportion of shared alleles at seven microsatellite loci (Chawia, 17 individuals; Ngangao, 54 individuals; Mbololo, 80 individuals). Three curved slashes (N, M, and C) across branches identify three population clusters. The letters on branch tips are sampling locations (i.e., population names); asterisks on branch tips represent putative immigrants (e.g., three immigrants from Ngangao into Chawia, top of figure). Modified from Pritchard *et al.* (2000).

continuities) are located, we can test whether the boundaries are concordant with some environmental gradient, a phenotypic trait distribution, or some ecological or landscape feature (e.g., a river or temperature gradient). This approach of associating population genetic boundaries with landscape or environmental features is a **landscape genetics** approach (Section 15.6, Manel *et al.* 2003).

These approaches can provide misleading results if samples of individuals are not collected evenly across space. In a continuously distributed population (see Figure 9.8), we might wrongly infer a genetic disconti-

nuity (barrier) between sampling locations if clusters of individuals are sampled from distant locations (Frantz *et al.* 2009). For example, Schwartz and McKelvey (2009) simulated a population of continuously distributed individuals, and then used common approaches to sample individuals from ten sampling locations. An individual-based STRUCTURE analysis was conducted, which suggested that many individuals could be assigned to the location from which they were sampled (Figure 16.13). This result could be interpreted to indicate the existence of discrete subpopulations, even though individual genotypes

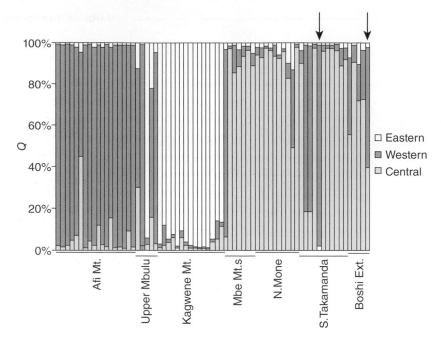

Figure 16.12 Bayesian clustering of individual Cross River gorilla based on genotypes at 11 microsatellite loci from forest fragments. Each fragment locality is named below the figure. Individuals are represented across the x-axis by a vertical bar that may be divided vertically into shaded segments that represent the individual's probability of originating (Q) from the Eastern (white), Western (black), or Central (gray) regions, computed using STRUCTURE (Pritchard *et al.* 2000). The left arrow indicates one of four putative migrants. The right arrow indicates an individual with approximately half of its genes (ancestry) from the Western regions and approximately half of its genes from the Central region. From Bergl and Vigilant (2007).

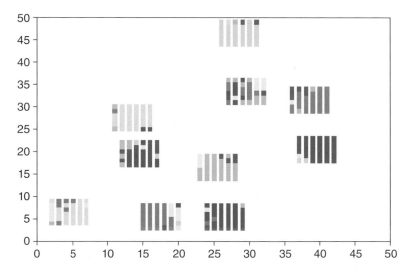

Figure 16.13 Potential problems when using individual-based methods to identify discrete subpopulations in the case of isolation by distance. Assignments using STRUCTURE (Pritchard *et al.* 2000) when 360 individuals are sampled in ten clusters of individuals from simulations of a continuously distributed population. Each of the ten sample locations consists of six vertical bars representing six individuals within each bar (36 individuals per sampling location). Each square represents an individual. Each square's shading indicates the cluster to which STRUCTURE assigned the individual, based on maximum Q. The values on the x and y axis are cell references in the 50 × 50 grid containing 250 individuals total. Modified from Schwartz and McKelvey (2009). See Color Plate 7.

were from a continuously distributed population having no discrete subpopulations or genetic discontinuities. In addition, this analysis suggested that there were some long-distance migrants, which is incorrect because a simple neighbor-mating process was used to generate the genotypes. Schwartz and McKelvey (2009) provide useful guidance to avoid these types of misinterpretations.

A potential problem with individual-based methods is that they can still yield uncertain results if genetic differentiation among populations is not substantial. Also the performance and reliability of individual-based methods has not been thoroughly evaluated (but see Evanno *et al.* 2005 for a performance-based analysis of the individual-based clustering method of Pritchard *et al.* 2000). Thus, it seems useful and prudent to use both individual-based and population-based methods.

16.4.3 Phylogeography

Phylogeography is the assessment of the correspondence between phylogenetic patterns and geographic patterns of distribution among taxa (Avise 2009). We expect to find phylogeographic structuring among populations with long-term geographic isolation. Substantial isolation for hundreds of generations is generally required for new mutations to arise locally, and to preclude their spread beyond local populations. Phylogeographic structure is expected in species with limited dispersal capabilities, with **philopatry**, or with distributions that span strong barriers to gene flow (e.g., mountains, rivers, roads and human development). In conservation biology, detecting phylogeographic structuring is important because it helps to identify long-isolated populations that might have distinct gene pools and local adaptations. Long-term reproductive isolation is one major criterion widely used to identify population units for conservation (see Section 16.5.2).

Intraspecific phylogeography was pioneered initially by J. Avise and colleagues (Avise *et al.* 1987). In a classic example, Avise *et al.* (1979a) analyzed mtDNA from 87 pocket gophers from across their range in southeastern US. The study revealed 23 different mtDNA genotypes, most of which were localized geographically (Figure 16.14). A major discontinuity in the maternal phylogeny clearly distinguished eastern and western populations. A potential conservation application of such results is that eastern and western populations of pocket gopher appear to be highly divergent with long-term isolation and thus potentially adaptive differences; this could warrant recognition as separate conservation units. However, additional data, including nuclear loci and nongenetic information, should be considered before making conservation management decisions (e.g., Section 16.6).

Phylogeographic studies can help to identify **biogeographic** provinces containing distinct flora and fauna worth conserving as separate geographic units in nature reserves. For example, multispecies phylogeographic studies in the southwestern US (Avise 1992) and northwest Australia (Moritz and Faith 1998) have revealed remarkably concordant phylogeographic patterns across multiple different species. Such multispecies concordance can be used to identify major biogeographic areas that can be prioritized as separate conservation units, and to identify locations to create nature reserves (Figure 16.15).

A widely used but controversial phylogeographic approach is **nested clade phylogeographic analysis** (NCPA, Templeton 1998). There has been substantial debate over the usefulness of NCPA (Knowles and Maddison 2002, Knowles 2008, Templeton 2008, Garrick *et al.* 2008). A shortfall of NCPA is that it does not incorporate error or uncertainty. This is the same problem with many phylogenetic approaches. For example, NCPA does not consider interlocus variation, as do coalescent-based population genetic models (see Section A10, Figure A12). Thus, NCPA might provide the correct inference about phylogeographic history, but we cannot easily quantify the probability of it being correct. Another limitation is that NCPA is somewhat ad hoc in using an inference key in order to distinguish between different historic processes, such as range expansion and population fragmentation. A recent study by Panchal and Beaumont (2007) evaluated NCPA by developing software to automate the NCPA procedure. Using simulations of random-mating populations, Panchal and Beaumont reported a high frequency of false positives for the detection of range expansions and for inferences of isolation by distance. The NCPA story illustrates the importance of conducting extensive performance evaluations (e.g., see Section A9) before using any novel computational approach.

Fortunately, the emerging field of **statistical phylogeography** promises to combine the strengths of NCPA with formal model-based approaches and statistical tests to test alternative hypotheses to explain phylogeographic patterns (Knowles and Maddison 2002).

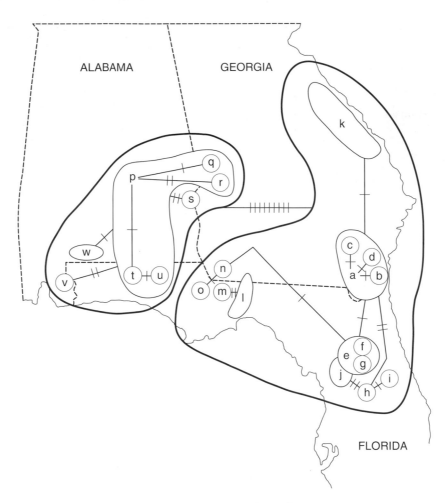

Figure 16.14 Mitochondrial DNA phylogenetic network for 87 pocket gophers. MtDNA genotypes are represented by lower-case letters and are connected by branches in a parsimony network. Slashes across branches are substitutions. Nine mutations separate the two major mtDNA clades encircled by heavy lines. From Avise (1994).

As formal modeling and validated statistical phylogeography approaches are becoming available, it would be prudent not to use NCPA, or use it only in combination with well-evaluated model-based phylogeographic approaches (Turner *et al.* 2000, Beaumont *et al.* 2010).

16.5 UNITS OF CONSERVATION

It is crucial to identify species and units within species to help guide management, monitoring, other conservation efforts, and to facilitate application of laws to conserve taxa and their habitats, as described in the first three paragraphs of this chapter. In this section, we consider the issues of identifying species and intraspecific conservation units.

16.5.1 Species

Identification of species is often problematic, even for some well-known taxa. One problem is that biologists cannot even agree on the appropriate criteria to define a species. In fact, more than two dozen species concepts have been proposed over the last few decades. Darwin

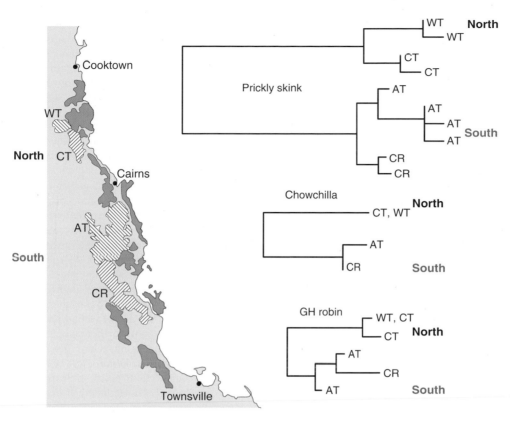

Figure 16.15 Neighbor-joining trees based on mtDNA sequence divergence for three species sampled from each of four areas (WT, CT, AT, and CR) from the tropical rainforests of northeastern Australia. Note the long branches separating the WT/CT in the north from the CR/AT populations in the south for all three species (prickly skink, chowchilla, and grey-headed robin). These results suggest long-term isolation for numerous species between the northern and southern rainforests. These regions merit recognition as separate conservation units. From Moritz and Faith (1998).

(1859) wrote that species are simply highly differentiated varieties. He observed that there is often a continuum in the degree of divergence from between populations, to between varieties, species, and higher taxonomic classifications. In this view, the magnitude of differentiation that is required to merit species status can be somewhat arbitrary.

The **biological species concept** (BSC) of Mayr (1942, 1963) is the most widely used species definition, at least for animals. Ernst Mayr defined species as "groups of interbreeding natural populations that are reproductively isolated from other such groups". This concept emphasizes reproductive isolation and isolating mechanisms (e.g., pre- and postzygotic). Criticisms of this concept are that (1) it can be difficult to apply to allopatric organisms (because we cannot observe or

test for natural reproductive barriers in disjunct populations); (2) it cannot easily accommodate asexual species; and (3) it has difficulties dealing with introgression between distinct forms, which is common in nature. Further, an emphasis on 'isolating mechanisms' implies that selection counteracts gene flow. However, the BSC generally does not allow for interspecific gene flow, even for a few segments of the genome that can introgress between species (Wu 2001, Rieseberg 2011).

The **phylogenetic species concept** (PSC, Cracraft 1989) relies largely on monophyly such that all members of a species must share a single common ancestor. This concept has fewer problems dealing with asexual organisms (e.g., many plants, fish, etc.) and with allopatric forms. However, it does not work well under hybridization and it can lead to oversplitting, for example, as more and more characters are used, as with powerful DNA sequencing techniques, so more 'taxa' might be identified.

A problem using PSC can arise if biologists interpret fixed DNA differences (monophyly) between populations as evidence for species status. For example, around the world many species are becoming fragmented, and population fragments aré becoming fixed (monophyletic) for different DNA polymorphisms. Under the PSC, this could cause the proliferation of 'new species' if biologists strictly apply the PSC criterion of monophlyly for species identification. This could result in oversplitting and the waste of limited conservation resources. This potential problem of fragmentation-induced oversplitting is described in a paper titled *Cladists in wonderland* (Avise 2000). To avoid such oversplitting, multiple independent DNA sequences (i.e., not mtDNA alone) should be used, along with nongenetic characters when possible.

The **ecological species concept** identifies species based on a distinct ecological niche (Van Valen 1976). The **evolutionary species concept** is used often by paleontologists to identify species based on change within lineages through time, but without splitting (anagenesis) (Simpson 1961). The many different concepts overlap, but emphasize different types of information. Generally, it is important to consider many kinds of information or criteria when identifying and naming species. If most criteria or species concepts give the same classification or conclusion, then we are more confident in the conclusion (e.g., species status is warranted).

A **unified species concept**, or more general concept, has been proposed (De Queiroz 2007, Hart 2010, Hausdorf 2011). Generality or unification is achieved by treating "separately evolving metapopulation lineages" and isolation processes as the only necessary properties of species (Hart 2010). This author reported that a conceptual shift in focus is occurring away from species diagnoses based on published species definitions, and towards analyses of the processes acting on lineages of metapopulations that will likely lead to different recognizable species (De Queiroz 2007). A possible advantage of this approach is that a species is recognized from **emergent properties** of measurable, ongoing population-level processes.

Similarly, a generalization of the genic concept (Wu 2001) has been proposed that defines species as groups of individuals that are characterized by features that would have negative fitness effects in other groups, if cross-breeding occurred between groups (Hausdorf 2011). This differential fitness concept has benefits, such as classifying groups that maintain differences and keep on differentiating despite occasional interbreeding, and it is not restricted to specific mutations or mechanisms causing speciation. In addition, it can be applied to the whole spectrum of organisms from biparentals to uniparentals (i.e., species with transmission of genotypes from only one parental type to all progeny).

African cichlid fishes illustrate some of the difficulties with the different species concepts. Approximately 1500 species of cichlids have recently evolved a diverse array of morphological differences (mouth structure, body color) and ecological differences (feeding and behaviors such as courtship). Morphological differences are pronounced among cichlids. However, the degree of genetic differentiation among cichlids is relatively low compared with other species, due to the recent radiation of African cichlid species (less than 1–2 million years ago). Further complicating efforts to identify species using molecular markers is the fact that reproductive isolation can be transient. For example, some cichlid species are reproductively isolated due to mate choice based on fixed color differences between species. However, this isolation breaks down when murky water prevents visual color recognition and leads to temporary interspecific gene flow (Seehausen *et al.* 1997).

16.5.1.1 Cryptic species

Molecular genetic data can help to identify species, especially cryptic species that have similar phenotypes (see also Section 22.1). For example, the neotropical skipper butterfly was recently identified as a complex of at least ten species, in part by the sequencing of a standard gene region for taxon identification (DNA 'barcoding', Section 22.1.1). The ten species have only subtle differences in adults and are largely sympatric (Hebert *et al.* 2004). However, they have distinctive caterpillars, different caterpillar food plants, and a relatively high genetic divergence (3%) in the mitochondrial gene cytochrome *c* oxidase I (COI). Section 22.1 discusses barcoding more extensively (see also Valentini *et al.* 2009a).

Molecular data can also help to identify taxa that are relatively well studied. For example, a recent study of African elephants used molecular genetic data to detect previously unrecognized species. Elephants from tropical forests are morphologically distinct from savannah elephants. Roca *et al.* (2001) biopsy-dart sampled 195 free-ranging elephants from 21 populations. Three populations were forest elephants in central Africa, 15 were savannah elephants (located north, east, and south of the forest populations), and three were unstudied and thus unclassified populations. DNA sequencing of 1732 base pairs from four nuclear genes revealed 52 nucleotide sites that were phylogenetically informative (i.e., at least two individuals shared a variant nucleotide).

All savannah elephant populations were closer genetically to every other savannah population than to any of the forest populations, even in cases where the forest population was geographically closer to savannah populations (Roca *et al.* 2001). Phylogenetic analyses revealed five fixed site differences between the forest and savannah (Figure 16.16a). By comparison, nine fixed differences exist between Asian and African elephants. Hybridization was considered to be "extremely limited", although the number of individuals sampled was only moderate, and one savannah individual apparently contained one nucleotide diagnostic for the forest elephants. The genetic data (see also Comstock *et al.* 2002, Figure 16.16b), combined with the morphological and habitat differences, suggests that species status is warranted.

This conclusion has been further supported by a recent study that sequenced nearly 40,000 bp of the nuclear genome in all living elephant taxa, as well as the extinct American mastodon and the woolly mammoth (Rohland *et al.* 2010). The recognition of the forest elephant as a distinct taxon has influenced conservation strategies, making it more urgent to protect and manage these increasingly endangered taxa separately.

16.5.1.2 Oversplitting

Genetic data may show that recognized species are not supported by reproductive relationships. Some authors have recognized the black sea turtle (*Chelonia* spp) as a distinct species on the basis of skull shape, body size, and color (Pritchard 1999). However, molecular analyses of mtDNA and three independent nuclear DNA fragments suggest that reproductive isolation does not exist between the black and green forms (Karl and Bowen 1999). Over the years, taxonomists have proposed more than a dozen species for different *Chelonia*

populations – with oversplitting occurring in many other taxa as well. Nevertheless for conservation purposes, it is clear that black turtles are distinct and could merit recognition as an intraspecific conservation unit (e.g., Section 16.5.2) that possesses potential local adaptations. Unfortunately, populations are declining, and additional data on adaptive differences are needed (e.g., food sources and feeding behavior).

Another example of oversplitting is the little brown bat from Alberta, Canada, and Montana, US, where strong mtDNA sequence divergence exists between bats from the two geographic areas. Lausen *et al.* (2008) tested the hypothesis that the divergence represented two species using ten nuclear microsatellite markers, and found a lack of differentiation between these two groups. In addition, the lack of geographic and morphologic boundaries suggested that no line can be drawn between these two groups. This study emphasizes the importance of investigating multiple nuclear loci to assess cryptic genetic species (see also Guest Box 9). It also has important implications for applications of DNA barcoding that use only mtDNA (Section 22.1.1).

16.5.2 Evolutionarily significant units

An **evolutionarily significant unit** (ESU) can be defined broadly as a population or group of populations that merit separate management or priority for conservation because of high distinctiveness both genetically and ecologically. The first use of the term ESU was by Ryder (1986). He used the example that five extant subspecies of tigers exist, but there is not space in zoos or captive breeding programs to maintain viable populations of all five. Thus sometimes we must choose which subspecies to prioritize for conservation action, and perhaps maintain only one or two global breeding populations (each perhaps consisting of more than one named subspecies). Since Ryder (1986), the term ESU has been used in a variety of frameworks for identifying conservation units (Box 16.3).

There is considerable confusion and controversy in the literature associated with the term ESU. For example, the ESA lacks any definition of a distinct population segment (DPS, Box 16.1). Waples (1991) suggested that a population or group of populations of salmon would be a DPS if it is an ESU. This has led to some confusion because some biologists equate a DPS and an ESU. We will use the term DPS when referring

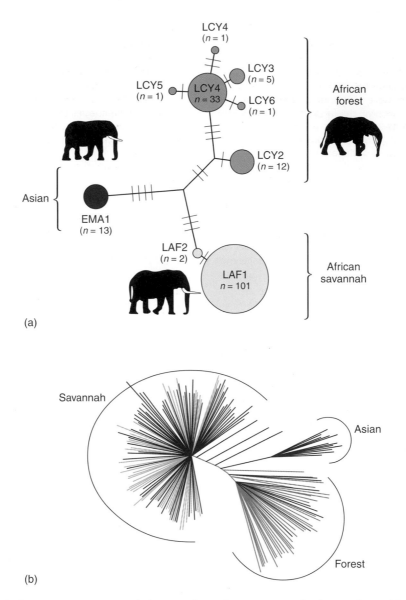

Figure 16.16   (a) Minimum spanning network showing relationships among nine haplotypes observed for the X-linked BNG gene for African forest, savannah, and Asian elephants. Each slash mark along branches separating each haplotypes represents one nucleotide difference (from Roca *et al.* 2001). (b) Neighbor-joining tree of 189 African elephants and 14 Asian elephants based on proportion of shared alleles (Dps) at 16 microsatellite loci. From Comstock *et al.* (2002).

to officially recognized 'species' under the ESA, and the term ESU in the more generally accepted sense.

It can be difficult to provide a single concise, detailed definition of the term ESU because of the controversy and different uses and definitions of the term in the literature. This ESU controversy is analogous to that surrounding the different species concepts mentioned above. The controversy is not surprising considering the problems surrounding the definition of species, and the fact that identifying intraspecific units is

---

Box 16.3   Proposed definitions of evolutionarily significant units (ESUs)

Ryder (1986): Populations that actually represent significant adaptive variation based on concordance between sets of data derived by different techniques. Ryder (1986) clearly argued that this subspecies problem is "considerably more than taxonomic esoterica". (Main focus: zoos for potential *ex situ* conservation of gene pools of threatened species.)

Waples (1991): Populations that are reproductively separate from other populations (e.g., as inferred from molecular markers) and that have distinct or different adaptations and that represent an important evolutionary legacy of a species. (Main focus: integrating different data types, and providing guidelines for identifying 'distinct population segments' of Pacific salmon which are given 'species' status for protection under the US Endangered Species Act.)

Dizon *et al.* (1992): Populations that are distinctive based on morphology, geographic distribution, population parameters and genetic data. (Main focus: concordance across some different data types, but always requiring some degree of genetic differentiation.)

Moritz (1994): Populations that are reciprocally monophyletic (see Figure 16.17) for mtDNA alleles and that show significant divergence of allele frequencies at nuclear loci. (Main focus: defining practical criteria for recognizing ESUs based on population genetics theory, while considering that variants providing adaptation to recent or past environments may not be adaptive (or might even retard the response to natural selection) in future environments.)

USFWS and NOAA (1996b) (US policy for recognition of discrete population segments, DPSs): (1) discreteness of the population segment in relation to the remainder of the species to which it belongs; and (2) the significance of the population segment to the species to which it belongs. This DPS policy is a further clarification of Waples (1991) Pacific salmon ESU policy that applies to all species under the US ESA.

Crandall *et al.* (2000): Populations that lack (1) "ecological exchangeability" (i.e., they have different adaptations or selection pressures (e.g., life histories, morphology, QTL variation, habitat, predators, etc.) and different ecological roles within a community); and (2) "genetic exchangeability" (e.g., they have had no recent gene flow, and show concordance between phylogenetic and geographic discontinuities). (Main focus: emphasizing adaptive variation and combining molecular and ecological criteria in a historical timeframe. Suggests returning to the more holistic or balanced two-part approach of Waples.)

Fraser & Bernatchez (2001): A lineage that demonstrates highly restricted gene flow from other such lineages within the higher organizational level (lineage) of the species. (Main focus: a context-based framework for delineating attempts to resolve conflicts among previous ESU definitions. Recognizes that different criteria will work better than others in some circumstances and can be used alone or in combination depending on the situation.)

---

generally more difficult than identifying species (Waples 1991). It is also not surprising considering the different rates of evolution that often occur for different molecular markers and phenotypic traits used in ESU identification. Different evolutionary rates lead to problems analogous to those in the classification of birds as a taxonomic class separate from reptiles due to the rapid evolution of birds, when in fact the class Aves is monophyletic within the class Reptilia (Figure 16.3).

In practice, an understanding of the underlying principles and the criteria used in the different ESU frameworks will help when identifying ESUs. The main criteria for several different ESU concepts are listed in Box 16.3, and synthesized at the end of this section (see also Fraser and Bernatchez 2001). Here we discuss

some details about three widely used ESU frameworks, each with somewhat different criteria as follows: (1) reproductive isolation and adaptation (Waples 1991); (2) **reciprocal monophyly** (Moritz 1994); and (3) "exchangeability" of populations (Crandall *et al.* 2000). This will provide background on principles and concepts, as well as a historical perspective of the controversy surrounding the different frameworks for identifying units of conservation.

16.5.2.1 Isolation and adaptation

Waples (1991) was the first to provide a detailed framework for ESU identification. His framework included the following two main requirements for an ESU: (1)

long-term reproductive isolation (generally hundreds of generations) so that an ESU represents a product of unique past evolutionary events that is unlikely to re-evolve, at least on an ecological timescale, and (2) ecological or adaptive uniqueness such that the unit represents a reservoir of genetic and phenotypic variation likely important for future evolutionary potential. This second part requiring ecological and adaptive distinctiveness was termed the "evolutionary legacy" of a species by Waples (1991). This framework has become the official policy under the ESA (USFWS and NOAA 1996b).

Waples (2005b) has argued that ESU identification is often most helpful if an intermediate number of ESUs are recognized within each species, with the goal of preserving a number of genetically distinct populations. Waples (2005b) reviewed other published ESU concepts and criteria (e.g., Box 16.3) and concluded that they often identified only a single ESU or a large number (hundreds) of ESUs in Pacific salmon species. This was a tentative conclusion based on the published criteria for other ESU concepts, many of which are sub-jective or qualitative. There is a need for more empirical examples in which multiple ESU concepts are applied to common problems, as in Waples 2005b).

16.5.2.2 Reciprocal monophyly

Moritz (1994) offered simple and thus readily applicable molecular criteria for recognizing an ESU: "ESUs should be reciprocally monophyletic for mtDNA (in animals) and show significant divergence of allele frequencies at nuclear loci". Mitochondrial DNA is widely used in animals because it has a rapid rate of evolution and lacks recombination, and thus facilitates phylogeny reconstruction. Cytoplasmic markers are often used in plants as they also lack recombination. "Reciprocally monophyletic" means that all DNA lineages within an ESU share a more recent common ancestor with each other than with lineages from other ESUs (Figure 16.17). These molecular criteria are relatively quick and easy to apply in most taxa because the necessary molecular markers (e.g., 'universal' PCR primers) and data analysis software have become widely available. Further,

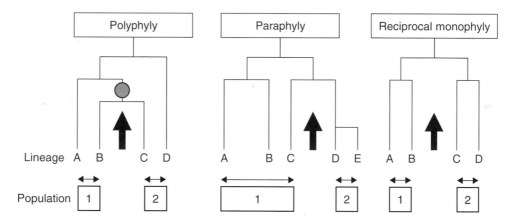

Figure 16.17 Development of phylogenetic relationships of alleles or lineages (A, B, C, D, E) in sister taxa (1 and 2). After a population splits into two because of the development of a barrier to reproduction (indicated by the large vertical arrows), the phylogenetic relationship of the alleles in the two sister populations usually proceeds from polyphyly through paraphyly to reciprocal monophyly. When two populations (1 and 2) first become isolated, they both will have some alleles that are more closely related to alleles in the other population (**polyphyly**). The filled circle at the root of the B and C branches indicates the most recent common ancestor between B and C (in the polyphyly example). After many generations of isolation, one population might become monophyletic for some alleles (D and E in population 2 in the **paraphyly** example; see also the black duck, Figure 16.7). But the other population (1) might maintain an allele (C) that is more related to an allele in the other population. After approximately four N_e generations, both sister populations are expected to be monophyletic with respect to each other (**reciprocal monophyly**) at nuclear loci. Modified from Moritz (1994).

speed is often important in conservation where management decisions may have to be made quickly, and before thorough ecological studies can be conducted.

An occasionally cited advantage of the Moritz (1994) monophyly criterion is that it can employ population genetics theory to infer the time since population divergence. For example, it takes a mean of $4N_e$ generations for a newly isolated population to coalesce to a single gene copy and therefore become reciprocally monophyletic through drift and mutation at a nuclear locus (Neigel and Avise 1986). This means that if a population splits into two daughter populations of size $N_e = 1000$, it would take an expected 1000 generations to become reciprocally monophyletic for mtDNA. For mtDNA to become monophyletic it requires fewer generations because the effective population size is approximately four times smaller for mtDNA than for nuclear DNA; thus lineage sorting is faster (see Section 9.6). Here it is important to recall that adaptive differentiation can occur in a much shorter time period than does monophyly.

A disadvantage of the Moritz ESU concept is it generally ignores adaptive variation, unlike the two-step approaches that incorporate the "evolutionary legacy" of a species (Waples 1991). The framework of Moritz is based on a cladistic phylogenetic approach (see Section 16.2) using neutral loci. Thus, unfortunately, the Moritz approach makes it likely that small populations (e.g., bottlenecked populations) could be identified as ESUs when in reality the populations are not evolutionarily distinct; small populations can quickly become monophyletic due to drift or lineage sorting. Worse perhaps, natural selection can lead to rapid adaptation, especially in large populations in which selection is efficient. Consequently, the strict Moritz framework often could fail to identify ESUs that have substantial adaptive differences.

One limitation of using only molecular information is that a phylogenetic tree might not equal the true population tree. This is analogous to the 'gene tree vs. species tree' problem discussed in Section 16.3.2 (Example 16.4). This problem of population trees not equaling gene trees is worse at the intraspecific level because there is generally less time since reproductive isolation at the intraspecific level and thus more problems caused by lineage sorting and paraphyly. Consequently, problems of gene trees not matching population trees will be relatively common at the intraspecific level. Unfortunately, in the conservation literature, mtDNA data are often used alone to attempt to identify ESUs. This should occur less often as nuclear DNA markers become more readily available.

16.5.2.3 Exchangeability

Crandall *et al.* (2000) suggested that ESU identification be based on the concepts of ecological and genetic "exchangeability". The idea of exchangeability is that individuals can be moved between populations and can occupy the same niche, and can perform the same ecological role as resident individuals, without any fitness reduction due to outbreeding depression. If we can reject the hypothesis of exchangeability between populations, then those populations represent ESUs. Ideally, exchangeability assessment would be based on heritable adaptive quantitative traits. Strengths of this approach are that it integrates genetic and ecological (adaptive) information, and that it is hypothesis-based.

Exchangeability can be tested using common-garden experiments and reciprocal transplant experiments. For example, if two plant populations from different locations have no reduced fitness when transplanted between locations, they might be exchangeable and would not warrant separate ESU status (see also Section 2.5, Figure 2.12, Figure 8.1).

The main problem with this approach is it is not generally practical, in that it is difficult to test the hypothesis of exchangeability in many species. For example, it is difficult to move a rhinoceros (or most endangered species) from one population to another and then to measure its fitness and the fitness of its offspring. Such studies are especially problematic in endangered species where experiments are often not feasible. Although difficult to test, exchangeability is a worthy concept to consider when identifying ESUs. Even when we cannot directly test for exchangeability, we might consider surrogate measures of exchangeability, such as life history differences, the degree of environmental differentiation, or the number of functional genes showing signatures of adaptive differentiation (e.g., see Section 16.6.1). Surrogates are often used when applying Waples' ESU definition.

16.5.2.4 Synthesis

Substantial overlap in criteria exists among different ESU concepts. Several concepts promote a two-pronged approach involving isolation and adaptive divergence. The main common principles and criteria are the

Example 16.4   Lack of concordance between mtDNA and nuclear genes in white-eyes

Degnan (1993) compared dendrograms from mtDNA and nuclear DNA for two species of white-eye birds from Australia. The mtDNA data yielded a single gene tree that does not reflect the organismal tree based on phenotypic characters. In contrast, the two nuclear DNA loci revealed phylogeographic patterns consistent with the traditional classification of the two species (Figure 16.18). The author concluded that the discordance between the mtDNA and nuclear DNA (and phenotype) likely results from past hybridization between the two species of white-eye and mtDNA introgression. Evidence for hybridization might have been lost in nuclear genes through recombination.

This study provides a clear empirical demonstration that single gene genealogies cannot be assumed to

accurately represent the true organismal phylogeny. Further, it emphasizes the need for analyses of multiple independent DNA sequences when inferring phylogeny and identifying conservation units. This is especially true for populations within species where relatively few generations have passed. For example, if we were trying to identify ESUs (or species) in this study by using mtDNA alone, we might identify three ESUs (corresponding to the three mtDNA haplogroups in Figure 16.18); however, these three are not concordant with the two groups identified by phenotype, nuclear DNA, and geography.

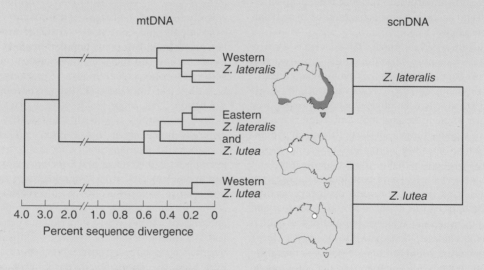

Figure 16.18   UPGMA dendrograms for mtDNA haplotypes (left) and scnDNA (single copy nuclear DNA) genotypes (right) in silver-eyes (*Zosterops lateralis*) and yellow white-eyes (*Z. lutea*). Distribution of silver-eye haplotypes in solid black, and yellow white-eye in outline circles. Note the middle map shows that the yellow white-eye samples from western Australia group in the mtDNA tree with silver-eye samples from eastern Australia, but group with the yellow white-eye in the scnDNA tree. Modified from Degnan (1993).

following: reproductive isolation (no gene flow), adaptive differentiation, and concordance across multiple data types (e.g., genetic, morphologic, behavioral, life history, and geographic). The *longer the isolation and the more different the environment* (selection pressures), the

more likely populations are to represent distinct units that are worthy of preservation and separate management. We should not rely on any single criterion, such as reciprocal monophyly of mtDNA. In fact, the greater the number of different data types showing concordant differentiation between populations, the stronger the evidence for ESU status.

16.5.3 Management units

Management units (MUs) are populations that are demographically independent (Moritz 1994); that is, their population dynamics (growth rate) depend on local birth and death rates rather than on immigration. The identification of these units, similar to 'stocks' recognized in fisheries biology, is useful for short-term management, such as delineating hunting or fishing areas, setting local harvest quotas, and monitoring habitat and population status (Palsbøll *et al.* 2007).

MUs often are subpopulations within a major metapopulation that represents an ESU. For example, fish populations are often structured on hierarchical levels, such as small streams (as MUs) that are nested within a major river drainage (ESU). MUs, unlike ESUs, generally do not show long-term independent evolution or strong adaptive differentiation. MUs should represent populations that are important for the long-term persistence of an ESU or species. The conservation of multiple populations, not just one or two, is critical for insuring the long-term persistence of species (Hughes *et al.* 1997, Hobbs and Mooney 1998).

Waples and Gaggiotti (2006) have suggested that, based on Hastings (1993), the transition from demographic dependence to independence generally occurs when the fraction of immigrants in a subpopulation falls below 10%. Hastings (1993) identified the 10% threshold as the point where population dynamics in two patches transitions from behaving independently to behaving as a single population (see Section 15.4.1).

Moritz (1994) originally defined the term "management unit" (MU) as a population that has substantially divergent allele frequencies at nuclear or mtDNA loci.

This view was supported by Bentzen (1998) who said that significant genetic differences and departure from panmixia are strong evidence that the populations are demographically independent, and should be considered as separate MUs. However, Palsbøll *et al.* (2007) found that significant genetic differentiation can be detected consistently, even with migration rates greater than 20%, if many highly variable genetic markers are used.

We saw in Chapter 9 that genetic divergence is largely a function of the number of migrants (mN). However, demographic independence is primarily a function of the number of migrants (N). Thus, allele frequency differentiation (e.g., F_{ST}) should not be used by itself to identify MUs (Figure 15.8). For example, large populations experience little drift (and little allele frequency differentiation) and thus can be demographically independent even if allele frequencies are similar. The same mN (and hence F_{ST}) can result in different migration rates (m) for different population sizes (N). As N goes up, m goes down for the same F_{ST} (Table 16.4). So in a large population, the number of migrants can be very small, and the population could be demographically independent, yet have a relatively low F_{ST}.

A related difficulty is determining whether migration rates would be sufficient for recolonization on an ecological timescale, for example, if a MU became extinct or overharvested. Allele frequency data can be used to estimate migration rates (mN), but at moderate to high rates of migration ($m > 0.01–0.10$), genetic estimators are notoriously imprecise (Faubet *et al.* 2007), such that confidence intervals on the mN estimate might include infinity (Waples 1998).

16.5.3.1 Oversplitting and undersplitting

Two general errors can occur in MU diagnosis, as with ESU diagnosis. First, identification of too few units could lead to underprotection, which could lead to the

Table 16.4 Inferring demographic independence of populations by using genetic differentiation data (F_{ST}) requires knowledge of the effective population size (N_e). Here, the island model of migration was assumed to compute mN_e (number of migrants) and m (proportion of migrants) from the F_{ST} (as in Figure 9.8). Recall that the effective population size is generally far less than the census size in natural populations (see Section 7.10).

F_{ST}	N_e	m	mN_e	Demographic independence
0.06	50	0.080	4	Unlikely
0.06	100	0.040	4	Likely
0.06	1000	0.004	4	Yes

reduction or loss of local populations. This problem could arise, for example, if statistical power is too low to detect genetic differentiation when differentiation is biologically significant. For example, too few MUs (and underprotection) could result if only one MU is identified when the species is actually divided into five demographically independent units. Consider that the sustainable harvest rate is 2% per year on the basis of total population, but that all the harvest comes from only one of the five MUs. Then the actual harvest rate for the single harvested MU is 10% (assuming equal sizes for the five MUs). This high harvest rate could result in overexploitation and perhaps extinction of the one harvested MU population. For example, if the harvested population's growth rate is only 4% per year and the harvest rate is 10%, overexploitation would be a problem (Taylor and Dizon 1999).

Here, undersplitting could result from either a lack of statistical power (e.g., due to too few data), or to the misidentification of population boundaries (e.g., due to cryptic population substructure). To help avoid misplacement of boundaries, researchers should sample many individuals that are widely distributed spatially, and use recently developed, individual-based statistical methods (see Section 16.4.2 and caveats therein; see also Manel *et al.* 2003).

Second, diagnosing too many MUs (oversplitting) could lead to unnecessary waste of conservation management resources. This error could occur if, for example, populations are designated as MUs because they have statistically significant differences in allele frequencies, but this differentiation is not associated with important biological differences. This becomes a potential problem as more and more molecular markers are used that are highly polymorphic and thus statistically powerful.

16.6 INTEGRATING GENETIC, PHENOTYPIC, AND ENVIRONMENTAL INFORMATION

Many kinds of information should be integrated, including life history differences, environmental characteristics, phenotypic divergence, and patterns of gene flow for the identification of conservation units (Figure 16.19, Guest Box 16). For example, if two geographically distant populations (or sets of populations)

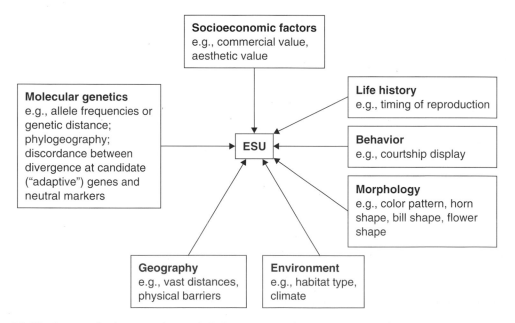

Figure 16.19 Sources of information that can help diagnose a population (or set of populations) as an evolutionarily significant unit. Modified from Moritz *et al.* (1995).

show large molecular differences that are concordant with life history (e.g., flowering time) and morphologic (e.g., flower shape) differences, we would be relatively confident in designating them as two geographic or population units important for conservation.

Researchers should always consider whether the environment or habitat type of different populations has been different for many generations, because this could lead to adaptations (even in the face of high gene flow) that are important for the long-term persistence of species. The more kinds of independent information that are concordant, the more certain one can be that a population merits recognition as a distinct conservation unit. The principle of considering multiple data types and testing for concordance is critical for identifying conservation units.

Difficulties arise when concordance is lacking among data types. For example, imagine that two populations show morphological differences in size or color of individuals, but show evidence of extensive recent gene flow. This scenario has arisen occasionally in studies that measure phenotypic traits from only small samples or nonrepresentative samples of individuals from each population (e.g., only 5–10 individuals of largely different sexes or ages from each population). In this example, taxonomic oversplitting results from biased or limited sampling, and conservation status as distinct units is

not warranted. This hypothetical example relates to the green/black turtle 'species' dilemma described above, where more extensive sampling and studies of life history and adaptive traits would be helpful.

16.6.1 Adaptive genetic variation

The incorporation of adaptive gene markers and gene expression studies can augment our understanding of conservation units (Funk *et al.* 2012). For example, adaptive and 'neutral' molecular variation can be integrated (Gebremedhin *et al.* 2009) by considering two separate axes in order to identify populations with high distinctiveness for both adaptive and neutral diversity (Figure 16.20).

Adaptive gene markers often give the same patterns of population relationships as neutral markers (e.g., Coop *et al.* 2009, Luikart *et al.* 2011). A recent comparison of assumed neutral and putatively selected alleles in over 640,000 autosomal SNPs in humans concluded that average allele frequency divergence at neutral loci is highly predictive of adaptive divergence and that neutral processes (migration and genetic drift) exert powerful influences over the geographic distribution of selected alleles (Coop *et al.* 2009). This result suggests that neutral loci can provide useful

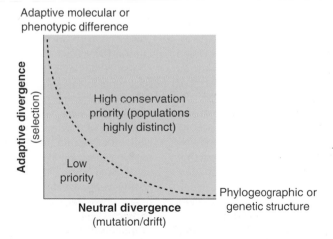

Figure 16.20 Adaptive information could be integrated with information from neutral markers and information on long-term isolation. Such an approach could help identify the most appropriate source population to translocate individuals into small, declining populations that require supplementation. This approach could also help rank or prioritize populations for conservation management. From Luikart *et al.* (2003).

descriptions of the patterns of divergence at adaptive loci at a relatively large spatial scale (regional to global) in this study.

Similarly, a study using neutral microsatellites revealed similar patterns to a study of gene expression and functional genetic divergence among populations (Hansen 2010). In the study, Tymchuk *et al.* (2010) used microarray hybridization (Chapter 4) with 16,000 gene targets (16K cDNA microarray) and RNA extracted from whole fry of 12 Atlantic salmon populations to examine patterns of gene expression. They found concordant patterns of divergence at seven microsatellite loci and gene expression. These results suggest that patterns of divergence at neutral loci reflect patterns of adaptive variation in gene expression.

Alternatively, studies with adaptive loci might suggest different population relationships compared with neutral loci. For example, as mentioned in Section 9.7, Wilding *et al.* (2001) genotyped 290 AFLP loci in an intertidal snail that has two different morphotypes adapted to two different habitats. When studying 15 putatively adaptive loci, populations with similar morphotypes and habitat type clustered together (Figure 16.21), but when the study excluded the 15 adaptive loci, the populations grouped together based on geographic proximity (Figure 16.21, Section 9.7.3). This study illustrates the potential complexity of using adaptive markers for delineating units of conservation. Other studies have also found that adaptive loci can yield substantially different inferences about population genetic relationships (Landry *et al.* 2002, Giger *et al.* 2006, Eckert *et al.* 2010, Hancock *et al.* 2011).

How should we identify and manage conservation units in these scenarios? Pitfalls arise when focusing on a set of adaptive loci rather than neutral patterns or genome-wide averages, largely because selection can be extremely complex and a complete understanding of adaptive divergence is unattainable (Luikart *et al.* 2003, Allendorf *et al.* 2010). Genes important for contemporary or past adaptations might not be those that will be crucial for adaptation in future environments. In addition, a focus on conserving variation at certain detectable adaptive genomic regions could result in loss of important genetic variation at other adaptive or neutral regions. Moreover, even when the same genomic regions are implicated in, for example, local adaptation across populations, the particular alleles involved may be different and perhaps even result in outbreeding depression when combined through admixture. Finally, much effort has been devoted to **genome-wide association studies** for detecting the genetic basis of complex traits using large samples of individuals and genetic markers, yet often a large proportion of the heritability remains unexplained (Frazer *et al.* 2009).

In summary, caution must be used when using adaptive loci for the identification of conservation units. Genetic patterns at neutral markers largely reflect the historic interaction of gene flow and genetic drift that are expected to affect the amount of genome-wide genetic variation within, and genetic divergence among, populations. These patterns are the foundation upon which natural selection operates to bring about adaptive differences among populations. Loci under selection generally should be used (a) as a supplement or complement to neutral loci, and (b) as one source of adaptive information that should be interpreted in combination with other data on phenotypes, life history, and the environment (e.g., Figure 16.21).

16.7 COMMUNITIES

Identification of groups of multiple co-distributed species might help to delineate conservation units and the spatial boundaries between units. One example is the discovery of congruent genealogical patterns among many species distributed across the southeastern US. Numerous species have phylogeographically concordant patterns across the Coastal Plain of the southeastern US, such that a phylogeographic break (Avise 1996) occurs for many taxa near the Apalachicola River Basin in western Florida. This drainage system is the range limit and/or contact zone for reptiles, amphibians, fishes, mammals, spiders, trees, and estuarine macroinvertebrates (e.g., Engle and Summers 2000); it is also a phylogeographic barrier within many species (Figure 16.14, Avise 1992, 1996). Pauly *et al.* (2007) found that the flatwoods salamander has a phylogeographic break (species boundary) near the same river basin. The authors suggested that these results emphasize that in the absence of taxon-specific data, the existence of concordant multitaxa spatial distributions can provide strong predictions for locations of management units for unstudied species of conservation concern.

Similarly, if interacting species have concordant phylogeographic structure due to coevolution, then the phylogeographic distinctiveness of populations of one

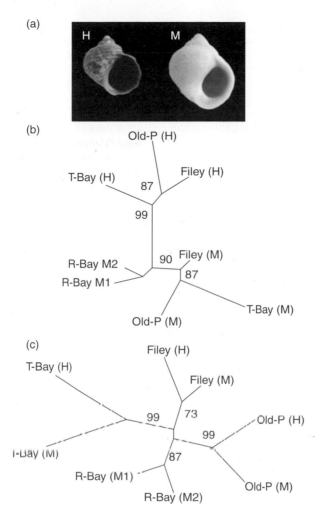

Figure 16.21   The effect of adaptive loci on the pattern of genetic similarity among intertidal snail populations. (a) The two intertidal snail morphotypes are thin shell with wide aperture (H), and thick shell with narrow aperture (M). (b) Neighbor-joining tree based on allele frequencies at all 290 AFLP loci examined. (c) Neighbor-joining tree based on 275 loci, with the 15 outlier loci removed (see Figure 9.14). Note the high bootstrap support for both trees. The two morphotypes cluster together when all 290 loci are used (a). However, when the outlier loci are removed, the geographically adjacent populations cluster together (b). This suggests that there is greater gene flow among adjacent populations, but that the differences in morphology and at the 15 outlier loci are maintained by local adaptation. From Luikart *et al.* (2003).

species could support the distinctiveness among populations within the other species. Criscione and Blouin (2007) found congruence in phylogeographic patterns between ESU boundaries of four Pacific salmonid species and of a trematode parasite. The authors suggested that the pattern from the trematode supports delineation of salmon ESU boundaries as biologically reasonable, as does the phylogeographic concordance among the four salmonid species.

Biek *et al.* (2006) used a rapidly evolving virus to help identify cougar population boundaries and to assess population connectivity. Their study suggested that virus population genetic structure could help to identify cougar MUs useful for conservation. Studies of multiple parasite species could further help to identify MUs or ESUs. For example, Whiteman *et al.* (2007) compared the phylogeographic and population genetic structure of three parasites of an endangered Galápagos hawk. This and other studies indicate that phylogeographic patterns of parasites can be predicted by the biogeographic history of their hosts, and that parasites can provide independent support for ESU boundary delineation.

Microbial community assemblages could also help to diagnose conservation units. Recently developed 'community phylogeny' and **metagenomics** approaches could identify distinct bacterial assemblages inferred by combining sequence data and ecological data to identify communities that represent both evolutionarily and ecologically distinct units (Cohan 2006). Microbial metagenomic spatial patterns could help to delineate distinct environments and geographic areas useful to help delineate conservation units for animal and plant species (e.g., Coleman and Chisholm 2010). Metagenomics is leading to new ways of thinking about microbial ecology and community ecology that supplant the concept of species or distinct populations in the sense of ESUs and MUs as discussed above. In this new paradigm, communities are becoming the units of evolutionary and ecological study among bacteria, archaea, and perhaps protists and fungi (Doolittle and Zhaxybayeva 2010).

Guest Box 16   Identifying units of conservation in a rich and fragmented flora
David J. Coates

The flora in the southwestern Australian biodiversity hotspot shows a diverse array of evolutionary patterns and exceptionally high species diversity (Hopper *et al.* 1996). This evolutionary complexity can be attributed to a number of unusual features of this region that combine an ancient, stable landscape, with no large-scale extinction episodes associated with glaciation since at least the late Cretaceous, and widespread climatic and habitat instability during the Pleistocene, leading to cyclic expansion and contraction of the mesic and arid zones. These processes have resulted in a high number of taxa with geographically restricted and fragmented or disjunct distributions (Hopper *et al.* 1996). Many are likely to be relictual and probably had wider, more continuous distributions during favourable climatic regimes, but now persist in disjunct remnants particularly through the semiarid transitional rainfall zone. As a consequence, substantial genetic differentiation between populations is typical of many species and is particularly evident in rare and geographically restricted species. Relatively high levels of population differentiation have been reported for 22 animal-pollinated, mainly outcrossing taxa with disjunct populations systems. These taxa cover a range of southwestern Australian genera, including long-lived woody shrubs and trees, and herbaceous perennials (Coates 2000).

Diversity at both the population and species levels presents a major challenge in the development of appropriate conservation strategies for this flora. To be effective these strategies should not only aim to preserve current levels of species diversity, but should also consider intraspecific variation and the evolutionary and ecological processes associated with the generation and maintenance of that variation.

Four case studies are presented in Figure 16.22 involving the animal-pollinated rare and threatened species: *Lambertia orbifolia* (Coates and Hamley 1999, Byrne *et al.* 1999), *Acacia anomala* (Coates 1988), *Banksia brownii* (unpublished), and *Stylidium nungarinense* (Coates *et al.* 2003). The partitioning of genetic variation within and among geographically disjunct population clusters is used to identify conservation units. This is carried out by combining significant divergence in allele frequencies at nuclear loci with phylogenetic analysis of gene frequency data to identify substantial genetic structure and delimit the conservation units. Three of these studies use allozyme data; phylogeographic relationships based on cpDNA variation were also investigated in *L. orbifolia*. Microsatellite data were used to delimit the conservation units in *B. brownii*. In some cases, such as *A. anomala*, the conservation units can also be distinguished by mode of reproduction.

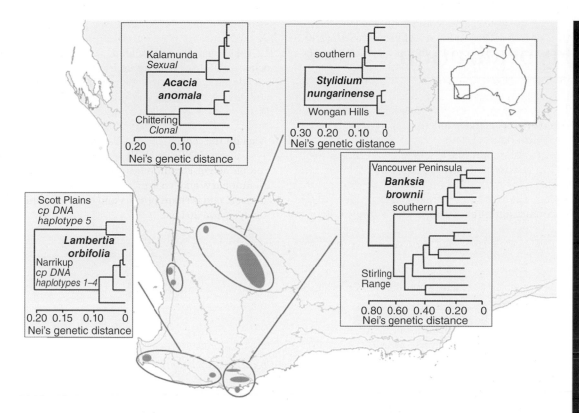

Figure 16.22 Geographic distributions of four plant species from southwestern Australia. Dendrograms show genetic relationships constructed using UPGMA and pairwise genetic distances between sample locations calculated from allele frequencies at allozyme loci (*A. anomala, L. orbifolia* and *S. nungarinense*), and microsatellite loci (*B. brownii*). The geographically separate population groups are recognized as separate conservation units and can be viewed as separate ESUs. Modified from Byrne *et al.* (1999), Coates (1988), Coates and Hamley (1999), and Coates *et al.* (2003).

In each case, we believe that the structure observed is the result of local extinction of intervening populations, and extended isolation of the remaining population groups resulting from the long-term effects of Pleistocene climatic instability in the southwest. These examples illustrate the different spatial scales at which population genetic structure can occur within species in this region, and that it does not readily correspond to topographic or biogeographic features.

The identification and characterization of conservation units, based on population genetic struc-

ture and phylogeographic patterns within species, provide a useful basis upon which more general conservation principles can be developed for the maintenance of these processes. Determining conservation units not only defines the appropriate units on which these strategies should be based, but also the geographic scale for management, which in some cases may involve a variety of land tenures such as national parks, nature reserves, private land, and local government land.

CHAPTER 17

Hybridization

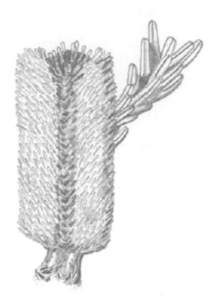

Banksia, *Section 17.2*

Hybridization, with or without introgression, frequently threatens populations in a wide variety of plant and animal taxa because of various human activities.

Judith M. Rhymer and Daniel Simberloff (1996)

Botanists have long viewed introgression as a potent evolutionary force that promoted the development and acquisition of novel adaptations.

Loren H. Rieseberg (2011)

Chapter Contents

Conservation and the Genetics of Populations, Second Edition. Fred W. Allendorf, Gordon Luikart, and Sally N. Aitken.
© 2013 Fred W. Allendorf, Gordon Luikart and Sally N. Aitken. Published 2013 by Blackwell Publishing Ltd.

Rates of **hybridization** and **introgression** have increased dramatically worldwide because of widespread intentional and incidental translocations of organisms and habitat modifications by humans (see Guest Box 17). Hybridization has contributed to the extinction of many species through direct and indirect means (Levin *et al.* 1996, Allendorf *et al.* 2001). The severity of this problem has been underestimated by conservation biologists (Rhymer and Simberloff 1996). The increasing pace of the three interacting human activities that contribute most to increased rates of hybridization (introductions of plants and animals, fragmentation, and habitat modification) suggests that this problem will become even more serious in the future (Kelly *et al.* 2010). For example, increased turbidity in Lake Victoria, Africa, has reduced color perception of cichlid fishes and has interfered with the mate choice that produced reproductive isolation among species (Seehausen *et al.* 1997). Similarly, increased turbidity because of land development and forest harvesting has led to increased hybridization among stickleback species in British Columbia, Canada (Wood 2003).

On the other hand, hybridization is a natural part of the evolutionary process. Hybridization has long been recognized as playing an important role in the evolution of plants (Rieseberg 2011, Figure 17.1). In addition, recent studies have found that hybridization has also played an important role in the evolution of animals (Arnold 1997, Dowling and Secor 1997, Grant and Grant 1998). Several reviews have emphasized the creative role that hybridization may play in adaptive evolution and speciation (e.g., Grant and Grant 1998, Seehausen 2004). Many early conservation policies generally did not allow protection of hybrids. However, increased appreciation of the important role of hybridization as an evolutionary process has caused a re-evaluation of these policies. Determining whether hybridization is natural or anthropogenic is crucial for conservation, but it is often difficult (Allendorf *et al.* 2001).

Hybridization provides an exceptionally tough set of problems for conservation biologists (Ellstrand *et al.* 2010). The issues are complex and controversial, beginning with the seemingly simple task of even defining hybridization (Harrison 1993). Hybridization has sometimes been used to refer to the interbreeding between species (e.g., Grant and Grant 1992b). However, we believe that this taxonomically restrictive use of hybridization can be problematic (especially since it is sometimes difficult to agree on what is a

species!). We have adopted the more general definition of Harrison (1990) that includes matings between "individuals from two populations, or groups of populations, which are distinguishable on the basis of one or more heritable characters".

The term 'hybrid' itself sometimes has a negative connotation, especially when used in conjunction with its opposite 'purebred'. In the US, a policy has been proposed to treat hybrids and hybridization under the ESA using the term "intercross" (suggested by John Avise) and "intercross progeny" to avoid the connotations of the term hybrid (USFWS and NOAA 1996a).

Detection of hybridization can be difficult, although it is becoming much easier through the application of various molecular techniques over the last two decades. Despite improved molecular data that can be collected with relative ease, interpreting the evolutionary significance of hybridization and determining the role of hybrid populations in developing conservation plans is more difficult than often appreciated. According to one review: "It is an understatement to say that hybridization is a complex business!" (Stone 2000).

In this chapter, we first consider the role that natural hybridization has played in the process of evolution. We next consider the possible harmful effects of anthropogenic hybridization and the fitness of hybrid individuals and populations. We also present and discuss genetic methods for detecting and evaluating hybridization. Finally, we consider the possible use of hybridization as a tool in conservation.

17.1 NATURAL HYBRIDIZATION

Consideration of the role of hybridization in systematics and evolution goes back to Linnaeus and Darwin (see discussion in Arnold 1997, p. 6). Botanical and zoological workers have tended to focus on the two opposing aspects of hybridization. Botanists have generally accepted hybridization as a pervasive and important aspect of evolution (e.g., Grant 1963, Stebbins 1959). They demonstrated that many plant taxa have hybrid origins and demonstrated that hybridization is an important mechanism for the production of new species and novel adaptations (Mallet 2007). In contrast, early evolutionary biologists working with animals were very interested in the evolution of reproductive isolation leading to speciation (Mayr 1942, Dobzhansky 1951). They emphasized that hybrid

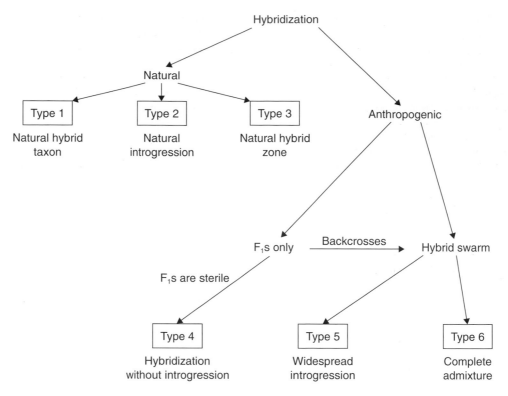

Figure 17.1 Framework to categorize hybridization. Each type should be viewed as a general descriptive classification used to facilitate discussion rather than a series of strict, all-encompassing divisions. Types 1–3 represent hybridization events that are a natural part of the evolutionary legacy of taxa; these taxa should be eligible for protection. Types 4–6 divide anthropogenic hybridization into three categories that have different consequences from a conservation perspective. From Allendorf *et al.* (2001).

offspring were often relatively unfit, and that this led to the development of reproductive isolation, and eventually speciation.

17.1.1 Intraspecific hybridization

Intraspecific hybridization in the form of gene flow among populations has several important effects. It has traditionally been seen as the cohesive force that holds species together as units of evolution (Mayr 1963). This view was challenged by Ehrlich and Raven (1969), who argued that the amount of gene flow observed in many species is too low to prevent differentiation thorough genetic drift or local adaptation.

The resolution to this conflict is the recognition that even very small amounts of gene flow can have a major cohesive effect. We saw in Chapter 9 that an average of one migrant individual per generation with the island model of migration is sufficient to make it likely that all alleles will be found in all populations. That is, populations may diverge quantitatively in allele frequencies, but qualitatively the same alleles will still be present. We saw in Chapters 9 and 12 that just one migrant per generation can greatly increase the local effective population size.

Rieseberg and Burke (2001) have presented a model of species integration that considers the effects of the spread of selectively advantageous alleles. They have shown that new mutations that have a selective advantage will spread across the range of a species much faster than selectively neutral mutations with low amounts of gene flow. They have proposed that it is the relatively rapid spread of highly advantageous alleles that holds a species together as an integrated unit of evolution.

High gene flow can reduce fitness and restrict the ability of populations to adapt to local conditions (migration load; see Box 13.1). Genetic swamping occurs when gene flow causes the loss of locally adapted alleles or genotypes (Lenormand 2002). This effect may be greatest in low density populations in which gene flow tends to be from densely populated areas (Garcia-Ramos and Kirkpatrick 1997). In such cases, the continued immigration of locally unfit genotypes reduces the mean fitness of a population and potentially could lead to what has been called a hybrid sink effect. This is a self-reinforcing process in which immigration produces hybrids that are unfit, which reduces local density and increases the immigration rate (Lenormand 2002).

Riechert *et al.* (2001) have provided evidence that gene flow in a desert spider has caused **genetic swamping** and the reduction in fitness of local populations. Riparian habitats favor spiders with a genetically determined nonaggressive phenotype in comparison with adjacent arid habitats, in which a competitive aggressive phenotype is favored. Nearly 10% of the matings of riparian spiders are with an arid-land partner. The resulting offspring have reduced survival in the riparian habitat compared with matings between riparian spiders. Modeling has shown that cessation of gene flow between spiders in different habitat types is expected to quickly result in the divergence in the frequency of aggressive and nonaggressive phenotypes in the two habitats (Figure 17.2).

17.1.2 Interspecific hybridization

Hybridization and introgression between species may occur more often than usually recognized (see Guest

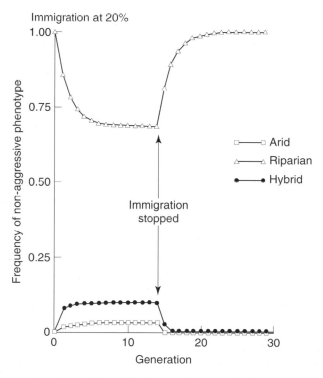

Figure 17.2 Predicted response to cessation of gene flow between desert spiders living in riparian and arid habitat patches, as measured by the frequency of the non-aggressive phenotype that has higher fitness in riparian but not arid environments. From Riechert *et al.* (2001).

Box 17). Whitney *et al.* (2010) reviewed the literature and concluded that interspecific hybridization is widespread among plants; on average, nearly 10% of all plant species have been described to produce hybrids. Interestingly, Mallet (2005) concluded that 6–10% of animal species hybridize. Bush (1994) defined speciation as a process of divergence of lineages that are sufficiently distinct from one another to follow independent evolutionary paths. Many independent lineages are capable of hybridizing and exchanging genes (introgression) for quite long periods without losing their phenotypic identities.

Interspecific hybridization can be an important source of genetic variation for some species. Grant and Grant (1998) studied two species of Galápagos finches on the volcanic island of Daphne Major for over 30 years. They found that hybridization between the two species (medium ground finch and cactus finch) has been an important source of genetic variation for the rarer cactus finch species; they have also suggested that their results may apply to many species. Interspecific introgression, or leakage between species, may cause a major shift in the way we think about species (Zimmer 2002).

Such introgression is especially important for island populations in which the effective population size is restricted because of isolation and the amount of available habitat. Two species of land snails (*Partula*) occur sympatrically on the island of Moorea in French Polynesia. In spite of being markedly different both phenotypically and ecologically, estimates of genetic distance based on molecular markers between some sympatric populations of these species are lower than is typical for conspecific comparisons for these taxa on different islands. Clarke *et al.* (1998) concluded that this apparent paradox was best explained by "molecular leakage, the convergence of neutral and mutually advantageous genes in two species through occasional hybridization".

Molecular leakage can be an important source of introgression for adaptive genes. For example, Song *et al.* (2011) found that German house mice carry a segment of DNA from Algerian mice which contains an allele that produces a blood clotting protein that provides resistance to warfarin, an anticoagulant used as a rodent poison. This allele is found at high frequency in German house mice where the two species overlap, apparently due to molecular leakage accompanied by strong selection for warfarin resistance. Interestingly, Abi-Rached *et al.* (2011) recently suggested that

modern humans acquired an MHC allele important for disease resistance from introgression with archaic humans called Denisovans, a likely sister group to the Neandertals.

Organelle DNA seems particularly prone to introgression and molecular leakage (Ballard and Whitlock 2004). There are many examples of cases where the mtDNA molecule of one species has completely replaced the mtDNA of another species in some populations without any evidence of nuclear introgression (mtDNA capture, Good *et al.* 2008). For example, the mtDNA in a population of brook trout in Lake Alain in Québec is identical to the Québec Arctic char genotype, yet the brook trout are morphologically indistinguishable from normal brook trout and have diagnostic brook trout alleles at nuclear loci (Bernatchez *et al.* 1995). A similar pattern of geographic haplotype-sharing occurs for chloroplast DNA in some congeneric plant species. For example, in Tasmania, 14 of 17 different species of Eucalypts studied were shown to share identical cpDNA haplotypes in the same geographic area (McKinnon *et al.* 2001).

Roca *et al.* (2005) have found that mtDNA genotypes can be misleading in describing the relationship between populations and species of elephants because of one-way introgression. Recurrent backcrossing of female hybrids between savannah and forest elephants to savannah elephant males results in populations that have the nuclear genome of savannah elephants with forest elephant mtDNA.

17.1.2.1 Hybrid zones

An interspecific hybrid zone is a region where two species are sympatric and hybridize to form at least partially fertile progeny. Hybrid zones usually result from secondary contact between species that have diverged in allopatry. Recent molecular analysis of plants and animals has revealed that hybrid zones occur widely in many taxa (Harrison 1993). Barton and Hewitt (1985) reviewed 170 reported hybrid zones and concluded that hybrids were selected against in most hybrid zones that have been studied. Nevertheless, some hybrid zones appear to be stable and persist over long periods of time through a balance between dispersal of parental types and selection against hybrids (Harrison 1993). Hybrid zones may act as selective filters that allow introgression of only selectively advantageous alleles between species (Martinsen *et al.* 2001).

Arnold (1997) proposed three models to explain the existence of a stable hybrid zone without genetic swamping of one or both of the parental species. In the Tension Zone Model, first- and second-generation hybrids are less fit than the parental types, but a balance between dispersal into the hybrid zone and selection against hybrids produces an equilibrium with a persistent, narrow hybrid zone containing F_1 individuals but few or no hybrids of F_2 or beyond. This model does not depend upon the ecological differences between habitats of the two parental types. In the Bounded Hybrid Superiority Model, hybrids are fitter than either parental species in environments that are intermediate to the parental habitats, but are less fit than the parental species in their respective native habitats (Example 17.1). The Mosaic Model is similar to the Bounded Hybrid Superiority Model, but the parental habitats are patchy rather than there being an environmental gradient between two spatially separated parental habitats. Under both models, theory predicts that hybridization and backcrossing would occur for many generations, creating introgressed populations containing individuals varying in their proportions of genetic material from the parental species.

17.1.3 Hybrid taxa

Approximately one-half of all plant species have been derived from **polyploid** ancestors, and many of these polyploid events involved hybridization between species or between populations within the same species (Stebbins 1950). Evidence suggests that all vertebrates went through an ancient polyploid event that might have involved hybridization (Lynch and Conery 2000). Other major vertebrate taxa have gone through additional polyploid events. For example, all salmonid fishes (trout, salmon, char, whitefish, and grayling) went through an ancestral polyploid event some 25–50 million years ago (Allendorf and Waples 1996).

Some hybrid taxa of vertebrates are unisexual. For example, unisexual hybrids between the northern redbelly dace and the finescale dace occur across the northern US (Angers and Schlosser 2007). Reproduction of such unisexual species is generally asexual or semisexual, and they are often regarded as evolutionary dead-ends. However, it appears that some tetraploid bisexual taxa had their origins in a unisexual hybrid (e.g., all salmonid fish).

Asexual and hybrid taxa provide some interesting challenges to conservation. For example, some species

---

Example 17.1   Genetic analysis of a hybrid zone between Sitka and white spruce

Sitka spruce is native to the mild, wet temperate rainforests of the Pacific Northwest. White spruce is abundant in North America's boreal forests, and is more drought and cold hardy than Sitka spruce. These species meet and hybridize extensively in the transition zone from the mild, wet maritime climate to the cold, drier continental climate in British Columbia and Alaska.

A hybrid index based upon genotypes of individuals at both neutral genetic markers (Bennuah et al. 2004) and SNPs in candidate genes (Hamilton et al. in press) was estimated to reflect the relative contribution of Sitka spruce and white spruce. The proportion of Sitka spruce ancestry is high in wet coastal areas, but drops rapidly where annual precipitation is below 1,000 mm farther inland (Hamilton et al. in press). Hybrid index is also correlated with the difference between mean summer and mean winter temperatures. All individuals sampled from hybrid populations are later-generation hybrids (beyond F_1 and F_2), and this hybrid zone is likely at least several thousand years old.

This hybrid zone fits expectations of the Bounded Hybrid Superiority model (Arnold 1997). Hybrid superiority in transitional, but not parental habitats, may result from combining the higher drought and cold hardiness of white spruce with the higher growth potential of Sitka spruce. Hybrids show some evidence for transgressive segregation of cold hardiness phenotypes, with hybrid individuals withstanding slightly colder temperatures than either parental species under some testing conditions (Hamilton et al. in press). Given the relatively steep clines observed, local adaptation should be managed through limiting seed transfer for reforestation along temperature and moisture gradients within the hybrid zone. The high levels of genetic diversity in these hybrid populations may facilitate rapid adaptation to climate change (Aitken et al. 2008).

of corals appear to be long-lived first-generation hybrids that primarily reproduce asexually (Vollmer and Palumbi 2002). However, even rare back-crossing and introgression between hybrid corals and their parental species can blur species boundaries (Miller and van Oppen 2003, Vollmer and Palumbi 2007). The morphology of intraspecific corals can be amazingly variable, and polyphyly of morphologically defined coral species appears to be common (Forsman *et al.* 2010, Pinzón and LaJeunesse 2010). Understanding reproductive boundaries is essential for setting conservation priorities for corals.

17.1.4 Transgressive segregation

Hybridization sometimes produces phenotypes that are extreme or outside the range of either parental type. This has been called **transgressive segregation**. Rieseberg *et al.* (2003) have shown that sunflower species found in extreme habitats tend to be ancient interspecific hybrids. They argue that new genotypic combinations resulting from hybridization have led to the ecological divergence and success of these species. A review of many hybrid species concluded that transgressive phenotypes are generally common in plant populations of hybrid origin (Rieseberg *et al.* 1999).

17.2 ANTHROPOGENIC HYBRIDIZATION

The increasing pace of introductions of plants and animals and habitat modifications has caused increased rates of hybridization among plant and animal species. The introduction of plants and animals outside of their native range clearly provides the opportunity for hybridization among taxa that were reproductively isolated. However, it is sometimes not appreciated just how much habitat modifications have increased rates of hybridization.

In many cases, it is difficult to determine whether hybridization is 'natural' or the direct or indirect result of human activities. In some cases, authors have referred to hybridization events resulting from habitat modifications as natural since they do not involve the introduction of species outside of their native range. Decline in abundance itself because of anthropogenic changes also promotes hybridization among species because of the greater difficulty in finding mates. In both of these cases, however, we believe that hybridization should be considered the indirect result of human activities.

Wiegand (1935) was perhaps the first to suggest that introgressive hybridization is observed most frequently in habitats disturbed by humans. The creation of extensive areas of modified habitats around the world has the effect of breaking down mechanisms of isolation between species (Rhymer and Simberloff 1996). For example, two native *Banksia* species in western Australia hybridize only in disturbed habitats where more vigorous growth has extended the flowering seasons of both species and removed asynchronous flowering as a major barrier to hybridization (Lamont *et al.* 2003). In addition, taxa that can adapt quickly to new habitats may undergo adaptive genetic change very quickly. It now appears that many of the most problematic invasive plant species have resulted from hybridization events (Ellstrand and Schierenbeck 2000, Gaskin and Schaal 2002, Blair and Hufbauer 2010). This topic is considered in more detail in Chapter 20.

Increased turbidity in aquatic systems because of deforestation, agricultural practices, and other habitat modifications has increased hybridization among aquatic species that use visual clues to reinforce reproductive isolation (Wood 2003). This has threatened sympatric species on the western coast of Canada (Kraak *et al.* 2001) and cichlid fish species in Lake Victoria (Seehausen *et al.* 1997). It is estimated that nearly half of the hundreds of species in Lake Victoria have gone extinct in the last 50 years, primarily because of the introduction of the Nile perch in the 1950s (Goldman 2003). The waters of this lake have grown steadily murkier, in part due to algal blooms resulting from the decline of cichlids. Mating between species now appears to be widespread and the loss of this classic example of adaptive radiation is now threatened (Goldman 2003).

Many other forms of habitat modification can lead to hybridization (Rhymer and Simberloff 1996, Seehausen *et al.* 2008). For example, the modification of patterns of water flow may bring species into contact that have been previously geographically isolated. It is likely that hybridization will continue to be more and more of a problem in conservation. Global environmental change may further increase the rate of hybridization between species in cases where it allows geographic range expansion. Kelly *et al.* (2010) have suggested that the melting of polar ice could cause increased frequency of hybridization among polar species. Seehausen (2006) suggested that loss of environmental heterogeneity causes a loss of biodiversity through increased hybridization (i.e., **reverse speciation)**.

Hybridization can contribute to the decline and eventual extinction of species in two general ways. In the case of sterile or partially sterile hybrids, hybridization results in loss of reproductive potential and may reduce the population growth rate below that needed for replacement (demographic swamping). In the case of fertile hybrids, genetically distinct populations may be lost through genetic mixing.

17.2.1 Hybridization without introgression

Many interspecific hybrids are sterile so that introgression (i.e., gene flow between populations whose individuals hybridize) does not occur. For example, matings between horses and donkeys produce mules which are sterile because of chromosomal pairing problems during meiosis. Sterile hybrids are evolutionary deadends. Nevertheless, the production of these hybrids reduces the reproductive potential of populations and can contribute to the extinction of species.

Hybridization represents a demographic threat to native species even without the occurrence of genetic admixture through introgression (type 4, Figure 17.1). In this case, hybridization is not a threat through genetic mixing, but wasted reproductive effort could pose a demographic risk. For example, females of the European mink hybridize with males from the introduced North American mink. Embryos are aborted so that hybrid individuals are not detected, but wastage of eggs through hybridization has accelerated the decline of the European species (Rozhnov 1993).

The presence of primarily F_1 hybrids should not jeopardize protection of populations affected by type 4 hybridization. However, care should be taken to determine conditions that favor the native species to protect and improve its status and reduce the wasted reproductive effort of hybridization. In extreme (and expensive) cases, it may be possible to selectively remove all of the hybrids and the non-native species to recover a native population.

Bull trout in Crater Lake National Park, Oregon, occur in a single stream, Sun Creek (Buktenica 1997). Introduced brook trout far outnumbered bull trout in this stream in the early 1990s and threatened to completely replace the native bull trout. The US National Park Service undertook a plan to recover this population by removing the brook trout. Brook trout were identified by visual observation and removed. Putative bull trout and hybrids were sampled by fin clips for genetic analysis and held until genetic testing revealed the identity of each individual (Spruell *et al.* 2001). Bull trout were then held in a small fishless stream and a hatchery. Sun Creek was then chemically treated to remove all fish from the stream. After treatment, the bull trout were placed back into Sun Creek. This population is currently increasing in abundance and distribution (Buktenica *et al.* in press).

17.2.2 Hybridization with introgression

In many cases, hybrids are fertile and may displace one or both parental taxa through the production of **hybrid swarms** (populations in which all individuals are hybrids by varying numbers of generations of backcrossing with parental types and mating among hybrids). This phenomenon has been referred to by many names (genetic assimilation, genetic extinction, or genomic extinction). The term 'genetic assimilation', which has been used in the literature (e.g., Cade 1983), should not be used in order to avoid confusion. Waddington (1961) used this phrase to mean a process in which phenotypically plastic characters that were originally 'acquired' become converted into inherited characters by natural selection (Pigliucci and Murren 2003).

Genomic extinction is a more appropriate term than the phrase genetic extinction; it is not genes or single locus genotypes that that are lost by hybridization; it is combinations of genotypes over the entire genome that are irretrievably lost. Genomic extinction results in the loss of the legacy of an evolutionary lineage; that is, the genome-wide combination of alleles and genotypes that have evolved over evolutionary time will be lost by genetic swamping through introgression with another lineage.

Perhaps surprisingly, introgression and admixture can spread even if hybridized individuals have reduced fitness. Population models (Epifanio and Philipp 2001) indicate that introgression may spread even when hybrids have severely reduced fitness (e.g., just 10% that of the parental taxa). This occurs because the production of hybrids is unidirectional, a sort of **genomic ratchet** (the Epifanio-Philipp effect). That is, all of the progeny of a hybrid will be hybrids. Thus, the frequency of hybrids within a local population may increase even when up to 90% of the hybrid progeny do not survive. The increase in the proportion of hybridized individuals in the population can occur even when the proportion of admixture in the population (i.e., the proportion of alleles in a hybrid swarm that come from each of the hybridizing taxa) is constant.

Hybridization can also spread rapidly when hybrids have reduced reproductive success, if hybrid individuals are more likely to disperse than non-hybrids. For example, Boyer *et al.* (2008) found that hybrids between native cutthroat trout and introduced rainbow trout were more likely to disperse than native cutthroat trout. Shine *et al.* (2011) have referred to this process as "**spatial sorting**".

17.2.3 Hybridization between wild species and their domesticated relatives

Hybridization between wild species and their domesticated relatives can be especially problematic because it can be difficult to detect the hybrids. Ellstrand *et al.* (1999) found that 12 of the world's 13 most valuable food crops hybridize with their wild relatives somewhere in the world. Such hybridization can be harmful because of reduced fitness of the wild population, as well as genetic swamping. Hybridization with domesticated species has increased risk of extinction in several wild species, including wild relatives of two of the world's 13 most important crops (rice and cotton seed). Such hybridization is also a concern because introgression between a crop and a related weedy taxon can produce a more aggressive weed (Ellstrand *et al.* 2010). Gene flow from a crop to a wild relative has been implicated in enhanced weediness in wild relatives of 7 of the world's 13 most important crops (Ellstrand *et al.* 1999).

Hybridization between wild animal species and their domesticated relatives is also a conservation concern (e.g., Brisbin and Peterson 2007). Randi (2008) has reviewed the effects of such introgression in European populations of wolves, wildcats, rock partridges, and red-legged partridges. He has concluded that introgressive hybridization is sometimes very common. He has recommended a number of steps to preserve the integrity of the gene pools of wild populations, including assessing the extent of hybridization, placing high conservation priority on non-hybridized populations, and

enforcing strict controls on the genetics of game species raised in captivity and released for restocking.

Hybridization between wild and hatchery populations of fish is a major conservation problem for many species (Araki and Schmid 2010). For example, up to two million Atlantic salmon are estimated to escape from salmon farms each year. Hybrids (F_1, F_2, and backcrosses) between farm and wild fish all show reduced survival compared with wild salmon (McGinnity *et al.* 2003). Nevertheless, farm and hybrid salmon show faster growth rates as juveniles and therefore may displace juvenile wild salmon. The repeated escapes of farmed salmon present a substantial threat to remaining wild populations of Atlantic wild salmon through accumulation of fitness depression by introgression. This issue is considered in more detail in Guest Box 4 and Chapter 19.

17.3 FITNESS CONSEQUENCES OF HYBRIDIZATION

Hybridization may have a wide variety of effects on fitness (Arnold and Martin 2010). In the case of **heterosis**, or **hybrid vigor**, hybrids have enhanced performance or fitness relative to either parental taxa. In the case of **outbreeding depression**, the hybrid progeny have lower performance or fitness than either parent (Lynch and Walsh 1998). Both heterosis and outbreeding depression have many possible causes, and the overall fitness of hybrids results from an interaction among these different effects. To further complicate matters, much of the heterosis that is often detected in F_1 hybrids is lost in subsequent generations, so that a particular cross may result in heterosis in the first generation and outbreeding depression in subsequent generations (Figure 17.3).

There are two primary mechanisms that may reduce the fitness of hybrids. The first mechanism is genetic incompatibilities between the hybridizing taxa; this has been referred to as both intrinsic outbreeding depression and endogenous selection. Outbreeding depression may also result from reduced adaptation to environmental conditions by hybrids; this has been referred to as extrinsic outbreeding depression and also as exogenous selection. With endogenous selection, fitness effects are independent of environments, while with exogenous selection, hybrids may have lower fitness than parental types in some environments, but higher fitness than parental types in other environments.

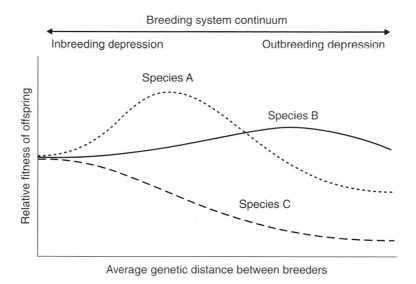

Figure 17.3 Heuristic model for visualizing the balance between inbreeding depression, hybrid vigor (heterosis), and outbreeding depression. Individual species exhibit different optimal levels of outcrossing, as illustrated by the plot of fitness relative to average genetic distance among breeders. For example, species A shows considerable inbreeding depression and also outbreeding depression. Species B exhibits little inbreeding depression and hybrid vigor. Redrawn from Waples (1995).

17.3.1 Hybrid superiority

Heterosis occurs when hybrid progeny have higher fitness than either of the parental types. In many regards, heterosis is the opposite of inbreeding depression. Therefore, the underlying causes of heterosis are the same as the causes of inbreeding depression: increased homozygosity and reduced heterozygosity (Crow 1993). The primary cause of heterosis is the sheltering of deleterious recessive alleles in hybrids. In addition, increased heterozygosity will increase the fitness of hybrid individuals for loci at where the heterozygotes have a selective advantage over homozygous genotypes. Heterosis is greatest in the F_1 hybrids. Heterosis will be diminished in subsequent generations when heterozygosity is reduced and homozygosity is increased because of Mendelian segregation.

The best example of heterosis has come from crossing inbred lines of corn to produce high-yielding hybrid corn. Virtually all agricultural corn grown today in the US is hybrid, compared with less than 1% of the corn planted in 1933 (Sprague 1978). A large number of self-fertilized lines of corn have been established from highly polymorphic populations. The yield of each inbred line decreases as homozygosity increases. Many lines are discontinued because their performance is so low. Inbred lines are expected to become homozygous for different combinations of deleterious recessive alleles. These deleterious recessive alleles will be sheltered in hybrids by heterozygosity. In addition, many alleles resulting in increased yield are dominant. The hybrids between two inbred lines are superior to both the inbred lines as well as corn from the original highly polymorphic populations. Many combinations of inbred lines are tested so that the combinations that produce the most desirable hybrids can be used.

Subdivision of natural populations (see Chapter 9) can provide the appropriate conditions for heterosis. Different deleterious recessive alleles will drift to relatively high frequencies in different populations. Therefore, progeny produced by matings between immigrant individuals are expected to have greater fitness than resident individuals. This effect is expected to result in a higher effective migration rate because immigrant alleles will be present at much higher frequencies than predicted by neutral expectations (Whitlock *et al.* 2000, Ingvarsson and Whitlock 2000, Morgan 2002).

Experiments involving immigration into inbred, laboratory populations of African satyrine butterflies have revealed surprisingly strong heterosis (Saccheri and Brakefield 2002). Immigrants were, on average, over 20 times more successful in contributing descendants to the fourth generation than were inbred non-immigrants. The mechanism underlying this rapid spread of immigrant alleles was found to be heterosis. The disproportionately large impact of some immigrants suggests that rare immigration events may be very important in evolution, and that heterosis may drive their fitness contribution.

Hybrids can have a more lasting fitness advantage because they possess advantageous traits from both parental populations (see Example 17.1). Lewontin and Birch (1966) suggested many years ago that hybridization can provide new variation that allows adaptation to new environments. This may be an important mechanism for adaptation to rapidly changing environments, e.g., anthropogenic climate change. As we saw in Section 17.1.4, Rieseberg *et al.* (2003) found that hybridization between sunflower species produced progeny that are adapted to environments very different from those occupied by the parental species. This was associated with the hybrids possessing new combinations of genetic traits. Choler *et al.* (2004) found hybrids between two subspecies of an alpine sedge that occurred only in marginal habitats for the two parental subspecies.

17.3.2 Intrinsic outbreeding depression

Intrinsic outbreeding depression results from genetic incompatibilities between hybridizing taxa.

17.3.2.1 Chromosomal

Reduced fitness of hybrids can result from heterozygosity for chromosomal differences between populations or species (see Chapter 3). Differences in chromosomal number or structure may result in the production of aneuploid gametes that result in reduced survival of progeny. We saw in Table 3.4 that hybrids between races of house mice with different chromosomal arrangements produced smaller litters in captivity. Hybrids between chromosomal races of the threatened owl monkey from South America show reduced fertility in captivity (De Boer 1982).

17.3.2.2 Genic

Reduced fitness of hybrids can also result from genetic interactions between genes originating in different taxa (Whitlock *et al.* 1995). Dobzhansky (1948) first used the word "coadaptation" to describe reduced fitness in hybrids between different geographic populations of the fruit fly *Drosophila pseudoobscura*. This term became controversial (and somewhat meaningless) following Mayr's (1963) argument that most genes in a species are coadapted because of the integrated functioning of an individual.

Reduced fitness of hybrids can potentially occur because of the effects of genotypes at individual loci. Perhaps the best example is that of the direction of shell coiling in snails (Johnson 1982). Shells of some species of snails coil either to the left (sinistral) or to the right (dextral). Variation in shell coiling direction occurs within populations of snails of the genus *Partula* on islands in the Pacific Ocean. Many species in this genus are now threatened with extinction because of the introduction of other snails (Mace *et al.* 1998). The variation in shell coiling in many snail species is caused by two alleles at a single locus (Sturtevant 1923, Johnson 1982). Snails that coil in different directions find mating difficult or impossible. Thus, the most common phenotype (sinistral or dextral) in a population will generally be favored, leading to the fixation of one type or the other. Hybrids between sinistral and dextral coiling populations may have reduced fitness because of the difficulty of mating with snails of the other type (Johnson *et al.* 1990).

Outbreeding depression may result from genic interactions between alleles at multiple loci (**epistasis**; Whitlock *et al.* 1995). That is, alleles that enhance fitness within their parental genetic backgrounds may reduce fitness in the novel genetic background produced by hybridization. Such interactions between alleles are known as Dobzhansky-Muller incompatibilities, because they were first described by these two famous Drosophila geneticists (Johnson 2000). Dobzhansky referred to these interactions as coadapted gene complexes. Such interactions are thought to be responsible for the evolution of reproductive isolation and eventually speciation.

There are few empirical examples of specific genes that show such Dobzhansky-Muller incompatibilities. Rawson and Burton (2002) have presented an elegant example of functional interactions between loci that code for proteins involved in the electron transport system of mitochondria in an intertidal copepod *Tigriopus californicus*. A nuclear gene encodes the enzyme cytochrome *c* (CYC), while two mtDNA genes encode subunits of cytochrome oxidase (COX). CYC proteins isolated from different geographic populations each had significantly higher activity in combination with the COX proteins from their own source population. These results demonstrate that proteins in the electron transport system form coadapted combinations of alleles, and that disruption of these coadapted gene complexes leads to functional incompatibilities that may lower the fitness of hybrids.

Self-fertilization in plants has long been recognized as potentially facilitating the evolution of adaptive combinations of alleles at many loci. Many populations of primarily self-fertilizing plants are dominated by a few genetically divergent genotypes that differ at multiple loci. Parker (1992) has shown that hybrid progeny between genotypes of the highly self-fertilizing hog peanut have reduced fitness (Figure 17.4). These genotypes naturally co-occur in the same habitats and the hybrid progeny have reduced fitness in a common garden. Thus, the reduced fitness of hybrids apparently results from Dobzhansky-Muller incompatibilities between genotypes.

17.3.3 Extrinsic outbreeding depression

Extrinsic outbreeding depression results from the reduced fitness of hybrids because of loss of adaptation by ecologically mediated selection.

The escape gaits of hybrids between the closely related whitetail and mule deer provide an interesting example of outbreeding depression (Lingle 1992). Whitetail deer gallop to escape rapidly from predation. In contrast, mule deer 'stott' by using long high bounds to escape predators. F_1 and other generation hybrids, between these species have an intermediate gait that is not as effective as either of the gaits of the parental species in making a quick escape.

Increased susceptibility to diseases and parasites is an important potential source of outbreeding depression because of the importance of disease in conservation and the complexity of immune systems and their associated gene complexes. Sage *et al.* (1986) studied a hybrid zone between two species of mice in Europe (*Mus musculus* and *M. domesticus*). Hybrids had significantly greater loads of pinworm (nematodes) and tapeworm (cestodes) parasites than either of the parental taxa (Figure 17.5). A total of 93 mice were examined within the hybrid zone. Fifteen of these mice had exceptionally high numbers of nematodes (>500) while 78 mice had 'normal' numbers of nematodes (<250). Fourteen of the 15 mice with high nematode loads were hybrids, while 37 of the 78 mice with normal loads were hybrids (P < 0.005). Cestode infections showed a similar pattern in hybrid and parental mice.

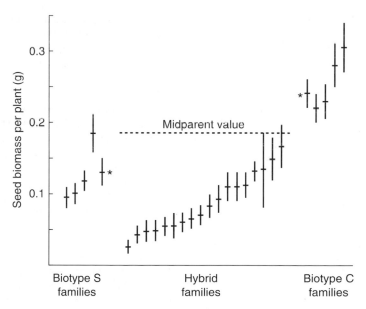

Figure 17.4   Fitness as measured by lifetime seed biomass of parental genotypes (biotypes S and C) and their hybrids in the highly self-fertilizing hog peanut. The two parental families marked with asterisks are the parents of the hybrids. From Parker (1992).

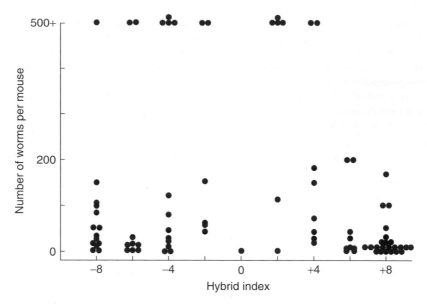

Figure 17.5 Nematode burdens (number of worms per mouse) in hybrid and parental mice. The hybrid index is based upon four diagnostic allozyme loci. Pure *M. musculus* has an index of −8 and pure *M. domesticus* has an index of +8. From Sage *et al.* (1986).

Currens *et al.* (1997) found that hybridization with introduced hatchery rainbow trout native to a different geographic region increased the susceptibility of wild native rainbow trout to myxosporean parasites. Similarly, Goldberg *et al.* (2005) found that hybrid largemouth bass from two genetically distinct subpopulations were more susceptible to largemouth bass virus. Parris (2004) found that hybrid frogs show increased susceptibility to emergent pathogens compared with the parental species.

17.3.4 Long-term fitness effects of hybridization

There is evidence that hybridization can increase fitness in the long term (tens of generations), even in cases where there is substantial outbreeding depression (Templeton 1986, Carney *et al.* 2000). For example, Hwang *et al.* (2011) found substantial reduction in fitness for many replicates of interpopulation hybrids of the marine copepod *Tigriopus californicus* in the first several generations. However, two of four long-term replicates showed equal fitness to parentals and two showed greater fitness than the parentals. Thus, in

some cases the increase in genetic variation from hybridization can lead to greater fitness in the long term, even when there is substantial outbreeding depression. In rapidly changing environments, the increased genetic variation from hybridization might facilitate long-term adaptation. However, in some situations the reduction in fitness can persist for long periods of time (Johnson *et al.* 2010).

17.4 DETECTING AND DESCRIBING HYBRIDIZATION

The detection of hybrid individuals relied upon morphological characteristics until the mid-1960s. However, not all morphological variation has a genetic basis, and the amount of morphological variation within and among populations is often greater than recognized (Campton 1987). The detection of hybrids using morphological characters generally assumes that hybrid individuals will be phenotypically intermediate to parental individuals (Smith 1992). This is often not the case, because hybrids sometimes express a mosaic of parental phenotypes. Furthermore, individuals from hybrid swarms that contain most of their genes from one of the parental taxa are often morphologically indistinguishable from that parental taxon (Leary *et al.* 1996, Brisbin and Peterson 2007). Morphological characters do not allow one to determine whether an individual is a first-generation hybrid (F_1), a backcross, or a later generation hybrid. These distinctions are crucial because if a population has not become a hybrid swarm and still contains a reasonable number of parental individuals, it could potentially be recovered by removal of hybrids or by a captive breeding program.

Genetic analysis of hybrids and hybridization is based upon loci at which the parental taxa have different allele frequencies. **Diagnostic loci** that are fixed or nearly fixed for different alleles in two hybridizing populations are the most useful, although hybridization can also be detected using multiple loci at which the parental types differ in allele frequency (Example 17.2, Cornuet *et al.* 1999).

The use of molecular genetic markers greatly simplifies the identification and description of hybridized populations. This procedure began with the development of protein electrophoresis (allozymes) in the mid-1960s. Recent advances in molecular techniques, especially the development of the polymerase chain reaction (PCR), have greatly increased the number of loci that can be used to detect hybridization.

Figure 17.7 outlines the use of diagnostic loci to analyze hybridization. First-generation (F_1) hybrids

---

Example 17.2   Hybridization between the threatened Canada lynx and bobcat

The Canada lynx is a wide-ranging felid that occurs in the boreal forest of Canada and Alaska (Schwartz *et al.* 2004). The southern distribution of native lynx extends into the northern contiguous US from Maine to Washington State. Lynx are also located in Colorado where a population was introduced in 1999. The Canada lynx is listed as Threatened under the US ESA. Canada lynx are elusive animals and their presence has routinely been detected by genetic analysis of mtDNA from hair and fecal samples (Mills *et al.* 2001). Samples of hair and feces confirmed that Canada lynx were present in northern Minnesota, after a ten-year suspected absence from the state. In 2001 a trapper was prosecuted for trapping a lynx. The trapper thought it was a bobcat, while the biologist registering the pelt and the enforcement officer processing the case thought it was a lynx. Initial analysis based on mitochondrial DNA showed the sample was a lynx. However, recognizing that mitochondrial DNA could only determine the female parent of the cat, Schwartz *et al.* (2004) designed an assay that could detect hybridization between bobcats and lynx. Hybridization between these species had never been confirmed in the wild.

The controversial sample was a hybrid. In addition, one of the other samples from a carcass and one hair sample collected on a putative lynx backtrack were also identified as hybrids using microsatellite analysis. The hybrids were identified as having one lynx-diagnostic allele and one bobcat-diagnostic allele (Figure 17.6). A heterozygote with one allele from each parental species is expected in a F_1 hybrid (although some F_2 hybrids will also be heterozygous for species-diagnostic alleles at some loci). The species-diagnostic alleles were identified (at two loci) by analyzing microsatellites in 108 lynx and 79 bobcats across North America, far away from potential hybridization zones between the two species. In addition, mitochondrial DNA analysis revealed that all hybrids had lynx mothers (i.e., lynx mtDNA). Therefore all three hybrid samples were produced by matings between female Canada lynx and male bobcats. After these results were published, researchers from Maine and New Brunswick requested that some of their study animals be screened using the hybrid test. Four additional hybrids were discovered – two in Maine and two in New Brunswick. However, in a screening of hundreds of lynx samples from the Rocky Mountains (another area where the two species co-occur), no hybrids were discovered.

These data have important conservation implications. First, bobcat trapping is legal, while it is illegal to trap lynx anywhere in the coterminous US. The trapping of bobcats in areas where Canada lynx are present could be problematic because both lynx and lynx–bobcat hybrids can be incidentally taken from extant populations. On the other hand, any factors that may favor bobcats in lynx habitat may lead to the production of hybrids and thus be potentially harmful to lynx recovery. Efforts need to be undertaken to describe the extent, rate, and nature of hybridization between these species, and to understand the ecological context in which hybridization occurs.

(Continued)

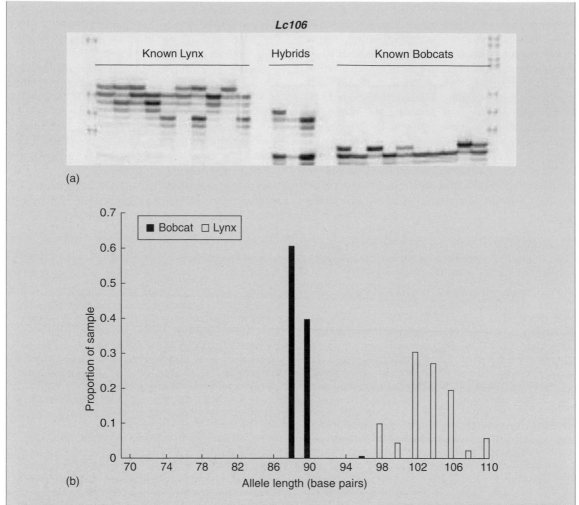

Figure 17.6   (a) Microsatellite gel image showing genotype profiles (locus *Lc106*) for 10 lynx, 10 bobcat, and three putative hybrids. Dark bands represent alleles; lighter bands are 'stutter' bands. Outer lanes show size standards. (b) Allele frequencies for locus *Lc106* in bobcat and lynx. This locus is diagnostic because the allele size ranges do not overlap between species. From Schwartz *et al.* (2004).

will be heterozygous for alleles from the parental taxa at all diagnostic loci. Later generation hybrids may result either from matings between hybrids or backcrosses between hybrids and one of the parental taxa (Example 17.3). The absence of such genotypes resulting in later generation hybrids suggests that the F_1 hybrids are sterile or have reduced fertility (Example 17.4). These two examples with different pairs of trout species demonstrate the contrasting results depending

upon whether or not the F_1 hybrids are fertile (Example 17.3) or sterile (Example 17.4).

17.4.1 Multiple loci and gametic disequilibrium

The distribution of gametic disequilibria (*D*) between pairs of loci is helpful to describe the distribution of

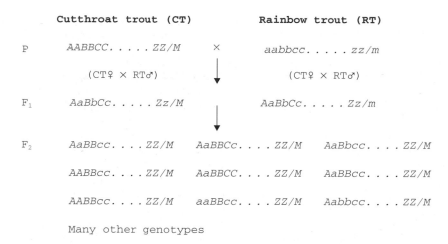

Figure 17.7 Outline of genetic analysis of hybridization between two species. Alleles present in one species at diagnostic nuclear loci are designated by capital letters, and the alleles in the other species by lower case letters. The parental (P) mtDNA haplotypes are designated by M and m, respectively.

Example 17.3 Hybrid swarms of cutthroat trout and rainbow trout

The loss of native cutthroat trout by hybridization with introduced rainbow trout has been recognized as a major threat for over 75 years in the western US (Allendorf and Leary 1988). The westslope cutthroat trout is one of four major subspecies of cutthroat trout (Allendorf and Leary 1988). The geographic range of westslope cutthroat trout is the largest of all subspecies and includes the Columbia, Fraser, Missouri, and Hudson Bay drainages of the US and Canada. The westslope cutthroat trout is genetically highly divergent at both nuclear and mitochondrial genes from the three other major subspecies: coastal, Yellowstone, and Lahontan cutthroat trout. For example, 10 out of 46 nuclear allozyme loci are fixed or nearly fixed for different alleles between westslope and Yellowstone cutthroat trout. This amount of divergence is far beyond that usually seen within a single species.

Introgressive hybridization with introduced rainbow trout and Yellowstone cutthroat trout occurs throughout the range of the westslope cutthroat trout. Hybridization of cutthroat and rainbow trout generally results in the formation of random mating populations in which all individuals are hybrids by varying numbers of generations of backcrossing with parental types and mating among hybrids (i.e., hybrid swarms).

Table 17.1 shows genotypes at eight diagnostic nuclear loci between native westslope cutthroat trout and Yellowstone cutthroat trout introduced into Forest Lake, Montana, in a representative sample of 15 individuals. All but one of these 15 fish are homozygous for both westslope and Yellowstone alleles at different loci. Each individual in this sample appears to be a later generation hybrid (see Figure 17.7). Thus, the fish in this lake are a hybrid swarm.

Muhlfeld et al. (2009) used parentage analysis to estimate the reproductive success (fitness) of individuals in a hybrid swarm between non-native rainbow and native westslope cutthroat trout. They found that small amounts of admixture markedly reduced fitness, causing reproductive success to decline by approximately 50%, with only 20% individual admixture for both females and males (Figure 17.8). The exception to this observation was first-generation (F_1) hybrids that showed much greater fitness than predicted on the basis of their proportion of admixture (Figure 17.8). This suggests that the sheltering of deleterious recessive alleles overcame the apparent extrinsic outbreeding depression in this population (see Section 17.3.1).

(Continued)

Table 17.1 Genotypes at eight diagnostic nuclear allozyme loci in a sample of westslope cutthroat trout, Yellowstone cutthroat trout, and their hybrids from Forest Lake, Montana (Allendorf and Leary 1988). Heterozygotes are *WY* while individuals homozygous for the westslope cutthroat trout allele are indicated as *W* and individuals homozygous for the Yellowstone cutthroat trout allele are indicated as *Y*. All individuals in this sample are later generation hybrids; thus, the fish in this lake are a hybrid swarm.

		Nuclear encoded loci							
No.	mtDNA	*Aat1*	*Gpi3*	*Idh1*	*Lgg*	*Me1*	*Me3*	*Me4*	*Sdh*
1	YS	W	W	WY	W	W	W	W	Y
2	YS	W	WY	WY	WY	Y	W	WY	Y
3	WS	WY	Y	Y	W	Y	WY	Y	WY
4	WS	Y	W	WY	WY	W	Y	W	WY
5	YS	Y	Y	Y	WY	WY	WY	Y	Y
6	YS	WY	Y	W	WY	W	W	W	Y
7	WS	WY	WY	Y	W	WY	W	W	W
8	WS	WY	Y	WY	WY	Y	W	Y	Y
9	WS	Y	Y	WY	WY	W	WY	WY	W
10	WS	WY	Y	WY	WY	WY	Y	W	Y
11	YS	Y	W	W	WY	W	Y	W	Y
12	WS	W	WY	Y	WY	W	WY	WY	Y
13	YS	W	Y	W	Y	W	WY	W	W
14	YS	Y	Y	WY	WY	WY	WY	WY	W
15	WS	WY	Y	WY	Y	W	Y	WY	W

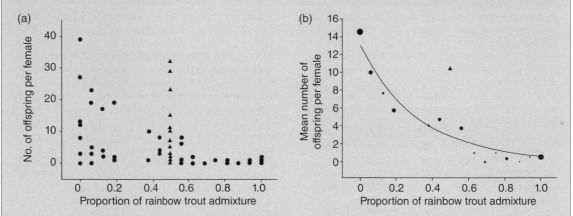

Figure 17.8 Effect of hybridization on fitness of native cutthroat trout. (a) Number of offspring per female versus the proportion of nonnative rainbow trout admixture. The plot includes 61 mothers and 397 juvenile assignments. Circles represent westslope cutthroat trout, rainbow trout, and later generation hybrids, and triangles represent first-generation (F₁) hybrids. (b) Bubble plot of the mean number of offspring per female plotted against the proportion of rainbow trout admixture. The mean value for first-generation hybrids is shown as a triangle, but these points were not included in the regression curve fitted through the points. From Muhlfeld *et al.* (2009).

Example 17.4 Near-sterility in hybrids between bull and brook trout

Bull trout are legally protected as threatened in the US under the Endangered Species Act. Hybridization with introduced brook trout is potentially one of the major threats to the persistence of bull trout. Bull trout and brook trout have no overlap in their natural distribution, but secondary contact between these species has occurred as a result of the introduction of brook trout into the bull trout's native range.

Leary *et al.* (1993b) described a rapid and almost complete displacement of bull trout by brook trout in which the initial phases were characterized by frequent hybridization. In the South Fork of Lolo Creek in the Bitterroot River drainage, Montana, brook trout first invaded in the late 1970s. In the initial sample collected in 1982, bull trout (44%) were the most abundant, followed by hybrids (36%) and brook trout (20%), and matings appeared to be occurring at random. By 1990, however, brook trout (65%) were more abundant than bull trout (24%) and hybrids (12%).

Table 17.2 shows genotypes at eight diagnostic nuclear loci between native bull trout and brook trout introduced into Mission Creek, Montana, in a sample of 15 individuals that were selected to be genetically analyzed because they appeared to be hybrids. Eleven of the 15 fish in this sample contained alleles from both species indicating that they are indeed hybrids. However, in striking contrast to Example 17.3, 10 of the 11 hybrids were heterozygous at all eight loci, suggesting that they are F_1 hybrids. It is extremely unlikely that a later generation hybrid would be heterozygous at all loci. For example, there is a 0.50 probability that an F_2 hybrid will be heterozygous at a diagnostic locus. Thus, there is a probability of $(0.50)^8 = 0.004$ that an F_2 will be heterozygous at all eight loci. The F_1 hybrids in this sample have both bull and brook trout mtDNA, indicating that both reciprocal crosses are resulting in hybrids.

This general pattern has been seen throughout the range of bull trout. Almost all hybrids appear to be F_1s with very little evidence of F_2 or backcross individuals (Kanda *et al.* 2002). The near-absence of progeny from hybrids of bull and brook trout may result from the sterility of the hybrids, their lack of mating success, the poor survival of their progeny, or combinations of these factors. Over 90% of the F_1 hybrids are male, suggesting some genetic incompatibility between these two genomes.

Table 17.2 Genotypes at eight diagnostic nuclear loci in a sample of bull trout, brook trout, and their hybrids from Mission Creek, Montana (Kanda *et al.* 2002). Heterozygotes are *LR* while individuals homozygous for the bull trout allele are indicated as *L* and individuals homozygous for the brook trout allele are indicated as *R*. Individuals are identified in the Status column as parental types bull trout (BL), brook trout (BR), or hybrids on the basis of their genotype.

No.	mtDNA	Nuclear encoded loci								Status
		Aat1	*Ck-A1*	*IDDH*	*sIDHP-2*	*LDH-A1*	*LDH-B2*	*MDH-A2*	*sSod-1*	
1	L	LR	LR	LR	LR	LR	LR	LR	LR	F1
2	L	LR	LR	LR	LR	LR	LR	LR	LR	F1
3	L	R	R	LR	LR	LR	LR	LR	R	F1xBR
4	L	L	L	L	L	L	L	L	L	BL
5	L	LR	LR	LR	LR	LR	LR	LR	LR	F1
6	R	LR	LR	LR	LR	LR	LR	LR	LR	F1
7	L	LR	LR	LR	LR	LR	LR	LR	LR	F1
8	R	LR	LR	LR	LR	LR	LR	LR	LR	F1
9	R	LR	LR	LR	LR	LR	LR	LR	LR	F1
10	L	LR	LR	LR	LR	LR	LR	LR	LR	F1
11	R	LR	LR	LR	LR	LR	LR	LR	LR	F1
12	R	LR	LR	LR	LR	LR	LR	LR	LR	F1
13	R	R	R	R	R	R	R	R	R	BR
14	R	R	R	R	R	R	R	R	R	BR
15	R	R	R	R	R	R	R	R	R	BR

hybrid genotypes and to estimate the 'age' of hybridized populations (see Table 10.2). Recently hybridized populations will have high D because they will contain parental types and many F_1 hybrids. By contrast, genotypes will be randomly associated among loci in hybrid swarms that have existed for many generations. This will occur rather quickly for unlinked loci because D will decay by one-half in each generation (equation 10.5). However, nonrandom association of alleles at different loci will persist for many generations at pairs of loci that are closely linked. Barton (2000) has provided single measures of gametic disequilibrium that can provide a meaningful measure to compare the amount of gametic disequilibrium at a number of unlinked loci in hybrid swarms.

Genetic data must be interpreted at both the individual and population level to understand the history of hybridization in populations (Barton and Gale 1993). Hybrid individuals can be first-generation (F_1) hybrids, second-generation hybrids (F_2), backcrosses to one of the parental taxa, or later generation hybrids. Parental types and F_1 hybrids can be reliably identified if many loci are examined. However, it is very difficult to distinguish between F_2 hybrids, backcrosses, and later generation hybrids, even if many loci are examined (Boecklen and Howard 1997).

New statistical approaches for assigning individuals to their population of origin based upon many highly polymorphic loci are especially valuable for identifying hybrids (Hansen *et al.* 2000). These techniques may be useful even when putative 'pure' (parental) populations are not available to provide baseline information. Example 17.5 presents genetic analysis of a particularly difficult hybridization situation in which known parental types were not available to determine the genetic composition of the parental taxa.

17.5 HYBRIDIZATION AND CONSERVATION

Hybridization is a natural part of evolution. Taxa that have arisen through natural hybridization should be eligible for protection. Nevertheless, increased anthropogenic hybridization is causing the extinction of many taxa (species, subspecies, and locally adapted populations) by both replacement and genetic mixing. Conservation policies should be designed to reduce anthropogenic hybridization. Nevertheless, developing policies to deal with the complex issues associated with hybridization has been difficult (Allendorf *et al.* 2001, Ellstrand *et al.* 2010).

17.5.1 Intentional hybridization

Some populations of listed taxa are small or have gone through a recent bottleneck, and therefore they contain little genetic variation. In some cases, it might be advisable to increase genetic variation in these populations through intentional hybridization. Under what circumstances should genetic rescue (see Section 15.5) by purposeful hybridization be used as a tool in conservation?

In extreme cases, some taxa might only be recovered through the use of intentional hybridization. However, the very characteristics of the local populations that make them unusual or exceptionally valuable could be lost through this purposeful introgression. In addition, such introductions could cause the loss of local adaptations and lower the mean fitness of the target population. The most well-known example of this dilemma is the decision to bring in panthers from Texas to reduce the apparent effects of inbreeding depression in Florida panthers. Another example is the artificial hybridization of the threatened species American chestnut. This species has been decimated by the introduced disease chestnut blight. The American Chestnut Foundation has hybridized susceptible American chestnuts with disease-resistant Chinese chestnuts, and then repeatedly backcrossed resistant individuals with American chestnuts.

Intentional hybridization should be used only after careful consideration of potential harm. Intentional hybridization would be appropriate when the population has lost substantial genetic variation through genetic drift and the detrimental effects of inbreeding depression are apparent (e.g., reduced viability or an increased proportion of obviously deformed or asymmetric individuals). Populations from as similar an environment as possible (that is, the greatest ecological exchangeability) should be used as the donor population (Crandall *et al.* 2000). In these situations, even a small amount of introgression might sufficiently counteract the effects of reduced genetic variation and inbreeding depression without disrupting local adaptations (Ingvarsson 2001).

Hybridization is least likely to result in outbreeding depression when there is little genetic divergence

Example 17.5 Genetic mixture analysis of Scottish wildcats

The Scottish wildcat has had full legal protection since 1988 (Beaumont *et al.* 2001). The presence of feral domestic cats and the possibility of hybridization have, however, made this protection ineffective because it has been impossible to unambiguously distinguish wildcats from hybrids. The amount of hybridization between domestic cats and existing wildcats is unknown. Some believe that there has been little hybridization until recently. However, the behavioral similarity between cats in a study area containing morphologically domestic and morphologically wildcat individuals suggests that hybridization may have had a substantial impact on the genetic composition of wildcats in Scotland.

Beaumont *et al.* (2001) studied nine microsatellite loci in 230 wild-living Scottish cats (including 13 museum skins) and 74 house cats from England and Scotland. In addition, pelage characteristics of the wild-living cats were recorded (Figure 17.9). Hybridization between Scottish wildcats and domestic cats was tested for to identify hybrid populations to guide conservation management.

The genetic mixture analysis method of Pritchard *et al.* (2000) using a Markov Chain Monte Carlo approach (see Section A.5) was used without specifying the allele frequencies of the source populations to estimate q (the proportion an individual's genome that comes from the wildcat population). The distribution of q is integrated over all possible gene frequencies in the two parental populations weighted by their posterior density to obtain a posterior density for q that is independent of the parental frequencies.

This analysis suggested the presence of two primary groups of wild-living cats: wildcats and domestic cats (Figure 17.10). The proportion of wildcats was much greater than domestic cats. There were also

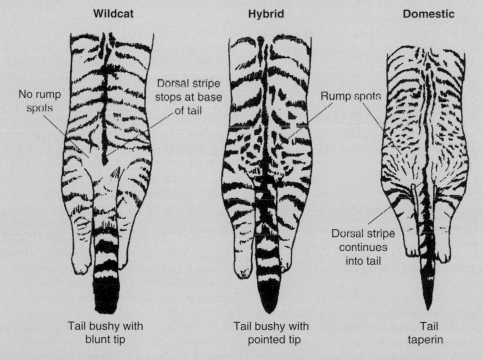

Figure 17.9 Diagram showing three (tail shape, dorsal stripe, and rump spots) of the five morphologic diagnostic characters used to distinguish wildcats and domestic cats. Tail tip color and paw color are not illustrated. From Beaumont *et al.* (2001).

(Continued)

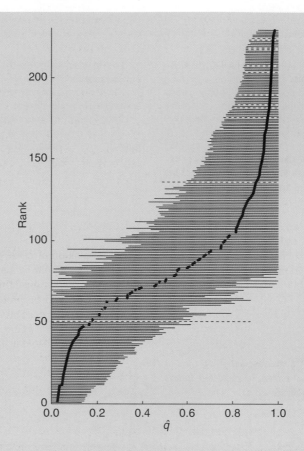

Figure 17.10 Genetic mixture analysis of hybridization between domestic cats and Scottish wildcats. $\hat{q}$ is the proportion of an individual's genome that comes from the wildcat population. This figure illustrates the ranked distribution of $\hat{q}$ among individuals. Also shown are lines giving the 95% equal-tail posterior probability intervals for each individual. The dotted lines are from museum specimens. Individuals with low $\hat{q}$ values are likely to be domestic cats, and individuals with high $\hat{q}$ values are likely to be wildcats. From Beaumont *et al.* (2001).

intermediate individuals that tended to have very wide probability limits for $\hat{q}$. There also was a strong correlation between the proportion of an individual's genome derived from wildcats and the morphology of individuals based upon the five morphological characters. The authors concluded that most of the wildcats have not experienced recent introgression from domestics. However, morphologic and genetic data suggest that earlier introgression from domestic cats has also occurred. The authors conclude there is strong evidence of a population of individuals that are different from domestic cats that may be worthy of legal protection. However, this will be difficult becausethere is no diagnostic test of a true wildcat that containsno domestic cat ancestry. In fact the evidencesuggests that such cats may not exist.

between the populations. Thus, it is most appropriate in the case of intraspecific hybridization as with the Florida panther example. Intrinsic outbreeding depression is probably not a major concern in most circumstances of intraspecific hybridization. However, in some circumstances genetic exchange between intraspecific populations could result in extrinsic outbreeding depression through the loss of important local adaptations that are crucial for the viability of local populations. This is more probable as the amount of genetic divergence between populations increases at molecular markers. Thus, populations that are genetically similar at molecular markers and are similar for a wide range of adaptive traits are the best candidates for intentional hybridization.

17.5.2 Protection of hybrids

Protection of hybrids under the US ESA has had a controversial history (Haig and Allendorf 2006). In May 1977, the US Department of the Interior's Office of the Solicitor issued a statement that:

> . . . because it defines 'fish or wildlife' to include any offspring without limitation, the Act's plain meaning dictates coverage of hybrids of listed animal species. The legislative history buttresses this conclusion for animals and also makes clear its applicability to plants.

However, response from the US Fish and Wildlife Service (July 1977) indicated:

> . . . since the Act was clearly passed to benefit endangered species, . . . it must have meant the offspring of two listed species and was not meant to protect a hybrid where that protection would in fact cause jeopardy to the continued existence of a species.

The Solicitor responded (August 1977 and reaffirmed in 1983) stating that "hybrids of listed species are not protected under the ESA" because he had learned there was the potential for a listed species to be harmed by hybridization. Overall, the US Fish and Wildlife Service's early position was to "discourage conservation efforts for hybrids between taxonomic species or subspecies and their progeny because they do not help and could hinder recovery of endangered taxon".

This series of correspondences and decisions that denied ESA protection for organisms with hybrid ancestry became known as the "Hybrid Policy" (O'Brien and Mayr 1991). O'Brien and Mayr pointed out that we would lose invaluable biological diversity if the ESA did not protect some subspecies or populations that interbreed (e.g., the Florida panther), or taxa derived from hybridization (e.g., the red wolf). Further, Grant and Grant (1992b) pointed out that few species would be protected by eliminating protection for any species interbreeding, since so many plant and animal species interbreed to some extent. Discussions such as

these and the Florida panther situation contributed to the US Fish and Wildlife Service suspending the Hybrid Policy in December 1990.

A proposed policy on hybrids was published in 1996 (USFWS and NOAA 1996a). This "Intercross Policy" was scheduled to be finalized one year later, but has never been approved. Thus, no official policy provides guidelines for dealing with hybrids under the ESA (Allendorf *et al.* 2001). The absence of a policy probably results from the difficulty in writing a hybrid policy that would be flexible enough to apply to all situations, but which would still provide helpful recommendations.

17.5.2.1 How much admixture is acceptable?

What proportion of cattle genes must be present in a 'bison' before it is no longer a bison? This is a difficult issue. Some have argued that only 'pure' bison with no admixture should be protected as bison (see Marris 2009). Moreover, most bison herds have been found to be admixed with cattle so that potentially valuable genetic variation could be lost if these herds were no longer protected as bison (Halbert and Derr 2007). Some have argued that a 'small' proportion of admixture should not disqualify populations from being protected. In addition, local knowledge could be lost if admixed populations are replaced with nonadmixed populations. For example, replacement nonadmixed bison might not be able to find water holes at certain times of year that are currently used by hybrid populations.

This issue has both a philosophical and practical aspect to it. Should an individual with both bison and cattle genes be considered a bison? If so, what proportion of cattle genes is acceptable? 5%? 10%? 25%? 50%? Practically speaking, admixture might also result in individuals with lower fitness. For example, there is some evidence that male bison with cattle mtDNA are smaller and have lower reproductive success than bison with bison mtDNA (James Derr, personal communication).

The amount of admixture that precludes protection is situation-specific. Setting some arbitrary limit of admixture below which a population will be considered 'pure' is problematic. First, estimating the proportion of admixture precisely is difficult because of a limited number of diagnostic markers. In addition, it is often hard to distinguish between a small proportion of admixture (e.g. <5%) and natural polymorphisms that might exist in some populations. Finally, setting an arbitrary threshold could give way to further erosion of the genetic integrity of the parental taxon by constantly lowering the definition of 'pure'. If 5% is acceptable, why not 6% or 10%?

Several factors need to be considered when assessing the potential value of a population hybridized because of human activities. One factor is how many pure populations of the taxon remain. The smaller the number of pure populations, the greater the conservation and restoration value of any hybridized populations. In addition, the greater the phenotypic (behavior, morphology, etc.) differentiation between the hybridized population and remaining pure populations, the greater the conservation value of the hybridized population. Another factor to consider is whether the continued existence of hybridized populations poses a threat to remaining pure populations. The greater the perceived threat, the lower the value of the hybridized population. Finally, if harmful ecosystem-level effects result from hybridization, the value of hybrids is clearly much lower (Schweitzer et al. 2011).

The creation of hybrid swarms between native cutthroat and introduced rainbow trout is widespread in the western US. For example, most local populations of native westslope cutthroat trout are now hybrid swarms with rainbow trout (see Example 17.3). What proportion of admixture must be present before a population should no longer be considered 'westslope cutthroat trout'? Some have argued that only nonhybridized populations should be included as westslope cutthroat trout in the unit to be considered for listing under the US ESA (Allendorf et al. 2004). Westslope cutthroat trout are a monophyletic lineage that has been evolutionarily isolated from other taxa for 1–2 million years (Allendorf and Leary 1988). This period of isolation and the amount of genetic divergence corresponds to that usually seen between congeneric species of fish. Only nonhybridized populations that still contain the westslope cutthroat trout genome that has evolved in isolation are likely to possess the local adaptations important for long-term persistence.

This recommendation has been supported by recent work showing that even small amounts of admixture reduce fitness (see Figure 17.8). Unfortunately, the USFWS (2003) finding on westslope cutthroat trout was that populations with less than 20% admixture from rainbow trout would be considered westslope cutthroat trout for listing under the US ESA.

17.5.3 Prediction of outbreeding depression

Outbreeding depression is much less predictable than inbreeding depression. The intensity of inbreeding depression differs, but all populations have some deleterious recessive alleles that will reduce fitness when they become homozygous because of inbreeding. However, the effects of matings between genetically distinct populations depend upon a combination of factors. Fitness can increase because of the sheltering of deleterious recessive alleles (heterosis). On the other hand, fitness can decrease because of genetic incompatibilities (intrinsic outbreeding depression) or loss of local adaptations (extrinsic outbreeding depression).

The predictability of outbreeding depression is an important practical issue because increasing habitat loss has led to reduced local population size and increased isolation of many local populations. The resulting loss in genetic variation can lead to increased probability of extinction (see Chapter 15). Establishing gene flow (genetic rescue, see Section 15.5) can reverse these effects, but managers are rightly concerned about possible outbreeding depression. Edmands (2007) cautiously recommended human-mediated hybridization only for populations that are clearly suffering inbreeding depression, and suggested testing for the effects of hybridization for at least two generations. There are two problems with this recommendation. First, it is difficult to detect inbreeding depression in wild populations (see Chapter 13), and second, it is impractical to test for the effects of hybridization in most species of conservation concern.

Frankham et al. (2011) have recently addressed this issue and have provided useful guidelines and a decision tree for guiding the use of gene flow to rescue isolated populations. In general, they argue that the concerns of managers (and Edmands 2007) have been overly excessive. They argue that outbreeding depression is unlikely between two populations within the same species if they have the same karyotype, have been isolated for less than 500 years, and they occupy similar environments. We agree with the recommendations of Frankham et al. (2011), except for the 500 years criterion. Genetic divergence is a function of the number of generations in isolation, not calendar years. For example, 500 years would correspond to 10 generations in tuatara (Table 7.6) and 500 generations in an annual plant. For most species, the 500-year recommendation would correspond to 50–100 generations in isolation.

We can apply the criteria of Frankham *et al.* (2011) to the case of the northern and southern populations of fisher in California described in Guest Box 9. Molecular genetic analysis suggests that these populations have been isolated for thousands of years; this would correspond to hundreds of generations. Therefore, some outbreeding depression is a significant risk if genetic exchange between these populations is mediated by managers.

Guest Box 17   Hybridization and the conservation of plants
Loren H. Rieseberg

Hybridization is a common feature in vascular plants. Estimates from eight well-studied floras suggest that approximately 9% of plant species hybridize (Whitney *et al.* 2010) and that close to one-quarter of hybridizing species are rare or endangered (Carney *et al.* 2000). In most instances, hybridization will not harm the rare taxon. Species with strong premating barriers, for example, or that have coexisted naturally for thousands of generations, are unlikely to be threatened by hybridization. Only when premating barriers are weak or when rare species come into contact with non-native species (or native species that have recently become aggressive due to human-induced habitat disturbance) is hybridization likely to cause genomic extinction. Because loss of rare populations may occur quickly, contact between native species and recently introduced or newly aggressive congeners requires swift assessment and action (Buerkle *et al.* 2003).

Recent theoretical analyses of contact between native and invading populations suggest that, under some circumstances, introgression will mostly be in the direction of the invading population (Currat *et al.* 2008). Thus, Currat *et al.* (2008) argue that the risk incurred by the native population when confronted with an invading taxon is primarily demographic rather than genetic. However, the theoretical models upon which this conclusion is based assume that the invader is at extremely low frequency when it first comes into contact with the native, and it becomes essentially 100% introgressed within just a few generations. These conditions are likely to be uncommon in natural situations.

Plant species from islands or other isolated floras are particularly vulnerable to hybridization because premating barriers are often weak and geographic ranges are small. Perhaps the best-studied example is the Catalina Island mahogany, whose population size has dwindled to six pure adult trees (Rieseberg and Gerber 1995). This distinctive species is restricted to Wild Boar Gully on the southwest side of Santa Catalina Island off the coast of California. When the population was first discovered in 1897, it consisted of more than 40 trees, but it has declined rapidly over the past century. Two factors appear to have caused this decline: grazing and rooting by introduced herbivores, and interspecific hybridization with its more abundant congener, mountain mahogany. Although the mountain mahogany is not found in Wild Boar Gully, hybridization between the two mahogany species appears to be frequent. In addition to the six pure Catalina mahogany trees in the gully, five other adult trees and at least 7% of newly established seedlings are of hybrid origin. Presumably, wind pollination allows mountain mahogany trees from nearby canyons to sire hybrid plants in Wild Boar Gully.

Other vulnerable island species include the Haleakala greensword and several Canary Island species in the genus *Argyranthemum*. In fact, the Haleakala greensword is now extinct; two hybrids contain the only known remnants of its genome (Carr and Medeiros 1998).

The wild relatives of crops are also often vulnerable to hybridization; indeed, 22 of the 25 most important crops are known to hybridize with wild relatives (Ellstrand 2003). Examples of ongoing introgression include the California black walnut, which hybridizes with the cultivated walnut, and the common sunflower. Populations of common sunflowers along cultivated sunflower fields consist entirely of crop–wild hybrids (Figure 17.11, Linder *et al.* 1998), a finding consistent with computer

(Continued)

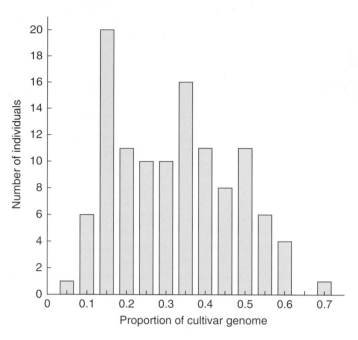

Figure 17.11 The proportion of cultivar genome carried by individuals from three 'wild' sunflower populations that are sympatric with cultivated sunflower. Note that all 'wild' individuals are actually crop-wild hybrids. From Linder *et al.* (1998).

simulations indicating that wild plants were likely to be replaced by crop–wild hybrids in less than 20 generations (Wolf *et al.* 2001).

Although most examples of genomic extinction by hybridization represent island endemics or crop relatives, even abundant mainland species may be at risk if faced with an aggressive congener. The native cordgrass (*Spartina foliosa*) in the San Francisco Bay is threatened by invading cordgrass (*S. alternifolia*) because the invader produces 21-fold more viable pollen than the native, and hybrids are strong and vigorous (Antilla *et al.* 1998). Simulations predict that native cordgrass could be extinct in 3–20 generations (Wolf *et al.* 2001).

Recent genomic studies show that many plant and animal species exhibit the footprints of past hybridization and introgression (e.g., Scascitelli *et al.* 2010). Even modern humans from Europe and Asia (but not Africa) share approximately 3% of their nuclear DNA with Neandertals (Green *et al.* 2010). Also, hybridization can sometimes lead to the origin of new adaptations or new species (e.g., Rieseberg *et al.* 2003). Thus, hybridization and introgression should be viewed as a natural part of the evolutionary process and may sometimes have positive rather than negative effects on biodiversity. Only when human activities lead to substantial changes in hybridization rates, and thereby threaten rare populations, is it sensible to implement management solutions. These include the elimination of the less-desired species from the area of hybridization, or the transplantation of the rare population to a remote location where the other hybridizing taxon does not occur.

CHAPTER 18

Exploited populations

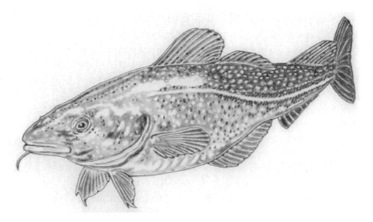

Atlantic cod, Guest Box 18

I suggest that minimizing the impact of sport hunting on the evolution of hunted species should be a major preoccupation of wildlife managers.

Marco Festa-Bianchet (2003, p. 191)

Evidence is mounting that fish populations won't necessarily recover even if overfishing stops. Fishing may be such a powerful evolutionary force that we are running up a Darwinian debt for future generations

Natasha Loder (2005)

Chapter Contents

Conservation and the Genetics of Populations, Second Edition. Fred W. Allendorf, Gordon Luikart, and Sally N. Aitken.
© 2013 Fred W. Allendorf, Gordon Luikart and Sally N. Aitken. Published 2013 by Blackwell Publishing Ltd.

Humans have **harvested** animals and plants from the wild since the beginning of our species. There is mounting evidence that overexploitation has led to the direct demographic extinction of many populations and species (Burney and Flannery 2005). Genetic changes brought about by **exploitation** pose a less obvious threat than direct extinction. Nevertheless, such genetic changes can greatly increase the complexity of managing populations in order to maintain sustainability (Ratner and Lande 2001, Walsh *et al.* 2006, Proaktor *et al.* 2007, Law 2007, Festa-Bianchet and Lee 2009, Palkovacs 2011).

Many natural resource managers have been reluctant to accept the potential for harvest to cause genetic change, and many are doubtful that any such changes are harmful (Harris *et al.* 2002, Heimer 2004, Conover 2007, Hutchings and Fraser 2008). However, intense and prolonged mortality caused by exploitation will inevitably result in genetic change. Harvest need not be selective to cause genetic change; uniformly increasing mortality independent of phenotype will select for earlier maturation (Law 2007). Genetic changes caused by exploitation can increase extinction risks and reduce recovery rates of over-harvested populations (Walsh *et al.* 2006, Olsen *et al.* 2004a).

Most of the concern in the literature about genetic changes caused by exploitation has focused on marine and freshwater finfish and hunted ungulate populations. However, an incredible variety of wild animal and plant populations are exploited by humans: terrestrial game birds, waterfowl, whales, snakes, turtles, land snails, a wide range of marine invertebrates (anemones, sea urchins, sponges, sea cucumbers, jellyfish), marine birds, kangaroos, forest primates, trees, cycads, orchids, and so on. The same concerns of genetic change elicited by harvest apply to all of these species. For example, the size at sexual maturity in rock lobsters off the west coast of Australia has declined substantially over the last 35 years (Figure 18.1). This change apparently is partially an evolutionary response to extremely high annual exploitation rates of adults (~75%), combined with a required minimum carapace length of 76 mm in harvested animals.

Understanding the genetic changes and evolutionary responses of exploited populations is crucial for the design of management aimed at sustainable exploitation of natural biological resources. In this chapter, we consider how harvest is likely to affect the genetics of wild populations. We also consider how the effects of both purposeful and inadvertent harvesting can affect wild populations. Finally, we consider the management and recovery of exploited populations.

18.1 LOSS OF GENETIC VARIATION

Harvesting can reduce effective population size and cause the loss of genetic variation both by reducing population size directly and by reducing the number of migrants into a local population (Section 18.3). The magnitude of local N_e is determined by demographic factors including N_C (census population size), sex ratio, and the mean and variance of lifetime number of progeny produced by males and females. Harvest often targets specific sex or age classes and thereby can reduce the effective population size, and increase the rate of loss of genetic variation. This effect is often exacerbated by ongoing habitat loss resulting in decreased population size and greater isolation. Many recent papers report reduced levels of genetic diversity in a wide variety of exploited species (see Table 18.1). For example, harvesting for rodent poison and making ropes has accelerated the rate of loss of genetic variation in a cycad species endemic to Cuba (Pinares *et al.* 2009).

Bishop *et al.* (2009) reported that uncontrolled hunting of Nile crocodiles in the Okavango Delta, Botswana, in the mid to late 20th century substantially reduced the census and effective population size. They estimated that the current census size is less than 10% of the historic population size, and inferred a recent loss of genetic variation using the bottleneck test of Piry *et al.* (1999) with seven microsatellite loci. They also estimated that the contemporary N_e/N_C ratio is 0.05. Simulations indicated that ongoing removal of adults will cause continued loss of allelic diversity and heterozygosity in this population.

Management operates in calendar time (e.g., years), whereas knowing N_e allows the prediction of the loss of heterozygosity per generation. When considering loss of variation over calendar time, a small N_e can be compensated for by a large generation interval (G) (see Section 7.9). Therefore, consideration of the effects of management on loss of genetic variation over time should not be restricted, as they often are, to N_e alone because effects of G are equally important (Sæther *et al.* 2009). For example, Ryman *et al.* (1981) found that different harvest regimes for moose can have strong effects on both effective population size and gen-

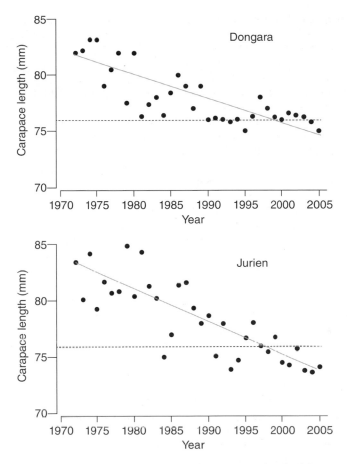

Figure 18.1 Observed decline in mean carapace length of rock lobsters captured in the fishery at two locations off the coast of Western Australia from 1972 to 2005 (Melville-Smith and de Lestang 2006). Only animals with a carapace length of greater than 76 mm (dotted line) can be legally harvested. This decline apparently is partially an evolutionary response to extremely high annual exploitation rates of adults (~75%), combined with a required minimum carapace length of 76 mm in harvested individuals. From Allendorf *et al.* (2008).

eration interval (Figure 18.2). Populations with smaller N_e tended to lose heterozygosity at a slower rate because those effects of hunting that reduced N_e (e.g., harvesting young animals) also tended to increase the generation interval. That is, hunted populations with relatively smaller N_e and a longer generation interval would lose genetic variation over time (not generations) more slowly than some populations with large N_e and shorter generation interval.

Harvest regulations can reduce the N_e/N_c ratio and thereby increase the rate of loss of heterozygosity without having any detectable effect on census population size. Male biased, or male only, harvest is practiced in many species of ungulates and some marine crustaceans (e.g., lobsters), and a skewed sex ratio amongst breeders might severely reduce effective population size. Sex ratios of less than one adult male for 10 adult females are not uncommon in hunted populations of deer and elk (Scribner *et al.* 1991, Noyes *et al.* 1996), and are probably common in other species where males are selectively hunted. For example, males comprised less than 1% of all adult elk in the Elkhorn Mountains of Montana in 1985 (Lamb 2010). In addition, harvest regulations can also increase the variance in reproductive success. For example, female brown bear, moose, and wild boar are protected by

Table 18.1 Examples of loss of genetic variation in exploited populations

Species	Observation
African elephant	Intense hunting in the early 1900s combined with slow post-bottleneck recovery and lack of gene flow into Addo Elephant National Park (South Africa) is associated with reduced microsatellite heterozygosity and allelic diversity. In contrast, the Krueger National Park population recovered faster due to immigration after a similar hunting-induced bottleneck, and has nearly double the heterozygosity and allelic diversity. (Nystrom *et al.* 2006)
Arctic fox	The Arctic fox population in Scandinavia probably numbered 10,000 historically, but heavy hunting pressure associated with a profitable fur trade in the early 20th century rapidly reduced the population to a few hundred individuals. Analysis of ancient DNA revealed that this population has lost approximately 25% of the microsatellite alleles and four out of seven mtDNA haplotypes. (Larson *et al.* 2002)
New Zealand snapper	Microsatellite heterozygosity and alleles per locus declined between 1950 and 1988 after commencement of a fishery on this population, in spite of an estimated standing population of well over 3 million fish. (Hauser *et al.* 2002)
Sea otter	Analysis of ancient DNA reveals that all the current populations examined exhibit considerably lower heterozygosities at microsatellite loci than samples pre-dating the population size bottleneck caused by extensive fur trading in the 18th and 19th centuries. (Larson *et al.* 2002)
Sika deer	Three out of seven mitochondrial DNA haplotypes in Hokkaido, Japan, were lost during a 200-year bottleneck caused by heavy hunting reinforced by heavy snow in two winters. (Nabata *et al.* 2004)
Tule elk	The Tule elk of the Central Valley of California, US, dwindled in 50 years from about half-a-million down to fewer than 30 animals in 1895 through habitat loss, hunting, and poaching following the gold rush. Approximately 60% of heterozygosity was lost, and the present population exhibits little genetic variation. (McCullough *et al.* 1996)
Red deer	Deer in both open and fenced hunted Spanish populations have lower levels of microsatellite heterozygosity than deer from protected areas. (Martinez *et al.* 2002)
White seabream	Mediterranean populations in areas protected from fishing have significantly less microsatellite allelic richness than those from non-protected areas. (Perez-Rusafa *et al.* 2006)

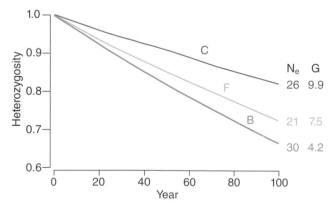

Figure 18.2 Expected decline of heterozygosity under three different sets of regulations in a population of moose in Sweden with a census size of 100 adults following the hunting season (Ryman *et al.* 1981). The effective population size and generation interval for each hunting regime is indicated on the right. In hunting regime B, all adults experience identical mortality rates, but calves (less than 1-year-old) are protected and are not hunted. In C, only calves are hunted. In F, adult females with calves are protected so that the risk of mortality of an adult female is reduced as a function of the number of calves (0, 1, or 2) with her at the beginning of hunting season. The regime (B) with the largest N_e is expected to lose heterozygosity at nearly twice the rate of the regime (C) with a smaller N_e, which has a longer generation interval. Redrawn from Ryman *et al.* (1981).

regulations in Sweden when accompanied by sub-adults. These policies will also result in the individuals surviving the hunting season being more closely related than expected by chance, thereby further decreasing N_e (e.g., Ryman *et al.* 1981).

Marine fish and invertebrates generally have much larger census and effective population sizes than terrestrial vertebrates. However, heterozygosity can be lost even in populations with large census population sizes because N_e is often much smaller than the census size in many marine species (Waples 1998). For example, N_e in New Zealand snapper was estimated to be approximately 100 based on monitoring temporal changes in allele frequency and losses of heterozygosity (Example 7.5). The minimum estimated population size (N_C) during this period was 3.3 million fish; thus, N_e/N_C was on the order of 0.0001. This suggests that even very large exploited marine fish populations might be in danger of losing genetic variation.

Loss of allelic diversity can have harmful effects in large, exploited marine populations where the loss of heterozygosity due to harvest is minimal (Ryman *et al.* 1995b). Allelic diversity is more sensitive than heterozygosity to dramatic reductions in population size, and N_e is a poor predictor of the rate of loss of allelic diversity. That is, populations with the same N_e can lose allelic diversity at very different rates. The reason for this effect is that allelic diversity is affected not only by N_e, but also by N_C (Crow and Kimura 1970, p. 455). Thus, reducing N_C from, for example, millions down to thousands, might have no effect on heterozygosity, but could result in a decline in allelic diversity. In contrast, greater harvest of males through hunting in ungulates could have a limited effect on allelic diversity while reducing heterozygosity, because N_C can remain large even when N_e is reduced due to increasingly skewed sex ratios favoring females (Section 7.2).

Tropical tree species often occur at low population densities in species-diverse forests, and are subjected to size-limited selective harvesting, where the largest trees are logged periodically. Degen *et al.* (2006) simulated the genetic and demographic effects of more- and less-intensive harvesting over four centuries on genetic diversity in four tropical tree species in French Guiana. The higher-intensity harvesting scenario reduced population sizes in three of four species, but the effects on genetic diversity were small due to overlapping generations and high seed and pollen dispersal. However, they did not evaluate whether this diameter-limit logging would result in selection for slower growth.

The loss of genetic variation will also be influenced by gene flow among subpopulations that comprise a metapopulation. Estimating the effective population size of a metapopulation is extremely complex (see Section 15.3). In addition, harvesting might have unexpected effects on the overall N_e of a metapopulation. For example, Hindar *et al.* (2004) found that small subpopulations within a metapopulation of Atlantic salmon contribute more per spawner to the overall effective population size than large subpopulations, and harvesting of the subpopulations jointly in mixed-stock fisheries has a relatively larger demographic effect on small than large populations. Consequently, the mixed-fishery harvest could reduce metapopulation N_e far more than expected.

18.2 UNNATURAL SELECTION

Fisheries, wildlife, and forest management that does not incorporate evolutionary considerations is at risk of reducing productivity in wild populations because exploitation removes phenotypes that might be those most favored by natural and sexual selection in the wild. Accounting for selection that acts counter to natural adaptive processes is therefore an important component of a comprehensive and effective sustainable management strategy. For example, Hanlon (1998) has concluded that commercial harvest of squid on the spawning grounds could impose unnatural sexual selection that could reduce recruitment and affect the long-term sustainability of these fisheries.

The reduction in the frequency of the silver morph in the red fox between 1834 and 1933 in eastern Canada was perhaps the first documented change over time resulting from selective harvest (Elton 1942). J.B.S. Haldane used these data to provide one of the first estimates of the strength of selection in a wild population, using his then recently developed mathematical models of the effects of selection on a single locus (Haldane 1942). The fur of the homozygous silver morph (*rr*) was worth approximately three times as much as the red fur of the heterozygous cross (*Rr*) or homozygous (*RR*) fox to the furrier, and, therefore, was more likely to be pursued by hunters.

The frequency of the desirable silver morph declined from 16% in 1830 to 5% in 1930 (Figure 18.3). Haldane concluded that this trend could be explained by a slightly greater harvest rate of the silver than the red and cross phenotypes. The lines in Figure 18.3

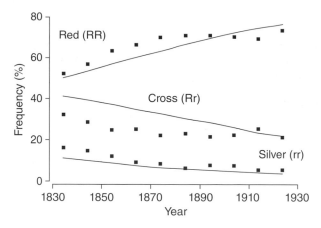

Figure 18.3 Reduction in frequency of the silver morph of the fox in eastern Canada resulting from the preferential harvest by hunters of the more valuable silver morph. The points represent data presented by Elton (1942). The lines represent the expected change in frequencies of the three phenotypes via selection at a single locus, assuming that the silver fox morph has a 3% survival disadvantage per generation relative to the red and cross morphs as modeled by Haldane (1942). The initial frequency of the R allele was 0.3 and the mean generation interval was 2 years. Modified from Allendorf and Hard (2009).

show the expected change in phenotypic frequencies, assuming that the relative fitness of the silver phenotype was 3% less than both the homozygous red and cross phenotypes, and the generation interval was two years. It is interesting to note the observed and expected decline in frequency of the heterozygous cross phenotype even though it had higher fitness than the silver phenotype. The favored cross phenotype is expected to decline in frequency because of the reduction in frequency of the r allele resulting from the lower fitness of the silver type.

Selective genetic changes within subpopulations resulting from exploitation are inevitable because increasing mortality will result in selection for earlier maturation, even if harvest is independent of phenotype (Figure 18.4). Moreover, harvesting of wild populations is inevitably phenotypically non-random (Law 2007); that is, individuals of certain phenotypes (e.g., sizes or behaviors) are more likely than others to be removed from a wild population by harvesting. Such selective harvest will bring about genetic changes in harvested populations if the favored phenotype has at least a partial genetic basis (i.e., is heritable). In addition, such changes are likely to both reduce the frequency of desirable phenotypes (Guest Box 11) and to reduce productivity.

The rate of genetic change caused by exploitative selection depends upon the amount of additive genetic variation for the trait (heritability) (equation 11.9):

$$R = H_N S$$

where H_N is the narrow-sense heritability, S is the selection differential (the difference in the phenotypic means between the selected parents and the whole population), and R is the response (the difference in the phenotypic means between the progeny generation and the whole population in the previous generation). In the case of exploitative selection, S will be affected both by the intensity of harvest (the proportion of the individuals harvested) and the phenotypic selectivity (e.g., the removal of only the largest individuals) of the harvest.

There are many examples in the literature of phenotypic changes that might be the result of exploitative selection (Table 18.2, Example 18.1). However, it has been difficult to determine whether observed phenotypic changes over time indicate genetic change, or are caused by other factors such as relaxing density-dependent effects on individual growth due to reductions in population density, or abiotic factors such as temperature affecting growth and development (Fenberg and Roy 2008). A recent review critically evaluated the observed evidence for a genetic basis of such phenotypic change and concluded that establishment of practices for routinely monitoring and sampling harvested fish stocks is vital for the detection and management of fisheries-induced evolution (Kuparinen and Merilä 2007). Guest Box 11 provides an excellent

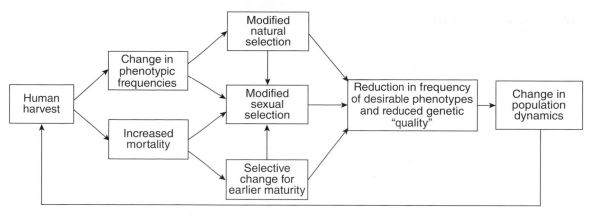

Figure 18.4 Human harvest can have a variety of direct and indirect genetic effects on populations, and it has the potential to affect the future yield and viability of exploited populations. From Allendorf and Hard (2009).

Table 18.2 Examples of phenotypic changes that could have resulted from exploitative selection

Species	Trait(s)	Observation
African elephant	Tusks	An increase in the proportion of tuskless females from 10% in 1969 to 38% in 1989 was directly attributed to illegal hunting in South Luangwa National Park, Zambia. (Jachmann *et al.* 1995)
American plaice	Body size Age at maturation	Three fish stocks with historically different levels of exploitation showed the same long-term shift towards maturation at younger ages and smaller sizes. This situation warrants further investigation to determine whether these stocks are truly demographically and genetically independent. (Barot *et al.* 2005)
Atlantic cod	Body size Age at maturation Growth rate	Survey data prior to the collapse of the fisheries in 1992 showed a significant genetic shift towards earlier maturation and at smaller sizes. In the 1970s fishing selection targeted slow-growing individuals; later in the 1980s the net mesh size was increased resulting in a bigger catch of large, faster-growing individuals. Application of a quantitative genetic model showed a reduction of length-at-age between cohorts of offspring and parents as a result of exploitative selection. (Swain *et al.* 2007)
Atlantic salmon	Time of spawning Body size	Earlier spawning fish experienced greater harvest by anglers. Allozyme and mitochondrial DNA data from four populations in Spain showed that the late-running individuals that escaped harvest were genetically distinct and significantly smaller. Catch records in Ireland extending back many decades and recent electronic counter data show a reduction in the abundance of early migrants and a decline in size of late migrants. (Consuegra *et al.* 2005)
European grayling	Age at maturation	Gill net fishing is suggested to have caused a constant reduction in the age and length at maturity in separate populations founded from the same common ancestors. (Haugen and Vollestad 2001)
North sea plaice	Age at maturation Body size	The reaction norms for age and length at maturation showed a significant trend towards younger age and shorter body length. (Grift *et al.* 2003)
Northern pike	Body size	Over a period of four decades, selective harvesting targeted large individuals and directional natural selection favors large body size. The result of these two opposing forces is stabilizing selection, but a reduction in overall fitness. (Edeline *et al.* 2007)
Red kangaroo	Body size	Hunters target the larger individuals in a group and there is evidence that average size has declined. (Croft 1999)

Example 18.1 Phenotypically selective harvest within and among local subpopulations

Phenotypically selective harvest of mixed populations comprising individuals from many contributing subpopulations can result in both exploitative selection within subpopulations and differential intensity of harvest on those subpopulations. For example, hundreds of reproductively isolated local subpopulations of sockeye salmon contribute to the Bristol Bay fishery in Alaska (Hilborn *et al.* 2003). There is a gillnet fishery in Bristol Bay that harvests these subpopulations before the salmon return to their home spawning grounds in freshwater. This mixed-stock fishery has the potential to harvest selectively depending upon run timing, body size, body shape, and life history (primarily age at sexual maturity).

Quinn *et al.* (2007) examined daily records in two fishing districts in Bristol Bay for evidence of temporally selective harvest over a 35-year period. They found that earlier migrants experienced lower capture rates in the fishery than later migrants. The timing of the run has become earlier over this period as expected in response to selection favoring individuals (and subpopulations) that arrive earlier. This observed phenotypic change apparently results from genetic changes both within subpopulations, resulting in earlier run timing of individuals, and among subpopulations with differential harvest intensity favoring earlier arriving subpopulations.

The gillnet fishery also results in selection within subpopulations and differential intensity of harvest on subpopulations, because the effectiveness of gillnets to capture migrating fish is dependent upon body size and shape (Hamon *et al.* 2000). Fish that are too small are able to escape by swimming through the mesh, and fish above the target size-class are too large to be wedged in the mesh. Therefore, capture by gillnets is likely to be selective on age at sexual maturity, size at age of sexual maturity, and body depth. Subpopulations that spawn in different habitat types show consistent differences for all of these characteristics (Quinn *et al.* 2007). For example, males from lake-spawning subpopulations generally have much deeper body depth than males from stream-spawning subpopulations.

Hamon *et al.* (2000) compared harvested fish with fish that escaped the fishery and returned to the spawning grounds to determine the relationship between morphology and fitness caused by selective capture in gillnets (Figure 18.5). They found that the effects of gillnet selectivity within subpopulations was strongly influenced by variability in age at reproduction. Subpopulations with mixed-aged fish at maturity experienced disruptive selection, with smaller and larger fish having the greatest fitness. In contrast, subpopulations predominantly of a single age-class experienced directional selection favoring smaller fish. In addition, differences in morphology among subpopulations resulted in large differences in harvest intensity. Some subpopulations experienced virtually no fishing mortality, while others sustained high mortality due to harvest (>70%).

Their analysis indicated that mixed age-classes are subject to disruptive selection, but that single age-class populations experienced directional selection. The effect of this selection depends upon cumulative selection pressures, which probably include natural and sexual selection on this trait. They concluded that selection by gillnets is a strong selective force affecting body size and shape within populations.

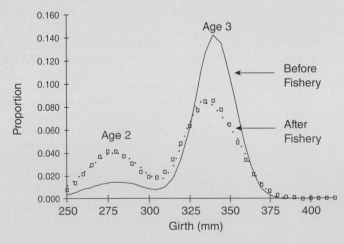

Figure 18.5 Distribution of girths before and after the fishery for female sockeye salmon in Little Togiak Lake, Alaska, in 1994. Solid lines represent the distributions of girths before the fishery. Squares indicate the distributions of girths after the fishery. Modified from Hamon *et al.* (2000).

example of disentangling the sources of phenotypic change to demonstrate effects of exploitative selection on horn length in bighorn sheep.

Many harvest regimes of trees, fish, and wildlife selectively remove larger individuals. Life history theory predicts that this should select for maturation at a younger age and smaller size (Marshall and Browman 2007). This prediction is concordant with the long-term trend towards earlier maturation that has been observed for many commercially exploited fish stocks. However, such trends might also be explained by **phenotypic plasticity** as a direct response to decreased population size, or long-term environmental changes.

Probabilistic maturation reaction norms (PMRNs) have been used to help disentangle genetic from plastic effects on maturation (Heino *et al.* 2002, Dieckmann and Heino 2007). Reconstructing PMRNs from historic data in exploited populations has provided evidence for fishery-induced selection. However, some have argued that because PMRNs do not fully account for physiological aspects of maturation, the observed shifts might reflect directional environmental effects on maturation rather than genetic changes (Marshall and McAdam 2007). It is impossible in most circumstances to completely disentangle genetic and plastic effects. Nevertheless, the use of PMRNs provides a useful method to determine whether genetic effects are at least partially responsible for an observed phenotypic change over time.

The selective harvesting of large, straight trees from native forests has occurred for centuries or millennia. The genetic effects of this historical logging have been suggested as the reason for the crooked stems and long branches of the Cedar of Lebanon, and for the poor form of Scots pine along waterways in Sweden and Finland, but these hypotheses have not been scientifically tested (Savolainen and Kärkkäinen 1992). Cornelius *et al.* (2005) evaluated the extent to which contemporary logging of natural mahogany populations might result in selection for growth or form. They concluded that, given the low heritability of size and form phenotypic traits in natural forests ($H_N \leq 0.1$), the phenotypic response to logging is likely to be less than 5% following a single logging event. While these effects may accumulate with multiple logging events and over generations, Cornelius *et al.* (2005) suggested that erosion of genetic diversity from reduced population sizes is a much greater concern in this tropical tree.

An interesting historical note is that Patrick Matthew published a book in 1831 that contained a forerunner of Darwin's concept of natural selection (Wells 1973). Matthew was aware of the great variation present within a single species of tree. He also pointed out that different species and varieties were well adapted for growing in certain habitats. He argued that foresters should use seed from the "largest, most healthy, and luxuriant growing trees" (Matthew 1831).

18.3 SPATIAL STRUCTURE

Virtually all species have separate local breeding groups (subpopulations) that are somewhat reproductively isolated. Harvest of wild populations can perturb genetic subdivision among populations within a species and reduce overall productivity. The primary problem is that harvesting a group of individuals that is a mixture of several subpopulations can result in the extirpation of one or more subpopulations. This will not be recognized unless the subpopulations are identified separately and individuals from population mixtures are assigned to subpopulations (see Section 22.5).

Extirpation of some subpopulations is likely to directly reduce overall population productivity. In addition, Schindler *et al.* (2010) have shown that productivity of certain subpopulations of sockeye salmon can change dramatically over time as environmental conditions change. Therefore, ensuring long-term productivity of the overall population depends upon conserving all subpopulations, including the currently less productive ones. In addition, reduction in the size or density of subpopulations might decrease the number of migrants among subpopulations and cause increased genetic drift and loss of genetic variation. Harvest can also increase the rate of gene flow into certain subpopulations and cause genetic swamping and loss of local adaptations. An understanding of this population genetic substructure at different points of a species life history is necessary to predict the potential effects of harvest on genetic subdivision.

In order to manage populations sustainably, we need to know what constitutes the harvested population and how it is genetically delineated (Palsbøll *et al.* 2007). If the harvested population is part of a wider geographic area connected by migration, then any effects of selective harvest might affect a larger geographic area than anticipated. Figure 18.6 shows different proportions of individuals from three loggerhead

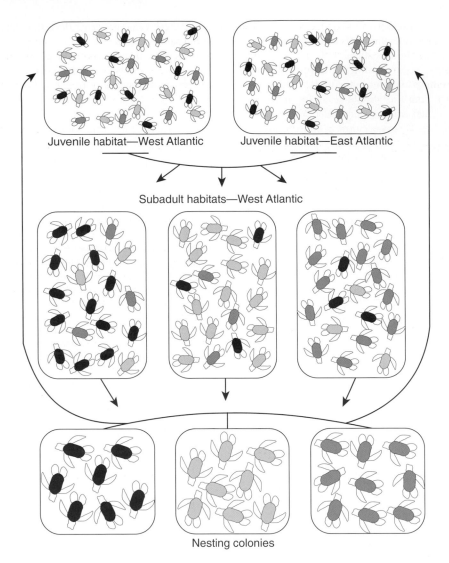

Juvenile habitat—West Atlantic   Juvenile habitat—East Atlantic

Subadult habitats—West Atlantic

Nesting colonies

Figure 18.6 Loggerhead turtle population structure in the North Atlantic for three hypothetical rookeries based on maternally inherited mtDNA. There is a stepwise increase in population structure through juvenile, subadult, and adult stages. In the juvenile stage, turtles from all three rookeries intermingle, and no population structure is apparent between eastern and western edges of the North Atlantic Gyre. In the subadult stage, turtles tend to move to feeding habitats in the vicinity of their natal rookery, inducing low but significant population structure. In the adults, females (and possibly males) have high site fidelity to breeding habitat, inducing strong population structure. From Bowen *et al.* (2005).

turtle spawning rookeries in the North Atlantic at three different life history stages. Harvest during the juvenile or subadult stages would affect all three rookeries.

The effect of this model of different population structure during different life history stages has been applied to hawksbill sea turtles in the Caribbean Sea, which are estimated to be on the order of 1% of the size of pre-exploitation populations (Bowen *et al.* 2007). The government of Cuba has argued that hawksbill turtles found in its waters are part of a closed system, and they have sought permission from

CITES to harvest 500 turtles a year. However, molecular genetic analysis of hawksbills collected on foraging grounds indicates that the harvest of turtles in Cuban waters would have potentially harmful effects on nesting colonies throughout the Caribbean because many turtles that breed outside of Cuba would be captured in Cuban waters.

Harvest of mixed populations is common in migratory waterfowl, marine mammals, ungulates, and many other species (see Example 18.2). For example,

---

Example 18.2 Genetic determination of composition of Canada goose harvests

Canada geese in eastern North America are managed to maintain the viability of individual breeding populations by managing harvest within acceptable numerical limits (Scribner *et al.* 2003). Estimating mortality rates for individual populations requires that they be sufficiently distinct geographically at some time to be monitored independently from other populations. Most harvest of Canada geese occurs after breeding in the fall and winter. Overlapping occurrence of birds from multiple breeding populations complicates harvest management, especially when management objectives differ among populations.

Estimating the proportional contribution of separate breeding populations in a mixed harvest is possible if there are genetic differences among breeding populations. In the extreme, each contributing breeding population may be characterized by a **diagnostic** marker (e.g., alleles or mitochondrial DNA haplotypes, such as the turtles in Figure 18.6). In such cases, each individual can be unambiguously assigned to a breeding population of origin. In the absence of fixed genetic differences, a statistical framework exists (e.g., assignment tests) for determining harvest composition and estimating statistical confidence based on population differences in allele or haplotype frequencies (see Section 22.5).

Scribner *et al.* (2003) used genotypes at five microsatellite loci to estimate proportions of giant and interior subspecies of Canada geese in harvest mixtures in the Great Lakes region of North America. They sampled baseline populations of Canada goose from both subspecies over broad geographic areas to describe genetic variation within and among breeding populations. Divergence among baseline populations was slight ($F_{ST} = 0.016$ over all breeding populations), and was slightly greater for pairs of populations from different subspecies.

The subspecies composition in harvests varied significantly among years and across early, regular, and late seasons within a year (Figure 18.7). Harvest composition varied spatially among management areas in different regions and between managed and private lands in close geographic proximity. Higher proportions of resident giant Canada geese were harvested during early hunting seasons and on private lands, relative to migratory interior Canada geese. Harvest estimates suggest that individuals from different subspecies and populations are differentially abundant or susceptible to harvest at different times of the season, during different years, and for populations across different geographic locations.

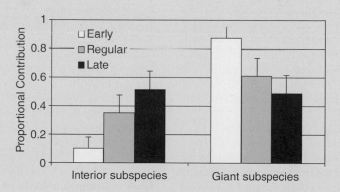

Figure 18.7 Mean-likelihood estimates (±SE) of proportional contributions based on five microsatellite loci of interior Canada goose and giant Canada goose subspecies to regular season harvests during 1994 in Michigan, US. From Scribner *et al.* (2003).

Pacific salmon are generally harvested in the ocean in mixed stocks that comprise many reproductively isolated subpopulations that spawn in freshwater (Schindler *et al.* 2010). Understanding which subpopulations are contributing to the harvest is essential to avoid overharvest and extirpation of some local subpopulations, while others experience very little harvest whatsoever (Example 18.2).

Extirpation of subpopulations through overharvesting has been observed in both marine and freshwater fish (Nelson and Soulé 1987, Dulvy *et al.* 2003). For example, the number of streams contributing substantially to production of four salmon species in southern British Columbia suffered a severe decline between 1950 and 1980 (Walters and Cahoon 1985). In general, the subpopulations that are less productive and the least resilient to exploitation have been the first to disappear (Loftus 1976). Moreover, stocks with the most desirable characteristics often experience the greatest exploitation. For example, "siskowet" lake trout, which were prized for their high fat content, were the first subpopulation to disappear from Lake Michigan (Nelson and Soulé 1987).

Some plant species used by humans have been overexploited and are threatened in the wild, yet are widely cultivated agriculturally. The orchid genus *Vanilla* provides an example of this. Aromatic species in this genus are widely used for the flavor of their pods. Approximately 95% of the world's commercial vanilla production uses *V. planifolia*, a species native to Mexico, Guatamala and Belize, and historically used by the Aztecs (Bory *et al.* 2008). This species is now cultivated and propagated around the world in tropical areas. However, remaining wild populations are extremely small and at very low population density, with one individual per 2–10 km² in two areas surveyed in Mexico. One expert spent 20 years searching in southern Mexico and only found 30 plants! These small populations are seriously threatened by deforestation and overharvesting, yet represent a valuable genetic resource to the vanilla industry from which to develop new varieties.

Exploitation can also increase gene flow or hybridization among subpopulations and potentially swamp local adaptations. Overexploitation could reduce the density of local subpopulations and allow for more immigration from nearby subpopulations less affected by exploitation. This could bring about the genetic swamping of the remnants of exploited subpopulations and thereby reduce fitness. Recent studies of red deer report that a change of fine-scale genetic structure appears to be associated with changes in harvest management (Nussey *et al.* 2005a, Frantz *et al.* 2008).

Examination of genetic samples collected over time (i.e., genetic monitoring, Section 22.6) is the most powerful way to detect genetic changes caused by harvest (see Guest Box 18). For example, the Flamborough Head population of North Sea Atlantic cod apparently went through a decline in genetic variation followed by genetic swamping between 1954 and 1998, based upon genetic variation at three microsatellite loci using otolith samples archived over this period (Hutchinson *et al.* 2003). Genetic diversity declined between 1954 and 1970, indicating reduced effective population size, apparently resulting from harvest. Genetic variation increased after this period because of increased immigration during a period of exceptionally high exploitation. Thus, the original genetic characteristics of the Flamborough Head population likely have been lost.

18.4 EFFECTS OF RELEASES

Large-scale exploitation of wild animals and plants through fishing, hunting, and logging often depends upon augmentation through releases of translocated or captive-raised individuals (Laikre *et al.* 2010b). Such releases are performed worldwide in vast numbers. Augmentation can be demographically and economically beneficial but can also cause four types of adverse genetic change to wild populations (Figure 18.8):
1 Loss of genetic variation.
2 Loss of adaptations.
3 Change in population composition.
4 Change in population structure.
While adverse genetic impacts are recognized and documented in fisheries, little effort is devoted to actually monitoring them. In forestry and wildlife management, genetic risks associated with releases are largely neglected. We outline the key features of programs to effectively monitor the consequences of such releases on natural populations.

18.4.1 Genetic effects of releases

Even releases that do not result in gene flow can have genetic consequences if they reduce local population

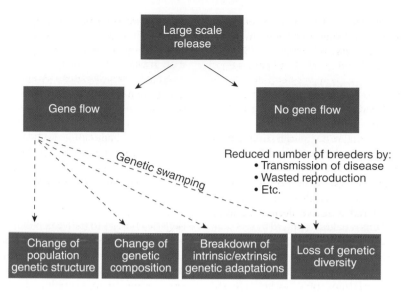

Figure 18.8 Primary pathways by which large-scale releases can change genetic characteristics between natural populations (population genetic structure) and within populations. From Laikre *et al.* (2010b).

size, for example, through competition or disease transmission, or through wasted reproductive effort by native individuals that mate with captive-bred individuals but do not produce viable offspring. The main concern in these cases is that changes to naturally existing genetic diversity within and among populations can reduce viability and productivity of exploited populations. This could be a problem both in the short term, by reducing individual fitness, and in the long term by reducing the capacity for populations to evolve and adapt to future conditions.

Introgression from introduced to native populations has been documented in a number of species subject to large-scale releases (Laikre *et al.* 2010b, p. 524). Although risks to native gene pools have been recognized for a variety of species, the most attention has focused on Pacific salmon, Atlantic salmon, and brown trout (Waples and Drake 2004). In Denmark, these concerns have resulted in a ban of all releases of salmonids originating from anything but the local population (Nielsen and Hansen 2008). However, this is the exception even for salmonids (but see Guest Box 4). Surprisingly little monitoring of the effects of mass releases has occurred in forestry and wildlife management, but examples are beginning to accumulate, such as Eucalypt populations used in Australian forestry

(Barbour *et al.* 2008) and exotic game birds in southern Europe (Barilani *et al.* 2005).

18.4.1.1 Loss of genetic variation

Wild populations might also lose genetic variation if their effective population size is reduced due to increased mortality caused by parasites or diseases transmitted by released individuals. A classic example refers to the effects of the unintended introduction of the parasite *Gyrodactylus salaris* (an ectoparasite living on the skin of Atlantic salmon) into Norway with juvenile salmon imported from Sweden in the early 1970s (Peeler *et al.* 2006). Substantial mortality has been observed in wild populations following spread of parasites through alien populations in salmon in Japan, red partridge in Spain, tortoises in the US, red deer in Italy, and rabbits in France and southern Europe (Daszak *et al.* 2001, Laikre *et al.* 2010b).

Individuals released for **supportive breeding** (see Section 19.7) can cause loss of genetic diversity by reducing N_e. Typically, relatively few parents are brought into captivity for reproduction; these parents often contribute disproportionately large numbers of genes to the next generation in the wild, potentially resulting in increased rates of inbreeding and genetic drift in the total population. Reduced genetic variation has been observed in populations of salmonid fishes subject to supportive breeding (Hansen *et al.* 2000).

18.4.1.2 Breakdown of adaptations

Releases can reduce adaptation by causing loss of extrinsic or intrinsic adaptation (Laikre et al. 2010b). Loss of fitness can occur when alleles that confer local adaptation are replaced by ones that are locally non-adaptive (extrinsic adaptation). Gene flow from a non-local source population can cause breakup of **coadapted gene complexes**, that is, alleles at multiple loci that work synergistically to increase fitness (intrinsic adaptation). Because this breakup is caused by recombination, loss of adaptation generally occurs only in the F_2 generation and beyond, and can be much more difficult to detect than loss of extrinsic adaptation. Empirical examples from wild populations show that both types of adaptation can be lost by gene flow from genetically divergent populations (Laikre *et al.* 2010b). Fitness effects can be insidious: in some studies, increased F_1 fitness due to heterosis has been followed by decreased fitness in F_2 or later generations as coadapted gene complexes are eroded (e.g., Muhlfeld *et al.* 2009, but see Hwang *et al.* 2011, Frankham *et al.* 2011).

A 37-year study of Atlantic salmon in Ireland found that naturally spawning farmed fish depress wild recruitment and disrupt the capacity of natural populations to adapt to higher water temperatures associated with climate change (McGinnity *et al.* 2009). Hansen *et al.* (2009) examined Danish populations of brown trout subject to hatchery supplementation for 60 years and found evidence for selection in the wild against alleles associated with non-native hatchery fish. Muhlfeld *et al.* (2009) showed that non-native rainbow trout that hybridized with native cutthroat trout had high F_1 reproductive success. However, in subsequent generations their fitness declined by nearly 50% (compared with fitness of native trout) following 20% introgression of non-native genes (see Example 17.3).

18.4.1.3 Change of population genetic structure

Most natural animal and plant species are structured into genetically distinct populations because of restricted gene flow, genetic drift, and local adaptation. Since large-scale releases can affect these microevolutionary processes, they can alter genetic structuring of natural populations, but such effects have rarely been monitored (Laikre *et al.* 2010b). Complete replacement of native gene pools of Mediterranean brown trout with introduced populations of Atlantic origin has occurred over large areas (Araguas *et al.* 2009). Slovenian populations of Adriatic grayling have been stocked with Danubian fish for over four decades, and levels of introgression are so high (40–50%) that few native individuals can be found (Sušnik *et al.* 2004).

18.4.2 Effects on species and ecosystem diversity

Genetic changes to native populations can have consequences that extend beyond the affected species. Evidence is accumulating from **community genetics** studies to show that genetic changes to one species can affect other species as well as entire communities and ecosystems (Whitham *et al.* 2008). For instance, genetic characteristics of individual plant populations can affect the composition of arthropods (Crutsinger *et al.* 2006). Foraging behavior of beavers is affected by the genetic makeup of the *Populus* species upon which they feed (Bailey *et al.* 2004).

New evidence indicates that high levels of genetic diversity increase resilience of species and ecosystems, and that genotypic diversity can complement the role of species diversity in a species-poor coastal ecosystem, and thus help buffer ecosystems against extreme climatic events. Genetic variation was positively correlated with recovery of seagrass ecosystems following overgrazing and climatic extremes (Worm *et al.* 2006). Reusch *et al.* (2005) conducted manipulative field experiments and found that increasing genotypic diversity of the cosmopolitan seagrass *Zostera marina* enhanced biomass production, plant density, and invertebrate faunal abundance, despite near-lethal water temperatures.

18.4.3 Monitoring of large-scale releases

Genetic monitoring (see Section 22.6) of releases should aim to provide answers to the following key questions:
1 What are the genetic characteristics of the natural population(s) prior to the release?
2 Do releases alter these characteristics?
3 What are the biological consequences?

Such monitoring should be included as a basic part of any program for commercial or other releases.

Assessments of risk–benefit tradeoffs are most effective if conducted prior to release activities. Waples and Drake (2004) outlined a framework for elements of comprehensive risk–benefit analysis that should be conducted prior to fish stock enhancement programs. Similarly, Barbour *et al.* (2008) discussed strategies for assessing risks of pollen-mediated gene flow from translocated species and hybrids of *Corymbia* and *Eucalyptus globulus* plantations into native populations. These studies showed that different risk–benefit assessment protocols are needed for different taxa and should be refined to fit particular species.

For releases that have already been carried out, an idealized monitoring design often cannot be followed. Sometimes, however, archived material such as fish scales, animal skins, plant leaves or DNA can help to address questions of genetic composition prior to release. Within forestry, so-called provenance trials have been used since the 19th century to identify populations with economically important characteristics. Such traditional tree-breeding programs are aimed at examining performance of trees from different geographic localities (provenances) to find the best sources of seed for selective breeding and planting. They have usually resulted in the use of relatively local populations for reforestation due to local adaptation. Geographic source materials for provenance trials are thus known, and existing trial stands can be used to study long-term effects of plantations, such as gene flow into neighboring, native populations (König *et al.* 2002).

18.5 MANAGEMENT AND RECOVERY OF EXPLOITED POPULATIONS

The most difficult political and economic decision in harvest management is to reduce the current harvest in order to increase the likelihood of long-term sustainability. This decision is especially difficult when taking actions to halt or reverse historic declines will come at the cost of economic hardship for dependent communities (Walters and Martell 2004). Management measures to reduce harmful long-term genetic effects are most likely to be adopted by managers if they also help to meet short-term management objectives. For example, maintaining large, old individuals within

populations provides both short- and long-term benefits (Birkeland and Dayton 2005).

The emphasis on disentangling genetic and plastic mechanisms of phenotypic change is crucial from a basic scientific perspective, but is less important from a strictly management perspective. It is not necessary to prove that an observed phenotypic shift in a wild population is an evolutionary response to harvest in order to apply evolutionary principles to management. Moreover, complete disentanglement of genetic and plastic responses is difficult, except in laboratory experiments, which have limited applicability to management of harvested wild populations (Hilborn 2006).

We recommend assuming that some genetic change due to harvest is inevitable, and to apply basic genetic principles combined with molecular genetic monitoring in order to develop management plans for harvested species. This approach can be especially powerful if archived samples that have been collected over time are available for analysis. Such archived samples are available for many species of fish (scales and otoliths), mammals (bones, teeth, and skin), birds (feathers and skin), and plants (leaves, stems, and reproductive parts).

The molecular genetic analysis of samples collected over a period of time has tremendous untapped potential to inform and guide management of exploited populations. Genetic monitoring can provide a window into the past, as the examples of genetic swamping of the Flamborough Head population of North Sea cod (Hutchinson *et al.* 2003) and the loss of genetic variation in New Zealand snapper illustrate (Hauser *et al.* 2002). Analysis of contemporary samples alone would not have uncovered these important consequences of past exploitation.

18.5.1 Loss of genetic variation

Small populations are most likely to be affected by the loss of genetic variation due to excessive harvest because of their smaller effective population size. Management actions that reduce effective population size below threshold values where loss of genetic variation might have harmful effects should be avoided. As we have seen, substantial loss of genetic variation can occur even when census population sizes are very large, because the genetically effective population size is often much smaller than the census size in many harvested species of marine fishes and invertebrates. The only way to detect such 'cryptic' loss of genetic variation of exploited populations is empirical observation of genetic variation over time (Luikart *et al.* 1998). Genetic monitoring programs can provide a powerful means to detect loss of genetic variation, if enough marker loci are used (Section 22.6).

18.5.2 Unnatural selection

Lowering rates of exploitation is the most direct way to reduce the effects of exploitative selection. Consideration should also be given to management approaches that spread the harvesting across the distribution of age and size classes, or target the intermediate-sized individuals by establishing an upper size limit on individuals to be harvested (especially for long-lived species). These actions will both reduce the long-term effects of exploitative selection and increase the number of older females that produce more and higher quality offspring in the short term (Birkeland and Dayton 2005). However, upper size limits might reduce N_e because individuals surviving to the size where they are 'safe' will contribute a disproportionately large number of progeny, and this is expected to increase the variance of family size (Ryman *et al.* 1981). This effect on N_e might be substantial in some cases, depending on the age distributions before and after introducing the limit. However, the expected effect on heterozygosity over calendar time would be more complicated because this harvest strategy could also lead to a longer generation interval (Figure 18.2).

The effects of selection can sometimes be reduced by harvesting fish after reproduction by changing either the time or location of harvesting. For example, the Northeast Arctic stock of Atlantic cod uses the Barents Sea for feeding, but spawns further south off the northwest coast of Norway (Heino 1998). Harvesting on breeding grounds in the Barents Sea rather than feeding grounds would avoid removing young fish before they can reproduce. In contrast, harvesting on the feeding grounds would select for early maturation because late-spawning fish might be harvested before they mature. However, there are challenges in making this biological solution socially palatable due to its potential economic impact on the fishing fleet and markets through increased seasonality of harvest and supply.

Recovery following relaxation, or even reversal, of exploitative selection will often be much slower than

the initial accumulation of harmful genetic changes (Heino 1998, Loder 2005, de Roos *et al.* 2006). This is because harvesting often creates strong selection differentials, while relaxation of this selective pressure will generally result in only mild selection in the reverse direction (however, see Conover *et al.* 2009). de Roos and colleagues (2006) used an age-structured fishery model to show that exploitation-induced evolutionary regime shifts can be irreversible under likely fisheries management strategies such as belated or partial fishery closure. This effect has been termed "Darwinian debt" (Loder 2005), and has been suggested to have general applicability (Tenhumberg *et al.* 2004). That is, the timescales of evolutionary recovery are likely to be much slower than those on which undesirable evolutionary changes occur. However, gene flow has the potential to accelerate the rate of recovery by restoring alleles or multiple-locus genotypes associated with the trait. For example, trophy-hunting might reduce or eliminate alleles for large horn size, but gene flow from national parks with no hunting might quickly restore alleles associated with large horn size.

18.5.3 Subdivision

The importance of individually managing reproductively isolated populations is obvious and has long been recognized in fisheries (Rich 1939). Nevertheless, application of this understanding is often complex and has proved difficult. For example, the concept of setting a maximum sustained yield in fisheries was developed to ensure long-term sustainability. However, if applied to a mixed-stock fishery, this policy is likely to result in a ratchet-like loss of the less productive local subpopulations (Larkin 1977).

There are two main approaches to this problem:
1 Harvest subpopulations individually.
2 Use genetic monitoring to determine the contribution of each subpopulation to a mixed harvest.

Genetic analysis of such mixed harvests can provide rapid and accurate estimates of the contribution of different subpopulations (Smith *et al.* 2005a). For example, samples of the Bristol Bay sockeye salmon fishery are analyzed shortly after capture and the results of a mixed-stock analysis are radioed to the fleet every other day so that the locations of harvest can be adjusted (Elfstrom *et al.* 2006, personal communication J.S. Seeb).

18.5.4 Protected areas

No-take protected areas have great potential for reducing the effects of both loss of genetic variation and harmful exploitative selection. Models of reserves in both terrestrial (Baskett *et al.* 2005) and marine (Kritzer and Sale 2004) systems support this approach for a wide variety of conditions. However, the effectiveness of such reserves on exploited populations outside of the protected area depends upon the amount of interchange between protected and non-protected areas, and upon understanding of the pattern of genetic subdivision (Palumbi 2003, Dawson *et al.* 2006). It has been suggested that as exploitation pressure intensifies outside protected areas, local protection could select for decreased dispersal distance, thereby increasing isolation and fragmentation, and potentially reducing the genetic capacity of organisms to respond to future environmental changes (Coltman 2007).

Guest Box 18   Long-term genetic changes in the Icelandic stock of Atlantic cod in response to harvesting
Guðrún Marteinsdóttir

Long-term changes in gene frequencies at the *Pantophysin* locus (*Pan* I) in the Icelandic stock of Atlantic cod provide clear evidence for genetic change caused by fishing. These fish have been managed historically as a single stock. There is now evidence for distinct types of cod that appear to have diverged in recent times and use different habitat niches. These different cod are exposed to exploitation in different ways (Pampoulie *et al.* 2008, Grabowski *et al.* 2011).

The Icelandic stock of Atlantic cod is distributed over the continental shelf all around Iceland (Begg and Marteinsdóttir 2003). Recovery of fish carrying data storage tags (DSTs) containing recordings of temperature and depth have shown that cod tagged on the same spawning ground displayed different patterns of behavior (Pálsson and Thorsteinsson, 2003).

Analysis of *Pan* I revealed that the coastal cod were more likely to have the genotype *AA* than *BB*, while the *BB* genotype was more common among the frontal deep-migrating cod (Pampoulie *et al.* 2008). Heterozygotes (*AB*) were detected within both behavior groups. Further analysis of the DST data showed that the different behavioral types occupied distinct seasonal thermal and bathymetric niches and demonstrated generally low levels of overlap throughout the year (Grabowski *et al.* 2011). Even while spawning and despite the fact that the different behavior types occupied the same spawning areas, they appeared to be isolated by fine-scale differences in habitat selection, such as depth and time. Additional evidence supporting these results has shown that cod with different genotypes also differ with respect to morphology (McAdam *et al.* in press) and life history (Jakobsdóttir *et al.* 2011). Indeed, *BB* cod were shown to have greater fin gaps, grow more slowly, and mature later than the *AA* cod.

Together with these data on the stock structure, analysis of genetic material from historically archived otoliths (1948–2002), also revealed unexpected results (Jakobsdóttir *et al.* 2011). During this period of intense fishing, long-term changes were observed in the frequencies of the *Pan* I genotypes, whereas no changes were observed at six microsatellite loci. The changes in genotype frequencies were quite dramatic, with the frequency of *BB* declining from 26% in the 1930s to 5% in 1990s, and the frequency of *AA* increasing to above 50% during the same time period (Figure 18.9). Supported by recent information on the life history and behaviour of the different types and their alignment with the *Pan* I genotypes, it is considered likely that the main cause for the long-term changes in the genotype frequencies stems from historic changes in exploitation patterns of the fishing fleet (Jakobsdóttir *et al.* 2011). Thus, following the termination of trawling within four nautical miles of the coast as well as the extension of the fishery jurisdiction to

(*Continued*)

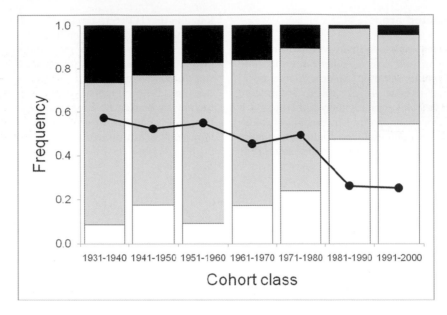

Figure 18.9   Long-term changes in frequencies of the *Pan* I genotypes of Icelandic stock of Atlantic cod (white: *AA*; gray: *AB*; black: *BB*) for different cohort classes (of 10 years each) collected in 1948–2002. The black line shows the frequency of the *B* allele. Adapted from Jakobsdóttir *et al.* (2011).

200 miles in 1976, larger ships with rapidly advancing fishing gear were redirected to deeper waters, thus targeting the *BB* genotypes with a progressively greater effort.

The Icelandic stock of Atlantic cod is one of the most studied commercial fish stocks in the world.

Due to this, it is quite intriguing to discover such a wealth of new information made possible by novel technologies both in terms of data loggers and genetic tools. At the same time, it is worrisome to imagine what we may be missing in terms of other stocks that have not received equal attention.

CHAPTER 19

Conservation breeding and restoration

Guam rail, Example 19.6

For nearly 3000 taxa of birds and mammals, conservation breeding may be the only possible way to avoid extinction.

Torbjörn Ebenhard (1995)

A major challenge of ex situ *conservation will be to ensure that sexually propagated samples of rare plants do not become museum specimens incapable of surviving under natural conditions.*

Spencer C.H. Barrett and Joshua R. Kohn (1989)

Chapter Contents

Conservation and the Genetics of Populations, Second Edition. Fred W. Allendorf, Gordon Luikart, and Sally N. Aitken.
© 2013 Fred W. Allendorf, Gordon Luikart and Sally N. Aitken. Published 2013 by Blackwell Publishing Ltd.

Captive breeding represents the last chance of survival for many species faced with imminent extinction in the wild (Conde *et al.* 2011, Conway 2011). The Guam rail (Example 19.6), black-footed ferret, and the kakapo (Example 19.1) would all almost certainly be extinct if the last few remaining individuals in the wild were not captured and brought into captivity where they have been bred successfully. Captive breeding has played a major role in the recovery of 17 of the 68 vertebrate species whose IUCN threat level has been reduced (Conway 2011). Less charismatic and well-known animal species have also avoided extinction by captive

---

Example 19.1   The kakapo: a conservation breeding challenge

The kakapo (night parrot) is one of the most unusual and rarest birds in the world (Cresswell 1996). It is a large (1.5–4 kg) flightless parrot that was once widespread throughout New Zealand. Kakapo are solitary birds that breed once every 2–5 years and live for many decades. They are the only flightless bird, the only New Zealand bird, and the only parrot in which lek behavior has been observed. Males construct tracks that lead to a shallow bowl on a prominent high point. The low-frequency booming of the males from the bowls to attract females travels up to 5 km and can go on every night for up to four months.

By the 1950s the only known kakapo consisted of a relict population in Fiordland on the South Island. The primary cause of decline was predation by introduced mammals (rats, cats, and stoats). Intensive investigation of this population in the 1970s revealed that it consisted only of a few males (Elliott *et al.* 2001). Another small population was discovered on Stewart Island in 1977. Some 61 kakapo were transferred to other islands because of high rates of predation by cats on Stewart Island. One male (named Richard Henry) from Fiordland, the last known surviving individual from mainland New Zealand, was transferred in 1975 and was part of the conservation breeding program until he died in 2010 at approximately 80 years of age.

The kakapo breeding program reached a low of 51 birds in 1995. More than one-half of all those currently living have been produced by the conservation breeding program. This program has faced a series of challenges associated with the unusual natural history of this bird. The lek breeding system has resulted in very high variance in male mating success. Approximately one-third of the birds born in the breeding program have been sired by a single male (named Felix). Supplemental feeding has been used in an attempt to increase the frequency of breeding. This did increase breeding success but not breeding frequency, but it also produced a significant excess of males (Clout *et al.* 2002). This effect is consistent with the general observation that polygynous birds produce an excess of the larger and more costly sex by females that are in good condition.

Little genetic variation has been found in the founding birds from Stewart Island (Robertson *et al.* 2009). In contrast, the single bird from the mainland, Richard Henry, was more genetically variable, and he was also substantially genetically divergent from the Stewart Island birds at these markers. Thus, the Stewart Island birds probably had a very small effective population size compared with birds on the South Island. Only 40% of eggs produced by females have hatched. This is an extremely low number for a bird species and might be caused by inbreeding depression associated with the low effective population size of birds on Stewart Island (Box 6.1).

A rapid increase in population size is essential for kakapo recovery since it is vulnerable to extinction because of its small population size and low reproductive rate. However, the near-absence of genetic variation in Stewart Island birds meant that it was essential that Richard Henry contribute progeny. However, he only produced three progeny with a single female. Thus, there is a conflict between the demographic needs of increasing the number of birds as soon as possible, and increasing the genetic variation in kakapo by incorporating the progeny of the single founder from the more genetically variable Fiordland population.

The management plan emphasizes the importance of increasing the contribution of the Fiordland population into the breeding population. Felix has been removed from the breeding population. A recent effort to find new birds from Fiordland was not successful. The last known individuals died in the late 1980s and the characteristic booming has not been heard since then. However, a hunter did report seeing what he thought to be a kakapo in 2004.

The kakapo is an icon of conservation in New Zealand. Every animal is named, and births and deaths are national news. Eleven chicks hatched in 2011 and have lived to become juveniles. The population of kakapo now totals 131 birds.

breeding programs. The white abalone became the first marine invertebrate to be listed under the Endangered Species Act (ESA) of the US in 2001. A captive breeding program was begun in 1999 to bring this species back from the brink of extinction and establish a self-sustaining population in the wild (USGS 2002).

Several plant species have also been rescued from extinction by similar intervention. *Kokia cookei* is one of Hawai'i's most beautiful and endangered plants (Mehrhoff 1996). It is a medium-sized tree with very large red and somewhat curved flowers. This species was discovered on the island of Molokai in 1871, and became extinct in the wild in 1919. Extinction resulted from habitat loss and predation by introduced species. The species is apparently adapted to bird pollination, and the loss of native nectar-feeding birds might have contributed to the decline of the species (USFWS 1998). Four seeds were collected from the last remaining tree in 1915. Only one mature tree resulted from these four seeds. This tree produced hundreds of progeny, but none of the progeny survived reintroduction. In 1976, a branch from the last remaining *Kokia cookei* was successfully grafted onto a closely related species. Twenty-eight grafted *Kokia cookei* were transplanted back to Molokai in 1991. Most of these transplants survived, but none have yet flowered (USFWS 2008).

The World Conservation Union (IUCN) has defined *ex situ* **conservation** as "the conservation of components of biological diversity outside their natural habitats" (IUCN 2002). There are a variety of *ex situ* (or **offsite**) techniques that are potentially valuable tools in the conservation of a wide variety of taxa that are threatened with extinction (e.g., captive breeding and germplasm banking). Approximately 20% of all bird and mammal species recognized with IUCN conservation status are maintained in zoos (Conde *et al.* 2011). Unfortunately, however, there are fewer than 50 individuals in zoos for just over half of all these species (Conde *et al.* 2011).

The Russian geneticist N. I. Vavilov initiated systematic collection of plant germplasm; collections have long been used to conserve genetic resources associated with plants used by humans (Frankel 1974). Eberhart *et al.* (1991) reviewed the long-term management of germplasm collections for the conservation of wild plant species. Li and Pritchard (2009) present a helpful overview of the role of *ex situ* populations in plant conservation.

We use the more general term **conservation breeding** to include efforts to manage the breeding of plant and animals species that do not necessarily involve captivity. For example, kakapo breeding is managed by moving groups of birds to predator-free islands, but they are not held in captivity (see Example 19.1).

Captive breeding has played a major role in the development of conservation biology. The first book on conservation biology (Soulé and Wilcox 1980) devoted five of 19 chapters to captive breeding. Modern conservation genetics had its beginnings in the use and application of genetic principles for the off-site preservation of plant genetic resources (Frankel 1974) and the development of genetically sound protocols for captive breeding programs in zoos (Ralls *et al.* 1979). Some early conservation biologists equated conservation genetics with captive breeding. Caughley (1994) concluded that there was little application of genetics in conservation other than captive breeding programs in zoos.

The maintenance of genetic diversity and demographic security are the primary goals for management of conservation breeding programs. These two goals are often compatible. However, there are situations in which maintaining the genetic characteristics of a population can reduce the population growth rate so that a conflict arises (see Example 19.1). This is most likely to occur when a species with only a few remaining individuals is brought into captivity in a last-ditch effort for survival. Demographic security will best be achieved by rapidly increasing the census size of the captive population.

Maintenance of genetic diversity generally requires maximizing effective population size by reducing variation in reproductive success among individuals (see Chapter 7). However, some individuals or pairs of individuals can be much more successful in captivity than others. Thus, maximizing the growth rate of a captive population can actually reduce the effective population size and result in more rapid erosion of genetic variation. In addition, allowing just a few founders to produce most of the captive population is expected to accelerate the rate of adaptation to captive conditions. Thus, maintaining the genetic characteristics of a captive population might come at the cost of reduced population growth rate.

Our goal in this chapter is to consider the genetic issues involved in conservation breeding and the introduction of individuals into the wild (Doremus 1999). When should captive breeding be considered as a conservation option? What are the potential problems with a conservation breeding program? What criteria should be used when choosing populations and individuals to introduce or move between populations? We also provide an overview of the genetic principles involved in managing captive populations. Interested readers should consult other sources that provide detailed instructions for genetic management of conservation breeding programs (e.g., Ballou and Foose 1996, Frankham 2008, Miller and Herbert 2010).

19.1 THE ROLE OF CONSERVATION BREEDING

There are three primary roles of offsite conservation breeding as part of a management or recovery program to conserve a particular species:

1 Provide demographic and genetic support for wild populations.
2 Establish sources for founding new populations in the wild.
3 Prevent extinction of species that have no immediate chance of survival in the wild.

The genetic objectives of these three roles are very different. Captive individuals used to provide demographic and genetic support for wild populations should be genetically matched to the wild population into which they will be introduced, so that they do not reduce the fitness of the population by outbreeding depression. In contrast, introduced new populations should have enough genetic variation present so that they can become adapted to their new environment by natural selection. In the last case, the initial concern of a captive breeding program is to insure that the species can be maintained in captivity (Midgley 1987). This might involve preferentially propagating individuals capable of reproducing in captivity and might result in adaptation to captivity.

Captive breeding has made many contributions to conservation other than just conservation breeding (e.g., public education, research, and professional training). The public display of species plays a very important role in conservation in providing opportunities for the public to come into contact with a wide variety of species that would otherwise just be names or pictures in books. The first author of this book became interested in biology because of visits to the Philadelphia Zoo as a child on class trips.

The goals of a display program are to establish an easily managed population that is well adapted to the captive environment (Frankham *et al.* 1986). These experiences provide an excellent opportunity for edu-

cation and also provide the setting for the public to develop affection and appreciation of a wide variety of species. Most people around the world will never have the opportunity to see a tiger, elephant, or a great ape in the wild. Zoos provide an important role in allowing the public to develop a first-hand connection to these species. People are more likely to support conservation efforts if they have knowledge, understanding, and appreciation of the species involved.

There is also a danger in this. Seeing elephants or tigers in the zoo can encourage the public and politicians to believe that these 'species' are now protected from extinction. However, a species is not just a collection of individuals that has been removed from the ecosystem in which they have survived and evolved for millions of years.

A condor is 5 percent feathers, flesh, blood, and bone. All the rest is place. Condors are soaring manifestations of the place that built them and coded their genes. (Devall and Sessions 1984, p. 317)

The ecologist David Barash (1973) said this in a somewhat different fashion:

Thus, the bison cannot be separated from the prairie, or the epiphyte from its tropical perch. Any attempt to draw a line between these is clearly arbitrary, so the ecologist studies the bison-prairie, acacia-bromeliad units.

Thus, the display of charismatic species to the public should be accompanied by educational efforts that emphasize that long-term species existence can only occur within the complex web of connections and interactions in their native ecosystems.

19.1.1 When is conservation breeding an appropriate tool for conservation?

This is an important and difficult question. Conservation breeding should be used sparingly because it is difficult and expensive, and worldwide resources are limited. In addition, directing resources to captive breeding and taking individuals into captivity might hamper efforts to recover species in the wild.

Captive breeding is perhaps too often promoted as a recovery technique. For example, Conservation Assessment and Management Plans under the Conservation Breeding Specialist Group of the IUCN have recommended captive breeding for 36% of the 3314 taxa considered (Seal *et al.* 1993). In the US, captive breeding has been recommended in 64% of 314 approved recovery plans for species listed under the ESA (Tear *et al.* 1993). The resources are not available to include captive breeding in the recovery plans of such a high proportion of species. It is important that it be used only for those species in which it can have the greatest effect.

Intensive field-based conservation can be an effective and cost-efficient alternative to captive propagation. Balmford *et al.* (1995) found that *in situ* management of well-protected reserves for large-bodied mammals resulted in comparable population growth rates and was consistently less expensive than captive propagation. These authors suggest that captive breeding is most cost-effective for smaller-bodied taxa and will only remain the best option for large mammals that are restricted to one or two vulnerable wild populations.

19.1.2 Priorities for conservation breeding

It is clear that only a relatively small proportion of the thousands of animal species that are threatened in the wild can be maintained in captivity because of constraints on space and other resources (Balmford *et al.* 1996, Snyder *et al.* 1996). It is generally assumed that a maximum of roughly 500 animal species could be maintained offsite in conservation breeding programs (IUDZG/CBSG 1993). As we have seen, however, captive breeding programs are often recommended for many taxa. Given this situation, what criteria should be used to determine which species should be maintained in conservation breeding programs?

Zoos have historically focused on large and charismatic species in breeding programs. Balmford *et al.* (1996) spelled out three general sets of criteria that should be considered in selecting candidate animal species for captive breeding:

1 Economic considerations. Which species can be conserved successfully in a captive breeding program most economically?

2 Biological suitability for captive breeding. Which species can be bred and raised successfully in captivity?

3 Likelihood of successful reintroduction. For which species is successful reintroduction to the wild a realistic option?

We suggest a fourth criterion: the potential effect on habit preservation. Will development of a captive breeding program increase or decrease the likelihood of habitat protection?

Invertebrates are generally better candidates for captive breeding than are large and charismatic vertebrates for which enormous resources have been used (Pearce-Kelly *et al.* 1998). Invertebrates have a relatively high probability of success for both the rearing and release phases. They also have small size and require relatively little space and cost. They typically have high reproductive potential and population size increases relatively rapidly in captivity and after release. Finally, there is a wealth of knowledge and techniques for rearing for numerous invertebrate species. For example, crickets, katydids, beetles, and butterflies have been widely and successfully raised in captivity.

A recent tragic accident with *Powelliphanta augusta*, a large, carnivorous land snail, demonstrates some of the dangers of captive breeding. This species was discovered in 1996 and was identified as a new species on the basis of molecular genetics and morphology (Walker *et al.* 2008). This species is only known from a small area on the South Island of New Zealand. The entire known habitat of this species was destroyed by coal mining by 2006. Some 2300 individuals were translocated to two other areas, where their survival suggests that the populations will not persist. Some 1600 snails were taken into captivity to develop a captive breeding program. However, a refrigerator malfunction resulted in 800 snails being frozen to death in November 2011.

Plants are generally better candidates for offsite breeding programs than animals for a variety of reasons (Templeton 1991, Li and Pritchard 2009). Many plants can be maintained for long periods as dormant seeds. This can be used to increase the generation interval and therefore reduce the rate of genetic change during offsite breeding. Other plants, such as trees, live a long time, so offsite breeding programs that take hundreds of years might only represent a handful of generations. This again will minimize the rate of genetic change by genetic drift and selection. Other problems with offsite breeding can be reduced because of the variety of modes of reproduction that are possible for plants with short generation intervals (e.g., selfing, apomixis, and clonal reproduction; Figure 19.1).

Guidelines for selecting candidate plants for **conservation collections** have been presented by the Center for Plant Conservation (1991). The decision to protect (or abandon) a particular population or species must be made within a larger framework of conservation. In addition, these guidelines are based on a natural genetic hierarchy: species, populations (or ecotypes),

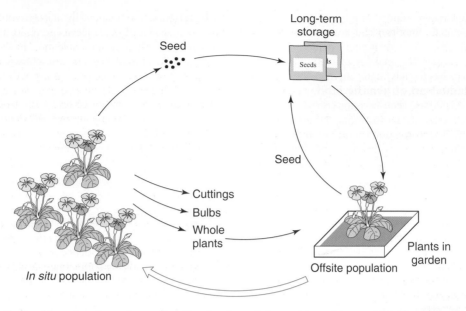

Figure 19.1 Possible modes of reproduction for offsite breeding of plants and possible interchange between offsite plants and *in situ* populations. Redrawn from Brown and Briggs (1991).

individuals, and alleles. The goal is to address diversity at several levels of organization rather than sampling a particular species without regard to genetic variation and future long-term viability. This approach includes five sampling decisions:

1 Which species should be collected?
2 How many populations within a species should be sampled?
3 How many individuals should be sampled per population?
4 How many propagules should be collected from each individual?
5 When should collections be made from multiple years?

19.1.3 Potential dangers of captive propagation

The use of captive breeding has been controversial. It is expensive, is sometimes ineffective, and can harm wild populations both indirectly and directly if not done correctly (Snyder *et al.* 1996). Perhaps the most serious criticism is that efforts directed towards captive breeding detract from grappling with the real problems (e.g., loss of habitat and protection). The dangers of

captive breeding are clearly demonstrated by the use of fish hatcheries to maintain stocks of Pacific salmon on the west coast of North America (Example 19.2).

19.2 REPRODUCTIVE TECHNOLOGIES AND GENOME BANKING

Reproductive technologies initially developed for agricultural species (e.g., cattle, sheep, and chickens) can be transferred to some related wild species to facilitate their conservation. These technologies include genome banking, cryopreservation, artificial insemination, and cloning.

Genome banking is the storage of sperm, ova, embryos, seeds, tissues, or DNA. Genome resource banking can help to move genetic material without moving individuals. It might, for example, allow for managed gene flow into isolated populations without the risks of translocating individuals. Genome banking also serves as an insurance against population or species extinction. It lengthens generation intervals and thereby reduces random genetic drift. It increases efficiency of captive breeding and reduces the number of individuals kept in captivity. Finally, banks are a source of tissue and DNA for basic and applied research.

Example 19.2 Who needs protection? We have hatcheries.

Fish hatcheries have a long and generally unsuccessful history in conservation efforts to protect populations of fish. Pacific salmon began a rapid decline on the west coast of the US in the late 1800s with the advent of the salmon-canning industry (Lichatowich 1999). The State of Oregon sought advice from the newly-created US Commission on Fish and Fisheries that was directed by Spencer Baird, a scientist with the Smithsonian Institution. In 1875 Baird (Lichatowich 1999, p. 112) recommended that:

> . . . instead of protective laws, which cannot be enforced except at very great expense and with much ill feeling, measures be taken, either by the joint efforts of the States and Territories interested or by the United States, for the immediate establishment of a hatching establishment on the Columbia River, and the initiation during the present year of the method of artificial hatching of these fish.

Unfortunately, this recommendation from the leading fisheries scientist of the US set in motion a paradigm for the conservation of salmon through hatcheries rather than facing the real problems of excessive fishing, dams that blocked spawning migrations, and habitat changes in the spawning rivers and streams. These efforts have failed profoundly (Meffe 1992). Some 26 different groups of Pacific salmon and anadromous rainbow trout (steelhead) are listed as threatened or endangered under the ESA at the time of writing this chapter. The role of hatcheries in salmon conservation continues to be controversial. There is current disagreement as to whether hatchery populations should be considered part of the distinct population segments that are listed and protected under the ESA (Myers *et al.* 2004, Box 16.1).

A number of recent studies have been performed to assess the possible genetic effects on wild populations of releasing hatchery fish into the wild. The consensus is clear: hybridization with hatchery fish has a dramatic harmful effect on the fitness of wild populations (Reisenbichler and Rubin 1999, McGinnity *et al.* 2003, Araki and Schmid 2010, Guest Box 4).

Reisenbichler and others have published several papers that compare the relative fitness of progeny of hatchery steelhead (the anadromous form of rainbow trout) to wild fish. Three primary results emerge from these studies. First, progeny from hatchery fish uniformly show reduced rates of survival. For example, Leider *et al.* (1990) found that the reproductive success of hatchery fish spawning in the wild relative to wild fish ranged from 5 to 15% in four successive year-classes. Second, progeny of hatchery fish have reduced survival at all life history stages between emergence from the gravel until returning from the ocean as adults (Figure 19.2, Reisenbichler and Rubin 1999). Finally, the decline of fitness observed in hatchery fish is proportional to the number of generations that the hatchery stock has been maintained in captivity. Chilcote *et al.* (2011) found that intrinsic measures of population productivity of steelhead, coho salmon, and Chinook salmon in Oregon declined as a function of the number of hatchery fish in the spawning population.

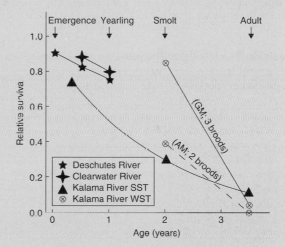

Figure 19.2 Results showing reduction in relative survival throughout the life cycle of progeny from hatchery steelhead spawning in the wild relative to the survival of progeny from wild fish. Data for the Kalama River winter steelhead (WST) are the geometric means (GM) for three year classes, or the arithmetic means (AM) for two year classes with an exceptional year class omitted. From Reisenbichler and Rubin (1999).

Genome banking is widely used for agricultural crop and farm animal preservation to help insure future agricultural productivity. Genome banking is also increasingly used for wild taxa. For example, for wild animals there exists a genome bank at the Smithsonian Institution's National Zoo where there are more than 1500 samples of frozen sperm or embryos from 69 species (including ~2% of mammalian species world-wide). Similarly, the San Diego Zoo maintains a 'frozen zoo' with samples (including cell lines and tissues) from more than 7000 species of endangered mammals, birds, and reptiles.

For wild plants there is increasing interest to establish seed banks. Researchers suggest that many tropical and rainforest species seeds can be banked. However, the ability to bank seeds is known for only about 4% of angiosperms (flowering plants). The FAO (Food and Agriculture Organization) reports that 6 million accessions exist in over 1300 seed banks around the world. However, less than about 10% are from wild plants.

Cryopreservation is the freezing and storage (often in liquid nitrogen at −180°C) of sperm, ova, embryos, seeds or tissues to manage and safeguard against loss of genetic variation in agricultural or wild populations. It is the principal storage method for animal material. In some plant species, seeds can be preserved dry at room temperature and can remain viable for 50–200 years. But for other species, longevity increases if seeds are kept at −20°C or colder.

Artificial insemination (AI) is a widely used and important technique for captive breeding. AI allows animals to breed that would not breed naturally, perhaps due to behavior problems such as aggression towards mates. Further, a genetically important male can still be used within a breeding program long after his death. Finally, instead of moving animals, sperm can be collected, cryopreserved, and shipped for AI (as mentioned above). AI has been used, for example, in breeding programs for the black-footed ferret and killer whales in the US, cheetah in Namibia, koalas in Australia, and gazelles in Spain and Saudi Arabia. AI was recently used successfully with corn snakes at the Henry Doorly Zoo in Omaha, Nebraska.

The use of cloning for conservation is controversial (Box 19.1). Cloning is generally conducted by (1) removing the nucleus from a donor egg cell of the

---

Box 19.1   How useful is cloning for animal conservation?

Cloning of endangered species might come to play a role in conservation (Bawa *et al.* 1997, Ryder and Benirschke 1997, Loi *et al.* 2001). Cloning could potentially allow resurrection of a recently extinct taxa (Holt *et al.* 2004, Pask *et al.* 2008). For example, Folch *et al.* (2009) cloned an individual from an extinct subspecies, Pyrenean ibex. DNA was taken from skin samples of the last-known animal of this subspecies which died in 2000. The DNA in eggs from domestic goats was replaced by DNA from the skin sample. The newborn ibex died shortly after birth due to physical defects in its lungs. Other cloned animals, including sheep, have been born with similar lung defects.

Cloning is very expensive, and it is technologically feasible for only a few species that are related to model research organisms (e.g., mice) or important in agriculture (e.g., cattle and sheep). Further, the success rate is very low: less than 0.1 to 5% of re-nucleated embryos lead to a live birth (Holt *et al.* 2004). It is generally agreed that long-extinct species, such as the woolly mammoth from the frozen Siberian permafrost, cannot be cloned because their DNA is fragmented.

Another potential advantage of cloning is to help bolster populations and avoid extinction of critically endangered species, such as the panda. However, the benefits of cloning compared with more traditional conservation breeding programs are questionable and the disadvantages are substantial.

Disadvantages are that cloned individuals and populations are genetically identical and thus would be equally susceptible to the same infectious diseases and have low adaptive potential to environmental change. Further, the money spent on cloning would often be better spent preserving habitat and conducting less expensive breeding programs. Extensive healthy habitats are necessary to ensure long-term persistence of any species.

Cloning should not be viewed as an alternative to habitat preservation and conservation breeding programs. In certain limited scenarios, perhaps cloning could be a last-resort approach in combination with habitat conservation and breeding programs to help insure species persistence and even recover extinct taxa. There are a few programs dedicated to storing frozen tissues in hopes that one day these can be used to produce living animals, for example, the Frozen Ark (http://frozenark.org/) and the Frozen Zoo (http://www.sandiegozooglobal.org).

animal that will carry the cloned embryo, and (2) injecting into the carrier's egg cell the nucleus from a cell of the animal to be cloned. For example, a nucleus from a tissue cell of a European wild sheep (mouflon) was injected into the nucleus-free cell of a close relative species, the domestic sheep. The resulting mouflon lamb was born and mothered by the domestic sheep. This is an example of cross-species cloning, which is more difficult than within-species cloning because of risks of incompatibility of mitochondrial genes from the donor egg and the nuclear genes from the animal to be cloned.

These technologies provide valuable opportunities for protecting species and increasing genetic variation within species on the brink of extinction. Nevertheless, it is essential that they be integrated so that they support ongoing conservation efforts rather than be used as alternatives.

19.3 FOUNDING POPULATIONS FOR CONSERVATION BREEDING PROGRAMS

Developing a captive breeding program begins with the selection of the founding individuals. In many situations when a species is on the brink of extinction, there is no choice involved because all remaining individuals are brought into captivity. In other cases, however, captive breeding programs are established when long-term survival of a species in the wild is unlikely, even though many individuals currently occur in the wild (e.g., tigers). In such cases, there are a variety of questions to be decided. Which subspecies or populations should be the source of individuals to be brought into captivity? How many subspecies or populations should be maintained? Should subspecies and populations be maintained separately or mixed together? How many individuals should there be in the founder population?

Confirming the taxonomic identification of individuals used in a captive breeding program is essential. There are many examples of captive breeding programs in which molecular analysis detected that hybrid individuals were unknowingly used in the captive breeding program: Przewalski's horse (Oakenfull and Ryder 1998), greater rhea (Delsuc *et al.* 2007), and cutthroat trout (Metcalf *et al.* 2007).

A recent study of 15 microsatellite loci in a breeding program for an endangered giant Galápagos tortoise from the island of Española discovered that one of 473 captive-bred tortoises released back to Española carried alleles at eight of 15 loci that were not present in any of the 15 founders (Milinkovitch *et al.* 2007). Bayesian clustering analysis of microsatellite genotypes at nine loci suggested that this anomalous individual was a hybrid between an Española tortoise and a tortoise from the island of Pinzón (see Figure 19.3). It is thought that a parent of this individual was transported by humans to Española before the initiation of the captive

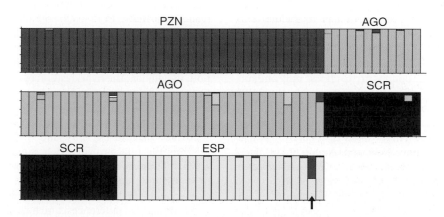

Figure 19.3 Inferred ancestry of giant Galápagos tortoises from the islands of Pinzón (*PZN*), Santiago (*AGO*), Santa Cruz (*SCR*), and Española (*ESP*) based upon nine microsatellite loci using the computer program STRUCTURE (Pritchard *et al.* 2000). The arrow indicates the hybrid individual found on Española that was apparently produced by a mating between Española and Pinzón parents. From Milinkovitch *et al.* (2007).

breeding program. Efforts are currently underway to identify individuals carrying genes from Pinzón tortoises and remove them from the Española population.

19.3.1 Source populations

Selecting the founding individuals and populations for a captive breeding program is an important and difficult problem for many species. Source populations should be selected in order to maximize genetic and ecological (adaptive) diversity. For example, there are currently four remaining subspecies of tigers in the wild. Recent genetic results have indicted substantial genetic divergence among these subspecies (Luo *et al.* 2004). A strong argument can be made that each of these subspecies represents a separate evolutionarily significant unit (ESU, see Chapter 16) and separate captive breeding programs should be established for each. However, space and other resources for captive breeding of tigers are limited. There are currently approximately 1000 spaces for tigers in captive breeding programs throughout the world. We then face a dilemma. How should we partition available captive breeding spaces among the four subspecies to enhance survival and retention of genetic variation?

Maguire and Lacy (1990) provided a very informative consideration of this problem. They identified three conservation goals: (1) to maximize the number of surviving subspecies; (2) to maximize genetic variation at the species level; and (3) to maximize genetic diversity at the subspecies level. They chose a timeframe of 200 years (32 tiger generations) to match recommendations for long-term conservation plans (Soulé *et al.* 1986). Their analysis also included consideration of the probabilities of persistence of the subspecies in the wild.

The two extreme options are to choose only one subspecies for captive breeding, or to divide the 1000 spaces equally among the four subspecies. They assume that the N_e/N_C ratio in captive tigers is 0.4 (Ballou and Seidensticker 1987). In the latter case, each of the four subspecies would have N_e of approximately 100 tigers (250×0.40). Using equation 6.7, we would expect to lose approximately 14% of the heterozygosity in each subspecies after 200 years ($t = 32$). General recommendations suggest a goal of retaining at least 90% of the heterozygosity after 200 years (Soulè *et al.* 1986). This would require an N_e of approximately 150, and an N_C of 375 for each subspecies. Maguire and Lacy

(1990) recommended devoting half of the available captive spaces to the *tigris* subspecies and dividing the remainder equally among the other three subspecies. See Example 22.5 for a consideration of how molecular genetic variation can be used to assign tigers already in captivity to their correct subspecies.

19.3.2 Admixed founding populations

Another option is to establish a captive population by hybridizing genetically divergent populations. For example, the State of Montana established a captive population of westslope cutthroat trout in 1985 to be used in a variety of restoration projects. Geographic populations of westslope cutthroat trout show substantial genetic divergence among populations, $F_{ST} = 0.32$ (Allendorf and Leary 1988). Space limitations required that only a single captive population could be maintained. The choice was to use a single representative population to establish the captive population, or to create a hybrid captive population by crossing individuals from a wide spectrum of native westslope cutthroat trout populations.

Do we choose one population to be brought into captivity or do we create a captive population by hybridizing individuals from different populations? The genetic choice that we face here is between genes and genotypes. We can maximize the allelic diversity of westslope cutthroat trout in the captive population by including fish from many streams in our founding population. However, hybridizing these populations will cause the loss of the unique combination of alleles (genotypes) that exists in each population. These genotypes might be important for local adaptation. These combinations of genes, and the resulting locally adapted phenotypes, will be lost through hybridization. In addition, the hybridization of different populations could result in outbreeding depression (see Section 17.3).

In some cases, genetically distinct populations have been brought into captivity and hybridized without realizing potential problems. For example, we saw in Table 3.3 that approximately 20% of orangutans born in captivity were hybrids between orangs captured in Borneo and Sumatra. These two populations are fixed for chromosomal differences, and it has been proposed they should be considered separate species. Current conservation breeding plans avoid the production and use of hybrids between these taxa.

There are no simple prescriptive answers to the best strategy in establishing a captive population. In the case of westslope cutthroat trout, the captive population was established by mixing from some 20 natural populations. There was some concern in this case about possible outbreeding depression caused by mixing together so many local populations. However, the alternative of using just one local population, which would contain such a small proportion of the total overall genetic variation, was considered less desirable.

19.3.2.1 Interspecific hybrid founding populations

Some have suggested that individuals from interspecific hybrid swarms could be used as founders to recover extinct species. The giant tortoise from the island of Floreana in the Galápagos became extinct within decades of Charles Darwin's visit, in which he described that massive numbers of these creatures were being removed to be stored in the hulls of ships to be used for food (Poulakakis *et al.* 2008). Examination of mtDNA and nuclear markers in historical museum specimens and extant tortoise from the nearby island of Isabela indicated the existence of a large number of individuals who are admixed and possess ancestry from both Floreana and Isabella populations. These admixed individuals could be used to found a captive breeding program in which screening for molecular markers could be used to increase the contribution from ancestors from Floreana.

19.3.3 Number of founder individuals

The number of founders recommended for establishing a captive population depends substantially upon the desired proportion of rare alleles to be captured, and upon the population growth rate expected in captivity. Approximately 30 diploid founders are required to have a 95% probability of sampling an allele at frequency 0.05. However, with 30 founders there is only approximately a 45% probability of including an allele of frequency 0.01 (see equation 6.8 and Figure 6.8). This probability increases to 63% if 50 founders are used; approximately 150 founders are needed to have a 95% probability of including an allele at a frequency of 0.01. Thus, we recommend more than 30 founders and preferably at least 50. Fifty founders will maintain approximately 98% of the original heterozygosity (see

equation 6.6). If the rate of population growth is low, additional founders or subsequent supplementation with additional individuals is recommended.

19.4 GENETIC DRIFT IN CAPTIVE POPULATIONS

A primary genetic goal of captive breeding programs is to minimize genetic change in captivity. Genetic changes in captive populations might reduce the ability of captive populations to reproduce and survive when returned to the wild. There are two primary sources of genetic change in captivity: genetic drift and natural selection.

19.4.1 Minimizing genetic drift

Genetic drift causes the loss of heterozygosity and allelic diversity. This reduced genetic diversity can have several consequences. First, inbreeding depression might limit population growth and lower the probability that the introduced population will persist. Second, reduced genetic diversity will limit the ability of introduced populations to evolve in their new or changing environments. In general, the effects of genetic drift can be minimized in captivity by managing the population to maximize the effective population size.

The primary method for minimizing genetic drift and maximizing effective population size is to equalize reproductive success among individuals. This is especially important for the founder individuals of a captive breeding program. We saw in Chapter 7 that the ideal population includes random variability in reproductive success. Under controlled captive conditions, it can be possible to reduce variance in reproductive success to near zero. In this case, the effective population might actually be nearly twice as great as the census population size (see equation 7.5). The most effective method to reduce variance in reproductive success depends upon the type of breeding scheme used in captivity (see Section 19.5).

19.4.2 Deleterious alleles and mutational meltdown

Deleterious alleles that are present at low frequencies in natural populations might drift to high frequencies

---

Example 19.3 Chondrodystrophy in California condors

The captive population of California condors was founded with the last remaining 14 individuals in 1987 (Ralls *et al.* 2000). California condors have bred well in captivity and the first individuals were reintroduced into the wild in 1992. However, nearly 5% of birds born in captivity have suffered from chondrodystrophy, a lethal form of dwarfism. This defect is apparently caused by a recessive allele that occurs at a frequency of 0.09 in the captive population.

Such deleterious alleles are likely to occur in any captive population founded by a small number of founders (Laikre 1999). What should be done? Ralls *et al.* (2000) considered three management options for this allele: (1) reduce its frequency by selection; (2) minimize its phenotypic frequency by avoiding matings between possible heterozygotes; and (3) ignore it.

Selective removal of this allele would require not using possible heterozygotes in the breeding program.

Under this scheme, over 50% of all birds would be eliminated from the breeding population. This is a very high cost to pay for elimination of a trait that affects less than 5% of all birds. In addition, it is likely that other traits caused by deleterious recessive alleles occur in this population. Selective removal of relatively low frequency alleles at multiple loci is generally not worth the cost of reducing the effective population size and further eroding genetic variation in the captive populations.

Ralls *et al.* (2000) recommended minimizing the phenotypic frequency of this trait by avoiding pairings between possible heterozygotes. They suggested that some selection would be feasible once the captive population has reached the carrying capacity in captivity. In addition, possible heterozygotes could be given a lower priority as candidates for introduction.

---

in captive populations because of the founder effect combined with relaxed natural selection (Example 19.3). Joron and Brakefield (2003) have suggested that relaxed natural selection in captivity can mask reduced fitness due to inbreeding. For example, wolves bred for conservation purposes in Scandinavia were found to have a high frequency of hereditary blindness apparently caused by an autosomal recessive allele (Laikre *et al.* 1993). Only six founders were originally brought into captivity (Figure 19.4). At least one of these founders apparently was heterozygous for a recessive allele associated with blindness. It is also possible that partial blindness might actually have some advantage in captivity for a wild animal such as a wolf.

Some populations of salmon and trout have high frequencies of null alleles at enzyme coding loci that are enzymatically nonfunctional (Section 5.4.2, Allendorf *et al.* 1984). Such alleles are not as deleterious in these fishes as in other species because gene duplication provided by their polyploid ancestry provides some redundancy (Allendorf *et al.* 1984). Nevertheless, developmental studies have found that these alleles do have harmful effects on developmental rate and developmental stability (Leary *et al.* 1993a).

Extensive surveys of natural and hatchery populations of trout and salmon indicate that enzymatically null alleles occur at high frequencies only in hatchery populations or natural populations that are restricted to lakes (Allendorf *et al.* 1984, Leary *et al.* 1993a). For example, a null allele at a lactate dehydrogenase (LDH) locus occurred at a frequency of 0.122 in a hatchery population of rainbow trout (Leary *et al.* 1993a). Homozygotes for this allele exhibited a 70% reduction in LDH activity in heart tissue. These hatchery populations usually have a large number of founders, so it is unlikely that the founder effect contributed to the high frequencies of these alleles.

In addition, new mildly deleterious mutations will occur in captive populations; these mutations might drift to high frequency in populations with a small N_e because natural selection is not effective in small populations (Chapter 8). Many of these new mutations with mild deleterious affects could accumulate in small populations and lead to so-called 'mutational meltdown' (Lande 1995, Chapter 14).

19.4.3 Inbreeding or genetic drift?

It is crucial to distinguish between the effects of inbreeding and genetic drift in captive populations. Some inbreeding (the mating of related individuals) will be unavoidable in small captive populations; this is

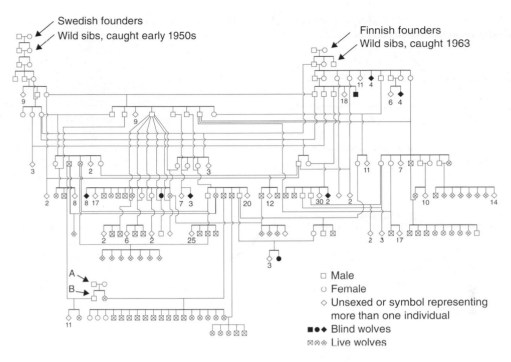

Figure 19.4 Pedigree of the captive population of 442 wolves bred in Scandinavian zoos as of January 1988. The numbers below the symbols indicate the number of individuals in a particular family. A, Russian founders, B, full-sibs imported in 1980. From Laikre and Ryman (1991).

the so-called inbreeding effect of small populations (see Chapter 6). In general, inbreeding should be avoided as much as possible in captive populations because the reduced fitness associated with inbreeding depression might threaten short-term persistence of the captive population.

However, the loss of genetic variation by genetic drift is a more serious and lasting effect than inbreeding. The harmful effects of inbreeding last for a single generation. That is, a mating between an inbred individual and an unrelated mate will produce a non-inbred progeny. However, the loss of alleles in a species through genetic drift is permanent. The long-term genetic viability of a captive population is more affected by the unequal representation of founders and effective population size than by matings between related individuals.

Schemes of mating with maximum avoidance of inbreeding will minimize the initial rate of loss of heterozygosity. However, perhaps surprisingly, there are often systems of mating that do a better job of retain-

ing heterozygosity in the long-term (Kimura and Crow 1963, Robertson 1964, Wright 1965b).

19.5 NATURAL SELECTION AND ADAPTATION TO CAPTIVITY

Natural selection will occur in captivity and bring about adaptation to captive conditions. Such changes will almost inevitably reduce the adaptiveness of the captive population to wild or natural conditions. For example, tameness in response to contact with humans is generally advantageous in captivity, but can have serious harmful effects in the wild.

The emphasis of captive breeding protocols has been primarily to reduce genetic drift by maximizing effective population size. This emphasis is appropriate for captive breeding programs of mammals and birds in zoos that have a relatively small number of individuals managed using pedigrees (Ballou and Foose 1996). However, increasing effective population size for some captive species (e.g., fish and plants) can increase the rate of adaptation to captive conditions.

19.5.1 Adaptation to captivity

Adaptation to captivity is probably the greatest threat in species that produce many offspring (e.g., insects, fish, amphibians, etc.). For example, females of many fish species produce thousands of eggs. Extremely strong natural selection can occur in the first few generations when founding a captive hatchery population of fish. Christie *et al.* (2012) found evidence for adaptation to hatchery conditions after only a single generation in captivity for anadromous rainbow trout (steelhead). Williams and Hoffman (2009) have provided a valuable review of the literature for captive conservation breeding programs that reported a strategy to minimize adaptation to captivity.

Darwin (1896) was very interested in the genetic changes brought about by selection during the process of domestication of animals bred in captivity. He attributed such changes to three mechanisms:

1 Systematic selection
2 Incidental (unintentional) selection
3 Natural selection

Systematic selection occurs when purposeful selection occurs for some desirable characteristics. For example, many hatchery populations of fish are selected for rapid growth rate. Incidental selection occurs when captive management favors a particular phenotype without being aware of their preference. For example, hatchery personnel might unconsciously favor a particular phenotype (e.g., large, colorful, etc.) when choosing fish to be mated. Finally, natural selection will act to favor those individuals who have characteristics that are favored under captive conditions. For example, many wild fish will not feed when brought into captivity. Therefore, natural selection for behaviors that permit feeding and surviving in captivity will be very strong.

This issue was raised many years ago by A. Starker Leopold (1944) in his consideration of the effects of release of 14,000 hybrid (wild × domestic) turkeys on the wild population of turkeys in southern Missouri, US. It was common practice throughout many parts of the US to release such hybrid turkeys in order to enhance wild populations that were hunted. Hybrid stocks were used because of the great difficulty in raising wild turkeys in captivity. He found that the hybrid birds were unsuccessful in the wild because of their tranquility, early breeding, and inappropriate behavior of chicks in response to the warning note of the hen:

> *Wild turkeys are wary and shy, which are advantageous characters in eluding natural and human enemies. They breed at a favorable time of the year. The hens and young automatically react to danger in ways that are self-protective . . . Birds of the domestic strain, on the other hand, are differently adapted. Many of the physiological reactions and psychological characteristics are favorable to existence in the barnyard but many preclude success in the wild. (Leopold 1944)*

Systematic selection and incidental selection can be greatly reduced in captivity by intensive effort. However, genetic divergence between wild and captive populations because of natural selection cannot be eliminated. Efforts are currently underway to reduce these effects in fish hatcheries by mimicking the natural environment. Nevertheless, it is impossible for a hatchery to simulate the complex and dynamic ecological heterogeneity of a natural habitat. In fact, any hatchery must create an environment that differs dramatically from the natural one to achieve its goal of producing more progeny per parent than occurs under natural circumstances. By definition, then, a goal of reducing mortality while retaining natural environmental conditions cannot be achieved; it is impossible to synthetically create conditions that are both identical to the natural ones and at same time provide a basis for increased survival.

19.5.2 Minimizing adaptation to captivity

Natural selection is most effective in large populations (see Section 8.5). Thus, rapid adaptation to captivity is expected to occur most rapidly in captive populations with a large N_e. Minimizing variance in reproductive success via pedigree management will also act to delay adaptation to captive conditions. However, pedigree management is probably not necessary or not practical for many species kept in captivity.

In contrast to this view, Bryant and Reed (1999) have suggested that the absence of any selection in captivity can lead to the deterioration in fitness and that captive programs should allow the alleles of less adapted individuals to be lost from the captive population. We agree with Lacy (2000a) that Bryant and

Reed overestimated the likely deterioration of fitness in this case, and they also overlook several other problems with the strategy that they propose.

In species with high fecundity (such as many fish, amphibians, and insects), rapid adaptation to captivity is most likely to occur because hundreds of progeny can be produced by single matings. Thus, natural selection can be very intense especially in the first few generations after being brought into captivity.

For example, the Apache trout, which is native to the southwestern US, is currently listed as threatened under the ESA. A single captive population, originating from individuals captured in the wild in 1983 and 1984, is the cornerstone of a recovery effort with an established goal of establishing 30 discrete populations within the native range of this species. Advances in culture techniques and the high fecundity of these fish have resulted in a program that spawns hundreds of mature fish and produces hundreds of thousands of fry per year for reintroduction.

The large number of spawners suggests that the effective population size of this population is very large, so the loss of genetic variation due to drift is not a concern. Nevertheless, these circumstances are ideal for natural selection to bring about rapid adaptation to captive conditions that would reduce the probability of successful establishment of reintroduced populations.

19.5.3 Interaction of genetic drift and natural selection

In many regards, actions taken to reduce genetic drift will also reduce the potential for natural selection. For example, minimizing variability in reproductive success among individuals will both maximize N_e and reduce the effects of natural selection (Allendorf 1993). However, as we saw in Chapter 8, natural selection is most effective in very large populations. Therefore, intermediate size populations would be large enough to avoid rapid genetic drift, but not so large that even weak natural selection could bring about adaptation to captive conditions.

Woodworth *et al.* (2002) tested these predictions with experimental populations of *Drosophila* to mimic captive breeding. They evaluated adaptation to captivity under benign captive conditions for 50 generations using effective population sizes of 25, 50, 100, 250, and 500. The small populations demonstrated reduced fitness after 50 generations due to inbreeding depression. The large populations demonstrated the most rapid adaptation to captive conditions. The least genetic change in captivity was observed in intermediate size populations as measured by moving the populations to simulated wild conditions (Figure 19.5). These authors suggested that adaptation to captivity

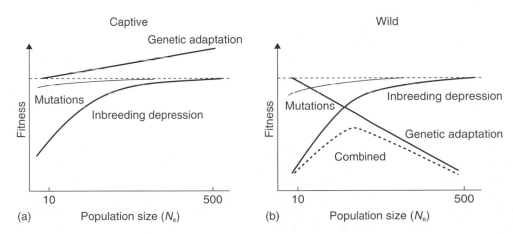

Figure 19.5 Expected relationship between fitness and population size (N_e) due to the inbreeding effect of small populations and genetic adaptation to captivity. The combined line represents the net effects of both factors. The effects are shown for populations maintained for approximately 50 generations under (a) benign captive conditions and (b) for these populations when introduced into the wild. Redrawn from Woodworth *et al.* (2002).

can be minimized by subdividing or fragmenting the captive population into a series of intermediate size populations. The effective population size of each population should be large enough to minimize the harmful effects of inbreeding and genetic drift, but small enough to minimize rapid adaptation to captive conditions.

19.6 GENETIC MANAGEMENT OF CONSERVATION BREEDING PROGRAMS

A primary genetic goal of captive breeding programs is to minimize genetic change caused by genetic drift and natural selection. Specific actions to achieve this goal depend upon the biology of the species. We first consider captive populations that are managed by keeping track of individual pedigrees (e.g., large mammals and birds). We then consider species for which large groups of individuals are held, but it is difficult or impractical to keep track of individuals (e.g., fishes and insects). Most of our examples concern animals, but the same underlying genetic principles hold for plants. The wide variety of possible modes of reproduction in plants (see Figure 19.1) makes it harder to provide general guidelines to apply these genetic principles. Guerrant (1996) provided an excellent review of maintaining offsite populations of plants for reintroduction.

19.6.1 Pedigreed populations

Genetic management by individual pedigrees is extremely powerful. It provides both maximum genetic information about the captive population and also maximum power to control the reproductive success of individuals chosen for mating. This approach is most appropriate for large mammals and birds. Most of the genetics literature dealing with management of captive populations deals with this situation.

Simply maximizing N_e might not be the best strategy for maintaining genetic variation in pedigreed populations (Ballou and Lacy 1995). Remember that genetic variation can be measured by either heterozygosity or allelic diversity. Maximizing N_e will minimize the loss of heterozygosity (by definition), but it might not be the best approach to retain allelic diversity. A strategy that uses all of the information contained in a pedigree can be developed to minimize the loss of heterozygosity and allelic diversity.

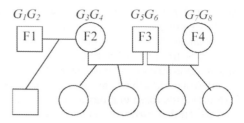

Figure 19.6   Hypothetical pedigree of a captive population founded by four individuals. We know that one allele at each locus has been lost from founder F1 because he left only one descendant in the captive population. Therefore, equalizing the contributions of these four founders in future generations would lead to an overrepresentation of alleles from F1.

Ballou and Lacy (1995) provided a lucid explanation of captive breeding strategies to maintain maximum genetic variation that is beyond the detail that we will consider here. This problem is extremely difficult because the pedigrees of captive populations are often extremely complicated (see Figure 19.4) and genetic planning is often not initiated until after the first few generations of captivity.

Simple rules of thumb such as equalizing the genetic contributions of founders to the captive population are not valid. We can see this in the hypothetical example presented in Figure 19.6, in which there are four founders of a captive population. What would be the result of a breeding strategy that equalized the genetic contributions of the founders? We can be absolutely certain that we have lost one of the two alleles carried by F1 at every locus since this founder only contributed one offspring to the captive population. However, there is some possibility that both alleles from the three other founders have been retained because they have contributed multiple progeny. Thus, we can maximize the retention of allelic diversity in the captive population by weighting the desirable contribution of each founder by the expected proportion of a founder's alleles retained (founder genome equivalents) (Lacy 1989).

Accurate calculations of kin relationships, inbreeding coefficients, and retention of founder alleles require complete knowledge of the pedigree. However, many pedigreed captive populations have some individuals with one or both parents unknown. Traditionally such individuals have been treated as founders unrelated to all nondescendant animals. In some circumstances, this can cause substantial errors in estimating genetic parameters (Ballou and Lacy 1995). Incorporation of molecular genetic information can often be used to resolve unknown relationships and can result in a substantially different view of the captive population (e.g., the whooping crane, Jones *et al.* 2002). Lacy (2012) has provided procedures to deal with missing information due to unknown or uncertain parentage in pedigrees.

Similarly, the founders of a population brought into captivity are generally assumed to be unrelated for pedigree analysis. However, this might not be the case. Incorrect assignment of founder relatedness will result in erroneous estimates of inbreeding coefficients, effective population size, and population viability. For example, the last remaining individuals in the wild might consist of just a few groups of sibs. This information should be taken into account along with molecular genetic analysis of relationships in order to maximize the retention of genetic variation in the captive population. Thus, correct classification of kin structure among founders is important for a captive breeding program.

19.6.2 Nonpedigreed populations

For many species held in captivity it is difficult or impractical to keep track of individuals and pedigrees. For example, a single female of the endangered Colorado pikeminnow can produce as many as 20,000 eggs each year. Other procedures, therefore, need to be developed to achieve the goal of minimizing genetic change by genetic drift or selection.

The large census population sizes at which some species are maintained in captivity should not be taken to mean that genetic drift is not a concern. For example, Briscoe *et al.* (1992) studied the genetics of eight captive populations of *Drosophila* held in populations with approximately thousands of individuals. All eight populations lost substantial heterozygosity at nine allozyme loci. Values of N_e estimated by the decline in heterozygosity were less than 5% of the census population size. Delpuech *et al.* (1993) reported similar results in their review of five species of insects held in captivity. Populations of all of these species had retained approximately 20% or less of their original heterozygosity at allozyme loci.

These results demonstrate the importance of genetic monitoring of populations (see Section 22.6). Regular examination of allele frequencies at molecular genetic loci should be used to detect the effects of genetic drift in captive populations in which individual reproductive success is not being monitored.

Adaptation to captive conditions is an even greater concern for large populations held in captivity. For example, Frankham and Loebel (1992) found that the average fitness in captivity of *Drosophila* doubled after being maintained for eight generations in captivity. Many other studies have found evidence for rapid adaptation to captivity in a variety of organisms (see discussion in Gilligan and Frankham 2003). The genes selected for in captivity are almost certain to decrease the fitness of individuals when they are returned to wild conditions. In addition, the strong selection in captivity will reduce the effective population size of the captive population. In fact, strong variance in reproductive success associated with this adaptation is the likely explanation of the small N_e/N_C ratios often found in captive populations.

A conceptual framework for minimizing the rate of adaptation to captivity (R) is provided by a modified form of the breeders' equation (see equation 11.9, Frankham and Loebel 1992):

$$R = \frac{H_N S(1-m)}{G} \tag{19.1}$$

where H_N is the narrow-sense heritability, S is the selection differential, m is the proportion of genes contributed from wild individuals, and G is the generation interval.

The goal is to minimize the rate of adaptation to captivity (R). Continued introduction of individuals from the wild (increased m) will slow the rate of adaptation to captivity. However, this will often not be possible. The generation interval (G) can be manipulated by increasing the average age of parents. For example, doubling the mean age of parents will double the generation interval and halve the rate of adaptation to captivity. However, increasing the age of parents will also slow the rate of population growth, so this approach is less feasible during the early stages of captivity before the population reaches carrying capacity.

Of course, reducing the intensity of selection (S) will slow adaptation to captivity. All efforts should be made to reduce differential survival and reproduction (fitness) in captivity. This can be done by minimizing mortality in captivity and by making the environmental conditions as close as possible to wild conditions.

Reducing differences in the number of progeny produced by individuals (family size) will also diminish the effects of selection in captivity (Allendorf 1993, Frankham *et al.* 2000). There will be no reproductive

Example 19.4   Supportive breeding of the world's largest freshwater fish

The Mekong giant catfish is a spectacular example of the potential problem with supportive breeding (Hogan *et al.* 2004). This is perhaps the largest species of fish found in freshwater. It grows up to 3 meters long and weighs over 300 kilograms! A century ago, this species was found throughout the entire Mekong River from Vietnam to southern China. This species began disappearing from fish markets in the 1930s, and efforts to find individuals in fish markets have failed in the last few years. Very few fish remain in the wild and the species is currently listed as endangered on the IUCN 'Red List' (critical World Distribution).

The Department of Fisheries of Thailand began a captive breeding program in 1984. Over 300 adult fish have been captured in the wild and brought into captivity over the last 20 years. However, this program further threatens this species because of the removal of adult fish from the wild and the release of large numbers of young fish from very few parents. For example, over 20 wild adults were sacrificed in 1999 to supply eggs and milt for artificial propagation. More than 10,000 of these fingerlings were released back into the wild in 2001. However, genetic analysis indicated that roughly 95% of these progeny were full-sibs produced by just two parents (Hogan *et al.* 2004).

differences between individuals if all individuals produce the same number of progeny. In this situation, natural selection will only operate through differences in relative survival of genotypes within families of full- or half-sibs. In a random mating population, approximately one-half of the additive genetic variance is within families and half is between families. Therefore, the rate of adaptation will be reduced by approximately 50% by equalizing family size. Equalizing family size will also increase N_e.

Using genome-wide genetic information to infer relatedness among individuals has the potential to minimize the loss of genetic variation in captive populations. de Cara *et al.* (2011) used simulations and concluded that using tens of thousands of SNPs spread throughout the genome generally performs better than using pedigrees information to maintain heterozygosity and allelic diversity within captive populations.

19.7 SUPPORTIVE BREEDING

Supportive breeding is the practice of bringing in a fraction of individuals from a wild population into captivity for reproduction and then returning their offspring into their native habitat where they mix with their wild counterparts (Ryman and Laikre 1991). The goal of these programs generally is to increase survival during key life stages in order to support the recovery of a wild population that is threatened with eminent extir-

pation. These programs would seem to pose a relatively small risk of causing genetic problems. Nevertheless, the favoring of only a segment of the wild population can also bring about changes in the wild population due to genetic drift and selection (Example 19.4).

19.7.1 Genetic drift and supportive breeding

Supportive breeding acts to increase the reproductive rate of one segment of the population (those brought into captivity). This will increase the variance in reproductive success (family size) among individuals and therefore potentially reduce effective population size. Demographic increases in population size can reduce the overall N_e and accelerate the loss of genetic variation. This effect is most likely to occur for species with high reproductive rates where large differentials in reproductive success are possible (e.g., fishes, amphibians, reptiles and insects).

Consider the situation where the breeding population consist of N_w effective parents that are reproducing in the wild, and N_c effective parents that are breeding in captivity, and their progeny are then released into the wild to supplement the wild population (Figure 19.7). Figure 19.8 presents the overall N_e as a function of the progeny that are produced in captivity. The overall effective size can be substantially smaller than the effective number of parents reproduc-

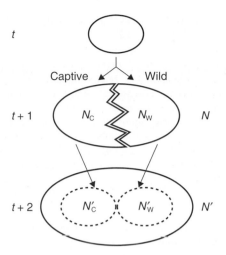

Figure 19.7 Schematic representation of supportive breeding. The total population of N individuals is divided into a captive (N_C) and a wild group (N_W) which reproduce in captivity and in the wild, respectively. The N'_C and N'_W offspring are mixed before breeding in generation $t + 2$. From Ryman *et al.* (1995a).

ing in the wild when the contribution of the captive population is high. For example, consider the case where the wild population consists of 22 effective parents, and that two of these parents are taken into captivity and then produce 50% of the total progeny. In this case, the total effective population size will be approximately 6 rather than the 22 that it would have been in the absence of a supportive breeding program.

Consideration of this problem has been extended to multiple generations (Wang and Ryman 2001, Duchesne and Bernatchez 2002). The effects of supportive breeding on N_e are complex. Moreover, the effects of supportive breeding on the inbreeding and variance effective population sizes (see Section 7.6) can differ. Nevertheless, supportive breeding, when carried out successfully over multiple generations, might increase not only the census but also the effective size of the supported population as a whole. If supportive breeding does not result in a substantial and continuous increase of the census size of the breeding population, however, it can be genetically harmful because of elevated rates of inbreeding and genetic drift.

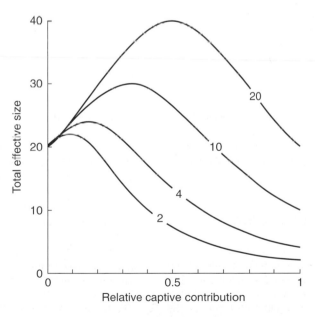

Figure 19.8 Total effective population size ($N_W + N_C$) when a natural population of 20 effective parents is supported by offspring from different numbers of captive parents, as indicated by the numbers on the different curves. The *x*-axis is the proportion of parents contributed by the captive parents. From Ryman and Laikre (1991).

19.7.2 Natural selection and supportive breeding

Supportive breeding can also have important genetic effects on supplemented populations because alleles that are harmful in the wild but advantageous in captivity might rise to high frequencies in captive populations (Lynch and O'Hely 2001). This genetic supplementation load will be especially severe when a captive population that is largely closed to import makes a contribution to the breeding pool of individuals in the wild. Moreover, theory indicates that this load might become substantial in a wild supplemented population when the captive breeders are always derived from the wild.

Many recent papers have modeled possible harmful genetic effects of supportive breeding programs of natural populations (e.g., Ford 2002, Duchesne and Bernatchez 2002, Theodorou and Couvet 2004). These populations can be managed to increase rather than decrease effective population size. Nevertheless, the effects of supportive breeding on adaptation of wild populations is more difficult to predict. Selection in captivity can substantially reduce the fitness of a wild population during supportive breeding. The continual introduction of wild individuals into the captive population can reduce but is not expected to eliminate this effect. These programs can reduce the probability of local extirpations, but it is essential that the genetic aspects of these programs are designed carefully.

19.8 REINTRODUCTIONS AND TRANSLOCATIONS

To insure a successful **reintroduction**, **introduction**, or supplemental **translocation**, we should consider several issues: (1) where to release the individuals, (2) how many populations to establish, (3) how many individuals to release, (4) age and sex of individuals to release, (5) which and how many source populations to use, and (6) how to monitor the population after the release of individuals. Genetics should play a role in all these issues (Sarrazin and Barbault 1996, Keller *et al.* 2012, Jamieson and Lacy 2012).

Monitoring after the release of individuals is crucially important to insuring the success of reintroductions and translocations. Unfortunately, post-release studies of genetic contribution or of population status are seldom conducted. Molecular markers can help monitor census population size, genetic diversity, effective population size, and reproductive contribution of released individuals. For example, if few founders actually reproduce, due to extreme polygamy, paternity analysis could detect the problem by identifying only a few males as fathers. If paternity analysis is not feasible, then monitoring for loss of alleles, rapid genetic change, and small effective population size could help to determine whether few founders reproduce (Example 19.5, Luikart *et al.* 1998).

Example 19.5   Rapid genetic decline in a translocated plant

The Corregin grevillea is one of the world's rarest plant species; only five plants were known in the wild in 2000 (Krauss *et al.* 2002). These plants occurred in degraded and isolated remnants of natural vegetation on road verges in Western Australia. In 1995, 10 plants were selected from the 47 plants known at the time to act as genetically representative founders for translocation into secure sites. Hundreds of ramets (tissue-cultured propagules of these 10 clones) were produced from these plants. By late 1998, 266 plants had been successfully translocated and were producing large numbers of seeds.

Krauss *et al.* (2002) used AFLPs to determine the genetic contribution of the 10 founders to this translocated population and their first-generation progeny.

They found that only 8 clones, not 10, were present in the translocated population. In addition, 54% of all plants were a single clone. They also found that F_1s produced between founders were on average 22% more inbred and 20% less heterozygous than their founders, largely because 85% of all seeds were the product of only four clones. They estimated that the effective population size of the translocated population was approximately two. That is, the loss in heterozygosity from the founders to the next generation was what would be expected if two founders had been used.

These results demonstrate the importance of genetic monitoring of translocation programs (see Section 22.6).

19.8.1 Reintroductions

Where to release individuals depends upon habitat suitability and availability. To maximize chances of a successful reintroduction, the habitat should be similar to that which the individuals to be released are adapted. Obviously, sufficient food, water, breeding habitat, and shelter or escape terrain should be available. Furthermore, the habitat should be free from exotic predators or competitive invasive species. For example, when threatened marsupials are reintroduced in Australia, exotic foxes and domestic cats should not be present because they are highly efficient at killing marsupials and preventing reestablishment of the population. In the case of the African rhino, there is abundant habitat, but little habitat free of human predators such as poachers (Section 16.1).

How many populations? At least two, and preferable several populations should be established and maintained. Populations should be independent demographically and environmentally to avoid a catastrophic species-wide decline due to severe weather, floods, fire, or disease epizootics, for example. Two or more populations should be established within each different environment or for each divergent genetic lineage, whenever divergent environments or lineages exist within a species' range. The preservation of multiple populations across multiple diverse environments can help insure long-term persistence of a species (Hilborn *et al.* 2003, see also Example 15.3). Genetics can help determine whether populations are independent demographically through genetic mark recapture to identify migrants. Molecular genetic markers are widely used to assess gene flow, an indicator of the degree of population independence or isolation.

How many individuals to release depends in part, on the breeding system, effective population size, and population growth rate after the reintroduction (Example 19.6). When feasible, at least 30–50 individuals should be reintroduced (Guest Box 19). More individuals will be required if the breeding system is strongly sex-biased (e.g., strong polygamy) or the effective population size is small compared with the census population size. Also, when population size does not increase above approximately 100–200 individuals within a few generations, more individuals should be released, when possible.

The sex and age of individuals can influence the success of the reintroductions and translocations. For example, it is often important to release more females

than males to maximize population growth, which limits demographic stochasticity and subsequent genetic drift. Many reintroductions of large game animals in western North America have used about 60–80% females. For supplemental translocations in polygynous species, it is better to reintroduce females than males if we want only limited gene flow, because a single male can potentially breed with many females, thereby swamping a population with introduced genes. Further, a male in a polygynous population might never breed, if he is not dominant, for example, making male-mediated gene flow highly variable and unpredictable. In territorial carnivores such as grizzly bears, it is often best to translocate females because males are more likely to fight for territory, sometimes to the death. Molecular genetic sexing can help to determine sex before translocation in some species (birds and reptiles), where sex is cryptic.

Age can influence the likelihood that a translocated individual remains in the location of release and integrates socially into the new population. In large mammals, young juvenile or yearling individuals are often more likely than adults to integrate socially and/ or not leave the release area. Currently there is no way to obtain age information for molecular genetic approaches, although the amount of telomere DNA on chromosomes is correlated with age and quantifying the amount might become feasible one day.

Which and how many source populations? For reintroductions and supplemental translocations, the source population generally should have high genetic diversity, genetic similarity, and environmental similarity when compared with the new or recipient population. Environmental similarity helps limit chances of maladaptation of the translocated individuals in the site of release. However, if populations have only recently become fragmented and differentiated, multiple differentiated source populations can help to maximize genetic diversity in reintroductions or translocations, with little risk of outbreeding depression. For example, a source population with greater genetic divergence from the recipient population will result in a greater increase in heterozygosity in the recipient population.

If no individuals are available from a similar environment, then individuals from several source populations could be mixed upon release to maximize diversity for natural selection to act upon. Mixing of individuals from multiple sources is less desirable in supplemental translocations where some locally adapted individuals still persist, because releasing a mixture of many

Example 19.6   Genetic management of a reintroduction: Guam rails

Over 50,000 Guam rails were estimated to be present on Guam in the 1960s. However, the introduction of the brown tree snake to Guam during the Second World War caused extinction or severe endangerment of all Guam's native forest birds. By 1986, Guam rails became extinct in the wild. However, 21 birds had been brought into captivity in 1983 and 1984 to initiate a captive breeding program (Haig *et al.* 1990).

The captive birds bred very successfully in captivity. By 1989, 113 birds were in the captive population and plans began to introduce Guam rails to the adjacent island of Rota. Environmental conditions on Rota are similar to Guam, except that the brown tree snake is not present. Initial plans were to introduce 90 birds to Rota. A number of factors were considered in designing an introduction program (e.g., behavior, demography, genetics, and the physical conditions of each animal; Griffith *et al.* 1989).

Haig *et al.* (1990) compared six possible genetic mating schemes to produce 90 chicks planned for introduction:
1 Select chicks at **random** from the captive population.
2 Select on the basis of **fitness**. That is, use chicks from those birds that produced the greatest number of progeny in captivity.
3 Select chicks that would maximize the heterozygosity of the introduced population at 23 **allozyme** loci.
4 Select chicks to **equalize founder contributions**.
5 Select chicks to maximize the **allelic diversity** in the introduced population.
6 Select chicks to maximize the **founder genome equivalents** (Section 19.6.1).

Each option was evaluated in terms of how well it would maintain genetic diversity in the introduced population using a gene drop analysis (Section 13.2).

The results indicated that some strategies would have done a poor job of maintaining genetic variation in the introduced population (Table 19.1). Selecting for reproductive fitness in captivity or heterozygosity at allozyme loci would have resulted in a substantial decline in allelic diversity in the introduced population. This shows the importance of minimizing differences in reproductive success among individuals in captivity.

The other three options (equalize founder contributions, maximize allelic diversity, and maximize founder genome equivalents) all performed fairly equally (Table 19.1). The founder genome equivalent strategy seems best because it would retain nearly as much allelic diversity, maintain more founder genome equivalents, and require fewer breeding pairs, which would make it logistically preferable.

Some authors have suggested that individuals should be chosen for breeding in captivity to increase genetic variation at certain loci, which can be examined with molecular techniques and might have particular adaptive importance (Wayne *et al.* 1986, Hughes 1991). However, the above results show that selecting for increased variation at a few detectable loci can reduce the effective population size and reduce genetic variation throughout the genome (Chapter 7).

Sixteen rails were reintroduced to Guam in 1998 to a 24-hectare enclosure surrounded by 2 m snake barrier. However, this population was extirpated by feral cats. Another introduction of 44 birds in 2003 had the same fate. Approximately 100 birds are released on the snake-free island of Rota annually in an effort to preserve this species.

Table 19.1 Comparison of six breeding options (see text for explanation) for creating a group of 90 Guam rails for an introduction program. From Haig *et al.* (1990).

Option	H_e	Number of alleles	Founder genome equivalents	Breeding pairs needed
Founders	1.00	42	21	–
Current population	0.98	31.5	10.5	–
1. Random (no selection)	0.95	24.1	9.4	23
2. Select for fitness	0.98	20.5	8.3	8
3. Maximize allozyme heterozygosity	1.00	19.9	7.1	13
4. Equalize founder contribution	0.98	27.2	13.4	8
5. Maximize allelic diversity	1.00	29.3	13.7	23
6. Founder genome equivalents	1.00	29.2	14.4	16

individuals could swamp the local gene pool and lead to loss of locally adapted genotypes.

19.8.2 Restoration of plant communities

These same genetic principles apply to developing sources to be used in restoration projects with plants (Example 19.7, Fenster and Dudash 1994, Lesica and Allendorf 1999, Li and Pritchard 2009). Restoration is an important tool for the preservation of native plant communities (Hufford and Mazer 2003, Kramer and Havens 2009). Restoration ecology is a synthesis of ecology and population genetics (see Guest Box 16).

In general, native local plants are the preferred source for restoration projects because of the potential importance of local adaptations (Linhart and Grant 1996). A variety of studies have found evidence that plants of relatively local origin are preferred as sources of reintroduction and restoration (Keller *et al.* 2000, Vergeer *et al.* 2004).

In some cases, local source populations might not be available. In addition, restoration projects might involve highly disturbed sites to which local genotypes are not adapted. In such cases, hybrids between populations, or mixtures of genotypes from different populations, might provide the best strategy (Figure 19.9, Guerrant 1996, Vergeer *et al.* 2004). Mixtures of genotypes from ecologically distinct populations or hybrids of these genotypes will possess high levels of genetic variation. Introduced populations with enhanced variation are more likely to rapidly evolve genotypes adapted to the novel ecological challenges of severely disturbed sites.

Strains of plants that have been selected for captive conditions are a common source of plants for restoration (Keller *et al.* 2000). Such **cultivars** are often readily available, and are much less expensive than acquiring progeny from wild seed sources. However, the widespread use of cultivars is likely to lead to the introduction of genes into the adjacent resident population through cross-pollination, although the degree of genetic introgression will depend upon the breeding system. Thus, widespread introductions of cultivars could alter the resident neutral gene pool. For these reasons, the use of cultivars should be restricted.

Example 19.7 Genetic management of a reintroduction: Mauna Kea silversword

The Mauna Kea silversword is a member of the silversword alliance, a group of Hawai'ian endemic plants that is one of the premier examples of adaptive radiation (Baldwin and Robichaux 1995). This plant is named for its mountain habitat and its striking rosette of dagger-shaped leaves covered with jewel-like silvery hairs (Robichaux *et al.* 1998). The Mauna Kea silversword historically was common in exposed subalpine and alpine habitats high on the 4205 m volcano on the island of Hawai'i. The introduction of sheep and other ungulates devastated this plant, presumably because of heavy browsing. By the 1970s, only a small remnant population confined to cliffs and rocks persisted.

Three plants from this remnant population of an estimated fewer than 100 plants flowered in 1973. Most Mauna Kea silverswords live up to 50 years and are monocarpic (i.e., they flower only once before dying). Seeds from two of these plants were removed, and over 800 plants resulted from outplanting seedlings from these seeds on Mauna Kea. Today over there are over 1500 plants in the reintroduced population that are first- or second-generation offspring of the two maternal founders. This intervention and sub-

sequent reintroduction dramatically increased the size of the silversword population on Mauna Kea.

This large reintroduced population went through a severe genetic bottleneck because it is based on just two maternal plants. Analysis of seven variable microsatellite loci indicated substantial loss of genetic variation in the outplanted population in comparison with the native population (Friar *et al.* 2000). Three of the seven loci that are variable in the native plants are fixed for a single allele. A total of 8 of the total of 21 alleles over all loci were not detected in the outplanted population. The expected average heterozygosity in the outplanted population (0.074) was 70% less than that the native population (0.250).

The greatest immediate genetic concern for recovery is the loss of allelic variation at the self-incompatibility locus in Mauna Kea silverswords (see Section 14.4.2, Robichaux *et al.* 1998). Loss of variation at this locus in the outplanted population might greatly reduce seed production and reduce the long-term chances for recovery of the species.

Efforts are underway to increase genetic variation by hand-transferring pollen from native plants that

(Continued)

flower into the outplanting program. This is not an easy task; collecting pollen often involves perching precariously on steep cliffs, because the remaining plants exist because they were out of the way of browsing ungulates. Two plants flowered in 1997, and a large number of seeds were produced by hand transfer of pollen. This doubled the number of founding maternal plants for the outplanting program.

More founders are expected to be added in the future. The program is currently concerned with bal-ancing the genetic contributions of founders by equalizing founder contributions. This might be difficult because so many plants have already been outplanted and so few plants flower in any given year. In addition, the long generation interval will make this even more challenging! This program is a clear example of the importance of taking genetic concerns into consideration in the recovery of species that have reached small numbers.

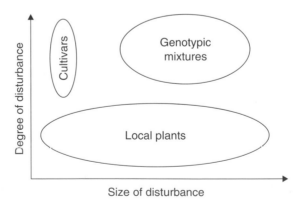

Figure 19.9 General relationship to degree and size of disturbance of three possible sources of plants to be used for restoration projects. In general, local plants should be preferred. However, cultivars may be appropriate in small but highly disturbed areas because they are more likely to establish themselves quickly. Hybrids between populations, or mixtures of genotypes from different populations, may provide the best strategy for highly disturbed sites to which local genotypes are not adapted. Introduced populations with enhanced variation are more likely to rapidly evolve genotypes adapted to the novel ecological challenges of severely disturbed sites. From Lesica and Allendorf (1999).

Guest Box 19   Understanding inbreeding depression: 25 years of experiments with *Peromyscus* mice
Robert C. Lacy

At first glance, inbreeding depression would seem to be well understood scientifically. It is a widespread consequence of matings between close relatives, and there is a simple mechanistic explanation – increased expression of recessive deleterious alleles in inbred individuals. However, when conservation geneticists began looking more closely at inbreeding depression, in order to make predictions about the vulnerability of populations and to provide sound management advice, the picture became much less clear.

For more than 25 years, my colleagues at the Brookfield Zoo and I have been studying the effects of inbreeding in some *Peromyscus* mice. We started with simple hypotheses and simple expectations, but have been led to conclude that inbreeding depression is a complex phenomenon that defies easy prediction. We first sought to show that populations of mice that were long isolated on small islands had already been purged of their deleterious alleles, leaving them able to inbreed with minimal further impact. Instead, we found that the island mice had low reproductive performance prior to inbreeding in the lab, and their fitness declined as fast or faster than in mice from more genetically diverse mainland populations when we forced inbred matings (Brewer *et al.* 1990).

We then tested whether we would get the same results if we repeated our measures of inbreeding depression on replicate laboratory stocks, each derived from the same wild populations. We found that different breeding stocks of a species could show very different effects of inbreeding (Lacy *et al.* 1996). The damaging effects of inbreeding showed up in different traits (e.g., in litter size, versus pup survival, versus growth rates), and to different extents. We found that the inbreeding depression was due largely to effects in the inbred descendants of some founders, while descendants of other founder pairs seemed to have little problem with inbreeding (see Section 13.5.3). This suggests that the effects of inbreeding are due mostly to a few alleles, and which animals carry these alleles is largely a matter of chance.

This situation might provide a good opportunity for selection to be effective at purging those deleterious alleles, when they are expressed in inbred homozygotes. When we examined our data to see whether inbreeding depression did become less through the generations (Lacy and Ballou 1998), we found purging in some subspecies but not in others (Figure 19.10). Thus, it might be that the cause of inbreeding depression (recessive alleles versus heterozygous advantage, few versus many loci, unconditional effects versus environmentally dependent effects) varies among natural populations, or it might be that natural selection is not efficient in the face of rapid drift in populations so small as to be subjected to high inbreeding. We hoped that the observed inbreeding depression might be a phenomenon of captive populations but not so prevalent in the less controlled wild environments. However, we found that inbred mice have much lower survival when released in natural habitats, with the inbreeding depression worse than in the lab (Jiménez *et al.* 1994).

Although predicting which populations would be most severely affected by inbreeding is difficult, the good news is that our most recent experiments verify that good genetic management of breeding programs can reduce the impacts of accumulated inbreeding. Experimental lineages of *Peromyscus* have been maintained for 20 generations, under three protocols for selecting breeders – selection for the most docile mice, pedigree-based breeding, and random-bred controls. As predicted, the pedigree-based management did retain more genetic variation, slowed the accumulation of inbreeding, and resulted in high reproductive success (increasing across generations, due to adaptation to captivity). The random-bred stocks lost more diversity, but have still bred as well, presumably because adaptation to captivity would be even faster. The populations of mice that were artificially selected for traits that we thought would adapt them to captivity lost variation the fastest and became more highly inbred because of the selection scheme, and then crashed because of declining reproductive success.

(Continued)

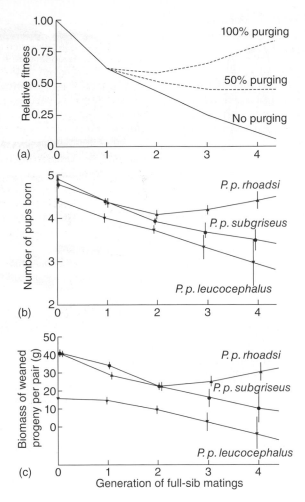

Figure 19.10   (a) Expected response of fitness to repeated generations of inbreeding, if the genetic load is due entirely to recessive lethal alleles (100% purging), overdominance (no purging), or half of each. (b) and (c) Observed responses to inbreeding in three subspecies of *Peromyscus polionotus* mice, measured as the depression in initial litter size and biomass of progeny weaned within 63 days, as predicted from regressions on inbreeding levels of litters and of prior generations. From Lacy and Ballou (1998).

CHAPTER 20

Invasive species

King brown snake eating a cane toad, Guest Box 20

Biological invaders are now widely recognized as one of our most pressing conservation threats.

Ingrid M. Parker et al. (2003)

The synergism arising from combining ecological, genetic, and evolutionary perspectives on invasive species may be essential for developing practical solutions to the economic and environmental losses resulting from these species.

Ann K. Sakai et al. (2001)

Chapter Contents

Conservation and the Genetics of Populations, Second Edition. Fred W. Allendorf, Gordon Luikart, and Sally N. Aitken.
© 2013 Fred W. Allendorf, Gordon Luikart and Sally N. Aitken. Published 2013 by Blackwell Publishing Ltd.

Invasion by **nonindigenous** (**alien**) species is recognized as second only to loss of habitat and landscape fragmentation as a major cause of loss of global biodiversity (Walker and Steffen 1997). The economic effect of these species is a major concern throughout the world. For example, an estimated 50,000 nonindigenous species established in the US cause major environmental damages and economic losses that total more than an estimated 125 billion dollars per year (Pimentel 2000). Management and control of nonindigenous species is perhaps the biggest challenge that conservation biologists will face in the next few decades.

A chapter on **invasive species** might at first seem out of place in a book on conservation genetics. However, we have chosen to include this chapter for several reasons. Molecular genetic analysis of introduced species can provide valuable information about the source and number of introduced populations (Le Roux and Wieczorek 2009). Also, understanding the ecological genetics of invasive species biology may provide helpful insights into developing methods of eradication or control. In addition, the study of species introductions offers exceptional opportunities to answer fundamental questions in population genetics that are important for the conservation of species. For example, how crucial is the amount of genetic variation present in introduced populations for their establishment and spread?

Molecular genetic analysis of introduced species (including diseases and parasites) can provide valuable information (Walker et al. 2003). Understanding the "epidemiology of invasions" (Mack et al. 2000) is crucial to controlling current invasions and preventing future invasions. Understanding the source of the introduced population, the frequency with which a species is introduced into an area, the size of each introduction, and the subsequent pattern of spread is important in order to develop effective mechanisms of control. However, observing such events is difficult, and assessment of the relative frequency of introductions or pattern of spread is extremely difficult. Molecular markers provide an important opportunity to answer these questions. In addition, many problematic diseases and parasites have been introduced and spread. Molecular markers are being used extensively to monitor and control these diseases (Criscione et al. 2005).

There is evidence that native species evolve and adapt to the presence of invasive species (see Guest Box 20). Native species sometimes change their response to predators, use different habitats, and have other adaptations that allow them to persist in invaded areas (Strauss et al. 2006). The ability of a population to respond to selection from invaders depends upon the effects of the invader, the presence of appropriate genetic variation, and the history of previous invasions. Adaptive change in natives can diminish the effects of invaders and potentially promote coexistence between invaders and natives.

An understanding of genetics may also help to predict which species are most likely to become invasive. There are two primary stages in the development of an invasive species (Figure 20.1). The first stage is the introduction, colonization, and establishment of a nonindigenous species in a new area. In other words, the introduced species must arrive, survive, and establish. The second stage is the spread and replacement of native species by the introduced species. The genetic principles that may help us predict whether or not a nonindigenous species will pass through these two stages to become invasive are the same principles that apply to the conservation of species and populations threatened with extinction: (1) genetic drift and the effects of small populations, (2) gene flow and hybridization, and (3) natural selection and adaptation.

In this chapter, we consider the possible importance of genetic change in the establishment and spread of invasive species. We also examine the significant role that hybridization may play in the development of invasive species. We consider ways in which genetic understanding may be applied to help predict which species are likely to be successful invaders and to help control invasive species. Finally, we discuss the use of genetic techniques to understand emerging diseases.

20.1 WHY ARE INVASIVE SPECIES SO SUCCESSFUL?

Not all introduced species become invasive (Sakai et al. 2001). A general observation is that only one out of every 10 introduced species becomes established, and only one out of every ten newly established species becomes invasive. Therefore, roughly only one out of every 100 introduced species becomes a pest. The next few sections consider what factors may influence whether a species becomes established and becomes invasive.

Invasive species provide an exceptional opportunity for basic research in the population biology and short-term evolution of species. Many of the best examples

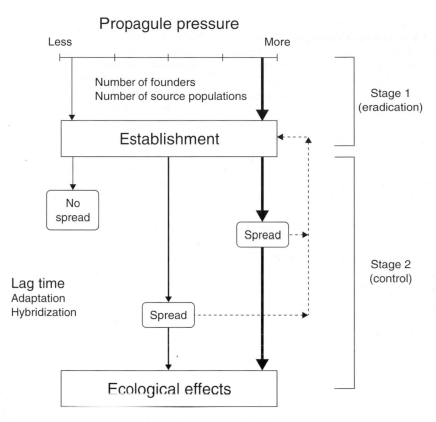

Figure 20.1 The two stages of invasion that generally coincide with different management responses. Propagule pressure is a continuum, with greater pressure leading to increased chance of establishment and spread with shorter lag times. If spread involves small groups of dispersing individuals, each group must be able to establish in a different area. Establishment or subsequent spread may be inhibited where groups reach the limits of particular environmental conditions. From Allendorf and Lundquist (2003).

of rapid evolutionary change come from the study of introduced populations (Lee 2002, Prentis *et al.* 2008). For example, *Drosophila subobscura* evolved a north–south cline in wing length just 20 years after introduction into North America that paralleled the pattern present in their native Europe of increased wing length with latitude (Huey *et al.* 2000). Similarly, two species of goldenrods evolved a cline in flowering time that resembled a cline in their native North America after being introduced into Europe (Weber and Schmid 1998).

Many unresolved central issues in the application of genetics to conservation – such as the inbreeding effects of small populations and the importance of local adaptation – can be much better experimentally addressed with introduced species. Two apparent para-

doxes emerge from comparison of our previous conclusions of the effects of small population size and local adaptation with the successful invasions by introduced species.

20.1.1 If population bottlenecks are harmful, then why are invasive species that have gone through a founding bottleneck so successful?

Much of the concern in conservation genetics relates to the potentially harmful effects of small population sizes. The loss of genetic variation through genetic drift and the inbreeding effect of small populations contribute to the increased extinction rate of small populations (e.g., Frankham and Ralls 1998). However, colonization of introduced species often involves a population bottleneck since the number of initial colonists is often small. Thus, a newly established population is likely to be much less genetically diverse than the population from which it is derived (Barrett and Kohn 1991).

The reduced genetic diversity can have two harmful consequences. First, inbreeding depression may limit population growth, and lower the probability that the population will persist. Second, reduced genetic diversity will limit the ability of introduced populations to evolve in their new environments. Thus we face a paradox: *if population bottlenecks are harmful, then why are invasive species that have gone through a founding bottleneck so successful?*

One answer to this paradox is that introduced species often have greater genetic variation than native species, because they are a mixture of several source populations (see Example 20.1, see Section 20.2.2).

Another solution is in the strong observed effect of **propagule pressure** on the invasiveness of species. That is, the clear association between the greater number of introduced individuals and the number of release events, and the probability of an introduced species becoming invasive, suggests that many invasive species are not as genetically depauperate as expected. In addition, plant species can avoid the reduction in genetic variation associated with colonization by their means of reproduction (Barrett and Husband 1990).

Many invasive plant species reproduce asexually by **apomixis**, which is the formation of seed without meiosis or fertilization, or through other forms of vegetative reproduction including layering, rhizomes, root suckering, and fragmentation (Baker 1995, Calzada *et al.* 1996). In all cases, the effects of inbreeding depression are avoided because the progeny are genetically identical to the parental plants. In addition, many invasive plant species are polyploids and can reproduce by self-pollination. In this situation, genetic variation is maintained the form of fixed heterozygosity because of genetic divergence between the genomes combined in the formation of the allopolyploid (Brown and Marshall 1981).

20.1.2 If local adaptation is important, then why are introduced species so successful at replacing native species?

The presence of local adaptations is often an important concern in the conservation of threatened species (McKay and Latta 2002). That is, adaptive differences between local populations are expected to evolve in response to selective pressures associated with different environmental conditions. The presence of such local adaptations in geographically isolated populations

---

Example 20.1   Genetic variation increases during invasion of a lizard

The brown anole is a small lizard that is native to the Caribbean (Kolbe *et al.* 2004). The brown anole has been introduced widely throughout the world (Hawaii, Taiwan, and the mainland US). Introduced populations often reach high population densities, show exponential range expansion, and are often competitively superior, as well as a predator of native lizards.

The brown anole first appeared in the Florida Keys in the late 19th century. Its range did not expand appreciably for 50 years, but widespread expansion throughout Florida began in the 1940s and increased in the 1970s.

Genetic analyses of mtDNA suggest that at least eight introductions have occurred from across this lizard's native range into Florida (Kolbe *et al.* 2004). This has resulted in an admixture from different geographic source populations and has produced populations that are substantially more genetically variable than native populations. Moreover, recently introduced brown anole populations around the world in turn originated from Florida, and some have maintained these elevated levels of genetic variation.

These authors suggest that one key to the invasive success of this species may be the occurrence of multiple introductions that transform among-population variation in native ranges to within-population variation in introduced areas. These genetically variable populations appear to be particularly potent sources for introductions elsewhere.

often plays an important role in the management of threatened species (Crandall *et al.* 2000).

When a species invades a new locality, it will almost certainly face a novel environment. However, many introduced species often outcompete and replace native species. For example, introduced brook trout are a serious problem in the western US where they often outcompete and replace ecologically similar native trout species. However, the situation is reversed in the eastern US where brook trout are native. They are in serious jeopardy because of competition and replacement from introduced rainbow trout that are native to the western US. Thus we face a second paradox: *if local adaptation is common and important, then why are introduced species so successful at replacing native species?*

A variety of explanations have been proposed to explain why introduced species often outperform indigenous species. First, some species may be intrinsically better competitors because they evolved in a more competitive environment. Second, the absence of enemies (e.g., herbivores in the case of plants) allows nonindigenous species to have more resources available for growth and reproduction and thereby outcompete native species. Siemann and Rogers (2001) found that an invasive tree species, the Chinese tallow tree, had evolved increased competitive ability in their introduced range. Invasive genotypes were larger than native genotypes and produced more seeds; however, they had lower quality leaves and invested fewer resources in defending them. Thus, there are a number of reasons why introduced species may fare well even though native species may be locally adapted.

In addition, local adaptation of native populations might only be essential during periodic episodes of extreme environmental conditions (e.g. winter storms, drought, or fire). Wiens (1977) has called these episodes "ecological crunches" and has suggested they are commonplace and limit the value of short-term studies of competition and fitness. For example, Rieman and Clayton (1997) have suggested that the complex life histories of some fish species (mixed migratory behaviors, etc.) are adaptations to periodic disturbances such as fire and flooding. Thus, introduced species may be able to outperform native species in the short term (a few generations) because the performance of native species in the short term is constrained by long-term adaptations that come into play every 50 or 100 years.

20.2 GENETIC ANALYSIS OF INTRODUCED SPECIES

Molecular genetic analysis of introduced species can provide valuable information about their origin. In addition, study of the amount and distribution of genetic variation in introduced species can provide valuable insight into the mechanisms of establishment and spread. In some cases, even identifying the species of invasive organisms may be difficult without genetic analysis. In other cases, populations of a native species may become invasive when introduced into a new ecosystem. Such populations would technically not be considered to be **alien** since conspecific populations were already present. Nevertheless, such populations may become invasive when introduced outside of their natural area (Genner *et al.* 2004). Current regulations dealing with invasive organisms are based upon species classification. However, recognizing biological differences between populations within the same species is important for control of invasive species.

20.2.1 Molecular identification of invasive species

In some cases, genetic identification may be necessary to identify the introduced species. For example, populations of Asian swamp eels (genus *Monopterus*) have been found throughout the southeastern US since 1994 (Collins *et al.* 2002). Swamp eels have a variety of characteristics that make them a potentially disruptive species. They are large predators (up to 1 m in length) that are capable of breathing out of water and dispersing over land. They also are extremely tolerant of drought because they produce large amounts of mucous that can prevent desiccation, and they can burrow when water levels drop.

The morphological similarity of swamp eels makes identification difficult. Collins *et al.* (2002) sampled four locations in Georgia and Florida to determine whether these eels were the result of a single introduction or multiple introductions. Examination of mtDNA revealed that introduced populations even in close proximity (<40 km) were genetically distinct, representing at least two and possibly three different species.

Genetics may also allow the detection of an invasive species that is conspecific with a native species. For example, Genner *et al.* (2004) used mtDNA to detect a non-native morph from Asia of the gastropod *Melanoides tuberulata* in Lake Malawi, Africa, which is sympatric with indigenous forms of the same species. This non-native morph was not present in historic collections and appears to be spreading rapidly and replacing the indigenous form.

20.2.2 Distribution of genetic variation in invasive species

Examination of published descriptions of the amount and patterns of genetic variation in introduced species reveals two contrasting patterns. In the first, introduction and establishment is often associated with a population bottleneck, or bottlenecks, so that introduced populations have less genetic variation than populations in the native range of the species. Under some circumstances, this reduced genetic variation may actually stimulate invasion (Example 20.2). In the second pattern, introduction and establishment is associated with admixture of more than one local population in the native range, so that populations in the introduced range have greater genetic variation. The bottleneck and admixture situations result in very different patterns of genetic variation in introduced species.

20.2.2.1 Bottleneck model

In many cases, introduced species may only have a few founders, so genetic variation is reduced by the founder effect. The land snail that we examined in Example 6.2 is an excellent example of this pattern (Johnson 1988). This species was introduced from Europe to Perth in Western Australia. The Perth population has reduced heterozygosity and allelic diversity compared with a population from France (see Figure 6.9). In addition, another population was founded on Rottnest Island by a limited number of founders from the Perth population. This second bottleneck further reduced heterozygosity and allelic diversity.

The rapa whelk, a predatory marine gastropod, presents an extreme example of this (Chandler *et al.* 2008). Sequences for two mtDNA genes (a total of 1292 bp) revealed 110 haplotypes in 178 individuals from eight populations in its native range of China, Korea, and Japan. However, all 106 individuals sampled from 12 introduced populations throughout Europe and North America had a single haplotype that was observed in four of the 178 native rapa whelk sequenced. Unfortunately, there are no nuclear gene data for this species.

Stone and Sunnucks (1993) have described the invasion of northern and western Europe of the gallwasp *Andricus quercusalicis* following human introduction of an obligate host plant, the Turkey oak from southeastern Europe. Populations further from the native range show reduced allelic diversity and heterozygosity at 13 allozyme loci (Figure 20.2). This suggests that this species has experienced a series of bottlenecks as it spread throughout Europe over the last 300–400 years. Patterns of allele frequency differentiation suggest that the invasion of this species followed a stepping-stone process rather than multiple introductions from its native range.

20.2.2.2 Admixture model

In contrast to the bottleneck model, many introduced species actually have greater variation in comparison with populations from the native range because their

---

Example 20.2   Loss of genetic variation in an introduced ant species promotes a successful invasion

Ants are among the most successful, widespread, and harmful invasive taxa. Highly invasive ants sometimes form unicolonial supercolonies in which workers and queens mix freely among physically separate nests. By reducing costs associated with territoriality, such unicolonial species can attain high worker densities, allowing them to achieve interspecific dominance.

Tsutsui *et al.* (2000) examined the behavior and population genetics of the invasive Argentine ant (*Linepithema humile*) in its native and introduced ranges.

They demonstrated with microsatellites that population bottlenecks have reduced the genetic diversity of introduced populations. This loss is associated with reduced intraspecific aggression among spatially separate nests, and leads to the formation of interspecifically dominant supercolonies. In contrast, native populations are more genetically variable and exhibit pronounced intraspecific aggression.

These findings provide an example of how a genetic bottleneck associated with introduction can lead to widespread ecological success.

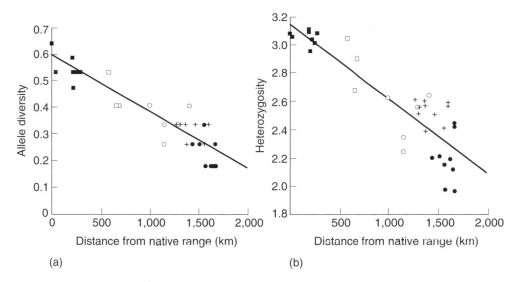

Figure 20.2   Relationships for an invasive gallwasp between distance from its native Hungary and (a) allelic diversity and (b) expected heterozygosity. Populations further from the native range show reduced genetic variation and patterns of allele frequency differentiation that suggest a stepping-stone invasion process rather than multiple introductions from its native range. The symbols indicate the country of origin of the sample. From Stone and Sunnucks (1993).

founders come from different local populations within the native range (Roman and Darling 2007). Admixing individuals from genetically divergent populations will increase genetic variation by converting genetic differences between populations to genetic variation between individuals within populations (Example 20.1).

We saw in Chapter 9 that the total heterozygosity (H_T) within a species can be partitioned into genetic variation within and between subpopulations:

$$H_T = H_S + D_{ST} \qquad (20.1)$$

where H_S is the average heterozygosity within subpopulations, and D_{ST} is the average gene diversity between subpopulations. D_{ST} is related to the more familiar F_{ST}:

$$D_{ST} = (F_{ST})(H_T) \qquad (20.2)$$

and

$$F_{ST} = \frac{D_{ST}}{H_T} \qquad (20.3)$$

Thus, D_{ST} is the proportion of the total heterozygosity due to genetic divergence between subpopulations (Nei 1987, p. 189).

For example, we saw in Example 9.2 that two separate demes of brown trout existed within a single Swedish lake and had substantial genetic differentiation at the *LDH A2* locus (and many other loci as well; Ryman *et al.* 1979):

$$H_T = H_S + D_{ST} = H_S + (F_{ST})(H_T)$$
$$= 0.128 + (0.728)(0.489) = 0.128 + 0.356 = 0.489$$

Thus, if we introduced an equal number of fish from each of these demes into a new lake and the fish mated at random, the expected heterozygosity in the newly founded population would be 0.489, nearly four times as great as in the original populations at this locus.

This effect can be seen in many introduced populations (see Roman and Darling 2007). For example, approximately 400 chaffinches were imported from England into New Zealand between 1862 and 1877. Overwintering birds from several populations on the European continent were included in the birds collected for introduction. Baker (1992) reported that chaffinches from eight populations in New Zealand have an average heterozygosity that is 38% greater (0.066 versus 0.048) than 10 native European populations at 42 allozyme loci. As expected, chaffinches in New Zealand have greatly reduced differentiation among subpopulations ($F_{ST} = 0.040$) compared with chaffinches in their native Europe ($F_{ST} = 0.222$).

20.2.3 Mechanisms of reproduction

Molecular genetic analysis can also be used to determine whether an introduced plant species is reproducing sexually or asexually, the breeding system, and the ploidy level of introduced plants. Further examination could determine how many different clonal lineages are present if an invader is reproducing asexually. This information, along with an understanding of genetic population structure, is essential for the development of effective control measures for invasive weed species (Chapman *et al.* 2004).

For example, most strains of the marine green alga *Caulerpa taxifolia* are not invasive. However, a small colony of *C. taxifolia* was introduced into the Mediterranean in 1984 from a public aquarium, spread widely and seriously reduced biological diversity in the northwestern Mediterranean (Jousson *et al.* 2000). The invasive strain differs from native tropical strains because it reproduces asexually, grows more vigorously, and is tolerant of lower temperatures. Colonies of *C. taxifolia* have recently been reported on the coast of California and have raised concerns about the danger of an invasion similar to that in the Mediterranean. Genetic analysis of the California alga has shown that it is the same strain as the one responsible for the Mediterranean invasion (Jousson *et al.* 2000). Thus, the rapid eradication of this introduced alga should receive high priority in order to reduce the probability of a new invasion.

Parthenogenesis (Greek for "virgin birth") is a form of asexual reproduction which occurs in animals without fertilization from a male. More than 80 taxa of fish, amphibians, and reptiles are now known to reproduce by parthenogenesis (Neaves and Baumann 2011). There are a variety of genetic mechanisms underlying parthenogenesis. Some result in completely homozygous progeny (automictic parthenogenesis). For example, automictic parthenogenesis has been reported in Komodo dragons (Watts *et al.* 2006), hammerhead sharks (Chapman *et al.* 2007), and zebra finches (Schut *et al.* 2008). This form of parthenogenesis is not likely to facilitate successful invasion because of the homozygosity of the progeny. However, apomictic parthenogenesis results in progeny that are genetically identical to their mother, and therefore has the potential to facilitate successful invasion.

North American crayfish have been successful invaders throughout the world, especially Europe where they have had major harmful effects on native crayfish. Buřič *et al.* (2011) recently discovered that females of the spiny-cheek crayfish (a successful invader of Europe) held in isolation are as reproductively successful as females held with males. All of the progeny from isolated females were genetically identical to the mothers at seven microsatellite loci; progeny from females held with males had genotypes indicating sexual reproduction. These authors have suggested that this reproductive plasticity might contribute to the great invasive success of these species.

20.2.4 Quantitative genetic variation

Although much information can be gained from molecular markers, characterization of the genetic variation controlling those life-history traits most directly related to establishment and spread is also crucial. These traits are likely to be under polygenic control with strong interactions between the genotype and the environment; they cannot be analyzed directly with molecular markers, although mapping quantitative trait loci (QTLs) affecting fitness, colonizing ability, or other traits affecting invasiveness may be possible (Barrett 2000). For example, variation in the number of rhizomes producing above-ground shoots, a major factor in the spread of the noxious weed johnsongrass, is associated with three QTLs (Paterson *et al.* 1995). This knowledge may provide opportunities for predicting the location of corresponding genes in other species and for growth regulation of major weeds.

Application of the methods of quantitative genetics could be useful for those species in which information can be obtained from progeny tests, parent–offspring comparisons, or use of the animal model (see Chapter 11). For example, one could compare the additive genetic variance/covariance structure of a set of life history traits of different populations to evaluate the role of genetic constraints on the evolution of invasiveness. Comparisons of the heritability of a trait could be made among different, newly established populations, or between invasive populations and the putative source population. Consideration of both the genetic and ecological context of these traits is critical, given the potentially strong interaction of genetic and environmental effects (Barrett 2000).

20.3 ESTABLISHMENT AND SPREAD OF INVASIVE SPECIES

One common feature of invasions is a lag between the time of initial colonization and the onset of rapid population growth and range expansion (Sakai *et al.* 2001) (see Figure 20.1). This lag time is often interpreted as an ecological phenomenon (the lag phase in an exponential population growth curve). Lag times are also expected if evolutionary change is an important part of the colonization process. This process could include the evolution of adaptations to the new habitat, the evolution of invasive life history characteristics, or the purging of genetic load responsible for inbreeding depression (Figure 20.3). It appears likely that in many cases there are genetic constraints on the probability of a successful invasion, and the lag times of successful invasions could be a result of the time required for adaptive evolution to overcome these genetic constraints (Ellstrand and Schierenbeck 2000, Mack *et al.* 2000).

20.3.1 Propagule pressure

Propagule pressure has emerged as the most important factor for predicting whether or not a nonindigenous species will become established (Kolar and Lodge 2001). Propagule pressure includes both the number of individuals introduced and the number of release events. Propagule pressure is expected to be an important factor in the establishment of introduced species on the basis of demography alone; it is unclear what role, if any, genetic effects may play in the effect of propagule pressure.

There are two primary ways in which the genetics of an introduced species may be affected by propagule pressure. First, a greater number of founding individuals would be expected to reduce the effect of any population bottleneck so that the newly established population would have greater genetic variation. Second, and perhaps most importantly, different releases may have different source populations. Therefore, hybridization between individuals from genetically divergent native populations may result in introduced populations having more genetic variation than native populations of the same species (Section 20.4).

20.3.2 Spread

Many recently established species often persist at low, and sometimes undetectable, numbers and then 'explode' to become invasive years or decades later (Sakai *et al.* 2001). Adaptive evolutionary genetic

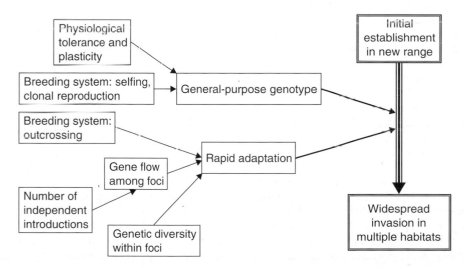

Figure 20.3 Factors that influence the process by which an introduced species moves from initial establishment in a new range to widespread invasion of multiple habitats. Two alternative, but not mutually exclusive, mechanisms are presented: rapid adaptation and the general-purpose genotype. Characteristics of the invading species (e.g., breeding system) or of the invasion process (e.g., number of introductions) that influence these two mechanisms are outlined. From Parker *et al.* (2003).

changes may explain the commonly observed lag time that is seen in many species that become invasive (García-Ramos and Rodríguez 2002). Many of the best examples of rapid evolutionary change come from the study of recently introduced populations (Lee 2002, Prentis *et al.* 2008).

20.4 HYBRIDIZATION AS A STIMULUS FOR INVASIVENESS

Hybridization might play an important role in introduced species becoming invasive (Example 20.3). As we have seen in the previous sections, many species become invasive only (1) after an unusually long lag time after initial arrival, and (2) after multiple introductions. Ellstrand and Schierenbeck (2000) have proposed that hybridization between species, or genetically divergent source populations, may serve as a stimulus for the evolution of invasiveness on the basis of these observations. They proposed four genetic mechanisms to explain how hybridization can stimulate invasiveness.

20.4.1 Evolutionary novelty

Hybridization can result in the production of novel genotypes and phenotypes that do not occur in either of the parental taxa. Evolutionary novelty can result either from the combination of different traits from both parents, or from traits in the hybrids that transgress the phenotypes of both parents (transgressive segregation; see Section 17.1.4).

20.4.2 Genetic variation

An increase in the amount of genetic variation may in itself be responsible for the evolutionary success of hybrids. That is, the greater genetic variation (heterozygosity and allelic diversity) in hybrid populations may provide more opportunity for natural selection to bring about adaptive evolutionary change.

20.4.3 Fixed heterosis

Many invasive plant species have genetic or reproductive mechanisms that stabilize first-generation hybridity and thus may fix genotypes at individual or multiple loci that demonstrate heterosis. These mechanisms include **allopolyploidy**, permanent translocation heterozygosity, **agamospermy**, and clonal reproduction. The increased fitness resulting from fixed heterozygosity may contribute to the invasiveness of many plant species.

Common cordgrass (*Spartina anglica*) has been identified as one of the world's worst invasive species by the World Conservation Union (Lowe *et al.* 2000). Common cordgrass is a perennial salt marsh grass that has been planted widely to stabilize tidal mud flats. Its invasion and spread leads to the exclusion of native plant species and the reduction of suitable feeding habitat for wildfowl and waders. This species originated by chromosome doubling of the sterile hybrid between the Old World *S. maritime* and the New World *S. alterniflora*. Genetic analysis has found an almost total lack of genetic differences among individuals. However, the

---

Example 20.3   Hybrid mosquitoes may spread West Nile virus by acting as a bridge between humans and birds

The rapid spread of West Nile virus in humans in North America provides an unusual example of the possible role of hybridization in the spread of invasive species (Couzin 2004). Mosquitoes in the *Culex pipens* complex are the primary vectors for the spread of West Nile virus to humans in North America and Europe (Fonseca *et al.* 2004). However, there have been few human outbreaks in Europe even though the virus, the birds that harbour it and the mosquitoes that spread it are all endemic.

Microsatellite analysis of *Culex* mosquitoes indicated there are several reproductively isolated taxa of mosquitoes that differ in biting behavior and physiology in northern Europe (Fonseca *et al.* 2004). In contrast, hybrids between these distinct taxa are found throughout the US. It appears that hybrid mosquitoes in North America bite both humans and birds, and as a result apparently serve as bridge vectors of the disease from birds to humans.

allopolyploid origin of this species has resulted in fixed heterozygosity at all loci for which these two parental species differed.

20.4.4 Reduction of genetic load

As we have seen, small isolated populations will accumulate deleterious recessive mutations so that mildly deleterious alleles become fixed, and this leads to a slow erosion of average fitness (Section 14.7). Hybridization between populations would reduce this mutational genetic load (Whitlock *et al.* 2000, Morgan 2002). Ellstrand and Schierenbeck (2000) have suggested that the increase in fitness of this effect may under some circumstances be sufficient to account for invasiveness.

20.5 ERADICATION, MANAGEMENT, AND CONTROL

Understanding the biology of invasive species is not necessary before taking action. Simberloff (2001) has described this as a policy of *"Shoot first, ask questions later"* (see also Ruesink *et al.* 1995). This recommendation is in agreement with basic population biology. The best way to reduce the probability that an introduced species becomes invasive is to eliminate it before it becomes abundant, widespread, and has had sufficient time to evolve any adaptations that may allow it to outcompete native species. Nevertheless, understanding population biology, genetics, and evolution may be helpful in the prediction of the potential for invasive species to evolve responses to management practices, and in the development of policy.

Genetics may play an important role in the potential of an established invader to evolve defenses against the effects of a control agent (e.g., evolution of resistance to herbicides or biological control agents). The rate of change in response to natural selection is proportional to the amount of genetic variation present (Fisher 1930). Therefore, the amount of heterozygosity or allelic diversity at molecular markers that are likely to be neutral with respect to natural selection may provide an indication of the amount of genetic variation at loci that potentially could be involved in response to a control agent.

The amount of molecular genetic variation may not be a reliable general indicator of the amount of heritable variation for adaptive traits (Frankham 1999, McKay and Latta 2002). However, molecular genetic variation is likely to be a reliable indicator for invasive species of the potential for adaptive change because of the genetic effects of recent colonization. For example, greatly reduced molecular variation in an invasive population relative to native populations of the same species is a good indicator of a small effective population during the founding event; this is expected to reduce the amount of variation at adaptive loci. In addition, greater molecular variation in an invasive population relative to native populations of the same species is a good indicator of introductions from multiple source populations. This may indicate that the invasive species has substantial amounts of adaptive genetic variation to escape the effects of a control agent.

Genetics should play a more central role in developing policy to manage and control invasive species. Regulations generally have not taken into account that some genotypes may be more invasive than others of the same species. According to the standards set by the International Plant Protection Convention, import cannot be restricted for species that are already widespread and not the object of an 'official' control program (Baskin 2002). For example, several well-known noxious range weeds (e.g., the yellow star thistle) are on the list of permitted imports in the State of Western Australia because they are widespread and the government is not officially trying to control them. However, they are subject to control attempts by landowners for which they are a problem. Allowing the future import of additional strains that could be more invasive seems unwise in situations such as this.

Allowing further introduction of individuals of established invasive species could also be a problem if the introduced species has already been affected by inbreeding depression. For example, the giant African land snail has been introduced throughout the world as a possible food resource, among other reasons, and is perhaps the most important snail pest in the world (Cowie 2001). It has been introduced to many islands in the Pacific and Indian Oceans, as well as to other regions. The giant African land snail often declines in abundance and size following initial success as an invader. It has been suggested that this decline could partially result from the loss of genetic variation at disease resistance loci (Civeyrel and Simberloff 1996) and inbreeding depression (Cowie 2001). Further introductions could restore lost genetic variation for disease resistance and provide genetic rescue from inbreeding depression.

20.5.1 Units of eradication

Eradication of introduced species is a potentially valuable tactic in restoration (Myers *et al.* 2000). The eradication of rats, mice, and other introduced mammals is becomingly increasingly common on oceanic islands and isolated portions of continental landmasses. The New Zealand Department of Conservation applied 120 metric tons (120,000 kg) of poison bait onto Campbell Island, a large subantarctic island (11,300 ha) approximately 700 km south of the New Zealand mainland. A survey of the island in 2003 found no trace of brown rats on the island, and an incredible recovery of bird and insect life (McLelland 2011).

Successful eradication requires a low risk of recolonization. Isolated island populations or populations limited to isolated 'habitat islands' have low risk of recolonization. However, other islands or regions that display no distinct geographic structure or barriers are more problematic. The eradication of a portion of a population, or a sink population within an unidentified source-sink dynamic, would result in rapid recolonization and a waste of resources (see examples in Myers *et al.* 2000). Fewster *et al.* (2011) have used the amount of genetic divergence between island and mainland populations to predict the likelihood of successful reinvasion of ship rats in New Zealand.

Genetic analysis can also provide important information about the source of reinvasion. For example, several species of mammalian pests were removed from the Rangitoto Island in the Hauraki Gulf of New Zealand in 2009 (Veale *et al.* 2012). A single stoat was found on the island in 2010, and it was not known whether this individual was a remnant from the eradicated populations or a migrant that swam at least 3 km from the mainland. Genetic analysis clearly determined that this individual was a new migrant, rather than a remnant from the eradicated population (Figure 20.4).

Genetics can be used to identify isolated reproductive units that are appropriate groups for eradication on the basis of patterns of genetic divergence (Calmet *et al.* 2001, Robertson and Gemmell 2004). Little genetic differentiation between spatially isolated populations is indicative of significant gene flow, while significant differentiation between adjacent populations indicates limited dispersal. Examination of the patterns of genetic variation can allow for the identification of dis-

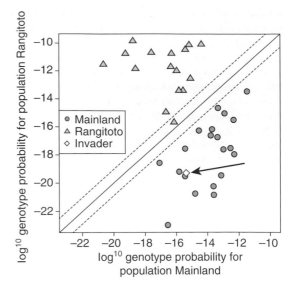

Figure 20.4 Molecular genetic analysis of a stoat found on Rangitoto Island one year after eradication. Log posterior probability plot for stoats from the pre-eradication population (*y*-axis) and the mainland population (*x*-axis). Triangles indicate pre-eradication stoats, circles indicate mainland stoats, and a stoat found on the island one year after eradication is indicated by a diamond and an arrow. Points below the solid diagonal line have greater posterior probability of belonging to the mainland than to Rangitoto. Points outside the dashed diagonal lines have over nine times greater posterior probability of belonging to one population than the other. The post-eradication stoat is approximately 10,000 times as likely to have originated from the mainland than from the pre-eradication population. From Veale *et al.* (2012).

tinct population units with negligible immigration. With appropriate care, these population units could be eradicated with little chance of recolonization. The identification of 'units of eradication' is an interesting analogue to the identification of units of conservation as seen in Chapter 16 (Robertson and Gemmell 2004). Genetic analysis would also allow distinction between an eradication failure (i.e., recovery by a few surviving individuals) and recolonization.

Robertson and Gemmell (2004) examined 18 microsatellite loci in two populations of brown rats separated by a glacier on South Georgia Island in the Antarctic. Rats were unintentionally introduced to South Georgia when commercial sealing started there in the late 1700s. The brown rats have devastated the island's avifauna. In addition, remaining rat-free areas on the island contain unusual assemblages of plants and animals that appear to be unable to sustain populations in the presence of rats.

The eradication of rats from the entire island is a daunting task because of its great size (400,000 ha). However, appropriate rat habitat is limited to coastal regions that are often separated by glaciers, permanent snow and ice, and icy waters. If such barriers preclude dispersal, then each discrete population could be considered as an eradication unit. Eradication of rats from South Georgia could then proceed sequentially with low risk of natural recolonization.

One population, Greene Peninsula, was earmarked for an eradication trial. Genetic diversity in 40 rats sampled from Greene Peninsula and a nearby population showed a pronounced genetic population differentiation that allowed individuals to be assigned to the correct population of origin (Robertson and Gemmell 2004). The results suggested limited or negligible gene flow between the populations, and that glaciers, permanent ice and icy waters restrict rat dispersal on South Georgia. Such barriers define eradication units that could be eradicated with low risk of recolonization, hence facilitating the removal of brown rats from South Georgia.

20.5.2 Genetics and biological control

Invasive species can undergo rapid adaptive evolution during the process of range expansion. Such evolutionary change during invasions has important implications for biological control programs (Wilson 1965). The degree to which such evolutionary processes might affect biological control efficacy remains largely unknown (Müller-Scharer et al. 2004).

The first applications of genetics in invasive species control have been in association with the sterile insect technique. One approach has been to introduce genotypes that could subsequently facilitate control or render the pest innocuous (Foster et al. 1972). Another approach was to release genotypes with chromosomal aberrations whose subsequent segregation would result in reduced fertility and damage the population (Foster et al. 1972). More recent efforts using the sterile insect technique have used transgenic insects homozygous for repressible female-specific lethal effects (Thomas et al. 2000). Release of males with this system may be an effective mechanism of control of some insect pests.

Hodgins et al. (2009) have modelled the use of a selfish genetic element called cytoplasmic male steril-ity (CMS) for controlling invasive weeds. CMS is caused by mutations in the mitochondrial genome that sterilize male reproductive organs. They found that the introduction of a CMS allele into plants can cause rapid population extinction, but only under a restricted set of conditions. They conclude that this approach would work only with species where pollen limitation is negligible, inbreeding depression is high, and the fertility advantage of females over hermaphrodites is substantial.

20.5.3 Pesticides and herbicides

Invasive species have the ability to evolve quickly in response to human control efforts (Example 20.4). Therefore, application of genetic principles is important for developing effective controls for invasive species. The evolution of resistance to insecticides and herbicides has increased rapidly in many species over the last 50 years (Denholm et al. 2002). Reducing the evolution of resistance to chemical control measures in invasive species will require an understanding of the origin, selection, and spread of resistant genes. Comparison of the genomes of insect species such as *Drosophila* and malaria vector mosquitoes should aid the development of new classes of insecticides, and should also allow the lifespan of current pesticides to be increased (Hemingway et al. 2002).

20.6 EMERGING DISEASES AND PARASITES

An emerging disease or parasite is one that has appeared in a population for the first time, or that existed previously but is rapidly increasing in prevalence or geographic range. Emerging parasites, including bacteria, viruses, protozoans, and helminthes are increasingly problematic for many wildlife and plant populations of conservation concern. An example is the bacterium *Yersinia pestis* that causes the plague in small mammals and humans worldwide. This disease has extirpated numerous prairie dog populations and consequently threatened the black-footed ferrets that depend on prairie dogs. Loss of prairie dogs and their influence in modifying landscapes could also reduce biodiversity of entire grassland ecosystems. Fortunately, researchers are advancing understanding of *Y. pestis* transmission through the use of DNA markers to help identify wildlife reservoir species such as

Example 20.4   Evolution of herbicide resistance in the invasive plant hydrilla

In the early 1950s, a female form of a dioecious strain of hydrilla was released into the surface water of Tampa Bay in Florida, and spread rapidly throughout the state (Michel *et al.* 2004). Today hydrilla is one of the most serious aquatic weed problems in the US. This invasive plant can rapidly cover thousands of contiguous hectares, displacing native plant communities and causing significant damage to the ecosystems.

Hydrilla has been controlled by the sustained use of the herbicide fluridone in lake water for several weeks. Fluridone is an inhibitor of phytoene desaturase (PDS), a rate-limiting enzyme in carotenoid biosynthesis. PDS is a nuclear-encoded protein and has activity in the chloroplasts, the site of carotenoid synthesis. Under high light intensities, carotenoids stabilize the photosynthetic apparatus by quenching the excess excitation energy. Inhibition of PDS decreases colored carotenoid concentration and causes photobleaching of green tissues.

An apparent decrease in the effectiveness of fluridone to control hydrilla has been observed in a number of lakes. Evolution of herbicide resistance was considered unlikely to occur in the absence of sexual reproduction. Nevertheless, a major effort was undertaken in 2001 and 2002 to test for herbicide-resistant hydrilla in 200 lakes throughout Florida. No within-site variation in fluridone resistance was detected. Approximately 90% of the lakes contained fluridone-sensitive hydrilla. However, three phenotypes of fluridone-resistant hydrilla populations were discovered in 20 water bodies of central Florida. A hydrilla phenotype with low-level resistance was found in eight lakes, the phenotype with intermediate resistance was found in seven lakes, and the most resistant phenotype was found in five lakes (Figure 20.5).

Sequencing of the phytoene desaturase locus (*pds*) indicated that the three fluridone-resistant types had different amino acid substitutions at codon 304 of the

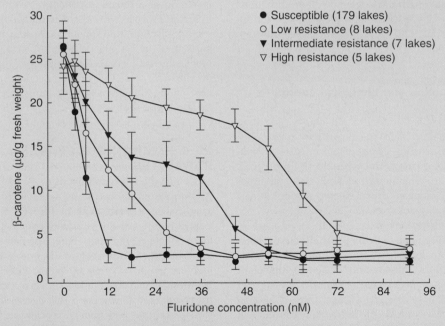

Figure 20.5   Differential response of hydrilla populations from lakes in Florida to the herbicide fluridone, which causes photobleaching of green tissues by decreasing β-carotene content. Low β-carotene content indicates high susceptibility to the effects of fluridone. Plots show mean and standard deviations of β-carotene content of hydrilla shoot apices following a 14-day laboratory exposure to fluridone at different concentrations. Four different phenotypes are shown: ●, susceptible (179 lakes); ○, low resistance (eight lakes); ▼, intermediate resistance (seven lakes); and ▽, high resistance (five lakes). From Michel *et al.* (2004.)

Rainbow River	cDNA/genomic DNA	ATTGCCTTAAAC⟨CGT⟩TTCCTTCAGGAA
	Amino acid	- I - - A - - L - - N - - **R** - - F - - L - - Q - - E -
Lulu	cDNA/genomic DNA	ATTGCCTTAAAC⟨AGT⟩TTCCTTCAGGAA
	Amino acid	- I - - A - - L - - N - - **S** - - F - - L - - Q - - E -
Pierce	cDNA/genomic DNA	ATTGCCTTAAAC⟨TGT⟩TTCCTTCAGGAA
	Amino acid	- I - - A - - L - - N - - **C** - - F - - L - - Q - - E -
Okahumpka	cDNA/genomic DNA	ATTGCCTTAAAC⟨CAT⟩TTCCTTCAGGAA
	Amino acid	- I - - A - - L - - N - - **H** - - F - - L - - Q - - E -

Figure 20.6 DNA sequences and their corresponding deduced amino-acid substitutions at codon 304 of the hydrilla *pds* gene that convert the susceptible (Rainbow River) into resistant biotypes: Lulu (low), Pierce (intermediate), and Okahumpka (high). From Michel *et al.* (2004).

sensitive form of PDS (Figure 20.6). The three PDS variants had specific activities similar to the sensitive form of the enzyme, but were two to five times less sensitive to fluridone. *In vitro* activity levels of the enzymes correlated with *in vivo* resistance of the corresponding populations.

It appears that fluridone resistance has arisen by somatic mutations that caused a single biotype quickly to become the dominant type within a lake. The establishment of herbicide-resistant biotypes as the dominant forms in these lakes was not anticipated.

Asexually reproducing plants are under strong uniparental constraints that limit their ability to respond to environmental changes (Holsinger 2000).

The future expansion of resistant biotypes poses significant environmental challenges in the future. Weed management in large water bodies relies heavily on fluridone, the only EPA-approved synthetic herbicide available for systemic treatments of lakes in the US. Current plans include regular monitoring to detect resistance and prevent the spread of these herbicide-resistant biotypes.

chipmunks, and to track the spread of this deadly bacterium among species and geographic areas (Example 20.5).

DNA markers offer enormous potential to understand the causes and consequences of parasite infection, including the emergence, spread, persistence, and evolution of infectious disease (Archie *et al.* 2009). Genetic markers are increasingly used to identify parasite species, quantify parasite abundance within individual animals or plants, trace back the origin of outbreaks, and predict transmission corridors and barriers (Section 15.6).

20.6.1 Detection of parasites

DNA markers are increasingly used to detect insect, fungal and microbial parasites in samples from the environment and from tissues or noninvasive samples from individual organisms (e.g., Beja-Pereira *et al.* 2009a). Detection is a crucial first step in determining prevalence of parasites and the effects of parasite infection on individual fitness and population persistence. For example, Lindner *et al.* (2011) developed a **quantitative PCR** (qPCR) diagnostic test to identify and track an emerging infectious disease called White-Nose Syndrome (WNS) that has caused major reductions in bat populations in western North America since 2006. Over 1,000,000 bats from six species in eastern North America have died from WNS since 2006, several species of bats may become endangered or extinct, and the disease is spreading rapidly (Foley *et al.* 2011). The qPCR test detects low-level amounts of the disease-causing fungal pathogen (*Geomyces destructans*) in bats and in soil samples collected from bat hibernacula. The qPCR test is crucial to allow assessment of effects of

Example 20.5   DNA markers identify wildlife reservoir for the plague

The plague, caused by the bacterium *Yersinia pestis*, has reduced or eliminated prairie dog populations throughout their range, and has contributed to the near-extinction of the endangered black-footed ferret. Lowell *et al.* (2009) isolated *Y. pestis* from infected humans and wildlife and their fleas in Colorado to help to identify wildlife reservoir species and focus disease surveillance efforts. Seventeen microsatellite loci were genotyped to construct a consensus tree by maximum parsimony analysis (assuming a stepwise mutation model) to identify genetically related *Y. pestis* isolates (Figure 20.7).

Results indicated that some human *Y. pestis* isolates were genetically similar to *Y. pestis* originating from the Colorado chipmunks and from the 'chipmunk flea' (see brackets on branch tips). Based on these data, the authors suggested that chipmunks are potential sources of human *Y. pestis* infection in Colorado and should be targeted more extensively in animal-based surveillance activities in Colorado and other states.

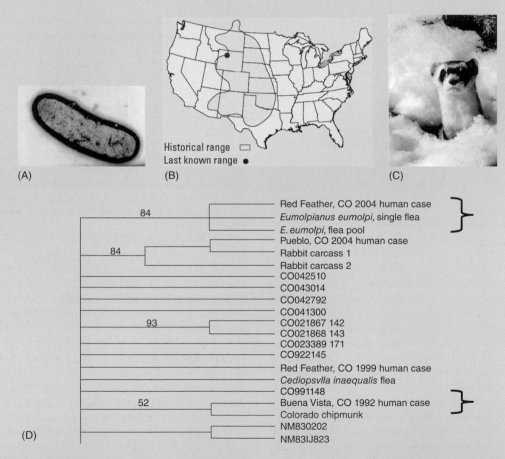

Figure 20.7   (A) *Yersinia pestis*. (B) Historical range of black-footed ferrets and last location in wild to become extirpated in the early 1980s. (C) Black-footed ferret. (D) Consensus maximum parsimony tree based on 17 microsatellite loci of *Y. pestis* plague isolates from humans, Colorado chipmunks, chipmunk fleas, and rabbits from Colorado or New Mexico (CO- or NM-labeled branch tips). The two numbers following the state abbreviation indicate the year collected. Numbers on the branches of the tree are jackknife support values. Brackets indicate closely related isolates found in humans and chipmunks. From Lowell *et al.* (2009).

the fungus on individual viability, characterization of the fungal lifecycle, and the tracking of spread across North America. Soon, in bats and many other host species, multiplex real-time PCR assays will permit the detection of multiple pathogens with high sensitivity and specificity (Singh *et al.* 2009).

20.6.2 Quantification of parasite abundance

Quantitative PCR can be used to quantify parasite abundance, which can help to understand and predict the effect of parasites on individuals and populations. For example, qPCR provides an estimate of parasite burden that is more sensitive than a blood smear and is predictive of clinical effects of malaria infection in pregnant women and newborns (Malhotra *et al.* 2005), suggesting the qPCR test would also be useful in wildlife. The qPCR test for detecting WNS in bats (previous section) also allows quantification of fungal load in bats.

20.6.3 Tracking origins of infectious disease outbreaks

DNA markers can help to determine the origin of disease outbreaks among wild and domestic animals. For example, a recent study illustrated the usefulness of DNA markers by genotyping bacteria that cause one of the world's most common bacterial **zoonoses**, brucellosis. Brucellosis causes abortions, inflammation of gonads, and can reduce reproductive success in wildlife populations. Brucellosis was introduced into North America with cattle imported from Europe in the early 1900s. Ironically, brucellosis in the US now exists only in wildlife around Yellowstone National Park, although recent outbreaks in domestic cattle have caused millions of dollars in lost revenue to the cattle industry and enormous controversy between wildlife enthusiasts and livestock growers.

Identifying the source of infectious disease outbreaks is difficult, especially for pathogens that infect multiple wildlife species, such as brucellosis. Beja-Pereira *et al.* (2009a) genotyped nine microsatellite loci in 56 *Brucella abortus* isolates from bison, elk, and cattle, to identify the wildlife species most likely to be the origin of recent outbreaks of brucellosis in cattle in the Greater Yellowstone Area. Isolates from cattle and elk were nearly identical to each other, but they were highly divergent from bison isolates of *B. abortus*. These data suggest that elk, not bison, are the reservoir species of origin for the recent cattle infections in Wyoming and Idaho. This illustrates the power of genetics to assess the origin of disease outbreaks, which are increasing worldwide following habitat fragmentation, climate change, and expansion of human and livestock populations.

Sudden oak death is a disease caused by the Oomycete *Phytophthora ramorum*. Since the mid-1990s, this pathogen has caused extensive mortality in several species of oaks, as well as tanoaks in coastal California and southern Oregon, and it infects a number of other tree and shrub taxa as well. It is not known where this pathogen originated, but it is thought to have been introduced to North America and Europe from different source populations, possibly in Asia. Genetic marker studies have shed considerable light on the population structure and the role of human activity in spreading the pathogen and causing this epidemic (Prospero *et al.* 2009).

The pathogen infecting forests along the west coast of the US represents three highly diverged clades, each likely derived from the introduction of a single clone, and all are currently reproducing clonally. One lineage is widespread in both forests and plant nurseries in North America (NA1), another has only been found in a few nurseries in California and Washington State (NA2), and the third is found primarily in Europe (EU1), but occurs sporadically in North American plant nurseries. These pathogens need to cross with a different mating type to reproduce sexually (A1 or A2). NA1 and NA2 are both mating type A2, and so are not compatible with each other. When EU1 samples of the A1 mating type were detected in North American nurseries, there was a fear that this would lead to sexual reproduction and recombination with NA1 or NA2, and great efforts were made to sanitize nurseries to prevent this from happening. It now appears that the EU1 lineage may be sexually incompatible with the North American lineages, from which it has long been separated. These population genetic studies have clearly shown that this pathogen has spread in North America through the long-distance transport of horticultural plants between nurseries and subsequent infection of native forests. This knowledge made it possible to establish quarantine zones and possibly prevent much broader dissemination of the pathogen.

In another example, the Tasmanian devil is threatened with extinction by an emerging disease called the devil facial tumor disease (DFTD) (Murchison *et al.* 2010, Guest Box 6). A recent study using microsatellites, mtDNA, transcriptome sequencing, and microRNA analyses suggested that the disease might have originated from peripheral nervous system protecting cells called Schwann cells (Murchison *et al.* 2010). DFTD was first detected in the mid-1990s and now threatens to extirpate wild Tasmanian devils within 25 to 35 years. The disease is characterized by large tumors on the face and mouth that lead to starvation and can metastasize to other parts of the body. It spreads when an infected Tasmanian devil bites an uninfected animal.

Murchison *et al.* (2010) genotyped 14 microsatellite loci on 25 tumor-host samples and on ten unaffected Tasmanian devil samples. The tumors all had the same genotype independent of the host's age, sex, or location. This genotype did not match other host tissues or samples from uninfected animals, suggesting the DFTD originated from one clonal cell line. Sequencing of 1180 bases of the mitochondrial control in 14 tumor-host pairs yielded similar results. Cloning and sequencing of 10 normal tissues and five tumor tissues identified 114 Tasmanian devil miRNAs. Again, tumor samples resembled one another but had distinct profiles compared with the tissue-specific miRNA patterns detected in other tissues. The miRNAs that were most highly expressed in DFTD included some known to be upregulated in other tumors. On the other hand, tumor suppressor miRNAs were found at low levels in DFTD samples.

Finally, transcriptome sequencing of tumor samples and normal Tasmanian devil testis tissue identified 31 transcripts that were significantly enriched in tumors compared with testis tissue. The researchers found that gene expression in the tumors most closely resembled that found in Schwann cells. Schwann cells produce the myelin sheath in the peripheral nervous system, while oligodendrocytes make myelin in the central system. This study represents one of the most thorough uses of genetic approaches to investigate the origins and spread of a recently emerged disease.

20.6.4 Assessing transmission routes

Molecular genetic analysis to infer pathogen transmission patterns (molecular epidemiology) is increasingly used to assess, predict, and manage risks of disease spread. For example, Biek *et al.* (2007) used phylogenetic analysis of raccoon rabies virus (RRV) to distinguish seven genetic lineages of the virus. Each of the seven lineages exhibited a general direction of spread relative to the first reported cases in West Virginia. The spread routes toward the southeast, southwest, and east were each associated with one particular viral lineage, whereas spread in the northwestern and northeastern directions was each marked by two different groups. Counties sampled 5–25 years after their first raccoon rabies cases consistently yielded viruses of the same genetic lineages that had colonized that area initially. This phylogeographic study illustrates the potential usefulness of sampling and sequencing years after an outbreak to help infer the routes of spread, which could help conservation geneticists understand, manage, and predict future disease transmission.

Guest Box 20   Rapid evolution of introduced cane toads and native snakes
Richard Shine

The cane toad is one of the largest toad species, with adults sometimes weighing more than 1 kg (Shine 2010). Native to South and Central America, these voracious predators have been introduced to more than 40 countries worldwide in misguided attempts at biological control of insect pests. The toads were brought to northeastern Australia in 1935, and released to control beetles that were attacking the sugarcane crop. The toads failed to control the beetles, but rapidly spread across tropical and subtropical Australia. In the process, they have killed many native predators. Toads possess distinctive defensive toxins, and the lack of native toads in Australia meant that most species of native predators had no evolutionary history of exposure to toad toxins. Many lizards, snakes, crocodiles, and marsupials have been killed when they tried to eat toads, and populations of some species have crashed dramatically (Shine 2010).

Rapid evolutionary changes have already occurred both in the toads and in their Australian predators. In the toads, the biggest changes have been in traits that increase rate of invasion. The cane toad invasion has been like a giant footrace across thousands of kilometers, and in every generation the fastest-dispersing toads inevitably will be the ones closest to the invasion front. These fast-dispersers breed with each other, producing off-spring that in some cases are even quicker than either parent – and so, the rate at which cane toads have invaded has increased steadily over the years. In the 1950s and 1960s, the toad front moved at about 10 km per year, but it has been accelerating ever since, and now averages around 50 to 60 km per year (Phillips *et al.* 2006).

Radio-tracking shows that toads at the invasion front often move more than 1 km per night in a highly directional way, whereas toads in established populations move much less, without any consistent dispersal direction (Alford *et al.* 2009). By raising offspring of toads from different populations under identical conditions, Phillips *et al.* (2010) showed that this dispersal shift is genetically based: offspring inherit the dispersal rates of their parents. Mathematical models suggest that all invasive species moving through suitable habitat will show a similar acceleration of dispersal rates due to rapid evolutionary change.

Over the same time period, native predators have adapted to the presence of the toxic toad. In areas where toads occur, frog-eating snakes exhibit modified behavior, physiology and morphology in ways that help them to coexist with the potentially fatal invader. For example, red-bellied blacksnakes are at great risk from toad invasion: they readily eat toads, and die as a result. But by comparing red bellied blacksnakes from areas with and without toads, Phillips and Shine (2004, 2006) showed that these snakes have adapted to toad presence by evolving a refusal to eat toads (they now discriminate between toads and native frogs, and only eat frogs), by greater physiological tolerance of the toads' poisons (so the snake is less likely to die if it eats a small toad, though a big toad would still be fatal), and by a reduction in head size relative to body size (because a small-headed snake can only eat a small toad compared with the predator's body size, again reducing the risk of a fatal overdose of toxin) (Figure 20.8). Thus, the ability for rapid adaptive response can help native fauna to survive the otherwise lethal impacts of an invasive species.

Managers need to consider the possibility of rapid evolutionary changes, both in invasive species and in the native fauna and flora that are affected by invaders (Ashley *et al.* 2003). For example, under-

standing the evolved acceleration of invasion fronts can help managers to predict how quickly the invading species will reach areas of specific conservation concern, and hence how quickly they need to institute management of those areas to combat invader impact. Understanding the possible adaptive response of native species is equally important because it affects the priority that managers need to apply to imperiled native taxa. A species that can adapt rapidly to the new challenge may warrant a lower priority for urgent management than a species that is unable to adapt, or adapts so slowly that local populations may disappear before an effective evolutionary response is mounted.

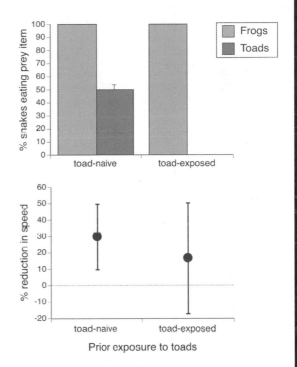

Figure 20.8 Rapid evolution of behavior and physiology in the red-bellied blacksnake enables it to better survive the invasion of the toxic cane toad. When toads first arrive in an area, most of the local red-bellied blacksnakes are fatally poisoned when they try to eat the highly toxic toads. Comparisons between toad-naive and toad-exposed populations of snakes show that within a few decades, the snakes evolve to ignore toads as prey (above), and they are more resistant to the toads' poisons as indicated by smaller reduction in swimming speed following toxin consumption (below). From Phillips and Shine (2006).

CHAPTER 21

Climate change

Whitebark pine, Section 21.7

Climate change is one of the largest threats to biodiversity of our times. Only when we, as a planet, adopt a socio-economic strategy that will allow organisms to adapt in pace with the changes in their environment can we prevent severe loss of species due to global climate change.

Marcel E. Visser (2008)

As climate changes in this century, the current distribution of climatic conditions will be rearranged on the globe; some climates will disappear entirely, and new (no-analogue) climates are expected in wide regions. For species to survive, the persistence of suitable climates is not sufficient. Species must also keep pace with climates as they move.

Scott R. Loarie et al. (2009)

Chapter Contents

Conservation and the Genetics of Populations, Second Edition. Fred W. Allendorf, Gordon Luikart, and Sally N. Aitken.
© 2013 Fred W. Allendorf, Gordon Luikart and Sally N. Aitken. Published 2013 by Blackwell Publishing Ltd.

The global distribution of plant and animal species is largely determined by climate. Genetic variation among populations within species for phenotypic traits involved in local adaptation is also frequently associated with climate. It has become abundantly clear in the last decade that climate is changing rapidly as a result of increasing greenhouse gases in the atmosphere (IPPC 2007). For those species whose ranges are determined primarily by climate, this means that suitable conditions are moving away from where species currently are found to new geographic areas (Ackerly *et al.* 2010). Novel climates comprising new combinations of temperature and precipitation may develop, and some climatic combinations may cease to exist. Climate change will present a great challenge to conserving populations and species at risk, and will test the abilities of populations to withstand rapid changes through phenotypic plasticity, adaptation to new conditions, or migration to favorable conditions elsewhere (see Example 15.2).

Some species may be quite resilient in the face of climate change, and established individuals may persist in locations outside of the climatic niche of that species. For example, some perennial plants such as trees that live for hundreds or even thousands of years can persist as adults through periods that might be climatically unsuitable for establishment or reproduction. Species that spread and reproduce clonally may be unable to reproduce sexually outside of their climatic niche, but may be able to persist clonally. For example, one trembling aspen clone in Utah has been estimated to be up to 10,000 years old, with root systems thought to have persisted through a wide range of past climates. Similarly, the alpine sedge *Carex curvula* was estimated to have clones as old as 2000 years in the Alps (Steinger *et al.* 1996). However, if conditions are too unfavorable for sexual reproduction in such environments, long-lived individuals and the populations they comprise may become functionally extirpated, unable to adapt and evolve.

Species will differ in their capacity to tolerate, adapt to, or migrate in the face of climate change. Interacting species may also migrate in different geographic directions as a result of differences in climatic niches determined by different climatic variables (Davis and Shaw 2001, Ackerly *et al.* 2010). This is likely to cause major shifts in interspecific interactions including competition, parasitism, predator–prey, and mutualism (Traill *et al.* 2010b, Van der Putten *et al.* 2010). Species may be limited in their migration capacity by the rate of migration of a symbiont (e.g., plants limited by the absence of mycorrhizae or pollinators). Hybrid zones may move geographically, and reproductively compatible species may come into or lose contact (Buggs 2007, Kelly *et al.* 2010). Hybridization can threaten the genetic integrity and persistence of a species, but may also provide the genetic variation required to adapt to new climatic conditions (see Chapter 17).

A population whose climatic niche is moving rapidly as a result of climate change faces four alternatives:

1 Individuals in the population might tolerate new conditions through phenotypic plasticity and persist.
2 The population might adapt to the new climatic conditions through natural selection.
3 Individuals in the population might move and track the climatic niche geographically.
4 The population may become unviable and become extirpated.

In general, two or more of these processes are likely to occur simultaneously (Davis and Shaw 2001). Only some combination of phenotypic plasticity and adaptation will prevent local extirpation of species facing environmental change (Gienapp *et al.* 2008).

In this chapter, we explore the alternative outcomes of climate change for populations, with a particular emphasis on phenotypic plasticity and adaptive responses. We also consider management options in the face of climate change, and discuss the implications of these actions for the conservation of populations.

21.1 PREDICTIONS AND UNCERTAINTY ABOUT FUTURE CLIMATES

Conservation strategies are usually focused on a species or ecosystem of concern within a fixed geographic context (Hansen *et al.* 2010). Populations targeted for conservation usually occur within defined geographic areas or regions that are often geopolitically bounded (i.e., within states, provinces, or countries). *In situ* conservation within parks, protected areas, and ecological reserves often forms the background of such strategies, with fixed land or aquatic areas maintained in a relatively natural state and managed to conserve plants and animal populations and their biological communities. If climate change results in protected areas no longer containing suitable habitat, population extirpations and ultimately species extinctions could result.

For example, Ackerly *et al.* (2010) modeled the future climates of over 500 protected areas in the San Francisco Bay Area of California. The model predicted that by 2100, only eight of these conservation areas will have temperatures within the current observed range. At a global scale, Loarie *et al.* (2009) predicted that current climatic conditions would persist within the boundaries of current protected areas for the next century in only 8% of protected areas globally. The need to consider climate change and the capacity for species to adapt or migrate as part of conservation planning is clear.

Both the paleoecological record and the recent history of contemporary populations provide evidence of species responses to climatic changes. While geologists and paleoecologists have documented large global and regional climatic shifts over geologic time, on a conservation timescale, climatic changes were not considered relevant to conservation until recently (Hansen *et al.* 2010). Biological responses to climate change have already been detected in natural populations (Parmesan 2006), and the extent of measured climatic changes over recent decades pales in comparison with modeled predictions for the coming century (IPPC 2007).

In order to interpret the implications of climate change for conservation, it is useful to understand the scientific sources of, and the uncertainty associated with, the predictions. Two types of models are combined to map the geographic distributions of suitable climatic conditions for species or populations: global climate models (GCMs) and bioclimatic envelope models (BEMs). **Global climate models** (GCMs) are complex mathematical models of global thermodynamics and atmospheric or oceanic circulation (IPPC 2007). These models are used to predict future climatic conditions for different levels of emissions of greenhouse gases. There is considerable uncertainty around predictions of future climates for two reasons: (1) GCMs differ in their scientific assumptions and methodologies, and thus produce different predictions; and (2) it is highly uncertain what the trajectory of greenhouse gas emissions will be over time, depending on how quickly humankind is able to curb carbon dioxide and other greenhouse gas emissions.

Species climatic niches are the range of climatic conditions over which the species can survive and reproduce. Climatic envelopes are described by multivariate combinations of temperature and precipitation-related variables that define the limits of the climatic niche.

Regional predictions of future climates from GCMs can be compared with the climatic conditions within an existing species range, to forecast the future distribution of species climatic niches in an approach called **bioclimate envelope modeling** (Pearson and Dawson 2003). Most bioclimate envelope models (BEMs) first determine the multivariate climatic tolerances of a species by analyzing their current distribution in conjunction with the distribution of temperature and moisture variables, and then predict where those conditions will occur under future climate scenarios.

A small subset of BEMs uses information on physiological climatic tolerances of species or populations, rather than using existing distributions to predict future distributions (Pearson and Dawson 2003). Called mechanistic or physiological bioclimate models, these models have the advantage of predictions being independent of current species distributions; however, they do not reflect the effects of interspecific competition and other types of species interactions. Range-based BEMs indirectly incorporate species interactions, as species do not fill that portion of climatic niche space in which they are outcompeted by other species.

The predictions from bioclimate envelope models are often presented as maps illustrating present observed species distributions and future predicted distributions of climate-based habitat. Such maps are valuable visual tools for education and conservation. However, they should be interpreted with caution. They usually depict areas where climatic conditions are predicted to be favorable without considering the biological capacity for populations to adapt to new conditions or migrate to areas that are climatically favorable. The non-climatic qualities of new potential habitats are also generally ignored. Locally adapted populations have smaller climatic niches than species, yet most bioclimatic models define climatic niches for entire species rather than for populations. Nevertheless, bioclimatic envelope model predictions for the coming century are a useful conservation planning tool, and have raised the alarm in the conservation biology community that climate change will threaten many species globally.

21.2 PHENOTYPIC PLASTICITY

Phenotypic plasticity occurs when the same genotype produces different phenotypes in different environments (Example 21.1). Recall from Chapters 2 and 11 that the phenotype, particularly for quantitative traits, is the product of both the genotype and the environment. If the environment changes, an individual's phenotype may change with no genetic change. Environmental conditions can affect phenotypes for **phenology**, growth, morphology, or reproductive traits. Phenotypic plasticity can be observed within an individual if phenotype changes in response to environmental changes within its lifespan, or between generations if genetic makeup does not change yet phenotypes change in response to environmental changes. Individuals and populations can vary genetically in their capacity for phenotypic plasticity.

One of the most commonly observed types of phenotypic plasticity related to climate is variation in phenology. The phenology of many species has already changed in response to shifts in climate over recent decades (Parmesan 2006). Many phenological traits, such as timing of bud break or flowering date in plants,

are based on the accumulation of heat sum in spring, and thus have inherent plasticity in responding to rising temperatures. Species with these traits may compensate for warmer conditions without genetic changes.

Changes in phenology as a result of global warming can be analyzed thanks to the long-standing tradition of recording the date of flowering or the date of leaf emergence from buds in some tree species in some locations. Chronologies of the flowering date of cherry trees in Japan over several centuries have shown that average flowering dates have advanced since 1952 (Menzel and Dose 2005). This shift in timing of flowering results from more rapid accumulation of the heat sum required for flower-containing buds to open.

Many animal behaviors and events are also temperature-dependent. Long-term records of migratory bird-arrival dates and egg-laying dates have been

---

Example 21.1 Phenotypic plasticity in the great tit

Phenological traits such as the timing of egg laying or spring migration in birds, or the timing of flowering in perennial plants, often occur as the result of a genetically determined response to temperature accumulation, or heat sum. These traits are quite plastic in a warming environment: as temperatures warm, heat sums accumulate more quickly, and these biological events are triggered earlier in spring. Temperature-initiated traits are expected to be more phenotypically plastic in the face of climate warming than phenological traits that are dependent on non-thermal cues, such as night length, and will require less genetic adaptation to keep pace with changes.

Phenological shifts, whether due to plasticity or adaptation, may occur at different rates in interacting species, resulting in asynchrony between species in key traits (Penuelas and Filella 2001). One of the most thoroughly studied systems illustrating the importance of phenotypic plasticity and genetic effects for species interactions in the face of climate change is the case of great tits and their caterpillar prey in the Netherlands (Nussey et al. 2005b). The mean laying date of great tits is determined by spring temperatures, and advanced by five to ten days from 1973 to 2006 in populations in the Netherlands and the United Kingdom (Husby et al. 2010) (Figure 21.1). This has

resulted in an asynchrony between the timing of egg hatching and the peak of caterpillar biomass (Nussey et al. 2005b). While both the mean date of egg laying and the peak of caterpillar biomass are highly temperature-dependent, and have advanced in the spring with warming temperatures in recent decades, these two peaks have become asynchronous as the critical thermal periods differ in timing between caterpillars and birds, resulting in an earlier peak in food availability than the peak period needed by the birds. This has resulted in an overall decline in the fitness of great tits.

Maternal parents vary in their degree of phenotypic plasticity for egg-laying, thus phenotypic plasticity itself shows genetic variation. Prior to 1983, there was no relationship between plasticity and fitness, whereas birds with the highest degree of phenotypic plasticity had the highest fitness from 1983 until 2002, suggesting directional selection for increased plasticity imposed by climate change. Furthermore, a comparison of a great tit population in a long-term study in the Netherlands with another in the UK revealed different levels and patterns of phenotypic variation for laying date and clutch size, suggesting that phenotypic plasticity itself shows population structure (Husby et al. 2010).

(Continued)

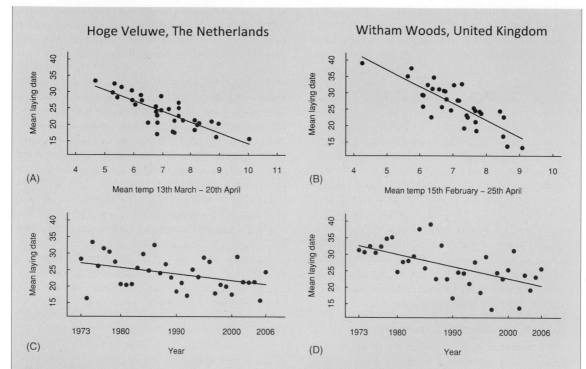

Figure 21.1 The effects of temperature on the laying date of populations of great tits in Hoge Veluwe National Park in the Netherlands (A and C), and in Wytham Woods near Oxford in the UK (B and D). The rate of increase of spring temperatures has been similar in the two locations, and phenotypic plasticity has resulted in earlier laying dates over time in both. However, the UK population has had a stronger phenotypic response to warming (−5.158 days °C⁻¹) than the Netherlands population (−3.256 days °C⁻¹). This suggests that the populations differ genetically in their capacity for phenotypic plasticity as a response to climate warming. From Husby *et al.* (2010), reproduced with permission of John Wiley & Sons, Ltd.

pivotal in understanding relationships between climate and life history events. For example, Gienapp *et al.* (2008) summarized data on egg-laying date from 15 studies and five bird species using records up to 70 years long. In all cases, birds had laid eggs earlier in later years by an average of 9.5 days. The primary cause is likely phenotypic plasticity, with birds responding to warmer conditions by laying eggs earlier.

Not all effects of climate warming are harmful for plant and animal species. In low temperature-limited environments, conditions are more favorable for growth for some species. Yellow-bellied marmots are large montane rodents that eat alpine vegetation. Between 1976 and 2008, the pre-hibernation weight of yellow-bellied marmots increased by 4–6% in a Colorado population, and the probability of survival increased with weight for both juvenile and mature

animals (Ozgul *et al.* 2010). The authors attributed the fatter pre-hibernation condition of the marmots primarily to two phenotypically plastic phenological shifts: earlier emergence from hibernation leading to a longer growing season, and earlier weaning of pups. A similar phenomenon has been recorded in long-term forestry provenance trials, where trees from northern locations planted in warmer climates have greater growth capacity than they do in their home provenances (Wang *et al.* 2010).

These shifts have been attributed primarily to phenotypic plasticity, although most of these studies of natural populations do not allow for the separation of plasticity from genetic adaptation (Gienapp *et al.* 2008, Visser 2008). Phenotypic plasticity is an effective mechanism for individuals to adjust to short-term environmental shifts, but is likely to be insufficient in the long term to allow populations to cope with predicted climate change and will not produce as great a shift in phenotype as a microevolutionary response in the longer term (Gienapp *et al.* 2008, Visser 2008).

21.3 MATERNAL EFFECTS AND EPIGENETICS

Characteristics of the environment during reproduction or development can affect reproductive traits or maternal care in ways that affect the phenotype of offspring. For example, in plants, the quality of the maternal environment often affects seed size, with plants producing larger seeds when conditions are more favorable for growth, and larger seeds in turn producing plants with more rapid juvenile growth due to the availability of additional nutritional resources in the seeds. Example 21.2 describes a maternal effect associated with timing of egg hatching that affects migration date in a migratory bird, and a maternal effect of maternal plant environment that affects the lifecycle of offspring.

Some maternal effects may result in **epigenetic** changes, which are heritable changes in gene expression and function that are not the result of changes in DNA sequence (Bossdorf *et al.* 2008, see Section 2.6). These effects include changes to the structure and configuration of chromatin, such as DNA methylation, that affect gene expression (Turner 2009). Epigenetic changes can be transmitted from mother to daughter cells through mitotic cell divisions, but more important to the adaptation to environments is transgenerational epigenetic inheritance, where the epigenetic state of DNA is transmitted from parents to offspring (Danchin *et al.* 2011). In laboratory rats, for example, it has been shown that the maternal care (e.g., licking and grooming behaviors) causes epigenetic changes in the promoter regions of target genes of offspring. These changes then affect gene expression and the behavior of offspring (Champagne 2011). Example 21.3 illustrates the potential adaptive significance of epigenetic inheritance for adaptation to temperature in a conifer.

Epigenetic effects on phenotypes could complement or supplant adaptation in rapidly changing climates for some populations or species. Indeed, some authors argue that such epigenetic mechanisms may have helped populations adapt rapidly to new environments during postglacial recolonization. However, Visser (2008) pointed out that in a rapidly changing environment, environments of offspring will differ from those of parents, particularly for long-lived organisms, and

---

Example 21.2 Maternal effects in blackcaps and the American bellflower

The phenotype of offspring can be modified by the maternal environment. This is a special type of phenotypic plasticity caused by the environment determined by or affected by the mother. Blackcaps are a migratory bird that provide a good example of such an effect. The autumn migration date of blackcaps is controlled by photoperiod, but this perception is modified by the photoperiod experienced by nestlings (Coppack *et al.* 2001, Pulido and Coppack 2004). The date of egg hatching is determined by the timing of egg-laying and by the environment prior to hatching. Blackcaps that are born later start their migration later than birds born earlier, but migrate at a slightly younger age, thus partially compensating for the later birth. A bird born a day later will initiate migration just half a day later.

The American bellflower provides an example of the effects of maternal environments on the germination of plant seeds. This species requires vernalization, a cold period to induce flowering. Plants can either germinate in the fall, overwinter, and then continue growing and flower the following summer as annuals, or germinate in the spring and flower the following spring as biennials. Galloway and Etterson (2007) found that seeds from mothers that were growing in the shade were more likely to germinate in spring, while seeds from mothers growing in full sun were more likely to germinate in the fall, a strategy more advantageous than spring germination in shady environments.

They concluded that these maternal effects are adaptive, in that seeds are more likely to disperse to germinate in environments similar to their mothers than to different environments, and that these maternal effects are predicted to increase fitness markedly. The challenge with rapid environmental change is that offspring will no longer inhabit environments similar to their mothers, especially for longer-lived plants and animals.

Example 21.3   Epigenetics in the Norway spruce

Epigenetic effects are difficult to study in non-laboratory species as they require experimental control over parental environments, and molecular genetic tools to confirm epigenetic structural changes to DNA such as methylation.

Norway spruce provides an example of apparently epigenetic effects that could be highly relevant to fitness in a rapidly changing environment. Trees were cloned through grafting to control genotype and planted in different locations or grown under different temperatures in controlled greenhouse experiments. The same mother tree genotypes produced seeds with markedly different phenotypes in different thermal environments (Johnsen *et al.* 2005b). On average, trees that produced seed under warmer conditions

had progeny that had longer growing seasons, delayed bud set, and lower cold-hardiness than progeny germinated from the seeds of genetically identical mother trees growing in cooler environments.

Johnsen and colleagues found that the change in phenotypes of seedlings due to maternal environment can be as great as that between natural populations separated by three to four degrees of latitude. This heritable effect of maternal environment on progeny phenology has been attributed to epigenetic effects of microRNAs on gene expression (Yakovlev *et al.* 2010). This research has also revealed genetic differences among families in the degree of epigenetic effect of maternal environment.

maternal effects including epigenetics may be insufficient for phenotypes to keep up with climate.

21.4 ADAPTATION

The capacity for adaptation of populations to new environmental conditions is substantial, as shown by the ability of many plant and animal species to adapt to local conditions during or since postglacial recolonization. As long as genetic variation exists for genes affecting phenotypic traits involved in adaptation to climate, selection imposed by new conditions should be able to shift average phenotypes in a population. However, the rates of anthropogenic climate change predicted for the coming century far outpace those seen during the Pleistocene. Here we first summarize the theoretical basis of adaptation to a changing environment, and then review examples of adaptive responses to global warming.

21.4.1 Theoretical predictions

Quantitative geneticists have developed theoretical predictions of the maximum rate of environmental change a population can withstand through adaptation. This rate is determined by the amount of genetic variation in the population, the fecundity and effective size of the population, environmental stochasticity, and the strength of selection. Beyond a threshold rate

of change, the population will be unable to adapt quickly enough to keep up with the necessary rate of change in optimum phenotype to maintain fitness. Over time, the population will lag increasingly behind its climate fitness optimum. When the **adaptational lag** between the average population phenotype and the optimum phenotype with the highest fitness becomes too great, extinction occurs.

Lynch and Lande (1993) developed an equation to predict this threshold rate of change (k_c, in units of phenotypic standard deviations, σ_p) based on the strength of stabilizing selection within populations (σ_w^2) and the maximum rate of population increase (r_{max}):

$$\frac{k_c}{\sigma_p} = \sigma_p \sqrt{2\, r_{max} / \sigma_w^2} \qquad (21.1)$$

This equation predicts that species with low fecundity ($r_{max} = 0.5$) can only tolerate a rate of environmental change corresponding to about 0.1 σ_p. However, for species with high fecundity, the maximum rate of climatic change that could be tolerated is much higher. For example, a pioneer tree species that can produce 10,000 seeds/tree would have an r_{max} of 9.2 and could theoretically tolerate a change of up to ~0.42 σ_p per generation (Aitken *et al.* 2008).

However, the maximum rate of climate change that a population can tolerate through adaptation is on a per generation basis. Trees and other long-lived perennial plants, some mammals, and some birds are experiencing a much higher rate of climate change in evolutionary terms than short-lived organisms like insects and annual plants.

Climatically imposed selection resulting from increased climatic stresses may shift demography in small populations from viable to non-viable by increasing mortality or decreasing reproduction. Bürger and Lynch (1995) expanded on the work of Lynch and Lande (1993) and developed a model to predict the effects of adaptational lag on population demography. They added demographic stochasticity to their model and found that random fluctuations in population size and fecundity can decrease genetic variation and adaptive potential through genetic drift and increased extinction risk. This effect is particularly relevant for conserving rare and endangered species, as it is greatest when population sizes are low ($N_e < 100$). Their sobering conclusion was that the threshold rate of environmental change for such species might be as low as 0.01 σ_p per generation. Populations for which this is true have very little chance of keeping pace with climate change through adaptation, even if they contain genetic variation for relevant genes and traits.

21.4.2 Empirical studies

A large number of studies have documented shifts in population phenotypes that correlate with recent climatic changes (Parmesan 2006), many of which invoke adaptation as an explanation (see Guest Box 21). However, few have conclusively demonstrated adaptation rather than phenotypic plasticity as the causal agent (reviewed by Gienapp *et al.* 2008). Detecting microevolution and adaptation conclusively and being able to attribute it to climatic trends is much more difficult than simply measuring and monitoring changes in phenotypes over time (Gienapp *et al.* 2008, Visser 2008). Phenotypic changes in a population may result from phenotypic plasticity, adaptation, epigenetic effects, or a combination of these processes. To prove adaptation has occurred, you need a common garden experiment or statistical control over environmental variation (Reale *et al.* 2003, Gienapp *et al.* 2008). As discussed in Chapter 11, the extent to which genetic variation underlies phenotypic variation also requires that related individuals be identified, either through known or inferred pedigrees or through marker-based estimates of relatedness.

Some scientists have attempted to show that adaptation is occurring in the short term, as illustrated by phenotypic changes occurring in the expected direction given climate-based selection. However, a substantial number of these studies have shown either no detectable changes, or changes in the opposite direction to that predicted. In some cases, phenotypic and genetic changes are in opposite directions. Gienapp *et al.* (2008) concluded that the primary reason for a lack of evidence of genetic responses to climate change is that genetic evidence is difficult to collect in natural populations.

In one of the few studies to conclusively show adaptation in response to warming environments, Reale *et al.* (2003) studied red squirrels in a pedigreed wild population in the Yukon, Canada. The average date of birth in the population advanced 18 days over a ten-year period (1989–1998). This amounted to a rate of change of 6 days per generation. More than half of this phenological shift (3.7 days) was attributed to phenotypic plasticity in response to an increase in availability of spruce seeds, while approximately 10% (0.6 to 0.8 days) was due to genetic causes (i.e., adaptation).

There are several explanations for the lack of adaptation to climate change observed:

1 The countergradient hypothesis: If environmental and genetic effects on phenotypes are in opposition (i.e., if genetic and environmental clines have opposing signs along a climatic gradient), then the effects of selection may be invisible as phenotypes will not change (see Section 2.4.1).

2 Lack of genetic variation: If a population lacks genetic variation for a phenotypic trait under selection, then no adaptive response can occur. This may be a major problem in small populations.

3 Antagonistic genetic correlations: If two traits under selection have an antagonistic genetic correlation (i.e., if selection increases fitness in one trait but decreases fitness in another due to pleiotropy or gametic disequilibrium), then evolutionary responses will be slow (Etterson and Shaw 2001, Hellmann and Pineda-Krch 2007).

4 Inbreeding: Small and declining populations are likely to have substantial levels of inbreeding, and this may also hamper adaptation.

5 Gene flow: If individuals or gametes (e.g., pollen) move between populations inhabiting and adapting to different climatic conditions, the extent to which gene flow facilitates or slows adaptation will depend on the source and recipient population climates as well as the size of the populations (i.e., the relative strength of gene flow from one population to

another) (Davis and Shaw 2001). Gene flow may slow adaptive responses unless the source populations of immigrants or gametes inhabit climates similar to the new climates faced by a population.

New genetic and genomic techniques will facilitate future studies of the capacity for rapid adaptation. The development of ecologically relevant genetic markers associated with key phenotypic traits and adaptation to climate will enable the study of changes in allele frequency resulting from selection without the need to control environmental variation. It will also help to determine amounts of standing genetic variation available for adaptive responses in wild populations. Experimental studies with model organisms in controlled environments will also provide needed information on the capacity of populations to adapt rapidly under a range of genetic, demographic, and environmental conditions.

21.5 SPECIES RANGE SHIFTS

The rate at which climates are predicted to shift with climate change varies geographically and topographically. As a result, the rate at which populations or species need to geographically shift to track their climatic habitat will vary. This ecological movement of species or populations is called migration, but should not be confused with genetic migration (i.e., gene flow). Loarie *et al.* (2009) estimated "velocities of climate change", defined as the horizontal rate of migration in km per year required to track temperature changes, for biomes globally. Their estimates averaged 0.42 km/yr, and ranged from 0.08 km/yr for tropical and subtropical coniferous forest biomes to 1.28 km/yr for flooded grassland, mangrove, and desert biomes. Predicted velocities were also lowest in mountainous terrain due to the rapid changes in temperature over short elevational distances. Population persistence and species survival will depend on their capacity to keep pace with these moving climates.

Species ranges can shift through population expansion and migration at the leading edge of the range, and population extirpation at the rear edge. Migration of populations is a function of population growth and dispersal coupled with the distribution and structure of available habitat (Higgins and Richardson 1999). The ability of populations to move to new habitat depends greatly on the life-history characteristics of the species including age of reproductive maturity,

fecundity, and mechanism of offspring dispersal, as well as the degree of fragmentation of suitable habitat. Also critical are the existence of barriers to dispersal such as mountain ranges, large water bodies, and deserts.

Parmesan (2006) reviewed evidence for changes in species ranges in response to contemporary climate change. The literature is already extensive, particularly for species at high latitudes or elevations. For example, evidence of range shifts is abundant for Antarctic bird species. Species dependent on sea ice, such as Adelie and emperor penguins, have shown marked shifts to higher latitudes. Open-water feeding penguins in the meantime have invaded more southern areas as ice shelves have contracted or collapsed. Many Arctic species have also shown range shifts, some resulting in major ecosystem transitions. For example, shrub species have been invading Arctic tundra in both North America and parts of Russia to the extent that fundamental changes in the albedo, carbon balance, and hydrology of those regions is predicted (Tape *et al.* 2006).

As trees and other plants are key habitat components for many other species, their migration rates may directly affect the migration potential of other species. Pollen records since the last glacial maximum have provided important evidence about the migration of plant populations. Cores can be extracted from bogs that contain a chronosequence of sediments, thereby preserving pollen records for common, wind-dispersed taxa such as many tree genera. The abundance of pollen of different taxa in sediments of known age can be used to infer past species distributions. However, genetic data have resulted in some large revisions of estimates from this paleobotanic approach. In eastern North America, maximum rates of tree migration based on pollen records have been cut in half based on the distribution of chloroplast haplotypes. The haplotype data suggest that pollen records for temperate angiosperm tree species do not capture pollen from small populations in more northern glacial refugia, and thus overestimate migration rates by assuming only southern refugia existed (McLachlan and Clark 2004). These studies indicate that tree migration rates during postglacial recolonization in North America averaged only about 100 m per year: an order of magnitude slower than that what would be needed to track climates spatially if climates warm to the extent predicted in the next century (McKenney *et al.* 2007).

The capacity for migration is likely to have a genetic component and to interact with adaptation (Davis *et al.* 2005). For example, poorly adapted phenotypes are likely to have lower fecundity, and lower fecundity produces fewer dispersants. Also, selection can act on phenotypic traits that affect dispersal ability, such as seed wing size or fruit characteristics for plants, or wing size for insects. Thus individuals that are less fit are less likely to migrate, and individuals with higher dispersal capabilities are more likely to migrate, both resulting in natural selection.

Invasive species often have short generation times and are well adapted for long-distance dispersal. Because of this, they may have a greater capacity than noninvasive species for migrating as the climate changes, which could lead to invasive species dominating more ecosystems in a rapidly changing climate. Many invasive species are early successional, and may become more problematic with climate warming, competing with and reducing the capacity of later-successional species to persist. This was demonstrated by Cole (2010), who studied vegetation records for the Grand Canyon from the Holocene and observed that those ecosystems became dominated by early successional plant species when climate change was rapid. These species typically have a greater migrational capacity as well as a shorter generation time.

21.6 EXTIRPATION AND EXTINCTION

If populations cannot tolerate rapid climate change through plasticity, adapt to new conditions, or migrate to track their climatic niche spatially, they will be extirpated. Species extinction may follow. To understand the relationship between climatic change and extinction, it is useful again to look back at the Pleistocene. Between 50,000 and 3000 years before present, 65% of large (>44 kg) mammal genera went extinct (Nogués-Bravo *et al.* 2010). There is some evidence that this high extinction rate resulted from climate changes in the late Quaternary, although the roles of climatic change versus anthropogenic factors such as hunting are hotly debated. Nogués-Bravo *et al.* (2010) analyzed paleorecords on climate and large mammal distributions over this period. They estimated 'climatic footprints' for this period based on the magnitude of climatic change. Their analysis revealed that continents other than South America that experienced higher rates of climate change, and had larger climatic

footprints, had higher extinction rates during this period.

Arctic and alpine populations are at high risk of species extirpation due to anthropogenic climate change. This is in part due to the greater rates of change predicted and observed, to date, for high-latitude areas, and in part due to a lack of adjacent habitat that remains climatically favorable. A low-vagility species inhabiting a mountain top has nowhere to go in a warming environment. Nine out of 25 recensused populations of American pika, a small alpine mammal in the Great Basin of the western US, were extirpated between the 1930s and 2007 (Beever *et al.* 2008, 2010). The populations that were extirpated occupied significantly lower-elevation habitats and experienced higher levels of heat stress than those that persisted, suggesting climate may have played a major role in extirpations.

Butterflies are one of the best documented groups of organisms used for monitoring contemporary range shifts and population extirpations in response to recent climate change (Parmesan 2006). Metapopulations of Apollo butterflies in France on plateaus less than 850 m in elevation have been extirpated, while those metapopulations above 900 m have remained relatively healthy (Parmesan 2006). This species appears to be dispersal-limited, and only those populations close to higher elevation habitats were able to disperse into and colonize those cooler areas. In the Sierra de Guadarrama of Spain, the lower elevation limit of Apollo butterflies also rose 300 m between 1967 and 2005, while mean annual temperature increased by 1.3°C over the same period (Wilson *et al.* 2005). This shift in elevation associated with climate warming has dramatically decreased the available habitat, since land area within elevational bands of fixed widths decreases rapidly with increasing elevation (Figure 21.2).

Predicting the risk of species extinction due to climate change is difficult. In general, species with larger geographic ranges have lower extinction risks. This is called extinction resistance. However, if a species contracts to only a few populations, then what will matter is threat tolerance of those remaining populations, not extinction resistance. A species that has a recently contracted range may not have any higher threat tolerance than a species that has always had a small range and relatively few populations (Waldron 2010).

Populations in different parts of a species range will face different degrees of threat from climate change,

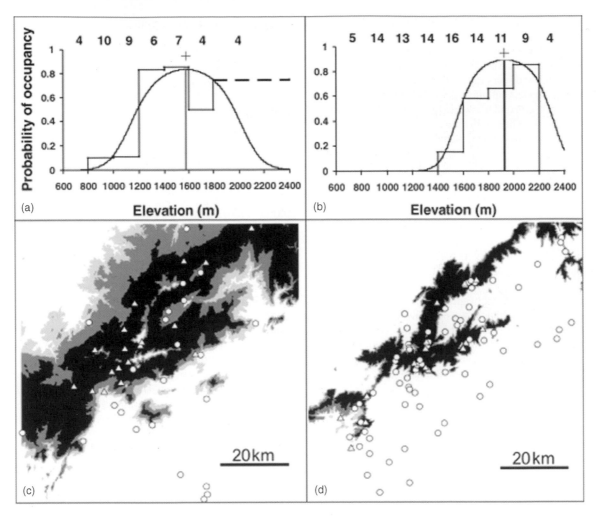

Figure 21.2 Modeled probability of occupancy at different elevations for the butterfly *Satyrus actaea* in central Spain. (a and b) Histograms of probability of occupancy in 200m intervals (bars), and probability of occupancy (*P*) modeled using logistic regression (curve) in (a) 1967–1973 and (b) 2004. Crosses show 'optimum' elevations with highest modeled probability of occupancy. Number of samples per 200m interval shown above each bar. In (a), the dashed line denotes the proportion of all four sites sampled above 1800m. (c and d) Distributions of suitable elevations based on equations in (a) and (b) respectively, for (c) 1967–1973 and (d) 2004. Black ≥ 50% probability of occupancy; dark gray ≥ 20%; pale gray ≥ 10%; white < 10%. Sample locations: triangles (occupied), circles (vacant). From Wilson *et al.* (2005).

depending on the position of the local environment within the species niche, population history, and contemporary patterns of gene flow across species ranges (Davis and Shaw 2001) (Figure 21.3). **Rear edge populations** (those at the warm or dry margin of a migrating species range) are at the greatest risk of extirpation during broad-scale shifts in species range.

These populations may already inhabit extreme environments at the margin of the species niche. They may also be older lineages with relatively high among-population diversity and high regional diversity as they may be located closer to glacial refugia (Hampe and Petit 2005). In contrast, theory suggests that **leading edge populations**, those at the cool margin of a

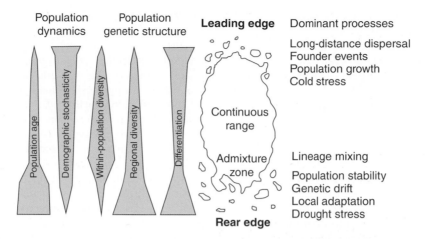

Figure 21.3 Population features and relevant processes at the leading and the rear edge of species ranges. The width of gray bars shown on the left hand indicates the quantity of features at the corresponding position within the range. From Hampe and Petit (2005).

species climatic niche that are likely to expand with warming, should experience increased fitness with climate change. Gibson *et al.* (2009) have made compelling arguments on conserving leading edge populations as they will play a key role in migration and species range shifts, potentially expanding into newly habitable areas. Hampe and Petit (2005) advocate for a conservation focus on rear edge populations as they are likely to be reservoirs of within-population genetic diversity and to show greater among-population divergence than younger populations.

If a species has a greater abundance in the center of the species range, population genetic theory has shown this is likely to result in greater gene flow from central to peripheral populations than the reverse (Kirkpatrick and Barton 1997, Davis and Shaw 2001). Populations at the leading edge of migration will receive gene flow from more central populations that carries alleles pre-adapted to the warmer conditions in the center of the species range. In a static environment, this means that northern populations occupy environments colder than optimal for their average genotypes. Warming will move the climate closer to optimal for the average phenotypes and genotypes of those leading edge populations. In contrast, rear edge populations are likely to have already inhabited climates warmer than optimal under pre-global warming conditions due to gene flow from more central populations in cooler environments. As those environments warm, the average genotypes and phenotypes of rear edge populations will become

less fit. The general pattern of populations from colder environments inhabiting colder climates than optimal for their genotypes, and populations from warmer climates inhabiting warmer climates than optimal for their genotypes, has been observed for lodgepole pine in a large, long-term reciprocal transplant study (Rehfeldt *et al.* 1999, Wang *et al.* 2006).

21.7 MANAGEMENT IN THE FACE OF CLIMATE CHANGE

The need to account for climate change in conservation planning is no longer in dispute, yet has only recently started to receive formalized attention (Glick *et al.* 2011). Povilitis and Suckling (2010) reported that while 124 of 1279 species listed under the US Endangered Species Act with recovery plans included climate change as a threat, plans for the vast majority of other listed species (1055 species) did not mention climate change. However, plans completed in recent years are much more likely to include consideration of climate change, indicating a growing awareness of this threat.

To maintain viable populations, conservation strategies should maintain or enhance the ability of populations to adapt to new conditions or migrate to new locations where conditions have become favorable. Hansen *et al.* (2010) proposed four basic principles for adapting *in situ* conservation strategies to a changing climate:

1 Protect adequate and appropriate space.

2 Reduce non-climatic stresses.

3 Use adaptive management to implement and test climate-change adaptation strategies.

4 Reduce the rate and extent of climate change to reduce overall risk.

Of these, the first two relate to genetic factors. Protected area reserve design will become increasingly important. Large protected areas that include a broad range of habitats and climates are more likely to contain suitable future habitat for extant species under new conditions than small or homogeneous reserves (Ackerly *et al.* 2010). Environmentally heterogeneous protected areas offer greater opportunities for migration over shorter distances into favorable microsites than more homogeneous areas. Large and heterogeneous protected areas are also more likely to contain numerous populations with correspondingly higher genetic diversity upon which selection can act.

Climate change may exacerbate other genetic risks to small and declining populations. Climate warming may reduce reproductive rates or increase mortality. If maladaptation due to climate change results in a decline in population size, then other genetic factors such as inbreeding and genetic drift will increasingly come into play, and further decrease the genetic variation present for an adaptive response to new climatic conditions. Monitoring threatened populations will become an even more important aspect of conservation.

21.7.1 Connectivity

Maintenance of genetic connectivity through gene flow has long been considered important for maintaining genetic diversity, but habitat for migration between reserves will become increasingly important for both adaptation and migration of populations. While gene flow can impede local adaptation under stable environmental conditions, in a rapidly changing environment, gene flow may be a source of genetic variation containing alleles pre-adapted to new conditions. Fragmented landscapes will pose a major barrier to climate-driven population migration. In some circumstances, **assisted gene flow** might be helpful for populations to adapt to future environments. For example, Bower and Aitken (2008) have recommended transfer of whitebark pine seeds from milder to colder climates in order to facilitate adaptation to future climates.

21.7.2 Assisted colonization

It is clear that many populations will not be able to migrate sufficiently rapidly to track their current climates spatially (Loarie *et al.* 2009). This has opened a hot debate among conservation biologists on the role of **assisted colonization**, also known as managed translocation, in conservation (Hoegh-Guldberg *et al.* 2008). Assisted colonization is the human relocation of individuals to new areas that have become climatically favorable. Proponents insist that this is a needed tool for preventing species extinctions. For example, Swarts and Dixon (2009) proposed assisted colonization as a necessary conservation tool for many terrestrial orchids. Opponents argue that this will lead to invasive species problems and the destabilization of ecosystems through the introduction of new species, and suggest the precautionary principle is in order (Ricciardi and Simberloff 2009). This process has also been called assisted migration (McLachlan *et al.* 2007), but we have used assisted colonization in order to distinguish it from moving genes among populations (assisted gene flow) and moving populations within existing species ranges (**population translocation**).

It is clear that decisions around assisted colonization need to evaluate both the risks of introducing species to new ecosystems and the risks of doing nothing. Unintentional assisted colonization is already common for garden plants in Europe (Van der Veken *et al.* 2008). Well-meaning individuals are also moving endangered species, perhaps without thorough scientific consideration. For example, a volunteer conservation group called the Torreya Guardians has already moved the endangered species Torreya 600 km without scientific assessment (Schwartz, M.W. *et al.* 2009).

The risks and benefits of assisted colonization depend on the scale of movement and the extent to which species in recipient ecosystems have previously coexisted with the species being introduced (McLachlan *et al.* 2007, Mueller and Hellmann 2008). Population translocation involves moving pre-adapted individuals from their native origins to locations with favorable climates within the existing species range. This is the least controversial form of assisted colonization, although it may have the potential to lead to outbreeding depression in cases where populations have previously been isolated. This is becoming a relatively widespread practice in reforestation of native tree species for which there is little evidence of outbreeding depression. **Assisted range expansion** is another type of assisted colonization. This involves moving individuals to regions adjacent to existing species ranges on the leading edge of migration. The ecological risk associated with assisted range expansion depends on whether species in the recipient ecosystem have previously been in contact on a relatively recent evolutionary timescale, or currently coexist in other areas. It also depends on the species being translocated and its ecological role and interactions. There is widespread agreement among conservation biologists that the **translocation of exotics** is a type of assisted colonization that carries much higher biological risks. This involves the intentional introduction of species from other bioregions within continents, or from other continents.

Population genetics has a role to play in informing assisted colonization in a conservation framework. Knowledge of genetics and patterns of local adaptation will inform conservationists about what source populations are more likely to provide more successful migrants for a recipient environment, how much genetic diversity will be available for adaptation, whether outbreeding depression is a likely outcome of population translocation, and whether translocated species can hybridize with congeneric species following assisted range expansion.

21.7.3 *Ex situ* conservation

Traditional *in situ* approaches to conservation in parks and protected areas may be inadequate for conserving populations of some species in a rapidly changing climate. Rear edge populations will be at greatest risk of population extirpation. *In situ* approaches to conservation are unlikely to be successful in these areas if populations are collapsing due to abiotic conditions or shifts in biotic interactions resulting from climatic changes. An alternative strategy for conserving the genetic diversity of these populations is *ex situ* conservation. For example, the vast Svalbard Global Seed Vault, sometimes nicknamed the Doomsday Vault, is dug into a mountain in Norway's Arctic and stores seeds for plant species (Charles 2006). Embryos or sperm of animals can similarly be cryopreserved.

21.7.4 Conclusion

Some target species may benefit from assisted gene flow, colonization, *ex situ* conservation, and other management actions. Nevertheless, it will be impossible to translocate entire ecosystems with the full suite of species that they contain. High fecundity, high vagility species are rarely the focus of conservation efforts, and these species are more likely to move and persist in a rapidly changing climate. Of course, the best conservation strategy for all species is to curb the production of greenhouse gases and slow climate change to a rate that species can tolerate through phenotypic plasticity in the shorter term, and adapt or move in response to in the longer term.

Guest Box 21 Rapid evolution of flowering time by an annual plant in response to climate fluctuation
S. J. Franks

Climate change is already affecting traits of many species, including altering the timing of reproduction in plants. For example, as the climate warms, many plants are flowering earlier in the spring. One important question is whether these changes are due to plasticity, meaning direct effects of environmental conditions on the traits, or to evolution, in which case climatic changes select for different trait values (such as earlier flowering) which are then passed on to new generations and increase in frequency in the population. Knowing the answer to this question would let us know how rapidly species can adapt to climate change.

To address this question, Franks and colleagues (2007) investigated natural populations of the annual plant *Brassica rapa* (field mustard) in southern California. For the five years from 1999–2004 there was very little rainfall, making this a period of extended drought. This drought was thus a climatic change, which could potentially cause evolution, particularly if it selected for plants that could flower early and escape the effects of the drought.

(Continued)

Fortunately, seeds of *B. rapa* had been collected from two populations in 1997, before the drought and after a series of wet years. Franks and collaborators grew these seeds, along with seeds collected in 2004, from the same two populations together under common conditions and crossed them within populations and years as well as across years, and examined traits in the offspring, using the 'resurrection' approach to examining evolution (Franks *et al.* 2008).

Plants derived from seeds collected in 2004 flowered earlier than the 1997 lines in both populations (Figure 21.4). This indicates rapid evolution of earlier flowering following drought, and was the first example of an evolutionary change in a natural plant population caused by a change in climate. The 'hybrids' (1997 × 2004 crosses) flowered at an intermediate time, which, along with heritability estimates from controlled crosses, showed that variation in flowering time is genetically based. When plants were subjected to a drought treatment, post-drought plants had higher fitness (greater survival) than pre-drought plants, demonstrating that the plants had adapted to drought conditions. Subsequent work showed that the drought caused the evolution of multiple life-history traits (Franks and Weis 2008), that plants that flowered early did so by beginning to flower at a smaller size and using water inefficiently, allowing rapid development (Franks 2011), and that the shift in the timing of reproduction could increase gene flow between populations (Franks and Weis 2009).

While this work documented a case of very rapid evolution in response to changes in climatic conditions, Franks and colleagues noted that this does not mean that all species will simply adapt to any changes in climate that occur. This example shows that under some circumstances, evolution can occur quickly, but it may not be fast enough for many species facing habitat loss and fragmentation, declining population size, as well as changing climatic conditions. A better understanding of evolution in natural populations can help us to conserve species as climates change.

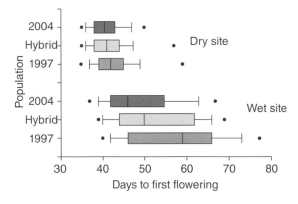

Figure 21.4 Box plots of flowering time of *Brassica rapa* plants from pre-drought (1997), post-drought (2004) and hybrid (1997 × 2004) lines grown under common conditions. The plants were from a dry site (Back Bay) and a relatively wetter site (Arboretum), both in southern California. Plants from the 2004 (post-drought) lines flowered earlier than the 1997 (pre-drought) lines in both populations, indicating the rapid evolution of earlier flowering following drought. From Franks *et al.* (2007).

CHAPTER 22

Genetic identification and monitoring

Red king crab, Section 22.5

Wildlife DNA forensics is essentially concerned with the identification of evidence items in order to determine the species, population, relationship or individual identity of a sample.

R. Ogden *et al*. (2009)

Genetic monitoring has been recognized in several international agreements and documents, and can be an important tool for the protection of biodiversity.

F.A. Aravanopoulos (2011)

Chapter Contents

Conservation and the Genetics of Populations, Second Edition. Fred W. Allendorf, Gordon Luikart, and Sally N. Aitken.
© 2013 Fred W. Allendorf, Gordon Luikart and Sally N. Aitken. Published 2013 by Blackwell Publishing Ltd.

Genetic identification is the use of molecular genetic analyses to identify the species, individual, relationship, or the population of origin of a sample. Genetic identification aids law enforcement, wildlife management, and biodiversity assessments. For example, molecular identification of an individual from its multilocus genotype or of a species from its mtDNA sequence can provide information to help convict poachers that illegally harvest protected plants and animals. Genetic identification can also help to monitor the presence of an endangered species in a nature reserve or national park, or of microbes (including pathogens) in environmental samples.

Applications of genetic identification in conservation genetics include the following:

1 Identification of the **species** of origin of tissues or products (e.g., whale meat or tiger bones sold on the open market) or of invasive species in environmental samples (e.g., sloughed cells or DNA from non-native mollusks, plants, or fish in water samples).

2 Assessment of **community composition** or diversity (e.g., as a first step in understanding processes driving speciation, adaptation, or ecosystem processes).

3 Identification of **individuals** or matching of tissue samples (e.g., matching blood stains in a national park to a trophy animal in a taxidermy shop).

4 Determination of the **relatedness or parentage** of individuals (e.g., paternity assignment of animals claimed to be born in captivity, but possibly taken from the wild).

5 Identification of the **population of origin** of a group of individuals (e.g., a boatload of fish or lobster), or of a single individual or tissue (e.g., an individual claimed to originate from a legal hunting area, rather than from a protected nature reserve).

Illegal 'trade' of wildlife, including fish and plants, represents the world's third largest type of illegal trafficking, after drugs and weapons (Clynes 2011). Poaching and trafficking are among the most serious threats to the persistence of many wild populations and species. Poaching and illegal trade threaten taxa ranging from plants (e.g., orchids and hardwood trees) to insects (exotic tropical beetles and butterflies), reptiles (snakes, turtles, and lizards), fish (sturgeon for making caviar), birds (parrots and canaries), and mammals (especially trophy-horned ungulates, large carnivores, primates, elephants, rhinos, and cetaceans). Among hardwood trees, illegal logging is so

extensive that in the year 2010, the harvest of contraband logs amounted to 0.24 million cubic meters (Clynes 2011).

The most important international treaty prohibiting the trade of endangered species is CITES (Convention on International Trade in Endangered Species), established in 1973 in association with the United Nations Environmental Program (UNEP). The main international program for monitoring wildlife trade is TRAFFIC – a network of dozens of staff and researchers across 20 countries jointly sponsored by the World Wide Fund for Nature and the World Conservation Union. Other organizations that work to control illegal wildlife trade include the international organization WildAid, headquartered in San Francisco, California, and PAW (Partnership for Action against Wildlife Crime) in the UK. Unfortunately, even with such programs, it is difficult to detect poaching and to enforce treaties and anti-poaching laws.

Genetic identification of wild taxa shares much in common with human **forensic** genetics (Jobling and Gill 2004, Kayser and Knijff 2011). However, wild taxa forensics often involves the identification of species rather than individuals as in human forensics (Ogden 2011). A US Fish and Wildlife Service forensics laboratory in Ashland, Oregon, is perhaps the only crime lab in the world dedicated entirely to wildlife. In 1991, the Wildlife Forensics DNA Laboratory at Trent University in Ontario, Canada, was the first lab to produce DNA evidence to be used in a North American court. It has been involved in over 50 cases a year, with convictions and fines ranging from $US1000–50,000. There are now more than 50 labs in North America, Europe, Australia, and other countries, that conduct forensics testing to help solve crimes such as illegal trafficking of plant and animal products, according to the recently formed Society for Wildlife Forensic Science.

In conservation and management, genetic identification of species or individuals is the first step in many applications of molecular markers. For example, individualization of samples is required in most **noninvasive** genetic studies, which use DNA from feces, shed hair, urine, saliva, sloughed skin, feathers, scent markings, menstrual fluid, and even mucus trails left by snails (Beja-Pereira *et al.* 2009b). Once individuals have been identified, we can estimate abundance of individuals (see Section 14.1), monitor their movements, identify immigrants, estimate sex ratios, and monitor population genetic parameters relating to neutral and adaptive molecular variation (see below).

Genetic monitoring is the quantification of temporal changes in either (1) a traditional ecological parameter, such as population size, (2) a population genetic metric, such as effective population size, or (3) natural selection and adaptive change, such as the increase in frequency of an adaptive allele following an environmental change. National and international organizations have established guidelines for monitoring biological diversity (Laikre *et al.* 2010b). However, most monitoring programs do not take full advantage of the enormous and growing potential afforded by molecular genetic markers, which can provide information on ecological and evolutionary timeframes. Genetic monitoring can sometimes cost less and be more reliable than traditional monitoring approaches that require the capture and recapture of individuals.

Molecular and statistical technologies are rapidly improving; together they provide enormous potential to facilitate conservation management and wildlife forensics. Nevertheless, this potential is not fully exploited. There is much room for further development and application of genetic approaches to help combat poaching and to improve wildlife management. We intend this chapter to encourage agency biologists, academic researchers, students, and funding organizations to improve and apply genetic identification and monitoring approaches wherever useful for biodiversity conservation.

22.1 SPECIES IDENTIFICATION

Many kinds of molecular genetic markers are used to identify species for both forensic and conservation applications. Mitochondrial DNA analysis is the most widely used molecular approach for animal species identification because many species have distinctive mtDNA sequences, and because 'universal' primers exist that work among taxa (mammals or even all vertebrates). Furthermore, mtDNA is relatively easy to extract from most tissues, including hair, elephant tusks, and old skins, because of its high copy number per cell (see Section 4.1).

Recently, nuclear DNA markers have been used for species identification. Nuclear markers are especially useful when interspecific hybridization is possible, because mtDNA analysis cannot detect male-mediated gene flow or introgression. New techniques have made it possible to identify fixed genetic differences between species or subspecies in order to understand the dynamics of hybridization and introgression. For example, Hohenlohe *et al.* (2011) used **next-generation sequencing** to identify nearly 3000 diagnostic SNPs between native westslope cutthroat trout and invasive rainbow trout.

Other nuclear markers include species-diagnostic microsatellites and AFLPs. Microsatellites have relatively high mutation rates, which can facilitate identification of species-diagnostic alleles between closely related species, such as sharks (Shivji *et al.* 2002) and mountain ungulates (Maudet *et al.* 2004). AFLPs have been used in detecting the trafficking of marijuana and endangered plant species (Miller Coyle *et al.* 2003).

Nuclear DNA single nucleotide polymorphisms (SNPs, Chapter 4) and quantitative PCR genotyping assays have been used to identify different species of crab harvested in commercial fisheries (Smith *et al.* 2005a). SNPs are also used to identify species of tropical hardwood trees harvested for trade worth hundreds of millions of US dollars annually (Example 22.1). SNPs are usually biallelic and easy to code in a database (0 or 1), they can be transferred between laboratories with little error in genotype scoring (unlike some microsatellites and AFLPs), and they can be genotyped using small quantities of partially degraded DNA thanks to PCR technology useful with short DNA fragments (<100 bp) (Campbell and Narum 2009). Thus, SNPs are especially useful for forensic applications or noninvasive sampling (DeJa-Pereti *et al.* 2009b) where degraded DNA is common and database errors and genotyping errors would be highly problematic.

22.1.1 DNA barcoding

DNA barcoding involves the analysis of short, standardized, and well-characterized DNA sequences for use as a tag for rapid and reliable species identification (Valentini *et al.* 2009b, Bucklin *et al.* 2011). Barcoding differs from phylogenetics (Section 16.3) in that the goal is not to determine evolutionary relationships, but to identify the species of origin of an unknown sample using a known taxonomic classification. Applications include the identification of insect larvae, which typically have few diagnostic morphological characters compared with adults, the assessment of the diet of an animal based on stomach contents or feces (Section 22.1.2), and the identification of products that are illegally traded such as wood, bone, horn, or fish products.

Example 22.1   Combating illegal logging with SNP genotyping to identify wood products from tropical tree species

Illegal logging contributes to deforestation and thus global warming and loss of habitats and biodiversity. Illegal logging in public lands alone causes losses in revenues estimated at billions of US dollars annually. Despite the enormous magnitude of the problem, few reliable legal means have existed to stop the illegal trade of forest products because the identification of illegally logged timber species has been technically difficult. However, recent developments in genetic identification technologies promise to aid law enforcement.

Ogden *et al.* (2008) developed a DNA test for the identification of the tropical hardwood tree ramin. Ramin was listed in CITES in 2004 shortly after the listing was requested by the Indonesian government. Ramin is illegally imported and traded in the European Union and the US in the form of finished products such as wooden tool handles that are difficult to iden-

tify to species anatomically. Ogden *et al.* (2008) sequenced **candidate genes** to identify SNPs diagnostic of ramin. Three candidate ramin-diagnostic SNPs were assessed for their suitability for genotyping with a SNP qPCR assay.

The authors designed a PCR assay (Chapter 4) for one diagnostic SNP and verified its reliability by genotyping DNA from different species and sample types, including worked products. More than one SNP (and SNP assay) should be genotyped to determine origins with certainty. Furthermore, many species of ramin (*Gonystylus* spp.) exist, and development of SNPs diagnostic of the main traded species (*G. bancanus*) would further help enforcement agencies (customs/border inspection) to monitor trade of ramin wood products. Nevertheless, this study shows how genetics research can be applied to facilitate law enforcement and help curb illegal logging and trade.

Barcoding is based on short DNA segments that have relatively low sequence variation within species, but high sequence divergence between species. Mitochondrial DNA has a relatively high mutation rate in most eukaryotic species, a relatively high divergence between species, and thus has been useful for barcoding. For example, Hebert *et al.* (2010) sequenced a 648 bp region of the cytochrome *c* oxidase I (COI) gene in more than 1300 lepidopteran species from eastern North America. The authors reported low intraspecific divergences averaging only 0.4% within species, whereas congeneric species revealed 18-fold higher mean divergences (7.7%). This study reported that 99.3% of the 1300 study species possessed diagnostic barcode sequences (with only nine cases of barcode-sharing among the 1327 species).

Barcoding has become an ambitious and somewhat controversial international initiative (Box 22.1) in which an mtDNA segment of ~650 base pairs is being sequenced for a large proportion of the world's species (Moritz and Cicero 2004). The goal is to develop a huge database of DNA sequences from a single gene (e.g., COI) for use in species identification and species discovery. This would facilitate biodiversity inventory, conservation, and the detection of illegal trafficking of wild taxa. Two increasingly common applications of bar-

coding are diet analysis from feces and environmental DNA analysis (Sections 22.1.2 and 22.1.3).

There are a number of potential problems with barcoding to consider (Taylor and Harris 2012). First, paraphyly and polyphyly are common (Section 16.3.2) and can lead to erroneous species identification. Second, mtDNA capture (introgression) is common between some taxa (Good *et al.* 2008). Third, insertions of mtDNA into the nuclear genome are also common in some taxa, which can also confound species identification (Song *et al.* 2008). Finally, problems with mtDNA arising from male-killing micro-organisms and cytoplasmic incompatibility-inducing symbionts (e.g., *Wolbachia* are) are somewhat common among insects (Hurst and Jiggins 2005). Such limitations suggest that the use of only mtDNA might not provide reliable species identification in some taxa.

Nevertheless, the proposed quality control, standardization, and scale of mtDNA sequencing should greatly improve species identification, discovery, and conservation (see the DNA Barcoding website and Box 22.1). New genetic technologies are helping to solve problems associated with using only a single short mtDNA fragment for barcoding and species identification (Box 22.1).

Chloroplast DNA markers are commonly used in plants for species identification because mitochondrial

Box 22.1   DNA barcoding: systems, challenges, and future perspectives

DNA barcoding provides a system for species identification based on digital characters, allowing for automated identifications, thereby improving the capacity to identify, monitor, and manage biodiversity. This system promises substantial societal and economic benefits. It also raises the possibility of identifying and tracking the spread of diseases and the pathogenic organisms themselves (Section 20.6).

The international Consortium for the Barcode of Life (CBOL) has over 200 member organizations in 50 countries. As of 2010, the barcoding database called BOLD, the Barcode of Life Data Systems, held barcode records for more than 850,000 specimens, representing approximately 100,000 species, which is 5.9% of the 1.7 million described species. Many individual organisms in BOLD are placed in museum collections, and their DNA is kept in a secure repository, so that they will be available for study by future technologies (Vernooy et al. 2010).

A substantial challenge for CBOL is the difficulty in obtaining tissue samples, DNA, or sequence data from certain countries with strict biodiversity exportation laws (e.g., India). Export laws exist to protect a country's **genetic resources** defined broadly as "genetic material of actual or potential value" from plants, animal, and microbes. This potential value is being realized in many fields, including pharmaceuticals, medicine, and plant- and animal-breeding.

Analysis of multiple loci could be considered as barcoding under a broad definition (Valentini et al.

2009b), such as follows: the use of any standardized DNA fragment(s) for identification of species or other taxonomic groups (genera or families). The identification of populations within a species using diagnostic markers or allele frequency differences among populations refers to assignment tests (Manel et al. 2005), and should not be considered as DNA barcoding (Valentini et al. 2009b).

Future barcoding systems will need to overcome the major limitation of using only a single, short, maternally inherited (mtDNA) locus (Taylor and Harris 2012). This will be facilitated by use of next-generation sequencing (Hajibabaei et al. 2011), gene-targeted **DNA enrichment** arrays (Mamanova et al. 2010), and DNA microarray chips (e.g., Gardner et al. 2010), which allow a multi-locus barcoding approach. For example, Timmermans et al. (2010) reported a protocol for sequencing multiple mtDNA genes using long-range PCR and 454 pyrosequencing to assess relationships among 30 beetle species. Hajibabaei et al. (2011) applied 454 pyrosequencing for environmental DNA (below) barcoding of benthic macroinvertebrates using the COI gene PCR products, which also could be extended to the sequencing of nuclear loci. Other researchers are developing protocols for sequencing many mtDNA genes and multiple nuclear genes simultaneously on pooled samples from 100 to 200 individuals, following DNA enrichment for the genes on a DNA arrays (Mamanova et al. 2010).

---

genes in plants have a low substitution rate. Universal primers for chloroplast noncoding regions are widely used (Taberlet et al. 1991). These primers work in a range of taxa from algae to gymnosperms and angiosperms. Hollingsworth et al. (2011) also identified a standard 2-locus DNA barcode for land plants. The 2-locus combination of **rbcL** + **matK** as the plant barcode is recognized by authors and the CBOL (Consortium for the Barcode of Life) Plant Working Group (2009). The CBOL group suggest that the 2-chloroplast loci provides a universal framework for routine use of DNA sequence data to identify taxa and discover cryptic species of land plants. Use of multiple loci, facilitated by new sequencing techniques, might become a common solution for species identification challenges in certain taxonomic groups (Box 22.1).

22.1.2 Diet analysis

Identification of species that have been consumed (diet analysis) by an endangered species can help to identify plant or animal resources important to protect or restore in order to conserve the endangered species (Cristobal-Azkarate and Arroyo-Rodriguez 2007, Valentini et al. 2009b). Also, the study of food webs is fundamental to understanding how the feeding habits of different species can influence the community and ecosystems. Thus, diet analysis of species coexisting in an area can improve our understanding of the functioning of ecosystems (e.g., Duffy et al. 2007).

Diet analysis has long been conducted by microhistological analysis of plant or animal parts identified in fecal samples. Recent DNA techniques can improve diet analysis from feces. DNA analysis is especially useful when the food items cannot be identified using morphological traits of partially digested organisms, or by observing the animal's feeding behavior.

DNA markers for diet analysis have been used on carnivores and herbivores including primates and birds, and even on historical and extinct animals using coprolites (reviewed in Valentini *et al.* 2009a). For example, Nyström *et al.* (2006) applied a PCR-based DNA analysis to identify ptarmigan species from bone remains in nests of the endangered gyrfalcons. The mtDNA-based identification was performed on 176 ptarmigan bones from 13 different breeding occasions occurring in five different territories. The results indicated that two ptarmigan species comprised ~93% of the gyrfalcon diet, and that rock ptarmigan was the most common prey during the breeding season. These data had implications for regulating hunting of the rock ptarmigan to help conserve the gyrfalcon.

In another study, short fragments (20–85 bp) of an intron of a gene encoding chloroplast transfer RNA was used to assess the diet of herbivorous animals (Valentini *et al.* 2009a). The PCR primers used are universal (for gymnosperms and angiosperms) and work with feces that contain degraded DNA. This barcoding system, combined with high-throughput parallel pyrosequencing, was used to analyze the diet of mammals, birds, mollusks, and insects. Such applications of DNA barcoding will likely become a valuable tool in conservation.

22.1.3 Environmental DNA

Environmental DNA (eDNA) is DNA in environmental samples, such as soil, water, and snow (Jerde *et al.* 2011). The uses of eDNA include the detection and surveillance of rare or elusive species difficult to study with traditional methods such as observations or trapping. Sampling eDNA can be differentiated from noninvasive DNA sampling because noninvasive sampling directly targets collection of tissues (hair) or feces or urine spots (Beja-Pereira *et al.* 2009b), whereas eDNA involves more diffuse samples where cells or free floating DNA are mixed in soil, water, or snow.

Environmental DNA samples can contain live microorganisms such as live bacteria, nematodes, larvae (e.g., from invasive mollusks), remains of microorganisms, or sloughed cells from vertebrates in the sampling area. Environmental samples have often been used to study microbial communities, using species-diagnostic

DNA sequences such as 16S or the internal transcribed spacer of rDNA (Herrera *et al.* 2007). With live organisms, DNA sequences are often thousands of base pairs long. However, environmental DNA is often degraded into fragments of less than 100–200 bp.

For secretive species, detecting their local distribution facilitates identification and protection of critical habitat to enhance survival or reproductive success. Fortunately, the environment can retain DNA from local species such that DNA analyses allow detection of secretive organisms without direct observation. For example, environmental water samples were PCR-amplified by Ficetola *et al.* (2008) using taxon-specific PCR primers that amplify short mitochondrial DNA sequences in order to track the presence of bullfrogs in natural wetlands. Such techniques can help to assess the distribution of species that are difficult to detect, especially during particular time periods or developmental stages, potentially biasing estimates of species diversity or distributions when using only traditional trapping or inventory methods.

Environmental DNA facilitates early detection of an invasion, which increases the feasibility of rapid responses to eradicate the species or contain its spread (Hoy *et al.* 2010). Environmental samples have been used for detection of invasive species in water from rivers and lakes. Invasive fish (Example 22.2) can be detected from water samples.

Use of eDNA requires careful control for errors when inferring species presence, due to high risks of false positives from contamination (Beja-Pereira *et al.* 2009b). Even if true positives are detected, it is difficult or impossible to estimate abundance or identify individuals. Finally, the occurrence of false negatives may be common but difficult to assess.

22.1.4 Forensic genetics

One of the most widely publicized forensic applications of wild population DNA analysis for conservation was the identification of illegally traded whale meat sold in Japanese and Korean markets (Baker *et al.* 1996). PCR-based analysis of mtDNA control region sequences revealed that about 50% of the whale meat sampled from markets had originated from protected species and not from the southern minke whale species that Japan is allowed to harvest under their scientific whaling program. For this study, the researchers were not allowed to export the tissue samples from Japan because many species of whale are protected by CITES, which

---

Example 22.2   Environmental DNA analyses detect underwater invasions

It is feared that two Asian carp species imported into the US decades ago by a fish farmer have recently breached a barrier and could endanger the Great Lakes ecosystem (Jerde *et al.* 2011). Both species, the silver and the bighead carp, can reach nearly 100 pounds and consume 20% of their weight in plankton per day. An electric barrier, with voltage pulsed at a certain frequency, was constructed to repel all fish and prevent them from moving upriver into Lake Michigan near Chicago, Illinois.

A study of eDNA suggested that carp had passed the barrier and might soon enter the Great Lakes, the largest freshwater ecosystem in the world. Jerde *et al.* (2011) developed diagnostic DNA markers for each species using publicly available sequences. The authors targeted short fragments of the mitochondrial control region to compare the carp markers with all available sequence data, including sequences from closely related species common to the Chicago area water system.

The authors collected approximately 1000, 2-liter surface water samples in 2009 and 2010. Each 2-liter sample was filtered onto a separate filter to capture sloughed cell material for DNA extraction. PCR tests detected DNA from both species of introduced carp in numerous water samples taken from several locations repeatedly over time, providing compelling genetic evidence that the carp have breached the electronic barrier and are within a few river-miles of Lake Michigan.

In June 2010 a commercial fisherman caught an adult bighead carp within 13 km of Lake Michigan, further supporting what the eDNA evidence suggested 8 months earlier. A comparison with traditional fisheries surveillance methods suggests greater sensitivity for eDNA (Jerde *et al.* 2011).

The use of only mtDNA in this study made it impossible to detect whether individuals were hybrids between these carp species (Lamer *et al.* 2010). In addition, individual identification with nuclear markers would help to determine how many carp have breached the barrier and how far individuals move.

---

forbids transportation of any tissues or DNA without a permit. Consequently, the researchers set up a portable PCR machine and amplified the mtDNA in a hotel in Japan. They subsequently transported the synthetic DNAs (not regulated by CITES) back to laboratories in the US and New Zealand for sequence analysis.

A particularly laudable application of genetics in species identification is in the Web-based *DNA Surveillance* software program (Ross *et al.* 2003). *DNA Surveillance* is a computer package that applies phylogenetic methods to the identification of species of whales, dolphins, and porpoises. One advantage of *DNA Surveillance* is it contains a database of validated, prealigned sequences with wide taxonomic and geographic representation, developed specifically for taxonomic identification (unlike GenBank, with limited sampling and variable data quality). The user typically pastes an mtDNA sequence (e.g., 400–500 bp of control region) of unknown origin into a data input window, and then receives back an alignment, genetic distance estimates, and an evolutionary tree (Example 22.3). This type of service is badly needed for other taxonomic groups. More recent studies have used microsatellite DNA genotyping to detect illegal international trade leading to closure of restaurants selling whale meat in the US and South Korea (Guest Box 22).

Penises from pinnipeds provide another example of using DNA analysis to detect illegal trade. Pinniped (seals, sea lions, fur seals, and walrus) penises are purchased in traditional Chinese medicine shops in Asia and North America. To investigate the trade of pinniped penises, researchers purchased 21 samples of unknown origin (labeled as pinnipeds) and sequenced 261 base pairs of the cytochrome *b* gene (Malik *et al.* 1997). One sample from Bangkok turned out to be from domestic cattle and six could not be identified because of lack of published reference sequences, although two were most similar to the African wild dog. The remaining samples were from seals. This study suggests that the lucrative market for pinniped penises may be encouraging the unregulated hunting of seals and other unidentified mammalian species. It also illustrates the importance of a large reference database. The trade of penises, bacula, and testes is lucrative and growing. For example, Australia exports nearly 5000 tons of domestic cattle penises to Chinese aphrodisiac markets each year (Malik *et al.* 1997).

Bird trafficking in Australia provides an example of how DNA analysis could help conserve threatened species. In 2006, the Australian Quarantine Inspection Service seized 23 bird eggs from a passenger who tried to smuggle the eggs through the Sydney

Example 22.3   Use of the *DNA Surveillance* website for taxonomic classification of unknown cetacean samples

The *DNA Surveillance* (Ross *et al.* 2003) website is useful for identifying the cetacean species of origin of a tissue sample of unknown origin. To use the program, you must obtain an mtDNA sequence (control region or cytochrome *b* fragment) from your individual tissue sample, and then simply cut and paste it into the 'Data Entry' window of the program.

Figure 22.1 shows a control region sequence of unknown origin ('Unknown1') pasted into the 'Data Entry' window. The sequence is from a meat sample purchased in Japanese markets in 1999 (see *DNA surveillance* website and Baker *et al.* 2002). Assuming it is from a baleen whale (Mysticetes), we simply click the circle under 'ctrl' and under 'Mysticetes', and then click on the 'submit' button in the DNA surveillance program window (Figure 22.1a). The program then aligns the user-submitted sequence against a set of validated sequences and outputs (to the computer screen) a cluster dendrogram (see Figure 22.1b). If we do not assume the sequence was from a baleen whale, we could click the circle 'All Cetacean' sequences available, but we would obtain a much larger tree (e.g., including dolphins etc.).

It turns out that 'Unknown1' (Figure 22.1b, *DNA surveillance* website) is an actual gray whale product purchased in Japan. This is presumably a 'Korean' western North Pacific gray whale, one of the most endangered populations of whales, unlike the 'California' or eastern North Pacific population. This finding was published as a likely infraction of international

Data Entry

>Unknown1
GAAAATATATATTGTACAATAACCACAAGGCCACAGTATTA
ATGTAACTTGTGCATGTATGTACTCCCACATAACCCATAGTA
TATGTATAATTGTGCATTCAATTATCTTCACTACGGAAGTTAA
AATATTTATTAATAGTACAATAGTACATGTTCTTATGCATCCT

Select a database:

Database	ctrl	cyt b
All Cetaceans	○	○
Mysticetes	◉	○
Odontocetes	○	○
Ziphiidae	○	○
Phocoenidae	○	○
Delphinidae (subgroups)		
Delphininae + Stenoninae	○	○
Globicephalinae + Orcininae	○	○
Lissodelphininae	○	○
Humpback Whale Populations	○	

Genomic regions:
ctrl = mtDNA control Region (= D-Loop);
cyt b = cytochrome b

(a)

Figure 22.1   (a) Data entry and database window in the *DNA Surveillance* web-based computer program. (b) Distance phenogram of mtDNA control region sequences from Mysticetes whales.

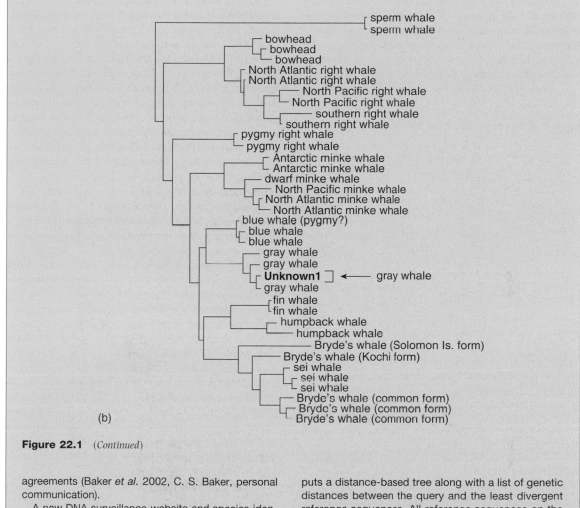

(b)

Figure 22.1 (*Continued*)

agreements (Baker *et al.* 2002, C. S. Baker, personal communication).

A new DNA surveillance website and species identification tool was recently established for carnivores (Chaves *et al.* 2011). This DNA Surveillance tool aligns an mtDNA sequence (provided by the user) and out-puts a distance-based tree along with a list of genetic distances between the query and the least divergent reference sequences. All reference sequences on the tree are named with scientific nomenclature followed by the GenBank accession number or the BOLD (Barcoding of Life Database) sample identification number.

International Airport from Thailand. The eggs were a quarantine risk for H5N1 (avian influenza) and were therefore euthanized and gamma-irradiated. Using DNA sequence data from two mitochondrial genes, the eggs were identified to be from seven species of parrots. One species, a cockatoo, was in the **CITES Appendix I**, which lists "species threatened with extinction" and whose "trade is permitted only in exceptional circum-stances" (CITES 2010, Johnson 2010). The defendant pleaded guilty to contravening the Commonwealth Environment Protection and Biodiversity Act 1999, the CITES laws, the Australian Customs Act 1901, and the Quarantine Act 1908. The birds were worth \$AU250,000 on the illegal black market. However, the defendant was fined only \$AU10,000 and sentenced to 2 years imprisonment. Illegal wildlife trafficking is an extremely lucrative crime with serious consequences for the targeted species and ecosystems, yet relatively low cost fines and few prosecutions occur. Authors have argued that high fines similar to the black market value (e.g., $AU250,000 in this case) are badly needed to help reduce poaching and trafficking.

22.1.5 Conservation and management

Species identification via DNA analysis is increasingly used for conservation management. Sequencing of mtDNA is often used to monitor for the presence of endangered species in nature reserves and in wildlife management areas. If an endangered species is detected in some area, that area might be granted protection from logging or development. For example, when the endangered long-footed potoroo (a small kangaroo) was first detected in forests of southeastern Australia, logging was halted in some areas. Potoroos, like many marsupials, are nocturnal, elusive, and their presence is difficult to detect. Thus, biologists detected potoroos by using field signs (e.g., soil diggings), feces and 'hair traps' (consisting of baited plastic tubes with sticky tape around the entrance to recover hairs). DNA analysis from hair or feces is necessary for species identification, because related potoroo species occur in sympatry (Luikart *et al.* 1997).

22.2 METAGENOMICS AND SPECIES COMPOSITION

Metagenomics involves simultaneous identification of many species through the analysis of many genomes or DNA sequences from a sample (Chapter 4). The approach is useful for characterizing the diversity of microbial communities, plant communities or animal communities. For example, metagenomics has been used on water and soil samples for comparing bacterial communities, permafrost for assessing past animal and plant communities, ice for assessing past animal and plant communities during glaciations, rodents' middens for describing past plant communities, and feces samples for health assessments or diet analysis (Section 22.1.2). One of the greatest advantages of metagenomics is that traditional bacterial or viral isolation or culture is not required to study microbes, many of which cannot be grown or detected using traditional culture or isolation approaches.

DNA microarrays (biochips) are being used to identify thousands of microbial species in complex mixed-species samples including water and soil. Biochips allow monitoring of known species and the discovery of new species. For example, Gardner *et al.* (2010) designed a pan-microbial microarray to detect all 2200 viruses and 900 bacteria for which full genome sequences were available. Microarrays can include species-specific probes and family-specific probes designed to tolerate some sequence variation to enable detection of divergent species and discovery of new species with homology to sequenced organisms, while having no significant matches to the human genome sequence.

The application of metagenomics to conservation is in its early stages, but shows promise. First, functional metagenomics of microbial communities provides a novel perspective on ecosystem processes, such as nutrient and energy flux. Although some studies have compared ecosystem functions across a broad scale of biomes (Dinsdale *et al.* 2008), similar comparative approaches may identify aspects of ecosystem function across sites within a habitat. Second, metagenomics allows assessment of physiological condition of individual organisms. For instance, Thurber *et al.* (2009) have found that shifts in the endosymbiont community (metazoans, protists, and microbes) of corals in response to stressors such as increased nutrients and temperature can profoundly shift the health status of the coral. Third, a metagenomic analysis of human fecal samples catalogued 3.3 million microbial genomes and found substantial differences in the microbial metagenome between healthy individuals and those with inflammatory bowel disease (Qin *et al.* 2010). It should be possible in the future to apply metagenomic techniques to fecal samples from wildlife species to assess health and physiological state, such as starvation stress.

Microbial community information can help evaluation of restoration and mitigation projects for conservation. For example, microbial community structure and function in restored and natural wetlands was compared to assess the success of wetland mitigation to replace ecosystem functions (Peralta *et al.* 2010). Microbial community composition was assessed using restriction fragment length polymorphisms (Chapter 4) targeting the 16S rRNA gene (total bacteria) and the nosZ gene (denitrifiers). Bacterial communities differed significantly between the restored and the reference wetlands, and denitrifier communities were similar among reference sites but highly variable among restored sites. This study indicated that wetland resto-

ration efforts did not successfully restore denitrification and that the differences in potential denitrification or community composition rates may be due to
distinct microbial assemblages in restored and natural
wetlands.

22.3 INDIVIDUAL IDENTIFICATION

Individual identification (DNA 'fingerprinting') is one
of the most widely used applications of molecular
markers in conservation genetics, forensics, and
molecular ecology. For example, in the lynx–bobcat
hybridization study in Example 17.2, researchers had
to individualize fecal samples to know the number of
different individuals sampled. For another example,
wildlife officers might need to match a tissue sample

(gut pile or blood stain) at the scene of a wildlife crime
to a trophy animal being transported through an
airport or highway checkpoint.

To match individual samples, we must first genotype
them with 10–20 highly polymorphic molecular
markers or with numerous moderately polymorphic
markers. We then compute a match probability (or
probability of identity, see below), by using allele frequencies estimated from the population of reference,
such as the national park or the geographic location
from which the sample at the 'crime' scene originated
(e.g., Tnah *et al.* 2010). If allele frequencies from the
reference population are not available, we can still estimate the match probability, but it requires additional
markers to achieve reasonably high power to resolve
individuals with high certainty (Menotte-Raymond
et al. 1997, see Box 22.2).

Box 22.2   Computation of the match probability (*MP*) for an individual sample (genotype) 'in hand'

Here we consider a scenario where we have one
sample in hand (e.g., a bloodstain at a wildlife crime
scene) and we want to compute the probability of
sampling a different individual that has an identical
multilocus genotype (in the same population). This is
often called the 'match probability' (*MP*).

I o compute the *MP*, consider two loci that each has
two alleles at the following frequencies: $p_1 = 0.50$,
$q_1 = 0.50$; and $p_2 = 0.90$, $q_2 = 0.10$, respectively. A
bloodstain from the scene of a wildlife crime (poaching
in a National Park) has a genotype that is heterozygous at both of these loci. What is the probability
that an individual sampled at random from the same
population has the same genotype (as the individual
whose bloodstain is 'in hand')?

First we compute each single locus *MP*:

Locus one: $2p_1q_1 = 2\,(0.50)\,(0.50) = 0.50$

Locus two: $2p_2q_2 = 2\,(0.90)\,(0.10) = 0.18$

Then, the multilocus *MP* is the product of the two
single locus probabilities: $0.50 \times 0.18 = 0.09$ (assuming independence between loci). We conclude there
is a 9% chance of sampling a different individual with
a double-heterozygous genotype identical to one 'in
hand'. Thus there is a 9% chance of matching the
bloodstain to the wrong individual. Clearly, many more
loci and perhaps more highly polymorphic loci are
needed to have a reasonably low chance (e.g.,
<1/10,000) of a match to the wrong individual. Recall
that here we are assuming unrelated individuals, no

substructure, and that the allele frequencies are
known for the population considered.

What if the wildlife crime occurs in a population
with no reference data (i.e., allele frequencies are
unknown)? How can we estimate the probability that
an individual sampled at random has the same double-
heterozygous genotype as the individual 'in hand'?
Here we could *assume* that the frequency of the
observed heterozygous genotype at each locus is high
(e.g., 0.50). This is the highest frequency possible
(assuming a biallelic locus and Hardy-Weinberg proportions), and gives the least power for individualization. Assuming that the heterozygote genotype
frequency is 0.50 is conservative and generally overestimates the true *MP* (Menotte-Raymond *et al.* 1997).
This is especially true if a locus is multiallelic because
the two alleles in a heterozygote 'in hand' could never
have a population frequency as high as 0.50.

We then could compute the multilocus match probability as follows: $0.50 \times 0.50 = 0.25$. Here, the estimated 25% chance of sampling this multilocus
genotype is much greater than the 9% chance estimated (above) by using the reference allele frequencies. This illustrates the power benefit of having
reference allele frequencies for the population. Note
that if two samples match and are homozygous, we
cannot use the locus, because we have no evidence
the locus is polymorphic (and thus informative) within
the population. Thus we need many more loci when
we do not know population allele frequencies, in order
to achieve a low *MP*.

Microsatellites have been the most widely used markers in forensics and genetic management because: (1) their short length (<300 bp) makes them relatively easy to PCR-amplify from partially degraded DNA (unlike like AFLP markers, for example, which are longer and more difficult to amplify; Section 4.3); (2) they are generally highly polymorphic; and (3) alleles from the same locus can be easily identified (unlike some AFLPs, and the multilocus DNA fingerprinting probes first used in human DNA forensic applications; Jeffreys *et al.* 1985). For human forensic investigations in the US and Britain, a standard set of 13 and 10 microsatellite loci are used, respectively (Watson 2000, Reilly 2001). These marker sets provide a chance of a match between two random people that is between about one in a million and one in a billion. For these marker sets, the genotyping of one individual sample costs approximately $US100 (Watson 2000). This is similar to the cost in some wildlife genetics laboratories.

An example of DNA-based individualization for wildlife management is the identification of problem animals. For example, when a wolf or bear attacks humans, kills livestock or steals a picnic basket, the problem animal will often be removed from the population. Removing the wrong bear could waste resources and eventually result in the removal of several animals before removing the correct one. This could negatively impact the population, especially if the individuals removed are reproductive females. Furthermore, knowing with certainty that the true problem individual was removed would satisfy some of the public. Thus, it is critical to identify the correct individual before removing it. Here, matching DNA from the scene of the 'crime' to an individual can help. For an example, see the description of DNA matching in the case of a family of grizzly bears from Glacier National Park, Montana (Example 22.4).

Other uses of individual identification in conservation management include **genetic tagging** for studying movements and estimating population census size from mark–recapture methods (see Section 14.1). Individualization also helps to identify clonal plants (genets) and animals (corals, anemone, and fishes). The identification of clones is required for accurate estimation of patterns and rates of gene flow, geographic distributions of clones, and inbreeding versus outcrossing rates.

22.3.1 Probability of identity

The statistical power of molecular markers to identify all individuals from their multilocus genotype is estimated as the average probability of identity (PI_{av}). PI_{av}

Example 22.4   DNA identifies a bear and her cubs

In 1998, a hiker was killed and partially consumed by a bear in Glacier National Park, USA. Evaluation of the evidence suggested that the hiker was chased a long distance and then attacked. A review panel decided that this was unnatural aggression and therefore warranted removal of the responsible bear. Park biologists noted that the killing occurred within the home range of a radio-collared grizzly bear female (female 235; Table 22.1), and hypothesized that she might have killed the hiker. They also were aware that this female had two 2-year-old cubs that might have been with her at the time of attack. Park biologists did not want to kill this grizzly and her cubs unless they could obtain conclusive evidence that she had killed and consumed the hiker.

Park biologists turned to DNA analysis of hair and fecal samples. Hair samples were taken from female

235 and one of the cubs (238) and sent to the Laboratory for Ecological and Conservation Genetics at the University of Idaho (Lisette Waits, personal communication). Park biologists then collected two bear hair and 11 fecal samples from the kill site. The laboratory evaluated: (1) did the suspect bear's genotype match that from the bear hair or feces found at the attack site; and (2) were there two cubs present at the kill site?

The lab conducted PCR of five highly variable microsatellite loci (Table 22.1, Paetkau *et al.* 1995). One sample matched the genotype of a cub and three samples matched the female. The probability of identity for the 3–5 loci that successfully amplified in different samples ranged from approximately one in 2000 to one in 40,000 (see equation 22.1). Thus, the park biologists attempted to capture all three bears.

Table 22.1 Genotypes at five microsatellite loci for the grizzly bear female (suspected mother), and one cub caught with the mother and another cub that was the suspected missing cub. Bold alleles are those that do not match either allele in the female. H and S samples represent hair and fecal samples found at the kill site.

| Sample | Locus | | | | |
	G1A	G10B	G10C	G1D	G10L
Female 235	189/189	155/155	102/110	171/175	153/155
Cub 238	189/**193**	155/**159**	**104**/110	175/**180**	153/155
Unknown cub	189/**193**	155/**159**	**104**/110	171/**180**	155/155
H-14	NS	155/**159**	**104**/110	175/**180**	153/155
S-37	189/189	155/155	102/110	NS	NS
S-34b	189/189	155/155	102/110	NS	153/155
S-3	NS	155/155	NS	171/175	153/155

NS, not scored.

The female and the cub 238 were captured quickly and removed from the park. But the second cub was not captured. A few weeks later a 2-year-old grizzly charged a group of people and was killed by park biologists in the general home range area of the female and her cubs. Park biologists believed this was the missing cub, so the lab genotyped this bear (Table 22.1). This cub shared one allele with the female at all loci which is consistent with the hypothesis that he is her offspring. Also, relatedness statistics (Queller and Goodnight 1989) reveal a pairwise relatedness of 0.70 between the mother and unknown cub, providing strong support that this bear was the missing cub.

is the probability of randomly sampling two individuals that have the same genotype (for the loci being studied). If we use highly polymorphic molecular markers, there is a low probability of two individuals sharing the same genotype at multiple loci. Thus, if we find two samples (e.g., bloodstains or tissues) with matching genotypes, we can determine with high probability that they come from the same animal or plant (e.g., found at a crime scene).

PI_{av} is computed using the following expression:

$$PI_{av} = \sum_{i=1}^{n} p_i^4 + \sum_{i>j=1}^{n} (2p_i p_j)^2 \qquad (22.1)$$

where p_i and p_j are the frequencies of ith and jth allele at the locus (Waits et al. 2001, Ayres and Overall 2004). Here, p_i^4 is simply the average probability of randomly sampling two homozygotes (e.g., aa), and $(2p_i p_j)^2$ is the average probability of randomly sampling two heterozygotes (e.g., Aa). This equation assumes Hardy-Weinberg proportions and that no substructure exists in the population. The multilocus PI_{av} is computed by using the product rule (i.e., 'multiplication rule', see Appendix Section A.1), and multiplying together the single-locus probabilities (see match probability Box 22.2). A reasonably low multilocus PI_{av} for forensics applications (e.g., matching blood from a wildlife crime scene to blood on a suspect's clothes) is approximately 1/10,000 to 1/100,000. Achieving this low a PI_{av} would require approximately 5–20 markers, depending on their polymorphism level (Figure 22.2).

PI_{av} is also often used to quantify the power of molecular markers for studies involving genetic tagging. A reasonably low PI_{av} for genetic tagging is approximately 1/100 (Waits et al. 2001). This is not as low as is needed for forensics, because it is less problematic to misidentify individuals in genetic tagging than in a law enforcement case where someone might be fined or imprisoned. To achieve a reasonably low PI_{av} for genetic tagging, approximately 5–10 highly polymorphic markers are often sufficient (see Box 22.2). For less

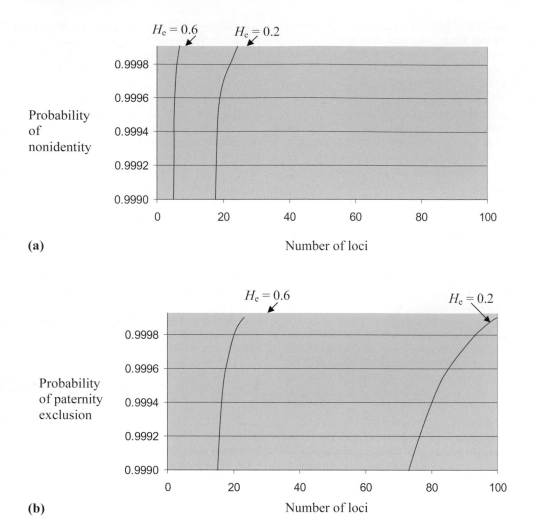

Figure 22.2 Relationship between the number of loci genotyped, and (a) the probability of nonidentity $(1 - PI_{av})$ and (b) the probability of paternity exclusion. The number of loci needed for paternity exclusion is approximately 3–5 fold higher than for individual identification. Also at low heterozygosity ($H_e = 0.2$; e.g., SNPs), 3–5 times more loci are needed compared with microsatellite loci ($H_e = 0.6$). The probability of identity (PI_{av}) was computed using equation 22.1 and allele frequencies that give a heterozygosity (H_e) of 0.6 representing microsatellites or 0.2 representing SNPs. For example, for loci with $H_e = 0.2$, two alleles have the frequency of 0.885 and 0.115 in the following equation: $H_e = 1 - \Sigma p_i^2$ (where p_i is the frequency of the ith allele). The probability of paternity exclusions was computed using the expression from Jamieson and Taylor (1997) for the case where genotypes are known for the mother and the offspring when testing to exclude a randomly sampled male that is not the true father. Modified from Morin *et al.* (2004).

polymorphic markers such as allozymes, SNPs, AFLPs (with heterozygosity typically from 0.20 to 0.40), more markers would be required. Power is lower for dominant markers like AFLPs than for codominant markers like microsatellites and SNPs (see Chapter 4). Interest-ingly, the power of a set of markers is better predicted by heterozygosity than allelic richness; loci with the same heterozygosity ($H = 0.6$) but a different number of alleles (3 or 10) have similar power to resolve indi-viduals (Waits *et al.* 2001).

It is important to note that equation 22.1 used to estimate PI_{av} assumes that individuals are unrelated (e.g., no siblings), sampled randomly, and that no substructure or gametic disequilibrium (i.e., independent loci) exists. These assumptions are often violated in natural populations. The violation of assumptions could cause underestimation of the true PI_{av}. For example, in datasets from wolves and bears, PI_{av} was underestimated by up to three orders of magnitude (e.g., 1/100,000 which underestimates the true value of 1/100; Waits et al. 2001). To avoid problems with underestimation, other PI_{av}-related statistics such as PI_{av}-sibs, should also be used to compute the probability of identity. Other PI_{av} statistics (e.g., accounting for potential substructure) can be computed with the user-friendly software API-CALC (Ayres and Overall 2004). It is confusing that sometimes the 'average probability of identity' (PI_{av}) is referred to as the 'average match probability' (MP), in the literature.

22.3.2 Match probability

The 'match probability' (MP) is a useful statistic related to PI_{av} (equation 22.1). While PI_{av} is the average probability of randomly sampling two individuals consecutively that have the same genotype, the MP is the actual probability of sampling one individual identical to the one already 'in hand' (i.e. sampled previously). PI_{av} is for computing the average power of a set of markers (considering all genotypes, homozygotes and heterozygotes, in a given study), whereas MP gives a probability of sampling the individual genotype in question, that was sampled (see Box 22.2). MP requires the same assumptions (no substructure, no gametic disequilibrium, and no siblings) as does PI_{av}, although more sophisticated MP statistics can correct for violations of these assumptions.

For example, in a recent study, Tnah et al. (2010) illustrated how DNA match probabilities can be computed for a tropical timber species across peninsular Malaysia. Match probabilities were assessed to help identify logs illegally removed from forest preserves. DNA typing was used to match stumps in forest preserves to logs sold or transported illegally. DNA matching of stumps to illegally trafficked logs can help to stop illegal deforestation (see also Example 22.1 on species identification). The authors genotyped 12 microsatellites from 30 populations and showed how effects of population substructure and inbreeding could be

incorporated into estimates of match probabilities. This estimation procedure, along with a large database of genotypes from many populations of the timber species, should help prosecute illegal loggers in Malaysia.

We have thus far considered only nuclear DNA markers for determining match probabilities and the average probability of identity. However, mtDNA can also be useful for individual identification. For example, the mtDNA control region in canids and felids has tandemly-repeated sequences (Fridez et al. 1999, Savolainen et al. 2000). These repeats are highly polymorphic and heteroplasmic (i.e., multiple clones with different repeat lengths are found within an individual). Thus, mtDNA analysis will occasionally be useful for individual differentiation because different individuals often have different mtDNA repeat profiles. But since mtDNA represents only one 'locus', it will provide much less certainty than multilocus nuclear DNA methods. An advantage of mtDNA is that it can be amplified from hair shafts, whereas nuclear DNA is only found in the hair root bulb (Watson 2000). Animal hairs are often found at poaching crime scenes and on people's clothing.

22.4 PARENTAGE AND RELATEDNESS

Determination of the mother (maternity) or father (paternity) of an individual can facilitate conservation in a variety of ways such as quantifying reproductive fitness of parents or verifying that individuals in pet stores originate from captive parents, as might be claimed by some pet-trade industry workers. An enormous problem for conservation is the illegal capture or collection of offspring from the wild, which occasionally involves killing of the wild parents (e.g., gorillas and orangutans). DNA typing of captive individuals to identify their parents could help detect the illegal capture of individuals from the wild, and thereby reduce threats to wild populations. In a similar way, estimating the relatedness of individuals could be used to identify wild caught animals, and also to assess relationships (siblings) among captive animals to help avoid inbreeding in captive breeding or reintroduction programs.

22.4.1 Parentage

Parentage analysis involves comparing genotypes of offspring to potential parents in order to identify the actual parents. This analysis generally depends on the fact that an offspring will have one allele per locus from each parent.

Two computational approaches exist for genetic parentage analysis: **parentage exclusion** and **parentage assignment** (e.g., Marshall *et al.* 1998, Taberlet *et al.* 2001). Exclusion involves the determination that both alleles at a locus in a candidate parent do not match either of the offspring's alleles, which leads to exclusion of that candidate parent as the true parent. Exclusion might be insufficient to resolve parentage unambiguously in paternity tests, when for example multiple candidate fathers cannot be excluded or when all potential fathers have not been sampled. Another deficiency with exclusion is that genotyping errors can cause the true father to be excluded erroneously.

When some males cannot be sampled or excluded or when genotyping errors exist, a statistically based method can help infer paternity. **Paternity assignment** is statistically based and involves the use of probabilities and likelihood computation. Paternity assignment is widely used to estimate the probability that a given male is the father (Slate *et al.* 2000; and see the CERVUS and PARANTE software, listed on this book's website). While statistical paternity inference is possible when not all potential fathers have been sampled, the power (or certainty) of assignment drops substantially when less than approximately 70–90% of males (potential fathers) are sampled (Marshall *et al.* 1998).

Parentage analysis requires more genetic markers than does individual identification (Section 22.3). For example, paternity exclusion requires approximately 10–15 microsatellites (Figure 22.2b), whereas individual identification requires only 5–10 loci (Figure 22.2a). This paternity exclusion example assumes that the mother's genotype is known, which often is the case because, for example, an offspring often associates closely with its mother (e.g., in mammals and birds) and thus maternity can be identified via observation.

Parentage analysis when neither parent is known requires even more loci (e.g., >20 microsatellites) than paternity analysis where the mother is known. Parentage analysis also requires more loci when heterozygosity is low, as for SNP loci or allozymes (biallelic loci). For example, approximately 40–60 SNPs (heterozygosity > 0.2) are required for paternity exclusion when the mother is known (Figure 22.2b). It is likely that parentage assignment with SNPs and unknown parents would often require 100 SNP loci.

To quantify the statistical power of molecular markers for parentage analysis (e.g., paternity exclusion), researchers can compute the expected paternity exclusion probability (*PE*), the probability of excluding a randomly chosen nonfather (e.g., Double *et al.* 1997). The power of a set of molecular markers for paternity exclusion (*PE*) can be quantified by plugging allele frequencies into the following equation (Jamieson and Taylor 1997):

$$PE = \sum_{i=1}^{n} p_i^2(1-p)^2 + \sum_{i>j=1}^{n} 2p_i p_j(1-p_i-p_j)^2 \qquad (22.2)$$

where p_i and p_j are the frequency of the ith and jth allele, respectively. For one locus, this expression gives the average probability of excluding (as father) a randomly sampled nonfather, when the *mother and offspring genotypes are both known*. To compute the multilocus *PE*, we multiply together the *PE* for each locus, assuming independence among loci. Often 10–15 highly variable markers (heterozygosity 0.50–0.60) are required to achieve a high probability of paternity exclusion (>99.9%), as mentioned above (Figure 22.2). Other expressions are available for estimating power for parent exclusion when neither parental genotype is known (Jamieson and Taylor 1997).

Another example comes from Australia where the owners of an adult female northern hairy-nosed wombat claimed that their juvenile wombat was the offspring of their adult female. The owners had a legal permit for the adult, but not for the young wombat. The owners claimed that their female must have been impregnated by a wild wombat somewhere near their backyard. Wildlife law enforcement officials questioned this story and conducted maternity testing in a genetics laboratory. The laboratory typed nine microsatellite loci on the mother and offspring and found no incompatibilities, that is, the mother had an allele at each locus compatible with the offspring (Andrea Taylor, personal communication). Thus there was no evidence that the offspring was taken from the wild, and the owners were allowed to keep the young wombat. The statistical certainty of a maternity (or paternity) assignment can be computed, based on allele frequencies at the loci (e.g., Slate *et al.* 2000).

Other applications of parentage analysis include understanding a species' mating system, estimating variance in reproductive success, and detecting multiple paternities. Such information is helpful for population management. Variance in reproductive success influences the effective population size and thus the rate of loss of genetic variation, inbreeding, and efficiency of selection. Knowing that variance in reproductive success is high, for example, can help biologists to predict that the N_e is much smaller than the census population size (see Section 7.10).

22.4.2 Relatedness

Relatedness estimation is useful in conservation of wild and captive populations to help understand animal social systems, to detect avoidance of inbreeding, and to optimize conservation breeding and translocation strategies (Gonçalves da Silva *et al.* 2010). Relatedness can be estimated using pedigrees (Chapter 13) or from genetic similarity between individuals estimated directly from DNA markers and allele sharing. For example, founders of captive populations are often assumed to be unrelated, which could bias breeding programs leading to mating between relatives, loss of genetic variation and reduced fitness, if closely related founders breed.

Gonçalves da Silva *et al.* (2010) used DNA markers to refine the captive breeding program for lowland tapirs. The authors used 10 microsatellite markers to genotype 49 captive individuals to evaluate relatedness using the metric of **mean kinship** (*mk*), which is a measure of how closely related each animal is to the population. It is an important measure of just how rare an individual's combination of genes is in the entire population. Animals with lower mean kinship values have relatively fewer genes in common with the rest of the population, and are therefore more genetically valuable in a breeding program (Ballou and Lacy 1995). Gonçalves da Silva *et al.* (2010) identified two individuals with low mean kinship (*mk* = 0.007), and thus of high genetic value for the population. These individuals might otherwise have been excluded from the breeding program, in the absence of empirical genetic data, because of their unknown origins. Common assumptions of individuals of unknown origin being highly related would have led to overestimates of mean kinship and underestimates of future heterozygosity, when compared with values found when genetic markers were used to inform kinship and breeding. This example shows how the computation of marker-based founder kinship coefficients could

improve captive management and the maintenance of genetic variation.

22.5 POPULATION ASSIGNMENT AND COMPOSITION ANALYSIS

Genetic markers can help to identify the population of origin of individuals, groups of individuals, or products made from plants and animals, such as from endangered forest trees or horn or bone from threatened wildlife (Waser and Strobeck 1998, Manel *et al.* 2002). Determining the population or geographic region of origin of wildlife products can help to identify populations being poached, and the trade routes used by traffickers selling illegally harvested individuals. Such information could help law enforcement officials to target geographic regions with poaching problems.

22.5.1 Assignment of individuals

Population assignment tests work by assigning an individual genotype to the population in which its genotype has the highest expected frequency (see Section 9.9.4 and Manel *et al.* 2005). Assignment tests generally require samples and multilocus genotype data from each candidate population of origin. **Assignment tests** can potentially determine the population of origin of individuals even when the F_{ST} is low ($F_{ST} = 0.02$), if many microsatellite loci are genotyped (Olsen *et al.* 2000).

Researchers recently used assignment tests to identify a wolverine in California. No wolverines had been detected in California for nearly 100 years (Moriarty *et al.* 2009). The researchers used 16 microsatellite loci to assign the sample to populations identified from 261 genetic samples, plus historic California samples. The assignment tests suggested the animal of unknown origin from California was not a remnant of a historic California population. Comparison with the available data revealed the individual was most closely related to populations from the western edge of the Rocky Mountains in Idaho. This represents the first evidence of movement from wolverine populations of the Rocky Mountains to the Sierra Nevada Mountains in California.

Assignment tests were also recently used along with genotypes from historical museum specimens to test whether individuals from a presumed extinct species might still exist in captivity. Russello *et al.* (2010)

genotyped nine microsatellites and mtDNA from museum specimens of extinct Galápagos tortoise species to assign captive tortoises to either the extinct or extant species. Galápagos tortoises live up to nearly 200 years, and many in captivity have unknown origins. The Bayesian assignment testing identified nine captive Galápagos tortoises that had previously been misassigned to extant species. Thus, the extinct species was discovered in captivity by assignment testing! The findings permitted captive breeding efforts to begin to breed the nine newly discovered individuals and to reestablish this extinct species of tortoise in the wild throughout its native range.

Similarly, assignment tests with microsatellites and mtDNA have been used to assign captive tigers to one of five extant subspecies (Example 22.5). This study illustrates the potential of assignment testing to identify genetic resources in captivity useful for conserving wild populations.

Another interesting example of assigning individuals for wildlife forensics came from a fishing competition. A fisherman claimed to have caught a large salmon in Lake Saimaa, Finland. However, the organizers of the competition questioned the origin of the salmon because of its unusually large size (5.5 kg). A genetic analysis of seven microsatellite loci was conducted on the large fish and on 42 fish from the tournament lake, Saimaa. A statistical analysis was conducted using the exclusion-simulation assignment test (Primmer *et al.* 2000). The exclusion test suggested that the probability of finding the large fish's genotype in Lake Saimaa was less than 1 in 10,000. Thus, the competition organizers excluded Lake Saimaa as the origin of the salmon. Subsequently, the fisherman confessed to having purchased the fish in a bait shop.

Population assignment in the fishing tournament example is based on **exclusion tests** and computer simulations to assess statistical confidence. In the exclusion–simulation approach, we 'assign' an individual to one population only if all other populations can be excluded with high certainty (e.g., P < 0.001). We exclude a population if the genotype in question is unlikely to occur in the population (P < 0.001), assuming Hardy-Weinberg proportions and gametic equilibrium (see Section 9.8).

An advantage of the exclusion–simulation approach is that it can be used when only one population (the putative source population) has been sampled; it does not assume that the true population of origin has been sampled. Other approaches, such as Bayesian (Pritchard *et al.* 2000) and likelihood ratio tests (Banks and Eichert 2000), generally require samples from at least two populations, and assume that the true population has been sampled. If the true population of origin has not been sampled, the assignment probabilities from Bayesian and likelihood ratio approaches could be misleading. It seems prudent to apply both the exclusion-based and Bayesian assignment approaches (Manel *et al.* 2002).

An advantage of the Bayesian approach is that it is generally more powerful than exclusion-based methods and other methods (Manel *et al.* 2002, Maudet *et al.* 2002). User-friendly software exists for all three (and other) approaches (Banks and Eichert 2000, Pritchard *et al.* 2000, Piry *et al.* 2004, Rannala and Mountain 1997).

A large-scale use of DNA analysis for assigning individuals to geographic populations is FishPopTrace, a €3.9 million research program to fight illegal fishing and fraudulent labeling of fish in European supermarkets (Stokstad 2010, Nielsen *et al.* 2012). Researchers estimate that fish fraud is a $US23 billion business worldwide. Fishing violations often involve harvest of fish from areas closed to fishing to allow the stocks to recover from overharvest. FishPopTrace used SNPs to assess the power for assignment in European hake, common sole, and Atlantic herring. Researchers sampled 50 fish from each of 20 populations around European waters, for each of the three species.

The FishPopTrace consortium created a fish SNP genotyping chip with 1536 SNPs for each species, and found that the 20 most informative SNPs could correctly identify the population of origin of >95% of fish within each species. This is impressive given the relatively high gene flow and low F_{ST} among marine fish populations. Associated with this, a European fishing law was passed in 2009 that explicitly mentions that genetic tests would be used as an enforcement tool to fight illegal harvest and fraudulent marketing. Researchers believe that random testing of fishing catch and markets will reduce fraud at a fraction of the cost of litigation (Stokstad 2010).

A final example of population assignment involves the reestablishment of migratory connectivity in a fragmented metapopulation of fish. Dams constructed without fish passage facilities have prevented migratory bull trout in Idaho and Montana, US, from returning to their natal spawning streams for nearly 100 years. To facilitate migration of the downstream fish back to their natal populations areas upstream, DeHaan

Example 22.5   Assignment tests identify origins of captive tigers and identify valuable individuals for captive breeding

Tigers are disappearing from the wild; fewer than 3000 remain (Luo *et al.* 2010). Three subspecies are extinct and one persists only in zoos. By contrast, captive tigers are flourishing, with 15,000–20,000 individuals worldwide, outnumbering their wild relatives by five fold. Researchers recently used genetic assignment tests to identify nonadmixed captive tigers that could be used for genetic management of subspecies in captivity and in the wild (Luo *et al.* 2008).

Subspecies-diagnostic genetic markers were developed by analyzing 134 reference voucher tigers from the five extant subspecies (Luo *et al.* 2004). Sequences of mtDNA were used to assign maternal ancestry to a subspecies. The authors also genotyped 30 microsatellite loci and used Bayesian assignment tests implemented in STRUCTURE (Pritchard *et al.* 2000, Section 16.4.2) to calculate the probability (q) that a tiger could be assigned to a single subspecies, or alternatively to quantify the extent of admixture of an individual from different subspecies. The genetic differentiation among subspecies, based on the 134 voucher specimens, was high (microsatellite R_{ST} = 0.314, mtDNA F_{ST} = 0.838, Luo *et al.* 2004).

The assignment tests assigned 49 tigers with high certainty ($q > 0.90$) to one of five subspecies (Amur, Sumatran, Malayan, Bengal, and Indochinese); 52 tigers had admixed subspecies origins (Figure 22.3). For example, 16 tigers had discordant mtDNA and microsatellite assignments and were therefore classified as admixed. Interestingly, 11 captive tigers previously thought to be 'purebred' (nonadmixed) were hybrids. However, most assignments (80%) of captives were consistent with the origins provided by owners, including 42 named as a specific subspecies and 41 suspected admixed (Luo *et al.* 2008).

The tested captive tigers contain substantial genetic diversity that has not been observed in their wild counterparts. For example, 46 new microsatellite alleles (36 in nonhybridized individuals and 10 in admixed tigers) were observed that were not present in the 134 voucher specimens.

This study suggests that genetic assignment tests can help us to identify captive individuals useful to supplement wild populations, and thereby provide insurance against extinction in the wild. Because captive and wild tigers today are consciously managed to avoid hybridization between subspecies, the discovery of 49 purebred tigers in a sample of 105 individuals (47%) has potentially important conservation implications. Assignment of subspecies ancestry offers a powerful tool to considerably increase the number of nonadmixed tigers suitable for conservation management.

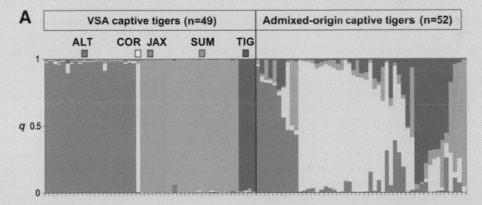

Figure 22.3   STRUCTURE population cluster analysis of 101 captive tigers. The left panel shows 49 tigers assigned to a single subspecies based on 134 reference tigers with verified subspecies ancestry (VSA). The right panel shows 52 tigers with apparent admixed origins. Each individual is represented by a thin vertical bar (defined by tick marks under colored regions on the *x* axis) partitioned into five colored segments representing individual ancestry (q) to the five indicated tiger subspecies. ALT, Amur; COR, Indochinese; JAX, Malayan; SUM, Sumatran; TIG, Bengal. From Luo *et al.* (2008). See Color Plate 8.

et al. (2011) genotyped 12 microsatellites. Based on genetic assignments for 259 migratory (downstream) adult bull trout collected below dams, 203 fish were manually transported upstream above one or more dams and released near their natal population inferred from the assignment tests. This reestablished connectivity and increased numbers of spawning adults in numerically depressed populations above the dams (DeHaan *et al.* 2011). In addition, in years following the upstream translocations, the authors used genetic parentage analysis (assignment of offspring to parents; Section 22.4.1) to document successful reproduction of many of the adults that had been transported upstream (S. Bernall, personal communication).

22.5.2 Assignment of groups

Assignment of *groups* of individuals to a population (or geographic region) of origin is also feasible. For example, the Alaska Department of Fish and Game confiscated a boatload of red king crab that they suspected was caught in an area closed to harvest near Bristol Bay in the Bering Sea (Seeb *et al.* 1989). The captain claimed that the crabs were caught near Adak Island in the Aleutian Islands, over 1500 km away from the closed area.

Thirteen populations of king crab from Alaskan waters had been previously examined at 42 allozyme loci (14 polymorphic loci) to describe the genetic population structure of red king crab (Seeb *et al.* 1989). Genetic data at eight loci indicated that the confiscated crabs could not have been caught near Adak Island (in the Aleutian Islands), the only area open to harvest. Allele frequencies at Adak Island significantly differed from the allele frequencies among the confiscated crab, as inferred using a chi-square test (e.g., $\chi^2 = 21.6$, P < 0.001; and $\chi^2 = 88.5$, P < 0.001, for the *Pghd* and *Alp* loci, respectively). Discriminate function analysis is a second statistical approach used to conclude that the confiscated crabs did not originate from the Adak Island area (see Seeb *et al.* 1989 for details). The allele frequencies in the confiscated sample matched the samples from further north in the Bering Sea (Figure 22.4). Based upon these results, the vessel owner and captain agreed to pay the State of Alaska a $US565,000 penalty for fishing violations.

Assigning a group of individuals can be easier (i.e., yield a higher statistical certainty) than assigning a single individual, because more information is available in a group of genotypes than in a single individual's genotype. A user-friendly software program for assigning groups (as well as individuals) to a population of origin is available in *GeneClass* 2.0 (Piry *et al.* 2004).

Wasser *et al.* (2007) used a novel DNA assignment method to determine the geographic origin(s) of large sets of elephant ivory seizures. They showed that a joint analysis of multiple tusks performs better than sample-by-sample methods in assigning sample clusters of known origin. They then used the joint assignment method to infer the geographic origin of the largest ivory seizure since the 1989 ivory trade ban. Authorities initially suspected that this ivory came from multiple locations across forest and savanna populations in Africa. However, DNA assignments showed that the ivory was entirely from savanna elephants, probably originating from an area centered on Zambia. Findings allowed law enforcement to focus their investigation on a few trade routes and led to changes within the Zambian government to improve antipoaching efforts. Such outcomes demonstrate the potential of genetic analyses to help combat the expanding wildlife trade by identifying origin(s) of large seizures, including multiple individuals in contraband ivory.

22.5.3 Population composition analysis

Many species are harvested in mixed populations such as mixed stock fisheries (salmon, marine mammals) and waterfowl (Example 18.2). Other species also migrate in mixed groups (neotropical song birds, butterflies, and others). Effective management of mixed stock fisheries and mixed populations requires that the populations or stocks that compose the mixture be identified and the extent of their contribution determined (Figure 22.5, Pella and Milner 1987). Stocks are generally analogous to management units (e.g., demographically independent populations), which were discussed in Section 16.5.

The fundamental unit of replacement or recruitment for anadromous salmon is the local breeding population because of homing (Rich 1939, Ricker 1972). That is, an adequate number of individuals for each local reproductive population are needed to ensure persistence of the many reproductive units that make up a fished stock of salmon. The homing of salmon to their natal streams produces a branching system of local reproductive populations that are demographically and genetically isolated. The demographic dynamics of a fish population are determined by the balance between reproductive potential (i.e., bio-

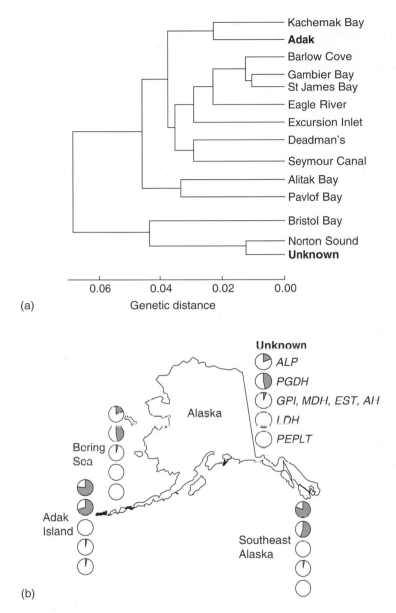

Figure 22.4 (a) Similarity clustering of king crab populations using allele frequency data. Note the 'unknown' confiscated crabs do not cluster with the Adak population, where the crab harvest was permitted (in the Aleutian Islands, see text). (b) Map of locations of populations (Adak and Bering Sea) in Alaska, and allele frequency pie charts for five allozyme loci. Modified from Seeb *et al.* (1989).

logical and physical limits to production) and losses due to natural death and fishing. "Population persistence requires replacement in numbers by the recruitment process" (Sissenwine 1984), so fishery scientists have focused on setting fishing intensity so that ade-

quate numbers of individuals 'escape' fishing to provide sufficient recruitment to replace losses.

The distinction between a local breeding population and a fished (harvested) stock is crucial (Beverton *et al.* 1984). A local breeding population is a local

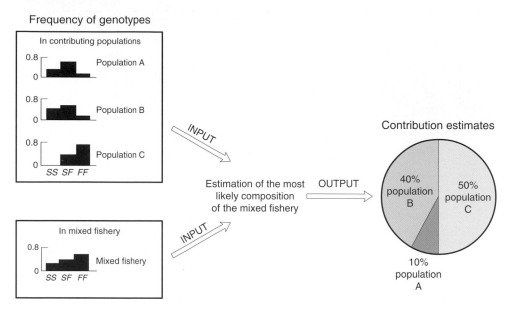

Figure 22.5   Outline of procedure for estimating the population (stock) composition on the basis of genetic data at a single locus with two alleles (S and F). The mixture is composed of three populations. From Milner *et al.* (1985).

population in which mating occurs. A stock is arbitrary and can refer to any recognizable group of population units that are fished (Larkin 1972, 1981). The literature has often been unclear on this distinction. In practice, it is extremely difficult to regulate losses to fishing on the basis of individual local breeding populations. Thousands of local breeding populations make up the US west coast salmon fishery, and many of these are likely to be intermingled in any particular catch. Nevertheless, the result of regulating fishing on a stock basis and ignoring the reproductive units that together constitute a stock is the disappearance or extirpation of some of the local breeding populations (Clark 1984).

The loss of local populations could lead to the crash or eventual extirpation of the entire metapopulation or species, with negative consequences for the larger ecosystem and regional economy (commercial fisheries, sports fisheries or hunting, ecotourism). The importance of maintaining numerous diverse local populations for insuring long-term metapopulation viability has been called the portfolio effect, and is illustrated in the study of salmon by Hilborn *et al.* (2003) (see Example 15.3) and Schindler *et al.* (2010).

An important application of mixed population analysis in conservation management is illustrated by a study of sockeye salmon in Alaska. Seeb *et al.* (2000)

genotyped 27 allozyme loci in all major spawning populations from the upper Cook Inlet and found substantial among-population differentiation (e.g., $F_{ST} = 0.075$, among nursery lakes). The salmon from these major populations are harvested in a mixed-stock aggregation that forms in upper Cook Inlet (Figure 22.6). A mixed stock analysis (based on maximum likelihood, see Manel *et al.* 2005) allowed estimation of the proportion of genes (and thus individuals) from each population in the pool of harvested fish. Impressively, the genotyping and statistical analysis can be conducted within 48 hours after harvest! This is critical because it allows real-time monitoring of harvest from each major population. It allows biologists to close the harvest if too many fish are harvested from any one major breeding population. This is critical to help prevent overfishing, longer-term closures of fishing, and the extinction of a major source population.

Population composition analysis differs from individual-based assignment tests in that composition analysis estimates the percentage of the gene pool (or alleles) that originates from each local breeding population, whereas individual-based assignment methods estimate the actual number and identify individuals originating from each breeding population (Manel *et al.* 2005). The Bayesian method programmed in the

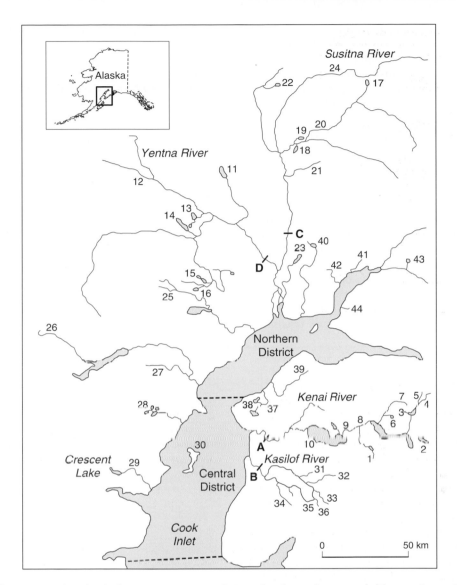

Figure 22.6 Locations (numbers) of major spawning populations of sockeye salmon sampled for genetic monitoring of harvest. The mixed stock fish harvest goes on in upper Cook Inlet, Central and Northern Districts. From Seeb *et al.* (2000).

STRUCTURE software (Pritchard *et al.* 2000) even computes the percentage of an individual's genome that originates from different breeding populations. The relative performance of the different assignment and composition analysis approaches depends on the question. More studies are needed to evaluate the relative performance of these analytical methods under different scenarios relevant to management and conservation.

22.6 GENETIC MONITORING

Genetic monitoring is the quantification of temporal changes in population genetic metrics (e.g., N_e or F_{ST}) or in demographic metrics (population census size, N_C) estimated using molecular genetic markers (Schwartz *et al.* 2007). Monitoring has a temporal dimension that is different from assessment, which quantifies a population characteristic at only a single time point. Monitoring of wild populations is increasingly feasible thanks to continual improvements in both molecular and statistical techniques. Nevertheless, genetic monitoring is not widely conducted despite the fact that many national and international organizations have established principles and promoted strategies for monitoring biological diversity (Laikre *et al.* 2010b).

Three categories of genetic monitoring can be delineated (Figure 22.7). Category I is the use of molecular markers for traditional ecological population monitoring through the identification of individuals and species and the estimation of population census size, which is often conducted using noninvasive sampling (Section 22.1). Category II includes the use of genetic markers to monitor population genetic parameters such as N_e or allelic diversity to detect potential population declines or genetic bottlenecks. Category III is the use of DNA markers in or near adaptive genes to assess effects of environmental change, disease epiz-

ootics, or other stresses on a population or its adaptive response.

22.6.1 Traditional ecological monitoring

Category I genetic monitoring is exemplified by the monitoring of DNA tags (DNA fingerprints) in lieu of physical tags or natural markings in order to identify individuals and estimate population census size (N_C). For example, Stetz *et al.* (2010) used encounter data from 379 grizzly bears identified genetically through bear rub surveys of hair samples to parameterize a series of mark–recapture simulations to assess the ability of noninvasive genetic sampling to detect declines in population size (N_C). They concluded that annual tree rub surveys would provide >80% power to detect a 3% annual decline within six or fewer sampling years. Estimates of the true population size (starting at $N_C = 765$) were unbiased, and became very precise within about four years. Thus, annual tree rub surveys and DNA-

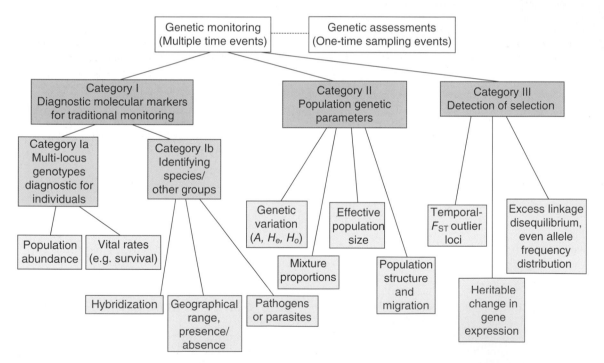

Figure 22.7 Categories of genetic monitoring. Category I includes the use of diagnostic molecular markers for traditional population monitoring through the identification of individuals and species, and the repeated (temporal) assessment of population size. Category II includes the use of genetic markers to monitor population genetic parameters. Category III involves use of DNA markers to detect changes in frequency of adaptive alleles or gene expression associated with environmental change. Modified from Schwartz *et al.* (2007).

based hair genotyping could provide a useful complement or alternative to traditional telemetry methods for monitoring trends in grizzly bear populations.

Another example of Category I monitoring involved noninvasive genetic monitoring of mortality and reproductive success of reintroduced otter populations in the Netherlands (Koelewijn *et al.* 2010). The study demonstrated that only a few dominant males successfully fathered offspring and thus the effective population size was small (Example 22.6).

Example 22.6   Genetic monitoring of an introduced otter population

Eurasian otters were extirpated from the Netherlands in 1989. From 2002 to 2008, 30 individuals were released into northern Netherlands. Post-reintroduction success was monitored using noninvasive genetic analyses. The founding individuals were genotyped along with feces collected in the release area. Researchers analyzed 1265 fecal samples (spraints) and anal secretions (jellies) with 7–15 microsatellite loci. Of the 1265 samples, 582 (46%) were successfully assigned to either a released or new genotype representing an offspring.

During the first three winters, seven microsatellites were sufficient for individual typing and parentage assessments. Subsequently, founder individuals died and relatedness increased, and 15 loci were required for parentage analyses, although the seven loci were still sufficient for individual identification (e.g., Figure 22.2a versus b). For example, during the final 2007/8 season the probability of identical genotypes in two random sibs (PI_{av}-sibs; Section 22.3) of the first set of seven loci was 2.1×10^{-3}, which is sufficient in small populations. When all 15 loci were used the PI_{av}-sibs decreased to 1.4×10^{-5}.

The researchers used genetic parentage assignment to identify 54 offspring (23 females and 31 males). The reproductive success among males was strongly skewed, with two dominant males fathering two-thirds of the offspring (Figure 22.8). One of the highly successful males was the son of the other. The effective population size was only about 30% of the detected number of individuals because of the large variance in reproductive success among males.

Genetic sex identification revealed that males had a higher mortality rate (22 out of 41 males (54%) vs. 9 out of 43 females (21%)), likely because most juvenile males dispersed to surrounding areas upon maturity. In contrast, juvenile females stayed inside the area next to the mother's territory. The main cause of mortality was traffic accidents.

This study demonstrates that noninvasive molecular methods can be used to monitor elusive species to reveal a comprehensive picture of population status. Nevertheless, the future of the Dutch otter population is unclear because it is new, small, and isolated, with a low effective population size. A connection with a second population in nearby wetlands is unlikely, and many animals are killed in traffic incidents when they move away from the current population. Furthermore, only a few males dominate reproductive process, which lowers the effective population size and increases relatedness and inbreeding.

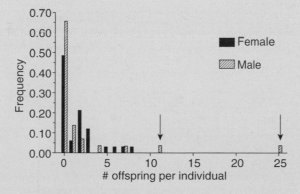

Figure 22.8   Reproductive success of introduced Eurasian otters during the period 2002–2008 in the Netherlands. The male with 11 offspring is the son of the male with 25 offspring (arrows). This high variance in male reproductive success greatly reduced the effective population size of this population. From Koelewijn *et al.* (2010).

22.6.2 Genetic parameters

Category II monitoring is exemplified by tests for loss of genetic variation or change of the effective population size (Figure 22.9). One of the best long-term examples of category II monitoring is from brown trout in Sweden (Palm *et al.* 2003). These authors monitored populations in small mountain lakes and creeks in the Hotagen Nature Reserve in central Sweden. Tissues for genetic analyses have been collected annually since the 1970s. The contemporary local effective population sizes of two natural brown trout populations were estimated using temporal changes in allele frequencies (i.e., a temporal F_{ST} or a related statistic, F_k), which is known as the temporal variance method or 'temporal method' (Waples 1989).

Point estimates of N_e fluctuated for individual populations, but the general trend indicated stability (Figure 22.9). N_e estimates were small considering that heterozygosity remained stable and relatively high. N_e was estimated to be 19 and 48 for the two populations. Heterozygosity for 17 polymorphic allozymes was 0.13 and 0.25 for localities I and II respectively, which were located above and below a waterfall.

How could the relatively high allozyme heterozygosity have been maintained in spite of the small N_e? The likely answer is that gene flow from other populations has prevented excessive loss of heterozygosity. The tem-poral method estimates the local contemporary effective size, whereas heterozygosity is related to the global (metapopulation) long-term effective size (Luikart *et al.* 2010). Recent research suggests that estimators of the local N_e are not sensitive to limited gene flow ($m < 0.05$), which can maintain heterozygosity yet not influence estimates of the local N_e (e.g., England *et al.* 2010).

22.6.3 Adaptive change

Monitoring to detect adaptive change can be accom-plished by focusing on molecular markers or on quantitative traits or phenotypes using the tools of quantitative genetics (Chapter 11) (Hansen *et al.* 2012). A challenge of monitoring quantitative traits is to dis-tinguish between adaptive genetic responses and **phe-notypic plasticity** (Gienapp *et al.* 2008). Monitoring molecular markers can avoid this problem, but raises another problem of how to associate genetic variation at a locus with variation in the selective agent acting on that variation (Vasemägi and Primmer 2005). We have already seen one example of monitoring for an adaptive response with Atlantic cod fish in which unnatural selection from harvest has shifted allele fre-quencies (see Guest Box 18).

Genetic monitoring to assess adaptive change at the molecular level often involves the use of population

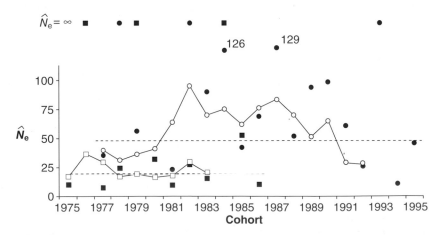

Figure 22.9   Point estimates of effective size ($\hat{N}_e$) for all pairs of consecutive cohorts (filled square and filled circle for localities I and II, respectively), and the corresponding (harmonic) mean ($\hat{N}_e$) (open symbols) obtained from moving averages for F_k (temporal variance in allele frequencies) over five consecutive cohorts (i.e., four cohort pairs). 'Cohort' on the *x*-axis represents the first cohort used for each N_e estimate. Note the broken *y*-axis, and that large N_e point estimates are given as numbers (∞ = infinity). From Palm *et al.* (2003).

genomics and outlier tests (Figure 9.14) to identify locus-specific effects (Section 9.7.3) including: (1) rapid loss of variation due to selective sweeps at a locus; (2) excessively high or low F_{ST}; (3) high linkage disequilibrium along a chromosome near a candidate locus; or (4) excessive deviation from mutation-drift equilibrium of the allele frequency distribution at a locus (Luikart *et al.* 2003). The tests for locus-specific outlier effects, such as F_{ST}, can be conducted in a monitoring framework, including the use of two samples separated by time from one population. The idea here is to test for locus-specific effects of a stress or challenge by genotyping a locus with an allele whose frequency would change rapidly during the stress event.

For example, a SNP or microsatellite near an immune system gene (MHC) could be monitored for excessive change in allele frequencies (compared with neutral loci) before and after a disease outbreak. In addition to this candidate gene approach, a genome-wide scan approach of genotyping hundreds of loci could be conducted to monitor for genetic signatures associated with disease outbreaks.

Hansen *et al.* (2012) define criteria needed to convincingly demonstrate adaptive evolutionary change for DNA markers. These include requirements that the monitored genes are relevant to the specific environ-mental stress, that selection is tested for by comparison with observed or expected genetic drift, and that shifts in allele frequencies coincide with the changes expected in response to the environmental change in question. Ideally, to further verify selection as the cause of evolutionary changes, researchers could conduct experiments such as reciprocal transplants or laboratory stress challenges to test whether locus-specific allele frequency changes occur as expected. For example, an allele for high-temperature tolerance increases in frequency during high-temperature challenges.

An example of Category III monitoring is the detection of an adaptive response to ongoing climate change in wild populations of *Drosophila melanogaster* and *D. subobscura* (Umina *et al.* 2005, Balanya *et al.* 2006). Each species displays a north–south cline in phenotypic traits influencing thermal adaptation and also in genetic markers involving well characterized genes and chromosome inversions associated with temperature tolerance. Laboratory selection experiments independently suggest the clines reflect temperature-related selection (Umina *et al.* 2005). This provides compelling evidence that the traits and gene polymorphisms are under selection caused by a specific environmental factor: temperature.

Guest Box 22   Genetic detection of illegal trade of whale meat results in closure of restaurants
C. Scott Baker

Molecular monitoring of commercial markets and genetic tracking of individual products are powerful tools for the control of legitimate and illegitimate trade in fisheries and wildlife (Baker 2008). By amplifying and sequencing short fragments of mitochondrial DNA, it is possible to identify the species origin of almost any wildlife or fisheries product, including meat, fur, feathers, horns, or bone, based on comparison with a reference database (Baker and Palumbi 1994). Multilocus genotyping, or 'DNA profiling', can then be used to identify the individual source of a product, or multiple products from the same individual (Baker *et al.* 2007), and to track their distribution from source to markets (Cipriano and Palumbi 1999, Wasser *et al.* 2007).

The ongoing hunting and sale of products from some species of whales provides an opportunity to apply these molecular tools as a 'mechanism of control' for international trade. All 13 species of whales regulated by the International Whaling Commission (IWC) are listed in Appendix I of the Convention on International Trade in Endangered Species (CITES). Species in Appendix I cannot be traded for commercial purposes unless a country has declared a 'reservation' to the listing. Japan, Iceland and Norway maintain reservations on the listing of some whales, allowing bilateral trade under certain circumstances, but these exceptions do not allow trade with other countries.

Given this prohibition on trade, it was a surprise when coincidental visits to Japanese-style sushi

(*Continued*)

restaurants in Santa Monica, California, and Seoul, South Korea, revealed menu items advertised as "whale" (Baker *et al.* 2010). To confirm that these items were, in fact, whale meat and to investigate their origin, products were purchased covertly and identified to species using mtDNA sequences and the web-based program *DNA surveillance* (Example 22.3). Two strips of raw meat purchased from the Santa Monica restaurant in October 2009 were identified as sei whale. This species is protected under the US Endangered Species Act but is hunted as part of Japan's controversial scientific whaling

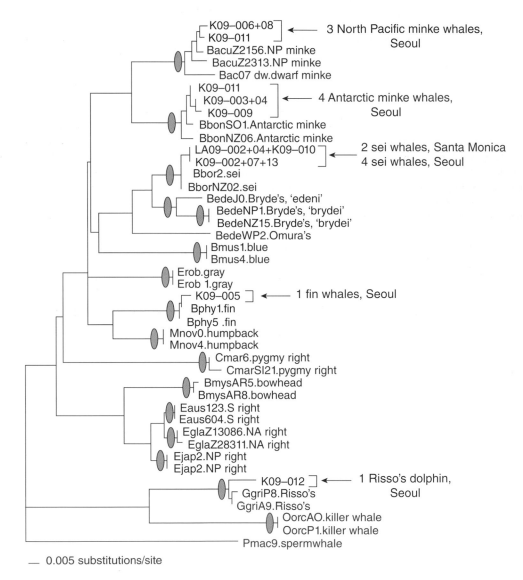

Figure 22.10 Phylogenetic identification of whale products from sushi restaurants in Santa Monica, California, and Seoul, South Korea, using sequences of the mtDNA control region and the web-based program *DNA surveillance*. Ovals show species-specific groupings supported by >90% of 1000 bootstrap simulations in a neighbor-joining reconstruction. From Baker *et al.* (2010).

program. A total of 13 products purchased from the Seoul restaurant during visits in June and September 2009 were found to include four species of whales and one species of dolphin: Antarctic minke whale, sei whale, North Pacific minke whale, fin whale, and Risso's dolphin (Figure 22.10). Like the sei whale, these other whale species are hunted in Japan's scientific whaling program and sold openly in markets and restaurants in Japan.

The circumstantial evidence pointed to Japan as the likely source of these whale products. However, other potential sources included illegal whaling and the sale of whales killed as 'bycatch' (Lukoschek *et al.* 2009). To establish a direct link in illegal trade between Korea and Japan, the microsatellite genotypes at seven loci of the fin whale purchased in Seoul were compared with microsatellite genotypes from fin whale products purchased during previous surveys of the Japanese market. This comparison was feasible because of the relatively small number of fin whales taken by the Japanese scientific whaling program (a total of 13 at the time) and the large size of the Japanese market survey (80 products representing 19 individual fin whales). The comparison revealed an exact match between the fin whale from the Seoul restaurant and a fin whale represented multiple products first purchased in Japan in September 2007. The probability of a match by chance at the 7 microsatellite loci was very low ($PI_{av} = 7 \times 10^{-11}$, see Section 22.2), providing strong evidence of illegal, international trade for the products from this individual whale. The results of the species identification were reported to relevant authorities in both countries. Subsequent investigation resulted in the closure of both restaurants.

Although species identification and genotype matching confirmed illegal trade between Japan and Korea, it was not possible to confirm with certainty that the products originated from the scientific whaling program. For this, it would be necessary to match the products to a 'DNA register', or electronic database, that includes the DNA profile of all individual whales taken in the hunt (Baker *et al.* 2010). Japan and Norway have both developed DNA registers for whales destined for commercial sale, including those killed in scientific whaling, commercial whaling and 'bycatch whaling' (e.g., IWC 1998). However, neither country has agreed to the sharing of the register with an independent third party, for the purposes of control of illegal trade. Without such an agreement, any mechanisms of control for future hunting and trade of whales will lack the critical requirement of transparency.

APPENDIX

Probability and statistics

> He has a lot of extremely abstruse, in fact almost esoteric mathematics. Mathematics, incidentally, of a kind which I certainly do not claim to understand. I am not a mathematician at all. My way of reading Sewall Wright's papers, which I think is perfectly defensible, is to examine the biological assumptions the man is making, and to read the conclusions he arrives at, and hope to goodness that what comes in between is correct.

Theodosius Dobzhansky (1962)

> Current research in population genetics employs advanced mathematical methods that are beyond the reach of most biology students.

James F. Crow (1986)

Appendix Contents

Conservation and the Genetics of Populations, Second Edition. Fred W. Allendorf, Gordon Luikart, and Sally N. Aitken.
© 2013 Fred W. Allendorf, Gordon Luikart and Sally N. Aitken. Published 2013 by Blackwell Publishing Ltd.

Box A1   Problems understanding sophisticated computational approaches

The senior author of this book was an Associate Editor in the initial days of the journal *Conservation Biology*. In the early 1990s, he handled a manuscript that applied some fairly sophisticated mathematical population genetics theory to a problem in conservation. He received the following review comments from a well-known population geneticist: "According to the Instructions to Reviewers for this journal, manuscripts should be understandable to conservation managers and gov-

ernment officials. It is not reasonable to expect either of these groups to understand stochastic theory of population genetics." This problem is much worse now than in the early 1990s because of the increasing sophistication of computational approaches as presented in this Appendix. It is becoming increasingly important to analyze empirical data with complex statistical approaches. However, it is also becoming increasingly difficult to evaluate the reliability of these analyses.

The gulf between mathematical population genetics and the understanding of most biologists has greatly increased over the last 25 years because of the introduction of a variety of new theoretical and computational approaches (Guest Box A). This can make it difficult for conservation geneticists to analyze datasets with recent computational approaches, and to publish their results in peer-reviewed conservation journals (Box A1).

The purpose of this appendix is to provide a basic understanding of the mathematical and statistical approaches used in this book. We have modeled this appendix after the appendices in Crow and Kimura (1970) and Crow (1986), with substantial use of Dytham (2011). We have not tried to provide mathematical rigor, but rather intend to make clear the general nature and limitations of the mathematical and statistical approaches used in this book. We aim to provide a conceptual understanding of different statistical approaches and show how to interpret results, rather than to teach details about how to actually conduct a certain statistical test or likelihood estimation.

The basic idea of statistical inference is simple: a sample is taken from a population, and you want to extrapolate from that sample to make general conclusions about the population from which the sample was taken. To do this, you must understand the relationship between a population **parameter** and a sample **statistic**. A sample statistic can be the mean, median, mode, or some other quantity that describes a characteristic of that particular sample. Statistics are computed from samples because all the items or individuals that constitute the entire population can seldom be collected and measured.

If you do measure the entire population and calculate a mean, though, this value is a parameter of the population, not a statistic. In most situations the parameter represents the 'truth', that is, the true population value that we want to estimate from the sample statistic. When a sample statistic is used to estimate the corresponding population parameter they are often written with a 'hat' (e.g., $\hat{F}_{ST} = 0.010$) to distinguish them from the population parameter ($F_{ST} = 0.015$). However, the 'hat' is not necessary if the context clearly describes a value (F_{ST}) as a statistic or an estimate. In this appendix, as in many texts, we do not use a hat to indicate a statistic; but rather we make it clear from context when we are discussing a statistic or estimate.

In this appendix, we first give a brief historic perspective and explain the major differences between the three approaches to statistical inference. Then, we present concepts of probability and basic statistics including hypothesis testing. Finally, we return to discuss in more detail the likelihood and Bayesian approaches, along with the coalescent and MCMC (Markov chain Monte Carlo) methods and their importance in conservation genetics.

A1 PARADIGMS

There are three main approaches or paradigms to statistical inference: frequentist, likelihood, and Bayesian approaches. Likelihood methods are sometimes classified within the frequentist approach (see Section A5).

The Bayesian philosophy and statistical approach to data analysis was developed in the 18th century by the Reverend Thomas Bayes. The classic frequentist approach was formalized later, during the early 1900s, by K. Pearson and R.A. Fisher (from England), as well as J. Neyman (from Poland); it quickly became dominant in science. Modern likelihood analysis was developed almost singlehandedly by R.A. Fisher between 1912 and 1922. A revival of the Bayesian approach has occurred during the last 20 years or so, thanks to advances in computer speed and simulation-based algorithms such as MCMC (see Section A6) that allow the analysis of complex probabilistic models containing multiple interdependent parameters such as genotypes, population allele frequencies, population size, migration rates, and variable mutation rates among loci.

The **frequentist** approach to statistical inference generally involves four steps: stating a hypothesis, collecting data, computing a summary statistic (e.g., $F_{ST} = 0.01$), and then inferring how frequently we would observe our statistic (0.01) by chance alone, if our null hypothesis (H_0) is true (e.g., H_0: $F_{ST} = 0.00$). If our statistic is so large (e.g., $F_{ST} = 0.10$) that we expect to observe it very *infrequently* by chance alone (e.g., only once per 100 independent experiments), we would reject the null hypothesis. The frequentist approach determines the expected long-term frequency of an observation or a summary statistic, if we were to repeat the experiment or observation many times. Frequentist approaches typically use the **moments** of the distribution (a summary statistic) and thus are called 'methods of moments'. The moments are the mean and variance, as well as skewness and kurtosis. These concepts are discussed below.

Likelihood approaches typically involve four steps: collecting data, developing a mathematical model with parameters (e.g., F_{ST}), plugging the raw data into the model (not a summary statistic), and computing the likelihood of the data for each of all possible parameter values, for example $F_{ST} = 0.00, 0.01, 0.02$, up to 1.00. This requires many computations or iterations. We then identify the parameter value that maximizes the likelihood of obtaining our actual data under the model. The main advantage of likelihood over frequentist (moments) approaches is that likelihood uses the raw data (e.g., allele counts at *each locus* separately) instead of a summary of it (e.g., F_{ST} *averaged across loci*: see Section A5). Thus, more information is used from the data (e.g., interlocus variation in F_{ST}), and therefore the estimates of parameters (and inference in general) often should be more accurate and precise (e.g., Williamson and Slatkin 1999).

The Bayesian approach is distinct in that (1) it can incorporate prior information (e.g., data from previous studies) to compute a probability estimate (i.e., a 'posterior probability'), and (2) it *directly* yields the probability that the hypothesis of interest is true (e.g., H_A: $F_{ST} > 0.00$). (Note: H_A in frequentist statistics is the alternative hypothesis, which is the hypothesis of interest.) Thus, Bayesian statistics more directly tests a hypothesis than frequentist methods that assess how frequently we expect to observe a summary statistic (e.g., $F_{ST} = 0.10$) if the null hypothesis is true. Recall that the null hypothesis is not the direct hypothesis of interest in the frequentist approach, but rather is the hypothesis we try to reject (see Section A4).

Bayesian methods combine a likelihood calculation with prior information to obtain a modified likelihood estimate called the posterior probability (see Section A6). Further, Bayesian approaches compute the probability (posterior probability) of the parameter given the data, whereas likelihood computes the probability of the data for a given parameter value in order to find the most likely value (maximum likelihood value). For example, when estimating N_e, the Bayesian approach outputs the (posterior) probability for different N_e values (e.g., for $N_e = 0$ to 500) given the data (see Section A6), whereas likelihood finds the parameter values that maximize the probability of the data (see Section A5).

Bayesian (and likelihood) approaches are *model-based*. It is important to define 'model-based' approaches because they "open doors for population geneticists and phylogeographers to the repertoire of likelihood-based analyses, including maximum likelihood estimation of model parameters and likelihood-ratio hypothesis tests" (Beaumont *et al.* 2010). Model-based approaches explicitly employ demographic models that include parameters such as population size and migration rates (e.g., Beaumont 1999). Model-based approaches have a goal of computing a likelihood function, that is, the probability of the data as a function of the parameters within a given model. An advantage is that a parameter estimated from two models (stable population versus declining population) can be compared to infer which model best fits your empirical data (e.g., to determine whether your population is declining).

We will return to model-based Bayesian and likelihood methods again, after considering the important concepts of probability, statistical distributions, and hypothesis testing. Such concepts will help explain the different methods of statistical inference and modeling.

A2 PROBABILITY

Probability was defined in 1812 by a French mathematician, Pierre Simon Laplace, as a number between 0 and 1 that measures the certainty of some event. A probability of 1.0 means the event is 100% certain to occur. An example is the probability of the A_1 allele being transmitted into a gamete by an A_1A_2 heterozygote; this probability is 0.5 (see Table 6.1). Probability concepts, including probability distributions such as posterior distributions, are important in statistics for using samples from a population to make inferences about the population, based on the sample characteristics (next section below).

Two important probability rules that we often use in genetics are the addition and the product rule. The **addition rule** is illustrated in Box 5.1. The addition rule (also known as the '**either/or**' probability rule, or the **sum rule**) is the probability of any of two or more mutually exclusive events occurring, which equals the **sum** of the separate probabilities of each event. In conservation genetics, we often study the probability of mutually exclusive events, such as being male or female, or of originating from population X, versus population Y or Z (see also Example 5.1). The sum of mutually exclusive events adds to one (1.0). For example, using Bayesian assignment tests (e.g., Sections 9.8 and 22.6), the estimated probability of an individual (multilocus genotype) originating from population X, versus Y or Z, might be 0.00, 0.01 or 0.99, respectively, all of which sum to a total probability of 1.0.

The **product rule** says that the probability of two independent events occurring simultaneously is equal to the product of the probabilities of the two events. The product rule (also called the '**both/and**' rule) is illustrated as follows: the probability a heterozygous parent will transmit **both** the A allele at a locus (Aa) **and** the B allele at another locus (Bb) is (0.50) (0.50) = 0.25, assuming independent loci. For an example application, consider a wildlife forensics case where the four-locus genotype from a bloodstain is $Aa/Bb/CC/dd$. What is the probability of randomly sampling a second individual with an identical genotype from this population, if the genotype frequencies are as follows: $Aa = 0.25$, $Bb = 0.50$, $CC = 0.10$, and $dd = 0.10$? Using the product rule (and assuming four independent loci), $P(AaBbCCdd) = (0.25)(0.50)(0.10)(0.10) = 0.00125$.

The probability of an event can be estimated from a large number of observations (e.g., flipping a coin hundreds of times and computing the long-term frequency of heads versus tails). This is called an **empirical probability** because it is obtained through empirical observations. This conceptual framework involving repeated events and their 'long-run' frequency is known as the 'frequentist approach' to probability and statistics.

The above concepts of probability are 'objective probabilities'. That is, there is no subjectivity, best-guess or intuition involved in computing the probability. For example, we know from Mendel's laws that each allele at a locus generally has an equal probability of being transmitted (e.g., a 50% chance). Furthermore, if we did not know the probability (0.50), we could empirically estimate the probability via repeated observations (e.g., repeated transmissions of alleles through genealogies or pedigrees).

A disadvantage of this frequentist approach is that it generally cannot give probability estimates for rare or infrequent events. Further, frequentist probability estimates cannot incorporate common sense or prior knowledge because the estimates are based only on a sample. For example, if you flip a coin ten times and obtain only three heads, your probability estimate will be 0.30. However, prior knowledge that unfair coins are rare would lead us to suspect that the estimate of 0.30 is too low (and should be close to 0.50). In this case, a more subjective approach to estimating probability could be used to incorporate all available information, and thereby obtain an estimate closer to 0.50.

'Subjective probability' is an important concept because it facilitates an alternative approach for describing probabilities. It can take into account previous knowledge, data, or 'best guesses'. For example, when computing the probability of extinction for a certain population, we can use input parameters in a population viability model (e.g., VORTEX, Chapter 14), which include 'best guesses' or intuitive predictions. When modeling population viability and the cost of inbreeding on population growth, we might use the average cost measured across mammals in captivity, if no data exist for our particular mammal species. The average cost of inbreeding is approximately a 30% reduction in juvenile survival of progeny by matings between full-sibs ($F = 0.25$) for mammals in captivity (see Section 13.5). This best guess of the cost is a somewhat 'subjective probability', if we do not measure the cost in the actual species and population being studied.

Another example of a 'subjective probability', and nonfrequentist approach is estimating the probability of a 1°C temperature increase due to global warming.

This type of computation is often conducted using a somewhat subjective model and parameter values (for example, including uncertainties inherent in the feedback processes that must be included in climate models).

Subjective probabilities are used in the Bayesian statistical approach (Section A6), which uses Bayes' theorem to incorporate prior information. The Bayesian approach uses a modifiable (or relativist) view of probability by using **prior probability** estimates (from prior knowledge) and then updating them with new data (from new observations) to give an 'improved' posterior probability estimate.

A2.1 Joint and conditional probabilities

We often must compute the probability of two events (E) occurring at the same time. This leads us to consider joint and conditional probabilities. For example, in order for inbreeding to increase the risk of population extinction, it is necessary that inbreeding reduces individual fitness ($E1$ = inbreeding depression) and that the reduced individual fitness also leads to reduced population performance ($E2$ = reduced population growth rate). Here, $P(E1\ and\ E2)$ is the joint probability of $E1$ and $E2$. Joint probabilities are important in the modeling of complex processes (e.g., Bayesian inference of processes) that have multiple sources of variation; for example, allele frequency changes are influenced by multiple sources of variation such as drift, selection, and migration (Beaumont and Rannala 2004).

A conditional probability is the probability of an event given that another event has happened. Conditional probabilities are used whenever considering events that are not independent. For example, if the effect of inbreeding on fitness increases with environmental stress, then we could compute the probability of inbreeding depression conditional upon a certain stress such as temperature change (resulting from global warming or an unusually hot summer). A conditional probability, the probability of $E2$ given $E1$ (i.e., conditioned on $E2$), is defined as follows:

$$P(E2\,|\,E1) = \frac{P(E1\,\&\,E2)}{P(E1)} \qquad (A1.1)$$

Note that conditioning on an independent event does not change the probability of the event: $P(E1\,|\,E2) = P(E1)$. Note that the '&' symbol means that E1 *and* E2 both occur.

Bayes' theorem is used to obtain a posterior probability conditioned on the data available from a sample. The posterior probability $P(E1\,|\,E2)$ uses the prior probability $P(E1)$ conditioned on the event $E2$ (the sample of data). Thus, the Bayesian approach computes revised ('updated') estimates of the probability of event $E1$ by conditioning on new data ($E2$), as data become available. A prior probability can be 'rectangular' and thus uninformative. For example, we could consider that microsatellite mutation rates range from 10^{-2} to 10^{-6}, with all values having an equal probability (a flat probability distribution). Alternatively, we could use a bell-shaped prior probability distribution with a higher probability for mutation rates between 10^{-3} and 10^{-4}, which is consistent with published observations suggesting that mutation rates are often near 10^{-3} or 10^{-4} (Section 12.1.2).

A2.2 Odds ratios and LOD scores

Another probability concept important in conservation genetics is that of 'odds'. The probability of an event can be expressed as the odds of an event. The odds ratio for an event E is computed as the probability that E will happen divided by the probability that E will not happen. For example, the probability 0.01 has the odds of 1 to 99 (or 1/99). Odds ratios (also called likelihood odds ratios) are used, for example, in paternity analysis to decide whether one candidate father is more likely than another candidate to be the true father (Marshall *et al.* 1998).

Odds ratios are also used in assignment tests to decide whether population X is more likely than population Z to be the origin of an individual (Banks and Eichert 2000). For example, we can compute the probability (expected genotype frequency, e.g., *2pq*, for a heterozygote) of a multilocus genotype originating (occurring) in Pop X versus Pop Z. If the logarithm of the ratio of the probabilities is very large (e.g., $\log_{10}\{P(\text{Pop }X)\,/\,P(\text{Pop }Z)\}$), we can conclude that Pop X is the origin of the individual. For example, we might decide to assign individuals to Pop X if the log of the odds (LOD) ratio is at least 2.0. In this case, with LOD = 2.0, we expect only 1/100 erroneous assignments where an individual assigned to Pop X actually originates from Pop Z. If the LOD score is 3.0, we expect only 1 in 1000 erroneous assignments (e.g., Banks and Eichert 2000).

A3 STATISTICAL MEASURES AND DISTRIBUTIONS

A **statistic** is a single measure of an attribute of a sample (e.g., its arithmetic mean value). A statistic is computed from a sample because the entire population usually cannot be collected or measured, as mentioned above. Five categories or kinds of statistical tests can be delineated based on the questions they address: descriptive statistics, tests for differences, tests for relationship, multivariate exploratory methods, and estimators of population parameters (Dytham 2011).

A3.1 Types of statistical descriptors or tests

Descriptive statistics are computed to describe and summarize sample data during the initial stages of data analysis, without fitting the data to a probability distribution or model (e.g., the normal distribution or model). Since no probability models are involved, descriptive statistics are not used to test hypotheses or to make testable predictions about the whole population. Nevertheless, computing descriptive statistics is an important part of data analysis that can reveal interesting features in the sample data. Examples of descriptive statistics are the mean and variance, which are described below in Section A3.2. Descriptive statistics are often called summary statistics (e.g., H_e, F_{IS}), as in approximately Bayesian computations (e.g., Section A7) (Tallmon $et\ al.$ 2008).

Tests for differences address questions such as 'do populations A and B have different heterozygosity?' Here, the null hypothesis is that A and B have the same heterozygosity. Tests for differences can also be used to compare distributions. For example, we might ask if the shape of the distribution of allele frequencies is different in populations A and B, or if the proportion of low-frequency alleles is less in population A than is expected in a large, stable (non-bottlenecked) population (Luikart $et\ al.$ 1998). There are many statistical tests for differences, including parametric and non-parametric tests (e.g., t-tests and signed-ranks tests) described below.

Tests for relationship ask questions like 'is fitness related to heterozygosity?' A null hypothesis might be: 'heterozygosity is not associated with juvenile survival'. Two classes of tests for relationships are correlation and regression. Correlation assesses the degree of association without implying a cause and effect. Regression fits a relationship (e.g., linear or curvilinear) between two variables so that one can be predicted from the other, implying a cause and effect relationship. The effect of inbreeding on fitness traits can be predicted via regression (see Section 13.5 and Figure 13.13). We could imagine a scenario where inbreeding is associated with reduced fitness, but inbreeding is not the direct cause. For example, if individuals from population A are more inbred, but also have poorer nutrition than individuals from population B, a correlation (between populations) for individual growth rate versus inbreeding could be caused by the environment, not genetics. A factor complicating the assessment of relationships is interactions (e.g., genetic by environment interactions). There are many ways to test for correlations, to compute regressions, and to account for interactions. One use of regressions is to compare many regression models to test for effects of environmental factors on genetic structure (as implemented in the computer program GESTE by Gaggiotti $et\ al.$ 2009).

Multivariate exploratory techniques ask questions such as 'what patterns exist in the data?', or 'can we assign individuals to groups based on multilocus genotypes?', or 'which factor (e.g., locus) is most useful (i.e., informative) when assigning individuals to groups?' Multivariate exploratory techniques can help identify hypotheses to test. In large datasets with multiple factors (e.g., many loci, morphological or environmental measurements), we might not initially test a specific hypothesis because so many potential hypotheses exist. Exploratory techniques are more appropriate for generating hypotheses than for formally testing them (i.e., they do not yield P-values, likelihoods or probability values). A wide range of statistical approaches exists, such as principal component analysis (PCA), frequency correspondence analysis (FCA), multidimensional scaling (MDS), or cluster analysis (see Section 9.7 and Section 16.4.2). Informative uses of PCA and potential misinterpretations of PCA outputs are described in Novembre and Stephens (2008).

Statistical estimators infer a population parameter by using data that are related to that parameter. For example, we could infer the effective population size (N_e) from data on the temporal change in allele frequencies between two generations. Change in allele frequencies is influenced by N_e, but might also be influenced by **sampling error** (see below), population structure, demographic status (expanding/declining), and selection or mutation rates. There are different approaches to statistical estimation (using the method of moments, maximum likelihood, Bayesian, and approximate Bayesian methods, see below).

Statistical tests can be divided into two classes: parametric and nonparametric. **Parametric** statistics assume that the data follow a known distribution – usually the normal distribution. Parametric distributions can be defined completely using very few parameters, such as the mean and variance, in a function or mathematical expression. Parametric statistical tests are generally more powerful than nonparametric tests, and thus are often preferred (see below). An example of a parametric test is the *t*-test, which assumes a normal distribution; it can be used to compare mean heterozygosity from two population samples, if the distribution of heterozygosity among loci is similar to the normal distribution (Archie 1985).

Nonparametric statistics require few or no assumptions about the distribution of the data or test statistic. Therefore, nonparametric statistics are called 'distribution-free' tests. Some of them are also called 'ranking tests' because they often involve ranking observations to generate an empirical cumulative distribution. These tests are generally less powerful, but more appropriate than parametric tests if the data might not follow a parametric distribution. An important example is Wilcoxon's signed-ranks test (a non-parametric version of the *t*-test), which is often used to test for lower mean heterozygosity in one population compared with another population (Luikart *et al.* 1999), or to test for a population bottleneck (Luikart and Cornuet 1998).

A3.2 Measures of location and dispersion

In statistics, the 'population' is defined as the totality of the individuals with some characteristic we are studying. The sample is the subset of our observations. We compute sample statistics to infer the population parametric value of a parameter (e.g., the mean). For any trait X, the general formulae for the population mean and sample mean are as follows:

$$\mu = \frac{\sum x_i}{N} \tag{A1.2}$$

$$\bar{x} = \frac{\sum x_i}{n} \tag{A1.3}$$

where i is the individual number, the bar over the x indicates the mean, and N and n are the population size and sample size, respectively.

The mean is a statistical measure of 'central value' or the central location (of a distribution). The arithmetic mean is given by expressions A1.2 and A1.3. Another kind of mean important in population genetics is the harmonic mean (see equation 7.8), which gives more weight to observations with small values.

The harmonic mean is used for computing the multigenerational effective population size from successive N_e estimates from each of several individual generations. An interesting controversy in conservation genetics results from, in part, confusing the arithmetic and harmonic mean when computing the ratio of N_e to N_C. The N_e, averaged across generations, is always computed as a harmonic mean, whereas N_C averaged across generations is often computed as an arithmetic mean (Frankham 1995). The harmonic mean is strongly influenced by low values, causing the (harmonic) mean N_e estimates to be lower than the (arithmetic) mean N_C. Thus, the estimates of N_e/N_C ratios (averaged across generations) can be biased low due to a statistical artifact of using the harmonic mean of N_e but the arithmetic mean of N_C (see Chapters 7, 15, and Kalinowski and Waples 2002).

Other familiar measures of central location are the median and mode. An advantage of the median is that it is less influenced than the mean by the skewness of the distribution of the statistic, so the median is resistant to extremely high or low outlier values. Thus, the median is said to be a relatively robust or resistant measure of central location.

A statistical measure of variability (or 'dispersion') of points around the mean is the variance. If all points have the same value, there is no dispersion and the variance is zero. If points have only very high and very low values, the variance would be high. The variance is the average of the squared deviations from the mean, and is computed as follows: the mean is subtracted from each observation point, this difference is squared, and finally the average of the squares is computed. The population variance (σ_x^2) and sample variance (s_x^2) are computed as follows.

$$\sigma_x^2 = \frac{\sum (x - \mu)^2}{N} \tag{A1.4}$$

$$s_x^2 = \frac{\sum (x - \bar{x})^2}{n - 1} \tag{A1.5}$$

where n (and N) is the number of sample (and population) observations, as above.

The standard deviation is another important measure of dispersion. It is computed as the square root of the variance ($s_x = \sqrt{s_x^2}$). We take the square root of the variance to avoid having to think in terms of squared measures, which are less interpretable (for example, it is easier to interpret the 'height' of individuals than the 'height squared'). Furthermore, recall that one standard deviation under the normal (bell-shaped) distribution encompasses 68% of the central area, while two standard distributions encompasses 95%, and finally three standard deviations contain 99% (99% fall between $\mu \pm 3\sigma$; see Section A3.3 below). Probability distributions such as the normal distribution, and their use for describing dispersion, are discussed more in the next section.

The **standard error** is a measure of the dispersion of a statistic (e.g., the sample mean, $\bar{x}$) computed from a sample. The standard error of the mean (SEM) is the standard deviation of a distribution of means for *repeated samples* from a population. The SEM should not be confused with the standard deviation, which describes the probability distribution of the underlying *population parameter* (μ) for the dataset. For example, the standard error describes the distribution of the SEM of heterozygosity in a dataset, whereas the standard deviation describes the probability distribution of the **population parametric** heterozygosity in the dataset (see Section A3.3 and Example A1 below). Probability distributions are discussed in the next section. For more information on standard errors see Dytham (2011, p. 56). Unfortunately, in publications, standard error and standard deviation are often confused or not well differentiated.

A3.3 Probability distributions

Probability distributions are crucial to understand because statistical tests and estimators require the use of a probability distribution. Different types of variables (height, temperature, locus-specific F_{ST}) have different probability distributions (see Figure A1).

Probability distributions are generally illustrated graphically as a curve or frequency histogram. The total area under a probability curve is 1.0. The probability of a rare or unusual observation is represented as a small area (e.g., 0.05) in the tail(s) of the distribution. We can obtain an empirical estimate of a probability distribution by plotting the relative frequency (histogram) of occurrence of each observation, for example the height of each individual, in a sample.

An example probability distribution is that from a common likelihood-based estimator of the parameter N_e (Figure A2b). A widely used Bayesian probability distribution in population genetics is that for the discrete variable k (number of demes), which is computed by program STRUCTURE (Pritchard *et al.* 2000). This distribution gives the probability of each of the possible values of k ($k = 1, 2, 3 \ldots$) and thus helps us to determine how many populations are represented by our sample of individuals from across a landscape, for example.

A3.3.1 Sampling distributions

Importantly, in frequentist statistics we use a sampling (probability) distribution of our sample statistic to obtain an estimate of the population parameter. The **sampling distribution** is defined as the distribution of an infinite number of samples of the same size as the sample in your study (Figure A2a). The point here is to realize that our sample is just one of a theoretically infinite number of samples we could have taken. Keeping the sampling distribution in mind, we understand that the statistic we computed from our sample is likely near the center of the sampling distribution, as most of the samples would likely have the statistic near the center.

Readers should not confuse probability distributions of sample statistics (e.g., mean heterozygosity of a sample) with probability distributions of the underlying population parameter (e.g., heterozygosity computed from the entire population) (see 'standard error' and 'SEM' above). Two important characteristics of sampling distributions are (1) they have lower variance than parameter distributions, simply because each sample mean includes multiple observations (and thus the probability distribution for a mean value is narrower than for individual observations), and (2) they approach the normal distribution when sample sizes are large, which is a surprising principle of the central limit theorem as mentioned above. The low variance and central limit theorem explain why we often see the normal distribution used in statistical tests and for computing confidence intervals.

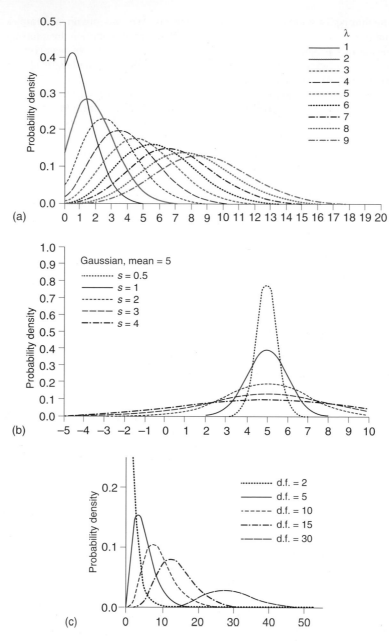

Figure A1 Probability distributions important in conservation genetics. (a) The Poisson distribution with a mean (λ) from 1 to 9 (nine curves). (b) Normal (Gaussian) distributions with variance (s) from 0.5 to 4, and mean 5. (c) The chi-square distribution (d.f. is the degrees of freedom). Figures modified from P. Bourke (personal communication).

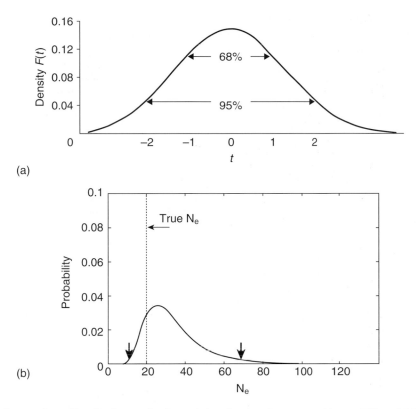

Figure A2   (a) A normal sampling distribution for the statistic t showing the upper and lower 68% and 95% confidence limits $[t\alpha/2, t1-\alpha/2]$, where the alpha (critical/threshold P value) equals 0.32 and 0.05, respectively (see text). (b) Probability distribution (likelihood curve) output from a likelihood-based estimation of N_e, and 95% support limits identified by thick vertical arrows, placing 2.5% of the area in each tail of the distribution. Modified from Berthier *et al.* (2002).

A3.3.2 Binomial distribution

An important probability distribution in genetics is the binomial distribution. The binomial is one of several theoretical probability distributions used for modeling (approximating) the distribution of observed data that occur in discrete classes, such as genotypes at a locus, as opposed to a continuous distribution of observations, such as height. The binomial is useful for modeling the proportion of binary events (male versus female births; transmission of allele A versus a; or survival versus death) that occur in a population sample of size n. Note that when more than two events are possible, we can use the multinomial distribution – a simple extension of the binomial.

The binomial distribution contains information on the number of times, x, an event with probability π occurs in a fixed number of observations n. The binomial distribution is defined as:

$$P(x = m) = \frac{n!}{(n-m)!m!}\pi^m(1-\pi)^{n-m} \tag{A1.6}$$

The factorials in the fraction give the number of ways m that positive outcomes (transmission of A) can occur out of n events (offspring). The binomial has a variance of:

$$V(x) = n\pi(1-\pi) \tag{A1.7}$$

For example, if the probability of transmitting the A allele is $\pi = 0.50$, then out of 100 transmissions (offspring), we expect a mean of $100 \times 0.50 = 50$ transmissions of the A allele, with a variance of

$100 \times 0.50 \times 0.50 = 25$ (standard deviation = 5.0). When the number of observations (*n*) becomes large, the binomial approaches the normal distribution.

A3.3.3 Poisson distribution

The Poisson distribution is another discrete distribution that is widely used in conservation genetics and ecology (Figure A1a). The Poisson distribution assumes an event is rare (relative to the maximum number of possible events), and that events are independent. Thus, the Poisson distribution is used to model rare and independent events that occur in a spatial or temporal sample. For example, in genetics, the Poisson distribution is used to model the probability of mutations through time (e.g., under the coalescent, see Section A10), because mutations are rare events that arise randomly (among individuals or lineages). The Poisson distribution is also used to model variance in family size (reproductive success), as in Section 7.3 (Figure 7.4). Ecologists use the Poisson to test whether the distribution of organisms over space is uniform versus random. For example, if the observed variance in distance between individuals is less than the mean distance, then the spacing is more uniform than random, because the mean equals the variance in a Poisson distribution.

An important property of the Poisson distribution is that the mean equals the variance. An example of this property comes from modeling a stable population. When using the Poisson to model a stable-sized (stationary) population, the mean family size (number of offspring per mating pair) equals two, as does the variance (Figure A1a, second curve from the left). This is called the **Wright-Fisher model**, which is widely used in modeling or simulating data for conservation genetics applications (e.g., Waples and Faulkner 2009). In such an ideal model, the effective population size (N_e; Chapter 7, Section 7.1) equals the census size (N_C). Although the Poisson is useful here, we know that in natural populations N_e is generally less than N_C because, for example, the variance in family size is often high (>2.0) (Figure 7.5). Thus, the Poisson is not always the most appropriate distribution for modeling N_e or variance in reproductive success in natural populations (Waples 1989).

Under the Poisson distribution, the probability of any number *x* of occurrences is:

$$P(x) = \frac{e^{-\mu}\mu^x}{x!}, \tag{A1.8}$$

where the mean number of occurrences μ equals N (the population size).

A3.3.4 Normal distribution

The normal (Gaussian) distribution is the most widely used continuous distribution: it is the famous symmetric 'bell-shaped' curve (Gauss 1809, Figure A1b). The binomial distribution approaches the normal distribution as sample sizes increase. Thus, for example, the shape of the distribution of the mean heterozygosity (*H*) approaches a smooth bell shape when sample size of loci approaches 30–50 loci.

The normal distribution is useful for modeling many observed variables (e.g., heterozygosity) because of the central limit theorem, which states that the distribution of the sample mean will approach the normal distribution as the sample size of observations increases (even if the observed variable itself is not normally distributed!). A normally distributed random variable is described by the following function:

$$P(x) = \frac{1}{\sigma 2\pi^2} e^{\frac{-(x-\mu)^2}{2\sigma^2}} \tag{A1.9}$$

For any continuously distributed variable, the probability distribution is defined as the probability of a random variable being less than or equal to a particular value $P(X \leq x) = P(x)$. Here, $P(x)$ is called the probability distribution function. The derivative of the probability distribution is called the probability density function (pdf). The area under any segment of a pdf curve is the probability of X being in a certain interval. Note that a pdf is the output of Bayesian analyses (posterior distribution) and also of maximum likelihood estimation (likelihood curve) where we estimate the probability of some parameter (e.g., N_e, F_{IS}, or mutation rate; see below).

The population probability distribution can be estimated empirically by computing the cumulative frequencies of observations in a sample, for example by plotting a histogram of cumulative frequencies of observations having values less than *x*. The accuracy of the empirical distribution (as an estimate of the population probability distribution) increases with large sample sizes.

A3.3.5 Chi-square distribution

The chi-square distribution is another continuous distribution widely used in statistics and in conservation genetics. It is asymmetric, unlike the normal distribution (Figure A1), and ranges from zero to infinity. The chi-square distribution is used to model and conduct tests comparing variance measures; thus, the chi-square probability distribution is used when studying, for example, the spatial variance in allele frequencies (F_{ST}) or the temporal variance in allele frequencies (F_k; Antao *et al.* 2011). The chi-square can be used to compute confidence intervals around F_{ST} or around N_e estimates that are computed from temporal variance in allele frequencies. Chi-square tests using the chi-square probability distribution are discussed in Section 5.3 and Examples 5.1 and 5.2.

We remind readers that chi-square tests use numbers (not proportions), and that if the 'expected number' in any class (e.g., genotype class) is less than approximately one (<1.0), we should consider using an exact multinomial test based on the multinomial probability distribution. **Exact tests** are explained in Example 5.3. Exact tests are performed by determining the exact probabilities of all possible sample outcomes, and then summing the probabilities of all equal and less proba-ble sample outcomes, to obtain the exact probability of the observed outcome.

Interval estimates are usually more useful than point estimates. In fact, without an interval estimate, a point estimate (e.g., mean H_e, F_{ST}, or N_e) is generally of little value. Two kinds of interval estimates often used in conservation genetics are confidence intervals (for frequentist approaches) and support limits or credible interval (for likelihood-based and Bayesian approaches).

Confidence intervals give the range of values within which the true population parameter (e.g., population mean) is likely to occur, with some chosen probability (usually 95% or 99%). Thus, confidence intervals (CIs) are measures of spread. Publications often report 95% CIs, which should span all but 5% of outcomes from repeated, independent sampling events. Note that error bars (e.g., on histograms) often report ±1 standard errors (±1 SE, or standard deviations of the mean), which represent 68% confidence intervals for normal/Gaussian distributed statistics (Example A1). Note also that 95% CIs are nearly twice as wide as 68% CIs, that is, a 95% CI represents approximately ±2 SE (Figure A3).

Example A1   Comparison of different types of error bars

Consider a hypothetical study where you discover a brain protein (LDE, language destroying enzyme) that causes people to utter strange words (Streiner 1996). You think LDE is in higher concentrations in administrators than in other people. You sample 25 administrators and 25 other people (as a control group) and compute the mean and standard deviation (Table A1). You present the data in a bar graph to make it more visually interpretable (Figure A3).

But how do you compute the error bars to extend above and below each histogram bar? In all studies it is important to report the standard deviation because this shows the dispersion of the actual raw data points. However, the reader generally also wants to know the sample to sample variation. For example, if we repeat this study 100 times, how much variation between the means of each study would we expect? Stated another way, how much confidence do we have in the estimation of the population mean from our sample mean? For this we must compute a standard error (i.e., a standard deviation of the mean).

Should we report one or two standard errors? We are generally interested in a range of values in which we are 95% certain. Thus we could report 2 SEs, which should contain *approximately* 95% of the study means (Figure A3). Furthermore, 2 SEs are used to compute exact 95% confidence intervals (assuming a normal distribution) when testing for statistically significant differences between populations means.

For example, using our table of the normal distribution, we find that 95% of the area falls between −1.96 and +1.96 SEs (SDs of the means, for this example). We compute 95% CIs as follows:

Table A1   Levels of LDE (language destroying enzyme) in the cerebrospinal fluid of administrators and controls (Streiner 1996).

Group	Number	Mean	SD
Administrators	25	25.83	5.72
Controls	25	17.25	4.36

(Continued)

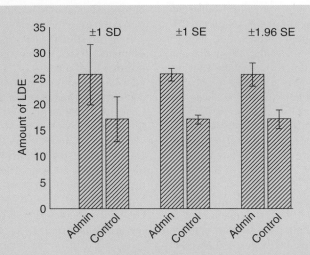

Figure A3 Computing error bars using standard deviations, standard errors (i.e., standard deviations of the mean), and 95% confidence intervals (assuming a normal distribution). Note that 1.96 SEs represent 95% confidence intervals. Because the error bars do not overlap for the ±1.96 SE, we can conclude with 95% confidence that the administrators and controls are significantly different. From Streiner (1996).

$$95\% \text{ CI} = \bar{x} \pm (1.96 \times \text{SE})$$

where $\bar{x}$ is the mean.

Of course, ±1.96 SDs of the mean nearly equals ±2 SDs of the mean. Confidence intervals show the range in which statistically significant differences exist between means. Showing 95% confidence intervals (or ±2 SEs) supports statistical testing (Section A4) and allows for an 'eyeball test' of significance. Note

that this eyeball approach does not work accurately when more than two groups are compared because of issues of multiple tests.

How do we interpret the error bar results? If the top of the lower bar (controls) and the bottom of the upper bar (administrators) do not overlap, then the difference between the groups is significant at the 5% level (see statistical testing below). We could then conclude that administrators have higher concentrations of LDE.

To compute a 95% CI, we choose an **alpha** (α) value of 0.05. Alpha is the critical threshold P-value used for rejecting the null hypothesis (e.g., if $P < 0.05$). For a sample statistic $t(x)$, we can compute a [$(1 - \alpha)$ 100%] confidence interval as CI[$t\alpha/2$, $t1-\alpha/2$], with lower and upper confidence limits of $t\alpha$ and $t1-\alpha/2$, respectively (where t_n is the nth quantile of the sampling distribution of the population parameter, T).

Support limits are used in likelihood and Bayesian approaches instead of CIs. Support limits can be computed, like confidence limits, such that the estimated sampling distribution (likelihood or posterior distribution) has cutoff points placing 2.5% of the probability density area in each tail. For an illustration, see Figure A2b. Support limits are generally reported with, and

plotted on, a probability curve (likelihood or posterior distribution), which allows visualization of the probability of different outcomes just by 'eyeballing' the curve (Figure A2b). This makes interpretation of probability estimates (from probability curves) more straightforward than frequentist CIs.

A4 FREQUENTIST HYPOTHESIS TESTING, STATISTICAL ERRORS, AND POWER

Hypothesis testing is widely used across scientific disciplines. It requires a formal statement called the null hypothesis (H_0), followed by a statistical test of the null hypothesis, which assesses the probability of null being true, by computing a P-value or a likelihood (probability) distribution. The null hypothesis is a negative statement that mirrors the alternative hypothesis. For example, a null hypothesis might be: 'population X is stable or growing'. The alternative hypothesis is: 'population X is declining'. Many researchers and funding agencies suggest that a well-written hypothesis should include a 'because phrase', such as 'population X is declining because of inbreeding depression', which helps researchers to develop predictions and focus on the testing of alternative hypotheses.

Errors in rejecting the null hypothesis can arise because we usually have only a small sample from an entire population, and because statistical tests only relate to the *probability* that the null hypothesis is false. Two kinds of errors, Type I and Type II, are possible when conducting a statistical test. A Type I error is rejecting the null when it is true. A Type II error is failing to reject the null when it is false (Table A2). The choice of the level α thus inevitably involves a compromise between significance and power, and consequently between the Type I error and the Type II error.

The lower the P-value, the more confident you are that the null hypothesis (H_0) is false. For example, if $P < 0.001$, you expect that in less than 1 of 1000 independent experiments you would observe an outcome (a test statistic) as unusual as the one observed. A P-value of 0.05 is often used in hypothesis testing as the threshold (α value) for rejecting the null hypothesis. When $P = 0.05$, we have 5 chances in 100 of rejecting the null when it is true (Type I error). The use of 0.05 is arbitrary and other α values can be used (0.10 or 0.01) depending on the importance of avoiding a Type I error.

A low Type I error rate (choosing a low critical α value) will increase the Type II error rate. Therefore choosing the appropriate α depends on the relative importance of avoiding a Type I versus Type II error. For example, consider the following null hypothesis: 'Population X is stable or growing'. An important question is, 'Would it be more risky to erroneously reject the null (wrongly accept or conclude that "population X is declining") or to erroneously fail to reject the null (wrongly conclude that "population X is stable or growing")?' If we wrongly conclude the population is stable (a Type II error), and it is actually declining, it could lead to extinction of the population or species.

In conservation biology it often is more risky to make a Type II error than to make a Type I error. Type I errors can be the more risky kind of error in other sciences, such as human medicine, where we must not reject the null when it is true. For example, we would not want to reject the following null hypothesis: 'medication X has no side-effects', unless we are highly certain (e.g., $P < 0.001$) the null hypothesis is false and there are no side-effects.

A4.1 One- versus two-tailed tests

Statistical tests can either be one- or two-tailed. In a *one*-tailed test, the alternative hypothesis (H_A) is a deviation in only *one* direction (Figure A4), for example, H_A: population X is declining. However in a *two*-tailed test, the alternative hypothesis would be H_A: population X is declining or growing (i.e., changing in size) (Figure A4a). Thus a two-tailed test checks for deviations in either of two directions. A one-tailed test is appropriate when (1) biological evidence suggests a deviation in one direction (e.g., a population has declined, so we conduct a one-tailed test for reduced allelic diversity), or (2) we only care about a deviation in one direction. For example, we might use a one-tailed test to detect reduced heterozygosity in a population that recently became isolated, if we care only about detecting a reduction of heterozygosity.

One-tailed tests generally have more power than two-tailed tests. Thus it is important to understand the difference between one- and two-tailed tests, and to use one-tailed tests when possible and appropriate. A one-tailed test (e.g., *t*-test) is more powerful, because more of the 'rejection region' (all 5%, not just 2.5%) is located in the one tail that we are interested in, making it easier to reject the null hypothesis (Figure A4 part (a) versus part (b)).

A4.2 Statistical power

An important consideration when choosing a statistical approach or test is its statistical power (see also

Table A2 Type I and Type II errors that can result when testing a null hypothesis (H_0).

	Accept H_0	Reject H_0
H_0 True	Correct	**Type I error**
H_0 False	**Type II error**	Correct

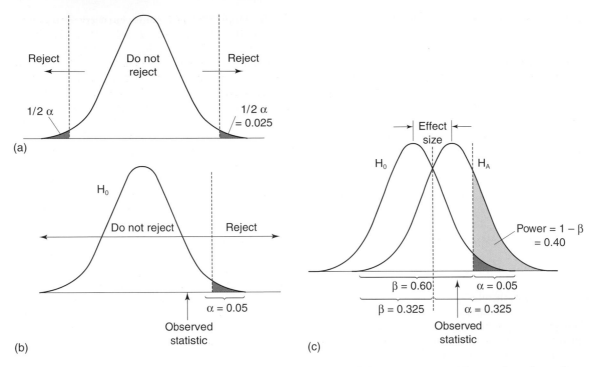

Figure A4   Two-tailed test (a) in contrast to a one-tailed test (b). A two-tailed test is appropriate when we do not know the direction of deviation expected (e.g., we do not expect H_e to be lower (or higher) in a certain population, for example. Panel (b) shows the conventional P < 0.05 (alpha = α = 0.05) as a threshold to reject the null hypothesis, whereas panel (c) shows a more balanced approach of choosing an alpha value leading to similar risk of Type I versus Type II errors. Note that the risk of a Type II error (beta, β) is 0.60 when alpha is 0.05. However, if we choose an alpha of 0.325, beta will also be 0.325. Further, note that the 'observed statistic' does not fall in the tail (right side of vertical dotted line for α = 0.5) in panel (b) (P > 0.05), so we would not reject the null hypothesis. However in panel (c), we would reject the null because the statistic is smaller than the threshold of rejection, α (P < 0.325). Modified from Taylor and Dizon (1999).

Section A9). Power is the probability of detecting an effect when the effect or phenomenon occurs. For example, the power of a statistical test for detecting a population decline (given that a decline occurs) is obviously important in conservation genetics. Power is also defined as the probability of rejecting the null hypothesis (H_0) when it is false.

Power is related to the Type II error rate as follows: Power = $1 - \beta$, where β is the Type II error rate (i.e., the false negative rate). Thus, the power of a test depends on the choice of β, such that choosing a small β leads to more power. Other factors that influence power, besides β, are the effect size (strength of the effect, e.g., severity of population decline) and the sample size (e.g., number of individuals and loci sampled).

Power is also influenced by the chosen statistical test itself. For example, we mentioned that parametric tests (t-test for loss of heterozygosity) are expected to be more powerful than nonparametric tests. A relevant example for conservation genetics is that the most powerful test for detecting a decline in heterozygosity is not the standard t-test, but rather a paired t-test. The paired test is more powerful because it treats each locus individually and thereby reduces the influence of interlocus variation that often is high. For example, different loci in a sample might have heterozygosity (H_e) ranging from 0.2 to 0.8, but the between-sample difference in mean H_e that we are testing might be only 0.6 versus 0.5 (e.g., in a large versus a small population). Interestingly, Wilcoxon's nonparametric test often is not less powerful than the parametric (paired) *t*-test when monitoring for loss of heterozygosity using two temporally spaced samples (Luikart *et al.* 1998).

A statistical power of 0.80 is often considered by statisticians as 'reasonably high' power for detecting the event of interest (e.g., population decline, migration, fragmentation, etc.), thus making it worth conducting the study of interest. A problem in science, and particularly in conservation biology, is the failure of researchers to compute the power of statistical tests. Fortunately, power analyses are becoming easier to conduct, thanks to the increasing availability of computer simulation programs that allow simulation of different population scenarios (e.g., population declines) and marker numbers and types (dominant, codominant) – see the simulation programs listed on this book's website.

A4.3 Problems with *P*-values

A problem with *P*-values and hypothesis testing via the frequentist approach is that *P*-values can be difficult to interpret (compared to Bayesian posterior probabilities, see below). A *P*-value should be interpreted as the chance, assuming the null is true, that you will get a similar or more extreme result if you repeat an experiment thousands of times. A value of P < 0.05 is sometimes misinterpreted to mean that there is 95% probability that the alternative hypothesis is true. This is different from the actual definition, given in the previous sentence.

P-values can overstate the strength of evidence, compared with Bayesian approaches. For example, Malakoff (1999) stated that a statistically significant trend in acid rain pollutants detected in some lakes by frequentist analyses disappeared upon a Bayesian reexamination. As explained in Gaggiotti (2010), the reason for this overstatement is that hypothesis testing based on *P*-values is appropriate only if the effects of all the various factors that influence the final result are minimized by randomization. This is impossible in typical ecological studies that rely on observational data. Thus, the rejection of the null hypothesis can be highly significant but does not necessarily mean that the alternative hypothesis can be considered as plausible. Bayesian approaches, on the other hand, can incorporate the uncertainty due to various factors, which can strongly reduce the support for the rejection of the null hypothesis.

Another problem of *P*-values often arises when the *P*-value is low, but not 'significant'. If $P = 0.06$, researchers might not 'reject the null' and subsequently conclude there is no effect (e.g., no evidence the population is declining). However, as mentioned above, the choice of $\alpha = 0.05$ is generally arbitrary, and in fact, $P = 0.06$ is suggestive of an effect (especially if the power of the test is low). Recall that if the effect size is small, we are unlikely to obtain a significant *P*-value (e.g., $P < 0.05$), unless sample sizes are very large and power is high (see Section A4.2 above).

Another problem with *P*-values is that 'negative results' ($P > 0.05$) are sometimes difficult to publish, and can lead to a bias in the scientific literature, and this leads to an underrepresentation of studies that find no 'significant' effect. For example, there might be more studies published that find a correlation between heterozygosity and fitness than do not, thereby leading to a biased proportion of (published) studies finding a correlation. This potential lack of publication of 'negative results' has been called the 'file drawer effect', because negative results might often end up in a file drawer, unpublished.

A5 MAXIMUM LIKELIHOOD

Likelihood is the probability of observing the data given some parameter value (e.g., $mN = 50$), under a certain statistical model (e.g., island model of migration). Maximum likelihood (ML) methods estimate the parameter value that maximizes the probability of obtaining the observed data under a given model. For example, we might compute the likelihood of each of many migration rates ($mN = 10, 11, 12 \ldots$ up to 500), and then choose the best (point) estimate of mN as the value that has the highest (maximum) likelihood (e.g., approximately $mN = 40$ in Figure A2b).

An advantage of likelihood analysis is that it is model-based and thus allows easy comparison of different models (even complex models), thereby improving inference about complex processes (e.g., different dispersal patterns, mutation models, stable versus declining population size) that might explain the data. Likelihood analysis is often used to test the fit of two different models by using the ratio of the MLE (maximum likelihood estimate) for one versus the other model. For example, if one model is far more likely (e.g., $\log_{10}(\text{MLE1}/\text{MLE2}) > 3$), we might reject the second model MLE2 (see Section A2.2). The two models might be, for example, a stable versus declining population, or alternatively the existence of two versus three subpopulations. Note that when '$\log_{10}(MLE1/MLE2) > 3$', the probability of MLE1 is generally 1000 times more likely that MLE2 (e.g., $P < 0.001$); when >2 the probability of MLE1 is considered to be 100 times more likely that MLE2 ($P < 0.01$).

Likelihood methods are sometimes classified as 'frequentist'. For example, when we compute the expected frequency in a large number of trials of a likelihood ratio (or a likelihood value), for example as part of a statistical test, this is a frequentist approach.

The main advantage of maximum likelihood approaches is they use 'all the data' in their raw form, and not some summary statistic (e.g., $\hat{H}_e$ or $\hat{F}_{IS}$). Because likelihood methods use a maximum of information from the data, they should, in theory, be more accurate and precise than moments-based methods (Luikart and England 1999). For example, likelihood-based methods use the raw date (number *and* genealogical divergence of each allele) to estimate N_e (or mN), and not a single summary statistic (e.g., H_e), as in classical moments-based estimators of N_e (or mN) (see Miller and Waits 2003).

Different datasets can give the same summary statistic (e.g., F_{ST}), whereas different datasets are less likely to yield the same ML estimates. For example, two independent sets of temporally spaced samples can have the same F_{ST} (temporal F_{ST}) even though they have different numbers of alleles. When using the summary statistic F_{ST} to estimate N_e (as in the classic 'temporal variance method': Waples 1989), we would not be using the information about the proportion of rare alleles, and thus might not achieve the most accurate or precise estimate of N_e. In another example, two independent metapopulations could have the same F_{ST}, but have different proportions of rare alleles. Information about the proportions of rare alleles can help to infer whether a metapopulation is stable, fragmenting, or is growing in size (e.g., Ciofi *et al.* 1999).

In practice, ML methods are often more accurate and precise than moment-based methods. For example, estimators of N_e based on likelihood provide tighter confidence intervals and less biased point estimates (Williamson and Slatkin 1999, Berthier *et al.* 2002). However, likelihood-based estimators generally require large sample sizes and can be biased and less precise than simpler summary statistics (moments-based methods), if sample sizes are small, for example, less than 40 or 50 individuals (e.g., see Lynch and Ritland 1999).

A6 BAYESIAN APPROACHES AND MCMC (MARKOV CHAIN MONTE CARLO)

Bayesian inference differs from classical frequentist statistics in two main ways. First, probabilities are defined and interpreted differently. In frequentist statistics, P-values (probability values) are interpreted as the average outcome of a repeated experiment in a large number of trials. P-values are interpreted as the probability of the test statistic being as extreme (or more extreme) than observed, if the null hypothesis is true. A frequentist test might yield $P = 0.05$, meaning there is 5% chance of observing the test statistic simply by chance alone.

Bayesian computations yield a more straightforward and direct probability answer that is easier to interpret than a P-value. For example, a Bayesian (posterior) probability might yield a probability of $P = 0.95$, meaning there is 95% probability that your hypothesis is true (e.g., that N_e is less than 100, for example). Recall that in the more complicated and less direct frequentist approach, we would construct a null hypothesis (e.g., H_0: N_e is greater than or equal to 100), and then reject the null if the P-value is low (e.g., $P < 0.05$); thereby finding support for the alternative hypothesis of interest 'N_e is less than 100'.

Furthermore, Bayesian posterior probability distributions (and support limits) are easier to interpret than confidence intervals because probability distributions show the probability visually as the area under a curve (e.g., in the tails of a probability distribution). We immediately get a feel for the width and degree of skewness of the probability distribution by observing the posterior distribution, which we cannot get from reading confidence limits. Thus, a probability distribution (posterior probability distribution) carries more information than a classical confidence interval and it gives a better feel for the relative probability of different parameter values (e.g., small versus large N_e, or Nm, or F_{IS}) (Ayres and Balding 1998).

Second, perhaps the main advantage of the Bayesian approach is the ability to include prior data or information when estimating the posterior probability that a hypothesis is correct. Bayes' theorem was developed to allow easy 'updating' of an existing estimation when presented with new data, such as observations from a new experiment. Classical frequentist statistics generally require each experiment to be totally independent and without reference to previous experiments. Prior

$$p(x)p(y|x) \propto p(x|y)$$

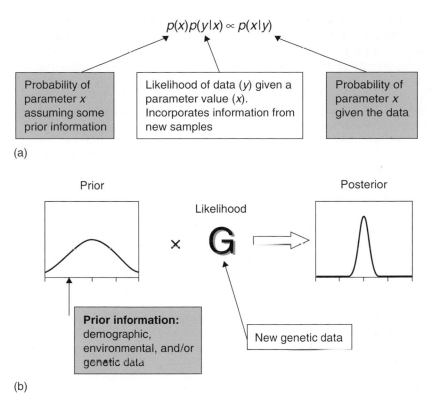

| Probability of parameter x assuming some prior information | Likelihood of data (y) given a parameter value (x). Incorporates information from new samples | Probability of parameter x given the data |

(a)

(b)

Figure A5 (a) Simplified Bayesian mathematical expression showing how the Bayesian approach allows us to combine the information from the data with prior information about the parameters of the model in order to obtain their **posterior** distribution (to estimate a parameter). (b) Graphical illustration of how prior information (in the prior probability distribution) is modified by the multiplication of it by the likelihood function (from the standard likelihood-based approach) to obtain a posterior probability distribution. Modified from O. Gaggiotti (personal communication).

information (previous data or even a hunch) can be incorporated into the computation of a probability (posterior probability) by multiplying the likelihood function by the prior information (Figure A5).

An example using of the Bayesian approach to incorporate prior information is estimating the effective population size (N_e) when the population census size is known (e.g., $N_C = 250$). Here, we can use the prior knowledge of N_C, and knowledge that N_e cannot be more than twice the census size ($N_e \leq 2N_C$; Chapter 7). Thus, the prior probability of N_e being greater than 500 equals zero ($P[N_e > 500] = 0.0$; as in Berthier *et al.* 2002 or Tallmon *et al.* 2008). Further, we know that N_e is often less than 50% of N_C (Frankham 1995, Kalinowski and Waples 2002). This information can be used to give more 'weight' to N_e estimates near or below 50% N_C (e.g., using a prior probability distribution, see below).

Another example use of prior information is in models that incorporate mutation dynamics. Published data suggest that most microsatellites have mutation rates between 10^{-2} and 10^{-5}. So, we might use a flat (rectangular) prior ranging between 10^{-2} and 10^{-5}, when modeling humans or other mammals. We also know that the average mutation rate is near 5×10^{-4}. Thus we might use a more informative prior (e.g., bell-shaped rather than flat) with a high probability peak near 5×10^{-4}. For an actual example, Beaumont (1999) used a prior mutation rate greater than zero for monomorphic loci, thereby allowing the use of monomorphic loci when testing for population bottlenecks. Other bottleneck inference tests do not use monomorphic loci (Luikart and Cornuet 1998). See Lewis (2001) for a simple example of Bayesian computation.

The Bayesian approach to incorporating prior information can be especially useful in conservation biology because it facilitates decision-making when data are few and we want to integrate all available knowledge. In conservation biology, we often must make decisions based on limited data. For example, wildlife managers often must decide if the size of a population is large enough to allow harvest, or alternatively if the population needs protection, monitoring, or supplementation. Interestingly, the US National Academy of Sciences panel recommended that fisheries scientists consider Bayesian methods to help estimate fish population status and guide management policies (Malakoff 1999). Harvest quotas could be more appropriate and flexible if the risks of population decline were calculated directly via Bayesian statistics incorporating prior information such as the probability that harvest actions might endanger a stock.

The main criticism of Bayesian approaches is they can be strongly influenced by prior information, and thus be less objective than classical approaches. For example, two different people could use different prior information and obtain different results. A counterargument is that we can quantify the effects of different priors (e.g., via sensitivity analysis using different priors); thus we can (and should) consider the magnitude of influence of the prior when making management decisions. Often prior information has little influence on the posterior, especially if data are extensive (Figure A6). Unfortunately, such sensitivity analysis is not always conducted. It seems reasonable to use both Bayesian and classical frequentist methods in many applications, such as the estimation of F_{ST}, N_e, or mN, especially when the performance of one or both methods has been poorly evaluated (Section A9).

An important general contribution of Bayesian approaches is that they allow for computations using complex models that that could not be achieved using other statistical approaches (Beaumont and Rannala 2004). Bayesian computation using complex models has been greatly facilitated by Markov Chain Monte Carlo (MCMC) computational methods.

A6.1 Markov chain Monte Carlo (MCMC)

MCMC is a simulation-based methodology to generate probability distributions that are difficult or impossible to obtain from analytical equations, including likelihood equations (Wu and Drummond 2011, Storz 2002). Analytical equations often cannot be developed to describe complex processes with many variables, such as population size, allele frequencies, and mutation rates. MCMC allows simulation of a special kind of stochastic process known as a **Markov chain**. A Markov chain generates a series of random variables whose future state depends only on the current state at any point in the chain (Beaumont and Rannala 2004).

MCMC allows us to obtain random samples from 'sample space', even when the sample space is enormous (e.g., billions of phylogenies or genealogies). MCMC combines (1) a Markov chain model, i.e., a model involving a random walk (chain of random steps) in which the next step is determined by the characteristics of the current or previous step, and (2) the 'Monte Carlo' process of drawing a random number that is necessary at each step of the random walk (Monte Carlo is a city famous for gambling, which also uses random events like the rolling of a dice).

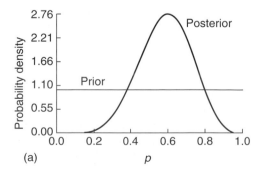
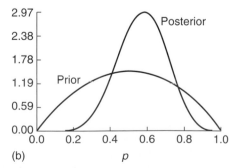

Figure A6   Effect of a flat prior (a) and an informative prior (b) on the posterior distribution. Here, the prior has little effect, as is the case when extensive data exists and the likelihood function (alone) is relatively informative. From Lewis (2001).

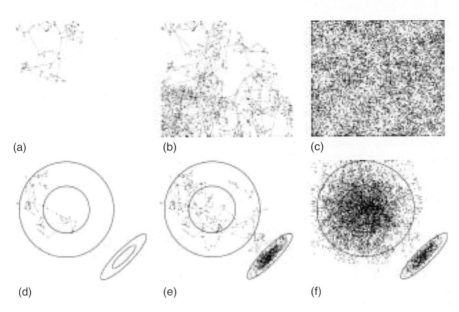

(a)   (b)   (c)

(d)   (e)   (f)

Figure A7 Illustration of the principles behind Markov chain Monte Carlo (MCMC) methods using a simple analogy of a 'random walk' in a square field (top panels) by a robot. The robot begins its walk in the upper left corner and continues for 100 steps (a), 1000 steps (b) and 10,000 steps (c), until nearly every portion of the field has been covered. Now suppose that two hills are present, represented by the concentric circles and smaller concentric ovals (d, e and f). The robot will take steps to points in proportion to their elevation, and thus higher points will be visited more often than lower ones. The proportion of time spent in any place approximates the probability of that location. From Lewis (2001).

MCMC is illustrated by an analogy of a robot taking a random walk in a square field (Figure A7). Each step of the robot varies in length and direction, randomly. Eventually, the robot visits every space within the field. However, the robot spends more time in spaces that are on hilltops at higher elevation (i.e., having higher probability). This is achieved by using a model with the following two main rules: (1) if a step takes the robot uphill, the robot will automatically take it; (2) if a step would take the robot downhill, the robot only takes the step with a probability depending on the elevation reduction (this probability can be computed several ways, e.g., via 'Metropolis' or 'Metropolis-Hastings' methods).

The first few steps or thousands of steps are called the 'burn-in', and are discarded to reduce influence of the starting point (bias). Once burn-in is achieved, the MCMC simulation has **converged**, that is, become independent of the starting point. The remaining steps, after convergence, give a good approximation of the landscape (probability space). This simulation of a random walk allows for estimation of the parts of the sample space with the highest probability (e.g., maximizing the probability of the data, given the model), as in maximum likelihood estimation (Section A5). Under the Bayesian approach (see below), MCMC simulation is often used to sample from the posterior distribution of a parameter in order to generate the posterior probability estimate of the parameter.

The main problem with MCMC approaches is sometimes we are not sure we have conducted a long-enough burn-in to achieve convergence and thus avoid bias. Also, MCMC simulation programs are generally difficult to write in computer code, and thus errors (bugs) are relatively likely to occur and can be difficult to detect. In addition, MCMC approaches are computationally slow and thus difficult to evaluate as to their performance among a range of scenarios (Section A9).

MCMC is primarily used within Bayesian approaches, but can also be used in maximum likelihood estimation. For example, some available software programs can use flat priors (or no priors) and give as output a likelihood (probability) curve or a posterior distribution if prior information is used (e.g., Beaumont 1999).

A7 APPROXIMATE BAYESIAN COMPUTATION (ABC)

ABC employs a Bayesian framework, incorporating prior information, to output an 'approximate' posterior probability distribution (Bertorelle *et al.* 2010). This posterior distribution is only an estimation of the full posterior, because all the raw data (e.g., full allelic distributions) are not used to compute the posterior. Instead, the posterior is usually approximated by summarizing the data using multiple different summary statistics (Tallmon *et al.* 2008, Sousa *et al.* 2009). For example, a full (exact) Bayesian approach might conduct MCMC simulations to obtain the exact posterior probability of the raw sample (allele number and frequency distribution), using each of thousands of simulated datasets (e.g., genealogies, for a population model under consideration, such as a stable population). Here, for example, we might consider population models (and simulations) with $N_e = 10$, 20, 30, and so on, if we were estimating the N_e for our observed data.

Unlike exact Bayesian computation, ABC would (1) replace (summarize) the raw observed data with multiple summary statistics of the data such as F_{ST}, H_e, and H_o, and then (2) compute the same summary statistics for each of thousands of simulated population datasets (for the population models under consideration), and finally (3) match the observed data summary statistics to those from simulated populations in order to choose the population parameter (N_e) estimate that best fits our data. This ABC approach is also called 'summary statistic matching' because we match our empirically observed summary statistics (F_{ST}, H_e, and H_o) to those from simulated population datasets (each with a different effective size, e.g., $N_e = 10$, 20, 30) to find the parameter estimate (N_e) that is most probable for our population, according to the matching of our empirical and simulated summary statistics.

ABC methods are becoming increasingly popular because they use nearly 'all the information' from the data (Beaumont *et al.* 2002), yet they are usually far less computationally demanding than fully Bayesian (MCMC) approaches. Thus, their performance can be evaluated thoroughly (Section A9), and they can be used with large datasets containing many loci or when conducting complex analyses with numerous parameters such as the population size, migration rate, and sex ratio. Finally, an experienced modeler can construct an ABC model in hours or days (e.g., Cornuet *et al.* 2008, Wegmann *et al.* 2010), whereas it can take weeks to construct a fully Bayesian MCMC model and the risks of programming errors can be far higher (M. Beaumont, personal communication).

A8 PARAMETER ESTIMATION, ACCURACY, AND PRECISION

Here we consider statistical frameworks (moments, likelihood, Bayesian) for inferring population parameters. To estimate a population parameter, such as the mean (μ), we usually compute a sample statistic ($\bar{x}$) from a sample of individuals. We can estimate a population parameter using different sample statistics such as the arithmetic mean, harmonic mean, median or mode. To further complicate things, to compute an estimator, such as the mode, we can use different approaches, including moment methods, maximum likelihood estimation, or Bayesian estimation.

The sample moments, for example $\bar{x}$, $\bar{x}^2$ and $\bar{x}^3$, are used to obtain estimates of location, variance (scale), and shape of the population distribution, respectively. Moment-based estimators are widely used (e.g., in classic frequentist statistics), but can yield biased estimates when the underlying population distribution is non-normal, especially when the 'higher' moments ($\bar{x}^2$, $\bar{x}^3$) are not considered. An example of such bias is the classical F_{ST}-based estimator of N_e, which is often biased because: (1) the underlying probability distribution of F_{ST} is often skewed with a long tail (unlike the normal distribution); and (2) the moment estimator (F_{ST}) incorporates information only from the first two moments, which do not contain information on skewness of the sampling distribution.

Maximum likelihood estimation (MLE) infers a parameter by finding the parameter value that maximizes the likelihood of obtaining the sample data (assuming some model such as Mendelian inheritance or a Wright-Fisher equilibrium population). MLE is increasingly used in population genetics for several reasons:

1 It yields probability distributions that are easy to interpret (see Section A5 and Figures A2b, A5b), rather than just a point estimate and confidence interval, as in moments methods.
2 MLE can help to evaluate and choose the best estimators (including moment-based estimators of the mean or variance, when data are normally distributed).
3 Faster computers and computer programs increasingly allow the computation of MLE estimates (see

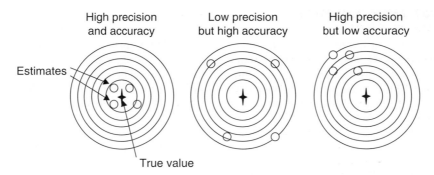

Figure A8   Illustration of the difference between accuracy and precision. Imagine these are archery targets with the bullseye in the middle (i.e., the true value).

LEMARK, MIGRATE, MsVAR and other computer programs on this book's webpage).

Which estimator and approach performs best? This is a critical question in conservation genetics that is often ignored or underappreciated. It is especially important in light of the many new methods and computer programs published in recent years. The performance of an estimator (accuracy, precision, and robustness, see Section A9) depends on the question, sample size, sample characteristics, and the parameter being estimated. For example, MLE approaches are generally most efficient (see Section A5) with large samples, but can be less efficient than moment methods when using small sample sizes, for example, fewer than 40 individuals. Efficiency refers to the ability to extract information from the data and to achieve high accuracy and precision in estimating the true population parameter.

Identifying the best estimator generally requires a **performance evaluation** comparing estimators. For examples of performance evaluations, see Section A9 (below) and publications such as Tallmon *et al.* (2004) and Wang (2002).

The accuracy and precision are critical concepts related to estimators of central tendency and dispersion, respectively. Accuracy of an estimator is its tendency to yield estimates near the true population parametric value. For example, if we compute an estimate of the mean heterozygosity (H_e) for each of 10 independent samples, the accuracy is good if 50% of estimates are high and 50% are low. If most of the independent estimates were low (or high) the estimator is considered to be biased.

If an estimator has poor precision, the 10 estimates will scatter widely from each other – often on both sides (e.g., high and low) of the true value. A precise statistical estimator will have relatively narrow confidence intervals, and the point estimates from independent estimations will cluster tightly together (see below). An estimator can have low precision but high accuracy, or vice versa (Figure A8).

Several different estimators should often be used whenever assessing a given question. For example, it is useful to estimate both the mean and median because if they are different we can infer that the distribution might be skewed. It is also useful to compute both moment-based and likelihood-based estimators, as we sometimes do not know which is most reliable or accurate. In general, when estimating parameters, it is prudent to use multiple methods and software programs, to avoid errors and to increase confidence in results.

Random and representative sampling is critical and often assumed without testing or discussing the assumption. If sampling is not random or not representative, the statistical estimate can be biased. For an extreme example, imagine that we sample only 10 individuals (F_1 offspring) from within only one family from a population containing hundreds of family groups. The sample is clearly not random or representative of the population. The allelic richness statistic we estimate often will be low compared with the true population value, simply because the individuals we sampled are closely related compared with individuals from a true population-wide sample (with random representation of all family groups).

A9 PERFORMANCE TESTING

Performance testing is the quantification of the accuracy (bias), precision, power and robustness of a statistical estimator or test. This includes quantifying the bias caused by violating assumptions (random sampling, no selection, etc.); such violations often occur in real datasets from natural populations.

Performance testing involves four main steps: (1) generate a test dataset (simulated or real) with a known parameter value for the parameter of interest (N_e, Nm, etc.); (2) estimate the parameter (e.g., with a confidence interval); (3) repeat both steps 1 and 2, 1000 times; and (4) compute the proportion of the 1000 estimates that give the true parameter most accurately and precisely (Figure A9).

Performance testing is critically important to allow conservation biologists to use statistical methods on real populations with minimal risk of making erroneous management decisions. Unfortunately, performance testing is rarely conducted thoroughly, but the growing availability of computer simulation programs (e.g., EASYPOP, METASIM; see this book's website) makes performance testing increasingly feasible, even for undergraduate students or as part of a PhD degree program.

Without performance evaluations, statistical methods are often used, and are later found to be biased. For example, some assignment tests and N_e estimators were shown to produce misleading or erroneous results (e.g., erroneous Type I error rates for assignment tests or underestimates of the true N_e), long after they were being used in natural populations (see Paetkau *et al.* 2004, England *et al.* 2006).

A10 THE COALESCENT AND GENEALOGICAL INFORMATION

To 'coalesce' means to fuse, unite, or come together. This refers to the process of tracing backward through time the joining of (coalescence of) homologous gene copies (haplotypes) from different individuals into the same parent or ancestor (Figure A10). The word **coalescent** is used in several ways in the genetics literature. Coalescent theory was developed (Kingman 1982) to model a genealogy of gene copies so that allele frequency patterns and genealogical patterns

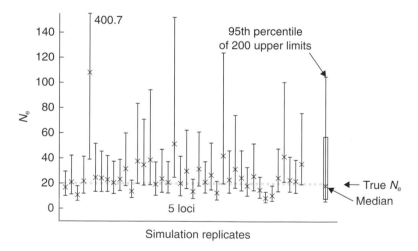

Figure A9   Example of power analysis where 200 independent populations with $N_e = 20$ were simulated, and then N_e (CI 90%) was estimated for each simulation replicate. Point estimates (×) of N_e, with confidence intervals (vertical lines) for each of 25 independent simulated populations are shown. The box plot graph on the far right summarizes the accuracy of point estimates by comparing the median of the many point estimates with the true N_e. The median is biased low (lower arrow). The box plot upper limit is the upper 95th percentile of the upper confidence interval limits (over 200 simulations). Modified from Berthier *et al.* (2002).

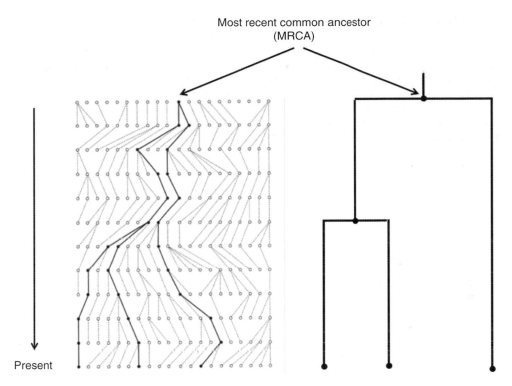

Most recent common ancestor
(MRCA)

Present

Figure A10 The coalescent approach for modelling the genealogy of individuals in a random mating population of 10 individuals for 11 generations. The complete genealogy is on the left; dark lines trace back through time (from bottom to top) the ancestries of three gene copies sampled in the present population. The 'subgenealogy' of the three sampled gene copies is on the right. The two gene copies on the left coalesced most recently. The coalescence times are proportional to branch lengths. The average (and distribution) of coalescence times provides information about the tree shape, which is used to make inferences about demographic history. Modified from Felsenstein (2004) and Rosenberg and Nordborg (2002).

(e.g., shapes of genealogies) could be used to infer parameters relating to demographic history such as population size, population growth, gene flow, time of divergence from another population, and selection.

The coalescent approach yields a the distribution of times to common ancestry among gene copies in a genealogy (Box A2) (Kuhner 2009). Coalescences are the points of common ancestry, or internal nodes, on a genealogy (Figure A10). The coalescent can be used with frequentist, maximum likelihood, or Bayesian statistical approaches, for example to generate the expected distribution of allele frequencies to test hypotheses and estimate parameters (Box A3).

The coalescent is a powerful modeling approach for analyzing population genetic data (Rosenberg and Nordborg 2002). It involves a different way of thinking

about population genetics compared with classic approaches. Classic approaches for modeling populations typically trace the inheritance of genes in a 'forward direction'. For example, individual parents are randomly mated to produce offspring; the offspring are eventually mated to produce the next generation, as in individual-based simulation modeling. In contrast, the coalescent approach looks backwards in time and traces gene copies back from offspring to parents, to grandparents, and eventually to a single most recent common ancestor.

The coalescent uses genealogical information from DNA sequences (i.e., information on the genealogical relationships among alleles at a locus). Classic statistical estimators in population genetics (e.g., F_{ST}) do not use genealogical information; they use only allele state

Box A2 Coalescent modeling

Coalescent modeling involves two main steps: first we generate a random genealogy of individuals backward through time. Here it helps to envision clonal individuals (or haploid chromosomes such as mtDNA). We start with a sample of clones and randomly connect them to parents, grandparents, great-grandparents, etc., until all clones coalesce into a single ancestor (the most recent common ancestor, MRCA; Figure A10). Going back in time, two lineages will coalesce whenever two clones are produced by the same parent. Going forward in time, lineages branch whenever a parent has two or more offspring, and branches end when no offspring are produced (i.e., lineage sorting, Section 7.8).

Second, we randomly place mutations on branches (e.g., using Monte Carlo simulations and a random number generator while considering the mutation rate). We start by assigning some allelic state to the MRCA and then 'drop' mutations along branches randomly moving forward. If a mutation is placed on a branch, then the allelic state (e.g., length for a microsatellite) must be determined by following rules of a model. For example, under a stepwise mutation model,

a mutation is equally likely to increase or decrease allele length by a single repeat unit (see Section 12.1.2).

Coalescent modeling is computationally efficient because we only simulate the sampled lineages ('subgenealogy' in Figure A10), and not the entire population as is done for individual-based forward models. Simulating only the subgenealogy requires less 'record-keeping' and saves computer time compared with the forward (individual-based) simulation modeling approach that requires record-keeping for all individuals including those not sampled.

In coalescent modeling, we often want to separate the two stochastic genealogical processes: (1) random neutral mutation and (2) random genetic drift (caused by random reproduction and population demography). These two processes determine the genetic make-up of the population of lineages. Separation of the two is important because we are often interested in the biological phenomena of demography and reproduction, but not mutation processes (Rosenberg and Nordborg 2002). For example, we are often interested in testing for population expansion or population subdivision, but not mutation dynamics.

Box A3 The coalescent used in frequentist, likelihood, and Bayesian approaches

The coalescent can be used for modeling or conducting statistical tests under different statistical frameworks including frequentist, likelihood, or Bayesian. For example, a frequentist coalescent approach might be used to test whether N_e is significantly smaller than 100. For this, we might (1) use the coalescent to simulate 1000 independent datasets for a population with $N_e = 100$; (2) compute an estimate of N_e for each simulated population (to obtain a distribution of N_e estimates consistent with a true N_e of 100); and (3) calculate how frequently (out of the 1000 datasets) we obtain a simulated N_e estimate as small as the empirical N_e estimate from our study population. If our population's N_e estimate is so small that it occurs only once in 1000 simulated datasets, then we would conclude that that our population's true (actual) N_e is significantly ($P < 0.001$) less than 100. This kind of approach was used in Funk *et al.* (1999) to test for small N_e in a salamander population.

In a maximum likelihood approach to test whether N_e is significantly smaller than 100, the coalescent

could be used to help compute the likelihood of $N_e = 1$, $N_e = 2$, $N_e = 3$ up to $N_e = 200$, given our raw data. Here, the coalescent could be used to simulate thousands of datasets for each N_e, and then compute the likelihood of each N_e ($N_e = 1$, $N_e = 2$, $N_e = 3$, etc.) given our real dataset. This would yield a probability (likelihood) distribution of N_e values (with $N_e = 1$, $N_e = 2$, $N_e = 3$ up to $N_e = 200$ on the x-axis). If all the area under the likelihood (probability) curve was less than 100 (i.e., did not include $N_e = 100$), we could conclude that our population's effective size is less than 100.

In a Bayesian approach, we would conduct the same computations as in the maximum likelihood approach just described, using the coalescent. However, we then would modify the resulting likelihood distribution by multiplying it by a prior distribution to obtain a posterior distribution, as illustrated in Figure A5.

This box illustrates how the coalescent can be used within different statistical frameworks to conduct statistical tests or estimate a population parameter.

and frequency information. With the advent of DNA sequencing (and restriction enzyme analysis, Chapter 4) most datasets contain information on relationships (i.e., amount of divergence between alleles). Microsatellites also contain genealogical information on the number of repeat unit differences between two alleles (assuming the stepwise mutation model discussed in Sections 4.2.1 and 12.1.2).

Genealogical methods, such as the coalescent, are generally not used to estimate evolutionary trees or infer a phylogeny (but see Carstens *et al.* 2005); rather they estimate parameters of the stochastic evolutionary processes that give rise to genealogies, such as gene flow rates, population size or population growth rates. For example, different population demographic histories yield differently shaped genealogies (Figure A11). Consequently, genealogical shape can be used to infer a population's demographic history (Emerson *et al.* 2001, Kuhner 2009).

Population growth yields star-like genealogies with many branches of similar length (Figures A11b and A12). Many long and similar-length branches are expected to arise during a long-term population expansion because new alleles (mutations) tend to persist a

long time because genetic drift (**lineage sorting**) is negligible in fast-growing populations. If a population is large and has expanded from a smaller size in the past, then coalescent events are relatively old (i.e., branch lengths are relatively long) and originate near the time the expansion began, compared with coalescent event times for a stable population (Figure A11a).

Random genealogical processes lead to many possible random genealogies for independent genes under a given demographic history (Figure A12). Therefore, we must study many genes to obtain accurate and precise estimates of demographic history. We can simulate thousands of random genealogies for each population history (e.g., a stable versus declining population) to test if one particular history best fits our empirical dataset. If one history (e.g., a decline) best fits our observed empirical data, then another history (stable population) might be rejected by comparing our observed data and the simulated genealogy data from the alternative population histories.

Natural selection can also cause distinctively shaped genealogies for a given locus. If the genealogy of one locus differs significantly from most other loci in the genome, we might infer that selection has influenced

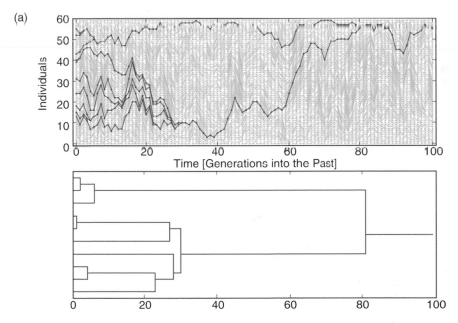

Figure A11   Three genealogies simulated with a coalescent model for a population with a mean effective population size of 60 individuals. (a) Constant sized population (random distribution of branch lengths).

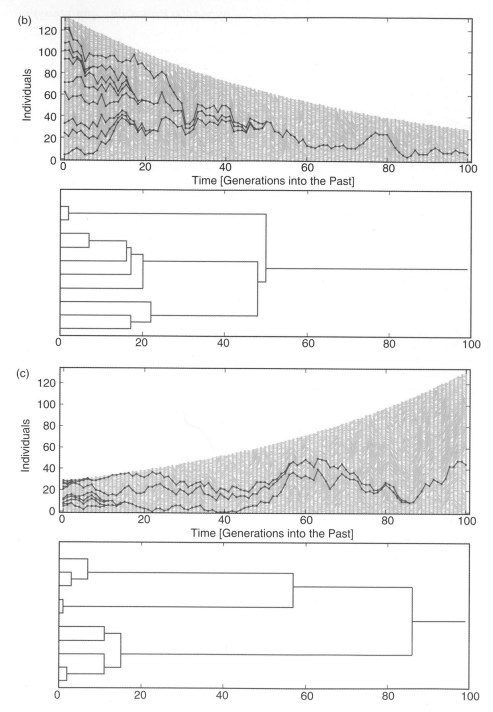

Figure A11 (*Continued*) (b) Growing population (multiple relatively long branches, less variable distribution of branch lengths, and more coalescent events relatively near the time of initiation of population growth many generations in the past). (c) Declining population (fewer long branches, distribution of branch lengths more variable than random as in the stable population, and most coalescences are relatively recent). Similarly to Figure A10, each complete genealogy is shown (top of each of the three panels) with dark zigzag lines tracing back through time (from left to right) the ancestries of ten gene copies sampled in the present population (time 0). The simplified 'subgenealogy' of the ten sampled gene copies is on the bottom of each panel. Population size is indicated by the gray area on each top panel. The horizontal axis (time) is the same (100 generations) for all three trees. Provided by P. Beerli.

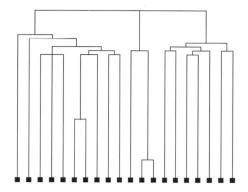 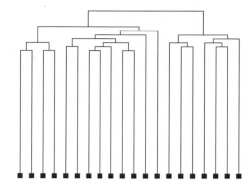

Figure A12 Two genealogies with the same demographic history – a rapidly growing population. Note that thousands of different genealogies are possible for the same identical demographic history. Thus, many independent genealogies must be examined to infer population history with precision. Consequently, distributions of many genealogies are used to infer (or exclude) different demographic histories. From Harpending *et al.* (1998).

the locus. Natural selection detection at a given locus requires the study of many independent loci because demography and selection can generate similar genealogies (Luikart *et al.* 2003).

For example, a selective sweep will remove many alleles (similar to a population decline) and subsequently lead to a star-shaped genealogy as new mutations arise in the population during the generations following the selective sweep. This single-locus star genealogy could resemble the star-shaped genealogies expected at all loci (genome-wide) following population growth. In a related way, balancing selection at a locus can mimic the effect of a population bottleneck by maintaining alleles at even frequencies at a single locus, similar to the way bottlenecks cause even frequencies at all loci genome-wide. Clearly, we must study many genes (loci) to obtain reliable estimates of demographic history and/or to test for selection at a single locus.

Guest Box A   Is mathematics necessary?
James F. Crow

Much of our understanding of the application of genetics to problems in conservation depends upon the field of population genetics. Population genetics used to consist of two quite different disciplines. One utilized observations of populations in nature or laboratory studies. These were often descriptive and involved no mathematics. This area is epitomized by the early work of Theodosius Dobzhansky, Ernst Mayr, and G. Ledyard Stebbins. At the same time a mathematical theory was being developed by J.B.S. Haldane, R.A. Fisher, and Sewall Wright. One of the earliest bridges was built in 1941 when Dobzhansky and Wright collaborated in a joint experimental paper with lots of theory.

Since that time, most work in population genetics has had some mathematical involvement. Almost every experiment or field observation now utilizes quantitative measurements, and that means statistics. The day is past when one can simply report results with no test of their statistical reliability. Increasingly, experiments are performed or observations are made based on some underlying theory. The person doing the experiments may develop the theory or make use of existing mathematical theory. Finally, there is the development of ever-deeper, more general, and more sophisticated theory. Much of this is being done by people with professional mathematics training.

(Continued)

We cannot all be mathematicians. But we can learn a minimum amount. Every population geneticist must know some mathematics and some statistics. I have done both experimental (usually driven by theory) and theoretical work. But my mathematics is limited and some of the research that I most enjoyed was done in collaboration with better mathematicians, notably Motoo Kimura.

There are two recent changes in the field. Computers have altered everything, and it is hardly necessary for me to mention that you need to know how to use them. It used to be that theoretical work was regularly stymied by insoluble problems. The computer has greatly broadened the range of problems that can be solved, not in the mathematical sense but numerically (e.g., MCMC and other simulation-based approaches), which is often what is wanted. At the same time, the mathematical theory itself is advancing as mathematicians enter the field.

The second change is the advent of molecular methods. Population genetics used to have a theory that was too rich for the data. That is no longer true. DNA analysis can yield mountains of data that call for improved, computerized analyses. Even in nonmodel species, datasets are becoming large enough that some sophisticated statistical methods can take days to conduct computations, and some analyses might not be feasible because, for example, computer programs take too long or do not converge.

If you are going to be an experimenter or analyze data with modern statistical tools, you need to know some mathematics and statistics, and be adept at computers. If you are going to develop theory (even if for application to natural populations and conservation), you usually need to be a *real* mathematician or collaborate with one.

Most readers of this book are primarily interested in understanding, but not contributing to the primary literature in population genetics. Much of the current literature in population genetics employs advanced mathematical methods that are beyond the reach of most biology students. Dobzhansky's method of reading and understanding the papers of Sewall Wright is one possible approach (see quote at the beginning of this appendix). Examining the biological assumptions being made is crucial, but not sufficient. However, a healthy amount of skepticism is probably a good thing. There was only one Sewall Wright!

Glossary

ABC *See* **approximate Bayesian computation**.

acrocentric Chromosomes and chromatids with a centromere near one end.

adaptation Evolutionary change resulting from natural selection that increases fitness, or a trait that increases fitness.

adaptational lag A population in a changing environment is sometimes unable to adapt quickly enough to keep up with the necessary rate of change to maintain fitness. Over time, the population can lag increasingly behind its climate fitness optimum. When the adaptational lag between the average population phenotype and the optimum phenotype with the highest fitness becomes too great, extinction occurs.

addition rule *See* **sum rule**.

additive genetic variation The portion of total genetic variation that is the average effect of substituting one allele, responsible for a phenotypic trait, for another. The proportion of variation that responds to natural selection.

admixture The formation of novel genetic combinations through hybridization of genetically distinct groups.

AFLP *See* **amplified fragment length polymorphism**.

agamospermy The asexual formation of seeds without fertilization in which mitotic division is sometimes stimulated by male gametes. Also called **apomixis**.

alien species A non-native species or any propagule of that species, such as eggs, spores, or tissue present in an ecosystem.

allele Alternative form of a gene.

allelic diversity A measure of genetic diversity based on the average number of alleles per locus present in a population.

allelic richness A measure of the number of alleles per locus; allows comparison between samples of different sizes by using various statistical techniques (e.g., rarefaction).

allopatric Species or populations that occur in geographically separate areas.

allopolyploid A polyploid originating through the addition of unlike chromosome sets, often in conjunction with hybridization between two species.

allozygous An individual whose alleles at a locus are descended from different ancestral alleles in the base population. Allozygotes may be either homozygous or heterozygous in state at this locus.

allozyme An allelic enzyme detected through protein electrophoresis used in many genetic applications such as hybrid identification and estimation of genetic variation.

AMOVA *See* **analysis of molecular variation**.

amplicon A segment of DNA formed by natural or artificial amplification events, for example by polymerase chain reactions.

amplified fragment length polymorphism (AFLP) A technique that uses PCR to amplify genomic DNA, cleaved by restriction enzymes, in order to generate DNA fingerprints; it is a combination of RFLP and arbitrary primer PCR. It does not require prior sequence knowledge.

amplify To use PCR to make many copies of a segment of DNA.

Conservation and the Genetics of Populations, Second Edition. Fred W. Allendorf, Gordon Luikart, and Sally N. Aitken.
© 2013 Fred W. Allendorf, Gordon Luikart and Sally N. Aitken. Published 2013 by Blackwell Publishing Ltd.

anagenesis Evolutionary changes that occur within a single lineage through time. *See* **cladogenesis**.

analysis of molecular variation (AMOVA) A statistical approach to partition the total genetic variation in a species into components within and among populations or groups at different levels of hierarchical subdivision. Analogous to ANOVA in statistics.

aneuploid A chromosomal condition resulting from either an excess or deficit of a chromosome or chromosomes so that the chromosome number is not an exact multiple of the typical haploid set in the species.

animal model A statistical model in which phenotypic variance is compartmentalized into environmental, genetic, and other causes, as in mixed models.

anneal The joining of single strands of DNA because of the pairing of complementary bases. In PCR, primers anneal to complementary target DNA sequences during cooling of the DNA (after DNA is made single-stranded by heating).

annual plant Those that germinate, flower, and mature seed in the same year, and survive winter or drought periods as seeds.

ANOVA Analysis of variance.

apomixis *See* **agamospermy**.

approximate Bayesian computation (ABC) A statistical framework using simulation modeling to approximate the Bayesian posterior distribution of parameters of interest (e.g., N_e, mN) often by using multiple summary statistics (H_e, number of alleles, F_{ST}). It is far faster computationally than fully Bayesian approaches but generally slightly less accurate and precise.

artificial selection Anthropogenic selection of phenotypes with a heritable genetic basis, to elicit a desired phenotypic change in succeeding generations.

ascertainment bias Selection of loci for marker development (e.g., SNPs or microsatellites) from an unrepresentative sample of individuals, or using a particular method, which yields loci that are not representative of the spectrum of allele frequencies in a population. For example, the choice of loci with high heterozygosity may bias assessments of allele frequency distributions in future studies using those loci such that alleles at low frequency (rare alleles) are underrepresented.

assignment tests A statistical method using multilocus genotypes to assign individuals to the population from which they most likely originated (i.e., in which their expected multilocus genotype frequency is highest).

assisted colonization Human relocation of individuals to sites where they do not currently occur or have not been known to occur in recent history.

assisted gene flow Human-mediated gene flow.

assisted range expansion Human movement of individuals to regions adjacent to existing species ranges on the leading edge of migration.

association mapping A method using gametic disequilibrium to associate phenotypes to genotypes in order to map QTLs. Also called linkage disequilibrium mapping.

associative overdominance An increase in fitness of heterozygotes at a neutral locus because it is in gametic disequilibrium at a locus that is under selection. Also known as **pseudo-overdominance**. Compare with **hitchhiking**.

assortative mating Preferential mating between individuals with a similar (or a different) phenotype is referred to as positive (or negative) assortative mating. *See also* **disassortative mating**.

autogamy Self-fertilization in a hermaphroditic species where the two gametes fuse in fertilization.

autosomal A locus that is located on an autosome (i.e., not on a sex chromosome).

autosomes Chromosomes that do not differ between sexes.

autozygosity A measure of the expected homozygosity where alleles are identical by descent.

autozygous Individuals whose alleles at a locus are identical by descent from the same ancestral allele.

balancing selection Diversifying selection that maintains polymorphism resulting from frequency-dependent selection, spatially heterogeneous selection, or heterozygous advantage.

Barr bodies Inactivated X-chromosomes in many female mammals that condense to form a darkly colored structure in the nuclei of somatic cells.

B chromosome *See* **supernumerary chromosome**.

binomial proportion A population will be in binomial proportions when it conforms to the binomial distribution so that the occurrence of a given event X, r_i times with a probability (p_i) of success, in a population of n total events is not significantly different than what would be expected based on random chance alone.

bioclimate envelope model Describes the multivariate climatic niche space occupied by a species, usually developed through correlating species occurrence with climatic variables.

biogeography The study of the geographic distribution of species and the principles and factors that influence these distributions.

biological species concept (BSC) Groups of naturally occurring interbreeding populations that are reproductively isolated from other such groups or species.

BLAST Basic Local Alignment Search Tool. Software to search for a DNA sequence similar to one in hand.

Bonferroni correction A correction used when several statistical tests are being performed simultaneously (since while a given α-value may be appropriate for each individual comparison, it is not for the set of all comparisons). In order to avoid a lot of spurious positives, the α-value needs to be adjusted to account for the number of comparisons being performed. Suppose we are testing for Hardy-Weinberg proportions at 20 loci. Instead of using the traditional 0.05 α-level, we would test at α of $0.05/20 = 0.0025$ level. This insures that the overall chance of making a Type I error is still less than 0.05.

bootstrap analysis A nonparametric statistical analysis for computing confidence intervals for a phylogeny or a point estimate (e.g., of F_{ST}). Re-sampling of a data set with replacement to estimate the proportion of times an event (such as the positioning of a node on a phylogenetic tree) appears.

bottleneck A special case of strong genetic drift where a population experiences a loss of genetic variation by temporarily going through a marked reduction in effective population size. In demography, a severe transient reduction in population size.

branch length Length of branches on a phylogenetic tree. Often proportional to the amount of genetic divergence between populations or species.

breeding value A measure of the value of an individual for breeding purposes, as assessed by the mean performance of its progeny.

broad-sense heritability (H_B) The proportion of phenotypic variation within a population that is due to genetic differences among individuals.

BSC *See* **biological species concept**.

candidate genes A gene that is thought to be more likely to be involved in the control of a trait in one species compared with a random gene from the genome, based on known functions in another species.

census population size The number of adult individuals in a population.

centromere A constricted region of a chromosome containing spindle microtubules responsible for chromosomal movement attach during mitosis and meiosis.

chi-square test A test of statistical significance based on the chi-squared statistic, which determines how closely experimental observed values fit theoretical expected values.

chloroplast DNA (cpDNA) A circular DNA molecule located in the chloroplasts.

chromosome A molecule of DNA in association with proteins (histones and non-histones) constituting a linear array of genes. In bacteria and archaea the circular DNA molecule contains the set of instructions necessary for the cell.

CITES Convention on International Trade in Endangered Species of Wild Fauna and Flora.

clade A species, or group of species that has originated and includes all the descendants from a common ancestor. A monophyletic group.

cladistics The classification of organisms based on phylogenies.

cladogenesis The splitting of a single evolutionary lineage into multiple lineages.

cladogram A diagram illustrating the relationship between taxa that is built using synapomorphies. Also called a **phylogeny**.

cline A gradual directional change in a character across a geographic or environmental gradient.

coadapted gene complex A multilocus genotype of alleles at several loci that results in high fitness and is commonly inherited as a unit because the loci are closely linked. See Guest Box 3.

coalescent The point at which the ancestry of two alleles converges at a common ancestral sequence.

coancestry Degree of relationship between two parents of a diploid individual.

codon A three nucleotide sequence on a strand of mRNA that gets translated into a specific amino acid within a protein.

common garden An experiment where individuals from different environments are reared in a common environment to facilitate the study of genetic differences in phenotypes. Term applies to both plants and animals.

community genetics The study of how genetic variation within a species affects interactions among species and community structure and diversity.

conservation breeding Efforts to control the breeding of plant and animal species for conservation.

conservation collections Living collections of rare or endangered organisms established for the purpose of contributing to the survival and recovery of a species.

conservation genomics The use of new genomic techniques to solve problems in conservation biology.

conservation unit A population of organisms that is considered distinct for purposes of conservation, such as a management unit (MU), distinct population segment (DPS), or evolutionarily significant unit (ESU).

conspecific A member of the same species.

continuous characters Phenotypic traits that are distributed continuously throughout the population (e.g., height or weight).

continuous distribution model of migration Individuals are continuously distributed across the landscape; neighborhoods of individuals exist that are areas within which panmixia occurs, and across which genetic differentiation occurs due to isolation by distance.

Convention on International Trade in Endangered Species of Wild Fauna and Flora (CITES) An international agreement that bans international trade or shipment of an agreed-upon list of endangered species, and that regulates and monitors trade in others that might become endangered.

converged The point where a MCMC simulation has become independent of starting parameter biases, or has been 'burnt in'. Typically, thousands of simulation steps are required (and discarded) before the MCMC simulation is used to estimate a parameter (e.g., N_e, mN, etc.).

countergradient variation Occurs when genetic effects on a trait oppose or compensate for environmental effects so that phenotypic differences across an environmental gradient among populations are minimized.

cpDNA *See* **chloroplast DNA**.

CU *See* **conservation unit**.

cultivars A human cultivated plant that was derived through anthropogenic selection and breeding.

cultural drift Stochastic fluctuations in the frequency of cultural traits in a population that can be especially strong in small populations.

cytogenetics A discipline of science combining cytology, the study of cells (their structure, function, and life history), and genetics.

cytoplasmic genes Genes located in cellular organelles such as mitochondria and chloroplasts.

degrees of freedom The total number of items in a dataset that are free to vary independently of each other. For example, in testing for Hardy-Weinberg proportions with a chi-square test, this is the number of possible genotypes minus the number of alleles.

deme A local group of individuals that mate at random.

demographic Topics relating to the structure and dynamics of populations, such as birth, death, and dispersal rates.

demographic rescue The decrease in the probability of extinction of an isolated population brought about by immigration that increases population size.

demographic stochasticity Differences in the dynamics of a population that are the effects of random events on individuals in the population.

dendrogram A tree diagram that serves as a visual representation of the relationships between populations within a species.

derived A derived character is one found only in a particular lineage within a larger group. For example, feathers are derived characters that distinguish birds from their reptile ancestors.

deterministic Events that have no random or probabilistic aspects but rather occur in a completely predictable fashion.

diagnostic locus A locus that is fixed, or nearly fixed for different alleles, allowing differentiation between parental species, populations, or their hybrids.

dioecious Varieties or species of plants that have separate male and female reproductive organs on unisexual individuals.

diploid The condition in which a cell or individual has two copies of every chromosome.

directional selection The increase in the frequency of a selectively advantageous allele, gene, or phenotypic trait in a population.

disassortative mating Preferential mating of individuals with different phenotypes.

discrete generations Generations that can be defined by whole integers and in which all individuals will breed only with individuals in their generation (e.g., pink salmon, or annual flowers without a seed bank).

dispersal In ecological literature, dispersal is the movement of individuals from one genetic population (or birthplace) into another. Dispersal is also known as **migration** in genetics literature.

distinct population segment (DPS) A level of classification under the US ESA that allows for legal protection of populations that are distinct, relatively reproductively isolated, and represent a significant evolutionary lineage to the species.

DNA Deoxyribonucleic acid.

DNA barcoding The use of a short DNA sequence from a standardized region of the genome to help discover, characterize, and distinguish species, and to assign unidentified individuals to species.

DNA enrichment A process through which one captures or enriches specific regions of the genome. Often used prior to next-generation sequencing for gene-targeted sequencing.

DNA fingerprinting Individual identification through the use of multilocus genotyping.

Dobzhansky–Muller incompatibilities Genic interactions between alleles at multiple loci in which alleles that enhance fitness within their parental genetic backgrounds may reduce fitness in the novel genetic background produced by hybridization.

dominance genetic variation The proportion of total genetic variation that can be attributed to the interactions of alleles at a locus in heterozygotes.

dominant An allele whose phenotypic effect is expressed in both homozygotes (AA) and heterozygotes (Aa).

DPS See **distinct population segment**.

drift load Reduction in average fitness of a population because of the increase in frequency of deleterious alleles caused by genetic drift in small populations.

ecosystem A community of organisms and its environment.

ecosystem services The products and services humans receive from functioning ecosystems.

ecotone The region that encompasses the shift between two biological communities.

eDNA See **environmental DNA**

effective number of alleles The number of equally frequent alleles that would create the same heterozygosity as observed in the population.

effective population size (N_e) The size of the ideal population that would experience the same amount of genetic drift as the observed population.

electrophoresis The movement of molecules through a medium across an electric field. Electro-phoresis is used to separate allelic enzymes (allozymes) on the basis of differences in charge, size, or shape, and DNA molecules on the basis of differences in size.

EM See **expectation maximization algorithm**.

emergent properties When several simple entities interact to form complex behaviors or structures as a collective. The complex structures or behaviors are not a property of any single entity and cannot be predicted from the behavior of a subset of entities.

empirical probability The ratio of the number of 'favorable' outcomes to the total number of trials in an actual sequence of experiments.

endemic Present exclusively in a particular area.

endonuclease An enzyme that cleaves either a single, or both strands of a DNA molecule. Bacterial endonucleases are used to split genomic DNA at specific sites for analysis. See **restriction enzyme**.

endosymbiont Any organism living inside the body or cells of another organism, such as algae inside of coral.

environmental DNA (eDNA) DNA collected from an environmental sample such as water or soil.

environmental stochasticity Random variation in environmental factors that influence population parameters affecting all individuals in that population.

EPA Environmental Protection Agency of the United States.

epidemiology The study of the spread and control of a disease in a population.

epigenetics Changes in gene expression that are stable through cell division but do not involve changes in the underlying DNA sequence. The most common example is cellular differentiation, but it is clear that environmental factors, such as maternal nutrition, can influence epigenetic programming.

epistasis In statistical genetics, this term refers to an interaction of different loci such that the multiple locus phenotype is different than predicted by simply combining the effects of each individual locus (i.e., there is a significant gene–gene interaction).

epistatic genetic variation The proportion of total genetic variation that can be attributed to the interaction between loci producing a combined effect different from the sum of the effects of the individual loci.

ESA Endangered Species Act of the United States.

ESPA Endangered Species Protection Act of Australia.

ESU *See* **evolutionarily significant unit.**

evolutionarily significant unit (ESU) A classification of populations that have substantial reproductive isolation which has led to adaptive differences so that the population represents a significant evolutionary component of the species. The original term used was "evolutionarily" (Ryder 1986). However, both evolutionarily and evolutionary are currently used in the literature.

exact tests An approach to compute the exact *P*-value for a statistical test rather than use an approximation such as the chi-square distribution.

exclusion tests A statistical genetic test to identify likely immigrants by excluding individuals as residents because their multilocus genotype is unlikely to occur in the focal population.

exon A coding portion of a gene that produces a functional gene product (e.g., a peptide).

expectation maximization (EM) algorithm An iterative mathematical algorithm used to estimate allele or gamete frequencies. EM alternates between the expectation (E) step, which computes the log-likelihood evaluated using the current estimates for the variables, and the maximization (M) step, which computes parameters maximizing the expected log-likelihood found in the *E* step. These parameter estimates are then used to determine the distribution of the variables in the next E step, until the values converge.

exploit To use a natural resource.

ex situ **conservation** The conservation of important evolutionary lineages of species outside the species natural habitat.

extant Currently living; not extinct.

extinction The disappearance of a species or other taxon so that it no longer exists anywhere.

extinction vortex The mutual reinforcement among biotic and abiotic processes that can drive small populations to extinction. For example, small population size causes increased inbreeding, which reduces individual fitness, which can further reduce the population size, leading to further increased inbreeding, and so on.

extirpation The loss of a species or subspecies from a particular area, but not from its entire range.

FCA Frequency correspondence analysis.

fecundity The potential reproductive capacity of an individual or population (e.g., the number of eggs or young produced by an individual per unit time).

fertility The ability to conceive and have offspring. Sometimes used for fecundity.

Fisher-Wright model *See* **Wright-Fisher population model.**

fitness The ability of an individual or genotype to survive and produce viable offspring. Quantified as the number or relative proportion of offspring contributed to the next generation.

fitness rebound Following an episode of inbreeding depression, successive generations of breeding may result in a rebound in fitness due to the selective decrease in frequency of deleterious alleles (purging). If inbreeding depression is due to deleterious recessive alleles (with negative fitness effects in a homozygous state), then successive generations of inbreeding may result in a rebound in fitness due to the selective decrease in frequency of deleterious alleles.

fixation index The proportional increase of homozygosity through population subdivision. F_{ST} is sometimes referred to as the fixation index.

fluctuating asymmetry (FA) Asymmetry in which deviations from symmetry are randomly distributed about a mean of zero. FA provides a simple measure of developmental precision or stability.

forensics The use of scientific methods and techniques, such as genetic fingerprinting, to solve crimes.

foundation species A foundation species is a dominant primary producer in an ecosystem both in terms of abundance and influence.

founder effect A loss of genetic variation in a population that was established by a small number of individuals that carry only a fraction of the original genetic diversity from a larger population. A special case of genetic drift.

founder-specific inbreeding coefficient The probability that an individual is homozygous (identical by descent) for an allele descended from a specific founder. The sum, across all founders, of the founder-specific partial inbreeding coefficients for an individual is equal to the inbreeding coefficient for that individual.

frequency-dependent selection Natural selection in which fitness varies as a function of the frequency of a phenotype.

frequentist A framework for making inferences from statistical samples. The result of a frequentist approach is either to accept or reject a hypothesis following a significance test or a conclusion if a confidence interval includes the true value.

fuzzy logic A mathematical logic that attempts to solve problems by assigning values to an imprecise spectrum of data in order to arrive at the most accurate conclusion possible.

gametic disequilibrium Non-random association of alleles at different loci within a population. Also known as **linkage disequilibrium**.

gametic equilibrium Random association of alleles at different loci within a population. Also known as **linkage equilibrium**.

gametogenesis The creation of gametes through meiosis.

gene A segment of DNA whose nucleotide sequence codes for protein or RNA, or regulates other genes.

gene drop Simulation of the transmission of alleles in a pedigree. Each founder is assigned two unique alleles, and the alleles are then passed on from parent to offspring, with each offspring receiving one allele chosen at random from each parent (modeling Mendelian segregation), until all individuals in the pedigree have an assigned genotype.

gene flow Exchange of genetic information between demes through migration.

gene genealogies The tracing of the history of inheritance of the genes in an individual. Gene genealogies are most easily constructed using non-recombining DNA such as mtDNA or the mammalian Y-chromosome.

generation interval The average age of parents when their progeny are born.

genet A genetically unique individual.

genetic assimilation A process in which phenotypically plastic characters that were originally 'acquired' become converted into inherited characters by natural selection. This term has also been applied to the situation in which hybrids are fertile and displace one or both parental taxa through the production of hybrid swarms (i.e., genomic extinction).

genetic correlation The genetic relationship between two quantitative traits, i.e., the proportion of genetic variation they share.

genetic distance matrix A pairwise matrix composed of distances between population (or individual) pairs that is calculated using a measure of genetic divergence, such as F_{ST}.

genetic divergence The evolutionary change in allele frequencies between populations.

genetic draft A stochastic process in which selective substitutions at one locus will reduce genetic diversity at neutral linked loci through hitchhiking.

genetic drift Random changes in allele frequencies in a population between generations due to sampling individuals that become parents and binomial sampling of alleles during meiosis. Genetic drift is more pronounced in small populations.

genetic engineering A process in which an organism's genes are artificially modified, often through splicing DNA fragments from different chromosomes or species, to achieve a desired result. Also called **genetic modification**.

genetic exchange *See* **gene flow**.

genetic hitchhiking The increase in frequency of a neutral or weakly selected mutation due to linkage with a positively selected mutation.

genetic linkage map A map of the relative locations of loci based on the amount of recombination that occurs between the loci.

genetic load The decrease in the mean fitness of individuals in a population compared with the theoretical mean fitness if all individuals had the most favored genotype. Caused by deleterious recessive alleles (mutation load), heterozygous advantage (segregation load), and several other types of load.

genetic modification *See* **genetic engineering**.

genetic monitoring The quantification of temporal change in population genetic metrics or other population data generated using genetic markers.

genetic rescue The recovery in the average fitness of individuals through gene flow into small populations, typically following a fitness reduction due to inbreeding depression.

genetic restoration The recovery of individual fitness and adaptive potential of populations through gene flow to increase neutral and adaptive genetic variants, as well as to eliminate the effects of deleterious alleles.

genetic swamping The loss of locally adapted alleles or genotypes caused by constant immigration and gene flow.

genetic variance The variance of phenotypic values caused used by genetic differences among individuals.

genetics The study of how genes are transmitted from one generation to the next and how those genes affect the phenotypes of the progeny.

genome-wide association studies (GWAS) The genotyping of many loci in different individuals to see whether any alleles or loci are associated with a particular trait.

genomic extinction   The situation in which hybrids are fertile and displace one or both parental taxa through the production of hybrid swarms so that the parental genomes no longer exist, even though the parental alleles are still present.

genomic ratchet   A process where hybridization producing fertile offspring will result in a hybrid swarm over time, even in the presence of outbreeding depression.

genomics   The study of the structure or function of large numbers of genes or markers in a genome.

genotype   An organism's genetic composition.

genotype-by-environment interaction   The phenotypic variation that results from interactions between genes and environments.

global climate model   Complex atmospheric and oceanic models based on fluid dynamics and thermodynamics, widely used for predicting climate change.

gnomics   The study of the complete genetic information of small creatures that live in the depths of the earth and guard buried treasure.

gynodioccy   The occurrence of female and hermaphroditic individuals in a population of plants.

haplogroup   A group of similar haplotypes that share a common ancestor.

haploid   The condition in which a cell or individual has one copy of every chromosome.

haplotype   The combination of alleles at loci that are found on a single chromosome or mtDNA molecule.

Hardy-Weinberg principle   The principle that allele and genotype frequencies will reach equilibrium in binomial proportions after one generation and remain constant in large random mating populations that experience no migration, selection, mutation, or nonrandom mating.

Hardy-Weinberg proportion   A state in which a population's genotypic proportions equal those expected with the binomial distribution.

harvest   To take or kill individuals (e.g. fish, deer, or trees) for food, sport, or other uses.

hemizygous   A term used to denote the presence of only one copy of an allele due to a locus being in a haploid genome, on a sex chromosome, or only one copy of the locus being present in an aneuploid organism.

heritability   The proportion of total phenotypic variation within a population that is due to individual genetic variation (H_B; broad-sense heritability). Heritability is more commonly referred to as the proportion of phenotypic variation within a population that is due to additive genetic variation (H_N; narrow-sense heritability).

hermaphrodite   An individual that produces both female and male gametes.

heterochromatin   Highly folded chromosomal regions that contain few functional genes. When these traits are characteristic of an entire chromosome, it is a heterochromosome or supernumerary chromosome.

heterogametic   The sex with different sex chromosomes (e.g., the male in mammals, XY, and female in birds, ZW).

heteromorphic   Having different forms.

heteroplasmy   The presence of more than one mitochondrial DNA haplotype in a cell or tissue.

heterosis   The case when hybrid progeny have higher fitness than either of the parental organisms. Also called **hybrid vigor.**

heterotic   A condition where loci exhibit heterozygous advantage.

heterozygosity   A measure of genetic variation that estimates either the observed, or expected proportion of individuals in a population that are heterozygotes.

heterozygosity-fitness correlations (HFC)   The observation that individuals with greater heterozygosity at marker loci have greater fitness. See Example 10.2.

heterozygote   An organism that has different alleles at a locus (e.g., *Aa*).

heterozygous advantage   A situation where heterozygous genotypes are more fit than homozygous genotypes. This fitness advantage can create a stable polymorphism. Also called **overdominance**.

heterozygous disadvantage   A situation where heterozygous genotypes are less fit than homozygous genotypes. Also called **underdominance**.

HFC   *See* **heterozygosity-fitness correlations.**

Hill-Robertson effect   An effect where selection at one locus will reduce the effective population size of linked loci, increasing the chance of genetic drift forming negative genetic associations that reduce the ability of associating loci to respond to selection. *See* **genetic draft**.

hitchhiking   The increase in frequency of a selectively neutral allele through gametic disequilibrium with a beneficial allele that selection increases in frequency in a population.

homogametic The sex that possesses the same sex chromosomes (e.g., the female in mammals (XX) and male birds (ZZ)).

homology Similarity in structural features such as genes or morphology that are derived from a shared ancestor by common descent.

homoplasmy The presence of a single mitochondrial DNA haplotype within a cell.

homoplasy Independent evolution or origin of similar traits, or gene sequences. At a locus, homoplasy can result from back-mutation or mutation to an existing allelic state.

homozygosity A measure of the proportion of individuals in a population that are homozygous, and is the reciprocal of heterozygosity.

homozygote An organism that has two or more copies of the same allele at a locus (*AA*).

horizontal gene transfer Incorporation of genetic material into an individual without vertical (parent–offspring) transmission. Common among single-celled organisms (e.g., transfer of genes for antibiotic resistance).

HW *See* **Hardy-Weinberg principle**.

hybridization Mating between individuals of two genetically distinct populations.

hybrid swarm A population of individuals that are all hybrids by varying numbers of generations of backcrossing with parental types and matings among hybrids.

hybrid sink The situation where immigration of locally unfit genotypes produces hybrids with low fitness that reduces local density and thereby increases the immigration rate.

hybrid vigor *See* **heterosis**.

hybrid zone An area of sympatry between two genetically distinct populations where hybridization occurs without forming a hybrid swarm in either parental population beyond the area of co-occurrence.

ideal population *See* **Wright-Fisher population model**.

identical by descent Alleles that are copies of the same allele from a common ancestor.

identity disequilibrium Nonrandom association of diploid genotypes in zygotes (e.g., more multiple-locus heterozygotes or homozygotes than expected based on single-locus heterozygosities).

inbreeding The mating between related individuals which results in an increase of homozygosity in the progeny because they possess alleles that are identical by descent.

inbreeding coefficient A measure of the level of inbreeding in a population, developed by Sewall Wright, that determines the probability that an individual possesses two alleles at a locus that are identical by decent. It can also be used to describe the proportion of loci in an individual that are homozygous.

inbreeding depression The relative reduction in fitness of progeny from matings between related individuals compared with progeny from unrelated individuals.

inbreeding effective number (N_{eI}) The size of the ideal population that loses heterozygosity at the same rate as the observed population.

***in situ* conservation** The conservation of a population or species in its natural habitat.

introduction The placement, or escape of a species, or individual into a novel habitat.

introgression The incorporation of genes from one population to another through hybridization that results in fertile offspring that further hybridize and backcross to parental populations.

intron A portion of a gene that produces a section of RNA strand that is cleaved prior to translation; a noncoding region between the exons.

invasive species An introduced alien species that is likely to cause harm to the natural ecosystem, the economy, or human health.

island model of migration A model of migration in which a population is subdivided into a series of demes, of equal size *N*, that randomly exchange migrants at a given rate, *m*.

isolation by distance In general, the case where genetic differentiation is greater the further that individuals (or populations) are from each other because gene flow decreases as geographic distance increases. Originally used when individuals are distributed continuously (e.g., coniferous tree species across boreal forests) and are not subdivided into discrete local populations by barriers to gene flow.

ISSR Inter-simple sequence repeat markers that use PCR with primers based on simple sequence repeats of microsatellites.

IUCN World Conservation Union (formerly International Union for Conservation of Nature).

karyotype The composition of the chromosomal complement of a cell, individual, or species.

landscape genetics The study of the influence of the landscape or environmental features on the genetics of populations.

landscape genomics Landscape genetics with many markers often in adaptive genes and with samples of individuals across environmental gradients to study both neutral and adaptive patterns and processes (e.g., adaptive differentiation and gene flow).

landscape resistance barrier A feature of the landscape (e.g., a river or land-cover type) that impedes movement and therefore reduces gene flow.

LE *See* **lethal equivalent**.

leading edge population A population at the expanding margin of a species range during a range shift.

least cost path Path with the least cumulative resistance to movement in landscape genetics. The least cost path is often longer in geographic distance than the direct straight-line path because the direct path often crosses poor habitat with high ecological cost to movement.

lek A specific area where the males of a population congregate and display for females.

lethal equivalent The number of deleterious alleles in an individual whose cumulative effect is the same as that of a single lethal allele. For example, four alleles, each of which would be lethal 25% of the time (or to 25% of their bearers), are equivalent to one lethal allele.

library Collection of DNA fragments from a given organism 'stored' in a virus or bacteria, or ligated to short DNA sequences (adaptors) used for next-generation sequencing.

likelihood statistics An approach for parameter estimation and hypothesis testing that involves building a model (i.e., a likelihood function) and the use of the raw data (not a summary statistic), which often provides more precision and accuracy than frequentist statistic approaches (method of moments). The parameter of interest is estimated as the member of the parameter space that maximizes the probability of obtaining your observed data. Likelihood approaches facilitate comparisons between different models (e.g., via likelihood ratio tests) and thus the testing of alternate hypotheses (e.g., stable versus declining population size).

lineage sorting A process where different gene lineages within an ancestral taxon are lost by drift or replaced by unique lineages evolving in a different derived taxon.

linkage The nonrandom segregation of two loci on the same chromosome. Parental combinations of alleles are found more frequently in offspring than nonparental combinations. Linkage is measured by the rate of recombination (r) between loci; $r < 0.5$ for linked loci, and 0.50 for unlinked loci.

linkage disequilibrium (LD) Non-random association of alleles at different loci within a population. Also known as **gametic disequilibrium**.

linkage equilibrium Random association of alleles at different loci within a population. Also known as **gametic equilibrium**.

local adaptation Greater fitness of individuals in their local habitats due to natural selection.

local scale The spatial scale at which individuals routinely interact with their environment.

locus The position on a chromosome of a gene or other marker.

LOD *See* **log of odds ratio**.

log of odds ratio (LOD) The odds ratio is the odds of an event occurring in one group to the odds of it occurring in another group. For example, if 80% of the individuals in a population are *Aa* and 20% are *AA*, then the odds of *Aa* over *AA* is four; there are four (4.0) times as many *Aa* as *AA* genotypes. The natural log of this ratio is often computed because it is convenient to work with statistically.

management unit (MU) A local population that is managed as a distinct unit because of its demographic independence.

marginal overdominance Greater overall fitness associated with heterozygous genotypes, which are not the most fit in any single environment, due to an organism's interactions with multiple environments that each favor different alleles.

Markov chain A mathematical system that undergoes transitions from one state to another, as a random process in which the next state depends only on the current state.

Markov chain Monte Carlo (MCMC) A tool or algorithm for sampling from probability distributions based on constructing a Markov chain. The state of the chain after many steps is then used as a sample from the desired distribution. Sometimes called a random walk Monte Carlo method.

match probability (MP) The probability of sampling an individual that has an identical multilocus genotype to the one already sampled ('in hand').

maternal effects The influence of the genotype or phenotype of the mother on the phenotype of the offspring. Maternal effects are not heritable because they have no genetic basis in the offspring.

maximum likelihood A statistical method of determining which of two or more competing alternative hypotheses (such as alternative phylogenetic trees) yields the best fit to the data.

maximum likelihood estimate (MLE) A method of parameter estimation that obtains the parameter value that maximizes the likelihood of the observed data.

MCMC *See* **Markov chain Monte Carlo**.

MDS *See* **multidimensional scaling**.

mean kinship (*mk*) A measure of how closely related each individual is to the entire population. Animals with lower *mk* have relatively fewer genes in common with the rest of the population, and are therefore more genetically valuable in a breeding program.

Mendelian segregation The random separation of paired alleles (or chromosomes) into different gametes.

meristic character A trait of an organism that can be counted using integers (e.g., fin rays or ribs).

metacentric A chromosome in which the centromere is centrally located.

metagenomics The analysis of DNA from the many species contained in an environmental sample, facilitated by high-throughput sequencing.

metapopulation A collection of spatially divided subpopulations that experience a certain degree of gene flow among them.

metapopulation scale The spatial scale at which individuals migrate between local subpopulations, often across habitat that is unsuitable for colonization.

microchromosomes Small chromosomes which, unlike heterochromosomes, carry functional genes.

microsatellite Tandemly repeated DNA consisting of short sequences of 1 to 6 nucleotides repeated approximately 5 to 100 times. Also known as **VNTRs** or **SSRs**.

migration The movement of individuals from one genetically distinct population to another, resulting in gene flow.

minimum viable population (MVP) The minimum population size at which a population is likely to persist over some defined period of time.

minisatellite Regions of DNA in which repeat units of 10–50 base pairs are tandemly arranged in arrays of 0.5–30 kb in length.

mitochondrial DNA (mtDNA) A small, circular, haploid DNA molecule found in the mitochondria cellular organelle of eukaryotes.

ML *See* **maximum likelihood**.

MLE *See* **maximum likelihood estimate**.

molecular clock The observation that mutations sometimes accumulate at relatively constant rates, thereby allowing researchers to estimate the time since two species shared a common ancestor.

molecular genetics The branch of genetics that studies the molecular structure and function of genes, or that (more generally) uses molecular markers to test hypotheses.

molecular mutations Changes to the genetic material of a cell, including single nucleotide changes, deletions, and insertions of nucleotides as well as recombinations and inversions of DNA sequences.

moments Quantitative measures of the shape of a distribution. The first moment of the distribution of the random variable X is the population mean. The second and third moments are the variance and skewness, respectively.

monoecious A plant in which male and female organs are found on the same plant but in different structures (for example, flowers in maize or cones in pines).

monomorphic The presence of only one allele at a locus, or the presence of a common allele at a high frequency (>95% or 99%) in a population.

monophyletic A group of taxa that include all species, ancestral and derived, from a common ancestor.

monophyly The presence of a monophyletic group.

monotypic A taxonomic group that encompasses only one taxonomic representative. The reptile family that contains tuatara (Sphenodontia) and the phylum containing ginkgo (Ginkgophyta) are currently monotypic.

morphology The study of the physical structures of an organism, including the evolution and development of these structures.

MP *See* **match probability**.

MRCA Most recent common ancestor.

mRNA Messenger ribonucleic acid.

MSD   Multiple factor sex determination.

mtDNA   *See* **mitochondrial DNA**.

MU   *See* **management unit**.

multidimensional scaling   A statistical graphing technique used to represent genetic distances between samples in two or three dimensions, and thereby visualizing similarities and differences between different groups or samples.

mutagenesis   The natural or intentional formation of mutations in a genome.

mutation   A change in the DNA sequence or chromosome in the transmission of genetic information from parent to progeny. *See also* **molecular mutations.**

mutational meltdown   The process by which a small population accumulates deleterious mutations, which leads to loss of fitness and decline of the population size, which leads to further accumulation of deleterious mutations. A population experiencing mutational meltdown is trapped in a downward spiral and eventually will go extinct.

mutualist species   A species in a symbiotic relationship in which both species benefit.

MVP   *See* **minimum viable population**.

narrow sense heritability (H_N)   The amount of individual phenotypic variation that is due to additive genetic variation.

native species   A species that was not introduced, and historically or currently occurs in a given ecosystem.

natural catastrophes   Natural events causing great damage to populations and that increase their probability of extinction.

natural selection   Differential contribution of genotypes to the next generation due to differences in survival and reproduction.

NCA   *See* **nested clade phylogeographic analysis**.

NCPA   *See* **nested clade phylogeographic analysis**.

nearly exact test   A method of using a nearly exact *P* value to test whether the observed test statistic deviates from the expected value under the null hypothesis. For example, a test of whether populations are in Hardy-Weinberg proportions by comparing the observed chi-squared value to the chi-squared values of random computer permutations of genotypes from the population's allele frequencies.

neighborhood   The area in a continuously distributed population that can be considered panmictic.

nested clade analysis (NCA)   *See* **nested clade phylogeographic analysis**.

nested clade phylogeographic analysis (NCPA)   A statistical approach to describe how genetic variation is distributed spatially within a species' geographic range. This method uses a haplotype tree to define a nested series of branches (clades), thereby allowing a nested analysis of the spatial distribution of genetic variation, often with the goal of resolving between past fragmentation, colonization, or range expansion events.

neutral allele   An allele that is not under selection because it has no effect on fitness.

next-generation sequencing   New DNA sequencing technologies that produce millions of short reads (from 25–500 bp) in a short time (1–5 days).

NMFS   US National Marine Fisheries Service.

NOAA   US National Oceanic Atmospheric Administration.

node   A branching point or end point on a phylogenic tree that represents either an ancestral taxon (internal node) or an extant taxon (external node).

nongenetic inheritance   Any effect on the phenotype of offspring brought about by factors other than DNA sequences from parents or more remote ancestors. *See* **maternal effects** and **epigenetics**.

nonindigenous species   Species present in a given ecosystem that were introduced and did not historically occur in that ecosystem.

noninvasive   Sampling DNA with no contact or skin break of a target organism. For example, collection of sloughed skin, shed feathers, feces, urine, or hair from a tree's bark or barbed wire fencing.

nuclear DNA (nDNA)   DNA that forms chromosomes in the cell nucleus of eukaryotes.

nuclear gene   A gene located on a chromosome in the nucleus of a eukaryotic cell.

nucleotides   The building blocks of DNA and RNA made up of a nitrogen-containing purine or pyrimidine base linked to a sugar (ribose or deoxyribose) and a phosphate group.

null allele   An allele that is not detectable either due to a failure to produce a functional product (e.g., allozymes) or a mutation in a primer site that precludes amplification during PCR analysis (e.g. for microsatellites).

Ockham's razor   The principle that the least complicated explanation (most parsimonious hypothesis) generally should be accepted to explain the data at hand.

offsite conservation *See ex situ* **conservation**.

outbreeding depression The reduction in fitness of hybrids compared with parental types.

outlier loci Loci that might be under selection (or in gametic disequilibrium with loci under selection) that are detected because they fall outside the range of the expected distribution for some summary statistic compared with that of neutral loci in a sample (e.g., extremely high or low values of F_{ST} or F_{IS}).

overdominance *See* **heterozygous advantage**.

overexploitation Use of a resource to the point of diminishing returns or destruction of the resource.

overlapping generations A breeding system where sexual maturity does not occur at a specific age, or where individuals breed more than once, causing individuals of different ages to interbreed in a given year.

paleoecology The study of ancient organisms and their interactions with environments, usually based on fossils.

panmictic Randomly mating.

paracentric inversion A chromosomal inversion that does not include the centromere because both breaks were on the same chromosomal arm.

parameter Numerical characteristic of a statistical population or model, often referring to the population parametric value computed from the total population of observations (not a sample).

paraphyletic A clade that does not include all of the descendants from the most recent common ancestor taxon. For example, reptiles are paraphyletic because they do not include birds.

parentage analysis The assessment of the maternity and/or paternity of a given individual.

parsimony The principle that the preferred phylogeny of an organism is the one that requires the fewest evolutionary changes; the simplest explanation.

parthenogenetic A form of asexual reproduction without fertilization.

PAW Partnership for Action against Wildlife Crime.

PCA Principal component analysis.

PCoA Principle coordinates analysis.

PCR *See* **polymerase chain reaction**.

pdf Probability density function.

PE Paternity exclusion (probability of).

pericentric inversion A chromosomal inversion that includes the centromere because the breaks were on opposite chromosomal arms.

phenetics Taxonomic classification solely based on overall similarity (generally of phenotypic traits) regardless of phylogeny.

phenogram A branching diagram or tree that is based on estimates of overall similarity between taxa derived from a suite of characters.

phenology The study of periodic plant and animal lifecycle events and how these are influenced by seasonal and interannual variations in climate.

phenotype The observable characteristics of an organism that are the product of the organism's genotype and environment.

phenotypic Relating to an aspect of an individual's phenotype.

phenotypic plasticity Phenotypic differences between individuals with similar genotypes or for an individual genotype over time due to differences in environmental factors during development.

philopatry A characteristic of reproduction of organisms where individuals faithfully home to natal sites. Individuals exhibiting philopatry are philopatric.

phylogenetic Evolutionary relationships between taxa or gene lineages. These relationships are often expressed visually in phylogenetic trees with nodes representing taxa or lineages (ancestral or derived), and branch lengths often corresponding to the amount of divergence between groups.

phylogenetic species concept (PSC) States that a species is a discrete lineage or recognizable monophyletic group.

phylogeny *See* **cladogram**.

phylogeography The assessment of the geographic distributions of the taxa of a phylogeny to understand the evolutionary history (e.g., origin and spread) of a given taxon.

phylopatry A characteristic of reproduction of organisms where individuals faithfully home to natal sites. Individuals exhibiting phylopatry are phylopatric.

PI *See* **probability of identity**.

pleiotropy The case where one gene affects more than one phenotypic trait.

PMRNs *See* **probabilistic maturation reaction norms**.

Poisson distribution A probability distribution, with identical mean and variance, that characterizes discrete events occurring independently of one another in time, when the mean probability of that event on any one trial is very small. Earthquake hazards, radioactive decay, and mutation events follow a Poisson distribution. The Poisson is a good approximation to the binomial distribution.

polygenic   A phenotype affected by more than one gene.

polymerase   A molecule that catalyzes the synthesis of DNA or RNA from a single-stranded template and free deoxynucleotides (e.g., during PCR).

polymerase chain reaction (PCR)   A technique to replicate a desired segment of DNA. PCR starts with primers that flank the desired target fragment of DNA. The DNA strands are first separated with heat, and then cooled allowing the primers to bind to their target sites. Polymerase then makes each single strand into a double strand, starting from the primer. This cycle is repeated multiple times creating a 10^6 increase in the gene product after 20 cycles and a 10^9 increase over 30 cycles.

polymorphic   The presence of more than one allele at a locus. Generally defined as when the most common allele is at a frequency less than 95% or 99%.

polymorphism   The presence of more than one allele at a locus. Polymorphism is also used as a measure of the proportion of loci in a population that are genetically variable or polymorphic (P).

polyphyletic   A group of taxa classified together that have descended from different ancestor taxa (i.e., taxa that do not all share the same recent common ancestor).

polyploid   Individuals whose genome consists of more than two sets of chromosomes (e.g., tetraploids).

polytomy   A node on a phylogenetic tree from which more than two branches emerge. Sometimes used either to indicate radiation events or ambiguities in knowledge about relationships.

population viability   The probability that enough individuals in a population will survive to reproductive age to prevent extirpation of the population.

population viability analysis (PVA)   The general term for the application of models that account for multiple threats facing the persistence of a population, to access the likelihood of the population's persistence over a given period of time.

primer   A small oligonucleotide (typically 18–22 base pairs long) that anneals to a specific single stranded DNA sequence to serve as a starting point for DNA replication (e.g., extension by polymerase during PCR).

prior probability   The prior probability distribution (i.e., 'prior') of an uncertain quantity k (e.g., the number of populations represented in a sample of individuals) is the probability distribution that would express one's uncertainty about k before the data (e.g., genotypes) are taken into account. It attributes uncertainty rather than randomness to the quantity.

private allele   An allele present in only one of many populations sampled.

probabilistic maturation reaction norm (PMRN)   A reaction norm is the set of phenotypes expressed by a single genotype across a range of environments reflecting phenotypic plasticity. A probabilistic maturation reaction norm describes an individual's probability of maturing at a given age as a function of size and other relevant phenotypic traits.

probability   The certainty of an event occurring. The observed probability of an event, r, will approach the true probability as the number of trials, n, approaches infinity.

probability of identity (*PI*)   The probability that two unrelated (randomly sampled) individuals would have an identical genotype. This probability becomes very small if many highly polymorphic loci are considered.

product rule   A statistical rule that states that the probability of n_i independent events occurring is equal to the product of the probability of each n independent event.

propagule   A dispersal vector. Any disseminative unit or part of an organism capable of independent growth (e.g., a seed, spore, mycelial fragment, sclerotium bud, tuber, root, or shoot).

propagule pressure   A measure of the introduction of nonindigenous individuals that includes the number of individuals (or propagules) introduced and the number of introductions.

proportion of admixture   The proportion of alleles in a hybrid swarm that come from each of the parental taxa.

protein   A polypeptide molecule.

PSC   *See* **phylogenetic species concept**.

pseudo-overdominance   *See* **associative overdominance**.

purging   The removal of deleterious recessive alleles from a population through inbreeding, which increases homozygosity, which in turn increases the ability of selection to act on recessive alleles.

PVA *See* **population viability analysis**.

qPCR *See* **quantitative PCR**.

QTL *See* **quantitative trait loci**.

quantitative PCR PCR that amplifies and simultaneously quantifies the number of copies of the targeted DNA molecule.

quantitative trait A phenotype (characteristic) that varies in degree due to polygenic effects (of two or more loci) and the environment.

quantitative trait loci Genetic loci that affect phenotypic variation (and potentially fitness), which are identified by a statistically significant association between genetic markers and measurable phenotypes. Quantitative traits are often influenced by multiple loci as well as environmental factors.

RADs Restriction-site associated DNA markers.

RAD sequencing Sequencing of the DNA segment that immediately flanks each side of a restriction enzyme site throughout the genome to discover or genotype SNPs.

RAPD Randomly amplified polymorphic DNA. A method of analysis in which PCR amplification using arbitrary oligonucleotide primers are used to create a multilocus band profile. RAPDs are no longer used because of their poor repeatability.

rarefaction A technique to compare allelic diversity in samples of different sizes. Rarefaction estimates the expected total number of alleles in a smaller sample drawn at random from a large pool. Rarefaction allows the comparison of the allelic diversity in samples of sizes.

rear edge population A population at the contracting or trailing edge of a species range during range migration.

reciprocal monophyly A genetic lineage is reciprocally monophyletic when all members of the lineage share a more recent common ancestor with each other than with any other lineage on a phylogenetic tree.

recombination The process that generates a haploid product of meiosis with a genotype differing from both the haploid genotypes that originally combined to form the diploid zygote.

reintroduction The introduction of a species or population into a historic habitat from which it had previously been extirpated.

relative fitness A measure of fitness that is the ratio of a given genotype's absolute fitness to the genotype with the highest absolute fitness. Relative fitness is used to model the effects of natural selection on allele frequencies.

resistance models Spatial models in which each location across a landscape is given a weight or 'resistance value' reflecting the influence of variables (e.g., land cover, slope, elevation) on movement by the species in question.

rescue effect When immigration into an isolated deme reduces the probability of the extinction of that deme either because of genetic or demographic effects. *See* **demographic rescue** and **genetic rescue**.

restriction enzyme An enzyme (*see* **endonuclease**), isolated from bacteria, that cleaves DNA at a specific nucleotide sequence. Over 3000 such enzymes exist that recognize and cut hundreds of different DNA sequences; used in RFLP, RAD, and AFLP analysis and to construct recombinant DNA (in genetic engineering).

restriction fragment length polymorphism (RFLP) A method of genetic analysis that examines polymorphisms based on differences in the number of fragments produced by the digestion of DNA with specific endonucleases. The variation in the number of fragments is created by mutations within restriction sites for a given endonuclease.

retroposon A mobile DNA sequence that can move to new locations through an RNA intermediate.

reverse mutation rate Back mutation rate. The rate at which a gene's ability to produce a functional product is restored. This rate is much lower than the forward mutation rate because there are many more ways to remove the function of a gene than restore it. Also used to describe mutation at microsatellite loci where (under the stepwise mutation model, for example) a back mutation yields an allele of length that already exists (i.e., homoplasy) in the population.

reverse speciation The loss of differences between species brought about by hybridization and admixture of two isolated gene pools into a single admixed population.

RFLP *See* **restriction fragment length polymorphism**.

ribonucleic acid (RNA) A polynucleotide similar to DNA that contains ribose in place of deoxyribose, and uracil in place of thymine. RNA is involved in the transfer of information from DNA, programming protein synthesis, and maintaining ribosome structure.

Robertsonian fission   An event where a metacentric chromosome breaks near the centromere to form two acrocentric chromosomes.

Robertsonian fusion   An event where two acrocentric chromosomes fuse to form one metacentric chromosome.

Robertsonian translocation   A special type of translocation where the break occurs near the centromere or telomere and involves the whole chromosomal arm so balanced gametes are usually produced.

sampling distribution   The sampling distribution of a statistic is the distribution of that statistic, considered as a random variable, when derived from a random sample.

SARA   *See* **Species at Risk Act of Canada**.

secondary contact zone   Contact between populations that had previously been geographically separate (i.e., allopatric).

selection coefficient   The reduction in relative fitness, and therefore genetic contribution to future generations, of one genotype compared with another.

selection differential   The difference between the mean value of a quantitative trait found in a population as a whole compared with the mean value of the trait in the selected breeding population.

selective sweep   The rapid increase in frequency by natural selection of an initially rare allele that also fixes (or nearly fixes) alleles at closely linked loci and thus reduces the genetic variation in a region of a chromosome.

sensitivity testing   A method used in population viability analyses where the effects of parameters on the persistence of populations are determined by testing a range of possible values for each parameter.

sequential Bonferroni correction   A method, similar to the Bonferroni correction, that is used to reduce the probability of a Type I statistical error when conducting multiple simultaneous tests.

sex chromosomes   Chromosomes that pair during meiosis but differ in the heterogametic sex.

sex-linked locus   A locus that is located on a sex chromosome.

sexual selection   Selection due to differential mating success either through competition for mates or mate choice.

shadow effect   A case usually caused by low marker polymorphism in mark–recapture studies in which a novel capture is labeled as a recapture due to identical genotypes at the loci studied.

SINEs   Short interspersed nuclear elements.

single nucleotide polymorphism (SNP)   A nucleotide site (base pair) in a DNA sequence that is polymorphic in a population and can be used as a marker to assess genetic variation within and among populations. Usually only two alleles exist for a SNP in a population.

SMM   *See* **stepwise mutation model**.

SNP   *See* **single nucleotide polymorphism**.

soft sweep   A selective substitution involving more than a single copy of the selected allele either taken from standing genetic variation or if new beneficial alleles arise by mutation during the spread to fixation.

spatial sorting   An evolutionary process that increases the frequency of phenotypes adept at rapid dispersal. This process assembles phenotypes through space, for example at an invasion front, rather than through time.

species   A group of organisms with a high degree of physical and genetic similarity, that naturally interbreed among themselves and can be differentiated from members of related groups of organisms.

Species at Risk Act of Canada (SARA)   Legislation (passed in 2002) to prevent wildlife species from becoming extinct and secure the necessary actions for their recovery. It provides for the legal protection of wildlife species and the conservation of critical habitat.

species concepts   The ideas of what constitutes a species, such as reproductive isolation (**Biological Species Concept**, BSC), or monophyly of a lineage (**Phylogenetic Species Concept**).

species scale   The spatial scale encompassing an entire species' distribution.

SSRs   Simple sequence repeats. *See* **microsatellite**.

stabilizing selection   Selection for a phenotype with a more intermediate state than either extreme states of the phenotype.

stable equilibrium   An equilibrium of allele frequencies to which the population returns if allele frequencies are perturbed away from the equilibrium.

stable polymorphism A polymorphism that is maintained at a locus through natural selection.

standard error The standard deviation of the sampling distribution of a statistic.

statistic A numerical measure of some attribute of a sample such as its mean or median.

statistical inference The process of drawing conclusions from data that are subject to random variation such as sampling variation.

stepping stone model of migration A model of migration that accounts for variation in the probability of migration over differing distances. This model is less effective at reducing genetic drift because of the decrease in gene flow between distant populations as a result of the lower probability of long-distance movements.

stepwise mutation model (SMM) A model of mutation in which the microsatellite allele length has an equal probability of either increasing, or decreasing (usually by a single repeat unit, as in the strict one-step SMM).

stochastic The presence of a random variable in determining the outcome of an event.

stock A term generally used in fisheries management that refers to a population that is demographically independent and often represents a subunit (e.g., MU) of an ESU.

STR Short tandem repeat. *See* **microsatellite**.

subfossil Remains not ancient enough to be considered true fossils, but not recent enough to be considered modern.

subpopulations Groups within a population delineated by reduced levels of gene flow with other groups.

subspecies A taxonomically defined subdivision within a species that is physically or genetically distinct, and often geographically separated.

sum rule A statistical rule that states that the probability of n_i mutually exclusive, independent events occurring is equal to the sum of the probabilities of each n event.

supergene Allelic combinations found at closely linked loci that affect related traits and are inherited together. An example of a supergene is the major histocompatibility complex (MHC), which in humans contains more than 200 genes adjacently located over several megabases of sequence on chromosome 6.

supernumerary chromosome A chromosome, often present in varying numbers, that is not needed for normal development, lacks functional genes, and does not segregate during meiosis. These small chromosomes, which are also called B chromosomes, are present in addition to the normal complement of functional chromosomes in an organism.

supportive breeding The practice of removing a subset of individuals from a wild population for captive breeding and releasing the captive-born offspring back into their native habitat to intermix with wild-born individuals and increase population size or persistence.

sympatric Populations or species that occupy the same geographic area.

synapomorphy A shared derived trait between evolutionary lineages. A homology that evolved in an ancestor common to all species on one branch of a phylogeny, but not common to species on other branches.

syntenic The location of two or more loci on the same chromosome.

Taq The bacterium *Thermus aquaticus* from which a heat-stable DNA polymerase used in PCR was isolated.

telomere A region of tandemly repeated segments of a short DNA sequence, one strand of which is G-rich and the other strand is C-rich, that forms the ends of linear eukaryotic chromosomes.

threshold The point at which environmental (or genetic) changes produce large phenotypic changes in an organism (or population). For example, there could be a threshold effect of inbreeding on fitness such that, after a certain level of inbreeding is reached, individual fitness declines increasingly rapidly.

threshold character A phenotypic character that contains a few discrete states controlled by many genes underlying continuous variation, which affects a character phenotypically only when a certain physiological threshold is exceeded.

TMRCA Time since the most recent common ancestor.

TRAFFIC An international wildlife trade monitoring network sponsored by the WWF and IUCN.

transgenic An individual or species with genes inserted from another species (e.g., a **genetically engineered** organism).

transgressive segregation Hybridization events that produce progeny that express phenotypic values outside the range of either parental phenotypic value. These differences are usually due to the disruption of polygenic traits.

transition A point mutation in which a purine base (A or G) is substituted for a different purine base, and a pyrimidine base (C or T) is substituted for a different pyrimidine base; for example, an A:T to G:C transition.

translocation   (1) The movement of individuals from one population (or location) to another that is usually intended to achieve either genetic or demographic rescue of an isolated population, or to allow adaptation to a rapidly changing climate. (2) A rearrangement occurring when a piece of one chromosome is broken off and joined to another chromosome.

transposable element   Any DNA sequence that can insert into a chromosome, exit, and relocate; includes insertion sequences, transposons, some bacteriophages, and controlling elements. A region of the genome, flanked by inverted repeats, a copy of which can be inserted at another place; also called a transposon or a jumping gene.

transposon   A mobile element of DNA that jumps to new genomic locations through a DNA intermediate and which usually carries genes other than those that encode for transposase proteins used to catalyze movement.

transversion   A point mutation in which a purine base is substituted for a pyrimidine base or vice versa; for example, an A:T to C:G transversion. *See* **transition**.

Type I statistical error   The probability of rejecting a true null hypothesis.

Type II statistical error   The probability of failing to reject a false null hypothesis.

underdominance *See* **heterozygous disadvantage**.

UNEP   United Nations Environmental Program.

unstable equilibrium   An equilibrium of allele frequencies in which allele frequencies move away from the equilibrium if they are perturbed.

UPGMA   Unweighted pair group method with arithmetic averages.

USFWS   US Fish and Wildlife Service.

variance   A statistical measure of variation calculated as the mean of the squared deviations from the arithmetic mean (the standard deviation squared). In quantitative genetics, phenotypic variance is partitioned into genetic and environmental components.

variance effective number (N_{ev})   The size of the ideal population that experiences changes in allele frequency at the same rate as the observed population.

viability   The probability of the survival of a given genotype to reproductive maturity (or of a population to persist through a certain time interval).

VNTRs   Variable number of tandem repeats. *See* **microsatellite** and **minisatellite.**

Wahlund effect   The deficit of heterozygotes compared with expected Hardy-Weinberg proportions because of the presence of two or more panmictic (random mating) demes.

Wright–Fisher population model   A constant-size population of size N in which the next generation is produced by drawing $2N$ genes at random from a large gamete pool to which all individuals contribute equally; therefore, selfing is possible. Allele frequencies change from generation to generation only by genetic drift.

WWF   World Wide Fund for Nature (formerly known as the World Wildlife Fund).

zoonoses   Any disease of nonhuman animals that can be transmitted to humans.

References

Abi-Rached, L., *et al.* 2011. The shaping of modern human immune systems by multiregional admixture with archaic humans. Science 334:89–94.

Ackerly, D. D., *et al.* 2010. The geography of climate change: implications for conservation biogeography. Diversity and Distributions 16:476–487.

Ackerman, M. W., C. Habicht, and L. W. Seeb. 2011. Single nucleotide polymorphisms (SNPs) under diversifying selection provide increased accuracy and precision in mixed stock analyses of sockeye salmon from Copper River, Alaska. Transactions of the American Fisheries Society 140:865–881.

Adams, J. M., G. Piovesan, S. Strauss, and S. Brown. 2002. The case for genetic engineering of native and landscape trees against introduced pests and diseases. Conservation Biology 16:874–879.

Adams, J. R., *et al.* 2011. Genomic sweep and potential genetic rescue during limiting environmental conditions in an isolated wolf population. Proceedings of the Royal Society B: Biological Sciences 278:3336–3344.

Adamski, P., and Z. J. Witkowski. 2007. Effectiveness of population recovery projects based on captive breeding. Biological Conservation 140:1–7.

Agashe, D. 2009. The stabilizing effect of intraspecific genetic variation on population dynamics in novel and ancestral habitats. American Naturalist 174:255–267.

Aitken, S. N., *et al.* 2008. Adaptation, migration or extirpation: climate change outcomes for tree populations. Evolutionary Applications 1:95–111.

Akçakaya, H. R., *et al.* 2000. Making consistent IUCN classifications under uncertainty. Conservation Biology 14:1001–1013.

Akey, J. M., *et al.* 2002. Interrogating a high-density SNP map for signatures of natural selection. Genome Research 12:1805–1814.

Alatalo, R. V., and A. Lundberg. 1986. Heritability and selection on tarsus length in the pied flycatcher (*Ficedula hypoleuca*). Evolution 40:574–583.

Alford, R. A., *et al.* 2009. Comparisons through time and space suggest rapid evolution of dispersal behaviour in an invasive species. Wildlife Research 36:23–28.

Allen, P. J., W. Amos, P. P. Pomeroy, and S. D. Twiss. 1995. Microsatellite variation in grey seals (*Halichoerus grypus*) shows evidence of genetic differentiation between two British breeding colonies. Molecular Ecology 4:653–662.

Allendorf, F. W. 1983. Isolation, gene flow, and genetic differentiation among populations. Pp. 51–65 *in* Schonewald-Cox, C., S. Chambers, B. MacBryde, and L. Thomas, eds. Genetics and Conservation. Benjamin/Cummings, Menlo Park, CA.

Allendorf, F. W. 1986. Genetic drift and the loss of alleles versus heterozygosity. Zoo Biology 5:181–190.

Allendorf, F. W. 1993. Delay of adaptation to captive breeding by equalizing family size. Conservation Biology 7:416–419.

Allendorf, F. W., and J. J. Hard. 2009. Human-induced evolution caused by unnatural selection through harvest of wild animals. Proceedings of the National Academy of Sciences USA 106:9987–9994.

Allendorf, F. W., and R. F. Leary. 1988. Conservation and distribution of genetic variation in a polytypic species: the cutthroat trout. Conservation Biology 2:170–184.

Allendorf, F. W., and L. L. Lundquist. 2003. Introduction: Population biology, evolution, and control of invasive species. Conservation Biology 17:24–30.

Allendorf, F. W., and S. R. Phelps. 1981. Use of allelic frequencies to describe population structure. Canadian Journal of Fisheries and Aquatic Sciences 38:1507–1514.

Allendorf, F. W., and N. Ryman. 2002. The role of genetics in population viability analysis. Pp. 50–85 *in* Beissinger, S. R.,

Conservation and the Genetics of Populations, Second Edition. Fred W. Allendorf, Gordon Luikart, and Sally N. Aitken.
© 2013 Fred W. Allendorf, Gordon Luikart and Sally N. Aitken. Published 2013 by Blackwell Publishing Ltd.

and D. R. McCullough, eds. Population Viability Analysis. University of Chicago Press, Chicago, Illinois.

Allendorf, F. W., and L. W. Seeb. 2000. Concordance of genetic divergence among sockeye salmon populations at allozyme, nuclear DNA, and mitochondrial DNA markers. Evolution 54:640–651.

Allendorf, F. W., and C. Servheen. 1986. Genetics and conservation of grizzly bears. Trends in Ecology & Evolution 1:88–89.

Allendorf, F. W., and R. S. Waples. 1996. Conservation and genetics of salmonid fishes. Pp. 238–280 *in* Avise, J. C., and J. L. Hamrick, eds. Conservation Genetics: Case Histories from Nature. Chapman Hall, New York.

Allendorf, F. W., K. L. Knudsen, and G. M. Blake. 1982. Frequencies of null alleles at enzyme loci in natural populations of ponderosa and red pine. Genetics 100:497–504.

Allendorf, F. W., G. Ståhl, and N. Ryman. 1984. Silencing of duplicate genes: a null allele polymorphism for lactate dehydrogenase in brown trout (*Salmo trutta*). Molecular Biology and Evolution 1:238–248.

Allendorf, F. W., W. A. Gellman, and G. H. Thorgaard. 1994. Sex-linkage of two enzyme loci in *Oncorhynchus mykiss* (rainbow trout). Heredity 72:498–507.

Allendorf, F. W., R. F. Leary, P. Spruell, and J. K. Wenburg. 2001. The problems with hybrids: setting conservation guidelines. Trends in Ecology & Evolution 16:613–622.

Allendorf, F. W., *et al.* 2004. Intercrosses and the US Endangered Species Act: Should hybridized populations be included as westslope cutthroat trout? Conservation Biology 18:1203–1213.

Allendorf, F. W., P. A. Hohenlohe, and G. Luikart. 2010. Genomics and the future of conservation genetics. Nature Reviews Genetics 11:697–709.

Ally, D., K. Ritland, and S. P. Otto. 2010. Aging in a long-lived clonal tree. PLoS Biol 8:e1000454.

Alstad, D. N. 2001. *Basic Populus Models of Ecology*. Prentice-Hall, Inc. Upper Saddle River, New Jersey.

Amato, G., R. DeSalle, O. A. Ryder, and H. C. Rosenbaum. 2009. Conservation Genetics in the Age of Genomics. Columbia University Press, New York.

Anderson, J. T., C.-R. Lee, and T. Mitchell-Olds. 2011. Life-history QTLs and natural selection on flowering time in *Boechera stricta*, a perennial relative of *Arabidopsis*. Evolution 65:771–787.

Angers, B., and I. J. Schlosser. 2007. The origin of *Phoxinus eos-neogaeus* unisexual hybrids. Molecular Ecology 16:4562–4571.

Anmarkrud, J., O. Kleven, L. Bachmann, and J. T. Lifjeld. 2008. Microsatellite evolution: Mutations, sequence variation, and homoplasy in the hypervariable avian microsatellite locus *HrU10*. BMC Evolutionary Biology 8:138.

Antao, T., A. Perez-Figueroa, and G. Luikart. 2011. Early detection of population declines: high power of genetic monitoring using effective population size estimators. Evolutionary Applications 4:144–154.

Antilla, C. K., C. C. Daehler, N. E. Rank, and D. R. Strong. 1998. Greater male fitness of a rare invader (*Spartina alterniflora*, Poaceae) threatens a common native (*Spartina foliosa*) with hybridization. American Journal of Botany 85:1597–1601.

Antolin, M. F. 1999. A genetic perspective on mating systems and sex ratios of parasitoid wasps. Researches on Population Ecology 41:29–37.

Araguas, R. M., *et al.* 2009. Role of genetic refuges in the restoration of native gene pools of brown trout. Conservation Biology 23:871–878.

Araki, H., and C. Schmid. 2010. Is hatchery stocking a help or harm? Evidence, limitations and future directions in ecological and genetic surveys. Aquaculture 308:S2–S11.

Aravanopoulos, F. A. 2011. Genetic monitoring in natural perennial plant populations. Botany 89:75–81.

Archie, E. A., G. Luikart, and V. O. Ezenwa. 2009. Infecting epidemiology with genetics: A new frontier in disease ecology. Trends in Ecology & Evolution 24:21–30.

Archie, J. W. 1985. Statistical analysis of heterozygosity data: Independent sample comparisons. Evolution 39:623–637.

Ardren, W. R., and A. R. Kapuscinski. 2003. Demographic and genetic estimates of effective population size (N_e) reveals genetic compensation in steelhead trout. Molecular Ecology 12:35–49.

Armbruster, P., and D. H. Reed. 2005. Inbreeding depression in benign and stressful environments. Heredity 95:235–242.

Armbruster, P., R. A. Hutchinson, and T. Linvell. 2000. Equivalent inbreeding depression under laboratory and field conditions in a tree-hole-breeding mosquito. Proceedings of the Royal Society B: Biological Sciences 267:1939–1945.

Armsworth, P. R., *et al.* 2007. Ecosystem-service science and the way forward for conservation. Conservation Biology 21:1383–1384.

Arnold, M. L. 1997. Natural Hybridization and Evolution. Oxford University Press, New York.

Arnold, M. L., and N. H. Martin. 2010. Hybrid fitness across time and habitats. Trends in Ecology & Evolution 25:530–536.

Arnold, S. J. 1981. Behavioral variation in natural populations. II. The inheritance of a feeding response in crosses between geographic races of the garter snake *Thamnophis elegans*. Evolution 35:510–515.

Ashley, M. V., *et al.* 2003. Evolutionarily enlightened management. Biological Conservation 111:115–123.

Aspinwall, N. 1974. Genetic analysis of North American populations of the pink salmon (*Oncorhynchus gorbuscha*), possible evidence for the neutral mutation-random drift hypothesis. Evolution 28:295–305.

Aubry, K. B., and J. C. Lewis. 2003. Extirpation and reintroduction of fishers (*Martes pennanti*) in Oregon: implications for their conservation in the Pacific states. Biological Conservation 114:79–90.

Avise, J. C. 1986. Mitochondrial DNA and the evolutionary genetics of higher animals. Proceedings of the Royal Society B: Biological Sciences 312:325–342.

Avise, J. C. 1990. Mitochondrial gene trees and the evolutionary relationship of mallard and black ducks. Evolution 44:1109–1119.

Avise, J. C. 1992. Molecular population structure and the biogeographic history of a regional fauna – a case history with lessons for conservation biology. Oikos 63:62–76.

Avise, J. C. 1994. Molecular Markers, Natural History, and Evolution. Chapman & Hall, New York.

Avise, J. C. 1996. Toward a regional conservation genetics perspective: Phylogeography of faunas in the southeastern united states. Pp. 431–470 in Avise, J. C., and J. L. Hamrick, eds. Conservation Genetics: Case Histories from Nature. Chapman & Hall, New York.

Avise, J. C. 2000. Cladists in wonderland. Evolution 54: 1828–1832.

Avise, J. C. 2004. Molecular Markers, Natural History, and Evolution, 2nd edn. Chapman & Hall, New York.

Avise, J. C. 2008. The history, purview, and future of conservation genetics. Pp. 5–15 in Carroll, S. P., and C. W. Fox, eds. Conservation Biology: Evolution in Action. Oxford University Press, New York.

Avise, J. C. 2009. Phylogeography: retrospect and prospect. Journal of Biogeography 36:3–15.

Avise, J. C. 2010. Conservation genetics enters the genomics era. Conservation Genetics 11:665–669.

Avise, J. C., R. A. Lansman, and R. O. Shade. 1979a. The use of restriction endonucleases to measure mitochondrial DNA sequence relatedness in natural populations. I. Population structure and evolution in the genus *Peromyscus*. Genetics 92:279–295.

Avise, J. C., et al. 1979b. Mitochondrial DNA clones and matriarchal phylogeny within and among geographic populations of the pocket gopher, *Geomys pinetis*. Proceedings of the National Academy of Sciences of the USA 76: 6694–6698.

Avise, J. C., et al. 1987. Intraspecific phylogeography: the mitochondrial DNA bridge between population genetics and systematics. Annual Review of Ecology and Systematics 18:489–522.

Awadalla, P., A. Eyrewalker, and J. M. Smith. 1999. Linkage disequilibrium and recombination in hominid mitochondrial DNA. Science 286:2524–2525.

Ayala, F. J., and J. R. Powell. 1972. Allozymes as diagnostic characters of sibling species of *Drosophila*. Proceedings of the National Academy of Sciences of the USA 69:1094–1096.

Ayre, D. J., R. J. Whelan, and A. Reid. 1994. Unexpectedly high levels of selfing in the Australian shrub *Grevillea barklyana* (Proteaceae). Heredity 72:168–174.

Ayres, K. L., and D. J. Balding. 1998. Measuring departures from Hardy-Weinberg: a Markov chain Monte Carlo method for estimating the inbreeding coefficient. Heredity 80:769–777.

Ayres, K. L., and A. D. J. Overall. 2004. API-CALC 1.0: a computer program for calculating the average probability of identity allowing for substructure, inbreeding and the presence of close relatives. Molecular Ecology Notes 4:315–318.

Baer, C. F., and J. Travis. 2000. Direct and correlated responses to artificial selection on acute thermal stress tolerance in a livebearing fish. Evolution 54:238–244.

Bahlo, M., and R. C. Griffiths. 2000. Inference from gene trees in a subdivided population. Theoretical Population Biology 57:79–95.

Bailey, J. K., et al. 2004. Beavers as molecular geneticists: A genetic basis to the foraging of an ecosystem engineer. Ecology 85:603–608.

Baillie, J. E. M., C. Hilton-Taylor, and S. N. Stuart. 2004. 2004 IUCN Red List of Threatened Species: A Global Species Assessment. IUCN Species Survival Commission, Gland, Switzerland, and Cambridge, UK.

Baird, N. A., et al. 2008. Rapid SNP discovery and genetic mapping using sequenced RAD markers. PLoS One 3:7.

Baker, A. J. 1992. Genetic and morphometric divergence in ancestral European and descent New Zealand populations of chaffinches (*Fringilla coelebs*). Evolution 46: 1784–1800.

Baker, C. S. 2008. A truer measure of the market: the molecular ecology of fisheries and wildlife trade. Molecular Ecology 17:3985–3998.

Baker, C. S., and S. R. Palumbi. 1994. Which whales are hunted? A molecular genetic approach to monitoring whaling. Science 265:1538–1539.

Baker, C. S., F. Cipriano, and S. R. Palumbi. 1996. Molecular genetic identification of whale and dolphin products from commercial markets in Korea and Japan. Molecular Ecology 5:671–687.

Baker, C. S., G. M. Lento, F. Cipriano, and S. R. Palumbi. 2000. Predicted decline of protected whales based on molecular genetic monitoring of Japanese and Korean markets. Proceedings of the Royal Society B: Biological Sciences 267:1191–1199.

Baker, C. S., M. L. Dalebout, G. M. Lento, and N. Funahashi. 2002. Gray whale products sold in commercial markets along the Pacific Coast of Japan. Marine Mammal Science 18:295–300.

Baker, C. S., et al. 2007. Estimating the number of whales entering trade using DNA profiling and capture–recapture analysis of market products. Molecular Ecology 16: 2617–2626.

Baker, C. S., et al. 2010. Genetic evidence of illegal trade in protected whales links Japan with the US and South Korea. Biology Letters 6:647–650.

Baker, H. G. 1995. Aspects of the genecology of weeds. Pp. 189–224 in Kruckeberg, A. R., R. B. Walker, and A. E. Leviton, eds. Genecology and Ecogeographic Races. Pacific Division of the American Association for the Advancement of Science, San Francisco.

Balanya, J., *et al.* 2006. Global genetic change tracks global climate warming in *Drosophila subobscura*. Science 313: 1773–1775.

Baldwin, B. G., and R. H. Robichaux. 1995. Historical biogeography and ecology of the Hawaiian silversword alliance (Asteraceae). Pp. 259–287 *in* Wagner, W. L., and V. A. Funk, eds. Hawaiian Biogeography: Evolution on a Hot Spot Archipelago. Smithsonian Institution Press, Washington, D.C.

Ballard, J. W. O., and M. C. Whitlock. 2004. The incomplete natural history of mitochondria. Molecular Ecology 13:729–744.

Ballou, J. 1983. Calculating inbreeding coefficients from pedigrees. Pp. 509–520 *in* Schonewald-Cox, C., S. Chambers, B. MacBryde, and L. Thomas, eds. Genetics and Conservation. Benjamin/Cummings, Menlo Park, CA.

Ballou, J. D. 1997. Ancestral inbreeding only minimally affects inbreeding depression in mammalian populations. Journal of Heredity 88:169–178.

Ballou, J. D., and T. J. Foose. 1996. Demographic and genetic management of captive populations. Pp. 263–283 *in* Kleiman, D. G., M. E. Allen, K. V. Thompson, and S. Lumpkin, eds. Wild Mammals in Captivity. University of Chicago Press, Chicago.

Ballou, J. D., and R. C. Lacy. 1995. Identifying genetically important individuals for management of genetic variation in pedigreed populations. Pp. 76–111 *in* Ballou, J. D., M. Gilpin, and T. J. Foose, eds. Population Management for Survival and Recovery. Columbia University Press, New York.

Ballou, J., and J. D. Seidensticker. 1987. The genetic and demographic characteristics of the 1983 captive population of Sumatran tigers (*Panthera tigris sumatrae*). Pp. 329–347 *in* Tilson, R. L., and U. S. Seal, eds. Tigers of the World: The Biology, Biopolitics, Management, and Conservation of an Endangered Species. Noyes Publications, Park Ridge, New Jersey.

Ballou, J. D., Gilpin, M., and Foose, T. J. 1994. Population Management for Survival and Recovery. Columbia University Press, New York.

Balloux, F. 2001. EASYPOP (Version 1.7): A computer program for population genetics simulations. Journal of Heredity 92:301–302.

Balloux, F. 2010. The worm in the fruit of the mitochondrial DNA tree. Heredity 104:419–420.

Balloux, F., N. Lugon-Moulin, and J. Hausser. 2000. Estimating gene flow across hybrid zones: how reliable are microsatellites? Acta Theriologica 45:93 101.

Balloux, F., W. Amos, and T. Coulson. 2004. Does heterozygosity estimate inbreeding in real populations? Molecular Ecology 13:3021–3031.

Balmford, A., N. Leader-Williams, and M. J. B. Green. 1995. Parks or arks: Where to conserve threatened mammals? Biodiversity and Conservation 4:595–607.

Balmford, A., G. M. Mace, and N. Leaderwilliams. 1996. Designing the ark: setting priorities for captive breeding. Conservation Biology 10:719–727.

Banks, M. A., and W. Eichert. 2000. WHICHRUN (Version 3.2): A computer program for population assignment of individuals based on multilocus genotype data. Journal of Heredity 91:87–89.

Barash, D. P. 1973. The ecologist as Zen master. The American Midland Naturalist 89:214–217.

Barbour, R. C., Y. Otahal, R. E. Vaillancourt, and B. M. Potts. 2008. Assessing the risk of pollen-mediated gene flow from exotic *Eucalyptus globulus* plantations into native eucalypt populations of Australia. Biological Conservation 141:896–907.

Barilani, M., *et al.* 2005. Detecting hybridization in wild (*Coturnix c. coturnix*) and domesticated (*Coturnix c. japonica*) quail populations. Biological Conservation 126:445–455.

Barker, J. S. F. 1994. Animal breeding and conservation genetics. Pp. 381–395 *in* Loeschcke, V., J. Tomiuk, and S. K. Jain, eds. Conservation Genetics. Birkhauser Verlag, Basel, Switzerland.

Barnosky, A. D., *et al.* 2011. Has the Earth's sixth mass extinction already arrived? Nature 471:51–57.

Barot, S., M. Heino, M. J. Morgan, and U. Dieckmann. 2005. Maturation of Newfoundland American plaice (*Hippoglossoides platessoides*): Long-term trends in maturation reaction norms despite low fishing mortality? ICES Journal of Marine Science 62:56–64.

Barrett, S. C. H. 2000. Microevolutionary influences of global changes on plant invasions. Pp. 115–139 *in* Mooney, H. A., and R. J. Hobbs, eds. Invasive Species in a Changing World. Island Press, Washington, DC.

Barrett, S. C. H., and J. R. Kohn. 1989. Quoted in "How to get plants into the conservationists' ark" by R. Lewin. Science 244:32–33.

Barrett, S. C. H., and J. R. Kohn. 1991. Genetic and evolutionary consequences of small population size in plants – implications for conservation. Pp. 3–30 *in* Falk, D. A., and K. E. Holsinger, eds. Genetics and Conservation of Rare Plants. Oxford University Press, New York.

Barrett, S. C. H., and B. C. Husband. 1990. The genetics of plant migration and colonization. Pp. 254–278 *in* Brown, A. H. D., M. T. Clegg, A. L. Kahler, and B. S. Weir, eds. Plant Population Genetics, Breeding, and Genetic Resources. Sinauer, Sunderland, MA.

Barton, N. H. 2000. Estimating multilocus linkage disequilibria. Heredity 84:373–389.

Barton, N. H. 2011. Estimating linkage disequilibria. Heredity 106:205–206.

Barton, N. H., and K. S. Gale. 1993. Genetic analysis of hybrid zones. Pp. 13–45 *in* Harrison, R. G., ed. Hybrid Zones and the Evolutionary Process. Oxford University Press, Oxford.

Barton, N. H., and G. M. Hewitt. 1985. Analysis of hybrid zones. Annual Review of Ecology and Systematics 16:113–148.

Barton, N. H., and P. D. Keightley. 2002. Understanding quantitative genetic variation. Nature Reviews Genetics 3:11–21.

Baskett, M. L., S. A. Levin, S. D. Gaines, and J. Dushoff. 2005. Marine reserve design and the evolution of size at maturation in harvested fish. Ecological Applications 15: 882–901.

Baskin, Y. 2002. A Plague of Rats and Rubbervines: The Growing Threat of Species Invasions. Island Press, Washington, DC.

Bass, R. A. 1979. Chromosomal polymorphism in cardinals, *Cardinalis cardinalis*. Canadian Journal of Genetics and Cytology 21:549–553.

Bataillon, T. 2003. Shaking the 'deleterious mutations' dogma? Trends in Ecology & Evolution 18:315–317.

Bateson, W. 1912. The Methods and Scope of Genetics. Cambridge University Press, Cambridge.

Battaglia, E. 1964. Cytogenetics of the B-chromosomes. Caryologia 8:205–213.

Bawa, K. S., S. Menon, and L. R. Gorman. 1997. Cloning and conservation of biological diversity: paradox, panacea, or Pandora's box? Conservation Biology 11:829–830.

Beaumont, M. A. 1999. Detecting population expansion and decline using microsatellites. Genetics 153:2013–2029.

Beaumont, M., and B. Rannala. 2004. The Bayesian revolution in genetics. Nature Reviews Genetics 5:251–261.

Beaumont, M., *et al.* 2001. Genetic diversity and introgression in the Scottish wildcat. Molecular Ecology 10:319–336.

Beaumont, M., W. Zhang, and D. Balding. 2002. Approximate Bayesian computation in population genetics. Genetics 162:2025–2035.

Beaumont, M. A., *et al.* 2010. In defence of model-based inference in phylogeography. Molecular Ecology 19: 436–446.

Beerli, P. 2006. Comparison of Bayesian and maximum-likelihood inference of population genetic parameters. Bioinformatics 22:341–345.

Beerli, P., and J. Felsenstein. 2001. Maximum likelihood estimation of a migration matrix and effective population sizes in *n* subpopulations by using a coalescent approach. Proceedings of the National Academy of Sciences of the USA 98:4563–4568.

Beever, E. A., *et al.* 2008. American pikas (*Ochotona princeps*) in northwestern Nevada: A newly discovered population at a low-elevation site. Western North American Naturalist 68:8–14.

Beever, E. A., C. Ray, P. W. Mote, and J. L. Wilkening. 2010. Testing alternative models of climate-mediated extirpations. Ecological Applications 20:164–178.

Begg, A. G., and G. Marteinsdóttir. 2003. Spatial and temporal partitioning of spawning stock biomass: effects of fishing on the composition of spawners. Fisheries Research 59:343–362.

Begun, D. J., and C. F. Aquadro. 1992. Levels of naturally occurring DNA polymorphism correlate with recombination rates in *D. melanogaster*. Nature 356:519–520.

Beissinger, S. R., and M. I. Westphal. 1998. On the use of demographic models of population viability in endangered species management. Journal of Wildlife Management 62:821–841.

Beja-Pereira, A., *et al.* 2009a. DNA genotyping suggests that recent brucellosis outbreaks in the Greater Yellowstone Area originated from elk. Journal of Wildlife Diseases 45:1174–1177.

Beja-Pereira, A., *et al.* 2009b. Advancing ecological understandings through technological transformations in noninvasive genetics. Molecular Ecology Resources 9:1279–1301.

Beletsky, L. D., and G. H. Orians. 1989. A male red-winged blackbird breeds for 11 years. Northwestern Naturalist 70:10–12.

Bellemain, E., *et al.* 2005. Estimating population size of elusive animals with DNA from hunter-collected feces: four methods for brown bears. Conservation Biology 19:150–161.

Bellen, H. J., *et al.* 1992. The Drosophila *couch potato* gene: an essential gene required for normal adult behavior. Genetics 131:365–375.

Belov, K. 2011. The role of the major histocompatibility complex in the spread of contagious cancers. Mammalian Genome 22:83–90.

Belovsky, G. E. 1987. Extinction models and mammalian persistence. Pp. 225–242 *in* Soulé, M. E., ed. Viable Populations for Conservation. Sinauer, Sunderland, MA.

Benirschke, K., and A. T. Kumamoto. 1991. Mammalian cytogenetics and conservation of species. Journal of Heredity 82:187–191.

Bennett, D. C., and M. L. Lamoreux. 2003. The color loci of mice – a genetic century. Pigment Cell Research 16:333–344.

Bennuah, S. Y., T. Wang, and S. N. Aitken. 2004. Genetic analysis of the *Picea sitchensis* X *glauca* introgression zone in British Columbia. Forest Ecology and Management 197:65–77.

Bensch, S., *et al.* 2006. Selection for heterozygosity gives hope to a wild population of inbred wolves. PLoS ONE 1:e72.

Bentzen, P. 1998. Seeking evidence of local stock structure using molecular genetic methods. *In* Hunt von Herbing, I., I. Kornfield, M. Tupper, and I. Wilson, eds. The Implications of Localized Fishery Stocks. Regional Agricultural Engineering Service, New York.

Bergl, R. A., and L. Vigilant. 2007. Genetic analysis reveals population structure and recent migration within the highly fragmented range of the Cross River gorilla (*Gorilla gorilla diehli*). Molecular Ecology 16:501–516.

Berlocher, S. H. 1984. Genetic changes coinciding with the colonization of California by the walnut husk fly, *Rhagoletis completa*. Evolution 38:906–918.

Bernatchez, L., H. Glemet, C. C. Wilson, and R. G. Danzmann. 1995. Introgression and fixation of Arctic char (*Salvelinus alpinus*) mitochondrial genome in an allopatric population of brook trout (*Salvelinus fontinalis*). Canadian Journal of Fisheries and Aquatic Sciences 52:179–185.

Berra, T. M., G. Alvarez, and F. C. Ceballos. 2010. Was the Darwin/Wedgwood dynasty adversely affected by consanguinity? BioScience 60:376–383.

Berry, O., and R. Kirkwood. 2010. Measuring recruitment in an invasive species to determine eradication potential. Journal of Wildlife Management 74:1661–1670.

Berteaux, D., D. Reale, A. G. McAdam, and S. Boutin. 2004. Keeping pace with fast climate change: can arctic life count on evolution? Integrative and Comparative Biology 44: 140–151.

Berthier, P., M. A. Beaumont, J.-M. Cornuet, and G. Luikart. 2002. Likelihood-based estimation of the effective population size using temporal changes in allele frequencies: a genealogical approach. Genetics 160:741–751.

Berthold, P. 1991. Genetic control of migratory behaviour in birds. Trends in Ecology & Evolution 6:254–257.

Berthold, P., and A. J. Helbig. 1992. The genetics of bird migration – stimulus, timing, and direction. Ibis 134: 35–40.

Bertorelle, G., A. Benazzo, and S. Mona. 2010. ABC as a flexible framework to estimate demography over space and time: Some cons, many pros. Molecular Ecology 19:2609–2625.

Berven, K. A., D. E. Gill, and S. J. Smith-Gill. 1979. Countergradient selection in the green frog *Rana clamitans*. Evolution 33:609–623.

Beverton, R. J. H., *et al.* 1984. Dynamics of single species: group report. Pp. 13–58 *in* May, R. M., ed. Dahlem Konferenzen. Springer, Berlin.

Biek R., A. Drummond, and M. Poss. 2006. A virus reveals population structure and recent demographic history of carnivore host. Science 311:538–541.

Biek, R., *et al.* 2007. A high-resolution genetic signature of demographic and spatial expansion in epizootic rabies virus. Proceedings of the National Academy of Sciences 104:7993–7998.

Bijlsma, R., J. Bundgaard, A. C. Boerema, and W. F. van Putten. 1997. Genetic and environmental stress, and the persistence of populations. Pp. 193–207 *in* Bijlsma, R., and V. Loeschcke, eds. Environmental Stress, Adaptation and Evolution. Birkhauser Verlag, Basel, Switzerland.

Bijlsma, R., J. Bundgaard, and W. F. van Putten. 1999. Environmental dependence of inbreeding depression and purging in *Drosophila melanogaster*. Journal of Evolutionary Biology 12:1125–1137.

Birkeland, C., and P. K. Dayton. 2005. The importance in fishery management of leaving the big ones. Trends in Ecology & Evolution 20:356–358.

Birky, C. W. Jr., T. Maruyama, and P. Fuerst. 1983. An approach to population and evolutionary genetic theory for genes in mitochondria and chloroplasts, and some results. Genetics 103:513–527.

Bishop, J. M., A. J. Leslie, S. L. Bourquin, and C. O'Ryan. 2009. Reduced effective population size in an overexploited population of the Nile crocodile (*Crocodylus niloticus*). Biological Conservation 142:2335–2341.

Bittner, T. D., and R. B. King. 2003. Gene flow and melanism in garter snakes revisited: a comparison of molecular markers and island vs. coalescent models. Biological Journal of the Linnean Society 79:389–399.

Black, F. L., and P. W. Hedrick. 1997. Strong balancing selection at HLA loci: evidence from segregation in South Amerindian families. Proceedings National Academy of Sciences of the USA 94:12452–12456.

Blair, A. C., and R. A. Hufbauer. 2010. Hybridization and invasion: one of North America's most devastating invasive plants shows evidence for a history of interspecific hybridization. Evolutionary Applications 3:40–51.

Blanchet, S., D. J. Paez, L. Bernatchez, and J. J. Dodson. 2008. An integrated comparison of captive-bred and wild Atlantic salmon (*Salmo salar*): Implications for supportive breeding programs. Biological Conservation 141:1989–1999.

Blanchong, J. A., *et al.* 2008. Landscape genetics and the spatial distribution of chronic wasting disease. Biology Letters 4:130–133.

Boakes, E. H., J. Wang, and W. Amos. 2007. An investigation of inbreeding depression and purging in captive pedigreed populations. Heredity 98:172–182.

Boecklen, W. J., and D. J. Howard. 1997. Genetic analysis of hybrid zones: numbers of markers and power of resolution. Ecology 78:2611–2616.

Boerner, M., and O. Krüger. 2009. Aggression and fitness differences between plumage morphs in the common buzzard (*Buteo buteo*). Behavioral Ecology 20:180–185.

Bogart, J. P. 1980. Evolutionary implications of polyploidy in amphibians and reptiles. Pp. 341–378 *in* Lewis, W., ed. Polyploidy: Biological Relevance. Plenum, New York.

Bollmer, J. L., F. H. Vargas, and P. G. Parker. 2007. Low MHC variation in the endangered Galapagos penguin (*Spheniscus mendiculus*). Immunogenetics 59:593–602.

Bonduriansky, R., and T. Day. 2009. Nongenetic inheritance and its evolutionary implications. Annual Review of Ecology Evolution and Systematics 40:103–125.

Bonfil, R., *et al.* 2005. Transoceanic migration, spatial dynamics, and population linkages of white sharks. Science 310:100–103.

Bonin, A., *et al.* 2007. Population adaptive index: a new method to help measure intraspecific genetic diversity and prioritize populations for conservation. Conservation Biology 21:697–708.

Bonneaud, C., J. Burnside, and S. V. Edwards. 2008. High speed developments in avian genomics. BioScience 58:587–595.

Bory, S., M. Grisoni, M.-F. Duval, and P. Besse. 2008. Biodiversity and preservation of vanilla: Present state of knowledge. Genetic Resources and Crop Evolution 55:551–571.

Bossdorf, O., C. L. Richards, and M. Pigliucci. 2008. Epigenetics for ecologists. Ecology Letters 11:106–115.

Boulanger, J., *et al.* 2008. Use of occupancy models to estimate the influence of previous live captures on DNA-based detection probabilities of grizzly bears. Journal of Wildlife Management 72:589–595.

Bouzat, J. 2010. Conservation genetics of population bottlenecks: The role of chance, selection, and history. Conservation Genetics 11:463–478.

Bowcock, A. M., et al. 1991. Drift, admixture, and selection in human evolution: A study with DNA polymorphisms. Proceedings National Academy of Sciences USA 88: 839–843.

Bowen, B. W. 1999. Preserving genes, species, or ecosystems? Healing the fractured foundations of conservation policy. Molecular Ecology 8:S5–S10.

Bowen, B. W., A. L. Bass, L. Soares, and R. J. Toonen. 2005. Conservation implications of complex population structure: lessons from the loggerhead turtle (Caretta caretta). Molecular Ecology 14:2389–2402.

Bowen, B. W., et al. 2007. Mixed-stock analysis reveals the migrations of juvenile hawksbill turtles (Eretmochelys imbricata) in the Caribbean Sea. Molecular Ecology 16:49–60.

Bower, A. D., and S. N. Aitken. 2008. Ecological genetics and seed transfer guidelines for Pinus albicaulis (Pinaceae). American Journal of Botany 95:66–76.

Boyer, M. C., C. C. Muhlfeld, and F. W. Allendorf. 2008. Rainbow trout (Oncorhynchus mykiss) invasion and the spread of hybridization with native westslope cutthroat trout (Oncorhynchus clarkii lewisi). Canadian Journal of Fisheries and Aquatic Sciences 65:658–669.

Bradley, K. M., et al. 2011. An SNP-based linkage map for zebrafish reveals sex determination loci. G3: Genes, Genomes, Genetics 1:3–9,

Bradshaw, H. D., and D. W. Schemske. 2003. Allele substitution at a flower colour locus produces a pollinator shift in monkeyflowers. Nature 426:176–178.

Bravington, M., and P. Grewe. 2007. A method for estimating the absolute spawning stock size of SBT, using close-kin genetics. Scientific Committee Report CCSBT-SC/0709/18, Commission for the Conservation of Southern Blue Fin Tuna.

Brenner, S. 2000. Genomics: the end of the beginning. Science 287:2173–2174.

Brewer, B. A., R. C. Lacy, M. L. Foster, and G. Alaks. 1990. Inbreeding depression in insular and central populations of Peromyscus mice. Journal of Heredity 81:257–266.

Brindley, D. C., and W. G. Banfield. 1961. Contagious tumor of hamster. Journal of the National Cancer Institute 26:949–957.

Brisbin, I. L., and A. T. Peterson. 2007. Playing chicken with red junglefowl: identifying phenotypic markers of genetic purity in Gallus gallus. Animal Conservation 10:429–435.

Briscoe, D. A., et al. 1992. Rapid loss of genetic variation in large captive populations of Drosophila flies: implications for the genetic management of captive populations. Conservation Biology 6:416–425.

Bromham, L., and P. H. Harvey. 1996. Behavioural ecology: Naked mole-rats on the move. Current Biology 6: 1082–1083.

Brook, B. W., et al. 2000. Predictive accuracy of population viability analysis in conservation biology. Nature 404:385–387.

Brookfield, J. F. Y. 1996. A simple new method for estimating null allele frequency from heterozygote deficiency. Molecular Ecology 5:453–455.

Brown, A. H. D. 1979. Enzyme polymorphisms in plant populations. Theoretical Population Biology 15:1–42.

Brown, A. H. D., and J. D. Briggs. 1991. Sampling strategies for genetic variation in ex situ collections of endangered plant species. Pp. 99–119 in Falk, D. A., and K. E. Holsinger, eds. Genetics and Conservation of Rare Plants. Oxford University Press, New York.

Brown, A. H. D., and D. R. Marshall. 1981. Evolutionary changes accompanying colonization in plants. Pp. 351–363 in Scudder, G. G. E., and J. L. Reveal, eds. Evolution Today, Proceedings of Second International Congress of Systematic and Evolutionary Biology. Hunt Institute for Botanical Documentation, Carnegie-Mellon University, Pittsburgh, PA.

Brown, A. H. D., and A. G. Young. 2000. Genetic diversity in tetraploid populations of the endangered daisy Rutidosis leptorrhynchoides and implications for its conservation. Heredity 85:122–129.

Brown, C. R., and M. B. Brown. 1998. Intense natural selection on body size and wing and tail asymmetry in cliff swallows during severe weather. Evolution 52:1461–1475.

Brown, J. H., and A. Kodric-Brown. 1977. Turnover rates in insular biogeography: effect of immigration on extinction. Ecology 58:445–449.

Brown, O. J. F. 2006. Tasmanian devil (Sarcophilus harrisii) extinction on the Australian mainland in the mid-holocene: Multicausality and ENSO intensification. Alcheringa: An Australasian Journal of Palaeontology 30, Supplement 1:49–57.

Brown, W. M., and J. W. Wright. 1979. Mitochondrial DNA analysis and the origin and relative age of parthenogenetic lizards (Genus Cnemidophorus). Science 203: 1247–1249.

Brown, W. M., M. George, and A. C. Wilson. 1979. Rapid evolution of animal mitochondrial DNA. Proceedings of the National Academy of Sciences USA 76:1967–1971.

Bruford, M. W., O. Hanotte, J. F. Y. Brookfield, and T. Burke. 1998. Multi-locus and single-locus DNA fingerprinting. Pp. 287–336 in Hoelzel, A. R., ed. Molecular Genetic Analysis of Populations. Oxford University Press, New York.

Brumfield, R. T., P. Beerli, D. A. Nickerson, and S. V. Edwards. 2003. The utility of single nucleotide polymorphisms in inferences of population history. Trends in Ecology & Evolution 18:249–256.

Brünner, H., and J. Hausser. 1996. Genetic and karyotypic structure of a hybrid zone between the chromosomal races Cordon and Valais in the common shrew, Sorex araneus. Hereditas 125:147–158.

Bryant, E. H., and D. H. Reed. 1999. Fitness decline under relaxed selection in captive populations. Conservation Biology 13:665–669.

Bucklin, A., D. Steinke, and L. Blanco-Bercial. 2011. DNA barcoding of marine metazoa. Annual Review of Marine Science 3:471–508.

Buerkle, C. A., D. E. Wolf, and L. H. Rieseberg. 2003. The origin and extinction of species through hybridization. Pp. 117–141 in Brigham, C. A., and M. W. Schwartz, ed. Population Viability in Plants: Conservation, Management, and Modeling of Rare Plants. Springer Verlag, Berlin.

Buggs, R. J. A. 2007. Empirical study of hybrid zone movement. Heredity 99:301–312.

Buktenica, M. W. 1997. Bull trout restoration and brook trout eradication at Crater Lake National Park, Oregon. Pp. 127–136 in Mackay, W. C., M. K. Brewin, and M. Monita, eds. Friends of the Bull Trout Conference Proceedings. Bull Trout Task Force (Alberta), Trout Unlimited Canada, Calgary.

Buktenica, M. W., et al. In press. Eradication of non-native brook trout using electrofishing and antimycin-A and the response of a remnant bull trout population. North American Journal of Fisheries Management.

Bull, J. J. 1983. Evolution of Sex Determining Mechanisms. Benjamin/Cummings, Menlo Park, CA.

Bunce, M., et al. 2003. Extreme reversed sexual size dimorphism in the extinct New Zealand moa Dinornis. Nature 425:172–175.

Bürger, R., and M. Lynch. 1995. Evolution and extinction in a changing environment: a quantitative-genetic analysis. Evolution 49:151–163.

Buřič, M., et al. 2011. A successful crayfish invader is capable of facultative parthenogenesis: A novel reproductive mode in decapod crustaceans. PLoS ONE 6.

Burney, D. A., and T. F. Flannery. 2005. Fifty millennia of catastrophic extinctions after human contact. Trends in Ecology & Evolution 20:395–401.

Bush, G. L. 1994. Sympatric speciation in animals: New wine in old bottles. Trends in Ecology & Evolution 9:285–288.

Bush, M., et al. 1996. Radiographic evaluation of diaphragmatic defects in gloden lion tamarins (Leontopithecus rosalia rosalia): implications for reintroduction. Journal of Zoo and Wildlife Medicine 27:346–357.

Byers, D. L., and D. M. Waller. 1999. Do plant populations purge their genetic load? Effects of population size and mating history on inbreeding depression. Annual Review of Ecology and Systematics 30:479–513.

Byers, J. A. 1997. American Pronghorn. University of Chicago Press, Chicago, IL.

Byrne, M., B. Macdonald, and D. Coates. 1999. Divergence in the chloroplast genome and nuclear rDNA of the rare Western Australian plant Lambertia orbifolia Gardner (Proteaceae). Molecular Ecology 8:1789–1796.

Cade, T. J. 1983. Hybridization and gene exchange among birds in relation to conservation. Pp. 288–310 in Schonewald-Cox, C., S. Chambers, B. MacBryde, and L. Thomas, eds. Genetics and Conservation. Benjamin/Cummings, Menlo Park, CA.

Calmet, C., M. Pascal, and S. Samadi. 2001. Is it worth eradicating the invasive pest Rattus norvegicus from Molene archipelago? Genetic structure as a decision-making tool. Biodiversity and Conservation 10:911–928.

Calzada, J. P. V., C. F. Crane, and D. M. Stelly. 1996. Botany – apomixis: the asexual revolution. Science 274:1322–1323.

Campbell, N., and S. Narum. 2009. Quantitative pcr assessment of microsatellite and SNP genotyping with variable quality DNA extracts. Conservation Genetics 10:779–784.

Campbell, R. K., and A. I. Sugano. 1993. Genetic variation and seed zones of Douglas-fir in the Siskiyou National Forest. USDA Forest Service Pacific Northwest Research Station Research Paper U1.

Campton, D. E. 1987. Natural hybridization and introgression in fishes: methods of detection and genetic interpretations. Pp. 161–192 in Ryman, N., and F. Utter, eds. Population Genetics and Fishery Management. University of Washington Press, Seattle, WA.

Capy, P., G. Gasperi, C. Biemont, and C. Bazin. 2000. Stress and transposable elements: co-evolution or useful parasites? Heredity 85:101–106.

Carney, S. E., K. A. Gardner, and L. H. Rieseberg. 2000. Evolutionary changes over the fifty-year history of a hybrid population of sunflowers (Helianthus). Evolution 54:462–474.

Caro, T. 2007. Behavior and conservation: a bridge too far? Trends in Ecology & Evolution 22:394–400.

Caro, T. M., and M. K. Laurenson. 1994. Ecological and genetic factors in conservation – a cautionary tale. Science 263:485–486.

Carr, D. E., and M. R. Dudash. 2003. Recent approaches into the genetic basis of inbreeding depression in plants. Philosophical Transactions of the Royal Society of London, Series B, Biological Sciences 358:1071–1084.

Carr, G. D., and A. C. Medeiros. 1998. A remnant greensword population from pu'u 'alaea, maui, with characteristics of Argyroxiphium virescens (Asteraceae). Pacific Science 52:61–68.

Carrington, M., et al. 1999. HLA and HIV-1: Heterozygote advantage and B*35-Cw*04 disadvantage. Science 283:1748–1752.

Carstens, B. C., J. D. Degenhardt, A. L. Stevenson, and J. Sullivan. 2005. Accounting for coalescent stochasticity in testing phylogcographical hypotheses: modelling Pleistocene population structure in the Idaho giant salamander Dicamptodon aterrimus. Molecular Ecology 14:255–265.

Carval, D., and R. Ferriere. 2010. A unified model for the coevolution of resistance, tolerance, and virulence. Evolution 64:2988–3009.

Case, R. A. J., et al. 2005. Macro- and micro-geographic variation in Pantophysin (Pan I) allele frequencies in NE Atlantic cod (Gadus morhua L.). Marine Ecology Progress Series 301:267–278.

Casellas, J., *et al.* 2008. Skew distribution of founder-specific inbreeding depression effects on the longevity of Landrace sows. Genetical Research 90:499–508.

Casellas, J., J. Piedrafita, G. Caja, and L. Varona. 2009. Analysis of founder-specific inbreeding depression on birth weight in Ripollesa lambs. Journal of Animal Science 87:72–79.

Castle, W. E. 1903. The laws of heredity of Galton and Mendel and some laws governing race improvement by selection. Proceedings of the American Academy of Arts and Sciences 39:223–242.

Castric, V., and X. Vekemans. 2004. Plant self-incompatibility in natural populations: a critical assessment of recent theoretical and empirical advances. Molecular Ecology 13:2873–2889.

Caughley, G. 1994. Directions in conservation biology. Journal of Animal Ecology 63:215–244.

Cavalli-Sforza, L. L., and A. W. F. Edwards. 1967. Phylogenetic analysis: models and estimation procedures. American Journal of Human Genetics 19:233–257.

CBOL Plant Working Group. 2009. A DNA barcode for land plants. Proceedings of the National Academy of Sciences USA 106:12794–12797.

Ceballos, G., and P. R. Ehrlich. 2002. Mammal population losses and the extinction crisis. Science 296:904–907.

Cenik, C., and J. Wakeley. 2010. Pacific salmon and the coalescent effective population size. PLoS ONE 5:e13019.

Center for Plant Conservation. 1991. Genetic sampling guidelines for conservation collections of endangered plants. Pp. 225–238 *in* Falk, D. A. and K. E. Holsinger, eds. Genetics and Conservation of Rare Plants. Oxford University Press, New York.

Chakraborty, R., and O. Leimar. 1987. Genetic variation within a subdivided population. Pp. 89–120 *in* Ryman, N., and F. Utter, eds. Population Genetics and Fishery Management. Washington Sea Grant Publications, University of Washington Press, Seattle.

Chambers, S. M. 1995. Spatial structure, genetic variation, and the neighborhood adjustment to effective population size. Conservation Biology 9:1312–1315.

Champagne, F. A. 2011. Maternal imprints and the origins of variation. Hormones and Behavior 60:4–11.

Chandler, E. A., J. R. McDowell, and J. E. Graves. 2008. Genetically monomorphic invasive populations of the rapa whelk, *Rapana venosa*. Molecular Ecology 17:4079–4091.

Chapman, D. D., *et al.* 2007. Virgin birth in a hammerhead shark. Biology Letters 3:425–427.

Chapman, H., B. Robson, and M. L. Pearson. 2004. Population genetic structure of a colonising, triploid weed, *Hieracium lepidulum*. Heredity 92:182–188.

Chapman, R. W., *et al.* 2011. The transcriptomic responses of the eastern oyster, *Crassostrea virginica*, to environmental conditions. Molecular Ecology 20:1431–1449.

Charles, D. 2006. Species conservation – a 'forever' seed bank takes root in the Arctic. Science 312:1730–1731.

Charlesworth, B. 1991. The evolution of sex chromosomes. Science 251:1030–1033.

Charlesworth, B. 1996. Background selection and patterns of genetic diversity in *Drosophila melanogaster*. Genetic Research 68:131–149.

Charlesworth, B., D. Charlesworth, and N. H. Barton. 2003. The effects of genetic and geographic structure on neutral variation. Annual Review of Ecology Evolution and Systematics 34:99–125.

Charlesworth, D. 2002. Plant sex determination and sex chromosomes. Heredity 88:94–101.

Charlesworth, D., and B. Charlesworth. 1987. Inbreeding depression and its evolutionary consequences. Annual Review of Ecology and Systematics 18:237–268.

Charlesworth, D., and J. H. Willis. 2009. The genetics of inbreeding depression. Nature Reviews Genetics 10:783–796.

Chaves, P. B., *et al.* 2011. DNA barcoding meets molecular scatology: Short mtDNA sequences for standardized species assignment of carnivore noninvasive samples. Molecular Ecology Resources 12:18–35.

Cheptou, P. O., and K. Donohue. 2011. Environment-dependent inbreeding depression: Its ecological and evolutionary significance. New Phytologist 189:395–407.

Cheung, W. W. L., T. J. Pitcher, and D. Pauly. 2005. A fuzzy logic expert system to estimate intrinsic extinction vulnerabilities of marine fishes to fishing. Biological Conservation 124:97–111.

Chilcote, M. W., K. W. Goodson, and M. R. Falcy. 2011. Reduced recruitment performance in natural populations of anadromous salmonids associated with hatchery reared fish. Canadian Journal of Fisheries and Aquatic Sciences 68:511–522.

Choler, P., *et al.* 2004. Genetic introgression as a potential to widen a species' niche: insights from alpine *Carex curvula*. Proceedings of the National Academy of Sciences of the USA 101:171–176.

Christie, M. R., M. L. Marine, R. A. French, and M. S. Blouin. 2012. Genetic adaptation to captivity can occur in a single generation. Proceedings of the National Academy of Sciences 109:238–242.

Ciofi, C., M. A. Beaumont, I. R. Swingland, and M. W. Bruford. 1999. Genetic divergence and units for conservation in the Komodo dragon *Varanus komodoensis*. Proceedings of the Royal Society B: Biological Sciences 266:2269–2274.

Cipriano, F., and S. R. Palumbi. 1999. Genetic tracking of a protected whale. Nature 397:307–308.

CITES. 2010. Convention on International Trade in Endangered Species of Wild Fauna and Flora. http://www.cites.org/

Civeyrel, L., and D. Simberloff. 1996. A tale of two snails: Is the cure worse than the disease? Biodiversity and Conservation 5:1231–1252.

Clark, A. G. 1988. The evolution of the Y chromosome with X-Y recombination. Genetics 119:711–20.

Clark, C. W. 1984. Strategies for multispecies management: objectives and constraints. Pp. 303–312 *in* May, R. M., ed. Dahlem Konferenzen. Springer, Berlin.

Clark, J. S. 2010. Individuals and the variation needed for high species diversity in forest trees. Science 327:1129–1132.

Clarke, B. 1979. The evolution of genetic diversity. Proceedings of the Royal Society B: Biological Sciences 205:19–40.

Clarke, B., M. S. Johnson, and J. Murray. 1998. How "molecular leakage" can mislead us about island speciation. Pp. 181–195 *in* Grant, P. R., ed. Evolution on Islands. Oxford University Press, Oxford.

Clarke, B. C., and L. Partridge. 1988. Frequency dependent selection: a discussion. Philosophical Transactions of the Royal Society of London B, Biological Sciences 319:457–640.

Clarke, G. M. 1993. Fluctuating asymmetry of invertebrate populations as a biological indicator of environmental quality. Environmental Pollution 82:207–211.

Clausen, J. 1951. Stages in the evolution of plant species. Cornell University Press, Ithaca, New York.

Clausen, J., D. D. Keck, and W. M. Hiesey. 1948. Experimental studies on the nature of species: III. Environmental responses of climatic races of *Achillea*. Carnegie Institution of Washington Publication 581:1–129.

Clegg, M. T. 1990. Molecular diversity in plant populations. Pp. 98–115 *in* Brown, A. H. D., M. T. Clegg, A. L. Kahler, and B. S. Weir, eds. Plant Population Genetics, Breeding, and Genetic Resources. Sinauer, Sunderland, MA.

Clegg, M. T., and M. L. Durbin. 2000. Flower color variation: a model for the experimental study of evolution. Proceedings of the National Academy of Sciences of the USA 97:7016–7023.

Clegg, M. T., and M. L. Durbin. 2003. Tracing floral adaptations from ecology to molecules. Nature Reviews Genetics 4:206–215.

Clements, G. R., C. J. A. Bradshaw, B. W. Brook, and W. F. Laurance. 2011. The safe index: Using a threshold population target to measure relative species threat. Frontiers in Ecology and the Environment 9:521–525.

Clobert, J., E. Danchin, A. A. Dhondt, and J. D. Nichols. 2001. Dispersal. Oxford University Press, Oxford.

Close, P. G., and G. Gouws. 2007. Cryptic estuarine gobiid fish larvae (Pisces: Gobfidae) identified as *Pseudogobius olorum* using allozyme markers. Journal of Fish Biology 71:288–293.

Clout, M. N., G. P. Elliott, and B. C. Robertson. 2002. Effects of supplementary feeding on the offspring sex ratio of kakapo: a dilemma for the conservation of a polygynous parrot. Biological Conservation 107:13–18.

Clynes, T. 2011. Confronting corruption. Conservation Magazine 11(4):24–33.

Coates, D. J. 1988. Genetic diversity and population structure in the rare Chittering grass wattle, *Acacia anomala*. Australian Journal of Botany 36:273–286.

Coates, D. J. 2000. Defining conservation units in a rich and fragmented flora: implications for the management of genetic resources and evolutionary processes in southwest Australian plants. Australian Journal of Botany 48:329–339.

Coates D. J., Carstairs S., and Hamley V. L. 2003. Evolutionary patterns and genetic structure in localized and widespread species in the *Stylidium caricifolium* complex (Stylidiaceae). American Journal of Botany 90:997–1008.

Coates, D. J., and V. L. Hamley. 1999. Genetic divergence and the mating system in the endangered and geographically restricted species, Lambertia orbifolia Gardner (Proteaceae). Heredity 83:418–427.

Cochran, R. G. 1954. Some methods for strengthening the common X^2 tests. Biometrics 10:417–451.

Cockerham, C. C., and B. S. Weir. 1977. Digenic descent measures for finite populations. Genetical Research 30:121–147.

Cohan, F. M. 2006. Toward a conceptual and operational union of bacterial systematics, ecology, and evolution. Proceedings of the Royal Society B: Biological Sciences 361:1985–1996.

Cohan, F. M., and A. A. Hoffmann. 1986. Genetic divergence under uniform selection. II. Different responses to selection for knockdown resistance to ethanol among *Drosophila melanogaster* populations and their replicate lines. Genetics 114:145–163.

Cole, K. L. 2010. Vegetation response to Early Holocene warming as an analog for current and future changes. Conservation Biology 24:29–37.

Coleman, M. L., and S. W. Chisholm. 2010. Ecosystem-specific selection pressures revealed through comparative population genomics. Proceedings of the National Academy of Sciences, USA 107:18634–18639.

Collins, T. M., J. C. Trexler, L. G. Nico, and T. A. Rawlings. 2002. Genetic diversity in a morphologically conservative invasive taxon: multiple introductions of swamp eels to the Southeastern United States. Conservation Biology 16:1024–1035.

Coltman, D. W. 2005. Testing marker-based estimates of heritability in the wild. Molecular Ecology 14:2593–2599.

Coltman, D. W. 2007. Molecular ecological approaches to studying the evolutionary impact of selective harvesting in wildlife. Molecular Ecology 16:221–235.

Coltman, D. W., J. G. Pilkington, J. A. Smith, and J. M. Pemberton. 1999. Parasite-mediated selection against inbred Soay sheep in a free-living, island population. Evolution 53:1259–1267.

Coltman, D. W., M. Festa-Bianchet, J. T. Jorgenson, and C. Strobeck. 2002. Age-dependent sexual selection in bighorn rams. Proceedings of the Royal Society of London B, Biological Sciences 269:165–172.

Coltman, D. W., *et al.* 2003. Undesirable evolutionary consequences of trophy hunting. Nature 426:655–658.

Comai, L., *et al.* 2004. Efficient discovery of DNA polymorphisms in natural populations by Ecotilling. The Plant Journal 37:778–786.

Comings, D. 1978. Mechanisms of chromosome banding and implications for chromosome structure. Annual Review of Genetics 12:25–46.

Comstock, K. E., *et al.* 2002. Patterns of molecular genetic variation among African elephant populations. Molecular Ecology 11:2489–2498.

Conde, D. A., *et al.* 2011. An emerging role of zoos to conserve biodiversity. Science 331:1390–1391.

Conner, M. M., and G. C. White. 1999. Effects of individual heterogeneity in estimating the persistence of small populations. Natural Resources Modeling 12:109–127.

Conover, D. O. 2007. Fisheries – Nets versus nature. Nature 450:179–180.

Conover, D. O., and E. T. Schultz. 1995. Phenotypic similarity and the evolutionary significance of countergradient variation. Trends in Ecology & Evolution 10:248–252.

Conover, D. O., and S. B. Munch. 2002. Sustaining fisheries yields over evolutionary time scales. Science 297:94–96.

Conover, D. O., S. B. Munch, and S. A. Arnott. 2009. Reversal of evolutionary downsizing caused by selective harvest of large fish. Proceedings of the Royal Society B: Biological Sciences 276:2015–2020.

Consuegra, S., C. G. DeLeaniz, A. Serdio, and E. Verspoor. 2005. Selective exploitation of early running fish may induce genetic and phenotypic changes in Atlantic salmon. Journal of Fish Biology 67:129–145.

Conway, W. G. 2011. Buying time for wild animals with zoos. Zoo Biology 30:1–8.

Cook, J. M., and R. H. Crozier. 1995. Sex determination and population biology in the Hymenoptera. Trends in Ecology & Evolution 10:281–286.

Cooke, F. 1987. Lesser snow geese: a long-term population study. Pp. 407–432 *in* Cooke, F. and P. A. Buckley, eds. Avian Genetics: A Population and Ecological Approach. Academic Press, London.

Coop, Graham, *et al.* 2009. The role of geography in human adaptation. PLoS Genetics 5:e1000500.

Cooper, D. W. 1968. The significance level in multiple tests made simultaneously. Heredity 23:614–617.

Coppack, T., and C. Both. 2002. Predicting life-cycle adaptation of migratory birds to global climate change. Ardea 90:369–378.

Coppack, T., F. Pulido, and P. Berthold. 2001. Photoperiodic response to early hatching in a migratory bird species. Oecologia 128:181–186.

Cornelius, J. P., C. M. Navarro, K. E. Wightman, and S. E Ward. 2005. Is mahogany dysgenically selected? Environmental Conservation 32:129–139.

Cornuet, J.-M., *et al.* 2008. Inferring population history with *DIY ABC*: A user-friendly approach to approximate Bayesian computation. Bioinformatics 24:2713–2719.

Cornuet, J.-M., *et al.* 1999. New methods employing multilocus genotypes for selecting or excluding populations as origins of individuals. Genetics 153:1989–2000.

Cosart, T., *et al.* 2011. Exome-wide DNA capture and next generation sequencing in domestic and wild species. BMC Genomics 12:347.

Cosentino, B., *et al.* 2012. Linking extinction-colonization dynamics to genetic structure in a salamander metapopulation. Proceedings of the Royal Society B: Biological Sciences 279:1575–1582.

COSEWIC. 2010. Canadian Wildlife Species at Risk. Committee on the Status of Endangered Wildlife in Canada. Available at http://www.cosewic.gc.ca/eng/sct0/rpt/rpt_csar_e.cfm [accessed 4 October 2010].

Cotgreave, P. 1993. The relationship between body size and population abundance in animals. Trends in Ecology & Evolution 8:244–248.

Court-Picon, M., C. Gadbin-Henry, F. Guibal, and M. Roux. 2004. Dendrometry and morphometry of *Pinus pinea* l. In Lower Provence (France): Adaptability and variability of provenances. Forest Ecology and Management 194:319–333.

Couzin, J. 2004. Hybrid mosquitoes suspected in West Nile virus spread. Science 303:1451.

Cowie, R. H. 2001. Can snails ever be effective and safe biocontrol agents? International Journal of Pest Management 47:23–40.

Cracraft, J. 1989. Speciation and its ontology: the empirical consequences of alternative species concepts for understanding patterns and processes of differentiation. Pp. 29–59 *in* Otte, D., and J. A. Endler, eds. Speciation and its Consequences. Sinauer Associates, Sunderland, MA.

Craig, J. K., and C. J. Foote. 2001. Countergradient variation and secondary sexual color: Phenotypic convergence promotes genetic divergence in carotenoid use between sympatric anadromous and nonanadromous morphs of sockeye salmon (*Oncorhynchus nerka*). Evolution 55:380–391.

Craighead, J. J., J. R. Varney, and F. C. Craighead Jr. 1973. A computer analysis of the Yellowstone grizzly bear population. Montana Cooperative Wildlife Unit, Missoula, Montana.

Craighead, J. J., J. S. Summner, and J. A. Mitchell. 1995. The Grizzly Bears of Yellowstone: Their Ecology in the Yellowstone Ecosystem, 1959–1992. Island Press, Washington, DC.

Crampe, H. 1883. Zuchtversuche mit zahmen Wanderratten. Landwirtschaftliches Jahrbuch 12:389–458.

Crandall, K. A., O. R. P. Binindaemonds, G. M. Mace, and R. K. Wayne. 2000. Considering evolutionary processes in conservation biology. Trends in Ecology & Evolution 15:290–295.

Cresswell, M. 1996. Kakapo recovery plan 1996–2005. Threatened species recovery plan No. 21. Department of Conservation, Wellington, New Zealand.

Criscione, C. D., and M. S. Blouin. 2007. Parasite phylogeographical congruence with salmon host evolutionarily significant units: implications for salmon conservation. Molecular Ecology 16:993–1005.

Criscione, C. D., R. Poulin, and M. S. Blouin. 2005. Molecular ecology of parasites: elucidating ecological and microevolutionary processes. Molecular Ecology 14:2247–2257.

Cristobal-Azkarate, J., and V. Arroyo-Rodriguez. 2007. Diet and activity pattern of howler monkeys (*Alouatta palliata*) in Los Tuxtlas, Mexico: Effects of habitat fragmentation and implications for conservation. American Journal of Primatology 69:1013–1029.

Crnokrak, P., and S. C. H. Barrett. 2002. Purging the genetic load: A review of the experimental evidence. Evolution 56:2347–2358.

Crnokrak, P., and D. A. Roff. 1999. Inbreeding depression in the wild. Heredity 83:260–270.

Croft, D. 1999. When big is beautiful: some consequences of bias in kangaroo culling. Pp. 70–73 in Wilson, M., ed. The kangaroo betrayed. Hill of Content Publishing, Melbourne, Australia.

Crombach, A., and P. Hogeweg. 2007. Chromosome rearrangements and the evolution of genome structuring and adaptability. Molecular Biology and Evolution 24:1130–1139.

Crow, J. F. 1948. Alternate hypotheses of hybrid vigor. Genetics 33:477–487.

Crow, J. F. 1954. Breeding structure of populations. II. Effective population number. Pp. 543–556 in Kempthorne, O., T. A. Bancroft, J. W. Gowen, and J. L. Lush, eds. Statistics and Mathematics in Biology. Iowa State College Press, Ames, Iowa.

Crow, J. F. 1957. Genetics of insect resistance to chemicals. Annual Review of Entomology 2:227–246.

Crow, J. F. 1970. Genetic loads and the cost of natural selection. Biomathematics: 1, Mathematical Topics in Population Genetics 128–177.

Crow, J. F. 1986. Basic Concepts in Population Genetics. Freeman, New York.

Crow, J. F. 1993. Mutation, mean fitness, and genetic load. Oxford Surveys in Evolutionary Biology 9:3–42.

Crow, J. F. 2001. The beanbag lives on. Nature 409:771.

Crow, J. F. 2010. Wright and Fisher on inbreeding and random drift. Genetics 184:609–611.

Crow, J. F., and K. Aoki. 1984. Group selection for a polygenic trait: estimating the degree of population subdivision. Proceedings of the National Academy of Sciences USA 81:6073–6077.

Crow, J. F., and C. Denniston. 1988. Inbreeding and variance effective population numbers. Evolution 42:482–495.

Crow, J. F., and M. Kimura. 1970. An Introduction to Population Genetics Theory. Burgess Publishing Company, Minneapolis, Minnesota.

Crozier, R. H. 1971. Heterozygosity and sex determination in haplo-diploidy. American Naturalist 105:399–412.

Crozier, R. H., and R. M. Kusmierski. 1994. Genetic distances and the setting of conservation priorities. Pp. 227–237 in Loeschcke, V., J. Tomiuk, and S. K. Jain, eds. Conservation Genetics. Birkhauser Verlag, Basel, Switzerland.

Crutsinger, G. M., et al. 2006. Plant genotypic diversity predicts community structure and governs an ecosystem process. Science 313:966–968.

Cruz, F., C. Vila, and M. T. Webster. 2008. The legacy of domestication: Accumulation of deleterious mutations in the dog genome. Molecular Biology and Evolution 25:2331–2336.

Currat, M., M. Ruedi, R. J. Petit, and L. Excoffier. 2008. The hidden side of invasions: massive introgression by local genes. Evolution 62:1908–1920.

Currens, K. P., et al. 1997. Introgression and susceptibility to disease in a wild population of rainbow trout. North American Journal of Fisheries Management 17:1065–1078.

Danchin, É., et al. 2011. Beyond DNA: Integrating inclusive inheritance into an extended theory of evolution. Nature Reviews Genetics 12:475–486.

Darimont, C. T., et al. 2009. Human predators outpace other agents of trait change in the wild. Proceedings of the National Academy of Sciences USA 106:952–954.

Darlington, C. D. 1969. The Evolution of Man and Society. Simon and Schuster, New York.

Darwin, C. 1859. The Origin of Species by Means of Natural Selection, or the Preservation of Favoured Races in the Struggle for Life. John Murray, London.

Darwin, C. 1876. The Effects of Cross and Self Fertilization in the Vegetable Kingdom. J. Murray and Co, London.

Darwin, C. 1896. The Variation of Animals and Plants under Domestication, Vol. II. D. Appleton and Co., New York.

Daszak, P., A. A. Cunningham, and A. D. Hyatt. 2000. Emerging infectious diseases of wildlife – threats to biodiversity and human health. Science 287:443–449.

Daszak, P., A. A. Cunningham, and A. D. Hyatt. 2001. Anthropogenic environmental change and the emergence of infectious diseases in wildlife. Acta Tropica 78:103–116.

Daugherty, C. H., A. Cree, J. M. Hay, and M. B. Thompson. 1990. Neglected taxonomy and continuing extinctions of tuatara (*Sphenodon*). Nature 347:177–179.

Davey, J. W., et al. 2011. Genome-wide genetic marker discovery and genotyping using next-generation sequencing. Nature Reviews Genetics 12:499–510.

Davies, T. J., et al. 2011. Extinction risk and diversification are linked in a plant biodiversity hotspot. PLoS Biol 9:e1000620.

Davis, M. B., and R. G. Shaw. 2001. Range shifts and adaptive responses to Quaternary climate change. Science 292:673–679.

Davis, M. B., R. G. Shaw, and J. R. Etterson. 2005. Evolutionary responses to changing climate. Ecology 86:1704–1714.

Dawson, L. 1982. Taxonomic status of fossil thylacines (Thylacinus, Thylacinidae, Marsupialia) from late quaternary deposits in eastern Australia. Pp. 527–536 in Archer, M., ed. Carnivorous Marsupials. Royal Zoological Society of N.S.W., Sydney.

Dawson, M. N., R. K. Grosberg, and L. W. Botsford. 2006. Connectivity in marine protected areas. Science 313:43–44.

De Boer, L. E. M. 1982. Karyological problems in breeding owl monkeys, *Aotus trivirgatus*. International Zoo Yearbook 22:119–124.

de Brabandere, L., and A. Iny. 2010. Scenarios and creativity: thinking in new boxes. Technological Forecasting and Social Change 77:1506–1512.

de Cara, M. A. R., J. Fernandez, M. A. Toro, and B. Villanueva. 2011. Using genome-wide information to minimize the loss of diversity in conservation programmes. Journal of Animal Breeding and Genetics 128:456–464.

Deeb, S. S. 2005. The molecular basis of variation in human color vision. Clinical Genetics 67:369–377.

Degen, B., *et al.* 2006. Impact of selective logging on genetic composition and demographic structure of four tropical tree species. Biological Conservation 131:386–401.

Degnan, S. M. 1993. The perils of single gene trees – mitochondrial versus single-copy nuclear DNA variation in white-eyes (Aves: Zosteropidae). Molecular Ecology 2:219–225.

DeHaan, P. W., *et al.* 2011. Use of genetic markers to aid in re-establishing migratory connectivity in a fragmented metapopulation of bull trout (*Salvelinus confluentus*). Canadian Journal of Fisheries and Aquatic Sciences 68:1952–1969.

de Jong, H. 2003. Visualizing DNA domains and sequences by microscopy: a fifty-year history of molecular cytogenetics. Genome 46:943–946.

de Leaniz, C. G., *et al.* 2007. A critical review of adaptive genetic variation in Atlantic salmon: implications for conservation. Biological Reviews 82:173–211.

DeLong, E. F. 2009. The microbial ocean from genomes to biomes. Nature 459:200–206.

Delpuech, J. M., Y. Carton, and R. T. Roush. 1993. Conserving genetic variability of a wild insect population under laboratory conditions. Entomologia Experimentalis et Applicata 67:233–239.

Delsuc, F., *et al.* 2007. Molecular evidence for hybridisation between the two living species of South American ratites: potential conservation implications. Conservation Genetics 8:503–507.

Demauro, M. M. 1993. Relationship of breeding system to rarity in the lakeside daisy (*Hymenoxys acaulis var. glabra*). Conservation Biology 7:542–550.

Dempster, A. P., N. M. Laird, and D. B. Rubin. 1977. Maximum likelihood estimation from incomplete data via the *EM* algorithm. Journal of the Royal Statistical Society, Series B 39:1–38.

de Nettancourt, D. 1977. Incompatibility in Angiosperms. Springer-Verlag, New York.

Denholm, I., G. J. Devine, and M. S. Williamson. 2002. Insecticide resistance on the move. Science 297:2222–2223.

De Queiroz, K. 2007. Species concepts and species delimitation. Systematic Biology 56:879–886.

de Roos, A. M., D. S. Boukal, and L. Persson. 2006. Evolutionary regime shifts in age and size at maturation of exploited fish stocks. Proceedings of the Royal Society B: Biological Sciences 273:1873–1880.

Devall, B., and G. Sessions. 1984. The development of natural resources and the integrity of nature. Environmental Ethics 6:293–322.

Devlin, R. H., and Y. Nagahama. 2002. Sex determination and sex differentiation in fish: an overview of genetic, physiological, and environmental influences. Aquaculture 208:191–364.

Diamond, J. 1997. Guns, Germs, and Steel. W.W. Norton, New York.

DiBattista, J. D., *et al.* 2009. Evolutionary potential of a large marine vertebrate: Quantitative genetic parameters in a wild population. Evolution 63:1051–1067.

Dieckmann, U., and M. Heino. 2007. Probabilistic maturation reaction norms: their history, strengths, and limitations. Marine Ecology Progress Series 335:253–269.

Dinerstein, E., and G. F. McCracken. 1990. Endangered greater one-horned rhinoceros carry high levels of genetic variation. Conservation Biology 4:417–422.

Dinsdale, E. A., *et al.* 2008. Functional metagenomic profiling of nine biomes. Nature 452:629–634.

Dizon, A. E., *et al.* 1992. Rethinking the stock concept – a phylogeographic approach. Conservation Biology 6:24–36.

Dobson, A. 2003. Metalife! Science 301:1488–1490.

Dobzhansky, T. 1948. Genetics of natural populations, XVIII. Experiments on chromosomes of *Drosophila pseudoobscura* from different geographical regions. Genetics 33:588–602.

Dobzhansky, T. 1951. Genetics and the origin of species, 3rd edn. Columbia University Press, New York.

Dobzhansky, T. 1962. Oral history memoir. Columbia University Press, New York (cited in Provine 1986).

Dobzhansky, T. 1970. Genetics of the Evolutionary Process. Columbia University Press, New York.

Dobzhansky, T., F. J. Ayala, G. L. Stebbins, and J. W. Valentine. 1977. Evolution. W.H. Freeman, San Francisco.

Doležel, J., and J. Bartoš. 2005. Plant DNA flow cytometry and estimation of nuclear genome size. Annals of Botany 95:99–110.

Doolittle, W. F., and O. Zhaxybayeva. 2010. Metagenomics and the units of biological organization. BioScience 60:102–112.

Doremus, H. 1999. Restoring endangered species: the importance of being wild. Harvard Environmental Law Review 23:3–92.

Double, M. C., A. Cockburn, S. C. Barry, and P. E. Smouse. 1997. Exclusion probabilities for single-locus paternity analysis when related males compete for matings. Molecular Ecology 6:1155–1166.

Dournon, C., A. Collenot, and M. Lauthier. 1988. Sex-linked peptidase-1 patterns in *Pleurodeles waltlii* Michah (Urodele amphibian): genetic evidence for a new codominant allele on the W sex chromosomes and identification of ZZ, ZW, and WW sexual genotypes. Reproduction, Nutrition, Development 28:979–987.

Dowling, D. K., U. Friberg, and J. Lindell. 2008. Evolutionary implications of non-neutral mitochondrial genetic variation. Trends in Ecology & Evolution 23:546–554.

Dowling, T. E., and C. L. Secor. 1997. The role of hybridization and introgression in the diversification of animals. Annual Review of Ecology & Systematics 28:593–619.

Dozier, H. L. 1948. Color mutations in the muskrat (*Ondatra z. macrodon*) and their inheritance. Journal of Mammalogy 29:393–405.

Dozier, H. L., and R. W. Allen. 1942. Color, sex ratio, and weights of Maryland muskrats. Journal of Wildlife Management 6:294–300.

Drake, J. W., B. Charlesworth, D. Charlesworth, and J. F. Crow. 1998. Rates of spontaneous mutation. Genetics 148:1667–1686.

Drechsler, M., *et al.* 2003. Ranking metapopulation extinction risk: from patterns in data to conservation management decisions. Ecological Applications 13:990–998.

Drew, R. E., *et al.* 2003. Conservation genetics of the fisher (*Martes pennanti*) based on mitochondrial DNA sequencing. Molecular Ecology 12:51–62.

Duchesne, P., and L. Bernatchez. 2002. An analytical investigation of the dynamics of inbreeding in multi-generation supportive breeding. Conservation Genetics 3:47–60.

Ducrest, A. L., L. Keller, and A. Roulin. 2008. Pleiotropy in the melanocortin system, coloration and behavioural syndromes. Trends in Ecology & Evolution 23:502–510.

Dudash, M. R. 1990. Relative fitness of selfed and outcrossed progeny in a self-compatible, protandrous species, *Sabatia angularis* L. (Gentianaceae): a comparison in three environments. Evolution 44:1129–1140.

Dudash, M. R., and C. B. Fenster. 2000. Inbreeding and outbreeding depression in fragmented populations. Pp. 35–53 *in* Young, A. G., and G. M. Clarke, eds. Genetics, Demography and Viability of Fragmented Populations. Cambridge University Press, Cambridge.

Dudley, J. W. 2007. From means to QTL: The Illinois long-term selection experiment as a case study in quantitative genetics. Crop Science 47:S20–S31.

Duffy, J. E., *et al.* 2007. The functional role of biodiversity in ecosystems: Incorporating trophic complexity. Ecology Letters 10:522–538.

Dulvy, N. K., Y. Sadovy, and J. D. Reynolds. 2003. Extinction vulnerability in marine populations. Fish and Fisheries 4:25–64.

Dumont, B. L., and B. A. Payseur. 2011. Genetic analysis of genome-scale recombination rate evolution in house mice. PLoS Genet 7:e1002116.

Dytham, C. 2011. Choosing and using statistics: a biologist's guide. 3rd edn. Wiley-Blackwell, Oxford.

Eanes, W. F. 1987. Allozymes and fitness: Evolution of a problem. Trends in Ecology & Evolution 2:44–48.

Ebenhard, T. 1995. Conservation breeding as a tool for saving animal species from extinction. Trends in Ecology & Evolution 10:438–443.

Eberhardt, L. L., and R. R. Knight. 1996. How many grizzlies in Yellowstone? Journal of Wildlife Management 60:416–421.

Eberhart, S. A., E. E. Roos, and L. E. Towill. 1991. Strategies for long-term management of germplasm collections. Pp. 135 *in* Falk, D. A., and K. E. Holsinger, eds. Genetics and Conservation of Rare Plants. Oxford University Press, New York.

Ebert, D., and R. Peakall. 2009. Chloroplast simple sequence repeats (cpssrs): Technical resources and recommendations for expanding cpssr discovery and applications to a wide array of plant species. Molecular Ecology Resources 9:673–690.

Echt, C. S., L. L. Deverno, M. Anzidei, and G. G. Vendramin. 1998. Chloroplast microsatellites reveal population genetic diversity in red pine, *Pinus resinosa* Ait. Molecular Ecology 7:307–316.

Eckert, A. J., *et al.* 2010. Back to nature: ecological genomics of loblolly pine (*Pinus taeda*, Pinaceae). Molecular Ecology 19:3789–3805.

Edeline, E., *et al.* 2007. Trait changes in a harvested population are driven by a dynamic tug-of-war between natural and harvest selection. Proceedings of the National Academy of Sciences USA 104:15799–15804.

Edmands, S. 2007. Between a rock and a hard place: evaluating the relative risks of inbreeding and outbreeding for conservation and management. Molecular Ecology 16:463–475.

Edwards, A. W. F. 2008. G. H. Hardy (1908) and Hardy-Weinberg equilibrium. Genetics 179:1143–1150.

Edwards, S. V., and P. W. Hedrick. 1998. Evolution and ecology of MHC molecules: from genomics to sexual selection. Trends in Ecology & Evolution 13:305–311.

Edwards, S. V., and W. K. Potts. 1996. Polymorphism of genes in the major histocompatibility complex (MHC): implications for conservation genetics of vertebrates. Pp. 214–237 *in* Smith, T. B. and R. K. Wayne, eds. Molecular Genetic Approaches in Conservation. Oxford University Press, New York.

Eggert, L. S., J. A. Eggert, and D. S. Woodruff. 2003. Estimating population sizes for elusive animals: the forest elephants of Kakum National Park, Ghana. Molecular Ecology 12:1389–1402.

Ehrlich, P. R., and P. H. Raven. 1969. Differentiation of populations. Science 165:1228–1232.

Eisner, T., *et al.* 1995. Building a scientifically sound policy for protecting endangered species. Science 268:1231–1233.

Elena, S. F., V. S. Cooper, and R. E. Lenski. 1996. Punctuated evolution caused by selection of rare beneficial mutations. Science 272:1802–1804.

Elfstrom, C. M., C. T. Smith, and J. E. Seeb. 2006. Thirty-two single nucleotide polymorphism markers for high-throughput genotyping of sockeye salmon. Molecular Ecology Notes 6:1255–1259.

Elias, J., D. Mazzi, and S. Dorn. 2009. No need to discriminate? Reproductive diploid males in a parasitoid with complementary sex determination. PLoS ONE 4:e6024.

Elle, E., S. Gillespie, S. Guindre-Parker, and A. L. Parachnowitsch. 2010. Variation in the timing of autonomous selfing among populations that differ in flower size, time to reproductive maturity, and climate. American Journal of Botany 97:1894–1902.

Ellegren, H. 1999. Inbreeding and relatedness in Scandinavian grey wolves Canis lupus. Hereditas 130:239–244.

Ellegren, H. 2000a. Microsatellite mutations in the germline: implications for evolutionary inference. Trends in Genetics 16:551–558.

Ellegren, H. 2000b. Evolution of the avian sex chromosomes and their role in sex determination. Trends in Ecology & Evolution 15:188–192.

Elliott, G. P., D. V. Merton, and P. W. Jansen. 2001. Intensive management of a critically endangered species: the kakapo. Biological Conservation 99:121–133.

Ellstrand, N. C. 2003. Dangerous Liaisons? When Cultivated Plants Mate with Their Wild Relatives. John Hopkins University Press, Baltimore, MD.

Ellstrand, N. C., and D. R. Elam. 1993. Population genetic consequences of small population size – implications for plant conservation. Annual Review of Ecology and Systematics 24:217–242.

Ellstrand, N. C., and K. A. Schierenbeck. 2000. Hybridization as a stimulus for the evolution of invasiveness in plants? Proceedings of the National Academy of Sciences USA 97:7043–7050.

Ellstrand, N. C., H. C. Prentice, and J. F. Hancock. 1999. Gene flow and introgression from domesticated plants into their wild relatives. Annual Review of Ecology and Systematics 30:539–563.

Ellstrand, N. C., et al. 2010. Crops gone wild: evolution of weeds and invasives from domesticated ancestors. Evolutionary Applications 3:494–504.

El Mousadik, A., and R. J. Petit. 1996. High level of genetic differentiation for allelic richness among populations of the argan tree [Argania spinosa (L.) Skeels] endemic of Morocco. Theoretical and Applied Genetics 92:832–839.

Elton, C. S. 1942. Voles, Mice, and Lemmings: Problems in Population Dynamics. Clarendon Press, Oxford.

Emerson, B. C., E. Paradis, and C. Thebaud. 2001. Revealing the demographic histories of species using DNA sequences. Trends in Ecology & Evolution 16:707–716.

Emerson, S. 1939. A preliminary survey of the Oenothera organensis population. Genetics 24:524–537.

Emerson, S. 1940. Growth of incompatible pollen tubes in Oenothera organensis. Botanical Gazette 101:890–911.

Engels, W. R. 2009. Exact tests for Hardy-Weinberg proportions. Genetics 183:1431–1441.

Engen, S., et al. 2007. Effective size of fluctuating populations with two sexes and overlapping generations. Evolution 61:1873–1885.

England, P. E., G. Luikart, and R. S. Waples. 2010. Early detection of population fragmentation using linkage disequilibrium estimation of effective population size. Conservation Genetics 11:2425–2430.

England, P. R., et al. 2003. Effects of intense versus diffuse population bottlenecks on microsatellite genetic diversity and evolutionary potential. Conservation Genetics 4:595–604.

England, P. R., et al. 2006. Estimating effective population size from linkage disequilibrium: severe bias in small samples. Conservation Genetics 7:303–308.

Engle, V. D., and J. K. Summers. 2000. Biogeography of benthic macroinvertebrates in estuaries along the Gulf of Mexico and western Atlantic coasts. Hydrobiologia 436:17–33.

Epifanio, J., and D. Philipp. 2001. Simulating the extinction of parental lineages from introgressive hybridization: the effects of fitness, initial proportions of parental taxa, and mate choice. Reviews in Fish Biology and Fisheries 10:339–354.

Epperson, B. K. 2005. Mutation at high rates reduces spatial structure within populations. Molecular Ecology 14:703–710.

Epperson, B. K. 2007. Plant dispersal, neighbourhood size and isolation by distance. Molecular Ecology 16:3854–3865.

Epps, C. W., et al. 2005. Highways block gene flow and cause a rapid decline in genetic diversity of desert bighorn sheep. Ecology Letters 8:1029–1038.

Epps, C. W., et al. 2007. Optimizing dispersal and corridor models using landscape genetics. Journal of Applied Ecology 44:714–724.

Erickson, G. M., P. J. Currie, B. D. Inouye, and A. A. Winn. 2006. Tyrannosaur life tables: An example of nonavian dinosaur population biology. Science 313:213–217.

Erwin, T. L. 1991. An evolutionary basis for conservation strategies. Science 253:750–752.

Escudero, A., J. M. Iriondo, and M. E. Torres. 2003. Spatial analysis of genetic diversity as a tool for plant conservation. Biological Conservation 113:351–365.

Estoup, A., and B. Angers. 1998. Microsatellites and minisatellites for molecular ecology: theoretical and empirical considerations. Pp. 55–86 in Carvalho, G. R., ed. Advances in Molecular Ecology. IOS Press, Amsterdam.

Etterson, J. R., and R. G. Shaw. 2001. Constraint to adaptive evolution in response to global warming. Science 294:151–154.

Evanno, G., S. Regnaut, and J. Goudet. 2005. Detecting the number of clusters of individuals using the software

STRUCTURE: a simulation study. Molecular Ecology 14: 2611–2620.

Ewens, W. J. 1982. On the concept of effective population size. Theoretical Population Biology 21:373–378.

Excoffier, L., and M. Slatkin. 1995. Testing for linkage disequilibrium in genotypic data using the Expectation–Maximization algorithm. Heredity 76:377–383.

Excoffier, L., P. E. Smouse, and J. M. Quattro. 1992. Analysis of molecular variance inferred from metric distances among DNA haplotypes – application to human mitochondrial DNA restriction data. Genetics 131:479–491.

Fabritius, H. 2010. Effective population size and the viability of the Siberian jay population of Suupohja, Finland. Master's thesis. University of Helsinki, Finland.

Fairbairn, D. J., and D. A. Roff. 1980. Testing genetic models of isozyme variability without breeding data: can we depend on the χ^2? Canadian Journal of Fisheries and Aquatic Sciences 37:1149–1159.

Faith, D. P. 2002. Quantifying biodiversity: a phylogenetic perspective. Conservation Biology 16:248–252.

Faith, D. P. 2008. Phylogenetic diversity and conservation. Pp. 99–115 in Carroll, S. P., and C. W. Fox, eds. Conservation Biology: Evolution in Action. Oxford University Press, New York.

Falconer, D. S., and T. F. C. Mackay. 1996. Introduction to Quantitative Genetics, 4th edn. Longman Scientific and Technical, Harlow, UK.

Fallour, D., B. Fady, and F. Lefevre. 1997. Study on isozyme variation in Pinus pinea L.: Evidence for low polymorphism. Silvae Genetica 46:201–207.

Faubet, P., R. S. Waples, and O. E. Gaggiotti. 2007. Evaluating the performance of a multilocus Bayesian method for the estimation of migration rates. Molecular Ecology 16:1149–1166.

Felsenstein, J. 1971. Inbreeding and variance effective numbers in populations with overlapping generations. Genetics 68:581–597.

Felsenstein, J. 1975. A pain in the torus – some difficulties with models of isolation by distance. American Naturalist 109:359–368.

Felsenstein, J. 2004. Inferring Phylogenies. Sinauer Assoctaes, Inc., Sunderland, MA.

Felsenstein, J. 2011. Theoretical Evolutionary Genetics. Seattle, WA. Available at http://evolution.gs.washington.edu/pgbook/pgbook.html.

Fenberg, P. B., and K. Roy. 2008. Ecological and evolutionary consequences of size-selective harvesting: How much do we know? Molecular Ecology 17:209–220.

Fenster, C. B., and M. R Dudash. 1994. Genetic considerations for plant population restoration and conservation. Pp. 34–62 in Bowles, M. L., and C. J. Whelan, eds. Restoration of Endangered Species. Cambridge University Press, Cambridge.

Fenster, C. B., X. Vekemans, and O. J. Hardy. 2003. Quantifying gene flow from spatial genetic structure data in a metapopulation of *Chamaecrista fasciculata* (Leguminosae). Evolution 57:995–1007.

Ferguson-Smith, M. A., and V. Trifonov. 2007. Mammalian karyotype evolution. Nature Reviews Genetics 8: 950–962.

Fernando, P., and D. J. Melnick. 2001. Molecular sexing eutherian mammals. Molecular Ecology Notes 1:350–353.

Festa-Bianchet, M. 2003. Exploitative wildlife management as a selective pressure for life-history evolution of large mammals. Pp. 191–207 in Festa-Bianchet, M., and M. Appollonio, eds. Animal Behavior and Wildlife Conservation. Island Press, Washington, DC.

Festa-Bianchet, M., and R. Lee. 2009. Guns, sheep and genes: when and why trophy hunting may be a selective pressure. Pp. 94–107 in B. Dickson, Hutton, and W. M. Adams, eds. Recreational Hunting, Conservation and Rural Livelihoods: Science and Practice. Blackwell Publishing, Oxford.

Feuk, L., et al. 2005. Discovery of human inversion polymorphisms by comparative analysis of human and chimpanzee DNA sequence assemblies. PLoS Genetics 1:e56.

Fewster, R. M., S. D. Miller, and J. Ritchie. 2011. DNA profiling a management tool for rat eradication. Pp. 426–431 in Veitch, C. R., M. N. Clout, and D. R. Towns, eds. Island Invasives: Eradication and Management. Proceedings of the International Conference on Island Invasives. Gland, Switzerland: IUCN and Auckland, New Zealand: CBB.

Ficetola, G. F., C. Miaud, F. Pompanon, and P. Taberlet. 2008. Species detection using environmental DNA from water samples. Biology Letters 4:423–425.

Fisher, R. A. 1918. The correlation between relatives on the supposition of Mendelian inheritance. Transactions Royal Society of Edinburgh 52:399–433.

Fisher, R. A. 1930. The Genetical Theory of Natural Selection. Clarendon Press, Oxford.

Fisher, R. A. 1935. The logic of inductive inference. Journal of the Royal Statistical Society, Series B 98:39–54.

Flather, C. H., G. D. Hayward, S. R. Beissinger, and P. A. Stephens. 2011. Minimum viable populations: Is there a 'magic number' for conservation practitioners? Trends in Ecology & Evolution 26:307–316.

Fleming, I. A., et al. 2000. Lifetime success and interactions of farm salmon invading a native population. Proceedings of the Royal Society B: Biological Sciences 267:1517–1523.

Folch, J., et al. 2009. First birth of an animal from an extinct subspecies (*Capra pyrenaica pyrenaica*) by cloning. Theriogenology 71:1026–1034.

Foley, J., D. et al. 2011. Investigating and managing the rapid emergence of white-nose syndrome, a novel, fatal, infectious disease of hibernating bats. Conservation Biology 25:223–231.

Foley, J. A., C. Monfreda, N. Ramankutty, and D. Zaks. 2007. Our share of the planetary pie. Proceedings of the National Academy of Sciences 104:12585–12586.

Foll, M., M. C. Fischer, G. Heckel, and L. Excoffier. 2010. Estimating population structure from AFLP amplification intensity. Molecular Ecology 19:4638–4647.

Foltz, D. W. 1986. Null alleles as a possible cause of heterozygote deficiencies in the oyster *Crassostrea virginica* and other bivalves. Evolution 40:869–870.

Foltz, D. W., B. M. Schaitkin, and R. K. Selander. 1982. Gametic disequilibrium in the self-fertilizing slug *Derocera laeve*. Evolution 36:80–85.

Fonseca, D. M., *et al.* 2004. Emerging vectors in the *Culex pipiens* complex. Science 303:1535–1538.

Foote, C. J., G. S. Brown, and C. W. Hawryshyn. 2004. Female colour and male choice in sockeye salmon: implications for the phenotypic convergence of anadromous and nonanadromous morphs. Animal Behaviour 67:69–83.

Forbes, S. H. 1990. Mitochondrial and nuclear genotypes in trout hybrid swarms: tests for gametic equilibrium and effects on phenotypes. PhD dissertation. University of Montana, Missoula.

Forbes, S. H., and F. W. Allendorf. 1991. Associations between mitochondrial and nuclear genotypes in cutthroat trout hybrid swarms. Evolution 45:1332–1349.

Forbes, S. H., and J. T. Hogg. 1999. Assessing population structure at high levels of differentiation: microsatellite comparisons of bighorn sheep and large carnivores. Animal Conservation 2:223–233.

Forbes, S. H., K. L. Knudsen, T. W. North, and F. W. Allendorf. 1994. One of two growth hormone genes in coho salmon is sex-linked. Proceedings of the National Academy of Sciences USA 91:1628–1631.

Ford, E. B. 1971. Ecological Genetics. Chapman & Hall, London.

Ford, M. J. 2002. Selection in captivity during supportive breeding may reduce fitness in the wild. Conservation Biology 16:815–825.

Formica, V. A., and E. M. Tuttle. 2009. Examining the social landscapes of alternative reproductive strategies. Journal of Evolutionary Biology 22.2395–2408.

Forsman, A., and V. Aberg. 2008. Variable coloration is associated with more northerly geographic range limits and larger range sizes in North American lizards and snakes. Evolutionary Ecology Research 10:1025–1036.

Forsman, A., J. Ahnesjo, S. Caesar, and M. Karlsson. 2008. A model of ecological and evolutionary consequences of color polymorphism. Ecology 89:34–40.

Forsman, Z. H., *et al.* 2010. Ecomorph or endangered coral? DNA and microstructures reveal Hawaiian species complexes: *Montipora dilatata/flabellata/turgescens & M. patula/verrilli*. PLoS ONE 5:e15021.

Foster, G. G., M. J. Whitten, T. Prout, and R. Gill. 1972. Chromosome rearrangement for the control of insect pests. Science 176:875–880.

Fournier-Level, A., *et al.* 2011. A map of local adaptation in *Arabidopsis thaliana*. Science 333:86–89.

Fowler, D. P., and D. T. Lester. 1970. Genetics of red pine. USDA Forest Service Research Paper WO-8, Washington, DC.

Fowler, D. P., and R. W. Morris. 1977. Genetic diversity in red pine: evidence for low genic heterozygosity. Canadian Journal of Forest Research 7:343–347.

Fox, C. W., and D. H. Reed. 2011. Inbreeding depression increases with environmental stress: An experimental study and meta-analysis. Evolution 65:246–258.

Fox, C. W., K. L. Scheibly, and D. H. Reed. 2008. Experimental evolution of the genetic load and its implications for the genetic basis of inbreeding depression. Evolution 62:2236–2249.

Fox, G. A. 2005. Extinction risk of heterogeneous populations. Ecology 86:1191–1198.

Frankel, O. H. 1970. Genetic conservation in perspective. Pp. 469–489 in Frankel, O. H., and E. Bennett, eds. Genetic Resources in Plants – Their Exploration and Conservation. Blackwell Scientific, Oxford.

Frankel, O. H. 1974. Genetic conservation: our evolutionary responsibility. Genetics 78:53–65.

Frankel, O. H., and Soulé, M. E. 1981. Conservation and Evolution. Cambridge University Press, Cambridge.

Frankham, R. 1995. Effective population size/adult population size ratios in wildlife: a review. Genetical Research 66:95–107.

Frankham, R. 1999. Quantitative genetics in conservation biology. Genetic Research 74:237–244.

Frankham, R. 2003. Genetics and conservation biology. Comptes Rendus Biologies 326:S22–S29.

Frankham, R. 2008. Genetic adaptation to captivity in species conservation programs. Molecular Ecology 17:325–333.

Frankham, R. 2010. Where are we in conservation genetics and where do we need to go? Conservation Genetics 11: 661–663.

Frankham, R., and D. A. Loebel. 1992. Modeling problems in conservation genetics using captive *Drosophila* populations – rapid genetic adaptation to captivity. Zoo Biology 11: 333–342.

Frankham, R., and K. Ralls. 1998. Conservation biology – inbreeding leads to extinction. Nature 392:441–442.

Frankham, R., *et al.* 1986. Selection in captive populations. Zoo Biology 5:127–138.

Frankham, R., *et al.* 1999. Do population size bottlenecks reduce evolutionary potential? Animal Conservation 2: 255–260.

Frankham, R., H. Manning, S. H. Margan, and D. A. Briscoe. 2000. Does equalization of family sizes reduce genetic adaptation to captivity? Animal Conservation 3:357–363.

Frankham, R., J. D. Ballou., and D. A. Briscoe. 2010. Introduction to Conservation Genetics, 2nd edn. Cambridge University Press, Cambridge.

Frankham, R., *et al.* 2011. Predicting the probability of outbreeding depression: critical information for managing fragmented populations. Conservation Biology 25:465–475.

Franklin, I. R. 1980. Evolutionary changes in small populations. Pp. 135–149 in Soulé, M. E., and B. A. Wilcox, eds. Conservation Biology: An Evolutionary–Ecological Perspective. Sinauer Associates, Sunderland, MA.

Franklin, I. R., and R. Frankham. 1998. How large must populations be to retain evolutionary potential? Animal Conservation 1:69–70.

Franks, S. J. 2011. Plasticity and evolution in drought avoidance and escape in the annual plant *Brassica rapa*. New Phytologist 190:249–257.

Franks, S. J., and A. E. Weis. 2008. A change in climate causes rapid evolution of multiple life-history traits and their interactions in an annual plant. Journal of Evolutionary Biology 21:1321–1334.

Franks, S. J., and A. E. Weis. 2009. Climate change alters reproductive isolation and potential gene flow in an annual plant. Evolutionary Applications 2:481–488.

Franks, S. J., S. Sim, and A. E. Weis. 2007. Rapid evolution of flowering time by an annual plant in response to a climate fluctuation. Proceedings of the National Academy of Sciences USA 104:1278–1282.

Franks, S. J., *et al.* 2008. The resurrection initiative: Storing ancestral genotypes to capture evolution in action. BioScience 58:870–873.

Frantz, A. C., J. L. Hamann, and F. Klein. 2008. Fine-scale genetic structure of red deer (*Cervus elaphus*) in a French temperate forest. European Journal of Wildlife Research 54:44–52.

Frantz, A. C., *et al.* 2009. Using spatial Bayesian methods to determine the genetic structure of a continuously distributed population: clusters or isolation by distance? Journal of Applied Ecology 46:493–505.

Fraser, D. J. 2008. How well can captive breeding programs conserve biodiversity? A review of salmonids. Evolutionary Applications 1:535–586.

Fraser, D. J., and L. Bernatchez. 2001. Adaptive evolutionary conservation: towards a unified concept for defining conservation units. Molecular Ecology 10:2741–2752.

Frazer, K. A., S. S. Murray, N. J. Schork, and E. J. Topol. 2009. Human genetic variation and its contribution to complex traits. Nature Reviews Genetics 10:241–252.

Freedberg, S., and M. J. Wade. 2001. Cultural inheritance as a mechanism for population sex-ratio bias in reptiles. Evolution 55:1049–1055.

Freeman, S., and Herron, J. C. 1998. Evolutionary Analysis, 1st edn. Prentice Hall, Upper Saddle River, NJ.

Frère, C. H., *et al.* 2010. Thar she blows! A novel method for DNA collection from cetacean blow. PLoS ONE 5:e12299.

Friar, E. A., T. Ladoux, E. H. Roalson, and R. H. Robichaux. 2000. Microsatellite analysis of a population crash and bottleneck in the Mauna Kea silversword, *Argyroxiphium sandwicense* ssp *sandwicense* (Asteraceae), and its implications for reintroduction. Molecular Ecology 9:2027–2034.

Fridez, F., S. Rochat, and R. Coquoz. 1999. Individual identification of cats and dogs using mitochondrial DNA tandem repeats? Science and Justice 39:167–171.

Fu, Y.-X., and W-H. Li. 1993. Maximum likelihood estimation of population parameters. Genetics 134:1261–1270.

Fu, Y.-X., and W-H. Li. 1999. Coalescing into the 21st century: an overview and prospects of coalescent theory. Theoretical Population Biology 56:1–10.

Funk, W. C., D. A. Tallmon, and F. W. Allendorf. 1999. Small effective population size in the long-toed salamander. Molecular Ecology 8:1633–1640.

Funk, W. C., J. K. McKay, P. A. Hohenlohe, and F. W. Allendorf. 2012. Harnessing genomics for delineating conservation units. Trends in Ecology & Evolution 27:489–496.

Funk, W. C., *et al.* 2005. Genetic basis of variation in morphological and life-history traits of a wild population of pink salmon. Journal of Heredity 96:24–31.

Gabriel, W., and R. Bürger. 1994. Extinction risk by mutational meltdown: synergistic effects between population regulation and genetic drift. Pp. 69–84 *in* Loeschcke, V., J. Tomiuk, and S. K. Jain, eds. Conservation Genetics. Birkhauser Verlag, Basel, Switzerland.

Gaggiotti, O. E. 2003. Genetic threats to population persistence. Annales Zoologici Fennici 40:155–168.

Gaggiotti, O. E. 2010. Preface to the special issue: advances in the analysis of spatial genetic data. Molecular Ecology Resources 10:757–759.

Gaggiotti, O. E., and M. Foll. 2010. Quantifying population structure using the F-model. Molecular Ecology Resources 10:821–830.

Gaggiotti, O. E., *et al.* 2009. Disentangling the effects of evolutionary, demographic, and environmental factors influencing genetic structure of natural populations: Atlantic herring as a case study. Evolution 63:2939–2951.

Galbusera, P., *et al.* 2000. Genetic variability and gene flow in the globally, critically-endangered Taita thrush. Conservation Genetics 1:45–55.

Galloway, L. F., and J. R. Etterson. 2007. Transgenerational plasticity is adaptive in the wild. Science 318:1134–1136.

Galtier, N., B. Nabholz, S. Glemin, and G. D. D. Hurst. 2009. Mitochondrial DNA as a marker of molecular diversity: a reappraisal. Molecular Ecology 18:4541–4550.

Gapare, W. J., and S. N. Aitken. 2005. Strong spatial genetic structure in peripheral but not core populations of sitka spruce [*Picea sitchensis* (Bong.) Carr.]. Molecular Ecology 14:2659–2667.

García-Ramos, G., and M. Kirkpatrick. 1997. Genetic models of adaptation and gene flow in peripheral populations. Evolution 51:21–28.

García-Ramos, G., and D. Rodriguez. 2002. Evolutionary speed of species invasions. Evolution 56:661–668.

Gardner, S., C. Jaing, K. McLoughlin, and T. Slezak. 2010. A microbial detection array (MDA) for viral and bacterial detection. BMC Genomics 11:668.

Garrick, R. C., R. J. Dyer, L. B. Beheregaray, and P. Sunnucks. 2008. Babies and bathwater: a comment on the premature obituary for nested clade phylogeographical analysis. Molecular Ecology 17:1401–1403.

Gaskin, J. F., and B. A. Schaal. 2002. Hybrid *Tamarix* widespread in US invasion and undetected in native Asian range. Proceedings of the National Academy of Sciences of the USA 99:11256.

Gauss, C. F. 1809. Theoria motus corporum coelestium in sectionis conicis solem ambientum. Hamburg, Germany.

Gebremedhin, B., *et al.* 2009. Frontiers in identifying conservation units: from neutral markers to adaptive genetic variation. Animal Conservation 12:107–109.

Gemmell, N. J., V. J. Metcalf, and F. W. Allendorf. 2004. Mother's curse: the effect of mtDNA on individual fitness and population viability. Trends in Ecology & Evolution 19:238–244.

Genner, M. J., *et al.* 2004. Camouflaged invasion of Lake Malawi by an Oriental gastropod. Molecular Ecology 13:2135–2141.

Gerardi, S., J. P. Jaramillo-Correa, J. Beaulieu, and J. Bousquet. 2010. From glacial refugia to modern populations: New assemblages of organelle genomes generated by differential cytoplasmic gene flow in transcontinental black spruce. Molecular Ecology 19:5265–5280.

Gerhardt, H. C., M. B. Ptacek, L. Barnett, and K. G. Torke. 1994. Hybridization in the diploid-tetraploid treefrogs *Hyla chrysoscelis* and *Hyla versicolor*. Copeia 1994:51–59.

Ghiselin, M. T. 1969. The Triumph of the Darwinian Method. University of Chicago Press, Chicago.

Gibbs, H. L., *et al.* 1991. Detection of a hypervariable locus in birds by hybridization with a mouse MHC probe. Molecular Biology and Evolution 8:433–446.

Gibson, S. Y., R. C. Van Der Marel, and B. M. Starzomski. 2009. Climate change and conservation of leading-edge peripheral populations. Conservation Biology 23:1369–1373.

Gienapp, P., *et al.* 2008. Climate change and evolution: disentangling environmental and genetic responses. Molecular Ecology 17:167–178.

Giger, T., *et al.* 2006. Life history shapes gene expression in salmonids. Current Biology 16:R281–2.

Gigord, L. D. B., M. R. Macnair, and A. Smithson. 2001. Negative frequency-dependent selection maintains a dramatic flower color polymorphism in the rewardless orchid *Dactylorhiza sambucina* (L.) Soo. Proceedings of the National Academy of Sciences of the USA 98:6253–6255.

Gilchrist, E. J., *et al.* 2006. Use of Ecotilling as an efficient SNP discovery tool to survey genetic variation in wild populations of *Populus* trichocarpa. Molecular Ecology 15:1367–1378.

Gillespie, J. H. 1992. The Causes of Molecular Evolution. Oxford University Press, Oxford.

Gillespie, J. H. 2001. Is the population size of a species relevant to its evolution? Evolution 55:2161–2169.

Gilligan, D. M., and R. Frankham. 2003. Dynamics of genetic adaptation to captivity. Conservation Genetics 4:189–197.

Gilpin, M. 1991. The genetic effective size of a metapopulation. Biological Journal of the Linnean Society 42:165–175.

Gilpin, M., and C. Wills. 1991. MHC and captive breeding – a rebuttal. Conservation Biology 5:554–555.

Gilpin, M. E., and M. E. Soulé. 1986. Minimum viable populations: processes of species extinction. Pp. 19–34 *in* Soulé, M. E., ed. Conservation Biology: The Science of Scarcity and Diversity. Sinauer, Sunderland, MA.

Gjoen, H. M., and H. B. Bentsen. 1997. Past, present, and future of genetic improvement in salmon aquaculture. ICES Journal of Marine Science 54:1009–1014.

Glick, P., B.A. Stein, and N.A. Edelson (editors). 2011. A Guide to Climate Change Vulnerability Assessment. National Wildlife Federation, Washington, D.C.

Godt, M. J. W., J. Walker, and J. L. Hamrick. 1997. Genetic diversity in the endangered lily *Harperocallis flava* and a close relative, *Tofieldia racemosa*. Conservation Biology 11:361–366.

Goetz, F. W., and S. MacKenzie. 2008. Functional genomics with microarrays in fish biology and fisheries. Fish and Fisheries 9:378–395.

Goldberg, T. L., *et al.* 2005. Increased infectious disease susceptibility resulting from outbreeding depression. Conservation Biology 19:455–462.

Goldman, E. 2003. Evolution: puzzling over the origin of species in the depths of the oldest lakes. Science 299:654–655.

Gompert, Z., *et al.* 2010. Secondary contact between *Lycaeides idas* and *L. melissa* in the Rocky Mountains: extensive admixture and a patchy hybrid zone. Molecular Ecology 19:3171–3192.

Gonçalves da Silva, A., *et al.* 2010. Genetic approaches refine ex situ lowland tapir (*Tapirus terrestris*) conservation. Journal of Heredity 101:581–590.

Good, J. M., *et al.* 2008. Ancient hybridization and mitochondrial capture between two species of chipmunks. Molecular Ecology 17:1313–1327.

Goodman, D. 1987. How do any species persist? Lessons for conservation biology. Conservation Biology 1:59–62.

Goodman, D. 2002. Predictive Bayesian population viability analysis: A logic for listing criteria, delisting criteria, and recovery plans. Pp. 447–469 *in* Beissinger, S. R. and D. R. McCullough, eds. Population Viability Analysis. University of Chicago Press, Chicago, IL.

Goodnight, K. F., and D. C. Queller. 1999. Computer software for performing likelihood tests of pedigree relationship using genetic markers. Molecular Ecology 8:1231–1234.

Gorman, G. C., and J. Renzi. 1979. Genetic distance and heterozygosity estimates in electrophoretic studies – effects of sample size. Copeia 1979:242–249.

Gouws, G., B. A. Stewart, and S. R. Daniels. 2004. Cryptic species within the freshwater isopod *Mesamphisopus capensis* (Phreatoicidea: Amphisopodidae) in the Western Cape, South Africa: allozyme and 12S rRNA sequence data and morphometric evidence. Biological Journal of the Linnean Society 81:235–253.

Grabowski, T. B., V. Thorsteinsson, B. J. McAdam, and G. Marteinsdóttir. 2011. Evidence of segregated spawning in a single marine fish stock: Sympatric divergence of ecotypes in Icelandic cod? PLoS ONE 6:e17528.

Grandbastien, M. A. 1998. Activation of plant retrotransposons under stress conditions. Trends Plant Science 3:181–187.

Grant, B. R., and P. R. Grant. 1998. Hybridization and speciation in Darwin's finches – the role of sexual imprinting on a culturally transmitted trait. Pp. 404–422 *in* Howard, D. J., and S. H. Berlocher, eds. Endless Forms. Oxford University Press, New York.

Grant, P. R. 1986. Ecology and Evolution of Darwin's Finches. Princeton University Press, Princeton, NJ.

Grant, P. R., and B. R. Grant. 1992a. Demography and the genetically effective sizes of two populations of Darwin finches. Ecology 73:766–784.

Grant, P. R., and B. R. Grant. 1992b. Hybridization of bird species. Science 256:193–197.

Grant, V. 1963. The Origin of Adaptations. Columbia University Press, New York.

Grant, W. S., and R. W. Leslie. 1993. Effect of metapopulation structure on nuclear and organellar DNA variability in semi-arid environments of southern Africa. South African Journal of Science 89:287–293.

Graur, D., and Li, W-H. 2000. Fundamentals of Molecular Evolution, 2nd edn. Sinauer Associates, Inc., Sunderland, MA.

Graves, J. A. M., and S. Shetty. 2001. Sex from W to Z: evolution of vertebrate sex chromosomes and sex determining genes. Journal of Experimental Zoology 290:449–462.

Gray, M. M., *et al.* 2009. Linkage disequilibrium and demographic history of wild and domestic canids. Genetics 181:1493–1505.

Green, D. M. 1991. Supernumerary chromosomes in amphibians. Pp. 333–357 *in* Green, D. M., and S. K. Sessions, eds. Amphibian Cytogenetics and Evolution. Academic Press, San Diego.

Green, R. E., *et al.* 2010. A draft sequence of the Neandertal genome. Science 328:710–722.

Greene, C. M., J. E. Hall, K. R. Guilbault, and T. P. Quinn. 2010. Improved viability of populations with diverse life-history portfolios. Biology Letters 6:382–386.

Greig, J. C. 1979. Principles of genetic conservation in relation to wildlife management in Southern Africa. South African Journal of Wildlife Research 9:57–78.

Griffith, B., J. W. Scott, J. W. Carpenter, and C. Reed. 1989. Translocation as a species conservation tool: status and strategy. Science 245:477–480.

Grift, R. E., *et al.* 2003. Fisheries-induced trends in reaction norms for maturation in North Sea plaice. Marine Ecology Progress Series 257:247–257.

Grinnell, J., Dixon, J. S., and Linsdale, J. M. 1937. Fur-bearing mammals of California: their natural history, systematic status, and relations to man. University of California Press, Berkeley, CA.

Gross, J. B., R. Borowsky, and C. J. Tabin. 2009. A novel role for *Mc1r* in the parallel evolution of depigmentation in independent populations of the cavefish *Astyanax mexicanus*. PLoS Genetics 5:e1000326.

Guerrant, E. O. 1996. Designing populations: Demographic, genetic, and horticultural dimensions. Pp. 171–207 *in* Falk, D. A., C. I. Millar, and M. Olwell, eds. Restoring Diversity. Island Press, Washington, DC.

Guillot, G., A. Estoup, F. Mortier, and J. F. Cosson. 2005. A spatial statistical model for landscape genetics. Genetics 170:1261–1280.

Gulisija, D., and J. F. Crow. 2007. Inferring purging from pedigree data. Evolution 61:1043–1051.

Gulisija, D., D. Gianola, K. A. Weigel, and M. A. Toro. 2006. Between-founder heterogeneity in inbreeding depression for production in Jersey cows. Livestock Science 104:244–253.

Gullion, G. W., and W. H. Marshall. 1968. Survival of ruffed grouse in a boreal forest. Living Bird 7:117–167.

Gutierrez-Espeleta, G. A., *et al.* 2001. Is the decline of desert bighorn sheep from infectious disease the result of low MHC variation? Heredity 86:439–450.

Gutschick, V. P., and H. BassiriRad. 2003. Extreme events as shaping physiology, ecology, and evolution of plants: toward a unified definition and evaluation of their consequences. New Phytologist 160:21–42.

Gyllensten, U., D. Wharton, A. Josefsson, and A. C. Wilson. 1991. Paternal inheritance of mitochondrial DNA in mice. Nature 352:255–257.

Haberl, H., *et al.* 2007. Quantifying and mapping the human appropriation of net primary production in earth's terrestrial ecosystems. Proceedings of the National Academy of Sciences 104:12942–12947.

Haig, S. M., and F. W. Allendorf. 2006. Hybirds and policy. Pp. 150–163 *in* Scott, J. M., D. D. Goble, and F. W. Davis, eds. The Endangered Species Act at Thirty: Conserving Biodiversity in Human-dominated Landscapes, Vol. 2. Island Press, Washington, DC.

Haig, S. M., J. D. Ballou, and S. R. Derrickson. 1990. Management options for preserving genetic diversity: reintroduction of Guam rails to the wild. Conservation Biology 4:290–300, 464.

Hailer, F., *et al.* 2012. Nuclear genomic sequences reveal that polar bears are an old and distinct bear lineage. Science 336:344–347.

Hajibabaei, M., *et al.* 2011. Environmental barcoding: A next-generation sequencing approach for biomonitoring applications using river benthos. PLoS ONE 6:e17497.

Halbert, N. D., and J. N. Derr. 2007. A comprehensive evaluation of cattle introgression into US federal bison herds. Journal of Heredity 98:1–12.

Haldane, J. B. S. 1942. The selective elimination of silver foxes in Canada. Journal of Genetics 44:296–304.

Hall, B. G. 2004. Phylogenetic Trees Made Easy: a How-to Manual. 2nd edn. Sinauer Associates, Sunderland, MA.

Hall, M. C., D. B. Lowry, and J. H. Willis. 2010. Is local adaptation in *Mimulus guttatus* caused by trade-offs at individual loci? Molecular Ecology 19:2739–2753.

Hall, P., L. C. Orrell, and K. S. Bawa. 1994. Genetic diversity and mating system in a tropical tree, *Carapa guianensis* (Meliaceae). American Journal of Botany 81:1104–1111.

Halliburton, R. 2004. Introduction to Population Genetics. Pearson Prentice Hall, Upper Saddle River, New Jersey.

Halligan, D. L., and P. D. Keightley. 2009. Spontaneous mutation accumulation studies in evolutionary genetics. Annual Review of Ecology Evolution and Systematics 40:151–172.

Hamede, R., *et al.* In review. Reduced impact of Tasmanian devil facial tumour disease at the current disease front. Conservation Biology.

Hamilton, A., *et al.* 2010. Quantifying uncertainty in estimation of tropical arthropod species richness. The American Naturalist 176:90–95.

Hamilton, J. A., C. Lexer, and S. N. Aitken. In press. Genomic and phenotypic architecture of a spruce hybrid zone (*Picea sitchensis* x *P. glauca*). Molecular Ecology.

Hamon, T. R., C. J. Foote, R. Hilborn, and D. E. Rogers. 2000. Selection on morphology of spawning wild sockeye salmon by a gill-net fishery. Transactions of the American Fisheries Society 129:1300–1315.

Hampe, A., and R. J. Petit. 2005. Conserving biodiversity under climate change: the rear edge matters. Ecology Letters 8:461–467.

Hamrick, J. L., and M. J. Godt. 1990. Allozyme diversity in plant species. Pp. 43–63 in Brown, A. H. D., M. T. Clegg, A. L. Kahler, and B. S. Weir, eds. Plant Population Genetics, Breeding, and Genetic Resources, Sinauer, Sunderland, MA.

Hamrick, J. L., and M. J. Godt. 1996. Conservation genetics of endemic plants. Pp. 281–304 in Avise, J. C. and J. L. Hamrick, eds. Conservation Genetics: Case Histories from Nature. Chapman and Hall, New York.

Hancock, A. M., *et al.* 2011. Adaptations to climate-mediated selective pressures in humans. PLoS Genet 7:e1001375.

Hankin, D. G., J. W. Nicholas, and T. W. Downey. 1993. Evidence for inheritance of age of maturity in chinook salmon (*Oncorhynchus tshawytscha*). Canadian Journal of Fisheries and Aquatic Sciences 50:347–358.

Hanlon, R. T. 1998. Mating systems and sexual selection in the squid *Loligo*: How might commercial fishing on spawning squids affect them? CalCOFI Reports 39:92–100.

Hanotte, O., T. Dessie, and S. Kemp. 2010. Time to tap Africa's livestock genomes. Science 328:1640–1641.

Hansen, B. D., and A. C. Taylor. 2008. Isolated remnant or recent introduction? Estimating the provenance of Yellingbo Leadbeater's possums by genetic analysis and bottleneck simulation. Molecular Ecology 17:4039–4052.

Hansen, B. D., P. Sunnucks, M. Blacket, and A. C. Taylor. 2005. A set of microsatellite markers for an endangered arboreal marsupial, Leadbeater's possum. Molecular Ecology Notes 5:796–799.

Hansen, B. D., D. K. P. Harley, D. B. Lindenmayer, and A. C. Taylor. 2009. Population genetic analysis reveals a long-term decline of a threatened endemic Australian marsupial. Molecular Ecology 18:3346–3362.

Hansen, L., J. Hoffman, C. Drews, and E. Mielbrecht. 2010. Designing climate-smart conservation: guidance and case studies. Conservation Biology 24:63–69.

Hansen, M. M. 2010. Expression of interest: transcriptomics and the designation of conservation units. Molecular Ecology 19:1757–1759.

Hansen, M. M., *et al.* 2000. Genetic monitoring of supportive breeding in brown trout (*Salmo trutta* L.), using microsatellite DNA markers. Canadian Journal of Fisheries and Aquatic Sciences 57:2130–2139.

Hansen, M. M., *et al.* 2007. Gene flow, effective population size and selection at major histocompatibility complex genes: brown trout in the Hardanger Fjord, Norway. Molecular Ecology 16:1413–1425.

Hansen, M. M., *et al.* 2012. Monitoring adaptive genetic responses to environmental change. Molecular Ecology 21:1311–1329.

Hanski, I. 1998. Metapopulation dynamics. Nature 396: 41–49.

Hanski, I., and M. Gilpin. 1991. Metapopulation dynamics – brief history and conceptual domain. Biological Journal of the Linnean Society 42:3–16.

Hanski, I., and M. E. Gilpin. 1997. Metapopulation Biology. Academic Press, San Diego, CA.

Hansson, B., S. Bensch, and D. Hasselquist. 2004. Lifetime fitness of short- and long-distance dispersing great reed warblers. Evolution 58:2546–2557.

Hansson, B., and L. Westerberg. 2002. On the correlation between heterozygosity and fitness in natural populations. Molecular Ecology 11:2467–2474.

Hardy, G. H. 1908. Mendelian proportions in a mixed population. Science 28:49–50.

Hardy, G. H. 1967. A Mathematician's Apology. Cambridge University Press, Cambridge.

Harpending, H. C., *et al.* 1998. Genetic traces of ancient demography. Proceedings of the National Academy of Sciences USA 95:1961–1967.

Harris, H. 1966. Enzyme polymorphism in man. Proceedings of the Royal Society B: Biological Sciences 164:298–310.

Harris, R. B., and F. W. Allendorf. 1989. Genetically effective population size of large mammals: an assessment of estimators. Conservation Biology 3:181–191.

Harris, R. B., W. A. Wall, and F. W. Allendorf. 2002. Genetic consequences of hunting: What do we know and what should we do? Wildlife Society Bulletin 30:634–643.

Harris, S. A., and R. Ingram. 1991. Chloroplast DNA and biosystematics – the effects of intraspecific diversity and plastid transmission. Taxon 40:393–412.

Harrison, R. G. 1990. Hybrid zones: windows on evolutionary processes. Oxford Surveys in Evolutionary Biology 7:69–128.

Harrison, R. G. 1993. Hybrids and hybrid zones: historical perspective. Pp. 3–12 in Harrison, R. G., ed. Hybrid Zones and the Evolutionary Process. Oxford University Press, Oxford.

Harrison, R. G., and S. M. Bogdanowicz. 1997. Patterns of variation and linkage disequilibrium in a field cricket hybrid zone. Evolution 51:493–505.

Harrison, S., J. Maron, and G. Huxel. 2000. Regional turnover and fluctuation in populations of five plants confined to serpentine seeps. Conservation Biology 14:769–779.

Hart, M. W. 2010. The species concept as an emergent property of population biology. Evolution 65:613–616.

Hartl, D. L., and Clark, A. G. 1997. Principles of Population Genetics, 3rd edn. Sinauer Assoc., Sunderland, MA.

Hartl, G. B., et al. 1992. Genetic diversity in the polish brown hare (Lepus europaeus, Pallas 1778) – implications for conservation and management. Acta Theriologica 37:15–25.

Hastings, A. 1981. Marginal underdominance at a stable equilibrium. Proceedings of the National Academy of Sciences USA 78:6558–6559.

Hastings, A. 1993. Complex interactions between dispersal and dynamics lessons from coupled logistic equations. Ecology 74:1362–1372.

Hauffe, H. C., and J. B. Searle. 1998. Chromosomal heterozygosity and fertility in house mice (Mus musculus domesticus) from northern Italy. Genetics 150:1143–1154.

Haugen, T. O., and L. A. Vollestad. 2001. A century of life-history evolution in grayling. Genetica 112–113: 475–491.

Hausdorf, B. 2011. Progress toward a general species concept. Evolution 65:923–931.

Hauser, L., and G. R. Carvalho. 2008. Paradigm shifts in marine fisheries genetics: ugly hypotheses slain by beautiful facts. Fish and Fisheries 9:333–362.

Hauser, L., et al. 2002. Loss of microsatellite diversity and low effective population size in an overexploited population of New Zealand snapper (Pagrus auratus). Proceedings of the National Academy of Sciences USA 99:11742–11747.

Haussler, D., et al. 2009. Genome 10K: A proposal to obtain whole-genome sequence for 10,000 vertebrate species. Journal of Heredity 100:659–674.

Hay, J. M., C. H. Daugherty, A. Cree, and L. R. Maxson. 2003. Low genetic divergence obscures phylogeny among populations of Sphenodon, remnant of an ancient reptile lineage. Molecular Phylogenetics and Evolution 29:1–19.

Hay, J. M., et al. 2010. Genetic diversity and taxonomy: a reassessment of species designation in tuatara (Sphenodon: Reptilia). Conservation Genetics 11:1063–1081.

Heard, W. R. 1991. Life history of pink salmon (Oncorhynchus gorbuscha). Pp. 119–230 in Groot, C., and L. Margolis, eds. Pacific Salmon Life Histories. UBC Press, Vancouver.

Heber, S. O. L., and J. V. Briskie. 2010. Population bottlenecks and increased hatching failure in endangered birds. Conservation Biology 24:1674–1678.

Hebert, P. D. N., et al. 2004. Ten species in one: DNA barcoding reveals cryptic species in the neotropical skipper butterfly

Astraptes fulgerator. Proceedings of the National Academy of Sciences USA 101:14812–14817.

Hebert, P. D. N., J. R. deWaard, and J.-F. Landry. 2010. DNA barcodes for 1/1000 of the animal kingdom. Biology Letters 6:359–362.

Hedgecock, D. 1994. Does variance in reproductive success limit effective population sizes of marine organisms? Pp. 122–134 in Beaumont, A. R., ed. Genetics and Evolution of Aquatic Organisms. Chapman and Hall, London.

Hedrick, P. W. 1987. Gametic disequilibrium measures: proceed with caution. Genetics 117:331–341.

Hedrick, P. W. 1994. Purging inbreeding depression and the probability of extinction: full-sib mating. Heredity 73:363–372.

Hedrick, P. W. 1999. Perspective: highly variable loci and their interpretation in evolution and conservation. Evolution 53:313–318.

Hedrick, P. W. 2002. Pathogen resistance and genetic variation at MHC loci. Evolution 56:1902–1908.

Hedrick, P. 2003. The major histocompatibility complex (MHC) in declining populations: an example of adaptive variation. Pp. 97–113 in Holt, W. V., A. R. Pickard, J. C. Rodger, and D. E. Wildt, eds. Reproductive Science and Integrated Conservation. Cambridge University Press, Cambridge.

Hedrick, P. 2005. 'Genetic restoration': a more comprehensive perspective than 'genetic rescue'. Trends in Ecology & Evolution 20:109.

Hedrick, P. 2007. Another view of conservation genetics. Trends in Ecology & Evolution 22:8–9.

Hedrick, P. W. 2011. Genetics of Populations, 4th edn. Jones and Bartlett Publishers, Sudbury, MA.

Hedrick, P. W., and M. E. Gilpin. 1997. Genetic effective size of a metapopulation. Pp. 165–181 in Hanski, I., and M. E. Gilpin, eds. Metapopulation Biology. Academic Press, San Diego, CA.

Hedrick, P. W., and S. T. Kalinowski. 2000. Inbreeding depression in conservation biology. Annual Review of Ecology and Systematics 31:139–162.

Hedrick, P. W., and P. S. Miller. 1992. Conservation genetics – techniques and fundamentals. Ecological Applications 2:30–46.

Hedrick, P. W., and K. Ritland. 2011. Population genetics of the white-phased "Spirit" black bear of British Columbia. Evolution 66:305–313.

Hedrick, P. W., R. C. Lacy, F. W. Allendorf, and M. E. Soulé. 1996. Directions in conservation biology: comments on Caughley. Conservation Biology 10:1312–1320.

Hedrick, P.W., and R. Fredrickson. 2010. Genetic rescue guidelines with examples from Mexican wolves and Florida panthers. Conservation Genetics 11:615–626.

Hedrick, P., R. Fredrickson, and H. Ellegren. 2001. Evaluation of d^2, a microsatellite measure of inbreeding and outbreeding, in wolves with a known pedigree. Evolution 55:1256–1260.

Hedrick, P. W., J. Gadau, and R. E. Page. 2006. Genetic sex determination and extinction. Trends in Ecology & Evolution 21:55–57.

Heimer, W. E. 2004. Inferred negative effects of "trophy hunting" in Alberta: the great Ram Mountain controversy. Proceedings of Biennial Symposium of the Northern Wild Sheep & Goat Council 14:193–209.

Heino, M. 1998. Management of evolving fish stocks. Canadian Journal of Fisheries and Aquatic Sciences 55:1971–1982.

Heino, M., U. Dieckmann, and O. R. Godo. 2002. Measuring probabilistic reaction norms for age and size at maturation. Evolution 56:669–678.

Heller, R., and H. R. Siegismund. 2009. Relationship between three measures of genetic differentiation G_{ST}, D_{EST} and G'_{ST}: how wrong have we been? Molecular Ecology 18: 2080–2083.

Hellmann, J. J., and M. Pineda-Krch. 2007. Constraints and reinforcement on adaptation under climate change: selection of genetically correlated traits. Biological Conservation 137:599–609.

Hemingway, J., L. Field, and J. Vontas. 2002. An overview of insecticide resistance. Science 298:96–97.

Hemmer-Hansen, J., E. E. Nielsen, J. Frydenberg, and V. Loeschcke. 2007. Adaptive divergence in a high gene flow environment: *Hsc70* variation in the European flounder (*Platichthys flesus* L.). Heredity 99:592–600.

Hendry, A. P. 2005. The power of natural selection. Nature 433:694–695.

Henrich, J. 2004. Demography and cultural evolution: why adaptive cultural processes produced maladaptive losses in Tasmania. American Antiquity 69:197–221.

Herrera, A., et al. 2007. Species richness and phylogenetic diversity comparisons of soil microbial communities affected by nickel-mining and revegetation efforts in New Caledonia. European Journal of Soil Biology 43:130–139.

Heslop-Harrison, J. S., and T. Schwarzacher. 2011. Organisation of the plant genome in chromosomes. Plant Journal 66:18–33.

Hey, J. 2000. Anticipating scientific revolutions in evolutionary genetics. Evolutionary Biology 32:97–111.

Hey, J. 2010. Isolation with migration models for more than two populations. Molecular Biology and Evolution 27:905–920.

Higgins, K., and M. Lynch. 2001. Metapopulation extinction caused by mutation accumulation. Proceedings of the National Academy of Sciences USA 98:2928–2933.

Higgins, S. I., and D. M. Richardson. 1999. Predicting plant migration rates in a changing world: The role of long-distance dispersal. American Naturalist 153:464–475.

Hilborn, R. 2006. Faith-based fisheries. Fisheries 31:554–555.

Hilborn, R., T. P. Quinn, D. E. Schindler, and D. E. Rogers. 2003. Biocomplexity and fisheries sustainability. Proceed-

ings of the National Academy of Sciences USA 100: 6564–6568.

Hill, W. G. 1974. Estimation of linkage disequilibrium in random mating populations. Heredity 33:229–339.

Hill, W. G. 1979. A note on the effective population size with overlapping generations. Genetics 92:317–322.

Hill, W. G. 1981. Estimation of effective population size from data on linkage disequilibrium. Genetical Research 38: 209–216.

Hill, W. G., and A. Robertson. 1966. The effect of linkage on limits to artificial selection. Genetical Research 8:269–294.

Hill, W. G., and A. Robertson. 1968. Linkage disequilibrium in finite populations. Theoretical and Applied Genetics 38:226–231.

Hindar, K., J. Tufto, L. M. Saettem, and T. Balstad. 2004. Conservation of genetic variation in harvested salmon populations. ICES Journal of Marine Science 61:1389–1397.

Hobbs, R. J., and H. A. Mooney. 1998. Broadening the extinction debate: Population deletions and additions in California and Western Australia. Conservation Biology 12:271–283.

Hodges, E., et al. 2007. Genome-wide in situ exon capture for selective resequencing. Nature Genetics 39:1522–1527.

Hodgins, K. A., L. Rieseberg, and S. P. Otto. 2009. Genetic control of invasive plants species using selfish genetic elements. Evolutionary Applications 2:555–569.

Hoegh-Guldberg, O., et al. 2008. Assisted colonization and rapid climate change. Science 321:345–346.

Hoekstra, H. E. 2006. Genetics, development and evolution of adaptive pigmentation in vertebrates. Heredity 97: 222–234.

Hoekstra, H. E., and M. W. Nachman. 2003. Different genes underlie adaptive melanism in different populations of rock pocket mice. Molecular Ecology 12:1185–1194.

Hoekstra, H. E., and T. Price. 2004. Parallel evolution in the genes. Science 303:1779–1781.

Hoekstra, H. E., et al. 2006. A single amino acid mutation contributes to adaptive beach mouse color pattern. Science 313:101–104.

Hoelzel, A. R. 1998. Molecular Genetic Analysis of Populations: A Practical Approach, 2nd edn. Oxford University Press, Oxford.

Hoffman, E. A., and M. S. Blouin. 2000. A review of colour and pattern polymorphisms in anurans. Biological Journal of the Linnean Society 70:633–665.

Hoffmann, A. A., R. J. Hallas, J. A. Dean, and M. Schiffer. 2003. Low potential for climatic stress adaptation in a rainforest *Drosophila* species. Science 301:100–102.

Hoffmann, A. A., and P. A. Parsons. 1997. Extreme environmental change and evolution. Cambridge University Press, Cambridge.

Hoffmann, A. A., and L. H. Rieseberg. 2008. Revisiting the impact of inversions in evolution: from population genetic markers to drivers of adaptive shifts and speciation? Annual Review of Ecology Evolution and Systematics 39:21–42.

Hoffmann, A. A., C. M. Sgro, and A. R. Weeks. 2004. Chromosomal inversion polymorphisms and adaptation. Trends in Ecology & Evolution 19:482–488.

Hoffmann, M., et al. 2010. The impact of conservation on the status of the world's vertebrates. Science 330: 1503–1509.

Hogan, Z. S., et al. 2004. The imperiled giants of the Mekong: ecologists struggle to understand and protect Southeast Asia's large migratory catfish. American Scientist 92: 228–237.

Hogbin, P. M., R. Peakall, and M. A. Sydes. 2000. Achieving practical outcomes from genetic studies of rare Australian plants. Australian Journal of Botany 48:375–382.

Hogg, J. T., S. H. Forbes, B. M. Steele, and G. Luikart. 2006. Genetic rescue of an insular population of large mammals. Proceedings of the Royal Society B: Biological Sciences 273:1491–1499.

Hohenlohe, P. A., et al. 2010a. Population genomic analysis of parallel adaptation in threespine stickleback using sequenced RAD tags. PLoS Genetics 6:e1000862.

Hohenlohe, P. A., P. C. Phillips, and W. A. Cresko. 2010b. Using population genomics to detect selection in natural populations: key concepts and methodological considerations. International Journal of Plant Sciences 171: 1059–1071.

Hohenlohe, P. A., et al. 2011a. Next-generation RAD sequencing identifies thousands of SNPs for assessing hybridization between rainbow and westslope cutthroat trout. Molecular Ecology Resources 11:117–122.

Hohenlohe, P. A., S. Bassham, M. Currey, and W. A. Cresko. 2012. Extensive linkage disequilibrium and parallel adaptive divergence across threespine stickleback genomes. Philosophical Transactions of the Royal Society of London, Series B, Biological Sciences 367:395–408.

Holderegger, R., and H. H. Wagner. 2008. Landscape genetics. BioScience 58:199–207.

Holliday, J. A., et al. 2008. Global monitoring of autumn gene expression within and among phenotypically divergent populations of Sitka spruce (*Picea sitchensis*). New Phytologist 178:103–122.

Holliday, J. A., K. Ritland, and S. N. Aitken. 2010. Widespread, ecologically relevant genetic markers developed from association mapping of climate-related traits in Sitka spruce (*Picea sitchensis*). New Phytologist 188:501–514.

Hollingsworth, P. M., S. W. Graham, and D. P. Little. 2011. Choosing and using a plant DNA barcode. PLoS ONE 6:e19254.

Holsinger, K. E. 2000. Reproductive systems and evolution in vascular plants. Proceedings of the National Academy of Sciences USA 97:7037–7042.

Holsinger, K. E., and B. S. Weir. 2009. Genetics in geographically structured populations: defining, estimating and interpreting FST. Nature Reviews Genetics 10:639–650.

Holsinger, K. E., R. J. Masongamer, and J. Whitton. 1999. Genes, demes, and plant conservation. Pp. 23–46 in Landweber, L. F., and A. P. Dobson, eds. Genetics and the Extinction of Species. Princeton University Press, Princeton, NJ.

Holt, W. V., A. R. Pickard, and R. S. Prather. 2004. Wildlife conservation and reproductive cloning. Reproduction 127:317–324.

Hopper, S. D., et al. 1996. The Western Australian biota as Gondwanan heritage, a review. In Hopper, S. D., J. A. Chappill, M. S. Harvey, and A. S. George, eds. Gondwanan Heritage: Past, Present and Future of the Western Australian Biota. Surrey Beatty & Sons, Chipping Norton.

Hosack, D. A., P. S. Miller, J. J. Hervert, and R. C. Lacy. 2002. A population viability analysis for the endangered Sonoran pronghorn, *Antilocapra americana sonoriensis*. Mammalia 66:207–229.

Houck, M. L., A. T. Kumamoto, D. S. Gallagher, and K. Benirschke. 2001. Comparative cytogenetics of the African elephant (*Loxodonta africana*) and Asiatic elephant (*Elephas maximus*). Cytogenetics and Cell Genetics 93:249–252.

Houle, D. 1992. Comparing evolvability and variability of quantitative traits. Genetics 130:195–204.

Howe G. T., et al. 2003. From genotype to phenotype: unraveling the complexities of cold adaptation in forest trees. Canadian Journal of Botany 81:1247–1266.

Hoy, M. S., K. Kelly, and R. J. Rodriguez. 2010. Development of a molecular diagnostic system to discriminate *Dreissena polymorpha* (zebra mussel) and *Dreissena bugensis* (quagga mussel). Molecular Ecology Resources 10:190–192.

Hudson, R. R. 1990. Gene genealogies and the coalescent process. Pp. 1–43 in Futuyma, D., and J. Antonovics, eds. Oxford Surveys in Evolutionary Biology, Volume 7. Oxford University Press, London.

Huey, R. B., et al. 2000. Rapid evolution of a geographic cline in size in an introduced fly. Science 287:308–309.

Hufford, K. M., and S. J. Mazer. 2003. Plant ecotypes: genetic differentiation in the age of ecological restoration. Trends in Ecology & Evolution 18:147–155.

Hughes, A. L. 1991. MHC polymorphism and the design of captive breeding programs. Conservation Biology 5: 249–251.

Hughes, J. B., G. C. Daily, and P. R. Ehrlich. 1997. Population diversity: its extent and extinction. Science 278:689–692.

Hurst, G. D. D., and F. M. Jiggins. 2005. Problems with mitochondrial DNA as a marker in population, phylogeographic and phylogenetic studies: the effects of inherited symbionts. Proceedings of the Royal Society B: Biological Sciences 272:1525–1534.

Husband, B. C., and H. A. Sabara. 2004. Reproductive isolation between autotetraploids and their diploid progenitors in fireweed, *Chamerion angustifolium* (Onagraceae). New Phytologist 161:703–713.

Husband, B. C., and D. W. Schemske. 1996. Evolution of the magnitude and timing of inbreeding depression in plants. Evolution 50:54–70.

Husby, A., *et al.* 2010. Contrasting patterns of phenotypic plasticity in reproductive traits in two great tit (parus major) populations. Evolution 64:2221–2237.

Hutchings, J. A., and D. J. Fraser. 2008. The nature of fisheries- and farming-induced evolution. Molecular Ecology 17:294–313.

Hutchinson, W. F., C. van Oosterhout, S. I. Rogers, and G. R. Carvalho. 2003. Temporal analysis of archived samples indicates marked genetic changes in declining North Sea cod (*Gadus morhua*). Proceedings of the Royal Society B: Biological Sciences 270:2125–2132.

Hutt, F. B. 1979. Genetics for Dog Breeders. W.H. Freeman and Company, San Francisco, CA.

Huynen, L., C. D. Millar, and D. M. Lambert. 2002. A DNA test to sex ratite birds. Molecular Ecology 11:851–856.

Huynen, L., C. D. Millar, R. P. Scofield, and D. M. Lambert. 2003. Nuclear DNA sequences detect species limits in ancient moa. Nature 425:175–178.

Huynh, L. Y., D. L. Maney, and J. W. Thomas. 2011. Chromosome-wide linkage disequilibrium caused by an inversion polymorphism in the white-throated sparrow (*Zonotrichia albicollis*). Heredity 106:537–546.

Hwang, A., *et al.* 2011. Long-term experimental hybrid swarms between moderately incompatible *Tigriopus californicus* populations: Hybrid inferiority in early generations yields to hybrid superiority in later generations. Conservation Genetics 12:895–909.

Ingman, M., H. Kaessmann, S. Paabo, and U. Gyllensten. 2000. Mitochondrial genome variation and the origin of modern humans. Nature 408:708–713.

Ingvarsson, P. K. 2001. Restoration of genetic variation lost – the genetic rescue hypothesis. Trends in Ecology & Evolution 16:62–63.

Ingvarsson, P. K., *et al.* 2008. Nucleotide polymorphism and phenotypic associations within and around the phytochrome B2 locus in European aspen (*Populus tremula*, Salicaceae). Genetics 178:2217–2226.

Ingvarsson, P. K., and M. C. Whitlock. 2000. Heterosis increases the effective migration rate. Proceedings of the Royal Society B: Biological Sciences 267:1321–1326.

Innocenti, P., E. H. Morrow, and D. K. Dowling. 2011. Experimental evidence supports a sex-specific selective sieve in mitochondrial genome evolution. Science 332:845–848.

IPPC. 2007. The Physical Science Basis. Contribution of Working Group I to the Fourth Assessment Report of the Intergovernmental Panel on Climate Change.

Isaac, N. J. B., *et al.* 2007. Mammals on the edge: Conservation priorities based on threat and phylogeny. PLoS ONE 2:e296.

IUCN. 2001. IUCN Red List Categories and Criteria: Version 3.1. IUCN Species Survival Commission, Gland, Switzerland, and Cambridge, UK.

IUCN. 2002. IUCN technical guidelines on the management of ex situ populations for conservation. Approved at the 14th Meeting of the Programme Committee of Council, Gland, Switzerland, 19 December 2002.

IUDZG/CBSG (IUCN/SSC). 1993. The World Zoo Conservation Strategy: The Roles of Zoos and Aquaria of the World in Global Conservation. Chicago Zoological Society, Brookfield, IL.

Ives, A. R., and M. C. Whitlock. 2002. Inbreeding and metapopulations. Science 295:454–455.

IWC (International Whaling Commission). 1998. Report of the Scientific Committee, Annex Q. Report of the working group on proposed specifications for a Norwegian DNA database register for minke whales. Report of the International Whaling Commission 49:287–289.

Jaari, S., M-H Li, and J. Merilä. 2009. A first-generation microsatellite-based genetic linkage map of the Siberian jay (*Perisoreus infaustus*): insights into avian genome evolution. BMC Genomics 10:1.

Jachmann, H., P. S. M. Berry, and H. Immae. 1995. Tusklessness in African elephants: A future trend. African Journal of Ecology 33:230–235.

Jacquard, A. 1975. Inbreeding: one word, several meanings. Theoretical Population Biology 7:338–363.

Jakóbsdottir, K. B., *et al.* 2011. Historical changes in genotypic frequencies at the Pantophysin locus in Atlantic cod (*Gadus morhua*) in Icelandic waters: evidence of fisheries-induced selection? Evolutionary Applications 4:562–573.

James, F. C. 1983. Environmental component of morphological differentiation in birds. Science 221:184–186.

James, F. C. 1991. Complementary descriptive and experimental studies of clinal variation in birds. American Zoologist 31:694–706.

James, J. W. 1971. The founder effect and response to artificial selection. Genetical Research 12:249–266.

Jamieson, A., and S. C. Taylor. 1997. Comparisons of three probability formulae for parentage exclusion. Animal Genetics 28:397–400.

Jamieson, I. G., and F. W. Allendorf. In press. How does the 50/500 rule apply to MVPs? Trends in Ecology & Evolution.

Jamieson, I. G., and R. C. Lacy. 2012. Managing genetic issues in reintroduction biology. Pp. 441–475 *in* Owen, J. G., D. P. Armstrong, K. A. Parker, and P. J. Seddon, eds. Reintroduction Biology: Integrating Science and Management. Wiley-Blackwell, New York.

Jamieson, I. G., and C. J. Ryan. 2000. Increased egg infertility associated with translocating inbred takahe (*Porphyrio hochstetteri*) to island refuges in New Zealand. Biological Conservation 94:107–114.

Jeffreys, A. J., V. Wilson, and S. L. Thein. 1985. Hypervariable 'minisatellite' regions in human DNA. Nature 314:67–73.

Jerde, C. L., A. R. Mahon, W. L. Chadderton, and D. M. Lodge. 2011. "Sight-unseen" detection of rare aquatic species using environmental DNA. Conservation Letters 4:150–157.

Jeukens, J., *et al.* 2010. The transcriptomics of sympatric dwarf and normal lake whitefish (*Coregonus clupeaformis* spp., Salmonidae) divergence as revealed by next-generation sequencing. Molecular Ecology 19:5389–5403.

Jiménez, J. A., *et al*. 1994. An experimental study of inbreeding depression in a natural habitat. Science 266:271–273.

Jobling, M. A., and P. Gill. 2004. Encoded evidence: DNA in forensic analysis. Nature Reviews Genetics 5:739–751.

Johnsen, O., O. G. Daehlen, G. Ostreng, and T. Skroppa. 2005a. Daylength and temperature during seed production interactively affect adaptive performance of *Picea abies* progenies. New Phytologist 168:589–596.

Johnsen, O., *et al*. 2005b. Climatic adaptation in *Picea abies* progenies is affected by the temperature during zygotic embryogenesis and seed maturation. Plant Cell and Environment 28:1090–1102.

Johnson, J., B. Fitzpatrick, and H. B. Shaffer. 2010. Retention of low-fitness genotypes over six decades of admixture between native and introduced tiger salamanders. BMC Evolutionary Biology 10:147.

Johnson, M. S. 1982. Polymorphism for direction of coil in *Partula suturalis*: behavioral isolation and positive frequency dependent selection. Heredity 49:145–151.

Johnson, M. S. 1988. Founder effects and geographic variation in the land snail *Theba pisana*. Heredity 61:133–42.

Johnson, M. S., and R. Black. 2006. Islands increase genetic subdivision and disrupt patterns of connectivity of intertidal snails in a complex archipelago. Evolution 60:2498–2506.

Johnson, M. S., B. Clarke, and J. Murray. 1990. The coil polymorphism in *Partula suturalis* does not favor sympatric speciation. Evolution 44:459–464.

Johnson, N. A. 2000. Speciation: Dobzhansky-Muller incompatibilities, dominance and gene interactions. Trends in Ecology & Evolution 15:480–482.

Johnson, R. 2010. The use of DNA identification in prosecuting wildlife-traffickers in Australia: Do the penalties fit the crimes? Forensic Science, Medicine, and Pathology 6:211–216.

Jones, A. G., E. Rosenqvist, A. Berglund, and J. C. Avise. 1999. Clustered microsatellite mutations in the pipefish *Syngnathus typhle*. Genetics 152:1057–1063.

Jones, J. C., M. R. Myerscough, S. Graham, and B. P. Oldroyd. 2004. Honey bee nest thermoregulation: diversity promotes stability. Science 305:402–404.

Jones, K. L., *et al*. 2002. Refining the whooping crane studbook by incorporating microsatellite DNA and leg-banding analyses. Conservation Biology 16:789–799.

Jones, M. E., D. Paetkou, E. Geffen, and C. Moritz. 2004. Genetic diversity and population structure of Tasmanian devils, the largest marsupial carnivore. Molecular Ecology 13:2197–2209.

Jones, M. E., *et al*. 2008. Life-history change in disease-ravaged Tasmanian devil populations. Proceedings of the National Academy of Sciences USA 105:10023–10027.

Jones, O. R., and J. Wang. 2010. Molecular marker-based pedigrees for animal conservation biologists. Animal Conservation 13:26–34.

Jones, R. N. 1991. B-chromosome drive. American Naturalist 137:430–442.

Jones, W. G., K. D. Hill, and J. M. Allen. 1995. *Wollemia nobilis*, a new living Australian genus and species in the Araucariaceae. Telopea 6:173–176.

Joron, M., and P. M. Brakefield. 2003. Captivity masks inbreeding effects on male mating success in butterflies. Nature 424:191–194.

Joshi, J., *et al*. 2001. Local adaptation enhances performance of common plant species. Ecology Letters 4:536–544.

Jost, L. 2008. G(ST) and its relatives do not measure differentiation. Molecular Ecology 17:4015–4026.

Jost, L. 2009. *D* vs. G_{ST}: response to Heller and Siegismund (2009) and Ryman and Leimar (2009). Molecular Ecology 18:2088–2091.

Jousson, O. J., *et al*. 2000. Invasive alga reaches California. Nature 408:157–158.

Jouventin, P., R. J. Cuthbert, and R. Ottvall. 2006. Genetic isolation and divergence in sexual traits: evidence for the northern rockhopper penguin *Eudyptes moseleyi* being a sibling species. Molecular Ecology 15:3413–3423.

Kahler, A. L., R. W. Allard, and R. D. Miller. 1984. Mutation rates for enzyme and morphological loci in barley (*Hordeam vulgare* L.). Genetics 106:729–734.

Kalinowski, S. T. 2009. How well do evolutionary trees describe genetic relationships among populations? Heredity 102:506–513.

Kalinowski, S. T., and P. W. Hedrick. 1999. Detecting inbreeding depression is difficult in captive endangered species. Animal Conservation 2:131–136.

Kalinowski, S. T., and P. W. Hedrick. 2001. Estimation of linkage disequilibrium for loci with multiple alleles: basic approach and an application using data from bighorn sheep. Heredity 87:698–708.

Kalinowski, S. T., and M. L. Taper. 2006. Maximum likelihood estimation of the frequency of null alleles at microsatellite loci. Conservation Genetics 7:991–995.

Kalinowski, S. T., and R. S. Waples. 2002. Relationship of effective to census size in fluctuating populations. Conservation Biology 16:129–136.

Kalinowski, S. T., P. W. Hedrick, and P. S. Miller. 2000. Inbreeding depression in the Speke's gazelle captive breeding program. Conservation Biology 14:1375–1384.

Kanda, N., R. F. Leary, and F. W. Allendorf. 2002. Evidence of introgressive hybridization between bull trout and brook trout. Transactions of the American Fisheries Society 131:772–782.

Kärkkäinen, K., *et al*. 1999. Genetic basis of inbreeding depression in *Arabis petraea*. Evolution 53:1354–1365.

Karl, S. A., and B. W. Bowen. 1999. Evolutionary significant units versus geopolitical taxonomy: molecular systematics of an endangered sea turtle (Genus *Chelonia*). Conservation Biology 13:990–999.

Karlsson, S., *et al*. 2011. Generic genetic differences between farmed and wild Atlantic salmon identified from a 7K SNP-chip. Molecular Ecology Resources 11:247–253.

Karron, J. D. 1991. Patterns of genetic variation and breeding systems in rare plant species. Pp. 87–98 *in* Falk, D. A., and K. E. Holsinger, eds. Genetics and Conservation of Rare Plants. Oxford University Press, New York.

Kaufmann, H. M., *et al.* 2002. Transplant tumor registry: Donor related malignancies. Transplantation 74:358–362.

Kaya, C. M. 1991. Rheotactic differentiation between fluvial and lacustrine populations of Arctic grayling (*Thymallus arcticus*), and implications for the only remaining indigenous population of fluvial "Montana grayling". Canadian Journal of Fisheries and Aquatic Sciences 48:53–59.

Kayser, M., and P. de Knijff. 2011. Improving human forensics through advances in genetics, genomics and molecular biology. Nature Reviews Genetics 12:179–191.

Keall, S. N., *et al.* 2001. Conservation in small places: reptiles on North Brother Island. New Zealand Journal of Zoology 28:367.

Kearsey, M. J. 1998. The principles of QTL analysis (a minimal mathematics approach). Journal of Experimental Botany 49:1619–1623.

Keightley, P. D., and M. Lynch. 2003. Toward a realistic model of mutations affecting fitness. Evolution 57:683–685.

Keller, L. F. 1998. Inbreeding and its fitness effects in an insular population of song sparrows (*Melospiza melodia*). Evolution 52:240–250.

Keller, L. F., and D. M. Waller. 2002. Inbreeding effects in wild populations. Trends in Ecology & Evolution 17:230–241.

Keller, L. F., *et al.* 1994. Selection against inbred song sparrows during a natural-population bottleneck. Nature 372:356–357.

Keller, L. F., A. B. Marr, and J. M. Reid. 2006. The genetic consequences of small population size: Inbreeding and loss of genetic variation. Pp. 113–137 *in* Smith, J. N. M., L. F. Keller, A. B. Marr, and P. Arcese, eds. Conservation and Biology of Small Populations. Oxford University Press, New York.

Keller, L. F., J. M. Reid, and P. Arcese. 2008. Testing evolutionary models of senescence in a natural population: Age and inbreeding effects on fitness components in song sparrows. Proceedings of the Royal Society B: Biological Sciences 275:597–604.

Keller, L. F., I. Biebach, S. R. Ewing, and P. E. A. Hoeck. 2012. The genetics of reintroductions: inbreeding and genetic drift. Pp. 360–394 *in* Owen, J. G., D. P. Armstrong, K. A. Parker, and P. J. Seddon, eds. Reintroduction Biology: Integrating Science and Management. Wiley-Blackwell, Chichester.

Keller, M., J. Kollmann, and P. J. Edwards. 2000. Genetic introgression from distant provenances reduces fitness in local weed populations. Journal of Applied Ecology 37:647–659.

Kelly, B. P., A. Whiteley, and D. Tallmon. 2010. The Arctic melting pot. Nature 468:891.

Kidwell, M. G. 2002. Keeping pace with opportunistic DNA. Science 295:2219–2220.

Kidwell, M. G., and D. R. Lisch. 1998. Hybrid genetics – transposons unbound. Nature 393:22–23.

Kimura, M. 1983. The Neutral Theory of Molecular Evolution. Cambridge University Press, Cambridge.

Kimura, M., and J. F. Crow. 1963. The measurement of effective population number. Evolution 17:279–288.

Kimura, M., and J. F. Crow. 1964. The number of alleles that can be maintained in a finite population. Genetics 49:725–738.

Kimura, M., and G. H. Weiss. 1964. The stepping stone model of population structure and the decrease of genetic correlation with distance. Genetics 49:561–576.

King, D. R., A. J. Oliver, and R. J. Mead. 1978. The adaptation of some western Australian mammals to food plants containing fluoroacetate. Australian Journal of Zoology 26:699–712.

King, M. 1993. Species Evolution: The Role of Chromosome Change. Cambridge University Press, Cambridge.

King, M. C., and A. C. Wilson. 1975. Evolution at two levels in humans and chimpanzees. Science 188:107–116.

Kingman, J. F. C. 1982. The coalescent. Stochastic Processes and Their Applications 13:235–248.

Kingsbury, N. 2009. A Plant Breeder's History of the World. University of Chicago Press, Chicago.

Kirby, G. C. 1975. Heterozygote frequencies in small populations. Theoretical Population Biology 8:31–48.

Kirkpatrick, M. 2010. How and why chromosome inversions evolve. PLoS Biol 8:e1000501.

Kirkpatrick, M., and N. H. Barton. 1997. Evolution of a species' range. American Naturalist 150:1–23.

Kirkpatrick, M., and N. Barton. 2006. Chromosome inversions, local adaptation and speciation. Genetics 173: 419–434.

Kirpichnikov, V. S. 1981. Genetic Bases of Fish Selection. Springer-Verlag, Berlin.

Klinka, D. R., and T. E. Reimchen. 2009. Adaptive coat colour polymorphism in the Kermode bear of coastal British Columbia. Biological Journal of the Linnean Society 98:479–488.

Knapp, S. 2011. Rarity, species richness, and the threat of extinction – are plants the same as animals? PLoS Biol 9:e1001067.

Knapton, R. W., and J. B. Falls. 1983. Differences in parental contribution among pair types in the polymorphic white-throated sparrow. Canadian Journal of Zoology 61:1288–1292.

Knaus, B. J., *et al.* 2011. Mitochondrial genome sequences illuminate maternal lineages of conservation concern in a rare carnivore. BMC Ecology 11:10.

Knowles, L. L. 2008. Why does a method that fails continue to be used? Evolution 62:2713–2717.

Knowles, L. L., and W. P. Maddison. 2002. Statistical phylogeography. Molecular Ecology 11:2623–2635.

Kocher, T. D., *et al.* 1989. Dynamics of mitochondrial DNA evolution in animals: Amplification and sequencing with conserved primers. Proceedings of the National Academy of Sciences USA 86:6196–6200.

Koelewijn, H., *et al.* 2010. The reintroduction of the Eurasian otter (*Lutra lutra*) into the Netherlands: hidden life revealed by noninvasive genetic monitoring. Conservation Genetics 11:601–614.

Koerper, H. C., and E. G. Stickel. 1980. Cultural drift: A primary process of culture change. Journal of Anthropological Research 36:463–469.

Kohn, B., *et al.* 1999. Feline adult beta-globin polymorphism reflected in restriction fragment length patterns. Journal of Heredity 90:177–181.

Kojima, K.-I. 1971. Is there a constant fitness value for a given genotype? No! Evolution 25:281–285.

Kolar, C. S., and D. M. Lodge. 2001. Progress in invasion biology: predicting invaders. Trends in Ecology & Evolution 16:199–204.

Kolbe, J. J., *et al.* 2004. Genetic variation increases during biological invasion by a Cuban lizard. Nature 431:177–181.

Konig, A. O., *et al.* 2002. Chloroplast DNA variation of oaks in western Central Europe and genetic consequences of human influences. Forest Ecology and Management 156:147–166.

Korpelainen, H. 2002. A genetic method to resolve gender complements investigations on sex ratios in *Rumex acetosa*. Molecular Ecology 11:2151–2156.

Kraak, S. B. M., B. Mundwiler, and P. J. B. Hart. 2001. Increased number of hybrids between benthic and limnetic three-spined sticklebacks in Enos Lake, Canada; the collapse of a species pair? Journal of Fish Biology 58: 1458–1464.

Kramer, A. T., and K. Havens. 2009. Plant conservation genetics in a changing world. Trends in Plant Science 14:599–607.

Krauss, S. L., B. Dixon, and K. W. Dixon. 2002. Rapid genetic decline in a translocated population of the endangered plant *Grevillea scapigera*. Conservation Biology 16:986–994.

Kreitman, M. E. 1983. Nucleotide polymorphism at the alcohol dehydrogenase locus of *Drosophila melanogaster*. Nature 304:412–417.

Kreitman, M. 2000. Methods to detect selection in populations with applications to the human. Annual Review of Genomics and Human Genetics 1:539–559.

Krimbas, C. B., and S. Tsakas. 1971. The genetics of *Dacus oleae*. V. Changes of esterase polymorphism in a natural population following insecticide control selection or drift? Evolution 25:454–460.

Kristensen, T. N., K. S. Pedersen, C. J. Vermeulen, and V. Loeschcke. 2010. Research on inbreeding in the 'omic' era. Trends in Ecology & Evolution 25:44–52.

Kritzer, J. P., and P. F. Sale. 2004. Metapopulation ecology in the sea: from Levins' model to marine ecology and fisheries science. Fish and Fisheries 5:131–140.

Krüger, O., J. Lindstrom, and W. Amos. 2001. Maladaptive mate choice maintained by heterozygote advantage. Evolution 55:1207–1214.

Krutovsky, K. V., *et al.* 2004. Comparative mapping in the Pinaceae. Genetics 168:447–461.

Kruuk, L. E. B. 2004. Estimating genetic parameters in natural populations using the 'animal model'. Philosophical Transactions of the Royal Society of London Series B Biological Sciences 359:873–890.

Kruuk, L. E. B., *et al.* 2000. Heritability of fitness in a wild mammal population. Proceedings of the National Academy of Sciences USA 97:698–703.

Kubo, T., and T. Mikami. 2007. Organization and variation of angiosperm mitochondrial genome. Physiologia Plantarum 129:6–13.

Kucera, T. E. 1991. Genetic variability in tule elk. California Fish and Game 77:70–78.

Kuhner, M. K. 2009. Coalescent genealogy samplers: windows into population history. Trends in Ecology & Evolution 24:86–93.

Kuparinen, A., and J. Merilä. 2007. Detecting and managing fisheries-induced evolution. Trends in Ecology & Evolution 22:652–659.

Kurian, V., and A. J. Richards. 1997. A new recombinant in the heteromorphy 'S' supergene in *Primula*. Heredity 78:383–390.

Lachish, S., *et al.* 2011. Evidence that disease-induced population decline changes genetic structure and alters dispersal patterns in the Tasmanian devil. Heredity 106:172–182.

Lack, D. 1947. Darwin's Finches. Cambridge University Press, Cambridge.

Lacy, R. C. 1988. A report on population genetics in conservation. Conservation Biology 2:245–247.

Lacy, R. C. 1989. Analysis of founder representation in pedigrees: founder equivalents and founder genome equivalents. Zoo Biology 8:111–123.

Lacy, R. C. 1997. Importance of genetic variation to the viability of mammalian populations. Journal of Mammalogy 78:320–335.

Lacy, R. C. 2000a. Should we select genetic alleles in our conservation breeding programs? Zoo Biology 19: 279–282.

Lacy, R. C. 2000b. Structure of the VORTEX simulation model for population viability analysis. Ecological Bulletins 48:191–203.

Lacy, R. C. 2012. Extending pedigree analysis for uncertain parentage and diverse breeding systems. Journal of Heredity 103:197–205.

Lacy, R. C., and J. D. Ballou. 1998. Effectiveness of selection in reducing the genetic load in populations of *Peromyscus polionotus* during generations of inbreeding. Evolution 52:900–909.

Lacy, R. C., A. Petric, and M. Warneke. 1993. Inbreeding and outbreeding in captive populations of wild animal species. Pp. 352–374 *in* Thornhill, N. W., ed. The Natural History of Inbreeding and Outbreeding. University of Chicago Press, Chicago, IL.

Lacy, R. C., G. Alaks, and A. Walsh. 1996. Hierarchical analysis of inbreeding depression in *Peromyscus polionotus*. Evolution 50:2187–2200.

Lacy, R. C., M. Borbat, and J. P. Pollak. 2009. VORTEX: A Stochastic Simulation of the Extinction Process. Version 9.99. Chicago Zoological Society, Brookfield, IL.

Laikre, L. 1999. Conservation genetics of Nordic carnivores: lessons from zoos. Hereditas 130:203–216.

Laikre, L., and N. Ryman. 1991. Inbreeding depression in a captive wolf (*Canis lupus*) population. Conservation Biology 5:33–40.

Laikre, L., N. Ryman, and E. A. Thompson. 1993. Hereditary blindness in a captive wolf (*Canis lupus*) population – frequency reduction of a deleterious allele in relation to gene conservation. Conservation Biology 7:592–601.

Laikre, L., *et al.* 2010a. Neglect of genetic diversity in implementation of the Convention on Biological Diversity. Conservation Biology 24:86–88.

Laikre, L., M. K. Schwartz, R. S. Waples, Nils Ryman, and GeM Working Group. 2010b. Compromising genetic diversity in the wild: unmonitored large-scale release of plants and animals. Trends in Ecology & Evolution 25:520–529.

Lalueza-Fox, C., *et al.* 2007. A *Melanocortin 1 Receptor* allele suggests varying pigmentation among Neanderthals. Science 318:1453–1455.

Lamb, L. 2010. Living up to its name: How controversial hunting regulations restored big bulls to the Elkhorn Mountains. Montana Outdoors 41 (Nov–Dec):14–19.

Lamer, J. T., *et al.* 2010. Introgressive hybridization between bighead carp and silver carp in the Mississippi and Illinois rivers. North American Journal of Fisheries Management 30:1452–1461.

Lamont, B. B., *et al.* 2003. Anthropogenic disturbance promotes hybridization between *Banksia* species by altering their biology. Journal of Evolutionary Biology 16:551–557.

Lande, R. 1979. Effective deme sizes during long-term evolution estimated from rates of chromosomal rearrangement. Evolution 33:234–251.

Lande, R. 1988. Genetics and demography in biological conservation. Science 241:1455–1460.

Lande, R. 1994. Risk of population extinction from fixation of new deleterious mutations. Evolution 48:1460–1469.

Lande, R. 1995. Mutation and conservation. Conservation Biology 9:782–791.

Lande, R. 1996. The meaning of quantitative genetic variation in evolution and conservation. Pp. 27–40 *in* Szaro, R. C. and D. W. Johnston, eds. Biodiversity in Managed Landscapes: Theory and Practice. Oxford University Press, New York.

Lande, R. 1999. Extinction risks from anthropogenic, ecological, and genetic factors. Pp. 1–22 *in* Landweber, L. F. and A. P. Dobson, eds. Genetics and the Extinction of Species. Princeton University Press, Princeton, NJ.

Lande, R., and G. F. Barrowclough. 1987. Effective population size, genetic variation, and their use in population management. Pp. 87–124 *in* Soulé, M. E., ed. Viable Populations for Conservation. Cambridge University Press, Cambridge.

Lande, R., and D. W. Schemske. 1985. The evolution of self-fertilization and inbreeding depression in plants. I. Genetic models. Evolution 39:24–40.

Landry, P.-A., M. T. Koskinen, and C. R. Primmer. 2002. Deriving evolutionary relationships among populations using microsatellites and $(d\mu)^2$: All loci are equal, but some are more equal than others. Genetics 161:1339–1347.

Landweber, L. F., and A. P. Dobson. 1999. Genetics and the Extinction of Species. Princeton University Press, Princeton, NJ.

Larkin, P. A. 1972. The stock concept and management of Pacific salmon. Pp. 11–15 *in* Simon, R. C. and P. A. Larkin, eds. The Stock Concept in Pacific Salmon. University of British Columbia, Vancouver.

Larkin, P. A. 1977. An epitaph for the concept of maximum sustained yield. Transactions of the American Fisheries Society 106:1–11.

Larkin, P. A. 1981. A perspective on population genetics and salmon management. Canadian Journal of Fisheries and Aquatic Sciences 38:1469–1475.

Larsen, L.-K., *et al.* 2011. Temporal change in inbreeding depression in life-history traits in captive populations of guppy (*Poecilia reticulata*): evidence for purging? Journal of Evolutionary Biology 24:823–834.

Larson, S., *et al.* 2002. Microsatellite DNA and mitochondrial DNA variation in remnant and translocated sea otter (*Enhydra lutris*) populations. Journal of Mammalogy 83:893–906.

Larsson, L. C., J. Charlier, L. Laikre, and N. Ryman. 2009. Statistical power for detecting genetic divergence-organelle versus nuclear markers. Conservation Genetics 10:1255–1264.

Latter, B. D. H., J. C. Mulley, D. Reid, and L. Pascoe. 1995. Reduced genetic load revealed by slow inbreeding in *Drosophila melanogaster*. Genetics 139:287–297.

Lausen, C. L., I. Delisle, R. M. R. Barclay, and C. Strobeck. 2008. Beyond mtDNA: nuclear gene flow suggests taxonomic oversplitting in the little brown bat (*Myotis lucifugus*). Canadian Journal of Zoology 86:700–713.

Law, R. 2007. Fisheries-induced evolution: Present status and future directions. Marine Ecology Progress Series 335:271–277.

Leamy, L. J., and C. P. Klingenberg. 2005. The genetics and evolution of fluctuating asymmetry. Annual Review of Ecology Evolution and Systematics 36:1–21.

Leary, R. F., and F. W. Allendorf. 1989. Fluctuating asymmetry as an indicator of stress: implications for conservation biology. Trends in Ecology & Evolution 4:214–217.

Leary, R. F., F. W. Allendorf, and K. L. Knudsen. 1984. Major morphological effects of a regulatory gene: *Pgm1-t* in rainbow trout. Molecular Biology and Evolution 1:183–194.

Leary, R. F., F. W. Allendorf, and K. L. Knudsen. 1993a. Null allele heterozygotes at two lactate dehydrogenase loci in rainbow trout are associated with decreased developmental stability. Genetica 89:3–13.

Leary, R. F., F. W. Allendorf, and S. H. Forbes. 1993b. Conservation genetics of bull trout in the Columbia and Klamath River drainages. Conservation Biology 7:856–865.

Leary, R. F., W. R. Gould, and G. K. Sage. 1996. Success of basibranchial teeth in indicating pure populations of rainbow trout and failure to indicate pure populations of westslope cutthroat trout. North American Journal of Fisheries Management 16:210–213.

Leberg, P. L. 1990. Influence of genetic variability on population growth – implications for conservation. Journal of Fish Biology 37:193–195.

Leberg, P. L. 1998. Influence of complex sex determination on demographic stochasticity and population viability. Conservation Biology 12:456–459.

Leberg, P. L., and B. D. Firmin. 2008. Role of inbreeding depression and purging in captive breeding and restoration programmes. Molecular Ecology 17:334–343.

Lee, C. E. 2002. Evolutionary genetics of invasive species. Trends in Ecology & Evolution 17:386–391.

Lee, C. E., and G. W. Gelembiuk. 2008. Evolutionary origins of invasive populations. Evolutionary Applications 1:427–448.

Leider, S. A., P. L. Hulett, J. J. Loch, and M. W. Chilcote. 1990. Electrophoretic comparison of the reproductive success of naturally spawning transplanted and wild steelhead trout through the returning adult stage. Aquaculture 88:239–252.

Leinonen, T., R. B. O'Hara, J. M. Cano, and J. Merilä. 2008. Comparative studies of quantitative trait and neutral marker divergence: a meta-analysis. Journal of Evolutionary Biology 21:1–17.

Lenormand, T. 2002. Gene flow and the limits to natural selection. Trends in Ecology & Evolution 17:183–189.

Lenski, R. E., and P. D. Sniegowski. 1995. "Adaptive mutation": the debate goes on. Science 269:285–286.

Leopold, A. S. 1944. The nature of heritable wildness in turkeys. Condor 46:133–197.

Le Roux, J., and A. M. Wieczorek. 2009. Molecular systematics and population genetics of biological invasions: towards a better understanding of invasive species management. Annals of Applied Biology 154:1–17.

Les, D. H., J. A. Reinartz, and E. J. Esselman. 1991. Genetic consequences of rarity in *Aster furcatus* (Asteraceae), a threatened, self-incompatible plant. Evolution 45:1641–1650.

Lesica, P., and F. W. Allendorf. 1992. Are small populations of plants worth preserving? Conservation Biology 6:135–139.

Lesica, P., and F. W. Allendorf. 1999. Ecological genetics and the restoration of plant communities: mix or match? Restoration Ecology 7:42–50.

Lessios, H. A. 1992. Testing electrophoretic data for agreement with Hardy-Weinberg expectations. Marine Biology 112:517–523.

Levene, H. 1949. On a matching problem arising in genetics. Annals of Mathematical Statistics 20:91–94.

Levin, D. A., J. Francisco Ortega, and R. K. Jansen. 1996. Hybridization and the extinction of rare plant species. Conservation Biology 10:10–16.

Levin, D. A., K. Ritter, and N. C. Ellstrand. 1979. Protein polymorphism in the narrow endemic *Oenothera organensis*. Evolution 33:534–542.

Levins, R. 1969. Some demographic and genetic consequences of environmental heterogeneity for biological control. Bulletin of the Entemological Society of America 15:237–240.

Levins, R. 1970. Extinction. Pp. 77–107 *in* Gerstenhaber, M., ed. Some Mathematical Problems in Biology. American Mathematical Society, Providence, RI.

Lewis, P. O. 2001. Phylogenetic systematics turns over a new leaf. Trends in Ecology & Evolution 16:30–37.

Lewontin, R. C. 1964. The interaction of selection and linkage. I. General considerations; heterotic models. Genetics 49:49–67.

Lewontin, R. C. 1965. The Theory of Inbreeding. Sir Ronald A. Fisher. Academic Press, New York, 2nd edn, 1965. viii + 150 pp. Science 150:1800–1801.

Lewontin, R. C. 1974. Genetic Basis of Evolutionary Change. Columbia University Press, New York.

Lewontin, R. C. 1988. On measures of gametic disequilibrium. Genetics 120:849–852.

Lewontin, R. C. 1991. 25 years ago in genetics – electrophoresis in the development of evolutionary genetics: milestone or millstone. Genetics 128:657–662.

Lewontin, R. C. 1999. Evolutionary Genetics from Molecules to Morphology. Cambridge University Press, Cambridge.

Lewontin, R. C. 2000. The Triple Helix: Gene, Organism, and Environment. Harvard University Press, Cambridge, MA.

Lewontin, R. C., and L. C. Birch. 1966. Hybridization as a source of variation for adaptation to new environments. Evolution 20:315–336.

Lewontin, R. C, and C. C. Cockerham. 1959. The goodness-of-fit test for detecting natural selection in random mating populations. Evolution 13:561–564.

Lewontin, R. C., and J. Felsenstein. 1965. The robustness of homogeneity tests in 2 X N tables. Biometrics 21:19–33.

Lewontin, R. C., and K. Kojima. 1960. The evolutionary dynamics of complex polymorphisms. Evolution 14:450–472.

Lewontin, R. C., L. R. Ginzburg, and S. D. Tuljapurkar. 1978. Heterosis as an explanation for large amounts of genetic polymorphism. Genetics 88:149–169.

Li, C. C. 1967. Castle's early work on selection and equilibrium. American Journal of Human Genetics 19:70–74.

Li, D.-Z., and H. W. Pritchard. 2009. The science and economics of *ex situ* plant conservation. Trends in Plant Science 14:614–621.

Li, M. H., and J. Merilä. 2010. Extensive linkage disequilibrium in a wild bird population. Heredity 104:600–610.

Li, W.-H. 1978. Maintenance of genetic variability under the joint effect of mutation, selection and random drift. Genetics 90:349–382.

Li, Y. C., et al. 2002. Microsatellites: genomic distribution, putative functions and mutational mechanisms: a review. Molecular Ecology 11:2453–2465.

Liao, W., and D. H. Reed. 2009. Inbreeding–environment interactions increase extinction risk. Animal Conservation 12:54–61.

Lichatowich, J. A. 1999. Salmon without Rivers. Island Press, Washington, DC.

Lincoln, F. C. 1930. Calculating waterfowl abundance on the basis of banding returns. United States Department of Agriculture Circular 118:1–4.

Linder, C. R., et al. 1998. Long-term introgression of crop genes into wild sunflower populations. Theoretical and Applied Genetics 96:339–347.

Lindner, D. L., et al. 2011. DNA-based detection of the fungal pathogen Geomyces destructans in soils from bat hibernacula. Mycologia 103:241–246.

Lindqvist, C., et al. 2010. Complete mitochondrial genome of a Pleistocene jawbone unveils the origin of polar bear. Proceedings of the National Academy of Sciences 107:5053–5057.

Lingle, S. 1992. Escape gaits of white-tailed deer, mule deer and their hybrids: gaits observed and patterns of limb coordination. Behaviour 122:153–181.

Linhart, Y. B., and M. C. Grant. 1996. Evolutionary significance of local genetic differentiation in plants. Annual Review of Ecology and Systematics 27:237–277.

Lively, C. M., C. Craddock, and R. C. Vrijenhoek. 1990. The Red Queen hypothesis supported by parasitism in sexual and clonal fish. Nature 344:864–866.

Loarie, S. R., et al. 2009. The velocity of climate change. Nature 462:1052–1111.

Loder, N. 2005. Point of no return. Conservation in Practice 6(3):29–34.

Loftus, K. H. 1976. Science for Canada's fisheries rehabilitation needs. Canadian Journal of Fisheries and Aquatic Sciences 33:1822–1857.

Loi, P., et al. 2001. Genetic rescue of an endangered mammal by cross-species nuclear transfer using post-mortem somatic cells. Nature Biotechnology 19:962–964.

Louis, E. J., and E. R. Dempster. 1987. An exact test for Hardy-Weinberg and multiple alleles. Biometrics 43:805–811.

Lowe, C. B., G. Bejerano, S. R. Salama, and D. Haussler. 2010. Endangered species hold clues to human evolution. Journal of Heredity 101:437–447.

Lowe, S. J., M. Browne, and S. Boudjelas. 2000. 100 of the World's Worst Invasive Alien Species. IUCN/SSC Invasive Species Specialist Group (ISSG), Auckland, New Zealand.

Lowe, W. H., and F. W. Allendorf. 2010. What can genetics tell us about population connectivity? Molecular Ecology 19:3038–3051.

Lowell, J. L., et al. 2009. Colorado animal-based plague surveillance systems: relationships between targeted animal species and prediction efficacy of areas at risk for humans. Journal of Vector Ecology 34:22–31.

Lowry, D. B., and J. H. Willis. 2010. A widespread chromosomal inversion polymorphism contributes to a major life-history transition, local adaptation, and reproductive isolation. PLoS Biology 8:e1000500.

Lowther, J. K. 1961. Polymorphism in the white-throated sparrow, Zonotrichia albicollis (Gmelin). Canadian Journal of Zoology 39:281–292.

Luck, G. W., G. C. Daily, and P. R. Ehrlich. 2003. Population diversity and ecosystem services. Trends in Ecology & Evolution 18:331–336.

Luikart, G., and J.-M. Cornuet. 1998. Empirical evaluation of a test for identifying recently bottlenecked populations from allele frequency data. Conservation Biology 12:228–237.

Luikart, G., and P. R. England. 1999. Statistical analysis of microsatellite DNA data. Trends in Ecology & Evolution 14:253–256.

Luikart, G., et al. 1997. Characterization of microsatellite loci in the endangered long-footed potoroo Potorous longipes. Molecular Ecology 6:497–498.

Luikart, G., W. B. Sherwin, B. M. Steele, and F. W. Allendorf. 1998. Usefulness of molecular markers for detecting population bottlenecks via monitoring genetic change. Molecular Ecology 7:963–974.

Luikart, G., J.-M. Cornuet, and F. W. Allendorf. 1999. Temporal changes in allele frequencies provide estimates of population bottleneck size. Conservation Biology 13:523–530.

Luikart, G. H., et al. 2003. The power and promise of population genomics: from genotyping to genome-typing. Nature Reviews: Genetics 4:981–994.

Luikart, G., et al. 2010. Estimation of census and effective population sizes: the increasing usefulness of DNA-based approaches. Conservation Genetics 11:355–373.

Luikart, G., et al. 2011. High connectivity among argali sheep from Afghanistan and adjacent countries: Inferences from neutral and candidate gene microsatellites. Conservation Genetics 12:921–931.

Lukoschek, V., et al. 2009. The rise of commercial 'by-catch whaling' in Japan and Korea. Animal Conservation 12:398–399.

Luo, S.-J., et al. 2004. Phylogeography and genetic ancestry of tigers (Panthera tigris). PLoS Biology 2:2275–2293.

Luo, S.-J., et al. 2008. Subspecies genetic assignments of worldwide captive tigers increase conservation value of captive populations. Current Biology 18:592–596.

Luo, S.-J., W. E. Johnson, J. L. D. Smith, and S. J. O'Brien. 2010. What is a tiger? Genetics and phylogeography. Pp. 35-51, in Tilson, R., and P. J. Nyhus, eds. Tigers of the World, 2nd edn: The Science, Politics and Conservation of Panthera tigris. Academic Press, London.

Lush, J. L. 1937. Animal Breeding Plans. Iowa State University Press, Ames, IA.

Lynch, M. 1996. A quantitative-genetic perspective on conservation issues. Pp. 471–501 in Avise, J. C. and J. L. Hamrick, eds. Conservation Genetics: Case Histories from Nature. Chapman and Hall, New York.

Lynch, M. 2001. The molecular natural history of the human genome. Trends in Ecology & Evolution 16:420–422.

Lynch, M., and J. S. Conery. 2000. The evolutionary fate and consequences of duplicate genes. Science 290:1151–1155.

Lynch, M., and W. Gabriel. 1990. Mutation load and the survival of small populations. Evolution 44:1725–1737.

Lynch, M., and R. Lande. 1993. Evolution and extinction in response to environmental-change. Pp. 234–250 in Kareiva, P. M., J. G. Kingsolver, and R. B. Huey, eds. Biotic Interactions and Global Change. Sinauer Associates, Sunderland, MA.

Lynch, M., and R. Lande. 1998. The critical effective size for a genetically secure population. Animal Conservation 1:70–72.

Lynch, M., and M. O'Hely. 2001. Captive breeding and the genetic fitness of natural populations. Conservation Genetics 2:363–378.

Lynch, M., and K. Ritland. 1999. Estimation of pairwise relatedness with molecular markers. Genetics 152:1753–1766.

Lynch, M., and Walsh, B. 1998. Genetics and analysis of quantitative traits. Sinauer Assoc., Sunderland, MA.

Lynch, M., R. Burger, D. Butcher, and W. Gabriel. 1993. The mutational meltdown in asexual populations. Journal of Heredity 84:339–344.

Lynch, M., J. Conery, and R. Burger. 1995. Mutational meltdowns in sexual populations. Evolution 49:1067–1080.

Lynch, M., et al. 1999. Perspective: spontaneous deleterious mutation. Evolution 53:645–663.

MacCluer, J. W., B. VandeBerg, B. Read, and O. A. Ryder. 1986. Pedigree analysis by computer simulation. Zoo Biology 5:147–160.

Mace, G. M., P. Pearcekelly, and D. Clarke. 1998. An integrated conservation programme for the tree snails (Partulidae) of Polynesia: a review of captive and wild elements. Journal of Conchology 89–96.

Mace, G. M., J. L. Gittleman, and A. Purvis. 2003. Preserving the tree of life. Science 300:1707–1709.

Mack, R. N., et al. 2000. Biotic invasions: causes, epidemiology, global consequences, and control. Ecological Applications 10:689–710.

Mackay, T. F. C. 2001. Quantitative trait loci in Drosophila. Nature Reviews Genetics 2:11–20.

Mackay, T. F. C. 2010. Mutations and quantitative genetic variation: Lessons from Drosophila. Philosophical Transactions of the Royal Society B: Biological Sciences 365:1229–1239.

Mackenzie, S., and L. McIntosh. 1999. Higher plant mitochondria. The Plant Cell 11:571–586.

Madsen, T., R. Shine, M. Olsson, and H. Wittzell. 1999. Conservation biology – restoration of an inbred adder population. Nature 402:34–35.

Maguire, L. A., and R. C. Lacy. 1990. Allocating scarce resources for conservation of endangered subspecies: partitioning zoo space for tigers. Conservation Biology 4:157–166.

Majerus, M. E. N., and N. I. Mundy. 2003. Mammalian melanism: natural selection in black and white. Trends Genetics 19:585–588.

Makela, M. E., and R. H. Richardson. 1977. The detection of sympatric sibling species using genetic correlation analysis. I. Two loci, two gamodemes. Genetics 86:665–678.

Malakoff, D. 1999. Bayes offers a 'new' way to make sense of numbers. Science 286:1460–1464.

Malhotra, I., et al. 2005. Real-time quantitative PCR for determining the burden of Plasmodium falciparum parasites during pregnancy and infancy. Journal of Clinical Microbiology 43:3630–3635.

Malik, S., et al. 1997. Pinniped penises in trade: a molecular-genetic investigation. Conservation Biology 11:1365–1374.

Mallet, J. 2005. Hybridization as an invasion of the genome. Trends in Ecology & Evolution 20:229–237.

Mallet, J. 2007. Hybrid speciation. Nature 446:279–283.

Mamanova, L., et al. 2010. Target-enrichment strategies for next-generation sequencing. Nature Methods 7:111–118.

Mammadov, J., et al. 2010. Development of highly polymorphic SNP markers from the complexity reduced portion of maize [Zea mays; L.] genome for use in marker-assisted breeding. Theoretical and Applied Genetics 121:577–588.

Manel, S., P. Berthier, and G. Luikart. 2002. Detecting wildlife poaching: identifying the origin of individuals using Bayesian assignment tests and multi-locus genotypes. Conservation Biology 16:650–657.

Manel, S., M. K. Schwartz, G. Luikart, and P. Taberlet. 2003. Landscape genetics: the combination of landscape ecology and population genetics. Trends in Ecology & Evolution 18:189–197.

Manel, S., O. E. Gaggiotti, and R. S. Waples. 2005. Assignment methods: Matching biological questions with appropriate techniques. Trends in Ecology & Evolution 20:136–142.

Manel, S., et al. 2010. Perspectives on the use of landscape genetics to detect genetic adaptive variation in the field. Molecular Ecology 19:3760–3772.

Maree, S., and C. Faulkes. 2008. Heterocephalus glaber. IUCN Red List of Threatened Species. Version 2011.1. Available at www.iucnredlist.org [accessed 13 August 2011].

Marker, L. L., et al. 2008. Molecular genetic insights on cheetah (Acinonyx jubatus) ecology and conservation in Namibia. Journal of Heredity 99:2–13.

Marr, A. B., et al. 2006. Interactive effects of environmental stress and inbreeding on reproductive traits in a wild bird population. Journal of Animal Ecology 75:1406–1415.

Marris, E. 2009. The genome of the American west. Nature 457:950–952.

Marshall, C. T., and H. I. Browman. 2007. Disentangling the causes of maturation trends in exploited fish populations. Marine Ecology Progress Series 335:249–251.

Marshall, C. T., and B. J. McAdam. 2007. Integrated perspectives on genetic and environmental effects on maturation can reduce potential for errors of inference. Marine Ecology Progress Series 335:301–310.

Marshall, T. C., J. Slate, L. E. B. Kruuk, and J. M. Pemberton. 1998. Statistical confidence for likelihood-based paternity inference in natural populations. Molecular Ecology 7:639–655.

Martinez, J. G., J. Carranza, J. L. Fernandez-Garcia, and C. B. Sanchez-Prieto. 2002. Genetic variation of red deer populations under hunting exploitation in southwestern Spain. Journal of Wildlife Management 66:1273–1282.

Martinsen, G. D., T. G. Whitham, R. J. Turek, and P. Keim. 2001. Hybrid populations selectively filter gene introgression between species. Evolution 55:1325–1335.

Matayoshi, T., et al. 1987. Heterochromatic variation in Cebus apella (Cebidae, Platyrrhini) of different geographical regions. Cytogenetics and Cell Genetics 44:158–162.

Mather, K. 1953. Genetical control of stability in development. Heredity 7:297–336.

Matsuki, Y., Y. Isagi, and Y. Suyama. 2007. The determination of multiple microsatellite genotypes and DNA sequences from a single pollen grain. Molecular Ecology Notes 7:194–198.

Matthew, P. 1831. On Naval Timber and Arboriculture. Adam Black, Edinburgh.

Mattson, D. J., and T. Merrill. 2002. Extirpations of grizzly bears in the contiguous United States, 1850–2000. Conservation Biology 16:1123–1136.

Mattson, D. J., K. C. Kendall, and D. P. Reinhart. 2001. Whitebark pine, grizzly bears, and red squirrels. Pp. 121–136 in Tomback, D. F., S. F. Arno, and R. E. Keane, eds. Whitebark Pine Communities: Ecology and Restoration. Island Press, Washington, DC.

Maudet, C., G. Luikart, and P. Taberlet. 2001. Development of microsatellite multiplexes for wild goats using primers designed from domestic Bovidae. Genetics Selection Evolution 33:S193–S203.

Maudet, C., et al. 2002. Microsatellite DNA and recent statistical methods in wildlife conservation management: applications in Alpine ibex [Capra ibex (ibex)]. Molecular Ecology 11:421–436.

Maudet, C., et al. 2004. A standard set of polymorphic microsatellites for threatened mountain ungulates (Caprini, Artiodactyla). Molecular Ecology Notes 4:49–55.

Maxted, N. 2003. Conserving the genetic resources of crop wild relatives in European Protected Areas. Biological Conservation 113:411–417.

May, B. 1998. Starch gel electrophoresis of allozymes. Pp. 1–28 in Hoelzel, A. R., ed. Molecular Genetic Analysis of Populations. Oxford University Press, Oxford.

May, R. M. 1990. Taxonomy as destiny? Nature 347:129–130.

May, R. M. 2004. Uses and abuses of mathematics in biology. Science 303:790–791.

Mayr, E. 1942. Systematics and the Origin of Species. Columbia University Press, New York.

Mayr, E. 1963. Animal Species and Evolution. Belknap Press, Cambridge, MA.

Mayr, E. 1981. Biological classification: toward of synthesis of classification. Science 214:510–516.

Mayr, E. 1982. The Growth of Biological Thought: Diversity, Evolution and Inheritance. Belknap Press, Cambridge, MA.

Melville-Smith, R., and S. de Lestang. 2006. Spatial and temporal variation in the size of maturity of the western rock lobster Panulirus cygnus George. Marine Biology 150:183–195.

McAdam, B. J., T. B. Grabowski, and G. Marteinsdóttir. In press. Identification of stock subunits using morphological markers. Fisheries Research.

McCallum, H., and M. Jones. 2012. Infectious cancers in wildlife. In Aguirre, A. A., R. S. Ostfeld, and P. Daszak, eds. Conservation Medicine: Applied Cases of Ecological Health. Oxford University Press, Oxford.

McCauley, D. E. 1991. Genetic consequences of local population extinction and recolonization. Trends in Ecology & Evolution 6:5–8.

McCauley, D. E. 1994. Contrasting the distribution of chloroplast DNA and allozyme polymorphisms among local populations of Silene alba: Implications for studies of gene flow in plants. Proceedings of the National Academy of Sciences USA 91:8127–8131.

McCauley, D. E., and J. R. Ellis. 2008. Recombination and linkage disequilibrium among mitochondrial genes in structured populations of the gynodioecious plant Silene vulgaris. Evolution 62:823–832.

McCauley, D. E., M. F. Bailey, N. A. Sherman, and M. Z. Darnell. 2005. Evidence for paternal transmission and heteroplasmy in the mitochondrial genome of Silene vulgaris, a gynodioecious plant. Heredity 95:50–58.

McClintock, B. 1984. The significance of responses of the genome to challenge. Science 226:792–801.

McCullough, D. R. 1978. Population dynamics of the Yellowstone grizzly bear. Pp. 173–196 in Fowler, C. W., and T. D. Smith, eds. Dynamics of Large Mammal Populations. John Wiley and Sons, New York.

McCullough, D. R., J. K. Fischer, and J. D. Ballou. 1996. From bottleneck to metapopulation: recovery of the tule elk in California. Pp. 375–403 in McCullough, D. R., ed. Metapopulations and Wildlife Conservation. Island Press, Covelo, CA.

McDermott, R., et al. 2009. Monoamine oxidase A gene (MAOA) predicts behavioral aggression following provocation. Proceedings of the National Academy of Sciences 106:2118–2123.

McDonald, J. H. 1994. Detecting natural selection by comparing geographic variation in protein and DNA polymorphisms. Pp. 88–100 *in* Golding, B., ed. Non-neutral Evolution: Theories and Molecular Data. Chapman and Hall, New York.

McGinnity, P., *et al.* 2003. Fitness reduction and potential extinction of wild populations of Atlantic salmon, *Salmo salar*, as a result of interactions with escaped farmed salmon. Proceedings of the Royal Society B: Biological Sciences 270:2443–2450.

McGinnity, P., *et al.* 2009. Impact of naturally spawning captive-bred atlantic salmon on wild populations: Depressed recruitment and increased risk of climate-mediated extinction. Proceedings of the Royal Society B: Biological Sciences 276:3601–3610.

McKay, J. K., and R. G. Latta. 2002. Adaptive population divergence: markers, QTL and traits. Trends in Ecology & Evolution 17:285–291.

McKenney, D. W., *et al.* 2007. Potential impacts of climate change on the distribution of North American trees. BioScience 57:939–948.

McKinnon, G. E., R. E. Vaillancourt, H. D. Jackson, and B. M. Potts. 2001. Chloroplast sharing in the Tasmanian eucalypts. Evolution 55:703–711.

McLachlan, J. S., and J. S. Clark. 2004. Reconstructing historical ranges with fossil data at continental scales. Forest Ecology and Management 197:139–147.

McLachlan, J. S., J. J. Hellmann, and M. W. Schwartz. 2007. A framework for debate of assisted migration in an era of climate change. Conservation Biology 21:297–302.

McLelland, P. J. 2011. Campbell Island – pushing the boundaries of rat eradication. Pp. 204–207 *in* Veitch, C. R., M. N. Clout, and D. R. Towns, eds. Island Invasives Eradication and Management. Proceedings of the International Conference on Island Invasives. Gland, Switzerland IUCN and Auckland, New Zealand CBB.

McNeely, J. A., *et al.* 1990. Conserving the world's biological diversity. IUCN, World Resources Institute, Conservation International, WWF-US and the World Bank, Washington, DC.

Meffe, G. K. 1992. Techno-arrogance and halfway technologies – salmon hatcheries on the Pacific coast of North America. Conservation Biology 6:350–354.

Mehrhoff, L. A. 1996. Reintroducing endangered Hawaiian plants. Pp. 101–120 *in* Falk, D. A., C. I. Millar, and M. Olwell, eds. Restoring Diversity. Island Press, Washington, DC.

Meine, C. D. 1998. Moving mountains: Aldo Leopold and a Sand County almanac. Wildlife Society Bulletin 26: 697–706.

Meirmans, P. G., and P. W. Hedrick. 2011. Assessing population structure: FST and related measures. Molecular Ecology Resources 11:5–18.

Melampy, M. N., and H. F. Howe. 1977. Sex ratio in the tropical tree *Triplaris americana* Polygonaceae. Evolution 31:867–872.

Mendel, G. 1865. Versuche uber Pflanzen-hybriden. Verhandlungen des naturforschenden Vereines in Brünn 4:3–57.

Menges, E. S. 2000. Population viability analyses in plants: challenges and opportunities. Trends in Ecology & Evolution 15:51–56.

Menotte-Raymond, M., V. A. David, J. C. Stephens, and S. J. O'Brien. 1997. Genetic individualization of domestic cats using feline STR loci for forensic analysis. Journal of Forensics Science 42:1039–1051.

Mensack, M. M., *et al.* 2010. Evaluation of diversity among common beans (*Phaseolus vulgaris* l.) from two centers of domestication using 'omics' technologies. BMC Genomics 11.

Menzel, A., and V. Dose. 2005. Analysis of long-term time series of the beginning of flowering by Bayesian function estimation. Meteorologische Zeitschrift 14:429–434.

Merilä, J., and P. Crnokrak. 2001. Comparison of genetic differentiation at marker loci and quantitative traits. Journal of Evolutionary Biology 14:892–903.

Meselson, M., and R. Yuan. 1968. DNA restriction enzyme from *E. coli*. Nature 217:1110–1114.

Metcalf, J. L., *et al.* 2007. Across the great divide: genetic forensics reveals misidentification of endangered cutthroat trout populations. Molecular Ecology 16:4445–4454.

Michalakis, Y., and L. Excoffier. 1996. A generic estimation of population subdivision using distances between alleles with special reference for microsatellite loci. Genetics 142:1061–1064.

Michel, A., *et al.* 2004. Somatic mutation-mediated evolution of herbicide resistance in the nonindigenous invasive plant hydrilla (*Hydrilla verticillata*). Molecular Ecology 13: 3229–3237.

Midgley, M. 1987. Keeping species on ice. Pp. 55–65 *in* McKenna, V., W. Travers, and J. Wray, eds. Beyond the Bars. The Zoo Dilemma. Thorsons, Wellingborough, UK.

Milinkovitch, M., *et al.* 2007. Giant Galapagos tortoises: molecular genetic analyses identify a trans-island hybrid in a repatriation program of an endangered taxon. BMC Ecology 7:2.

Millar, C. D., *et al.* 2008. New developments in ancient genomics. Trends in Ecology & Evolution 23:386–393.

Miller, C. R., and L. P. Waits. 2003. The history of effective population size and genetic diversity in the Yellowstone grizzly (*Ursus arctos*): implications for conservation. Proceedings of the National Academy of Sciences USA 100:4334–4339.

Miller, D. J., and M. J. H. van Oppen. 2003. A 'fair go' for coral hybridization. Molecular Ecology 12:805–807.

Miller, E. J., and C. A. Herbert. 2010. Breeding and genetic management of captive marsupial populations. Pp. 5–32 *in* Deakin, J. E., P. D. Waters, and J. A. Marshall Graves, eds. Marsupial Genetics and Genomics. Springer, Netherlands.

Miller, P. S., and P. W. Hedrick. 1991. MHC polymorphism and the design of captive breeding programs – simple solutions are not the answer. Conservation Biology 5:556–558.

Miller, P. S., and Lacy, R. C. 2005. VORTEX: A Stochastic Simulation of the Extinction Process. Version 9.50 User's

Manual. Conservation Breeding Specialist Group (SSC/IUCN), Apple Valley, MN.

Miller, W., *et al.* 2011. Genetic diversity and population structure of the endangered marsupial Sarcophilus harrisii (Tasmanian devil). Proceedings of the National Academy of Sciences early edition:1–6.

Miller Coyle, H., *et al.* 2003. A simple DNA extraction method for marijuana samples used in amplified fragment length polymorphism (AFLP) analysis. Journal of Forensic Science 48:343–347.

Mills, L. S., and F. W. Allendorf. 1996. The one-migrant-per-generation rule in conservation and management. Conservation Biology 10:1509–1518.

Mills, L. S., *et al.* 1996. Factors leading to different viability predictions for a grizzly bear data set. Conservation Biology 10:863–873.

Mills, L. S., *et al.* 2000. Estimating animal abundance using noninvasive DNA sampling: promise and pitfalls. Ecological Applications 10:283–294.

Mills, L. S., K. Pilgrim, M. K. Schwartz, and K. McKelvey. 2001. Identifying lynx and other North American felids based on mtDNA analysis. Conservation Genetics 1:285–289.

Milner, G. B., D. J. Teel, F. M. Utter, and G. A. Winans. 1985. A genetic method of stock indentification in mixed populations of Pacific salmon, *Oncorhynchus* spp. Marine Fisheries Review 47:1–8.

Mimura, M., and S. N. Aitken. 2007. Adaptive gradients and isolation-by-distance with postglacial migration in *Picea sitchensis*. Heredity 99:224–232.

Mitton, J. B. 1997. Selection in Natural Populations. Oxford University Press, New York.

Modiano, D., *et al.* 2001. Haemoglobin C protects against clinical *Plasmodium falciparum* malaria. Nature 414:305–308.

Moore, J. A., B. D. Bell, and W. L. Linklater. 2008. The debate on behavior in conservation: New Zealand integrates theory with practice. BioScience 58:454–459.

Morell, V. 1993. Evidence found for a possible 'aggression gene'. Science 260:1722–1723.

Morgan, J. W. 1999. Effects of population size on seed production and germinability in an endangered, fragmented grassland plant. Conservation Biology 13:266–273.

Morgan, M. T. 2002. Genome-wide deleterious mutation favors dispersal and species integrity. Heredity 89:253–257.

Morgan, T. J., *et al.* 2005. Molecular and quantitative genetic divergence among populations of house mice with known evolutionary histories. Heredity 94:518–525.

Moriarty, K. M., *et al.* 2009. Wolverine confirmation in California after nearly a century: Native or long-distance immigrant? Northwest Science 83:154–162.

Morin, P. A., G. Luikart, R. K. Wayne, and the SNP workshop group. 2004. SNPs in ecology, evolution and conservation. Trends in Ecology & Evolution 19:208–216.

Morin, P. A., K. K. Martien, and B. L. Taylor. 2009. Assessing statistical power of SNPs for population structure and conservation studies. Molecular Ecology Resources 9:66–73.

Moritz, C. 1994. Defining 'Evolutionarily Significant Units' for conservation. Trends in Ecology & Evolution 9:373–375.

Moritz, C., and C. Cicero. 2004. DNA barcoding: promise and pitfalls. PLoS Biology 2: e354.

Moritz, C., and D. P. Faith. 1998. Comparative phylogeography and the identification of genetically divergent areas for conservation. Molecular Ecology 7:419–429.

Moritz, C., S. Lavery, and R. Slade. 1995. Using allele frequency and phylogeny to define units for conservation and management. American Fisheries Society Symposium 17:249–262.

Morizot, D. C., J. C. Bednarz, and R. E. Ferrell. 1987. Sex linkage of muscle creatine kinase in Harris' hawks. Cytogenetics and Cell Genetics 44:89–91.

Morton, N. E., J. F. Crow, and H. J. Muller. 1956. An estimate of the mutational damage in man from data on consanguineous marriages. Proceedings of the National Academy of Sciences USA 42:855–863.

Mosher, J. A., and C. J. Henny. 1976. Thermal adaptiveness of plumage color in screech owls. Auk 93:614–619.

Mosseler, A., D. J. Innes, and B. A. Roberts. 1991. Lack of allozymic variation in disjunct Newfoundland populations of red pine (*Pinus resinosa*). Canadian Journal of Forest Research 21:525–528.

Mosseler, A., K. N. Egger, and G. A. Hughes. 1992. Low levels of genetic diversity in red pine confirmed by random amplified polymorphic DNA markers. Canadian Journal of Forest Research 22:1332–1337.

Mousseau, T. A., and D. A. Roff. 1987. Natural selection and the heritability of fitness components. Heredity 59:181–198.

Mueller, J. M., and J. J. Hellmann. 2008. An assessment of invasion risk from assisted migration. Conservation Biology 22:562–567.

Muhlfeld, C. C., *et al.* 2009. Hybridization rapidly reduces fitness of a native trout in the wild. Biology Letters 5:328–331.

Muir, C. C., B. M. F. Galdikas, and A. T. Beckenbach. 2000. mtDNA sequence diversity of orangutans from the islands of Borneo and Sumatra. Journal of Molecular Evolution 51:471–480.

Muller, H. J. 1950. Our load of mutations. American Journal of Human Genetics 2:111–176.

Müller-Scharer, H., U. Schaffner, and T. Steinger. 2004. Evolution in invasive plants: implications for biological control. Trends in Ecology & Evolution 19:417–422.

Mundy, N. I., *et al.* 2004. Conserved genetic basis of a quantitative plumage trait involved in mate choice. Science 303:1870–1873.

Müntzing, A. 1966. Accessory chromosomes. Bulletin of the Botanical Society of Bengal 20:1–15.

Murchison, E. P. 2009. Clonally transmissible cancers in dogs and Tasmanian devils. Oncogene 27:S19–S30.

Murchison, E. P., *et al.* 2010. The Tasmanian devil transcriptome reveals Schwann cell origins of a clonally transmissible cancer. Science 327:84–87.

Murgia, C., *et al.* 2006. Clonal origin and evolution of a transmissible cancer. Cell 126:477–487.

Murray, B. G., and A. G. Young. 2001. Widespread chromosome variation in the endangered grassland forb *Rutidosis leptorrhynchoides* F. Muell. (Asteraceae: Gnaphalieae). Annals of Botany 87:83–90.

Myers, J. H., D. Simberloff, A. M. Kuris, and J. R. Carey. 2000. Eradication revisited: dealing with exotic species. Trends in Ecology & Evolution 15:316–320.

Myers, N., and A. H. Knoll. 2001. The biotic crisis and the future of evolution. Proceedings National Academy of Sciences USA 98:5389–5392.

Myers, R. A., *et al.* 2004. Hatcheries and endangered salmon. Science 303:1980.

Nabata, D., R. Masuda, and O. Takahashi. 2004. Bottleneck effects on the sika deer *Cervus nippon* population in Hokkaido, revealed by ancient DNA analysis. Zoological Science 21:473–481.

Nachman, M. W., and S. L. Crowell. 2000. Estimate of the mutation rate per nucleotide in humans. Genetics 156:297–304.

Nachman, M. W., and J. B. Searle. 1995. Why is the house mouse karyotype so variable? Trends in Ecology & Evolution 10:397–402.

Nachman, M. W., H. E. Hoekstra, and S. L. D'Agostino. 2003. The genetic basis of adaptive melanism in pocket mice. Proceedings of the National Academy of Sciences USA 100:5268.

Nadeau, N. J., and C. D. Jiggins. 2010. A golden age for evolutionary genetics? Genomic studies of adaptation in natural populations. Trends in Genetics 26:484–492.

Nagaraj, S. H., R. B. Gasser, and S. Ranganathan. 2007. A hitchhiker's guide to expressed sequence tag (EST) analysis. Briefings in Bioinformatics 8:6–21.

Nagy, E. S., and K. J. Rice. 1997. Local adaptation in two subspecies of an annual plant: implications for migration and gene flow. Evolution 51:1079–1089.

Nakajima, M., N. Kanda, and Y. Fujio. 1991. Fluctuation of gene frequency in sub-populations originated from one guppy population. Bulletin of the Japanese Society of Scientific Fisheries 57:2223–2227.

Narum, S. R., N. R. Campbell, C. C. Kozfkay, and K. A. Meyer. 2010. Adaptation of redband trout in desert and montane environments. Molecular Ecology 19:4622–4637.

Nauta, M. J., and F. J. Weissing. 1996. Constraints on allele size at microsatellite loci: Implications for genetic differentiation. Genetics 143:1021–1032.

Neaves, W. B., and P. Baumann. 2011. Unisexual reproduction among vertebrates. Trends in Genetics 27:81–88.

Nei, M. 1965. Variation and covariation of gene frequencies in subdivided populations. Evolution 19:256–258.

Nei, M. 1972. Genetic distance between populations. American Naturalist 106:283–292.

Nei, M. 1975. Molecular Population Genetics and Evolution. American Elsevier, New York.

Nei, M. 1977. F-statistics and analysis of gene diversity in subdivided populations. Annals of Human Genetics 41:225–233.

Nei, M. 1978. Estimation of average heterozygosity and genetic distance from a small number of individuals. Genetics 89:583–590.

Nei, M. 1987. Molecular Evolutionary Genetics. Columbia University Press, New York.

Nei, M. 2005. Selectionism and neutralism in molecular evolution. Molecular Biology and Evolution 22:2318–2342.

Nei, M., and W-H. Li. 1973. Linkage disequilibrium in subdivided populations. Genetics 75:213–219.

Nei, M., and F. Tajima. 1981. Genetic drift and estimation of effective population size. Genetics 98:625–640.

Nei, M., T. Maruyama, and R. Chakraborty. 1975. The bottleneck effect and genetic variability in populations. Evolution 29:1–10.

Neigel, J. E. 2002. Is F_{ST} obsolete? Conservation Genetics 3:167–173.

Neigel, J. E., and J. C. Avise. 1986. Phylogenetic relationships of mitochondrial DNA under various demographic models of speciation. Pp. 515–534 *in* Nevo, E., and S. Karlin, eds. Evolutionary Processes and Theory. Academic Press, New York.

Neigel, J. E., and J. C. Avise. 1993. Application of a random walk model to geographic distributions of animal mitochondrial DNA variation. Genetics 135:1209–1220.

Nelson, K., and M. Soulé. 1987. Genetic conservation of exploited fishes. Pp. 345–368 *in* Ryman, N., and F. Utter, eds. Population Genetics and Fishery Management. Washington Sea Grant Program, Seattle, WA.

Nevo, E., A. Bailes, and R. Ben-Shlomo. 1984. The evolutionary significance of genetic diversity: ecological, demographic and life history correlates. Pp. 13–213 *in* Levin, S., ed. Lecture Notes in Biomathematics. Vol. 53: Evolutionary Dynamics of Genetic Diversity, G. S. Mani (ed.). Springer-Verlag, Berlin.

Newhouse, A., *et al.* 2007. Transgenic American elm shows reduced Dutch elm disease symptoms and normal mycorrhizal colonization. Plant Cell Reports 26:977–987.

Newman, D., and D. Pilson. 1997. Increased probability of extinction due to decreased genetic effective population size: experimental populations of *Clarkia pulchella*. Evolution 51:354–362.

Newman, D., and D. A. Tallmon. 2001. Experimental evidence for beneficial fitness effects of gene flow in recently isolated populations. Conservation Biology 15:1054–1063.

Nichols, R. 2001. Gene trees and species trees are not the same. Trends in Ecology & Evolution 16:358–364.

Nichols, R. A., M. W. Bruford, and J. J. Groombridge. 2001. Sustaining genetic variation in a small population: evidence from the Mauritius kestrel. Molecular Ecology 10:593–602.

Nichols, R. A., and K. L. M. Freeman. 2004. Using molecular markers with high mutation rates to obtain estimates of relative population size and to distinguish the effects of gene flow and mutation: a demonstration using data from endemic Mauritian skinks. Molecular Ecology 13:775–787.

Nielsen, E. E., et al. 2012. Gene-associated markers provide tools for tackling illegal fishing and false eco-certification. Nature Communications 3:851.

Nielsen, E. E., and M. M. Hansen. 2008. Waking the dead: the value of population genetic analyses of historical samples. Fish and Fisheries 9:450–461.

Nielsen, R., D. K. Mattila, P. J. Clapham, and P. J. Palsbøll. 2001. Statistical approaches to paternity analysis in natural populations and applications to the North Atlantic humpback whale. Genetics 157:1673–1682.

Nielsen, E. E., B. R. MacKenzie, E. Magnussen, and D. Meldrup. 2007. Historical analysis of Pan I in Atlantic cod (Gadus morhua): temporal stability of allele frequencies in the southeastern part of the species distribution. Canadian Journal of Fisheries and Aquatic Sciences 64:1448–1455.

Nielsen, E. E., et al. 2009. Genomic signatures of local directional selection in a high gene flow marine organism: the Atlantic cod (Gadus morhua). BMC Evolutionary Biology 9:276.

Nogués-Bravo, D., R. Ohlemuller, P. Batra, and M. B. Araujo. 2010. Climate predictors of late Quaternary extinctions. Evolution 64:2442–2449.

Normandeau, E., J. A. Hutchings, D. J. Fraser, and L. Bernatchez. 2009. Population-specific gene expression responses to hybridization between farm and wild Atlantic salmon. Evolutionary Applications 2:489–503.

Novembre, J., and M. Stephens. 2008. Interpreting principal component analyses of spatial population genetic variation. Nature Genetics 40:646–649.

Noyes, J. H., et al. 1996. Effects of bull age on conception dates and pregnancy rates of cow elk. Journal of Wildlife Management 60:508–517.

Nunney, L. 1997. The effective size of a hierarchically structured population. Evolution 53:1–10.

Nunney, L. 2000. The limits to knowledge in conservation genetics – the value of effective population size. Evolutionary Biology 32:179–194.

Nunney, L. 2002. The effective size of annual plant populations: The interaction of a seed bank with fluctuating population size in maintaining genetic variation. American Naturalist 160:195–204.

Nunney, L., and K. A. Campbell. 1993. Assessing minimum viable population size – demography meets population genetics. Trends in Ecology & Evolution 8:234–239.

Nussey, D. H., et al. 2005a. Rapidly declining fine-scale spatial genetic structure in female red deer. Molecular Ecology 14:3395–3405.

Nussey, D. H., E. Postma, P. Gienapp, and M. E. Visser. 2005b. Selection on heritable phenotypic plasticity in a wild bird population. Science 310:304–306.

Nyström, J., et al. 2006. Effect of local prey availability on gyrfalcon diet: DNA analysis on ptarmigan remains at nest sites. Journal of Zoology 269:57–64.

Oakenfull, E. A., and O. A. Ryder. 1998. Mitochondrial control region and 12S rRNA variation in Przewalski's horse (Equus przewalskii). Animal Genetics 29:456–459.

O'Brien, S. J., and J. F. Evermann. 1988. Interactive influence of infectious disease and genetic diversity in natural populations. Trends in Ecology & Evolution 3:254–259.

O'Brien, S. J., and E. Mayr. 1991. Bureaucratic mischief: recognizing endangered species and subspecies. Science 251:1187–1188.

O'Brien, S. J., et al. 1983. The cheetah is depauperate in genetic variation. Science 221:459–462.

O'Brien, S. J., et al. 1985. Genetic basis for species vulnerability in the cheetah. Science 227:1428–1434.

O'Brien, S. J., Menninger, J. C., and Nash, W. G. 2006. Atlas of Mammalian Chromosomes. John Wiley & Sons, Inc., Hoboken, NJ.

O'Donald, P. 1987. Polymorphism and sexual selection in the Arctic skua. Pp. 433–452 in Cooke, F., and P. A. Buckley, eds. Avian Genetics: A Population and Ecological Approach. Academic Press, London.

O'Donald, P., and J. W. F. Davis. 1959. The genetics of the colour phases of the Arctic skua. Heredity 13:481–486.

O'Donald, P., and J. W. F. Davis. 1975. Demography and selection in a population of Arctic skuas. Heredity 35:75–83.

Ogden, R. 2011. Unlocking the potential of genomic technologies for wildlife forensics. Molecular Ecology Resources 11:109–116.

Ogden, R., et al. 2008. SNP-based method for the genetic identification of ramin Gonystylus spp. timber and products: Applied research meeting CITES enforcement needs. Endangered Species Research 9:225–261.

Ogden, R., N. Dawnay, and R. McEwing. 2009. Wildlife DNA forensics – bridging the gap between conservation genetics and law enforcement. Endangered Species Research 9:179–195.

O'Grady, J. J., et al. 2006. Realistic levels of inbreeding depression strongly affect extinction risk in wild populations. Biological Conservation 133:42–51.

Ohta, T. 1971. Associative overdominance caused by linked detrimental mutations. Genetical Research 18:277–286.

Ohta, T., and M. Kimura. 1969. Linkage disequilibrium due to random drift. Genetical Research 13:47–55.

Ohta, T., and M. Kimura. 1973. A model of mutation appropriate to estimate the number of electrophoretcially detectable alleles in a finite population. Genetical Research 22:201–204.

Olsen, E. M., *et al.* 2004a. Maturation trends indicative of rapid evolution preceded the collapse of Northern cod. Nature 428:932–935.

Olsen, J. B., *et al.* 2000. Microsatellites reveal population identity of individual pink salmon to allow supportive breeding of a population at risk of extinction. Transactions of the American Fisheries Society 129:232–242.

Olsen, J. B., C. Habicht, J. Reynolds, and J. E. Seeb. 2004b. Moderately and highly polymorphic microsatellites provide discordant estimates of population divergence in sockeye salmon, *Oncorhynchus nerka*. Environmental Biology of Fishes 69:261–273.

O'Malley, K. G., M. J. Ford, and J. J. Hard. 2010. Clock polymorphism in Pacific salmon: evidence for variable selection along a latitudinal gradient. Proceedings of the Royal Society B: Biological Sciences 277:3703–3714.

Oostermeijer, J. G. B., S. H. Luijten, and J. C. M. den Nijs. 2003. Integrating demographic and genetic approaches in plant conservation. Biological Conservation 113:389–398.

Otto, S. P., and J. Whitton. 2000. Polyploid incidence and evolution. Annual Review of Genetics 34:401–437.

Ouborg, N. J., Y. Piquot, and J. M. Vangroenendael. 1999. Population genetics, molecular markers and the study of dispersal in plants. Journal of Ecology 87:551–568.

Ouborg, N. J., *et al.* 2010. Conservation genetics in transition to conservation genomics. Trends in Genetics 26:177–187.

Ovenden, J. R., and R. W. G. White. 1990. Mitochondrial and allozyme genetics of incipient speciation in a landlocked population of *Galaxias truttaceus*. Genetics 124:701–716.

Ozgul, A., *et al.* 2010. Coupled dynamics of body mass and population growth in response to environmental change. Nature 466:482–485.

Packer, L., and R. Owen. 2001. Population genetic aspects of pollinator decline. Conservation Ecology 5:4.

Paetkau, D., W. Calvert, I. Stirling, and C. Strobeck. 1995. Microsatellite analysis of population structure in Canadian polar bears. Molecular Ecology 4:347–354.

Paetkau, D., *et al.* 1997. An empirical evaluation of genetic distance statistics using microsatellite data from bear (Ursidae) populations. Genetics 147:1943–1957.

Paetkau, D., *et al.* 1998. Variation in genetic diversity across the range of North American brown bears. Conservation Biology 12:418–429.

Paetkau, D., R. Slade, M. Burden, and A. Estoup. 2004. Genetic assignment methods for the direct, real-time estimation of migration rate: a simulation-based exploration of accuracy and power. Molecular Ecology 13:55–65.

Painter, T. S. 1933. A new method for the study of chromosome rearrangements and the plotting of chromosome maps. Science 78:585–586.

Palkovacs, E. P. 2011. The overfishing debate: An eco-evolutionary perspective. Trends in Ecology & Evolution 26:616–617.

Palm, S., L. Laikre, P. E. Jorde, and N. Ryman. 2003. Effective population size and temporal genetic change in stream resident brown trout (*Salmo trutta*, L.). Conservation Genetics 4:249–264.

Palmer, A. R., and C. Strobeck. 1986. Fluctuating asymmetry: measurement, analysis, patterns. Annual Review of Ecology and Systematics 17:391.

Palmer, A. R., and C. Strobeck. 1997. Fluctuating asymmetry and developmental stability: Heritability of observable variation vs. heritability of inferred cause. Journal of Evolutionary Biology 10:39–49.

Palsbøll, P. J., *et al.* 1997. Genetic tagging of humpback whales. Nature 388:767–769.

Palsbøll, P. J., M. Berube, and F. W. Allendorf. 2007. Identification of management units using population genetic data. Trends in Ecology & Evolution 22:11–16.

Palsbøll, P. J., M. Z. Peery, M. T. Olsen, S. R. Beissinger, and M. Bérubé. In press. Inferring recent historic abundance from current genetic diversity. Molecular Ecology.

Pálsson, O. K., and V. Thorsteinsson. 2003. Migration patterns, ambient temperature, and growth of icelandic cod (*Gadus morhua*): Evidence from storage tag data. Canadian Journal of Fisheries and Aquatic Sciences 60:1409–1423.

Palstra, F. P., and D. E. Ruzzante. 2008. Genetic estimates of contemporary effective population size: what can they tell us about the importance of genetic stochasticity for wild population persistence? Molecular Ecology 17: 3428–3447.

Palumbi, S. R. 2003. Population genetics, demographic connectivity, and the design of marine reserves. Ecological Applications 13:S146–S158.

Pamilo, P., and S. Pálsson. 1998. Associative overdominance, heterozygosity and fitness. Heredity 81:381–389.

Pampoulie, C., *et al.* 2006. The genetic structure of Atlantic cod (*Gadus morhua*) around Iceland: insight from microsatellites, the *Pan* I locus, and tagging experiments. Canadian Journal of Genetics and Cytology 63:2660–2674.

Pampoulie, C., P. Steingrund, M. O. Stefansson, and A. K. Danielsdottir. 2008. Genetic divergence among East Icelandic and Faroese populations of Atlantic cod provides evidence for historical imprints at neutral and non-neutral markers. Ices Journal of Marine Science 65:65–71.

Panchal, M., and Beaumont M. A. 2007. The automation and evaluation of nested clade phylogeographic analysis. Evolution 61:1466–1480.

Pardini, A. T., *et al.* 2001. Sex-biased dispersal of great white sharks – In some respects, these sharks behave more like whales and dolphins than other fish. Nature 412:139–140.

Park, L. 2011. Effective population size of current human population. Genetics Research 93:105–114.

Parker, I. M., J. Rodriguez, and M. E. Loik. 2003. An evolutionary approach to understanding the biology of invasions: local adaptation and general-purpose genotypes in the weed *Verbascum thapsus*. Conservation Biology 17:59–72.

Parker, M. A. 1992. Outbreeding depression in a selfing annual. Evolution 46:837–841.

Parmesan, C. 2006. Ecological and evolutionary responses to recent climate change. Annual Review of Ecology, Evolution, and Systematics 37:637–669.

Parris, M. J. 2004. Hybrid response to pathogen infection in interspecific crosses between two amphibian species (Anura: Ranidae). Evolutionary Ecology Research 6: 457–471.

Pask, A. J., R. R. Behringer, and M. B. Renfree. 2008. Resurrection of DNA function *in vivo* from an extinct genome. PLoS ONE 3:e2240.

Patarnello, T., and B. Battaglia. 1992. Glucosephosphate isomerase and fitness – effects of temperature on genotype dependent mortality and enzyme activity in 2 species of the genus *Gammarus* (Crustacea, Amphipoda). Evolution 46:1568–1573.

Paterson, A. H., *et al.* 1995. The weediness of wild plants: molecular analysis of genes influencing dispersal and persistence of johnsongrass, *Sorghum halepense* (L.). Proceedings of the National Academy of Sciences USA 92:6127–6131.

Paterson, S., K. Wilson, and J. M. Pemberton. 1998. Major histocompatibility complex variation associated with juvenile survival and parasite resistance in a large unmanaged ungulate populations (*Ovis aries L.*). Proceedings of the National Academy of Sciences USA 95: 3714–3719.

Pauly, G. B., O. Piskurek, and H. B. Shaffer. 2007. Phylogeographic concordance in the southeastern United States: the flatwoods salamander, Ambystoma cingulatum, as a test case. Molecular Ecology 16:415–429.

Peakall, R., *et al.* 2003. Comparative genetic study confirms exceptionally low genetic variation in the ancient and endangered relictual conifer, *Wollemia nobilis* (Araucariaceae). Molecular Ecology 12:2331–2343.

Pearce-Kelly, P., *et al.* 1998. The captive rearing of threatened Orthoptera: a comparison of the conservation potential and practical considerations of two species' breeding programmes at the Zoological Society of London. Journal of Insect Conservation 2:201–210.

Pearse, A.-M., and K. Swift. 2006. Transmission of devil facial-tumour disease. Nature 439:549.

Pearson, R. G., and T. P. Dawson. 2003. Predicting the impacts of climate change on the distribution of species: Are bioclimate envelope models useful? Global Ecology & Biogeography 12:361–371.

Peeler, E., M. Thrush, L. Paisley, and C. Rodgers. 2006. An assessment of the risk of spreading the fish parasite *Gyrodactylus salaris* to uninfected territories in the European union with the movement of live Atlantic salmon (*Salmo salar*) from coastal waters. Aquaculture 258:187–197.

Pella, J. J., and G. B. Milner. 1987. Use of genetic analysis in stock composition analysis. Pp. 247–276 *in* Ryman, N. and F. Utter, eds. Population Genetics and Fisheries Management. University of Washington Press, Seattle, WA.

Pemberton, J. 2004. Measuring inbreeding depression in the wild: the old ways are the best. Trends in Ecology & Evolution 19:613–615.

Pemberton, J. M. 2008. Wild pedigrees: the way forward. Proceedings of the Royal Society B: Biological Sciences 275:613–621.

Penuelas, J., and I. Filella. 2001. Phenology – responses to a warming world. Science 294:793–795.

Peralta, A. L., J. W. Matthews, and A. D. Kent. 2010. Microbial community structure and denitrification in a wetland mitigation bank. Applied and Environmental Microbiology 76:4207–4215.

Perez-Rusafa, A., *et al.* 2006. Effects of fishing protection on the genetic structure of fish populations. Biological Conservation 129:244–255.

Perkel, J. 2008. SNP genotyping: Six technologies that keyed a revolution. Nature Methods 5:447–453.

Petit, R. J., *et al.* 1997. Chloroplast DNA footprints of postglacial recolonization by oaks. Proceedings of the National Academy of Sciences USA 94:9996–10001.

Petit, R. J., A. El Mousadik, and O. Pons. 1998. Identifying populations for conservation on the basis of genetic markers. Conservation Biology 12:844–855.

Petit, R. J., *et al.* 2005. Comparative organization of chloroplast, mitochondrial and nuclear diversity in plant populations. Molecular Ecology 14:689–701.

Phillips, B. L., and R. Shine. 2004. Adapting to an invasive species: Toxic cane toads induce morphological change in Australian snakes. Proceedings of the National Academy of Sciences USA 101:17150–17155.

Phillips, B. L., and R. Shine. 2006. An invasive species induces rapid adaptive change in a native predator: cane toads and black snakes in Australia. Proceedings of the Royal Society B: Biological Sciences 273:1545–1550.

Phillips, B. L., G. P. Brown, J. K. Webb, and R. Shine. 2006. Invasion and the evolution of speed in toads. Nature 439:803–803.

Phillips, B. L., G. P. Brown, and R. Shine. 2010. Evolutionarily accelerated invasions: the rate of dispersal evolves upwards during the range advance of cane toads. Journal of Evolutionary Biology 23:2595–2601.

Phillips, P. C. 2008. Epistasis – the essential role of gene interactions in the structure and evolution of genetic systems. Nature Reviews Genetics 9:855–867.

Piálek, J., H. C. Hauffe, and J. B. Searle. 2005. Chromosomal variation in the house mouse. Biological Journal of the Linnean Society 84:535–563.

Pickup, M., and A. G. Young. 2008. Population size, self-incompatibility and genetic rescue in diploid and tetraploid races of *Rutidosis leptorrhynchoides* (Asteraceae). Heredity 100:268–274.

Pigliucci, M., and C. J. Murren. 2003. Perspective: Genetic assimilation and a possible evolutionary paradox: can macroevolution sometimes be so fast as to pass us by? Evolution 57:1455–1464.

Pimentel, D. 2000. Biological control of invading species. Science 289:869.

Pimm, S. L., *et al.* 2001. Environment – can we defy nature's end? Science 293:2207–2208.

Pimm, S. L., H. L. Hones, and J. M. Diamond. 1988. On the risk of extinction. American Naturalist 132:757–785.

Pinares, A., *et al.* 2009. Genetic diversity of the endangered endemic *Microcycas calocoma* (Miq.) A. DC (Zamiaceae, Cycadales): Implications for conservation. Biochemical Systematics and Ecology 37:385–394.

Pinzon, J. H., and T. C. LaJeunesse. 2010. Species delimitation of common reef corals in the genus *Pocillopora* using nucleotide sequence phylogenies, population genetics and symbiosis ecology. Molecular Ecology 20:311–325.

Piry, S., G. Luikart, and J. M. Cornuet. 1999. BOTTLENECK: A computer program for detecting recent reductions in the effective population size using allele frequency data. Journal of Heredity 90:502–503.

Piry, S., *et al.* 2004. GENECLASS2: a software for genetic assignment and first-generation migrant detection. Journal of Heredity 95:536–539.

Pogson, G. H., and S. E. Fevolden. 2003. Natural selection and the genetic differentiation of coastal and Arctic populations of the Atlantic cod in northern Norway: a test involving nucleotide sequence variation at the pantophysin (*PanI*) locus. Molecular Ecology 12:63–74.

Poissant, J., *et al.* 2009. Genome-wide cross-amplification of domestic sheep microsatellites in bighorn sheep and mountain goats. Molecular Ecology Resources 9:1121–1126.

Pollak, E. 1983. A new method for estimating the effective population size from allele frequency changes. Genetics 104:531–548.

Porter, C. A., and J. W. Sites. 1987. Evolution of *Sceloporus grammicus* complex (Sauria: Iguanidae) in central Mexico. II. Studies of nondisjunction and the occurrence of spontaneous chromosomal mutations. Genetica 75:131–144.

Postma, E., *et al.* 2011. Disentangling the effect of genes, the environment and chance on sex ratio variation in a wild bird population. Proceedings of the Royal Society B: Biological Sciences 278:2996–3002.

Poulakakis, N., *et al.* 2008. Historical DNA analysis reveals living descendants of an extinct species of Galapagos tortoise. Proceedings of the National Academy of Sciences USA 105:15464–15469.

Povilitis, A., and K. Suckling. 2010. Addressing climate change threats to endangered species in US recovery plans. Conservation Biology 24:372–376.

Powell, J. R. 1994. Molecular techniques in population genetics: a brief history. Pp. 131–156 *in* Schierwater, B., B. Streit, G. P. Wagner, and R. DeSalle, eds. Molecular Ecology and Evolution: Approaches and Applications. Burkhauser Verlag, Basel, Switzerland.

Powell, R. A. 1993. The Fisher: Life History, Ecology, and Behavior. University of Minnesota Press, Minneapolis.

Prentis, P. J., *et al.* 2008. Adaptive evolution in invasive species. Trends in Plant Science 13:288–294.

Primmer, C. R. 2009. From conservation genetics to conservation genomics. Annals of the New York Academy of Sciences 1162:357–368.

Primmer, C. R., M. T. Koskinen, and J. Piironen. 2000. The one that did not get away: individual assignment using microsatellite data detects a case of fishing competition fraud. Proceedings of the Royal Society B: Biological Sciences 267:1699–1704.

Pritchard, J. K., M. Stephens, and P. Donnelly. 2000. Inference of population structure using multilocus genotype data. Genetics 155:945–959.

Pritchard, P. C. H. 1999. Status of the black turtle. Conservation Biology 13:1000–1003.

Proaktor, G., T. Coulson, and E. J. Milner-Gulland. 2007. Evolutionary responses to harvesting in ungulates. Journal of Animal Ecology 76:669–678.

Prospero, S., N. J. Grünwald, L. M. Winton, and E. M. Hansen. 2009. Migration patterns of the emerging plant pathogen *Phytophthora ramorum* on the West Coast of the United States of America. Phytopathology 99:739–749.

Prout, T. 1973. Appendix to: population genetics of marine pelecypods. III. Epistasis between functionally related isoenzymes of *Mytilus edulis*. Genetics 73:493–496.

Provan, J., W. Powell, and P. M. Hollingsworth. 2001. Chloroplast microsatellites: new tools for studies in plant ecology and evolution. Trends in Ecology & Evolution 16:142–147.

Provine, W. B. 1986. Sewall Wright and Evolutionary Biology. University of Chicago Press, Chicago.

Provine, W. B. 2001. The Origins of Theoretical Population Genetics with a New Afterword. University of Chicago Press, Chicago.

Prum, R. O. 2003. Palaeontology: dinosaurs take to the air. Nature 421:323–324.

Ptacek, M. B., H. C. Gerhardt, and R. D. Sage. 1994. Speciation by polyploidy in treefrogs: multiple origins of the tetraploid, Hyla versicolor. Evolution 48:898–908.

Pulido, F., P. Berthold, G. Mohr, and U. Querner. 2001. Heritability of the timing of autumn migration in a natural bird population. Proceedings of the Royal Society B: Biological Sciences 268:953–959.

Pulido, F., and T. Coppack. 2004. Correlation between timing of juvenile moult and onset of migration in the blackcap, *Sylvia atricapilla*. Animal Behaviour 68:167–173.

Punnett, R. C. 1904. Merism and sex in "*Spinax niger*". Biometrika 3:313–362.

Pyle, P. 1997. Identification Guide to North American Birds – Part I. Slate Creek Press, Bolinas, CA.

Qin, J., *et al.* 2010. A human gut microbial gene catalogue established by metagenomic sequencing. Nature 464: 59–65.

Queller, D. C., and K. F. Goodnight. 1989. Estimating relatedness using genetic markers. Evolution 43:258–275.

Quinn, T. P., *et al.* 2007. Directional selection by fisheries and the timing of sockeye salmon (*Oncorhynchus nerka*) migrations. Ecological Applications 17:731–739.

Raberg, L., A. L. Graham, and A. F. Read. 2009. Decomposing health: Tolerance and resistance to parasites in animals. Philosophical Transactions of the Royal Society of London B 364:37–49.

Radwan, J., A. Biedrzycka, and W. Babik. 2010. Does reduced mhc diversity decrease viability of vertebrate populations? Biological Conservation 143:537–544.

Raeymaekers, J. A. M., *et al.* 2007. Divergent selection as revealed by P_{ST} and QTL-based F_{ST} in three-spined stickleback (*Gasterosteus aculeatus*) populations along a coastal-inland gradient. Molecular Ecology 16:891–905.

Ralls, K., and J. Ballou. 1983. Extinction: lessons from zoos. Pp. 164–184 *in* Schonewald-Cox, C., S. Chambers, B. MacBryde, and L. Thomas, eds. Genetics and Conservation. Benjamin/Cummings, Menlo Park, CA.

Ralls, K., K. Brugger, and J. Ballou. 1979. Inbreeding and juvenile mortality in small populations of ungulates. Science 206:1101–1103.

Ralls, K., J. D. Ballou, and A. Templeton. 1988. Estimates of lethal equivalents and the cost of inbreeding in mammals. Conservation Biology 2:185–193.

Ralls, K., J. D. Ballou, B. A. Rideout, and R. Frankham. 2000. Genetic management of chondrodystrophy in California condors. Animal Conservation 3:145–153.

Ralls, K., S. R. Beissinger, and J. F. Cochrane. 2002. Guidelines for using population viability analysis in endangered-species management. Pp. 521–550 *in* Beissinger, S. R. and D. R. McCullough, eds. Population Viability Analysis. University of Chicago Press, Chicago, IL.

Randi, E. 2008. Detecting hybridization between wild species and their domesticated relatives. Molecular Ecology 17:285–293.

Rannala, B., and J. L. Mountain. 1997. Detecting immigration by using multilocus genotypes. Proceedings of the National Academy of Sciences USA 94:9197–9201.

Ratner, S., and R. Lande. 2001. Demographic and evolutionary responses to selective harvesting in population with discrete generations. Ecology 82:3093–3104.

Rawson, P. D., and R. S. Burton. 2002. Functional coadaptation between cytochrome c and cytochrome c oxidase within allopatric populations of a marine copepod. Proceedings of the National Academy of Sciences USA 99:12955.

Reale, D., A. G. McAdam, S. Boutin, and D. Berteaux. 2003. Genetic and plastic responses of a northern mammal to climate change. Proceedings of the Royal Society B: Biological Sciences 270:591–596.

Rebbeck, C. A., *et al.* 2009. Origins and evolution of a transmissible cancer. Evolution 63:2340–2349.

Reed, D. H., and E. H. Bryant. 2000. Experimental tests of minimum viable population size. Animal Conservation 3:7–14.

Reed, D. H., and R. Frankham. 2001. How closely correlated are molecular and quantitative measures of genetic variation? A meta-analysis. Evolution 55:1095–1103.

Reed, D. H., A. C. Nicholas, and G. E. Stratton. 2007. Genetic quality of individuals impacts population dynamics. Animal Conservation 10:275–283.

Reed, T. E., *et al.* 2010. Phenotypic plasticity and population viability: the importance of environmental predictability. Proceedings of the Royal Society B: Biological Sciences 277:3391–3400.

Reed, T. E., D. E. Schindler, and R. S. Waples. 2011. Interacting effects of phenotypic plasticity and evolution on population persistence in a changing climate. Conservation Biology 25:56–63.

Reeve, H. K., *et al.* 1990. DNA "fingerprinting" reveals high levels of inbreeding in colonies of the eusocial naked mole-rat. Proceedings of the National Academy of Sciences 87:2496–2500.

Reeves, P. A., and C. M. Richards. 2007. Distinguishing terminal monophyletic groups from reticulate taxa: performance of phenetic, tree-based, and network procedures. Systematic Biology 56:302–32.

Regan, H. M., M. Colyvan, and M. A. Burgman. 2000. A proposal for fuzzy International Union for the Conservation of Nature (IUCN) categories and criteria. Biological Conservation 92:101–108.

Rehfeldt, G. E., C. C. Ying, D. L. Spittlehouse, and D. A. Hamilton. 1999. Genetic responses to climate in *Pinus contorta*: niche breadth, climate change, and reforestation. Ecological Monographs 69:375–407.

Reich, D. E., *et al.* 2001. Linkage disequilibrium in the human genome. Nature 411:199–204.

Reid, J. M., P. Arcese, and L. F. Keller. 2003. Inbreeding depresses immune response in song sparrows (*Melospiza melodia*): Direct and inter-generational effects. Proceedings of the Royal Society B: Biological Sciences 270:2151–2157.

Reid, J. M., *et al.* 2005. Hamilton and Zuk meet heterozygosity? Song repertoire size indicates inbreeding and immunity in song sparrows (*Melospiza melodia*). Proceedings of the Royal Society B: Biological Sciences 272:481–487.

Reid, J. M., *et al.* 2007. Inbreeding effects on immune response in free-living song sparrows (*Melospiza melodia*). Proceedings of the Royal Society B: Biological Sciences 274:697–706.

Reilly, P. 2001. Legal and public policy issues in DNA forensics. Nature Reviews Genetics 2:313–317.

Reinartz, J. A., and D. H. Les. 1994. Bottleneck-induced dissolution of self-incompatibility and breeding system consequences in *Aster furcatus* (Asteraceae). American Journal of Botany 81:446–455.

Reisenbichler, R. R., and S. P. Rubin. 1999. Genetic changes from artificial propagation of Pacific salmon affect the productivity and viability of supplemented populations. ICES Journal of Marine Science 56:459–466.

Remington, D. L., and D. M. O'Malley. 2000. Evaluation of major genetic loci contributing to inbreeding depression for survival and early growth in a selfed family of *Pinus taeda*. Evolution 54:1580–1589.

Resco, V., J. Hartwell, and A. Hall. 2009. Ecological implications of plant's ability to tell the time. Ecology Letters 12:583–592.

Reusch, T. B. H., A. Ehlers, A. Hammerli, and B. Worm. 2005. Ecosystem recovery after climatic extremes enhanced by genotypic diversity. Proceedings of the National Academy of Sciences USA 102:2826–2831.

Rhymer, J. M., and D. Simberloff. 1996. Extinction by hybridization and introgression. Annual Review of Ecology and Systematics 27:83–109.

Ricciardi, A., and D. Simberloff. 2009. Assisted colonization is not a viable conservation strategy. Trends in Ecology & Evolution 24:248–253.

Rice, W. R. 1989. Analyzing tables of statistical tests. Evolution 43:223–225.

Rice, W. R. 1996. Evolution of the Y sex chromosome in animals. Bioscience 46:331–343.

Rich, W. H. 1939. Local populations and migration in relation to the conservation of Pacific salmon in the western states and Alaska. Contributions No. 1, Fish Commission of Oregon, Salem, Oregon.

Richard, F., C. Messaoudi, M. Lombard, and B. Dutrillaux. 2001. Chromosome homologies between man and mountain zebra (*Equus zebra hartmannae*) and description of a new ancestral synteny involving sequences homologous to human chromosomes 4 and 8. Cytogenetics & Cell Genetics 93:291–296.

Richards, A. J. 1986. Plant Breeding Systems. Allen and Unwin, London.

Richards, C. L., O. Bossdorf, and M. Pigliucci. 2010. What role does heritable epigenetic variation play in phenotypic evolution? BioScience 60:232–237.

Richards, C. L., *et al.* 2008. Plasticity in salt tolerance traits allows for invasion of novel habitat by japanese knotweed s. L. (*Fallopia japonica* and *F. bohemica*, Polygonaceae). American Journal of Botany 95:931–942.

Richards, C. M. 2000. Genetic and demographic influences on population persistence: gene flow and genetic rescue in *Silene alba*. Pp. 271–291 *in* Young, A. and G. Clarke, eds. Genetics, Demography and Viability of Fragmented Populations. Cambridge University Press, Cambridge.

Ricker, W. E. 1972. Hereditary and environmental factors affecting certain salmonid populations. Pp. 27–160 *in* Simon, R. C., and P. A. Larkin, eds. The Stock Concept in Pacific Salmon. University British Columbia, Vancouver.

Ricker, W. E. 1981. Changes in the average body size and average age of Pacific salmon. Canadian Journal of Fisheries and Aquatic Sciences 38:1636–1656.

Ricklefs, R. E., and G. L. Miller. 2000. Ecology, 4th edn. W.H. Freeman and Co., New York.

Ridley, M. 2004. The Agile Gene: How Nature Turns on Nurture. Harper Collins Publishers, New York.

Riechert, S. E., F. D. Singer, and T. C. Jones. 2001. High gene flow levels lead to gamete wastage in a desert spider system. Genetica 112–113:297–319.

Rieman, B., and J. Clayton. 1997. Wildlife and native fish: issues of forest health and conservation of sensitive species. Fisheries 22(11):6–15.

Rieman, B. R., *et al.* 1993. Consideration of extinction risks for salmonids. Fish Habitat Relationships Technical Bulletin (US Forest Service) 14:1–12.

Rieseberg, L. H. 2001. Chromosomal rearrangements and speciation. Trends in Ecology & Evolution 16:351–358.

Rieseberg, L. 2011. Adaptive introgression: The seeds of resistance. Current Biology: CB 21:R581–R583.

Rieseberg, L. H., and J. M. Burke. 2001. A genic view of species integration. Journal of Evolutionary Biology 14:883–886.

Rieseberg, L. H., and D. Gerber. 1995. Hybridization in the Catalina Island mountain mahogany (*Cercocarpus traskiae*): RAPD evidence. Conservation Biology 9:199–203.

Rieseberg, L. H., M. A. Archer, and R. K. Wayne. 1999. Transgressive segregation, adaptation and speciation. Heredity 83:363–372.

Rieseberg, L. H., *et al.* 2003. Major ecological transitions in wild sunflowers facilitated by hybridization. Science 301:1211–1216.

Rising, J. D., and G. F. Shields. 1980. Chromosomal and morphological correlates in two new world sparrows (Emberizidae). Evolution 34:654–662.

Ritland, K. 1996. Inferring the genetic basis of inbreeding depression in plants. Genome 39:1–8.

Ritland, K. 2000. Detecting inheritance with inferred relatedness in nature. Pp. 187–199 *in* Mousseau, T. A., B. Sinervo, and J. A. Endler, eds. Adaptive Genetic Variation in the Wild. Oxford University Press, New York.

Ritland, K. 2002. Extensions of models for the estimation of mating systems using *n* independent loci. Heredity 88:221–228.

Ritland, K., C. Newton, and H. Marshall. 2001. Inheritance and population structure of the white-phased "Kermode" black bear. Current Biology 11:1468–1472.

Ritzema-Bos, J. 1894. Untersuchungen uber die Folgen der Zucht in engster Blutverwandtschaft. Biologisches Centralblatt 14:75–81.

Roach, J. C., *et al.* 2010. Analysis of genetic inheritance in a family quartet by whole-genome sequencing. Science 328:636–639.

Roberge, C., S. Einum, H. Guderley, and L. Bernatchez. 2006. Rapid parallel evolutionary changes of gene transcription profiles in farmed Atlantic salmon. Molecular Ecology 15:9–20.

Roberge, C., et al. 2008. Genetic consequences of inter-breeding between farmed and wild Atlantic salmon: insights from the transcriptome. Molecular Ecology 17: 314–324.

Robertson, A. 1960. A theory of limits in artificial selection. Proceedings of the Royal Society B: Biological Sciences 153B:234–249.

Robertson, A. 1962. Selection for heterozygotes in small populations. Genetics 47:1291–1300.

Robertson, A. 1964. The effect of non-random mating within inbred lines on the rate of inbreeding. Genetical Research 5:164–167.

Robertson, A. 1965. The interpretation of genotypic ratios in domestic animal populations. Animal Production 7:319–324.

Robertson, A. 1967. The nature of quantitative genetic variation. Pp. 265–280 in Brink, R. A., and E. D. Styles, eds. Heritage from Mendel. University of Wisconsin Press, Madison.

Robertson, B. C., and N. J. Gemmell. 2004. Defining eradication units in pest control programmes. Journal of Ecology 41:1042–1048.

Robertson, B. C., and N. J. Gemmell. 2006. PCR-based sexing in conservation biology: wrong answers from an accurate methodology? Conservation Genetics 7:267–271.

Robertson, B. C., et al. 2009. Thirty polymorphic microsatellite loci from the critically endangered kakapo (Strigops habroptilus). Molecular Ecology Resources 9:664–666.

Robichaux, R. H., et al. 1998. Restoring Mauna Kea's crown jewel. Endangered Species Bulletin 23:21–25.

Robinson, T. J., and F. E. B. Elder. 1993. Cytogenetics – its role in wildlife management and the genetic conservation of mammals. Biological Conservation 63:47–51.

Roca, A. L., N. Georgiadis, J. Pecon-Slattery, and S. J. O'Brien. 2001. Genetic evidence for two species of elephant in Africa. Science 293:1473–1477.

Roca, A. L., N. Georgiadis, and S. J. O'Brien. 2005. Cytonuclear genomic dissociation in African elephant species. Nature Genetics 37:96–100.

Roff, D. A. 1997. Evolutionary Quantitative Genetics. Chapman & Hall, New York.

Roff, D. 2003. Evolutionary quantitative genetics: Are we in danger of throwing out the baby with the bathwater? Annales Zoologici Fennici 40:315–320.

Roff, D. A. 2007. A centennial celebration for quantitative genetics. Evolution 61:1017–1032.

Roff, D. A., and P. Bentzen. 1989. The statistical analysis of mitochondrial DNA polymorphisms: chi-square and the problem of small samples. Molecular Biology and Evolution 6:539–545.

Roff, D. A., and T. A. Mousseau. 1987. Quantitative genetics and fitness: lessons from Drosophila. Heredity 58: 103–118.

Rohland, N., et al. 2010. Genomic DNA sequences from mastodon and woolly mammoth reveal deep speciation of forest and Savanna elephants. PLoS Biol 8:e1000564.

Roman, J., and J. A. Darling. 2007. Paradox lost: genetic diversity and the success of aquatic invasions. Trends in Ecology & Evolution 22:454–464.

Roman, J., and S. R. Palumbi. 2003. Whales before whaling in the North Atlantic. Science 301:508–510.

Romanov, M. N., et al. 2009. The value of avian genomics to the conservation of wildlife. BMC Genomics 10:S10.

Romanov, M. N., J. B. Dodgson, R. A. Gonser, and E. M. Tuttle. 2011. Comparative BAC-based mapping in the white-throated sparrow, a novel behavioral genomics model, using interspecies overgo hybridization. BMC Research Notes 4:211.

Romm, R. 2011. All his children: A sperm donor discovers his rich, unsettling legacy. The Atlantic 308(5):22–24.

Rompler, H., et al. 2006. Nuclear gene indicates coat-color polymorphism in mammoths. Science 313:62.

Rosenberg, N. A., and M. Nordborg. 2002. Genealogical trees, coalescent theory and the analysis of genetic polymorphisms. Nature Reviews Genetics 3:380–390.

Ross, H. A., et al. 2003. DNA Surveillance: Web-based molecular identification of whales, dolphins, and porpoises. Journal of Heredity 94:111–114.

Ross Gillespie, A., M. J. O'Riain, and L. F. Keller. 2007. Viral epizootic reveals inbreeding depression in a habitually inbreeding mammal. Evolution 61:2268–2273.

Roy, M. S., et al. 1994. Patterns of differentiation and hybridization in North American wolflike canids, revealed by analysis of microsatellite loci. Molecular Biology and Evolution 11:553–570.

Royer, D. L., L. J. Hickey, and S. L. Wing. 2003. Ecological conservatism in the "living fossil" ginkgo. Paleobiology 29:84–104.

Rozhnov, V. V. 1993. Extinction of the European mink: ecological catastrophe or a natural process? Lutreola 1:10–16.

Rubin, C. J., et al. 2010. Whole-genome resequencing reveals loci under selection during chicken domestication. Nature 464:587–591.

Ruesink, J. L., I. M. Parker, M. J. Groom, and P. M. Kareiva. 1995. Reducing the risks of nonindigenous species introductions – guilty until proven innocent. BioScience 45: 465–477.

Ruiz-Pesini, E., et al. 2000. Human mtDNA haplogroups associated with high or reduced spermatozoa motility. American Journal of Human Genetics 67:682–696.

Russello, M. A., et al. 2010. DNA from the past informs ex situ conservation for the future: An "extinct" species of Galapagos tortoise identified in captivity. PLoS ONE 5:e8683.

Ryder, O. A. 1986. Species conservation and systematics: the dilemma of subspecies. Trends in Ecology & Evolution 1:9–10.

Ryder, O. A. 1987. Conservation action for gazelles: an urgent need. Trends in Ecology & Evolution 2:143–144.

Ryder, O. A., and K. Benirschke. 1997. The potential use of "cloning" in the conservation effort. Zoo Biology 16:295–300.

Ryder, O. A., and L. G. Chemnick. 1993. Chromosomal and mitochondrial DNA variation in orangutans. Journal of Heredity 84:405–409.

Ryder, O. A., A. T. Kumamoto, B. S. Durrant, and K. Benirschke. 1989. Chromosomal divergence and reproductive isolation in diks-diks. Pp. 208–225 in Otte, D., and J. A. Endler, eds. Speciation and Consequences. Sinauer Associates, Sunderland, MA.

Ryder, O. A., et al. 2000. Ecology – DNA banks for endangered animal species. Science 288:275–277.

Rye, M., and B. Gjerde. 1996. Phenotypic and genetic parameters of body composition traits and flesh colour in Atlantic salmon, *Salmo salar* L. Aquaculture Research 27:121–133.

Ryman, N. 1994. Supportive breeding and effective population size – differences between inbreeding and variance effective numbers. Conservation Biology 8:888–890.

Ryman, N., and L. Laikre. 1991. Effects of supportive breeding on the genetically effective population size. Conservation Biology 5:325–329.

Ryman, N., and O. Leimar. 2008. Effect of mutation on genetic differentiation among nonequilibrium populations. Evolution 62:2250–2259.

Ryman, N., and O. Leimar. 2009. G_{ST} is still a useful measure of genetic differentiation – a comment on Jost's *D*. Molecular Ecology 18:2084–2087.

Ryman, N., F. W. Allendorf, and G. Ståhl. 1979. Reproductive isolation with little genetic divergence in sympatric populations of brown trout (*Salmo trutta*). Genetics 92:247–262.

Ryman, N., R. Baccus, C. Reuterwall, and M. H. Smith. 1981. Effective population size, generation interval, and potential loss of genetic variability in game species under different hunting regimes. Oikos 36:257–266.

Ryman, N., P. E. Jorde, and L. Laikre. 1995a. Supportive breeding and variance effective population size. Conservation Biology 9:1619–1628.

Ryman, N., F. Utter, and L. Laikre. 1995b. Protection of intraspecific biodiversity of exploited fishes. Reviews in Fish Biology and Fisheries 5:417–446.

Ryman, N., et al. 2006. Power for detecting genetic divergence: differences between statistical methods and marker loci. Molecular Ecology 15:2031–2045.

Räikkönen, J., et al. 2009. Congenital bone deformities and the inbred wolves (*Canis lupus*) of Isle Royale. Biological Conservation 142:1025–1031.

Saccheri, I. J., and P. M. Brakefield. 2002. Rapid spread of immigrant genomes into inbred populations. Proceedings of the Royal Society B: Biological Sciences 269:1073–1078.

Saccheri, I., et al. 1998. Inbreeding and extinction in a butterfly metapopulation. Nature 392:491–494.

Saether, B. E., S. Engen, and E. J. Solberg. 2009. Effective size of harvested ungulate populations. Animal Conservation 12:488–495.

Sage, R. D., D. Heyneman, K-C. Lim, and A. C. Wilson. 1986. Wormy mice in a hybrid zone. Nature 324:60–63.

Saito, Y., K. Sahara, and K. Mori. 2000. Inbreeding depression by recessive deleterious genes affecting female fecundity of a haplo-diploid mite. Journal of Evolutionary Biology 13:668–678.

Sakai, A. K., et al. 2001. The population biology of invasive species. Annual Review of Ecology and Systematics 32:305–332.

Sala-Torra, O., et al. 2006. Evidence of donor-derived hematologic malignancies after hematopoietic stem cell transplantation. Biology of Blood and Marrow Transplantation 12:511–517.

Salemi, M., and Vandamme, A-M. 2003. The Phylogenetic Handbook: A Practical Approach to DNA and Protein Phylogeny. Cambridge University Press, Cambridge.

Salmon, A., M. L. Ainouche, and J. F. Wendel. 2005. Genetic and epigenetic consequences of recent hybridization and polyploidy in *Spartina* (Poaceae). Molecular Ecology 14:1163–1175.

Sarrazin, F., and R. Barbault. 1996. Reintroduction: challenges and lessons for basic ecology. Trends in Ecology & Evolution 11:474–478.

Sarre, S. D., and A. Georges. 2009. Genetics in conservation and wildlife management: a revolution since Caughley. Wildlife Research 36:70–80.

Savolainen, O., and K. Kärkkäinen. 1992. Effect of forest management on gene pools. New Forest 6:329–345.

Savolainen, O., and T. Pyhajarvi. 2007. Genomic diversity in forest trees. Current Opinion in Plant Biology 10:162–167.

Savolainen, P., L. Arvestad, and J. Lundeberg. 2000. mtDNA tandem repeats in domestic dogs and wolves: mutation mechanism studied by analysis of the sequence of imperfect repeats. Molecular Biology and Evolution 17:474–488.

Savolainen, O., T. Pyhajarvi, and T. Knurr. 2007. Gene flow and local adaptation in trees. Annual Review of Ecology Evolution and Systematics 38:595–619.

Scascitelli, M., et al. 2010. Genome scan of hybridizing sunflowers from Texas (*Helianthus annuus* and *H. debilis*) reveals asymmetric patterns of introgression and small islands of genomic differentiation. Molecular Ecology 19:521–541.

Schaal, B. B., and K. M. Olsen. 2000. Gene genealogies and population variation in plants. Proceedings of the National Academy of Sciences USA 97:7024–7029.

Schindler, D. E., et al. 2010. Population diversity and the portfolio effect in an exploited species. Nature 465:609–612.

Schlager, G., and M. M. Dickie. 1971. Natural mutation rates in the house mouse: estimates for five specific loci and dominant mutations. Mutation Research 11:89–96.

Schlötterer, C. 1998. Microsatellites. Pp. 237–262 *in* Hoelzel, A. R., ed. Molecular Genetic Analysis of Populations. Oxford University Press, New York.

Schlötterer, C. 2004. The evolution of molecular markers – just a matter of fashion? Nature Reviews Genetics 5:63–69.

Schluter, D., and G. L. Conte. 2009. Genetics and ecological speciation. Proceedings of the National Academy of Sciences 106 (Supplement 1):9955–9962.

Schnable, P. S., and R. P. Wise. 1998. The molecular basis of cytoplasmic male sterility and fertility restoration. Trends in Plant Science 3:175–180.

Schoettle, A. W., and R. A. Sniezko. 2007. Proactive intervention to sustain high-elevation pine ecosystems threatened by white pine blister rust. Journal of Forest Research 12:327–336.

Schoettle, A., J. Klutsch, M. Antolin, and S. Field. 2011. A population genetic model for high-elevation five-needle pines: Projecting population outcomes in the presence of white pine blister rust. *In* Keane, R. E., D. F. Tomback, M. P. Murray, and C. M. Smith, eds. The future of high-elevation five-needle white pines in America: Proceedings of the High Five Symposium. Proceedings RMRS-P-63, Missoula, MT.

Schonewald-Cox, C. M., S. M. Chambers, B. MacBryde, and W. L. Thomas. 1983. Genetics and Conservation. Benjamin/Cummings, Menlo Park, CA.

Schut, E., N. Hemmings, and T. R. Birkhead. 2008. Parthenogenesis in a passerine bird, the Zebra Finch *Taeniopygia guttata*. Ibis 150:197–199.

Schwartz, M. K., and K. S. McKelvey. 2009. Why sampling scheme matters: the effect of sampling scheme on landscape genetic results. Conservation Genetics 10:441–452.

Schwartz, M. K., D. A. Tallmon, and G. Luikart. 1998. Review of DNA-based census and effective population size estimators. Animal Conservation 1:293–299.

Schwartz, M. K., et al. 2002. DNA reveals high dispersal synchronizing the population dynamics of Canada lynx. Nature 415:520–522.

Schwartz, M. K., et al. 2004. Hybridization between Canada lynx and bobcats: Genetic results and management implications. Conservation Genetics 5:349–355.

Schwartz, M. K., G. Luikart, and R. S. Waples. 2007. Genetic monitoring as a promising tool for conservation and management. Trends in Ecology & Evolution 22:25–33.

Schwartz, M. K., et al. 2009. Wolverine gene flow across a narrow climatic niche. Ecology 90:3222–3232.

Schwartz, M. W., J. J. Hellmann, and J. S. McLachlan. 2009. The precautionary principle in managed relocation is misguided advice. Trends in Ecology & Evolution 24:474–474.

Schweitzer, J., et al. 2011. Forest gene diversity is correlated with the composition and function of soil microbial communities. Population Ecology 53:35–46.

Scribner, K. T., M. H. Smith, R. A. Garrott, and L. H. Carpenter. 1991. Temporal, spatial, and age-specific changes in genotypic composition of mule deer. Journal of Mammalogy 72:126–137.

Scribner, K. T., et al. 2003. Genetic methods for determining racial composition of Canada goose harvests. Journal of Wildlife Management 67:122–135.

Seal, U. S., S. A. Ellis, T. J. Foose, and A. P. Byers. 1993. Conservation assessment and management plans (CAMPs) and global action plans (GCAPs). Conservation Breeding Specialist Group Newsletter 4(2):5–10.

Searle, J. B. 1986. Meiotic studies of Robertsonian heterozygotes from natural populations of the common shrew. Cytogenetics and Cell Genetics 41:154–162.

Seeb, J. E., C. E. Pascal, R. Ramakrishnan, and L. W. Seeb. 2009. SNP genotyping by the 5'-nuclease reaction: advances in high throughput genotyping with non-model organisms. Pp. 277–292 *in* Komar, A., ed. Methods in Molecular Biology, Single Nucleotide Polymorphisms, 2nd edn. Humana Press, New York.

Seeb, J. E., et al. 2011. Single-nucleotide polymorphism (SNP) discovery and applications of SNP genotyping in nonmodel organisms. Molecular Ecology Resources 11:1–8.

Seeb, L. W., J. E. Seeb, G. H. Kruse, and R. G. Weck. 1989. Genetic structure of red king crab populations in Alaskan facilitates enforcement of fishing regulations. Pp. 491–502 *in* Proceedings of the International Symposium on King and Tanner Crabs, Anchorage, Alaska, USA. Alaska Sea Grant College Program, Fairbanks.

Seeb, L. W., et al. 2000. Genetic diversity of sockeye salmon of Cook Inlet, Alaska, and its application to management of populations affected by the Exxon Valdez oil spill. Transactions of the American Fisheries Society 129:1223–1249.

Seehausen, O. 2004. Hybridization and adaptive radiation. Trends in Ecology & Evolution 19:198–207.

Seehausen, O. 2006. Conservation: Losing biodiversity by reverse speciation. Current Biology 16:R334–R337.

Seehausen, O., J. J. M. Vanalphen, and F. Witte. 1997. Cichlid fish diversity threatened by eutrophication that curbs sexual selection. Science 277:1808–1811.

Seehausen, O., G. Takimoto, D. Roy, and J. Jokela. 2008. Speciation reversal and biodiversity dynamics with hybridization in changing environments. Molecular Ecology 17:30–44.

Seuanez, H. N. 1986. Chromosomal and molecular characterization of the primates: its relevance in the sustaining of the primate populations. Pp. 887–910 *in* Benirschke, K., ed. Primates: The Road to Self Sustaining Populations. Springer-Verlag, New York.

Severns, P. M., and A. Liston. 2008. Intraspecific chromosome number variation: a neglected threat to the conservation of rare plants. Conservation Biology 22:1641–1647.

Sgrò, C. M., A. J. Lowe, and A. A. Hoffmann. 2011. Building evolutionary resilience for conserving biodiversity under climate change. Evolutionary Applications 4:326–337.

Shaffer, M. L. 1978. *Determining minimum viable population sizes: a case study of the grizzly bear (*Ursus arctos *L.).* PhD dissertation, Duke University.

Shaffer, M. L. 1981. Minimum population sizes for species conservation. BioScience 31:131–134.

Shaffer, M. 1987. Minimum viable populations: coping with uncertainty. Pp. 69–86 *in* Soulé, M. E., ed. Viable Populations for Conservation. Sinauer, Sunderland, MA.

Shaffer, M. L., and F. B. Samson. 1985. Population size and extinction: A note on determining critical population sizes. American Naturalist 125:144–152.

Shaffer, M., L. H. Watchman, W. J. Snape III, and I. K. Latchis. 2002. Population viability analysis and conservation policy. Pp. 123–142 *in* Beissinger, S. R., and D. R. McCullough, eds. Population Viability Analysis. University of Chicago Press, Chicago, IL.

Shaw, F. H., C. J. Geyer, and R. G. Shaw. 2002. A comprehensive model of mutations affecting fitness and inferences for *Arabidopsis thaliana*. Evolution 56:453–463.

Sherwin, W. B. 2010. Entropy and information approaches to genetic diversity and its expression: genomic geography. Entropy 12:1765–1798.

Shields, G. F. 1982. Comparative avian cytogenetics: a review. Condor 84:45–58.

Shields, W. M. 1993. The natural and unnatural history of inbreeding and outbreeding. Pp. 143–169 *in* Thornhill, N. W., ed. The Natural History of Inbreeding and Outbreeding. University of Chicago Press, Chicago, IL.

Shine, R. 2010. The ecological impact of invasive cane toads (*Bufo marinus*) in Australia. Quarterly Review of Biology 85:253–291.

Shine, R., G. P. Brown, and B. L. Phillips. 2011. An evolutionary process that assembles phenotypes through space rather than through time. Proceedings of the National Academy of Sciences USA 108:5708–5711.

Shivji, M., *et al.* 2002. Genetic identification of pelagic shark body parts for conservation and trade monitoring. Conservation Biology 16:1036–1047.

Siddle, H. V., *et al.* 2010. MHC gene copy number variation in Tasmanian devils: Implications for the spread of a contagious cancer. Proceedings of the Royal Society B: Biological Sciences 277:2001–2006.

Siemann, E., and W. E. Rogers. 2001. Genetic differences in growth of an invasive tree species. Ecology Letters 4:514–518.

Simberloff, D. 1988. The contribution of population and community biology to conservation science. Annual Review of Ecology and Systematics 19:437–511.

Simberloff, D. 2001. Biological invasions – how are they affecting us, and what can we do about them? Western North American Naturalist 61:308–315.

Simon, J-P., Y. Bergeron, and D. Gagnon. 1986. Isozyme uniformity in red pine (*Pinus resinosa*) in the Abitibi Region, Quebec. Canadian Journal of Forest Research 16:1133–1135.

Simpson, G. G. 1945. The principles of classification and a classification of Mammals. Bulletin of the American Museum of Natural History 85:1–350.

Simpson, G. G. 1961. Principles of animal taxonomy. Columbia University Press, New York.

Singh, S., P. Goswami, R. Singh, and K. J. Heller. 2009. Application of molecular identification tools for lactobacillus, with a focus on discrimination between closely related species: A review. LWT – Food Science and Technology 42:448–457.

Singh, R. C., R. C. Lewontin, and A. Felton. 1976. Genetic heterogeneity within electrophoretic "alleles" of xanthine dehydrogenase in *Drosophila pseudoobscura*. Genetics 84:609–626.

Sinnock, P. 1975. The Wahlund effect for the two-locus model. American Naturalist 109:565–570.

Sissenwine, M. P. 1984. Why do fish populations vary? Pp. 59–94 *in* May, R. M., ed. Dahlem Konferenzen. Springer, Berlin.

Skaug, H. J. 2001. Allele-sharing methods for estimation of population size. Biometrics 57:750–756.

Slate, J., and N. J. Gemmell. 2004. Eve 'n' Steve: Recombination of human mitochondrial DNA. Trends in Ecology & Evolution 19:561–563.

Slate, J., and J. M. Pemberton. 2007. Admixture and patterns of linkage disequilibrium in a free-living vertebrate population. Journal of Evolutionary Biology 20:1415–1427.

Slate, J., T. Marshall, and J. Pemberton. 2000. A retrospective assessment of the accuracy of the paternity inference program CERVUS. Molecular Ecology 9:801–808.

Slate, J., *et al.* 2004. Understanding the relationship between the inbreeding coefficient and multilocus heterozygosity: theoretical expectations and empirical data. Heredity 93:255–265.

Slatkin, M. 1977. Gene flow and genetic drift in species subject to frequent local extinctions. American Naturalist 12:253–262.

Slatkin, M. 1985. Rare alleles as indicators of gene flow. Evolution 39:53–65.

Slatkin, M. 1987. Gene flow and the geographic structure of natural populations. Science 236:787–792.

Slatkin, M. 1995. A measure of population subdivision based on microsatellite allele frequencies. Genetics 139:457–462.

Slatkin, M. 2008. Linkage disequilibrium – understanding the evolutionary past and mapping the medical future. Nature Reviews Genetics 9:477–485.

Slatkin, M., and N. H. Barton. 1989. A comparison of three indirect methods for estimating average levels of gene flow. Evolution 43:1349–1368.

Smales, I. J. 1994. The discovery of Leadbeater's possum, *Gymnobelideus leadbeateri* McCoy, resident in a lowland swamp woodland. Victorian Naturalist 111:178–182.

Smith, C. T., J. E. Seeb, P. Schwenke, and L. W. Seeb. 2005a. Use of the 5'-nuclease reaction for single nucleotide polymorphism genotyping in Chinook salmon. Transactions of the American Fisheries Society 134:207–217.

Smith, G. R. 1992. Introgression in fishes – significance for paleontology, cladistics, and evolutionary rates. Systematic Biology 41:41–57.

Smith, J. J., et al. 2005b. A comprehensive EST linkage map for tiger salamander and Mexican axolotl: enabling gene mapping and comparative genomics in Ambystoma. Genetics 171:1161–1171.

Smith, J. N. M., L. F. Keller, A. B. Marr, and P. Arcese. 2006. Conservation and Biology of Small Populations. Oxford University Press, New York.

Snyder, G. 1990. The Practice of the Wild. North Point Press, San Francisco.

Snyder, N. F. R., et al. 1996. Limitations of captive breeding in endangered species recovery. Conservation Biology 10:338–348.

Soliva, M., and A. Widmer. 1999. Genetic and floral divergence among sympatric populations of Gymnadenia conopsea SL (Orchideaceae) with different flowering phenology. International Journal of Plant Sciences 160:897–905.

Solmsen, N., J. Johannesen, and C. Schradin. 2011. Highly asymmetric fine-scale genetic structure between sexes of African striped mice and indication for condition dependent alternative male dispersal tactics. Molecular Ecology 20:1624–1634.

Soltis, D. E., and P. S. Soltis. 1989. Genetic consequences of autoploidy in Tolmiea (Saxifragaceae). Evolution 586–594.

Song, H., J. E. Buhay, M. Whiting, and K. A. Crandall. 2008. Many species in one: DNA barcoding overestimates the number of species when nuclear mitochondrial pseudogenes are coamplified. Proceedings of the National Academy of Sciences USA 105:13486–13491.

Song, Y., et al. 2011. Adaptive introgression of anticoagulant rodent poison resistance by hybridization between Old World mice. Current Biology: CB 21:1296–1301.

Sorensen, F. C. 1969. Embryonic genetic load in coastal Douglas fir, Pseudotsuga menziesii var. menziesii. American Naturalist 103:389–398.

Sorensen, F. C. 1999. Relationship between self-fertility, allocation of growth, and inbreeding depression in three coniferous species. Evolution 53:417–425.

Soulé, M. E. 1980. Thresholds for survival: maintaining fitness and evolutionary potential. Pp. 151–170 in Soulé, M. E., and B. M. Wilcox, eds. Conservation Biology: An Evolutionary-Ecological Perspective. Sinauer, Sunderland, MA.

Soulé, M. E. 1987a. Viable Populations for Conservation. Sinauer, Sunderland, MA.

Soulé, M. E. 1987b. Introduction. Pp. 1–10 in Soulé, M. E., ed. Viable Populations for Conservation. Sinauer, Sunderland, MA.

Soulé, M. E. 1987c. Where do we go from here? Pp. 175–183 in Soulé, M. E., ed. Viable Populations for Conservation. Sinauer, Sunderland, MA.

Soulé, M. E., and L. S. Mills. 1992. Conservation genetics and conservation biology: a troubled marriage. Pp. 55–65 in Sandlund, O. T., K. Hindar, and A. H. D. Brown, eds. Species and Ecosystem Conservation. Scandinavian University Press, Oslo.

Soulé, M. E., and L. S. Mills. 1998. No need to isolate genetics. Science 282:1658–1659.

Soulé, M. E., and B. M. Wilcox. 1980. Conservation Biology: An Evolutionary–Ecological Perspective. Sinauer, Sunderland, MA.

Soulé, M., M. Gilpin, W. Conway, and T. Foose. 1986. The millennium ark: how long a voyage, how many staterooms, how many passengers? Zoo Biology 5:101–113.

Sousa, V. C., M. Fritz, M. A. Beaumont, and L. S. Chikhi. 2009. Approximate Bayesian computation without summary statistics: The case of admixture. Genetics 181:1507–1519.

Spencer, C. N., B. R. McClelland, and J. A. Stanford. 1991. Shrimp stocking, salmon collapse, and eagle displacement: cascading interactions in the food web of a large aquatic ecosystem. BioScience 41:14–21.

Spencer, H. G., and R. W. Marks. 1993. The evolutionary construction of molecular polymorphisms. New Zealand Journal of Botany 31:249–256.

Spitze, K. 1993. Population structure in Daphnia obtusa – quantitative genetic and allozymic variation. Genetics 135:367–374.

Sprague, G. F. 1978. Introductory remarks to the session on the history of hybrid corn. Pp. 11–12 in Walden, D. B., ed. Maize Breeding and Genetics. John Wiley & Sons, New York.

Spruell, P., et al. 1999a. Inheritance of nuclear DNA markers in gynogenetic haploid pink salmon (Oncorhynchus gorbuscha). Journal of Heredity 90:289–296.

Spruell, P., et al. 1999b. Genetic population structure within streams: microsatellite analysis of bull trout populations. Ecology of Freshwater Fish 8:114–121.

Spruell, P., M. L. Bartron, N. Kanda, and F. W. Allendorf. 2001. Detection of hybrids between bull trout (Salvelinus confluentus) and brook trout (Salvelinus fontinalis) using PCR primers complementary to interspersed nuclear elements. Copeia 2001:1093–1099.

Spruell, P., et al. 2003. Conservation genetics of bull trout: geographic distribution of variation at microsatellite loci. Conservation Genetics 4:17–29.

Stapley, J., et al. 2010. Adaptation genomics: The next generation. Trends in Ecology & Evolution 25:705–712.

Stebbins, G. L. 1938. Cytological characteristics associated with the different growth habits in the dicotyledons. American Journal of Botany 25:189–198.

Stebbins, G. L. 1950. Variation and Evolution in Plants. Columbia University Press, New York.

Stebbins, G. L. 1959. The role of hybridization in evolution. Proceedings of the American Philosophical Society 103:231–251.

Steinberg, E. K., *et al.* 2002. Rates and patterns of microsatellite mutations in pink salmon. Molecular Biology and Evolution 19:1198–1202.

Steinger, T., C. Korner, and B. Schmid. 1996. Long-term persistence in a changing climate: DNA analysis suggests very old ages of clones of alpine *Carex curvula*. Oecologia 105:94–99.

Stephens, J. C., *et al.* 2001. Haplotype variation and linkage disequilibrium in 313 human genes. Science 293: 489–493.

Stetz, J. B., K. C. Kendall, and C. Servheen. 2010. Evaluation of bear rub surveys to monitor grizzly bear population trends. Journal of Wildlife Management 74:860–870.

Stevens, N. M. 1908. The chromosomes in *Diabrotica vittata*, *Diabrotica soror*, and *Diabrotica 12-punctata*. Journal of Experimental Zoology 5:453–470.

Stinchcombe, J. R., and H. E. Hoekstra. 2008. Combining population genomics and quantitative genetics: Finding the genes underlying ecologically important traits. Heredity 100:158–170.

Stockwell, C. A., and M. V. Ashley. 2004. Diversity – rapid adaptation and conservation. Conservation Biology 18: 272–273.

Stockwell, C. A., A. P. Hendry, and M. T. Kinnison. 2003. Contemporary evolution meets conservation biology. Trends in Ecology & Evolution 18:94–101.

Stokstad, E. 2010. To fight illegal fishing, forensic DNA gets local. Science 330:1468–1469.

Stone, G. 2000. Phylogeography, hybridization and speciation. Trends in Ecology & Evolution 15:354–355.

Stone, G. N., and P. Sunnucks. 1993. Genetic consequences of an invasion through a patchy environment – the cynipid gallwasp *Andricus queruscalicis* (Hymenoptera: Cynipidae). Molecular Ecology 2:251–268.

Storfer, A. 1996. Quantitative genetics: a promising approach for the assessment of genetic variation in endangered species. Trends in Ecology & Evolution 11:343–348.

Storz, J. F. 2002. Contrasting patterns of divergence in quantitative traits and neutral DNA markers: analysis of clinal variation. Molecular Ecology 11:2537–2551.

Strauss, S. Y., J. A. Lau, and S. P. Carroll. 2006. Evolutionary responses of natives to introduced species: What do introductions tell us about natural communities? Ecology Letters 9:357–374.

Streiner, D. L. 1996. Maintaining standards: differences between the standard deviation and standard error, and when to use each. Canadian Journal of Psychiatry 41:498–502.

Strickberger, M. W. 2000. Evolution, 3rd edn. Jones and Bartlett Publishers, Sudbury, MA.

Stuart, S. N., *et al.* 2010. The barometer of life. Science 328:177.

Sturtevant, A. H. 1923. Inheritance of shell coiling in *Limnaea*. Science 58:269–270.

Sturtevant, A. H., and Beadle, G. W. 1939. An Introduction to Genetics. Dover, New York.

Sunnucks, P. 2000. Efficient genetic markers for population biology. Trends in Ecology & Evolution 15:199–203.

Sušnik, S., *et al.* 2004. Genetic introgression between wild and stocked salmonids and the prospects for using molecular markers in population rehabilitation: the case of the Adriatic grayling (*Thymallus thymallus* L. 1785). Heredity 93:273–282.

Sutherland, B., D. Stewart, E. R. Kenchington, and E. Zouros. 1998. The fate of paternal mitochondrial DNA in developing female mussels, *Mytilus edulis*: implications for the mechanism of doubly uniparental inheritance of mitochondrial DNA. Genetics 148:341–347.

Sutherland, W. J. 1996. Why census? Pp. 1–10 *in* Sutherland, W. J., ed. Ecological Census Techniques, a Handbook. Cambridge University Press, Cambridge.

Swain, D. P., A. F. Sinclair, and J. M. Hanson. 2007. Evolutionary response to size-selective mortality in an exploited fish population. Proceedings of the Royal Society B: Biological Sciences 274:1015–1022.

Swarts, N. D., and K. W. Dixon. 2009. Terrestrial orchid conservation in the age of extinction. Annals of Botany 104:543–556.

Syvänen, A.-C. 2005. Toward genome-wide SNP genotyping. Nature Genetics 37:S5–S10.

Szulkin, M., N. Bierne, and P. David. 2010. Heterozygosity-fitness correlations: A time for reappraisal. Evolution 64: 1202–1217.

Taberlet, P., L. Gielly, G. Pautou, and J. Bouvet. 1991. Universal primers for amplification of three non-coding regions of chloroplast DNA. Plant Molecular Biology 17:1105–1109.

Taberlet, P., L. P. Waits, and G. Luikart. 1999. Noninvasive genetic sampling: look before you leap. Trends in Ecology & Evolution 14:323–327.

Taberlet, P., G. Luikart, and E. Geffen. 2001. Novel approaches for obtaining and analyzing genetic data for conserving wild carnivore populations. Pp. 313–334 *in* Gittleman, J. L., S. M. Funk, D. W. MacDonald, and R. K. Wayne, eds. Carnivore Conservation. Cambridge University Press, Cambridge.

Takezaki, N. 2010. Evolution of microsatellites. *In* Encyclopedia of Life Sciences. John Wiley & Sons,Chichester. DOI: 10.1002/9780470015902.a0022866.

Talbot, S. L., and G. F. Shields. 1996. Phylogeography of brown bears (*Usus arctos*) of Alaska and paraphyly within the Uidae. Molecular Phylogenetics and Evolution 5:477–494.

Tallmon, D. A., G. Luikart, and M. A. Beaumont. 2004. Comparative evaluation of a new effective population size estimator based on approximate Bayesian summary statistics. Genetics 167:977–988.

Tallmon, D. A., A. Koyuk, G. Luikart, and M. A. Beaumont. 2008. ONeSAMP: a program to estimate effective population size using approximate Bayesian computation. Molecular Ecology Resources 8:299–301.

Tallmon, D. A., *et al.* 2010. When are genetic methods useful for estimating contemporary abundance and detecting population trends? Molecular Ecology Resources 10:684–692.

Tanaka, M. M., R. Cristescu, and D. W. Cooper. 2009. Effective population size of koala populations under different population management regimes including contraception. Wildlife Research 36:601–609.

Tape, K., M. Sturm, and C. Racine. 2006. The evidence for shrub expansion in Northern Alaska and the Pan-Arctic. Global Change Biology 12:686–702.

Tarr, C. L., S. Conant, and R. C. Fleischer. 1998. Founder events and variation at microsatellite loci in an insular passerine bird, the Laysan finch (*Telespiza cantans*). Molecular Ecology 7:719–731.

Tave, D. 1984. Quantitative genetics of vertebrae number and position of dorsal spines in the velvet belly shark, *Etmopterus spinax*. Copeia 1984:794–797.

Taylor, A. C., W. B. Sherwin, and R. K. Wayne. 1994. Genetic variation of microsatellite loci in a bottlenecked species: the northern hairy-nosed wombat *Lasiorhinus kreffti*. Molecular Ecology 3:277–290.

Taylor, B. L., and A. E. Dizon. 1999. First policy then science: why a management unit based solely on genetic criteria cannot work. Molecular Ecology 8:S11–S16.

Taylor, E. B., C. J. Foote, and C. C. Wood. 1996. Molecular genetic evidence for parallel life-history evolution within a Pacific salmon (sockeye salmon and kokanee, *Oncorhynchus nerka*). Evolution 50:401–416.

Taylor, H. R., and W. E. Harris, 2012, An emergent science on the brink of irrelevance: A review of the past eight years of DNA barcoding. Molecular Ecology Resources 12:377–378.

Tear, T. H., J. M. Scott, P. H. Hayward, and B. Griffith. 1993. Status and prospects for success of the Endangered Species Act: a look at recovery plans. Science 262:976–977.

Templeton, A. R. 1982. Adaptation and the integration of evolutionary forces. *in* Milkman, R., ed. Perspectives on Evolution. Sinauer, Sunderland, MA.

Templeton, A. 1986. Coadaptation and outbreeding depression. Pp. 105–116 *in* Soule, M. E., ed. Conservation Biology: The Science of Scarcity and Diversity. Sinauer, Sunderland, MA.

Templeton, A. R. 1991. Off-site breeding of animals and implications for plant conservation strategies. Pp. 182–194 *in* Falk, D. A., and K. E. Holsinger, eds. Genetics and Conservation of Rare Plants. Oxford University Press, New York.

Templeton, A. R. 1998. Species and speciation – geography, population structure, ecology, and gene trees. Pp. 32–43 *in* Howard, D. J., and S. H. Berlocher, eds. Endless Forms. Oxford University Press, New York.

Templeton, A. R. 2002. The Speke's gazelle breeding program as an illustration of the importance of multilocus genetic diversity in conservation biology: response to Kalinowski *et al.* Conservation Biology 16:1151–1155.

Templeton, A. R. 2008. Nested clade analysis: an extensively validated method for strong phylogeographic inference. Molecular Ecology 17:1877–1880.

Templeton, A. R., and B. Read. 1984. Factors eliminating inbreeding depression in a captive herd of Speke's gazelle. Zoo Biology 3:177–199.

Templeton, A. R., and B. Read. 1994. Inbreeding: one word, several meanings, much confusion. Pp. 91–105 *in* Loeschcke, V., J. Tomiuk, and S. K. Jain, eds. Conservation Genetics. Birkhauser Verlag, Basel, Switzerland.

Templeton, A. R., and B. Read. 1998. Elimination of inbreeding depression from a captive population of Speke's gazelle: validity of the original statistical analysis and confirmation by permutation testing. Zoo Biology 17:77–94.

Templeton, A. R., R. J. Robertson, J. Brisson, and J. Strasburg. 2001. Disrupting evolutionary processes: the effect of habitat fragmentation on collared lizards in the Missouri Ozarks. Proceedings of the National Academy of Sciences USA 98:5426–5432.

Tenhumberg, B., A. J. Tyre, A. R. Pople, and H. P. Possingham. 2004. Do harvest refuges buffer kangaroos against evolutionary responses to selective harvesting? Ecology 85:2003–2017.

Thelen, G. C., and F. W. Allendorf. 2001. Heterozygosity-fitness correlations in rainbow trout: Effects of allozyme loci or associative overdominance? Evolution 55:1180–1187.

Theodorou, K., and D. Couvet. 2004. Introduction of captive breeders to the wild: harmful or beneficial? Conservation Genetics 5:1–12.

Thomas, D. D., C. A. Donnelly, R. J. Wood, and L. S. Alphey. 2000. Insect population control using a dominant, repressible, lethal genetic system. Science 287:2474–2476.

Thomas, J. W., *et al.* 2008. The chromosomal polymorphism linked to variation in social behavior in the white-throated sparrow (*Zonotrichia albicollis*) is a complex rearragement and suppressor of recombination. Genetics 179:1455–1468.

Thompson, G. G. 1991. Determining minimum viable populations under the Endangered Species Act. NOAA Technical Mememorandum NMFS F/NWC-198.

Thorneycroft, H. B. 1975. A cytogenetic study of the white-throated sparrow, *Zonotrichia albicollis* (Gmelin). Evolution 29:611–621.

Thornton, I. W. B. 1978. White tiger genetics: further evidence. Journal of Zoology 185:389–394.

Thoss, M., P. Ilmonen, K. Musolf, and D. J. Penn. 2011. Major histocompatibility complex heterozygosity enhances reproductive success. Molecular Ecology 20:1546–1557.

Thrall, P. H., C. M. Richards, D. E. McCauley, and J. Antonovics. 1998. Metapopulation collapse: the consequences of limited gene-flow in spatially structured populations. Pp. 83–104 *in* Bascompte, J., and R. V. Solé, eds. Modeling Spatiotemporal Dynamics in Ecology. Springer-Verlag, Berlin.

Thrower, F. P., and J. J. Hard. 2009. Effects of a single event of close inbreeding on growth and survival in steelhead. Conservation Genetics 10:1299–1307.

Thurber, R. V., et al. 2009. Metagenomic analysis of stressed coral holobionts. Environmental Microbiology 11:2148–2163.

Timmermans, M. J. T. N., et al. 2010. Why barcode? High-throughput multiplex sequencing of mitochondrial genomes for molecular systematics. Nucleic Acids Research 38:e197.

Tishkoff, S. A., et al. 2001. Haplotype diversity and linkage disequilibrium at human G6PD: Recent origin of alleles that confer malarial resistance. Science 293:455–462.

Tnah, L. H., et al. 2010. Forensic DNA profiling of tropical timber species in Peninsular Malaysia. Forest Ecology and Management 259:1436–1446.

Todd, C. R., and M. A. Burgman. 1998. Assessment of threat and conservation priorities under realistic levels of uncertainty and reliability. Conservation Biology 12:966–974.

Toews, D. P. L., and D. E. Irwin. 2008. Cryptic speciation in a holarctic passerine revealed by genetic and bioacoustic analyses. Molecular Ecology 17:2691–2705.

Tokarska, M., et al. 2009. Effectiveness of microsatellite and SNP markers for parentage and identity analysis in species with low genetic diversity: The case of European bison. Heredity 103:326–332.

Tolar, J., and J. P. Neglia. 2003. Transplacental and other routes of cancer transmission between individuals. Journal of Pediatric Hematology/Oncology 25:430–434.

Torres, T. T., M. Metta, B. Ottenwalder, and C. Schlötterer. 2008. Gene expression profiling by massively parallel sequencing. Genome Research 18:172–177.

Tortajada, A. M., M. Carmona, and M. Serra. 2009. Does haplodiploidy purge inbreeding depression in rotifer populations? PLoS One 4:e8195.

Traill, L. W., B. W. Brook, R. R. Frankham, and C. J. A. Bradshaw. 2010a. Pragmatic population viability targets in a rapidly changing world. Biological Conservation 143:28–34.

Traill, L. W., M. L. M. Lim, N. S. Sodhi, and C. J. A. Bradshaw. 2010b. Mechanisms driving change: altered species interactions and ecosystem function through global warming. Journal of Animal Ecology 79:937–947.

Travis, S. E., C. E. Proffitt, and K. Ritland. 2004. Population structure and inbreeding vary with successional stage in created Spartina alterniflora marshes. Ecological Applications 14:1189–1202.

Trimble, H. C., and C. E. Keeler. 1938. The inheritance of "high uric acid excretion" in dogs. Journal of Heredity 29:280–289.

Tsutsui, N. D., A. V. Suarez, D. A. Holway, and T. J. Case. 2000. Reduced genetic variation and the success of an invasive species. Proceedings of the National Academy of Sciences USA 97:5948–5953.

Tucker, J. M., et al. Submitted. Historical and contemporary DNA indicate fisher decline and isolation occurred prior to the European settlement of California.

Turner, B. M. 2009. Epigenetic responses to environmental change and their evolutionary implications. Philosophical Transactions of the Royal Society B: Biological Sciences 364:3403–3418.

Turner, J. R. G. 1985. Fisher's evolutionary faith and the challenge of mimicry. Oxford Surveys in Evolutionary Biology 2:159–196.

Turner, T. F., J. C. Trexler, J. L. Harris, and J. L. Haynes. 2000. Nested cladistic analysis indicates population fragmentation shapes genetic diversity in a freshwater mussel. Genetics 154:777–785.

Tuttle, E. M. 2003. Alternative reproductive strategies in the white-throated sparrow: behavioral and genetic evidence. Behavioral Ecology 14:425–432.

Tymchuk, W. V., et al. 2010. Conservation genomics of Atlantic salmon: Variation in gene expression between and within regions of the Bay of Fundy. Molecular Ecology 19:1842–1859.

Ujvari, B., M. Dowton, and T. Madsen. 2007. Mitochondrial DNA recombination in a free-ranging Australian lizard. Biology Letters 3:189–192.

Umina, P. A., et al. 2005. A rapid shift in a classic clinal pattern in Drosophila reflecting climate change. Science 308:691–693.

Unger, F. 1852. Versuch einer Geschichte der Pflanzenwelt. Braumuller, Wein.

USFWS. 1983. Endangered and threatenend species listing and recovery priority guidelines. Federal Register 48(184): 43098–43105.

USFWS. 1998. Recovery Plan for Kokia cookei. US Fish and Wildlife Service, Portland, OR.

USFWS. 2003. Endangered and threatened wildlife and plants: reconsidered finding for an amended petition to list the westslope cutthroat trout as threatened throughout its range. Federal Register 68:46989–47009.

USFWS. 2004. 12-month finding for a petition to list the West Coast distinct population segment of the fisher (Martes pennanti): proposed rule. Federal Register 69:18770–18792.

USFWS. 2008. Kokia cookei (Cooke's koko'o) 5-Year Review Summary and Evaluation. US Fish and Wildlife Service, Honolulu, Hawaii.

USFWS and NOAA. 1996a. Endangered and threatened wildlife and plants; proposed policy and proposed rule on the treatment of intercrosses and intercross progeny (the issue of "hybridization"); request for public comment. Federal Register 61(26):4710–4713.

USFWS and NOAA. 1996b. Policy Regarding the Recognition of District Vertebrate Population. Federal Register 61(26):4721–4725.

USGS (US Geological Survey). 2002. White abalone restoration. USGS Western Ecological Center Sheet.

US National Science Board Committee on International Science's Task Force on Global Biodiversity. 1989. Loss of biological diversity: A global crisis requiring international solutions. National Science Board Report 89–171, Washington, DC.

Utter, F. M. 2005. Farewell to allozymes(?) Journal of Irreproducible Results 49:35.

Utter, F., P. Aebersold, and G. Winans. 1987. Interpreting genetic variation detected by electrophoresis. Pp. 21–46 *in* Ryman, N., and F. Utter, eds. Population Genetics and Fishery Management. University of Washington Press, Seattle, WA.

Valdes, A. M., M. Slatkin, and N. B. Freimer. 1993. Allele frequencies at microsatellite loci: the stepwise mutation model revisited. Genetics 133:737–749.

Valentini, A., *et al.* 2009a. New perspectives in diet analysis based on DNA barcoding and parallel pyrosequencing: the TML approach. Molecular Ecology Resources 9:51–60.

Valentini, A., F. Pompanon, and P. Taberlet. 2009b. DNA barcoding for ecologists. Trends in Ecology & Evolution 24:110–117.

Van Aarde, R. J., and A. Van Dyk. 1986. Inheritance of the king coat color pattern in cheetahs *Acinonyx jubatus.* Journal of Zoology 209:573–578.

Van Asch, M., and M. E. Visser. 2007. Phenology of forest caterpillars and their host trees: the importance of synchrony. Annual Review of Entomology 52:37–55.

Van Bers, N. E. M., *et al.* 2010. Genome-wide SNP detection in the great tit parus major using high throughput sequencing. Molecular Ecology 19:89–99.

VanCamp, L. F., and Henny, C. J. 1975. The Screech Owl: Its Life History and Population Ecology in Northern Ohio. North American Fauna, No. 71, US Fish and Wildlife Service.

Van der Putten, W. H., M. Macel, and M. E. Visser. 2010. Predicting species distribution and abundance responses to climate change: why it is essential to include biotic interactions across trophic levels. Philosophical Transactions of the Royal Society B: Biological Sciences 365: 2025–2034.

Van der Veken, S., *et al.* 2008. Garden plants get a head start on climate change. Frontiers in Ecology and the Environment 6:212–216.

Vane-Wright, R. I., C. J. Humphries, and P. H. Williams. 1991. What to protect – Systematics and the agony of choice. Biological Conservation 55:235–254.

Van Noordwijk, A. J., and W. Scharloo. 1981. Inbreeding in an island population of the great tit. Evolution 35:674–688.

Van Valen, L. 1973. A new evolutionary law. Evolutionary Theory 1:1–30.

Van Valen, L. 1976. Ecological species, multispecies, and oaks. Taxon 25:233–239.

Vasemägi, A., and C. R. Primmer. 2005. Challenges for identifying functionally important genetic variation: the promise of combining complementary research strategies. Molecular Ecology 14:3623–3642.

Veale, A. J., M. N. Clout, and D. M. Gleeson. 2012. Genetic population assignment reveals a long-distance incursion to an island by a stoat (*Mustela erminea*). Biological Invasions 14:735–742.

Vega Thurber, R., *et al.* 2009. Metagenomic analysis of stressed coral holobionts. Environmental Microbiology 11:2148–2163.

Vekemans, X., and O. J. Hardy. 2004. New insights from fine-scale spatial genetic structure analyses in plant populations. Molecular Ecology 13:921–935.

Vendramin, G. G., *et al.* 2008. Genetically depauperate but widespread: The case of an emblematic Mediterranean pine. Evolution 62:680–688.

Vergeer, P., E. Sonderen, and N. J. Ouborg. 2004. Introduction strategies put to the test: Local adaptation versus heterosis. Conservation Biology 18:812–821.

Vernooy, R., *et al.* 2010. Barcoding life to conserve biological diversity: Beyond the taxonomic imperative. PLoS Biol 8:e1000417.

Vieira, C. P., and D. Charlesworth. 2002. Molecular variation at the self-incompatibility locus in natural populations of the genera *Antirrhinum* and *Misopates.* Heredity 88: 172–181.

Vinkey, R., *et al.* 2006. When reintroductions are augmentations: the genetic legacy of fishers (*Martes pennanti*) in Montana. Journal of Mammalogy 87:265–271.

Visser, M. E. 2008. Keeping up with a warming world: assessing the rate of adaptation to climate change. Proceedings of the Royal Society B: Biological Sciences 275:649–659.

Vithayasai, C. 1973. Exact critical values of the Hardy-Weinberg test statistic for two alleles. Communications in Statistics 1:229–242.

Vogel, F., and Motulsky, A. G. 1986. Human Genetics: Problems and Approaches, 2nd edn. Springer-Verlag, Berlin.

Voipio, P. 1950. Evolution at the population level with special reference to game animals and practical game management. Papers on Game Research (Helsinki) 5:1–175.

Vollmer, S. V., and S. R. Palumbi. 2002. Hybridization and the evolution of reef coral diversity. Science 296:2023–2025.

Vollmer, S. V., and S. R. Palumbi. 2007. Restricted gene flow in the Caribbean staghorn coral *Acropora cervicornis*: Implications for the recovery of endangered reefs. Journal of Heredity 98:40–50.

vonHoldt, B. M., *et al.* 2010. Genome-wide SNP and haplotype analyses reveal a rich history underlying dog domestication. Nature 464:898–902.

vonHoldt, B. M., *et al.* 2011. A genome-wide perspective on the evolutionary history of enigmatic wolf-like canids. Genome Research 21:1294–1305.

Vos, P., *et al.* 1995. AFLP: a new technique for DNA fingerprinting. Nucleic Acids Research 23:4407–4414.

Voss, S. R. 1995. Genetic basis of paedomorphosis in the axolotl, *Ambystoma mexicanum*: a test of the single-gene hypothesis. Journal of Heredity 86:441–447.

Voss, S. R., and H. B. Shaffer. 1997. Adaptive evolution via a major gene effect: paedomorphosis in the Mexican axolotl. Proceedings of the National Academy of Sciences USA 94:14185–14189.

Voss, S. R., and J. J. Smith. 2005. Evolution of salamander life cycles: a major effect quantitative trait locus contributes to discreet and continuous variation for metamorphic timing. Genetics 170:275–281.

Vrijenhoek, R. C. 1989. Genotypic diversity and coexistence among sexual and clonal forms of *Poeciliopsis*. Pp. 386–400 *in* Otte, D., and J. Endler, eds. Speciation and Its Consequences. Sinauer Associates, Sunderland, MA.

Vrijenhoek, R. C., and P. L. Leberg. 1991. Let's not throw the baby out with the bathwater – a comment on management for MHC diversity in captive populations. Conservation Biology 5:252–254.

Vrijenhoek, R. C., and S. Lerman. 1982. Heterozygosity and developmental stability under sexual and asexual breeding systems. Evolution 36:768–776.

Vrijenhoek, R. C., E. Pfeiler, and J. Wetherington. 1992. Balancing selection in a desert stream-dwelling fish, *Poeciliopsis monacha*. Evolution 46:1642–1647.

Waddington, C. H. 1961. Genetic assimilation. Advances in Genetics 10:257–290.

Wahlund, S. 1928. Zusammensetzung von Populationen und Korrelationserscheinungen vom Standpunkt der Vererbungslehre aus betrachtet. Hereditas 11:65–106.

Waits, J. L., and P. L. Leberg. 2000. Biases associated with population estimation using molecular tagging. Animal Conservation 3:191–199.

Waits, L. P., S. L. Talbot, R. H. Ward, and G. F. Shields. 1998. Mitochondrial DNA phylogeography of the North American brown bear and implications for conservation. Conservation Biology 12:408–417.

Waits, L. P., G. Luikart, and P. Taberlet. 2001. Estimating the probability of identity among genotypes in natural populations: cautions and guidelines. Molecular Ecology 10:249–256.

Wakeley, J. 2009. Coalescent Theory: An Introduction. Roberts and Company Publishers, Greenwood Village, CO.

Waldron, A. 2010. Lineages that cheat death: Surviving the squeeze on range size. Evolution 64:2278–2292.

Waldron, P. J., *et al.* 2009. Functional gene array-based analysis of microbial community structure in groundwaters with a gradient of contaminant levels. Environmental Science and Technology 43:3529–3534.

Walker, B., and W. Steffen. 1997. An overview of the implications of global change for natural and managed terrestrial ecosystems. Conservation Ecology [online]1(2): 2. Available at http://www.consecol.org/vol1/iss2/art2/

Walker, K. J., S. A. Trewick, and G. M. Barker. 2008. *Powelliphanta augusta*, a new species of land snail, with a description of its former habitat, Stockton Coal Plateau, New Zealand. Journal of the Royal Society of New Zealand 38:163–186.

Walker, N. F., P. E. Hulme, and A. R. Hoelzel. 2003. Population genetics of an invasive species, *Heracleum mantegazzianum*: implications for the role of life history, demographics and independent introductions. Molecular Ecology 12:1747–1756.

Wallace, A. R. 1892. Island Life. Macmillan and Co., London.

Wallace, A. R. 1923. Darwinism: An Exposition of the Theory of Natural Selection with Some of its Applications. Macmillan and Co., London.

Wallace, B. 1968. Topics in Population Genetics. Norton, New York.

Wallace, B. 1991. Fifty Years of Genetic Load: An Odyssey. Cornell University Press, Ithaca, NY.

Wallace, B., and Dobzhansky, T. 1959. Radiation, Genes, and Man. Holt, New York.

Waller, D. M., J. Dole, and A. J. Bersch. 2008. Effects of stress and phenotypic variation on inbreeding depression in *Brassica rapa*. Evolution 62:917–931.

Walsh, M. R., S. B. Munch, S. Chiba, and D. O. Conover. 2006. Maladaptive changes in multiple traits caused by fishing: Impediments to population recovery. Ecological Letters 9:142–148.

Walter, R., and B. K. Epperson. 2001. Geographic pattern of genetic variation in *Pinus resinosa*: area of greatest diversity is not the origin of postglacial populations. Molecular Ecology 10:103–111.

Walters, C. J., and P. Cahoon. 1985. Evidence of decreasing spatial diversity in British Columbia salmon stocks. Canadian Journal of Fisheries and Aquatic Sciences 42:1033–1037.

Walters, C. J., and Martell, S. J. D. 2004. Fisheries Ecology and Management. Princeton University Press, Princeton, NJ.

Wang, I. J., W. K. Savage, and H. B. Shaffer. 2009. Landscape genetics and least-cost path analysis reveal unexpected dispersal routes in the California tiger salamander (*Ambystoma californiense*). Molecular Ecology 18:1365–1374.

Wang, J. L. 2002. An estimator for pairwise relatedness using molecular markers. Genetics 160:1203–1215.

Wang, J. L., and N. Ryman. 2001. Genetic effects of multiple generations of supportive breeding. Conservation Biology 15:1619–1631.

Wang, T., *et al.* 2006. Use of response functions in selecting lodgepole pine populations for future climates. Global Change Biology 12:2404–2416.

Wang, T. L., G. A. O'Neill, and S. N. Aitken. 2010. Integrating environmental and genetic effects to predict responses of tree populations to climate. Ecological Applications 20:153–163.

Wang, W., and H. Lan. 2000. Rapid and parallel chromosomal number reductions in muntjac deer inferred from mitochondrial DNA phylogeny. Molecular Biology and Evolution 17:1326–1333.

Wang, Z. S., A. J. Baker, G. E. Hill, and S. V. Edwards. 2003. Reconciling actual and inferred population histories in the house finch (*Carpodacus mexicanus*) by AFLP analysis. Evolution 57:2852–2864.

Waples, R. S. 1987. A multispecies approach to the analysis of gene flow in marine shore fishes. Evolution 41:385–400.

Waples, R. S. 1989. A generalized approach for estimating effective population size from temporal changes in allele frequency. Genetics 121:379–391.

Waples, R. S. 1990. Conservation genetics of Pacific salmon: III. Estimating effective population size. Journal of Heredity 81:277–289.

Waples, R. S. 1991. Pacific salmon, *Oncorhynchus* spp., and the definition of "species" under the Endangered Species Act. Marine Fisheries Review 53:11–22.

Waples, R. S. 1995. Genetic effects of stock transfers of fish. Pp. 51–69 in Philipp, D., et al., eds. Proceedings of the World Fisheries Congress, Theme 3. Oxford and IBH Publishing Co. Pvt. Ltd., New Delhi.

Waples, R. S. 1998. Separating the wheat from the chaff: Patterns of genetic differentiation in high gene flow species. Journal of Heredity 89:438–450.

Waples, R. S. 2002. Definition and estimation of effective population size in the conservation of endangered species. Pp. 147–168 in Beissinger, S. R., and D. R. McCullough, eds. Population Viability Analysis. University of Chicago Press, Chicago, IL.

Waples, R. S. 2005a. Genetic estimates of contemporary effective population size: to what time periods do the estimates apply? Molecular Ecology 14:3335–3352.

Waples, R. S. 2005b. Identifying conservation units of Pacific salmon using alternative ESU concepts. Pp. 127–149 in Scott, J. M., D. D. Goble, and F. W. Davis, eds. The Endangered Species Act at Thirty: Conserving Biodiversity in Human-Dominated Landscapes. Island Press, Washington, DC.

Waples, R. S., and C. Do. 2008. LDNE: a program for estimating effective population size from data on linkage disequilibrium. Molecular Ecology Resources 8:753–756.

Waples, R. S., and C. Do. 2010. Linkage disequilibrium estimates of contemporary Ne using highly variable genetic markers: a largely untapped resource for applied conservation and evolution. Evolutionary Applications 3:244–262.

Waples, R. S., and J. Drake. 2004. Risk-benefit considerations for marine stock enhancement: a Pacific salmon perspective. Pp. 260–306 in Leber, K. M., S. Kitada, H. L. Blankenship, and T. Svasand, eds. Blackwell, Oxford.

Waples, R. S., and J. R. Faulkner. 2009. Modelling evolutionary processes in small populations: Not as ideal as you think. Molecular Ecology 18:1834–1847.

Waples, R. S., and O. Gaggiotti. 2006. What is a population? An empirical evaluation of some genetic methods for identifying the number of gene pools and their degree of connectivity. Molecular Ecology 15:1419–1439.

Waples, R. S., C. Do, and J. Chopelet. 2011. Calculating ne and ne/n in age-structured populations: A hybrid Felsenstein-Hill approach. Ecology 92:1513–1522.

Ward, R. D., D. O. F. Skibinski, and M. Woodwark. 1992. Protein heterozygosity, protein structure, and taxonomic differentiation. Evolutionary Biology 26:73–160.

Warwell, M. V., G. E. Rehfeldt, and N. L. Crookston. 2007. Modelling contemporary climate profiles of whitebark pine (*Pinus albicaulis*) and predicting responses to global warming. In Goheen, E., ed. Whitebark Pine: a Pacific Coast Perspective. USDA Forest Service R6-NR-FHP-2007-01.

Waser, P. M., and C. Strobeck. 1998. Genetic signatures of interpopulation dispersal. Trends in Ecology & Evolution 13:43–44.

Wasser, S. K., et al. 2007. Using DNA to track the origin of the largest ivory seizure since the 1989 trade ban. Proceedings of the National Academy of Sciences USA 104: 4228–4233.

Waters, J. M., L. H. Dijkstra, and G. P. Wallis. 2000. Biogeography of a southern hemisphere freshwater fish: how important is marine dispersal? Molecular Ecology 9:1815–1821.

Watson, A. 2000. New tools. A new breed of high-tech detectives. Science 289:850–854.

Watson, J. D., and F. A. C. Crick. 1953. Genetical implications of the structure of deoxyribonucleic acid. Nature 171:964–967.

Watt, W. B. 1995. Allozymes in evolutionary genetics: beyond the twin pitfalls of "neutralism" and "selectionism". Revue Suisse de Zoologie 102:869–882.

Watts, P. C., et al. 2006. Parthenogenesis in Komodo dragons. Nature 444:1021–1022.

Wayne, R. K., and P. A. Morin. 2004. Conservation genetics in the new molecular age. Frontiers Ecology Environment 2:89–97.

Wayne, R. K., et al. 1986. Genetic markers of zoo populations: morphological and electrophoretic assays. Zoo Biology 5:215–232.

Weber, E., and B. Schmid. 1998. Latitudinal population differentiation in two species of *Solidago* (Asteraceae) introduced into Europe. American Journal of Botany 85:1110–1121.

Wegmann, D., C. Leuenberger, S. Neuenschwander, and L. Excoffier. 2010. ABCtoolbox: A versatile toolkit for approximate Bayesian computations. BMC Bioinformatics 11:116.

Weinberg, W. 1908. Uber den Nachweis der Vererbung beim Menschen. Jahresh. Verein f. vaterl. Naturk. in Wurttemberg 64:368–382.

Weir, B. S., and W. G. Hill. 1980. Effect of mating structure on variation in linkage disequilibrium. Genetics 95: 477–488.

Weir, B. S. 1996. Genetic Data Analysis II. Methods for Discrete Population Genetic Data. Sinauer, Sunderland, MA.

Wells, K. D. 1973. The historical context of natural selection: The case of Patrick Matthew. Journal of the History of Biology 6:225–258.

Welsh, J., and M. McClelland. 1990. Fingerprinting genomes using PCR with arbitrary primers. Nucleic Acids Research 18:7213–7218.

Westemeier, R. L., *et al.* 1998. Tracking the long-term decline and recovery of an isolated population. Science 282: 1695–1698.

Wheeler, D. L., *et al.* 2000. Database resources of the National Center for Biotechnology Information. Nucleic Acids 28:10–14.

White, B. A., and J. B. Shaklee. 1991. Need for replicated electrophoretic analyses in multiagency genetic stock identification (GSI) programs – examples from a pink salmon (*Oncorhynchus gorbuscha*) GSI fisheries study. Canadian Journal of Fisheries and Aquatic Sciences 48:1396–1407.

White, M. J. D. 1973. Animal Cytology and Evolution, 3rd edn. Cambridge University Press, Cambridge.

White, M. J. D. 1978. Modes of Speciation. W.H. Freeman and Company, San Francisco.

Whiteman, N. K., R. T. Kimball, and P. G. Parker. 2007. Co-phylogeography and comparative population genetics of the threatened Galapagos hawk and three ectoparasite species: ecology shapes population histories within parasite communities. Molecular Ecology 16:4759–4773.

Whitham, T. G., *et al.* 2008. Extending genomics to natural communities and ecosystems. Science 320:492–495.

Whitlock, M. C. 2008. Evolutionary inference from Q_{ST}. Molecular Ecology 17:1885–1896.

Whitlock, M. C. 2011. G'ST and D do not replace FST. Molecular Ecology 20:1083–1091.

Whitlock, M. C., and F. Guillaume. 2009. Testing for spatially divergent selection: Comparing Q_{ST} to F_{ST}. Genetics 183:1055–1063.

Whitlock, M. C., and D. E. McCauley. 1999. Indirect measures of gene flow and migration: $F_{ST} \neq 1/(4Nm + 1)$. Heredity 82:117–125.

Whitlock, M. C., and S. P. Otto. 1999. The panda and the phage: compensatory mutations and the persistence of small populations. Trends in Ecology & Evolution 14: 295–296.

Whitlock, M. C., P. C. Phillips, F. B. G. Moore, and S. J. Tonsor. 1995. Multiple fitness peaks and epistasis. Annual Review of Ecology and Systematics 26:601–629.

Whitlock, M. C., P. K. Ingvarsson, and T. Hatfield. 2000. Local drift load and the heterosis of interconnected populations. Heredity 84:452–457.

Whitney, K. D., *et al.* 2010. Patterns of hybridization in plants. Perspectives in Plant Ecology, Evolution and Systematics 12:175–182.

Whitton, J., C. J. Sears, E. J. Baack, and S. P. Otto. 2008. The dynamic nature of apomixis in the angiosperms. International Journal of Plant Sciences 169:169–182.

Wiegand, K. M. 1935. A taxonomist's experience with hybrids in the wild. Science 81:161–166.

Wiens, J. A. 1977. On competition and variable environments. American Scientist 65:590–597.

Wiens, J. J., C. A. Kuczynski, and P. R. Stephens. 2010. Discordant mitochondrial and nuclear gene phylogenies in emydid turtles: Implications for speciation and conservation. Biological Journal of the Linnean Society 99:445–461.

Wilding, C. S., R. K. Butlin, and J. Grahame. 2001. Differential gene exchange between parapatric morphs of *Littorina saxatilis* detected using AFLP markers. Journal of Evolutionary Biology 14:611–619.

Willi, Y., J. Van Buskirk, and A. A. Hoffmann. 2006. Limits to the adaptive potential of small populations. Annual Review of Ecology Evolution and Systematics 37:433–458.

Williams, C. L., B. Lundrigan, and O. E. Rhodes. 2004. Microsatellite DNA variation in tule elk. Journal of Wildlife Management 68:109–119.

Williams, S. E., and E. A. Hoffman. 2009. Minimizing genetic adaptation in captive breeding programs: A review. Biological Conservation 142:2388–2400.

Williamson, E. G., and M. Slatkin. 1999. Using maximum likelihood to estimate population size from temporal changes in allele frequencies. Genetics 152:755–761.

Willis, J. H. 1993. Partial self-fertilization and inbreeding depression in two populations of *Mimulus guttatus*. Heredity 71:145–154.

Willis, J. H. 1999. The role of genes of large effect on inbreeding depression in *Mimulus guttatus*. Evolution 53:1678–1691.

Willis, K., and R. J. Wiese. 1997. Elimination of inbreeding depression from captive populations: Speke's gazelle revisited. Zoo Biology 16:9–16.

Wilson, A. C., G. L. Bush, S. M. Case, and M. C. King. 1975. Social structuring of mammalian populations and rate of chromosomal evolution. Proceedings National Academy Sciences USA 72:5061–5065.

Wilson, A. C., *et al.* 1985. Mitochondrial DNA and two perspectives on evolutionary genetics. Biological Journal of the Linnean Society 26:375–400.

Wilson, A. J., *et al.* 2006. Environmental coupling of selection and heritability limits evolution. Plos Biology 4: 1270–1275.

Wilson, E. O. 1984. Biophilia. Harvard University Press, Cambridge, MA.

Wilson, F. 1965. Biological control and the genetics of colonizing species. Pp. 307–325 *in* Baker, H. G., and G. L. Stebbins, eds. The Genetics of Colonizing Species. Academic Press, New York.

Wilson, G. A., and B. Rannala. 2003. Bayesian inference of recent migration rates using multilocus genotypes. Genetics 163:1177–1191.

Wilson, R. J., *et al.* 2005. Changes to the elevational limits and extent of species ranges associated with climate change. Ecology Letters 8:1138–1146.

Woinarski, J. C. Z., and A. Fisher. 1999. The Australian Endangered Species Protection Act 1992. Conservation Biology 13:959–962.

Wolf, D. E., N. Takebayashi, and L. H. Rieseberg. 2001. Predicting the risk of extinction through hybridization. Conservation Biology 15:1039–1053.

Wood, C. C., and C. J. Foote. 1996. Evidence for sympatric genetic divergence of anadromous and nonanadromous morphs of sockeye salmon (*Oncorhynchus nerka*). Evolution 50:1265–1279.

Wood, P. M. 2003. Will Canadian policies protect British Columbia's endangered pairs of sympatric sticklebacks? Fisheries 28(5):19–26.

Woodruff, D. S. 2001. Declines of biomes and the future of evolution. Proceedings National Academy Sciences USA 98:5471–5476.

Woodruff, R. C., and J. N. Thompson. 1992. Have premeiotic clusters of mutation been overlooked in evolutionary theory? Journal of Evolutionary Biology 5: 457–464.

Woodruff, R. C., H. Huai, and J. N. Thompson Jr. 1996. Clusters of identical new mutations in the evolutionary landscape. Genetica 98:149–160.

Woodworth, L. M., M. E. Montgomery, D. A. Briscoe, and R. Frankham. 2002. Rapid genetic deterioration in captive populations: Causes and conservation implications. Conservation Genetics 3:277–288.

Worm, B., et al. 2006. Impacts of biodiversity loss on ocean ecosystem services. Science 314:787–790.

Worthy, T. N., and Holdaway, R. N. 2002. The Lost World of the Moa: Prehistoric Life of New Zealand. Indiana University Press, Bloomington, IN

Wright, D. A., and C. M. Richards. 1983. Two sex-linked loci in the leopard frog, *Rana pipiens*. Genetics 103:249–261.

Wright, S. 1921. Systems of mating, I: The biometric relation between parent and offspring. Genetics 6:111–123.

Wright, S. 1922. Coefficients of inbreeding and relationship. American Naturalist 56:330–338.

Wright, S. 1931. Evolution in Mendelian populations. Genetics 16:97–159.

Wright, S. 1939. Statistical genetics in relation to evolution. Actualités scientifiques et industrielles. 802. Herman & Cie, Paris. Republished in Provine (ed.) 1986.

Wright, S. 1940. Breeding structure of populations in relation to speciation. American Naturalist 74:232–248.

Wright, S. 1943. Isolation by distance. Genetics 28: 114–138.

Wright, S. 1945. Tempo and mode in evolution: A critical review. Ecology 26:415–419.

Wright, S. 1946. Isolation by distance under diverse systems of mating. Genetics 32:303–324.

Wright, S. 1951. The genetical structure of populations. Annals of Eugenics 15:323–354.

Wright, S. 1960. On the number of self-incompatibility alleles maintained in equilibrium by a given mutation rate in a population of given size: A re-examination. Biometrics 16:61–85.

Wright, S. 1965a. The distribution of self-incompatibility alleles in populations. Evolution 18:609–619.

Wright, S. 1965b. The interpretation of population structure by F-statistics with special regard to systems of mating. Evolution 19:355–420.

Wright, S. 1969. Evolution and the Genetics of Populations. Vol. 2. The Theory of Gene Frequencies. University of Chicago Press, Chicago.

Wu, C.-H., and A. J. Drummond. 2011. Joint inference of microsatellite mutation models, population history and genealogies using transdimensional Markov Chain Monte Carlo. Genetics 188:151–164.

Wu, C.-I. 2001. The genic view of the process of speciation. Journal of Evolutionary Biology 14:851–865.

Xu, X., and U. Arnason. 1996. The mitochondrial DNA molecule of Sumatran and a molecular proposal for two (Bornean and Sumatran) species of orangutan. Journal of Molecular Evolution 43:431–437.

Yakovlev, I. A., C. G. Fossdal, and Ø. Johnsen. 2010. MicroRNAs, the epigenetic memory and climatic adaptation in Norway spruce. New Phytologist 187:1154–1169.

Young, A. G., and B. G. Murray. 2000. Genetic bottlenecks and dysgenic gene flow into re-established populations of the grassland daisy, *Rutidosis leptorrhynchoides*. Australian Journal of Botany 48:409–416.

Young, A. G., and M. Pickup. 2010. Low S-allele numbers limit mate availability, reduce seed set and skew fitness in small populations of a self-incompatible plant. Journal of Applied Ecology 47:541–548.

Young, A. G., et al. 2000a. Genetic erosion, restricted mating and reduced viability in fragmented populations of the endangered grassland herb *Rutidosis leptorrhynchoides*. Pp. 335–359 in Young, A. G., and G. M. Clarke, eds. Genetics, Demography and Viability of Fragmented Populations. Cambridge University Press, Cambridge.

Young, A. G., C. Millar, E. A. Gregory, and A. Langston. 2000b. Sporophytic self-incompatibility in diploid and tetraploid races of *Rutidosis leptorrhynchoides*. Australian Journal of Botany 48:667–672.

Yunis, J. J., and O. Prakash. 1982. The origin of man: a chromosomal pictorial legacy. Science 215:1525–1529.

Zakharov, V. M. 2001. Ontogeny and population: developmental stability and population variation. Russian Journal of Ecology 32:146–150.

Zapata, C. 2000. The D' measure of overall gametic disequilibrium between pairs of multiallelic loci. Evolution 54: 1809–1812.

Zapata, C., and G. Alvarez. 1992. The detection of gametic disequilbrium between allozyme loci in natural populations of *Drosophila*. Evolution 46:1900–1917.

Zayed, A., and L. Packer. 2005. Complementary sex determination substantially increases extinction proneness of haplodiploid populations. Proceedings National Academy Sciences USA 102:10742–10746.

Zeder, M. A., and B. Hess. 2000. The initial domestication of goats (*Capra hircus*) in the Zagros mountains 10,000 years ago. Science 287:2254–2257.

Zhi, L., *et al.* 1996. Genomic differentiation among natural populations of orang-utan (*Pongo pygmaeus*). Current Biology 6:1326–1336.

Zimmer, C. 2002. Darwin's avian muses continue to evolve. Science 296:633–635.

Index

Page numbers in *italics* refer to figures and tables, those in **bold** refer to guest boxes.

Conservation and the Genetics of Populations, Second Edition. Fred W. Allendorf, Gordon Luikart, and Sally N. Aitken.
© 2013 Fred W. Allendorf, Gordon Luikart and Sally N. Aitken. Published 2013 by Blackwell Publishing Ltd.